Dario Leister, PhD
Editor

Plant Functional Genomics

Pre-publication
REVIEWS,
COMMENTARIES,
EVALUATIONS . . .

"This is the most comprehensive book on plant functional genomics published to date. It is an impressive and timely overview that brings the reader up-to-date with recent techniques in plant functional genomics."

Stefan Jansson, PhD
Professor of Plant Cell and Molecular Biology,
Umeå Plant Science Centre,
Umeå University

"*Plant Functional Genomics* is a must for the plant molecular laboratory. It covers the full breadth of technologies involved in identifying and understanding plant gene function in the era of genomics. This mammoth volume does an outstanding job of explaining to researchers and newcomers how the technologies developed historically and how they work. The book also provides access to numerous high-profile case studies in which these technologies have been successfully employed. The virtues of various plant genetic systems ranging from cyanobacteria to angiosperms are covered—a particular strength that wil help researchers."

William Martin
Professor of Botany,
Heinrich-Heine-Universität,
Düsseldorf, Germany

NOTES FOR PROFESSIONAL LIBRARIANS AND LIBRARY USERS

This is an original book title published by Food Products Press®, an imprint of The Haworth Press, Inc. Unless otherwise noted in specific chapters with attribution, materials in this book have not been previously published elsewhere in any format or language.

CONSERVATION AND PRESERVATION NOTES

All books published by The Haworth Press, Inc. and its imprints are printed on certified pH neutral, acid-free book grade paper. This paper meets the minimum requirements of American National Standard for Information Sciences-Permanence of Paper for Printed Material, ANSI Z39.48-1984.

Plant Functional Genomics

FOOD PRODUCTS PRESS®

Crop Science

Amarjit S. Basra, PhD
Senior Editor

Bacterial Disease Resistance in Plants: Molecular Biology and Biotechnological Applications by P. Vidhyasekaran

Handbook of Formulas and Software for Plant Geneticists and Breeders edited by Manjit S. Kang

Postharvest Oxidative Stress in Horticultural Crops edited by D. M. Hodges

Encyclopedic Dictionary of Plant Breeding and Related Subjects by Rolf H. G. Schlegel

Handbook of Processes and Modeling in the Soil-Plant System edited by D. K. Benbi and R. Nieder

The Lowland Maya Area: Three Millennia at the Human-Wildland Interface edited by A. Gómez-Pompa, M. F. Allen, S. Fedick, and J. J. Jiménez-Osornio

Biodiversity and Pest Management in Agroecosystems, Second Edition by Miguel A. Altieri and Clara I. Nicholls

Plant-Derived Antimycotics: Current Trends and Future Prospects edited by Mahendra Rai and Donatella Mares

Concise Encyclopedia of Temperate Tree Fruit edited by Tara Auxt Baugher and Suman Singha

Landscape Agroecology by Paul A Wojkowski

Concise Encyclopedia of Plant Pathology by P. Vidhyasekaran

Molecular Genetics and Breeding of Forest Trees edited by Sandeep Kumar and Matthias Fladung

Testing of Genetically Modified Organisms in Foods edited by Farid E. Ahmed

Agrometeorology: Principles and Applications of Climate Studies in Agriculture by Harpal S. Mavi and Graeme J. Tupper

Concise Encyclopedia of Bioresource Technology edited by Ashok Pandey

Genetically Modified Crops: Their Development, Uses, and Risks edited by G. H. Liang and D. Z. Skinner

Plant Functional Genomics edited by Dario Leister

Immunology in Plant Health and Its Impact on Food Safety by P. Narayanasamy

Abiotic Stresses: Plant Resistance Through Breeding and Molecular Approaches edited by Muhammad Ashraf and Philip John Charles Harris

Multinational Agribusinesses edited by Ruth Rama

Crops and Environmental Change: An Introduction to Effects of Global Warming, Increasing Atmospheric CO_2 and O_3 Concentrations, and Soil Salinization on Crop Physiology and Yield by Seth G. Pritchard and Jeffrey S Amthor

Teaching in the Sciences: Learner-Centered Approaches edited by Catherine McLoughlin and Acram Taji

Plant Functional Genomics

Dario Leister, PhD
Editor

Food Products Press®
An Imprint of The Haworth Press, Inc.
New York • London • Oxford

Published by

Food Products Press®, an imprint of The Haworth Press, Inc., 10 Alice Street, Binghamton, NY 13904-1580.

Cover design by Lora Wiggins.

Rice photo: Takuji Sasaki, photographer.

Targeting, *Arabidopsis*, array, cluster, insertions, and proteomics photos, Dario Leister, photographer.

Physcomitrella cDNA illustration: Originally published in Egener, T., Granado, J., Guitton, M. C., Hohe, A., Holtorf, H., Lucht, J. M., Rensing, S., Schlink, K., Schulte, J., Schween, G., et al. (2002). High frequency of phenotypic deviations in *Physcomitrella patens* plants transformed with a gene-disruption library. *BMC Plant Biology* 2: 6.

Library of Congress Cataloging-in-Publication Data

Plant functional genomics / Dario Leister, editor.
p. cm.
Includes bibliographical references and index.
ISBN 1-56022-998-5 (hard : alk. paper)—ISBN 1-56022-999-3 (soft : alk. paper)
1. Plant molecular genetics. 2. Plant genome mapping. 3. Plant gene expression. 4. Plant genomes. I. Leister, Dario.
QK981.4.P579 2004
572.8'2—dc22

2004009819

CONTENTS

SECTION II: SPECIES EXAMPLES

SECTION V: PROTEIN FAMILIES

ABOUT THE EDITOR

Dario Leister, PhD, is head of the research group on functional genomics of *Arabidopsis* photosynthesis at the Department of Plant Breeding and Yield Physiology at the Max Planck Institute for Plant Breeding Research in Germany. His research interests include the genetic and genomic analysis of photosynthesis, chloroplast-nucleus signaling, genome evolution, and evolution of chloroplasts and mitochondria. He has been published in such venues as *Nature Genetics, Trends in Genetics, The Plant Journal, The Plant Cell, Trends in Plant Science, Plant Physiology,* and others.

CONTRIBUTORS

Kenji Akiyama is a technical staff employee at Plant Functional Genomics Research Group, RIKEN Genomic Sciences Center, Japan.

Pauline Bablak is a research scientist at the University of Wales, Aberystwyth, and holds an MPhil in plant physiology from the University of Leicester, England.

Søren Bak is an associate professor in the Department of Plant Biology at the Royal Veterinary and Agricultural University, Kopenhagen, Denmark, where he also received a PhD in plant biochemistry.

Frédéric Berger is a senior staff scientist at the École Normale Supérieure de Lyon, France, and holds a PhD in cell biology and genetics from the University of Lyon, France.

Hans J. Bohnert is a professor at the University of Illinois, Champaign/Urbana, in the Department of Plant Biology and the Department of Crop Sciences.

Karine Boivin is a postdoctoral fellow at the Plant Genomic Research Lab, INRA-Evry, France, and holds a PhD in plant biology from the University of Köln, Germany.

Piero Carninci is a senior scientist at Genome Science Laboratory, RIKEN (Japan), and holds a PhD in molecular biology from the University of Trieste, Italy.

Barry Causier is a postdoctoral fellow at the Centre for Plant Sciences, University of Leeds, United Kingdom. He obtained his PhD from the University of Leeds.

Françoise Daniel-Vedele is a senior staff scientist at and director of the Plant Nitrogen Nutrition Lab, INRA-Versailles, France, and holds a PhD in genetics from the INA-PG, Paris, France.

Brendan Davies is a senior lecturer at the University of Leeds, United Kingdom, and obtained his PhD at the Imperial Cancer Research Fund in London.

Koen Dekker is a research scientist at the Max Planck Institute for Plant Breeding Research, Köln, Germany, and holds a PhD in agricultural chemistry from Nagoya University, Japan.

Rachel M. Dent is a postgraduate researcher in the Department of Plant and Microbial Biology at the University of California, Berkeley, and holds a PhD in plant sciences from Cambridge University, England.

Marcelo Desimone is a group leader at the Center for Plant Molecular Biology, University of Tübingen, Germany, and holds a PhD from the University of Freiburg, Germany.

Martine Devic is a senior staff scientist at the University of Perpignan, France, and holds a PhD from the University Louis Pasteur, Strasbourg, France.

John Draper is a professor of plant pathology at the University of Wales, Aberystwyth, and holds a PhD in plant genetic engineering from the University of Nottingham, England.

Akiko Enju is a technical staff employee at Plant Functional Genomics Research Group, RIKEN Genomic Sciences Center, Japan.

Ulf-Ingo Flügge is a full professor at the Botanical Institute, University of Cologne, Germany.

Wolfgang Frank is an assistant professor at the Chair of Plant Biotechnology at the University of Freiburg, Germany, and holds a PhD in genetics from the University of Cologne, Germany.

Wolf B. Frommer is a senior staff scientist at the Carnegie Institution of Washington, Stanford, California, and was a professor at the Center for Plant Molecular Biology, University of Tübingen, Germany.

Miki Fujita is a postdoctoral fellow at Plant Functional Genomics Research Group, RIKEN Genomic Sciences Center, Japan, and holds a PhD in plant pathology from Kyoto University, Japan.

David W. Galbraith is a professor in the Department of Plant Sciences at the University of Arizona, Tucson, and holds a PhD in biochemistry from Cambridge University, England.

John Graham is a postdoctoral fellow at the Centre for Plant Sciences, University of Leeds, United Kingdom. He obtained his PhD from the University of Leeds.

Anna Haldrup is an associate professor in the Department of Plant Biology at the Royal Veterinary and Agricultural University, Kopenhagen, Denmark, and obtained a PhD in plant biotechnology at Danisco Biotechnology, also in Copenhagen.

Klaus Harter is a full professor at the Botanical Institute of the University of Cologne, Germany, and holds a PhD in plant physiology from the University of Freiburg, Germany.

Claudia Hass is a PhD student at the Botanical Institute of the University of Cologne, Germany, and holds a diploma in biology from the University of Bremen, Germany.

Robert Hasterok is a lecturer at the University of Silesia, Poland, and holds a PhD in plant biology from the same institution.

Yoshihide Hayashizaki is a project director at Genome Exploration Research Group, RIKEN Genomic Sciences Center, Japan; is a chief scientist at Genome Science Laboratory, RIKEN; and holds an MD and PhD from Osaka University Medical School, Japan.

Joshua L. Heazlewood is a postdoctoral research associate at the University of Western Australia, Perth, and holds a PhD in plant molecular biology from La Trobe University, Victoria, Australia.

Hauke Holtorf is a senior research assistant at the Chair of Plant Genetics of the Federal Institute of Technology (ETH), Zürich, Switzerland, and holds a PhD in plant genetics from the ETH Zürich.

Kei Iida is a technical staff employee at Plant Functional Genomics Research Group, RIKEN Genomic Sciences Center, Japan.

Junko Ishida is a technical staff employee at Plant Functional Genomics Research Group, RIKEN Genomic Sciences Center, Japan.

Glyn Jenkins is a lecturer in genetics at the University of Wales, Aberystwyth, and holds a PhD in plant genetics from the same institution.

Poul Erik Jensen is an associate professor in the Department of Plant Biology at the Royal Veterinary and Agricultural University, Kopenhagen, Denmark, where he also received a PhD in plant genetics.

Takakazu Kaneko is a staff scientist at Kazusa DNA Research Institute, Japan, and holds a PhD in plant biology from Nagoya University, Japan.

Jun Kawai is a team leader at Genome Exploration Research Group, RIKEN Genomic Sciences Center, Japan, and holds a PhD in science from Kyoto University, Japan.

Anne Krapp is a senior staff scientist at the Plant Nitrogen Nutrition Lab, INRA-Versailles, France, and holds a PhD in plant physiology from the University of Bayreuth, Germany.

Jörg Kudla is a professor at the Botanical Institute of the University of Münster, Germany, and holds a PhD from the University Halle/Wittenberg, Germany.

Reinhard Kunze is a group leader at the Botanical Institute, University of Cologne, Germany, and holds a PhD in genetics from the University of Cologne.

Heidi K. Ledford is a PhD student in the Department of Plant and Microbial Biology at the University of California, Berkeley, and holds a BS in biology from Duke University, North Carolina.

Loïc Lepiniec is a senior staff scientist at the Institut Jean-Pierre Bourgin de Versailles, France, and holds a PhD from the University Paris-XI, Orsay, France.

Xin Li is an assistant professor at the Biotechnology Laboratory at the University of British Columbia, Vancouver, and holds a PhD in plant pathology from Oklahoma State University, Stillwater.

Christina Lunde is a postdoctoral scientist at the University of Adelaide, Australia, and holds a PhD in molecular plant biology from the Royal Veterinary and Agricultural University, Kopenhagen, Denmark.

Joachim Messing is a university professor of molecular biology at Rutgers, the State University of New Jersey, New Brunswick; is the director of the Waksman Institute of Microbiology at Rutgers; and holds a PhD degree in biochemistry and pharmacy from the Ludwig Maximilian University of Munich, Germany.

Christian Meyer is a senior staff scientist at and deputy director of the Plant Nitrogen Nutrition Lab, INRA-Versailles, France, and holds a PhD from the INA-PG, Paris, France.

A. Harvey Millar is a QEII research fellow at the University of Western Australia, Perth, and holds a PhD in plant biochemistry from the Australian National University, Canberra.

Luis A. J. Mur is a lecturer in plant pathology at the University of Wales, Aberystwyth, and holds a PhD in plant pathology from the University of the West of England, Bristol.

Maiko Nakajima is a technical staff employee at Plant Functional Genomics Research Group, RIKEN Genomic Sciences Center, Japan.

Mari Narusaka is a technical staff employee at Plant Functional Genomics Research Group, RIKEN Genomic Sciences Center, Japan.

Krishna K. Niyogi is an associate professor in the Department of Plant and Microbial Biology at the University of California, Berkeley, and holds a PhD in biology from the Massachusetts Institute of Technology, Cambridge.

John B. Ohlrogge is a professor of plant biology at Michigan State University, East Lansing. He obtained a PhD in biochemistry from the University of Michigan, Ann Arbor.

Youko Oono is a PhD student at the University of Tsukuba, Japan.

Mathilde Orsel is a PhD student at the Plant Nitrogen Nutrition Lab, INRA-Versailles, France.

Ralf Reski is the chair of plant biotechnology at the University of Freiburg, Germany, and holds a PhD in general botany from the University of Hamburg, Germany.

Hélène Roussel was a postdoctoral fellow at the Plant Nitrogen Nutrition Lab, INRA-Versailles, France, and holds a PhD in plant physiology from the University of Dijon, France.

Sari A. Ruuska is a postdoctoral research associate in the Department of Plant Biology at Michigan State University, East Lansing, and holds a PhD in plant sciences from the Australian National University, Canberra.

Tetsuya Sakurai is a technical scientist at Plant Functional Genomics Research Group, RIKEN Genomic Sciences Center, Japan.

Takuji Sasaki is a director of the Genome Research Department of the National Institute of Agrobiological Sciences, Japan, and holds a PhD in biophysics and biochemistry from the University of Tokyo.

Shusei Sato is a staff scientist at Kazusa DNA Research Institute, Japan, and holds a PhD in plant biology from Nagoya University, Japan.

Masakazu Satou is a technical staff employee at Plant Functional Genomics Research Group, RIKEN Genomic Sciences Center, Japan.

Henrik Vibe Scheller is a professor in the Department of Plant Biology at the Royal Veterinary and Agricultural University, Copenhagen, Denmark. He holds a PhD in plant biochemistry from the Royal Veterinary and Agricultural University and a Dr. Scient. degree from the University of Copenhagen.

Motoaki Seki is a senior scientist at Plant Functional Genomics Research Group, RIKEN Genomic Sciences Center, Japan, and holds a PhD in science from Hiroshima University, Japan.

Allen Sessions is a senior staff scientist at Syngenta Biotechnology, Inc., Research Triangle Park, North Carolina, and holds a PhD in plant biology from The University of California at Berkeley.

Kazuo Shinozaki is a project director at Plant Functional Genomics Research Group, RIKEN Genomic Sciences Center, Japan; is a chief scientist at Riken's Laboratory of Plant Molecular Biology; and holds a PhD in science from Nagoya University, Japan.

Shauna Somerville is a staff scientist at the Department of Plant Biology, Carnegie Institution of Washington, Stanford, California, and an adjunct professor in the Department of Biological Sciences, Stanford University, Stanford, California. She received her PhD from the University of Illinois, Urbana-Champaign.

Satoshi Tabata is head of the Department of Plant Gene Research section at Kazusa DNA Research Institute, Japan, and holds a PhD in molecular biology from Kyoto University, Japan.

Christelle Thibault is a postdoctoral fellow at the IGBMC, Ilkirch, France, and holds a PhD in biology from the University of Poitiers, France.

Richard N. Trethewey is the chief scientific officer of Metanomics GmbH & Co. KGaA, Berlin, Germany, and holds a PhD in plant sciences from the University of Cambridge, United Kingdom.

Stefan Weinl is a PhD student in the Department of Molecular Botany at the University of Ulm, Germany.

Klaas J. van Wijk is an assistant professor in the Department of Plant Biology at Cornell University, Ithaca, New York. He received his PhD in biology from the University of Groningen, the Netherlands.

Kazuko Yamaguchi-Shinozaki is a senior researcher at Japan International Research Center for Agricultural Sciences (JIRCAS) and holds a PhD in life chemistry from the Tokyo Institute of Technology.

Yuelin Zhang is an assistant professor at the Biotechnology Laboratory at the University of British Columbia, Vancouver, and holds a PhD in biochemistry from Oklahoma State University, Stillwater.

Laurent Zimmerli is a postdoctoral scientist in the Department of Plant Biology, Carnegie Institution of Washington, Stanford, California, and received his PhD from the University of Fribourg, Switzerland.

Preface

Definition and Origin of Plant Functional Genomics

The expression *functional genomics* is nowadays widely used as an expansion of the term *genomics*. In the end of the 1980s, genomics arose as a term describing the mapping and sequencing of genomes, as well as the analysis of the information content present in genomic sequences. Functional genomics appeared ten years later when complete genome sequences were considered as a basis for systematic functional analyses. This led Hieter and Boguski to propose in 1997 the division of genomics into the two disciplines of structural genomics and functional genomics. Structural genomics corresponds to the initial phase of genome analysis resulting ultimately in the revelation of the complete DNA sequence of an organism, while functional genomics makes use of the genome sequence to assess, on a large scale, the functions of genes. As such, the birth of functional genomics goes back to the year 1996 when the genome sequence of the budding yeast *Saccharomyces cerevisiae* was completed. Only slightly later the main components of functional genomics had been already defined: the genome-wide analysis of (1) mutants, (2) RNA expression data, (3) proteins and their interactions, and (4) metabolites. When the complete, or partial, sequence of several genomes became available, genomics was expanded to consider the comparison of entire gene complements to assign (in an indirect way) functions to genes, to reconstruct the evolution of traits, as well as to understand the evolution of species and of their genomes (thus generating what is now termed *comparative genomics* and *evolutionary genomics*).

The current picture is such that also for plant biology the methodology of yeast functional genomics has been adapted, refining it, however, for its specific needs. With the completion of the sequencing of the genome of *Arabidopsis thaliana* in 2000, the first collection of all genes of a flowering plant has become available. The rice genome is also at the end of the pipeline; a number of draft and/or high-quality chromosomal sequences already exist for this species.

In this book, a cross section through plant functional genomics is provided. Section I presents breakthrough techniques; Section II provides an overview of the state of functional genomics for several plant and plant-related species (including *Synechocystis*, the model species for the cyano-bacterial endosymbiont giving rise to chloroplasts); Sections III and IV present and discuss the results of projects aiming to dissect particular plant functions; Section V deals with the genomic analysis of particular protein families. Comparative genomic analysis also has been considered here, in particular in Section II, and with the completion of the rice genome and of the genome of the green alga *Chlamydomonas*, exciting analyses can be expected to be published in the near future.

Abbreviations

ABA	abscisic acid
ABRC	*Arabidopsis* Biological Resource Center
ACCase	acetyl-CoA carboxylase
ACP	acyl carrier protein
AFGN	*Arabidopsis* Functional Genomics Network
AFLP	amplified fragment length polymorphism
AGI	Arabidopsis Genome Initiative
ARAMEMNON	Arabidopsis Membrane Protein Database
ArrayExpress	DNA array expression database maintained at EBI
BAC	bacterial artificial chromosome
BC	biotin carboxylase
BCCP	biotin carboxylase carrier protein
BIND	Biomolecular Interaction Network Database
BLAST	Basic Local Alignment Search Tool
bp	basepair(s)
CATMA	Complete Arabidopsis Transcriptome MicroArray
CBL	calcineurin B-like
CDPK	calcium-dependent protein kinase
CE	capillary electrophoresis
CHASE domain	cyclase/histidine kinase-associated sensing extracellular domain
CID	collision-induced disassociation
CIPK	CBL-interacting protein kinase
cM	centimorgan
CSIRO	Commonwealth Scientific and Industrial Research Organization (Australia)
cTP	chloroplast transit peptide
ddNTP	didesoxynucleotide triphosphate
DECAL	differential expression using customized amplification libraries
DHPLC	denaturing high-performance liquid chromatography
DNA	deoxyribonucleic acid
dNTP	desoxynucleotide triphosphate
DOP-PCR	degenerate oligonucleotide primer-PCR
EBI	European Bioinformatics Institute

EMS	ethyl methanesulfonate
ESI-MS	electrospray ionization mass spectrometry
EST	expressed sequence tag
FANCY	functional analyses by coresponse in yeast
FAS	fatty acid synthase
FCM	flow cytometric
FISH	fluorescence in situ hybridization
FNR	ferredoxin-NADP reductase
FPC	fingerprint contig
FST	flanking sequence tag
FTIC-MS	Fourier transform ion cyclotron resonance mass spectrometry
FTMS	Fourier transform mass spectrometry
GABI	German Plant Genome Analysis Project (Genomanalyse im biologischen System Pflanze)
GARNet	Genome Arabidopsis Resource Network
Gbp	gigabasepair(s)
GC	gas chromatography
GEO	Gene Expression Omnibus (DNA array expression database maintained at NCBI)
GFP	green fluorescene protein
GMOD	Generic Model Organism Database
GST	gene sequence tag
GUS	β-glucuronidase
HAPPY mapping	mapping procedure employing haploid equivalents of DNA and PCR
HATS	high-affinity transport system
HCA	hierarchical clustering analysis
HIP	highly iterated palindrome
HMM	hidden Markov model
HPLC	high-performance liquid chromatography
IAA	indole-3-acetic acid
ICAT	isotope-coded affinity tag
IEF	isoelectric focusing
IGBMC	Institut de Genetique et de Biologie Moleculaire et Cellulaire
ihpRNA	intron-containing hairpin RNA
InDel	insertion-deletion polymorphism
INRA	Institut National de la Recherché Agronomique
iPCR	inverse PCR
IPG	immobilized pH gradient
JGI	Joint Genome Institute

KAOS	Kazusa *Arabidopsis* data opening site
kbp	kilobasepair(s)
KDRI	Kazusa DNA Research Institute
λ-FLC vector	lambda full-length cDNA vector
LATS	low-affinity transport system
LC	liquid chromatography
LCM	laser-capture microdissection
LHCI	light-harvesting complex I
LHCII	light-harvesting complex II
LMS-PCR	ligand-mediated suppression PCR
LRR	leucin-rich repeat
lTP	(thylakoid) lumen transit peptide
MAGE	microarray and gene expression
MALDI-TOF	matrix-assisted laser desorption ionization—time of flight
MAPK	mitogen-activated protein kinase
MATDB	MIPS *Arabidopsis thaliana* Database
Mbp	megabasepair(s)
MCF	mitochondrial carrier family
MGED	microarray gene expression data
MIAME	minimum information about a microarray experiment
MIPS	Munich Information Center for Protein Sequences
MITE	miniature inverted-repeat transposable element
MPSS	massive parallel signature sequencing
MS	mass spectrometry
MTM Database	Maize Targeted Mutagenesis Database
mTP	mitochondrial transit peptide
MudPIT	multidimensional protein-identification technology
NAD(P)	nicotinamide adenine dinucleotide (phosphate)
NASC	Nottingham *Arabidopsis* Stock Center
NCBI	National Center for Biotechnology Information
NIAS	(Japanese) National Institute of Agrobiological Sciences
NiR	nitrite reductase
NMR	nuclear magnetic resonance
NPQ	nonphotochemical quenching
NR	nitrate reductase
OEE	oxygen-evolving enhancer
ORF	open reading frame
PAC	P1 artificial chromosome
PCA	principle component analysis
PcG protein	polycomb group protein

PCR	polymerase chain reaction
PDF	peptide deformylase
PFAM Database	Protein Families Database of Alignments and HMMs
PEP	phosphoenolpyruvate
PEPRK	PPCK-related kinase
PHA	polyhydroxyalkanoate
PHB	polyhydroxybutyrate
PK	protein kinase
PMF	peptide mass fingerprint
PP	protein phosphatase
PPCK	PEP carboxylase kinase
PQ	plastoquinone
PQH2	plastoquinol
PSI	photosystem I
PSII	photosystem II
PTGS	posttranscriptional gene silencing
qE	energy-dependent quenching
qI	photoinhibitory quenching
qT	state transition quenching
QTL	quantitative trait locus
Q-TOF	quadrupole time-of-flight
RAFL clone	Riken *Arabidopsis* full-length clone
RAPD	random amplification of polymorphic DNA
REGIA	Regulatory Gene Initiative in Arabidopsis
RFLP	restriction fragment length polymorphism
RISC	RNAi silencing complex
RLK	receptor-like kinase
RNA	ribonucleic acid
RNAi	RNA interference
RT-PCR	reverse transcription PCR
SAG	senescence-associated protein
SAGE	serial analysis of gene expression
SAIL	Syngenta *Arabidopsis* Insertion Library
SBE	single-base extension
SCF	Skp1/Cullin/F-box
SIGNAL	Salk Institute Genome Analysis Lab
SLAT	Sainsbury Laboratory *Arabidopsis* transposants
SMC	structural maintenance chromosome
SNF	sucrose-nonfermenting
SNP	single nucleotide polymorphism
SnRK	SNF-related kinase
SOS	salt overly sensitive

SPUTNIK	MIPS EST clustering and annotation system
SRP	signal recognition particle
SRP-RNC	SRP/ribosome/nascent chain complex
SSH	suppression subtractive hybridization
SSLLM	single-strand linker ligation method
SSR	simple sequence repeat
TAG	triacylglyceride
TAIL-PCR	thermal asymmetric interlaced PCR
TAIR	The *Arabidopsis* Information Resource
TAK	thylakoid-associated kinase
TCA	tricarboxylic acid
T-DNA	portion of the Ti (tumor-inducing) plasmid transferred to plant cells
TFC	tubulin-folding cofactors
TIGR	The Institute for Genomic Research
TILLING	targeting induced local lesions in genomes
Tn	transposon
TPT	triosephosphate translocator
TUC	tentative unique contig
TUS	tentative unique singlet
UTR	untranslated region
VDE	violaxanthin de-epoxidase
WAK	cell wall-associated kinase
YTH	yeast two-hybrid
ZE	zeaxanthin epoxidase
ZIGIA	Center for Functional Genomics in *Arabidopsis* (Zentrum zur Identifikation von Genfunktionen durch Insertionsmutagenese bei *Arabidopsis thaliana*)

SECTION I: BREAKTHROUGH TECHNIQUES

Chapter 1

Full-Length cDNAs for the Discovery and Annotation of Genes in *Arabidopsis thaliana*

Motoaki Seki
Masakazu Satou
Tetsuya Sakurai
Kenji Akiyama
Kei Iida
Junko Ishida
Maiko Nakajima
Akiko Enju
Mari Narusaka
Miki Fujita
Youko Oono
Kazuko Yamaguchi-Shinozaki
Piero Carninci
Jun Kawai
Yoshihide Hayashizaki
Kazuo Shinozaki

INTRODUCTION

Arabidopsis thaliana has been adopted as a model organism in the study of plant biology because of its small size, short generation time, and high efficiency of transformation (Meinke et al., 1998). To sequence its small genome (125 Mbp) (Arabidopsis Genome Initiative, 2000), scientists in Japan, Europe, and the United States collaborated in the *Arabidopsis* genome sequencing project (Bevan, 1997). Two of the five chromosomes (chromosomes 2 and 4, except for the nucleolar organizer regions and centromeres) were sequenced in 1999 (Lin et al., 1999; Mayer et al., 1999), and the re-

This work was supported by a grant for Genome Research from RIKEN, the Program for Promotion of Basic Research Activities for Innovative Biosciences, the Special Coordination Fund of the Science and Technology Agency, and a Grant-in-Aid from the Ministry of Education, Science and Culture of Japan to K. S. It was also supported in part by a Grant-in-Aid for Scientific Research on Priority Areas (C) "Genome Science" from the Ministry of Education, Science, Sports and Culture of Japan to M. S. It was also supported in part by a Research Grant for the RIKEN Genome Exploration Research Project from the Ministry of Education, Culture, Sports, Science and Technology of the Japanese Government to Y. H.

maining three chromosomes were sequenced in 2000 by the Arabidopsis Genome Initiative (AGI) (Arabidopsis Genome Initiative, 2000).

About 226,000 expressed sequence tags (ESTs) from *Arabidopsis* have been deposited in the EST database (dbEST) as of July 2003, including sequences from large-scale EST projects in France (Höfte et al., 1993; Cooke et al., 1996), United States (Newman et al., 1994; White et al., 2000), and Japan (Asamizu et al., 2000). These projects have produced EST data from different tissues, organs, seeds, and developmental stages (Höfte et al., 1993; Newman et al., 1994; Cooke et al., 1996; Asamizu et al., 2000; White et al., 2000). However, most of these EST projects are based on cDNA libraries in which most of the inserts are not full-length. ESTs are useful for making a catalog of expressed genes, but not for further study of gene function. Consequently, genome-scale collections of the full-length cDNAs of expressed genes are becoming important for the analysis of the structure and function of genes and their products in this functional genomics era.

We previously made full-length cDNA libraries using the biotinylated CAP trapper method (Carninci et al., 1996, 1997) from *Arabidopsis* plants (Seki et al., 1998). We have constructed *Arabidopsis* full-length cDNA libraries from plants grown under different conditions (Seki et al., 1998; Seki, Narusaka, Kamiya, et al., 2002) by the biotinylated CAP trapper method using trehalose-thermoactivated reverse transcriptase (Carninci et al., 1996, 1997, 1998). We isolated 155,144 RIKEN *Arabidopsis* full-length (RAFL) clones and they were clustered into 14,668 nonredundant cDNA groups, about 60 percent of all *Arabidopsis* predicted genes (Seki, Narusaka, Kamiya, et al., 2002). Haas et al. (2002) also reported isolation of 5,000 *Arabidopsis* full-length cDNAs (among them, ca. 4,000 clones overlap with the RAFL cDNA collection).

Full-length cDNAs also have many advantages for functional analysis of genes in the postsequencing era (Figure 1.1) (Seki, Narusaka, Abe, et al., 2001; Seki, Narusaka, Ishida, et al., 2002). Microarrays are powerful tools for systematically analyzing expression profiles of large numbers of genes, including stress-inducible gene expression, tissue-specific gene expression, and changes in the expression profiles of mutants and transgenic plants (Eisen and Brown, 1999; Richmond and Somerville, 2000) (see Chapter 4). We have used the full-length cDNAs for the microarray analysis of expression profiles of *Arabidopsis* genes under various stress conditions and in various mutants and transgenic plants (Figure 1.1) (Seki, Narusaka, Abe, et al., 2001; Seki, Narusaka, Ishida, et al., 2002). Full-length cDNAs are also useful resources for determining the three-dimensional structures of proteins by X-ray crystallography and NMR (nuclear magnetic resonance) spectroscopy and for studying the protein function by biochemical analy-

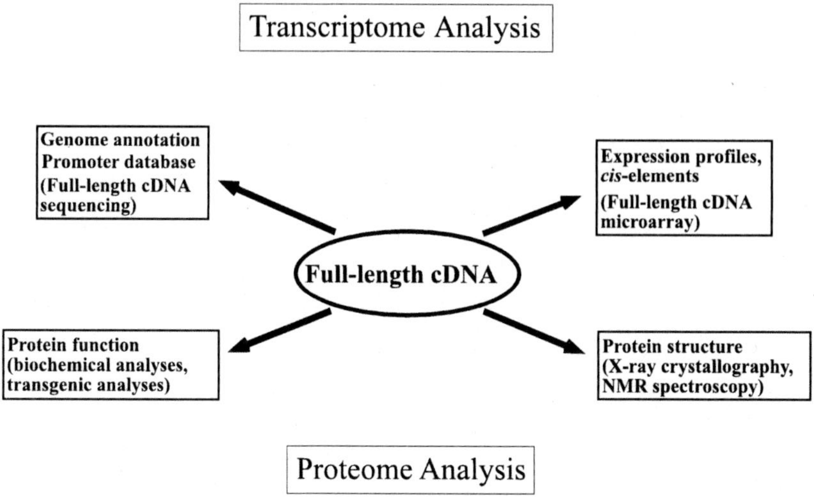

FIGURE 1.1. Application of full-length cDNAs to functional genomics.

ses and transgenic analyses (Figure 1.1) (Seki, Narusaka, Yamaguchi-Shinozaki, et al., 2001).

In this chapter, we summarize the present state and perspectives of the analysis of *Arabidopsis* full-length cDNAs, including their collection and annotation, and their application to the expression profiling, and the structural and functional analysis of plant proteins.

COLLECTION OF ARABIDOPSIS *FULL-LENGTH cDNAs*

We have constructed *Arabidopsis* full-length cDNA libraries from plants grown under different conditions as reported previously (Seki et al., 1998; Seki, Narusaka, Kamiya, et al., 2002) by the biotinylated CAP trapper method using trehalose-thermoactivated reverse transcriptase (Carninci et al., 1996, 1997, 1998). We used lambda-ZAP (Carninci et al., 1996; Seki et al., 1998) and lambda-FLC (Carninci et al., 2001) vectors for construction of the cDNA libraries. The lambda-FLC vectors accommodate cDNAs in a broad range of sizes and are useful for the high-efficiency cloning of long cDNA fragments (Carninci et al., 2001). The lambda-FLC vectors can also be bulk-excised by a Cre-*lox*-based system free of size bias to produce the plasmid libraries. In the construction of full-length cDNA libraries (RAFL 12, 13, 14, 15, 16, 17, 18, 19, and 21; Seki, Narusaka, Kamiya, et al., 2002), we used a single-strand linker ligation method (SSLLM) (Shibata et al., 2001) that uses DNA ligase to add a double-stranded (ds) DNA linker to single-stranded (ss) full-length cDNA. This SSLLM method simplifies sequencing and protein expression from the cDNA clones because, in the construction of the full-length cDNA libraries (RAFL 12, 13, 14, 15, 16, 17, 18, 19, and 21), no oligo (dG) tailing was used. Normalization and subtraction procedures (Carninci et al., 2000) were also introduced in the construction of full-length cDNA libraries (RAFL 11, 12, 13, 17, 18, 19, and 21; Seki, Narusaka, Kamiya, et al., 2002) to reduce the representation of highly expressed mRNAs in the library, and to remove cDNAs already categorized by using one-pass sequencing, respectively. The method is based on hybridization of the first-strand full-length cDNA with several RNA drivers, including starting mRNA as the normalizing driver and run-off transcripts from rearrayed clones as subtracting drivers. This method dramatically speeds up the discovery of new cDNAs. The overall strategy for preparing cDNA libraries, including standard, normalized, and subtracted libraries, has been described (Seki, Narusaka, Yamaguchi-Shinozaki, et al., 2001). We constructed 19 full-length cDNA libraries from *Arabidopsis* plants grown under various stress, hormone, and light conditions; plants at various developmental stages; and various plant tissues (Seki, Narusaka, Kamiya, et al., 2002).

We have performed single-pass sequencing of the RAFL cDNA clones from both ends and 224,530 RAFL cDNA clones have been isolated as of March 2004. Among them, we reported the results on the functional annotation of the 155,144 RAFL cDNA clones (Seki, Narusaka, Kamiya, et al., 2002). In this chapter, we summarize the results of our previous paper (Seki, Narusaka, Kamiya, et al., 2002). The 155,144 3'-ESTs were clustered and then mapped onto the *Arabidopsis* genome. Finally, 14,668 nonredundant RAFL cDNA clones were identified and mapped on the *Arabidopsis* genome (Seki, Narusaka, Kamiya, et al., 2002). The information of the 14,668 RAFL cDNA clones (the "RAFL cDNA" genes) is available at <http://pfgweb.gsc.riken.go.jp/> (Seki, Narusaka, Kamiya, et al., 2002). Assuming that the total number of *Arabidopsis* genes is about 26,000, the RAFL clones isolated should account for about 60 percent of all *Arabidopsis* genes. Our evaluation of 349 RAFL cDNA clones by single-pass sequencing showed that ca. 98 percent of the clones contained both start and stop codons. Thus the cDNA libraries constructed by the biotinylated CAP trapper contained a very high proportion of full-length cDNAs.

Haas et al. (2002) also reported the sequence of 5,016 full-length gene transcripts from three *Arabidopsis* size-fractionated full-length cDNA libraries. According to the paper of Haas et al. (2002), the starting material for cDNA library construction was polysomal RNA isolated from the topmost inflorescence tissues (Wassilewskija ecotype) and from roots (Landsberg *erecta* ecotype). The mRNA from the two ecotypes was mixed before library construction. To capture full-length cDNAs, an oligonucleotide was first attached to intact 5' ends. After first- and second-strand synthesis, the full-length cDNAs were selected, size fractionated, and cloned into pBluescript. After the 5' end sequences of ca. 120,000 clones were determined, the sequences were clustered and the clone with the longest 5' end in each cluster was selected for full-length sequencing. Finally, the full-length sequences of 5,016 clones were determined (Haas et al., 2002).

APPLICATION OF FULL-LENGTH cDNAs TO GENOME ANNOTATION

Algorithms for the Prediction of Transcription Units from Genomic Sequence Data and Their Limitations

Based on 115.4 Mbp of *Arabidopsis* genome sequence available, 25,498 putative coding genes were identified and annotated within the assembled sequences (Arabidopsis Genome Initiative, 2000). After the whole genome sequence was determined, several *Arabidopsis* genome analysis groups, such as the Institute for Genomic Research (TIGR), Kazusa DNA Research Institute, the Stanford Genome Technology Center, the Arabidopsis Information Resources (TAIR), and the Munich Information Center for Protein Sequences (MIPS) have reported genome annotation in *Arabidopsis* using several software programs (Cho and Walbot, 2001). The annotation databases include the description of which regions of the genomic DNA sequence are transcribed into messenger RNAs, the positions of the introns in the mRNA, and the translated protein sequences of the predicted genes. The annotation database is available at AGAD (TIGR Arabidopsis Genome Annotation Database, <www.tigr.org/tdb/edb2/ath1/htmls/annotation.htmlDAtD>), KAOS (Kazusa *Arabidopsis* data opening site, <http://www.kazusa.or.jp/kaos/>, DatA (Database of *Arabidopsis thaliana* Annotation at the Stanford Genome Technology Center (<http://baggage.stanford.edu/group/arabprotein/>; Palm et al., 2000), TAIR homepage <http://www.arabidopsis.org/home.html>; Garcia-Hernandez et al., 2002) and MATDB (MIPS *Arabidopsis thaliana* database, <http://mips.gsf.de/proj/ thal/db/index.html>; Schoof et al., 2002), respectively. Annotation of the *Arabidopsis* genome and functional characterization of all genes is in progress. TAIR, TIGR, and MIPS predict that the total number of *Arabidopsis* genes is, respectively, 26,184; 27,288; and 26,621 as of December 2002 in <ftp://

ftp.arabidopsis.org/home/tair/Genes/>, <ftp://ftp.tigr.org/pub/data/a_thaliana/>, and <ftp://ftpmips.gsf.de/cress/>.

Several programs for computational gene structure prediction have been developed (Pertea and Salzberg, 2002). Despite many improvements in recent years, the computational methods for genome annotation have several shortcomings that result in the following errors in annotation. The accuracy of gene prediction programs for predicting exon boundaries correctly is only about 80 percent, even for the most intensively studied organisms (Burge and Karlin, 1998; Lin et al., 1999; Pavy et al., 1999). Thus only $0.8^4 = 40$ percent of a gene with four exons will be completely correctly predicted, and genes with more exons will be even less likely to be correctly predicted. Gene prediction programs also tend to merge and split gene models. Thus, two real genes may be merged into one predicted transcript, or vice versa. In addition, programs to align genomic DNA to protein sequences often miss small exons, especially when the homologous proteins are not well conserved. Annotation protocols also tend to miss short genes (Haas et al., 2002). Alignment programs also make mistakes when genes occur in tandemly repeated copies. Furthermore, alignment of protein sequence to genomic DNA cannot predict untranslated regions (UTRs), and the leading gene prediction programs, such as Genscan (Burge and Karlin, 1997), GlimmerM (Salzberg et al., 1999), and Genemark.HMM (Lukashin and Borodovsky, 1998), are not designed for predicting UTRs; most of them predict only the coding portion of a transcript.

To improve the annotation, it is essential to isolate and analyze the full-length cDNAs (Figure 1.1). Sequencing the full-length cDNAs solves the following problems: First, it permits accurate identification of the 5' and 3' UTRs. Second, in comparison with complete genomic sequence, it enables one to identify the precise locations of all introns. Third, it aids in the discovery of new genes. Two reports (Seki, Narusaka, Kamiya, et al., 2002; Haas et al., 2002) on the application of the full-length cDNAs to genome annotation in *Arabidopsis* have been published so far. In this section, we briefly summarize these reports.

Improved Annotation of the **Arabidopsis** *Genome by RAFL cDNAs*

From the 5'-end sequences of mRNAs the promoter sequences can be obtained by comparison with the *Arabidopsis* genomic sequences (Figure 1.1). We obtained 5'-ESTs of 14,034 RAFL cDNA clones and constructed a promoter database (Seki, Narusaka, Kamiya, et al., 2002) using the plant *cis*-acting regulatory DNA elements (PLACE) database (Higo et al., 1999).

The *Arabidopsis* promoter database contains information on the genomic sequences 1000 bp upstream from the 5'-termini of each RAFL cDNA clone, and about 300 *cis*-acting elements known from plants (Seki, Narusaka, Kamiya, et al., 2002). The *Arabidopsis* promoter database constructed is available at <http://pfgweb.gsc.riken.go.jp/>. cDNA microarrays are powerful tools for the systematic analysis of expression profiles of large numbers of genes, including stress-inducible gene expression and changes in the expression profiles of mutants or transgenics (Eisen and Brown, 1999; Richmond and Somerville, 2000; Seki, Narusaka, Yamaguchi-Shinozaki, et al., 2001). One interesting type of application of microarray analysis is the identification of novel *cis*-elements that regulate the expression of genes in response to various experimental treatments. By identifying subsets of the genes that have a common expression profile, it might be possible to identify conserved motifs in the promoter regions. Our promoter database will be useful for systematic analysis of *cis*-acting elements in *Arabidopsis* (Figure 1.1) (Seki, Narusaka, Kamiya, et al., 2002; Seki, Satou, et al., 2002).

One significant class of genes missing from the existing genome annotation is the class that gives rise to non-protein-coding RNAs. In addition to their role in protein synthesis (ribosomal and transfer RNAs), noncoding RNAs have been implicated in control processes such as chromosomal silencing, transcriptional regulation, developmental control, and responses to stress (MacIntosh et al., 2001). Of the 14,668 RAFL cDNA clones mapped onto the *Arabidopsis* genome, 13,831 were matched to AGI identifiers, leaving 837 RAFL cDNA clones unmatched (Figure 1.2C) (Seki, Narusaka, Kamiya, et al., 2002). These 837 RAFL cDNAs have not yet been predicted by AGI and thus represent false negatives in the genome annotation. Among the 837 RAFL cDNAs, several putative noncoding RNAs exist.

In the postsequencing era, it is important to summarize all known expressed *Arabidopsis* genes. In this review, we report our previous summary on *Arabidopsis* expressed genes (Seki, Narusaka, Kamiya, et al., 2002). To analyze all known expressed *Arabidopsis* genes, we used data from:

1. 5,100 complete cDNAs that *Arabidopsis* researchers have sequenced and deposited in GenBank as of August 18, 2001 (http://www.ncbi.nlm.nih.gov:80/entrez/query.fcgi?db=Nucleotide);
2. 127,031 *Arabidopsis* ESTs identified as of May 22, 2001 (http://www.tigr.org/tdb/agi/release); and
3. 5,000 *Arabidopsis* full-length cDNAs that Ceres, Inc., released to TIGR on December 19, 2000 (http://www.tigr.org/tdb/e2k1/ath1/ceres/ceres.shtml).

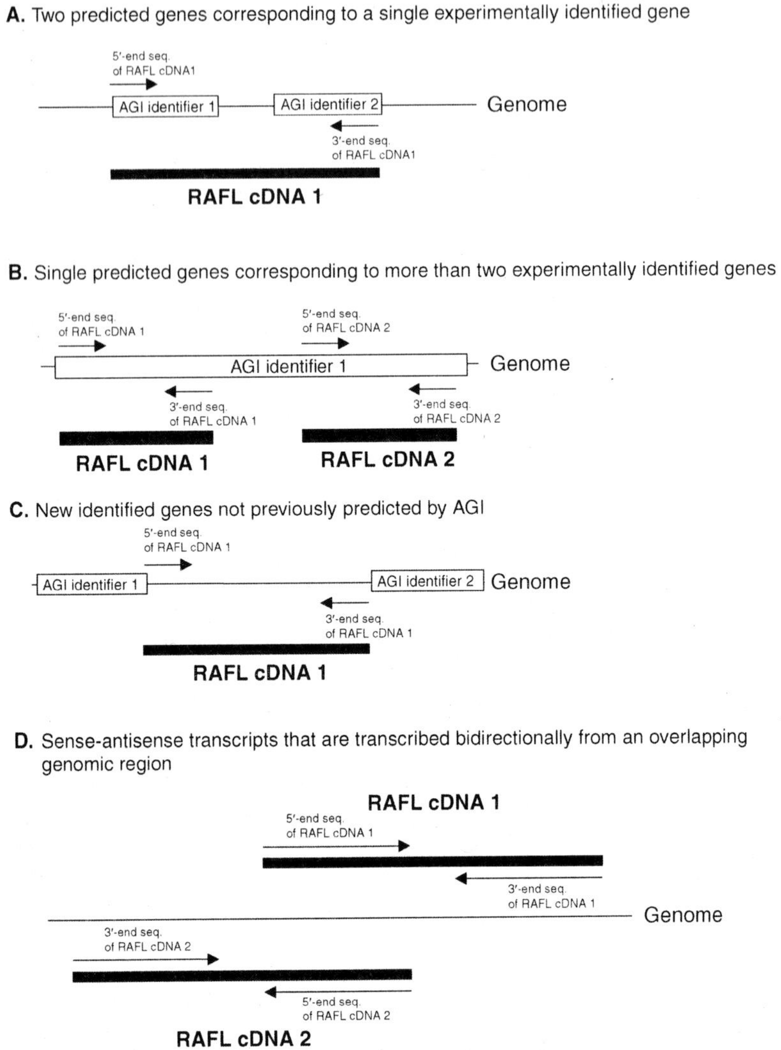

FIGURE 1.2. Examples of misannotation of genes predicted by AGI. (A) We identified a total of 201 RAFL cDNA clones whose gene covers more than two predicted genes (Seki, Narusaka, Kamiya, et al., 2002). The information of the 201 RAFL cDNA clones is available at <http://pfgweb.gsc.riken.go.jp/>. (B) We identified a total of 297 AGI identifiers whose predicted gene corresponds to more than two experimentally identified genes (Seki, Narusaka, Kamiya, et al., 2002). The information of the 297 AGI identifiers is available at <http://pfgweb.

gsc.riken.go.jp/>. (C) We identified a total of 837 newly identified genes not previously predicted by AGI (Seki, Narusaka, Kamiya, et al., 2002). The information of the 837 RAFL cDNA clones is available at <http://pfgweb.gsc.riken.go.jp/>. (D) In this study, a total of 308 pairs of sense-antisense transcripts overlapping by more than 20 bases. The information on the 308 pairs of sense-antisense transcripts is available at Supplemental Table 1 from <http://pfgweb.gsc.riken.go.jp/>. Note that the information described here is based on the previous data as of August 2001 (Seki, Narusaka, Kamiya, et al., 2002). After publication of our works in March 2002 (Seki, Narusaka, Kamiya, et al., 2002), we sent our data to the MIPS group. The MIPS group then renewed the MIPS annotation database; therefore, the MIPS annotation database is currently updated.

Altogether these genes (the "Reported EST or cDNA" genes) were subjected to a homology search as described previously (Seki, Narusaka, Kamiya, et al., 2002) against the sequence database of its corresponding AGI identifier using the BlastN program. The "Reported EST or cDNA" genes covered a total of 14,551 AGI identifiers (Seki, Narusaka, Kamiya, et al., 2002). Also, 2,437 of the RAFL cDNAs mapped to the AGI identifiers were novel genes not identified so far (Seki, Narusaka, Kamiya, et al., 2002). A total of 11,394 genes corresponded to both "Reported EST or cDNA" and "RAFL cDNA" genes. These results bring the total number of *Arabidopsis* genes whose expression has been experimentally confirmed to 17,956 (Seki, Narusaka, Kamiya, et al., 2002). In comparison, AGI lists 17,119 experimentally confirmed genes of which 16,999 were predicted (Seki, Narusaka, Kamiya, et al., 2002). The discrepancies are likely due to two predicted genes corresponding to a single experimentally identified gene (Figure 1.2A), or single predicted genes corresponding to more than two experimentally identified genes (Figure 1.2B) (Seki, Narusaka, Kamiya, et al., 2002). Some RAFL cDNA clones correspond to each of these circumstances (Seki, Narusaka, Kamiya, et al., 2002). We concluded that 9,286 predicted genes need further data to be confirmed as "expressed" genes or "unidentified" genes (Seki, Narusaka, Kamiya, et al., 2002). Because these "unidentified" genes have not been confirmed by any ESTs, some of the predicted genes may represent false positives or pseudogenes. Alternatively, these unidentified genes might have remained undetected by the EST approach because of their weak expression in specific tissues. Note that the compilation of the *Arabidopsis* expressed genes described previously is based on the data as of August 2001 (Seki, Narusaka, Kamiya, et al., 2002). After publication of our work in March 2002 (Seki, Narusaka, Kamiya, et al., 2002), we sent our data to the MIPS group. The MIPS group then renewed the MIPS annotation database; therefore, the MIPS annotation database is currently updated.

Recently, antisense RNA has been shown to play important roles such as inhibition of gene expression and genetic imprinting (Sleutels et al., 2002). We have searched for pairs of transcripts that are transcribed bidirectionally from an overlapping genomic region. We identified 308 pairs of sense-antisense transcripts overlapping by more than 20 bases (Figure 1.2D). A list of the genes is available as Supplementary Table 1 from <http://pfgweb.gsc.riken.go.jp/>. In many cases of the sense-antisense transcripts identified, the RAFL cDNAs existing in one strand of the *Arabidopsis* genomic regions were mapped on the AGI identifiers, but the RAFL cDNAs existing in the opposite strand were not mapped on the AGI identifiers.

The biological roles and biochemical functions of the RAFL cDNA clones were assigned by a homology search using the BLAST program (Seki et al., 2002a). The results showed that cDNA clones of some functional categories, such as energy production, protein synthesis, and ion homeostasis, were well represented in RAFL. More than 80 percent of cDNAs for genes involved in energy production, protein synthesis, and ionic homeostasis were found in RAFL, and ca. 70 percent of cDNAs for genes involved in metabolism, protein destination, cellular transport and transport mechanisms, and cellular organization were found in RAFL. It has been estimated that ca. 1,500 transcription factor (Riechmann et al., 2000) and about 1,000 protein kinase (http://plantsp.sdsc.edu/) genes existed in the *Arabidopsis* genome. The RAFL cDNA collection as of February 2002 included 1,087 transcription factor and 506 protein kinase genes (Seki, Narusaka, Kamiya, et al., 2002).

After we had summarized all reported expressed *Arabidopsis* genes in March 2002 (Seki, Narusaka, Kamiya, et al., 2002), we isolated around 3,500 new RAFL cDNA clones. We will summarize all expressed *Arabidopsis* genes, including these clones, in the future.

Improved Annotation of the Arabidopsis *Genome by Ceres/TIGR cDNAs*

Recently, Haas et al. (2002) also reported that full-length cDNAs are useful for improving genome annotation in *Arabidopsis*. They used sequences from ca. 5,000 full-length transcripts sequenced by Ceres, Inc., and released in the TIGR database (ftp://ftp.tigr.org/pub/data/a_thaliana/ceres) in March 2001. Haas et al. (2002) mapped and aligned the cDNA sequences to the *Arabidopsis* genome using the following four programs:

1. sim4 (Florea et al., 1998)
2. dds/gap2 (Huang et al., 1997)

3. est_genome (Mott, 1997)
4. GeneSeqer (Usuka et al., 2000)

The first three programs (sim4, dds/gap2, and est_genome) are general-purpose alignment algorithms designed to map EST or cDNA sequences to a genome. The GeneSeqer program is more specialized, in that it has particular subroutines designed to recognize splice sites in *Arabidopsis,* and it is able to find noncanonical AT-AC introns. The first three programs (sim4, dds/gap2, and est_genome) require only that an intron begin with the dinucleotide GT and end with AG. Note that none of these three programs (sim4, dds/gap2, and est_genome) is tuned to find noncanonical AT-AC introns (Tarn and Steitz, 1996) or any other introns with nonconsensus splice sites. Haas et al. (2002) evaluated how often the four programs produced identical alignments between the cDNA sequences and the genome. The analysis indicated that gap2 and sim4 performed the best when the entire cDNA was matched to the genome (Haas et al., 2002). The analysis also indicated that sim4, dds/gap2, and GeneSeqer were excellent in extending the alignment to cover the full length of the cDNA and that only est_genome consistently failed to extend the alignments (Haas et al., 2002). The analysis also showed that there were 4,918 cDNAs that yielded identical splice sites (and therefore identical introns) regardless of which programs were used to generate the alignment and that in 64 cases three programs agreed and only one disagreed, while in the remaining 32 cases there was no alignment shared by a majority of the programs. In summary, the comparison of the alignments indicated that the currently available programs to align the cDNA sequences to genome sequence work well, but leave little room for improvement.

Haas et al. (2002) used the alignments generated from the cDNA sequences to create new gene models for the corresponding genes in the *Arabidopsis* genome. They assumed that the protein-coding region is the longest open reading frame (ORF) on the forward strand, and required it to span at least 40 percent of the cDNA length. They updated the annotation of the genome as follows:

1. For the vast majority of genes, 5' and 3' untranslated regions (UTRs) had not been annotated previously, and these were added with the incorporation of the cDNA data.
2. Of the gene models, 2,978 contained identical protein-coding regions to what had already been annotated and required only UTR refinements, but 1,591 were adjusted, yielding more accurate protein sequences. Some of these contained very short "micro-exons" that are

typically missed by both gene-finding programs and alignment algorithms. Using new software protocols developed originally, Haas et al. (2002) found 47 micro-exons, ranging from 3 to 25 bp in length, distributed evenly across all five chromosomes.

3. There was the addition of 240 completely novel genes not previously included in the *Arabidopsis* genome annotation. Of the 240 novel genes, 92 had significant homology to known proteins, and the rest did not match any previously described proteins.

In summary, the comparison of full-length cDNAs from Ceres with previous genome annotation revealed that 62 percent of the matching gene models correctly predicted the exon-intron structure for the gene, 33 percent needed to be modified, and 5 percent represented previously undiscovered genes (Haas et al., 2002).

Analysis of cDNA sequences can help to estimate the frequency of alternative splicing. Alternative splicing appears to be relatively common in animals (Goldstrohm et al., 2001); in plants this phenomenon has been less frequently observed. Recently, some reports on the alternative splicing in plants have been published (Lazar and Goodman, 2000). Haas et al. (2002) examined the alignments of cDNAs to the genome, looking for cases where more than one cDNA aligned to overlapping locations on the same chromosome in such a way as to predict a different splicing pattern. Their working hypothesis was that if two cDNAs mapped to the same locus, but presented distinct sets of exons, this would constitute evidence of alternative splicing, or possibly another type of splicing anomaly. Haas et al. (2002) used all the complete cDNAs available from GenBank for the search. Finally, Haas et al. (2002) identified 158 genes with apparent splicing anomalies. They fell into many different classes, representing different genetic events as follows: 27 alignments indicated an alternative 3' acceptor site for an intron; 17 alignments indicated an alternative 5' donor site for an intron; and 23 alignments indicated that one or more introns remained unspliced. The unspliced transcripts may arise from nuclear rather than mature cytoplasmic mRNA sequences. Six alignments indicated that an internal exon was missing in one isoform; presumably the adjacent introns were spliced as a single intron containing the exon sequence. Whether the different spliced products are translated and whether the splicing differences reflect programmed developmental variation or simply splicing errors remains to be proved. However, incomplete splicing and consequential variants in plants have been noted previously to be associated with gene silencing and were postulated to reflect the regulated production of aberrant RNA products not destined to be translated (Metzlaff et al., 2000). The full-length cDNAs should also become a valuable resource for discovering more about alternative splicing and gene regulation.

APPLICATION OF FULL-LENGTH CDNAS TO FUNCTIONAL ANALYSIS

Microarrays are powerful tools for the systematic analysis of expression profiles of large numbers of genes, including stress-inducible gene expression, tissue-specific gene expression, and changes in the expression profiles of mutants and transgenic plants (Eisen and Brown, 1999; Richmond and Somerville, 2000). We have used the full-length cDNAs for the microarray analysis of expression profiles of *Arabidopsis* genes under various stress conditions, such as drought, cold and high-salinity stresses (Seki, Narusaka, Abe, et al., 2001; Seki, Narusaka, Ishida, et al. 2002), and high-light stress (Kimura et al., 2003); with various treatment conditions, such as ABA treatment (Seki, Ishida, et al., 2002) and proline treatment (Satoh et al., 2002); and in various mutants and transgenic plants (Seki, Narusaka, Abe, et al., 2001; Osakabe et al., 2002; Abe et al., 2003) (Figure 1.1). Our microarray studies showed that the full-length cDNA microarray is a useful material to analyze the expression pattern of *Arabidopsis* genes under various stress and hormone treatments, and to identify target genes of stress-related transcription factors and potential *cis*-acting DNA elements by combining the expression data with the genomic sequence data.

There are several advantages and disadvantages to using the RAFL cDNA microarrays for the systematic analysis of expression profiles of large numbers of genes. Using the full-length cDNA microarrays, it is easy to isolate full-length cDNAs for further functional analysis. Biochemical characteristics of the gene products are easily analyzed from the overexpression of the full-length cDNAs in bacteria, yeast, or wheat germ cell-free protein synthesis systems. Functions of the gene products *in planta* can be analyzed by the overexpression of full-length cDNAs in transgenic plants. Moreover, promoter sequences and putative *cis*-acting elements of each gene can be predicted by comparison of full-length cDNA sequences with the *Arabidopsis* genomic sequence. However, cross-hybridization between highly related sequences may occur in the full-length cDNA microarray (Richmond and Somerville, 2000). To avoid cross-hybridization problems, oligonucleotide microarrays, such as the Agilent Oligo Microarray (Agilent Technologies, Inc., Palo Alto, California) and Affymetrix GeneChip Array (Affymetrix, Inc., Santa Clara, California), are thought to be powerful tools for the future.

Full-length cDNAs are useful resources for determining the three-dimensional structures of proteins by X-ray crystallography and NMR spectroscopy (Seki, Narusaka, Yamaguchi-Shinozaki, et al., 2001) (Figure 1.1). We have started to determine the 3-D structures of plant proteins using

Arabidopsis full-length cDNAs by both methods in the RIKEN Structural Genomics Initiative (Yokoyama et al., 2000). In the RIKEN Structural Genomics Initiative, we are mainly using cell-free protein synthesis systems for protein expression. Cell-free in vitro systems have three advantages over conventional in vivo expression systems: (1) cell-free systems are suitable for automated, high-throughput expression, as proteins can be produced without the need for cloning genes into expression vectors; (2) milligram quantities of proteins can be obtained in several hours; and (3) proteins that are difficult for expression in vivo can be produced in vitro. Recently, Endo's group at Ehime University has modified the wheat germ cell-free protein synthesis system to produce milligram quantities of proteins (Madin et al., 2000; Sawasaki et al., 2002). We are applying this system for expression of plant proteins to study protein-protein and protein-DNA interactions (Figure 1.1).

Systematic approaches using transgenic plants, such as overexpression (Jaglo-Ottosen et al., 1998; Liu et al., 1998; Kasuga et al., 1999; Iuchi et al., 2001; Taji et al., 2002), antisense suppression (Nanjo et al., 1999), and double-stranded RNA interference (dsRNAi) (Chuang and Meyerowitz, 2000; Smith et al., 2000), are effective for functional analysis of plant proteins. Full-length cDNAs are also useful for these transgenic studies (Seki, Narusaka, Yamaguchi-Shinozaki, et al., 2001b) (Figure 1.1). The production of transgenic *Arabidopsis* plant lines that overexpress full-length cDNAs under the control of the CaMV 35S promoter or a chemical-inducible system (Zuo and Chua, 2000) is thought to be a useful alternative to activation tagging (Walden et al., 1994). This system may yield phenotypic changes in transgenics because of both overexpression and cosuppression.

CONCLUSIONS AND PERSPECTIVES

Although many algorithms have been developed to predict a transcription unit from genomic sequence data, the accuracy of their predictions is still limited. A more direct and efficient approach to identify coding sequences is to sequence full-length cDNAs (Figure 1.1). The results described demonstrate the power of full-length cDNA sequences for improving the quality of multiple aspects of genome annotation. We have been determining the full-length sequences of the RAFL cDNA clones in collaboration with the Arabidopsis SSP group of the United States. Complete sequences of the *Arabidopsis* full-length cDNAs will be used for gene annotation, such as UTRs, transcription start site, and alternative splicing. The RAFL cDNA clones are publicly available from the RIKEN Bioresource Center (http://www.brc.riken.go.jp/lab/epd/Eng/index.html).

Alignment of the full-length cDNA sequences to genome sequence can provide substantial evidence in favor of exon/intron assignment (Seki, Narusaka, Kamiya, et al., 2002; Haas et al., 2002; Xiao et al., 2002). After the complete *Arabidopsis* genome sequence was determined in 2000 (Arabidopsis Genome Initiative, 2000), several useful tools, such as Affymetrix Gene Chip (Lipshutz et al., 1999), were developed by use of the genome annotation database in *Arabidopsis*. However, as the original Affymetrix annotation table is outdated, it should be updated with the improved *Arabidopsis* genome annotation data (Ghassemian et al., 2001). Also, several neural network-based tools, such as ChloroP (Emanuelsson et al., 1999) and Target P (Emanuelsson et al., 2000), have been developed. The updated database of *Arabidopsis* genes/proteins should be systematically analyzed for their predicted targeting.

We have shown that the full-length cDNA microarray is a useful tool for analyzing the expression patterns of *Arabidopsis* genes under various stress and hormone treatment conditions, for identifying target genes of stress-related transcription factors, and for identifying potential *cis*-acting DNA elements by combining the expression data with the genomic sequence data. By accumulating data on gene expression by tissue types, developmental stages, stress and hormone treatments, and genetic backgrounds, it should be possible to identify the genes involved in many important processes of development and responses to environmental conditions in plants.

Full-length cDNAs will become a standard resource for the analysis of functions and structures of plant proteins, such as overexpression in transgenic plants, and for the analysis of protein expression for studying protein-protein interactions (see Chapter 6) and for determining protein structure by X-ray crystallography and NMR spectroscopy. Once all the *Arabidopsis* full-length cDNAs are functionally annotated, the database will be the Rosetta stone for understanding the network of gene functions in higher plants (Appels et al., 2002; Clarke et al., 2002). Our approaches using the *Arabidopsis* full-length cDNAs will also be applied to other model plants, such as rice, *Physcomitrella patens,* wheat, soybean, and poplar, in the near future.

REFERENCES

Abe, H., Urao, T., Ito, T., Seki, M., Shinozaki, K., Yamaguchi-Shinozaki, K. (2003). Arabidopsis AtMYC2 (bHLH) and AtMYB2 (MYB) function as transcriptional activators in abscisic acid signaling. *Plant Cell* 15: 63-78.

Appels, R., Francki, M., Chibbar, R. (2003). Advances in cereal functional genomics. *Funct. Integr. Genomics* 3: 1-24.

Arabidopsis Genome Initiative (2000). Analysis of the genome sequence of the flowering plant *Arabidopsis thaliana*. *Nature* 408: 796-815.

Asamizu, E., Nakamura, Y., Sato, S., Tabata, S. (2000). A large scale analysis of cDNA in *Arabidopsis thaliana:* Generation of 12,028 non-redundant expressed sequence tags from normalized and size-selected cDNA libraries. *DNA Res.* 7: 175-180.

Bevan, M. (1997). Objective: The complete sequence of a plant genome. *Plant Cell* 9: 476-478.

Burge, C., Karlin, S. (1997). Prediction of complete gene structures in human genomic DNA. *J. Mol. Biol.* 268: 78-94.

Burge, C.B., Karlin, S. (1998). Finding the genes in genomic DNA. *Curr. Opin. Struct. Biol.* 8: 346-354.

Carninci, P., Kvam, C., Kitamura, A., Ohsumi, T., Okazaki, Y., Itoh, M., Kamiya, M., Shibata, K., Sasaki, N., Izawa, M., et al. (1996). High-efficiency full-length cDNA cloning by biotinylated CAP trapper. *Genomics* 37: 327-336.

Carninci, P., Nishiyama, Y., Westover, A., Itoh, M., Nagaoka, S., Sasaki, N., Okazaki, Y., Muramatsu, M., Hayashizaki, Y. (1998). Thermostabilization and thermoactivation of thermolabile enzymes by trehalose and its application for the synthesis of full length cDNA. *Proc. Natl. Acad. Sci. USA* 95: 520-524.

Carninci, P., Shibata, Y., Hayatsu, N., Itoh, M., Shiraki, T., Hirozane, T., Watahiki, A., Shibata, K., Konno, H., Muramatsu, M., Hayashizaki, Y. (2001). Balanced-size and long-size cloning of full-length, cap-trapped cDNAs into vectors of the novel lambda-FLC family allows enhanced gene discovery rate and functional analysis. *Genomics* 77: 79-90.

Carninci, P., Shibata, Y., Hayatsu, N., Sugahara, Y., Shibata, K., Itoh, M., Konno, H., Okazaki, Y., Muramatsu, M., Hayashizaki, Y. (2000). Normalization and subtraction of cap-trapper-selected cDNAs to prepare full-length cDNA libraries for rapid discovery of new genes. *Genome Res.* 10: 1617-1630.

Carninci, P., Westover, A., Nishiyama, Y., Ohsumi, T., Itoh, M., Nagaoka, S., Sasaki, N., Okazaki, Y., Muramatsu, M., Schneider, C., Hayashizaki, Y. (1997). High efficiency selection of full-length cDNA by improved biotinylated cap trapper. *DNA Res.* 4: 61-66.

Cho, Y., Walbot, V. (2001). Computational methods for gene annotation: The Arabidopsis genome. *Curr. Opin. Biotechnol.* 12: 126-130.

Chuang, C.F., Meyerowitz, E. (2000). Specific and heritable genetic interference by double-stranded RNA in *Arabidopsis thaliana*. *Proc. Natl. Acad. Sci. USA* 97: 4985-4990.

Clarke, B., Lambrecht, M., Rhee, S.Y. (2003). *Arabidopsis* genomic information for interpreting wheat EST sequences. *Funct. Integr. Genomics* 3: 33-38.

Cooke, R., Raynal, M., Laudié, M., Grellet, F., Delseny, M., Morris, P.C., Guerrier, D., Giraudat, J., Quigley, F., Clabault, G., et al. (1996). Further progress towards a catalogue of all Arabidopsis genes: Analysis of a set of 5,000 non-redundant ESTs. *Plant J.* 9: 101-124.

Eisen, M.B., Brown, P.O. (1999). DNA arrays for analysis of gene expression. *Methods in Enzymology* 303: 179-205.

Emanuelsson, O., Nielsen, H., Brunak, S., Heijne, G.V. (2000). Predicting subcellular localization of proteins based on their N-terminal amino acid sequence. *J. Mol. Biol.* 300: 1005-1016.

Emanuelsson, O., Nielsen, H., Heijne, G.V. (1999). ChloroP, a neural network-based method for predicting chloroplast transit peptides and their cleavage sites. *Protein Science* 8: 978-984.

European Union Arabidopsis Genome Sequencing Consortium and Cold Spring Harbor, Washington University in St. Louis and PE Biosystems Arabidopsis Sequencing Consortium (1999). Sequence and analysis of chromosome 4 of the plant *Arabidopsis thaliana. Nature* 402: 769-777.

Florea, L., Hartzell, G., Zhang, Z., Rubin, G.M., Miller, W. (1998). A computer program for aligning a cDNA sequence with a genomic DNA sequence. *Genome Res.* 8: 967-974.

Garcia-Hernandez, M., Berardini, T.Z., Chen, G., Crist, D., Doyle, A., Huala, E., Knee, E., Lambrecht, M., Miller, N., Mueller, L.A., et al. (2002) TAIR: A resource for integrated *Arabidopsis* data. *Funct. Integr. Genomics* 2: 239-253.

Ghassemian, M., Waner, D., Tchieu, J., Gribskov, M., Schroeder, J.I. (2001). An integrated *Arabidopsis* annotation database for Affymetrix Genechip data analysis, and tools for regulatory motif searches. *Trends Plant Sci.* 6: 448-449.

Goldstrohm, A.C., Greenleaf, A.L., Garcia-Blanco, M.A. (2001). Co-transcriptional splicing of pre-messenger RNAs: Considerations for the mechanism of alternative splicing. *Gene* 277: 31-47.

Haas, B.J., Volfovsky, N., Town, C.D., Troukhan, M., Alexandrov, N., Feldmann, K.A., Flavell, R.B., White, O., Salzberg, S.L. (2002). Full-length messenger RNA sequences greatly improve genome annotation. *Genome Biol.* 3: 1-12.

Higo, K., Ugawa, Y., Iwamoto, M., Korenaga, T. (1999). Plant *cis*-acting regulatory DNA elements (PLACE) database: 1999. *Nucl. Acids Res.* 27: 297-300.

Höfte, H., Desprez, T., Amselem, J., Chiapello, H., Caboche, M., Moisan, A., Jourjon, M.F., Charpenteau, J.L., Berthomieu, P., Guerrier, D., et al. (1993). An inventory of 1,152 expressed sequence tags obtained by partial sequencing of cDNAs from *Arabidopsis thaliana. Plant J.* 4: 1051-1061.

Huang, X., Adams, M.D., Zhou, H., Kerlavage, A.R. (1997). A tool for analyzing and annotating genomic sequences. *Genomics* 46: 37-45.

Iuchi, S., Kobayashi, M., Taji, T., Naramoto, M., Seki, M., Kato, T., Tabata, S., Kakubari, Y., Yamaguchi-Shinozaki, K., Shinozaki, K. (2001). Regulation of drought tolerance by gene manipulation of 9-*cis*-epoxycarotenoid dioxygenase, a key enzyme in abscisic acid biosynthesis, in *Arabidopsis. Plant J.* 27: 325-333.

Jaglo-Ottosen, K.R., Gilmour, S.J., Zarka, D.G., Schabenberger, O., Thomashow, M.F. (1998). *Arabidopsis* CBF1 overexpression induces *cor* genes and enhances freezing tolerance. *Science* 280: 104-106.

Kasuga, M., Liu, Q., Miura, S., Yamaguchi-Shinozaki, K., Shinozaki, K. (1999). Improving plant drought, salt, and freezing tolerance by gene transfer of a single stress-inducible transcription factor. *Nature Biotechnol.* 17: 287-291.

Kimura, M., Yamamoto, Y.Y., Seki, M., Sakurai, T., Satou, M., Abe, T., Yoshida, S., Manabe, K., Shinozaki, K., Matsui, M. (2003). Identification of *Arabidopsis* genes regulated by high light stress using cDNA microarray. *Photochem. Photobiol.* 77: 226-233

Lazar, G., Goodman, H.M. (2000). The *Arabidopsis* splicing factor SR1 is regulated by alternative splicing. *Plant Mol. Biol.* 42: 571-581.

Lin, X., Kaul, S., Rounsley, S., Shea, T.P., Benito, M.I., Town, C.D., Fujii, C.Y., Mason, T., Bowman, C.L., Barnstead, M., et al. (1999). Sequence and analysis of chromosome 2 of *Arabidopsis thaliana. Nature* 402: 761-768.

Lipshutz, R.J., Fodor, S.P.A., Gingeras, T.R., Lockhart, D.J. (1999). High density synthetic oligonucleotide arrays. *Nat. Genet.* 21: 20-24.

Liu, Q., Kasuga, M., Sakuma, Y., Abe, H., Miura, S., Yamaguchi-Shinozaki, K., Shinozaki, K. (1998). The transcription factors, DREB1 and DREB2, with an EREBP/AP2 DNA binding domain separate two cellular signal transduction pathways in drought- and low-temperature-responsive gene expression, respectively, in *Arabidopsis. Plant Cell* 10: 1391-1406.

Lukashin, A.V., Borodovsky, M. (1998). GeneMark.hmm: New solutions for gene finding. *Nucleic Acids Res.* 26: 1107-1115.

MacIntosh, G.C., Wilkerson, C., Green, P.J. (2001). Identification and analysis of *Arabidopsis* expressed sequence tags characteristics of non-coding RNAs. *Plant Physiol.* 127: 765-776.

Madin, K., Sawasaki, T., Ogasawara, T., Endo, Y. (2000). A highly efficient and robust cell-free protein synthesis system prepared from wheat embryos: Plants apparently contain a suicide system directed at ribosomes. *Proc. Natl. Acad. Sci. USA* 97: 559-564.

Mayer, K., Schuller, C., Wambutt, R., Murphy, G., Volckaert, G., Pohl, T., Dusterhoft, A., Stiekema, W., Entian, K.D., Terryn, N., et al. (1999). Sequence and analysis of chromosone 4 of the plant *Arabidopsis thaliana. Nature* 402: 769-777.

Meinke, D.W., Cherry, J.M., Dean, C.D., Rounsley, S., Koornneef, M. (1998). *Arabidopsis thaliana:* A model plant for genome analysis. *Science* 282: 662-682.

Metzlaff, M., O'Dell, M., Hellens, R., Flavell, R.B. (2000). Developmentally and transgene regulated nuclear processing of primary transcripts of chalcone synthase A in petunia. *Plant J.* 23: 63-72.

Mott, R. (1997). EST_GENOME: A program to align spliced DNA sequences to unspliced genomic DNA. *Comp. Appl. Biosci.* 13: 477-478.

Nanjo, T., Kobayashi, M., Yoshiba, Y., Sanada, Y., Wada, K., Tsukaya, H., Kakubari, Y., Yamaguchi-Shinozaki, K., Shinozaki, K. (1999). Biological functions of proline in morphogenesis and osmotolerance revealed in antisense transgenic *Arabidopsis thaliana. Plant J.* 18: 185-193.

Newman, T., deBruijn, F.J., Green, P., Keegstra, K., Kende, H., McIntosh, L., Ohlrogge, J., Raikhel, N., Somerville, S., Thomashow, M., Retzel, E., Somerville, C. (1994). Genes galore: A summary of methods for accessing results from large-scale partial sequencing of anonymous *Arabidopsis* cDNA clones. *Plant Physiol* 106: 1241-1255.

Osakabe, Y., Miyata, S., Urao, T., Seki, M., Shinozaki, K., Yamaguchi-Shinozaki, K. (2002). Overexpression of *Arabidopsis* response regulators, ARR4/ATRR1/

IBC7 and ARR8/ATRR3, alters cytokinin responses differentially in the shoot and in callus formation. *Biochem. Biophys. Res. Commun.* 293: 806-815.

Palm, C.J., Federspiel, N.A., Davis, R.W. (2000). DatA: Database of *Arabidopsis thaliana* annotation. *Nucl. Acids Res.* 28: 102-103.

Pavy, N., Rombauts, S., Dehais, P., Mathe, C., Ramana, D.V., Leroy, P., Rouze, P. (1999). Evaluation of gene prediction software using a genomic data set: Application to *Arabidopsis thaliana* sequences. *Bioinformatics* 15: 887-899.

Pertea, M., Salzberg, S.L. (2002). Computational gene finding in plants. *Plant Mol. Biol.* 48: 39-48.

Richmond, T., Somerville, S. (2000). Chasing the dream: Plant EST microarrays. *Curr. Opin. Plant Biol.* 3: 108-116.

Riechmann, J.L., Heard, J., Martin, G., Reuber, L., Jiang, C., Keddie, J., Adam, L., Pineda, O., Ratcliffe, O.J., Samaha, R.R., et al. (2000). *Arabidopsis* transcription factors: Genome-wide comparative analysis among eukaryotes. *Science* 290: 2105-2110.

Salzberg, S.L., Pertea, M., Delcher, A.L., Gardner, M.J., Tettelin, H. (1999). Interpolated Markov models for eukaryotic gene finding. *Genomics* 59: 24-31.

Satoh, R., Nakashima, K., Seki, M., Shinozaki, K., Yamaguchi-Shinozaki, K. (2002). ACTCAT, a novel *cis*-acting element for proline- and hypoosmolarity-responsive expression of the *ProDH* gene encoding proline dehydrogenase in *Arabidopsis*. *Plant Physiol.* 130: 709-719.

Sawasaki, T., Ogasawara, T., Morishita, R., Endo, Y. (2002). A cell-free protein synthesis system for high-throughput proteomics. *Proc. Natl. Acad. Sci. USA* 99: 14652-14657.

Schoof, H., Zaccaria, P., Gundlach, H., Lemcke, K., Rudd, S., Kolesov, G., Arnold, R., Mewes, H.W., Mayer, K.F.X. (2002). MIPS *Arabidopsis thaliana* database (MAtDB): An integrated biological knowledge resource based on the first complete plant genome. *Nucl. Acids Res.* 30: 91-93.

Seki, M., Carninci, P., Nishiyama, Y., Hayashizaki, Y., Shinozaki, K. (1998). High-efficiency cloning of *Arabidopsis* full-length cDNA by biotinylated CAP trapper. *Plant J.* 15: 707-720.

Seki, M., Ishida, J., Narusaka, M., Fujita, M., Nanjo, T., Umezawa, T., Kamiya, A., Nakajima, M., Enju, A., Sakurai, T., et al. (2002). Monitoring the expression pattern of ca. 7000 *Arabidopsis* genes under ABA treatments using a full-length cDNA microarray. *Funct. Integr. Genomics* 2: 282-291.

Seki, M., Narusaka, M., Abe, H., Kasuga, M., Yamaguchi-Shinozaki, K., Carninci, P., Hayashizaki, Y., Shinozaki, K. (2001). Monitoring the expression pattern of 1300 *Arabidopsis* genes under drought and cold stresses using full-length cDNA microarray. *Plant Cell* 13: 61-72.

Seki, M., Narusaka, M., Ishida, J., Nanjo, T., Fujita, M., Oono, Y., Kamiya, A., Nakajima, M., Enju, A., Sakurai, T., et al. (2002). Monitoring the expression profiles of 7000 *Arabidopsis* genes under drought, cold, and high-salinity stresses using a full-length cDNA microarray. *Plant J.* 31: 279-292.

Seki, M., Narusaka, M., Kamiya, A., Ishida, J., Satou, M., Sakurai, T., Nakajima, M., Enju, A., Akiyama, K., Oono, Y., et al. (2002). Functional annotation of a full-length *Arabidopsis* cDNA collection. *Science* 296: 141-145.

Seki, M., Narusaka, M., Yamaguchi-Shinozaki, K., Carninci, P., Kawai, J., Hayashizaki, Y., Shinozaki, K. (2001). *Arabidopsis* encyclopedia using full-length cDNAs and its application. *Plant Physiol. Biochem.* 39: 211-220.

Seki, M., Satou, M., Sakurai, T., Shinozaki, K. (2002). RIKEN *Arabidopsis* full-length (RAFL) cDNA database. *Trends Plant Sci.* 7: 562-563.

Shibata, Y., Carninci, P., Watahiki, A., Shiraki, T., Konno, H., Muramatsu, M., Hayashizaki, Y. (2001). Cloning full-length, cap-trapper-selected cDNAs by using the single-strand linker ligation method. *Biotechniques* 30: 1250-1254.

Sleutels, F., Zwart, R., Barlow, D.P. (2002). The non-coding *Air* RNA is required for silencing autosomal imprinted genes. *Nature* 415: 810-813.

Smith, N.A., Singh, S.P., Wang, M.B., Stoutjesdijk, P.A., Green, A.G., Waterhouse, P.M. (2000). Total silencing by intron-spliced hairpin RNAs. *Nature* 407: 319-320.

Taji, T., Ohsumi, C., Iuchi, S., Seki, M., Kasuga, M., Kobayashi, M., Yamaguchi-Shinozaki, K., Shinozaki, K. (2002). Important roles of drought- and cold-inducible genes for galactinol synthase in stress tolerance in *Arabidopsis thaliana*. *Plant J.* 29: 417-426.

Tarn, W.Y., Steitz, J.A. (1996). A novel spliceosome containing U11, U12, and U5 snRNPs excises a minor class (AT-AC) intron in vitro. *Cell* 84: 801-811.

Usuka, J., Zhu, W., Brendel, V. (2000). Optimal spliced alignment of homologous cDNA to a genomic DNA template. *Bioinformatics* 16: 203-211.

Walden, R., Fritze, K., Hayashi, H., Miklashevichs, E., Harling, H., Schell, J. (1994). Activation tagging: A means of isolating genes implicated as playing a role in plant growth and development. *Plant Mol. Biol.* 26: 1521-1528.

White, J.A., Todd, J., Newman, T., Focks, N., Girke, T., Ilárduya, O.M.D., Jaworski, J.G., Ohlrogge, J.B., Benning, C. (2000). A new set of *Arabidopsis* expressed sequence tags from developing seeds: The metabolic pathway from carbohydrates to seed oil. *Plant Physiol.* 124: 1582-1594.

Xiao, Y.L., Malik, M., Whitelaw, C.A., Town, C.D. (2002). Cloning and sequencing of cDNAs for hypothetical genes from chromosome 2 of *Arabidopsis*. *Plant Physiol.* 130: 2118-2128.

Yokoyama, S., Hirota, H., Kigawa, T., Yabuki, T., Shirouzu, M., Terada, T., Ito, Y., Matsuo, Y., Kuroda, Y., Nishimura, Y., et al. (2000). Structural genomics projects in Japan. *Nature Struct. Biol.:* 943-945.

Zuo, J., Chua, N.H. (2000). Chemical-inducible systems for regulated expression of plant genes. *Curr. Opin. Biotechnol.* 11: 146-151.

Chapter 2

T-DNA Mutagenesis: From Tagging to Insertion Sequence Databases

Allen Sessions

CROWN GALLS, NEOPLASIA, AND AGROBACTERIUM

For centuries farmers have noticed the tumorous crown galls that grow at the base of certain fruit trees. The causative agent for this syndrome was identified in 1907 as *Agrobacterium tumefaciens* (Smith and Townsend, 1907). This soil bacterium infects a broad host range of dicotyledonous plants, causing neoplastic growth of plant tissues (reviewed recently by Binns, 2002; Tzfira and Citovsky, 2002; Zupan et al., 2000). Although *Agrobacterium* causes neoplastic growth of dicotyledonous plant tissue, this occurs only when a bacterial cell comes in contact with a wounded plant cell of a susceptible host. Studies in the 1930s and 1940s demonstrated that once the neoplasia is induced, the continuous presence of the bacterium is not necessary for sustained tumorous growth. This suggested that a tumor-inducing factor is transferred from *Agrobacterium* to the host plant (reviewed by Braun, 1982). In the early 1970s, a large extrachromosomal plasmid, termed the Ti (tumor-inducing) plasmid, was identified and shown to be essential for tumor induction by *Agrobacterium* (Zaenen et al., 1974; Van Larebeke et al., 1974). In 1977, Chilton et al. demonstrated that a portion of the Ti plasmid, termed the T-DNA (transfer-DNA), was transferred to the plant tissues that undergo neoplastic growth. Besides viruses, *Agrobacterium* is the only agent known to participate in the transkingdom

This chapter is dedicated to Ken Feldmann, a pioneer of T-DNA tagging and *Arabidopsis* reverse genetics, and to the researchers and studies that have elucidated the biology of *Agrobacterium* and harnessed it to good use (only part of this story has been mentioned in this chapter). It should be acknowledged that there are several other excellent "non T-DNA" resources for reverse genetics in *Arabidopsis,* and they are listed at <http://www.arabidopsis.org/links/insertion.html>, and, last, that T-DNA insertional mutagenesis is also an active area in rice research (Jeon et al., 2000).

transfer of genetic material, and it has also been shown to transfer genetic material to various fungal species (Bundock et al., 2002; Hanif et al., 2002) and human cells (Kunik et al., 2001), using a mechanism involving a type IV transporter that appears similar to bacterial conjugation (see references in Tzfira and Citovsky, 2002).

Subsequent studies in the past twenty-five years have further defined the specifics of the transfer process of the T-DNA from *Agrobacterium* into plant cells (reviewed recently by Binns, 2002; Tzfira and Citovsky, 2002; Zupan et al., 2000). Molecular and genetic studies identified genes on the T-DNA that encode enzymes which induce the overproduction of cell proliferation-causing phytohormones, resulting in neoplasia. Furthermore, there are genes on the T-DNA encoding biosynthetic enzymes that catalyze the production of specific amino acid derivatives, termed opines, that only *Agrobacterium* can metabolize. Other regions of the Ti plasmid outside of the T-DNA encode for enzymes that metabolize the specific opines synthesized by the T-DNA. The conclusion has been drawn that *Agrobacterium* transforms plant cells to proliferate and synthesize opines, to create a unique habitat full of a metabolite on which only it can grow. After the finding that *Agrobacterium* could transfer any DNA molecule positioned between the imperfect 25 bp repeats that are at the ends of the T-DNA (Wang et al., 1984), the road was paved to insert any sequence (i.e., a selectable marker) into a susceptible plant host genome.

AGROBACTERIA AS FRIEND NOT FOE

Following the discovery that agrobacteria transfer genetic material to plant genomes, efforts were initiated to harness this incredible tool. Besides the obvious utilities of transferring any DNA to a dicot plant cell, other efforts involved the creation of insertion mutants for the functional analysis of individual plant genes, primarily in *Arabidopsis*. In the late 1980s and early 1990s, large collections of *Agrobacterium*-transformed *Arabidopsis* were created and screened for mutant phenotypes in specific pathways and the T-DNA used to rescue the gene sequence. Forward genetics refers to the process of gene identification and functional characterization starting with a mutant phenotype and using it to identify the disrupted gene. More recently, with the availability of the complete *Arabidopsis* genome sequence, large collections of plants have been screened for disruptions in specific genes, starting with the gene sequence. The process of characterizing a gene's function by starting with its sequence and then identifying a loss- or gain-of-function allele that confers a mutant phenotype is referred to as reverse genetics (see also Chapters 3 and 10). This chapter focuses on the use of *Agrobacterium* and the T-DNA for reverse genetics studies aimed at the functional characterization of disrupted plant genes, primarily those of *Arabidopsis*. Since the inception of the idea to use T-DNA for insertional mutagenesis over twenty-five years ago (Feldmann and Marks, 1987), consistent improvement in *Agrobacterium* methodologies and reverse genetic technologies has occurred. Key advances with *Agrobacterium* have occurred with improved transformation technologies and the development of new selectable markers. Concurrent with these developments, *Drosophila* biologists have developed many techniques for performing reverse genetics on a large scale. These developments have been exploited by many researchers to generate large collections of T-DNA insertion mutants, some of which are listed in Table 2.1 and discussed in this chapter.

TABLE 2.1. T-DNA insertion collections with searchable flanking sequence tag databases.

Collection	Back-ground	T-DNA transformation vector	Border rescued	Border isolation	End sequences (5/04)	Reference
SIGNAL	Col-O	pROK-2	Left	Adapter ligation/ suppression PCR	141,486	<http://signal. salk.edu/cgibin/ tdnaexpress>
SAIL	Col-O	pDAP101, pCSA110	Left	Modified TAIL-PCR	~100,000	Sessions et al., 2002; <http://www. nadii.com/pages/ collaborations/ garlic_files/ GarlicDescription. html>
GABI-Kat	Col-O	pAC106 and pAC161 (activation tag)	Left and right	Adapter ligation PCR	~42,000	<http://www.mpiz-koeln.mpg.de/ GABI-Kat/>
Versailles Collection	Ws	pGKB5	Left	PCR-Walking (Balzergue et al., 2001)	~29,000	<http://flagdb-genoplante-info.infobiogen.fr/ projects/fst/ DocsIntro/Page_ accueil.html>; Samson et al., 2002
Biological Research Center T-DNA Insertion Library	Col-O	Various	Left and right	Modified TAIL-PCR (Szabados et al., 2002)	978	<http://www.szbk. u-szeged.hu/ ~arabidop/>; Szabados et al., 2000

DEVELOPMENT OF TRANSFORMATION PROTOCOLS FOR THE EFFICIENT GENERATION OF T-DNA INSERTION MUTANTS

The first demonstration of the transfer of recombinant DNA by *Agrobacterium* into plants and their progeny was reported in 1984 (De Block et al., 1984). The first general protocols for plant transformation (i.e., Horsch et al., 1985) by agrobacteria required cocultivation of plant extracts with cultures of *Agrobacterium* harboring a selectable marker-containing T-DNA. Any resultant resistant calli were regenerated into whole plants. This tissue culture step was slow and laborious and precluded the generation of large numbers of transformants. It would be another two years before the first reports of transformation of *Arabidopsis* (Lloyd et al., 1986) using the leaf disk transformation-regeneration method and hygromycin selection would appear. Using this method, primary transformants could be regenerated in three to five months. Following the development of kanamycin as a selectable marker, and the use of root explants as the transformed tissue, the time in tissue culture was reduced to two to three months (Valvekens et al., 1988). Aside from the laborious and time-consuming protocols with plant explant transformation, the cultivation of plant cells in tissue culture for these long periods frequently induced epigenetic changes in plant phenotypes that were not related to T-DNA insertions (e.g., see discussion in Feldmann and Marks, 1987).

In 1987, Feldmann and Marks developed a protocol for transforming *Arabidopsis* with *Agrobacterium* that obviated the tissue culture step. They reported that *Arabidopsis* seeds incubated with cultures of *Agrobacterium* containing a kanamycin resistance conferring T-DNA occasionally produced progeny that contained one to many T-DNAs and were resistant to kanamycin. It was initially hypothesized that *Agrobacterium* might diffuse into the seed and transform one of the few cells in the embryo that gives rise to the germ cell lineage. This would result in the majority of progeny transformants arising from a single transformed embryo being homozygous for all inserted T-DNAs. The surprising finding was that almost all of the primary transformants were hemizygous for inserted T-DNAs, implying that the transformations had to occur late in development in the plants produced from treated seeds, perhaps within gametophytes or the progeny zygotes (Feldmann and Marks, 1987). The seed transformation protocol developed by Feldmann and Marks was a key advance, allowing the creation of a large population of insertionally mutagenized plants for the functional analysis of plant genes. Aside from the rapidity at which transformants could be generated, this method had the advantage that plants did not need to be regenerated from callus and thus would lack phenotypic effects induced by tissue culture.

Populations created using this method (Feldmann, 1991) were used for forward genetic screens to identify genes involved in specific processes. Mutant screens were carried out for alterations in phenotype in a variety of growth conditions. The advantage of the T-DNA for inducing mutations and identifying genes, in contrast to other chemical and ionizing radiation mutagens at the time, was that the T-DNA allowed isolation of the disrupted gene by virtue of being engineered to contain plasmid sequences that facilitated the isolation of plant DNA–T-DNA junctions. A number of mutants identified in these screens led to isolation of the first genes of the modern plant molecular biology age controlling hormone responses, flower development, wax deposition, trichome development, phototropic responses, phytohormones response, plant size, and flowering time (see Feldmann, 1991). Investigators studying the genetic basis of any trait in *Arabidopsis* put screening for T-DNA mutants as their top priority. It largely changed the way plant biology was done. Limitations of the populations were that many genes were not identified by phenotypically mutant alleles, and that the population size was not saturating for mutations. Another finding was that many of the visible mutants were not associated with T-DNA insertions, and when they were, they frequently had complicated integration patterns that made plasmid rescue of flanking sequences difficult (Castle et al., 1993).

EVEN FASTER AND MORE EFFICIENT TRANSFORMATION

Bechtold and colleagues (1993) made another significant advance in transformation methodology with the development of the vacuum infiltration method of *in planta* transformation. Adult flowering plants that had been vacuum infiltrated with a culture of *Agrobacterium* containing a Basta resistance conferring T-DNA produced transformed progeny at a frequency of 1 to 4 percent. Aside from the increase in transformation efficiency, the use of Basta as a selection agent allowed transformants to be identified by simple foliar spraying of Basta on soil-grown plants, obviating the need for selection of seedlings on sterile agar plants with hygromycin or kanamycin. Bechtold and colleagues later identified the female gametophyte as the target of transformation (Bechtold et al., 2000), suggesting that the target cells of the seed treatment (Feldmann and Marks, 1987) transformation method are likely the female gametophytes, as originally postulated. Further improvements have been made to the *in planta* infiltration method (Clough and Bent, 1998). The increase in transformation efficiency and speed resulted in the generation of many large populations of *Arabidopsis* T-DNA insertion mutants (i.e., Krysan et al., 1999; Weigel et al., 2000; Galbiati et al., 2000; Sussman et al., 2000; Sessions et al., 2002).

To identify insertions within specific genes, *Arabidopsis* researchers followed the lead of *Drosophila* biologists (Ballinger and Benzer, 1989) and began screening pools of T-DNA mutagenized plants using PCR with T-DNA and gene-specific primers. This "reverse" genetics screening has the advantage that researchers can identify insertion mutants that are not found in simple forward genetic screens, such as those that do not have a visible phenotype, as well as lethals. Furthermore, with the sensitivity of PCR and the low frequency of insertion mutants in any specific sequence, mutant plants can be pooled in large numbers and screened together as a group to minimize the number of PCR reactions to perform. The first use of this "site-selected mutagenesis" using PCR (Kaiser and Goodwin, 1990) in plants was for the identification of mutations within the *Arabidopsis ACTIN* gene family (McKinney et al., 1995). In this case, McKinney and colleagues used T-DNA border primers and degenerate primers designed to a conserved *ACTIN* motif to screen 5,300 T-DNA insertion mutants as pools of 100 individuals. In this study, of the ten target *ACTIN* genes, insertions were identified in two of them within the 5,300 plants. This was followed by a screen for T-DNA insertions within 63 target genes in a population of 9,100 individuals screened as seven pools of ~1,300 lines (Krysan et al., 1996), and other screens in a variety of laboratories (reviewed by Bouche and Bouchez, 2001) using gene-specific, as well as degenerate, primers (Young et al., 2001).

Krysan and colleagues (1996) and others calculated the expected frequency of T-DNA insertion isolation for different size T-DNA populations and determined that a 95 percent chance of finding an insertion would occur with a population of 95,000 individuals. They established a population of ~60,000 transformants that can be screened using PCR for insertions in genes of interest (Krysan et al., 1999; Sussman et al., 2000). This collection has been developed into a public resource for PCR screening (Sussman et al., 2000) that is arrayed into 30 pools of ~2,000 individuals. Screening information can be found at <http://www.biotech.wisc.edu/Arabidopsis/>. The Arabidopsis Information Resource lists other collections of pools of genomic DNA available for screening (http://www.arabidopsis.org/).

FROM A POPULATION OF T-DNA INSERTION MUTANTS TO A FILE OF T-DNA FLANKING SEQUENCE TAGS

Arabidopsis researchers, now armed with the complete *Arabidopsis* genome sequence (Arabidopsis Genome Initiative, 2000), began using an idea originated by *Drosphila* biologists (Spradling et al., 1995) to isolate and sequence insertion borders as a way to prescreen mutant collections for all identifiable insertions. Using such a strategy, all identifiable T-DNA insertions can be mapped to a physical position in the genome. This reduces the screening step for identifying an insertion in a specific sequence or gene to a simple computer search. In plant systems, large numbers of insert flanking sequence tags were first generated for a population of 500 *Arabidopsis* plants containing newly mobilized *dissociation* elements (Parinov et al., 1999) and, later, for large collections of *Arabidopsis* T-DNA insertion mutants (<http://signal.salk.edu/cgibin/tdnaexpress>; Szabados et al., 2002; Samson et al., 2002; Sessions et al., 2002). At least five collections available now have insertion sequence tags from lines that were created by floral dip method and are searchable by gene sequence or gene name (Table 2.1). The four largest collections are summarized in the following sections.

The Salk Institute Genome Analysis Lab (SIGNAL) Collection

The SIGNAL collection is the largest population of T-DNA mutants and corresponding insertion flanking sequence tags (http://signal.salk.edu/cgi-bin/tdnaexpress). The population will eventually consist of ~150,000 Columbia (Col-O) individuals transformed with a kanamycin resistance encoding T-DNA (http://signal.salk.edu/cgi-bin/tdnaexpress). Insertion flanking sequences are generated from the T-DNA left border using an adapter ligation/suppression PCR method (Siebert et al., 1995) at a rate of ~6,000/month (http://signal.salk.edu/cgi-bin/tdnaexpress). Sequences are compared to the *Arabidopsis* genome using BLAST- and the top high-scoring similarity pair (HSP) for each line is identified (Alonso, 2002). BLAST hit and sequence information is made available for insertion mutant screens on a Web site (http://signal.salk.edu/cgi-bin/tdnaexpress). T3 seeds for mutant lines are distributed by the Arabidopsis Biological Resource Center (http://signal.salk.edu/cgi-bin/tdnaexpress). The SIGNAL group has plans to perform PCR-based screens on pools of these plants for insertions in specific genes not identified by the insertion flanking sequence tag data (Alonso, 2002).

The Syngenta Arabidopsis *Insertion Library (SAIL) Collection*

The SAIL collection was created as collaboration between industry and academic scientists and consists of ~100,000 Col-O individuals transformed with Basta resistance conferring T-DNAs (Sessions et al., 2002). Insertion flanking sequences have been rescued from the T-DNA left border using a modification of TAIL-PCR (Liu et al., 1995; Sessions et al., 2002). In the original TAIL-PCR protocol, three nested T-DNA border primers are each used with each of six arbitrary degenerate (AD) primers in three successive reactions where the PCR cycling is alternated between high and low annealing temperatures. Modifications used with SAIL included pooling four of the six AD primers and using only two nested reactions, thereby shortening the number of reactions per border from 18 to 2. The modified TAIL-PCR protocol typically generates 3.1 products that range in size from 100 to 1,500 bp. Similar to the SIGNAL collection, the SAIL reactions were sequenced without gel isolation of individual products. These sequences are frequently composed of spans of nucleotides that arise from distinct fragments in the reaction that are called into a single sequence of multiple tags. Mapping the tags back to the genome to definitive chromosomal sites involves a number of steps (Presting, 2003). The *Arabidopsis* genome is first split into overlapping 2 kbp fragments, compared to the mTAIL tags using BLAST, and the results filtered by sequence quality, BLAST score, and nucleotide diversity. Of the remaining associations, the fragment of the *Arabidopsis* genome that is identified in the top HSP for each span of a SAIL sequence is then associated with that SAIL line from which it derives in a table (http://www.nadii.com/pages/collaborations/garlic_files/PlantCell.html) that lists all predicted insertions. Using the coordinates of the alignment, a predicted insertion site can be calculated and primers can be designed that flank each predicted insertion site to be used for confirming predicted insertions by PCR. This table can be searched for insertions in genes of interest predicted to be present in the SAIL collection, or users can compare specific sequences to the collection using BLAST. Progeny seeds of primary transformants are provided to requestors who sign a material transfer agreement (see http://www.nadii.com/pages/collaborations/garlic_files/GarlicDescription.html). As of the end of 2003, ~9,800 lines have been distributed to labs worldwide. One limitation of the methods used to process the collection is that only 75 percent of the insertions predicted by BLAST can be confirmed by PCR. Analysis of the SAIL T-DNA insertion flanking sequences suggests that the collection contains at least nearly 85,000 insertions distributed throughout the genome.

The GenomAnalyse im Biologischen System Pflanze Kölner Arabidopsis T-DNA (GABI-Kat) Collection

The GABI-Kat collection consists of ~42,000 T-DNA insertion flanking sequences from 60,000 Col-O individuals transformed with a sulfadiazine resistance encoding T-DNA with an activation tag at the T-DNA right - border (Rosso et al., 2003). Sequences are generated from T-DNA left and right borders using an adapter-ligation PCR method and compared to the *Arabidopsis* genome using BLAST. Researchers can search for insertions on a Web site using gene names or sequences as a query. T2 seeds from lines containing predicted insertions can be ordered (http://www.mpiz-koeln.mpg.de/GABI-Kat/db/seed_request.php) in exchange for a signature on an MTA, and for a fee the user can confirm the presence of a T-DNA in the line requested. Similar to the SAIL collection, only ~75 percent of the T-DNA insertions predicted can be confirmed using PCR.

The Versailles Collection

This collection is composed of ~50,000 Wassilevskiya (Ws) individuals transformed with a kanamycin and Basta resistance encoding T-DNA with promoterless GUS at the T-DNA right border (http://flagdb-genoplante-info.infobiogen.fr/projects/fst/). Sequences are generated from T-DNA left and right borders using PCR-Walking (Balzergue et al., 2001) and compared to the *Arabidopsis* genome using BLAST (Samson et al., 2002). Researchers can search for insertions on a Web site using gene sequences as a query (http://flagdb-genoplante-info.infobiogen.fr/cgi-bin/projects/fst/blast/blastAccueil.pl). Seeds from lines containing predicted insertions can be ordered from the Nottingham Stock Center.

AMPLIFICATION OF T-DNA BORDERS AND THE DISTRIBUTION OF PREDICTED INSERTIONS

Many methods have been developed to amplify T-DNA flanking sequences (Liu et al., 1995; Balzergue et al., 2001; Spertini et al., 1999; Choi et al., 1999; Szabados et al., 2002; Sessions et al., 2002; Siebert et al., 1995). Several protocols (Balzergue et al., 2001; Sessions et al., 2002; <http://signal.salk.edu/cgibin/tdnaexpress>) and studies (Castle et al., 1993; Krysan et al., 1996) have had reasonable success at isolating or amplifying T-DNA left borders. However, T-DNA right borders are often positioned to adjacent T-DNAs, either in direct (right border–left border) or inverted (right border–right border) repeat structures, leading to a large number of

T-DNA-only, right border flanking sequences (discussed in Sessions et al., 2002). This has been found with TAIL-PCR strategies, as well as with adapter-ligation-PCR strategies.

The distribution of predicted T-DNA insertions in four large collections has been reported (Alonso, 2002; Krysan et al., 2002; Szabados et al., 2002; Sessions et al., 2002). Insertion hot spots have not been identified in any of the collections. Although the definitions of genic and intergenic vary in these reports, they are all in agreement with the finding that approximately 30-35 percent of the predicted insertions are within transcribed regions and that there appears to be a slight bias against identifying insertions within genes. This slight bias however does not detract from the utility of the T-DNA as an insertional mutagen (Krysan et al., 2002).

REFERENCES

Alonso, J. (2002). A sequence-indexed library of insertion mutations in the *Arabidopsis* genome. Presentation at the Twelfth International Conference on Arabidopsis Research.

Arabidopsis Genome Initiative (2000). Analysis of the genome sequence of the flowering plant *Arabidopsis thaliana*. *Nature* 408: 796-815.

Ballinger, G.B., Benzer, S. (1989). Targeted gene mutations in *Drosophila*. *Proc. Natl. Acad. Sci. USA* 86: 9402-9406.

Balzergue, S., Dubreucq, B., Chauvin, S., Le-Clainche, I., Le Boulaire, F., de Rose, R., Samson, F., Biaudet, V., Lecharny, A., Cruaud, C., et al. (2001). Improved PCR-walking for large scale isolation of plant T-DNA borders. *Biotechniques* 30: 496-504.

Bechtold, N., Ellis, J., Pelletier, G. (1993). *In planta Agrobacterium* mediated gene transfer by infiltration of adult *Arabidopsis thaliana* plants. *C. R. Acad. Sci. Paris, Life Sciences* 316: 1194-1199.

Bechtold, N., Jaudeau, B., Jolivet, S., Maba, B., Vezon, D., Voisin, R., Pelletier, G. (2000). The maternal chromosome set is the target of the T-DNA in the *in planta* transformation of *Arabidopsis thaliana*. *Genetics* 155: 1875-1887.

Binns, A.N. (2002). T-DNA of *Agrobacterium tumefaciens*: 25 years and counting. *Trends Plant Sci.* 7: 231-233.

Bouche, N., Bouchez, D. (2001). *Arabidopsis* gene knockout: Phenotypes wanted. *Curr. Opin. Plant Biol.* 4: 111-117.

Braun, A.C. (1982). A history of the crown gall problem. In Kahl, G., Schell, J. (Eds.), *Molecular Biology of Plant Tumors* (pp. 155-210). New York: Academic Press.

Bundock, P., van Attikum, H., den Dulk-Ras, A., Hooykaas, P.J. (2002). Insertional mutagenesis in yeasts using T-DNA from *Agrobacterium tumefaciens*. *Yeast* 19: 529-536.

Castle, L.A., Errampalli, D., Atherton, T.L., Franzmann, L.H., Yoon, E.S., Meinke, D.W. (1993). Genetic and molecular characterization of embryonic mutants identified following seed transformation in *Arabidopsis*. *Mol. Gen. Genet.* 241: 504-514.

Chilton, M.-D., Drummond, M.H., Merlo, D.J., Sciaky, D., Montoya, A.L., Gordon, M.P., Nester, E. (1977). Stable incorporation of plasmid DNA into higher plant cells: The molecular basis of crown gall tumorogenesis. *Cell* 11: 263-271.

Choi, Y.I., Noh, E.W., Han, M.S., Son, S.H. (1999). Adaptor-aided PCR to identify T-DNA junctions in transgenic plants. *Biotechniques* 27: 222-226.

Clough, S.J., Bent, A. (1988). Floral dip: A simplified method for *Agrobacterium*-mediated transformation of *Arabidopsis thaliana*. *Plant J.* 16: 735-743.

De Block, M., Herrera-Estrella, L., van Montagu, M., Schell, J., Zambryski, P. (1984). Expression of foreign genes in regenerated plants and in their progeny. *EMBO J.* 3: 1681-1689.

Feldmann, K.A. (1991). T-DNA insertion mutagenesis in *Arabidopsis:* Mutational spectrum. *Plant J.* 1: 71-82.

Feldmann, K.A., Marks, M.D. (1987). *Agrobacterium*-mediated transformation of germinating seeds of *Arabidopsis thaliana:* A non-tissue culture approach. *Mol. Gen. Genet.* 208: 1-9.

Galbiati, M., Moreno, M., Nadzan, G., Zourelidou, M., Dellaporta, S.L. (2000). Large-scale T-DNA mutagenesis in *Arabidopsis* for functional genomic analysis. *Funct. Integr. Genomics* 1: 25-34.

Hanif, M., Pardo, A.G., Gorfer, M., Raudaskoski, M. (2002). T-DNA transfer and integration in the ectomycorrhizal fungus Suillus bovines using hygromycin B as a selectable marker. *Curr. Genet.* 41: 183-188.

Horsch, R.B., Fry, J.E., Hoffmann, N.L., Eichholtz, D., Rogers, S.G., Fraley, R.T. (1985). A simple and general method for transferring genes into plants. *Science* 227: 1229-1231.

Jeon, J.S., Lee, S., Jung, K.H., Jun, S.H., Jeong, D.H., Lee, J., Kim, C., Jang, S., Yang, K., Nam, J., et al. (2000). T-DNA insertional mutagenesis for functional genomics in rice. *Plant J.* 22: 561-570.

Kaiser, K., Goodwin, S.F. (1990). "Site-selected" transposon mutagenesis of *Drosophila*. *Proc. Natl. Acad. Sci. USA* 87: 1686-1690.

Krysan, P.J., Young, J.C., Jester, P.J., Monson, S., Copenhaver, G., Preuss, D., Sussman, M.R. (2002). Characterization of T-DNA insertion sites in *Arabidopsis thaliana* and the implications for saturation mutagenesis. *OMICS* 6: 163-174.

Krysan, P.J., Young, J.C., Sussman, M.R. (1999). T-DNA as an insertional mutagen in *Arabidopsis*. *Plant Cell* 11: 2283-2290.

Krysan, P.J., Young, J.C., Tax, F., Sussman, M.R. (1996). Identification of transferred DNA insertions within *Arabidopsis* genes involved in signal transduction and ion transport. *Proc. Natl. Acad. Sci. USA* 93: 8145-8150.

Kunik, T., Tzfira, T., Kapulnik, Y., Gafni, Y., Dingwall C., Citovsky, V. (2001). Genetic transformation of HeLa cells by *Agrobacterium*. *Proc. Natl. Acad. Sci. USA* 98: 1871-1876.

Liu, Y., Mitsukawa, N., Oosumi, T., Whittier, R.F. (1995). Efficient isolation and mapping of *Arabidopsis thaliana* T-DNA insert junctions by thermal asymmetric interlaced PCR. *Plant J.* 8: 457-463.

Lloyd, A.M., Barnason, A.R., Rogers, S.G, Byrne, M.C., Fraley, R.T., Horsch, R.B. (1986). Transformation of *Arabidopsis thaliana* with *Agrobacterium tumefaciens*. *Science* 234: 464-466.

McKinney, E.C., Ali, N., Traut, A., Feldmann, K.A., Belostsky, D.A., McDowell, J.A., Meagher, R.B. (1995). Sequence-based identification of T-DNA insertion mutations in *Arabidopsis:* Actin mutants *act2*-1 and *act4-1*. *Plant J.* 8: 613-622.

Parinov, S., Sevugan, M., De, Y., Yang, W.C., Kumaran, M., Sundaresan, V. (1999). Analysis of flanking sequences from dissociation insertion lines: A database for reverse genetics in *Arabidopsis*. *Plant Cell* 11: 2263-2270.

Presting, G. G. (2003). Mapping multiple co-sequenced T-DNA insertion sites within the *Arabidopsis* genome. *Bioinformatics* 19: 579-586.

Rosso, M. G., Li, Y., Strizhov, N., Reiss, B., Dekker, K., Weisshaar, B. (2003). An *Arabidopsis thaliana* T-DNA mutagenized population (GABI-Kat). *Plant. Mol. Biol.* 53: 247-259.

Samson, F., Brunaud, V., Balzergue, D., Dubreucq, B., Lepiniec, L., Pelletier, G., Caboche, M., Lecharny, A. (2002). FLAGdb/FST: A database of mapped flanking insertion sites (FSTs) of *Arabidopsis thaliana* T-DNA transformants. *Nucleic Acids Res.* 30: 94-97.

Sessions, A., Burke, E., Presting, G., Aux, G., McElver, J., Patton, D., Dietrich, B., Ho, P., Bacwaden, J., Ko, C., et al (2002). A high-throughput *Arabidopsis* reverse genetics system. *Plant Cell* 14: 2985-2994.

Siebert, P.D., Chenchik, A., Kellogg, D.E., Kukyanova, K.A., Lukyano, S.A. (1995). An improved PCR method for walking in uncloned genomic DNA. *Nucleic Acids Res.* 23: 1087-1088.

Smith, E.F., Townsend, C.O. (1907). A plant tumor of bacterial origin. *Science* 25: 671-673.

Spertini, D., Beliveau, C., Bellemare, G. (1999). Screening of transgenic plants by amplification of unknown genomic DNA flanking T-DNA. *Biotechniques* 27: 308-314.

Spradling, A.C., Stern, D., Kiss, I., Roote, J., Laverty, T., Rubin, G.M. (1995). Gene disruptions using P transposable elements. *Proc. Natl. Acad. Sci. USA* 92: 10824-10830.

Sussman, M.R., Amasino, R.M., Young, J.C., Krysan, P.J., Austin-Phillip, S. (2000). The *Arabidopsis* knockout facility at the University of Wisconsin-Madison. *Plant Physiol.* 124: 1465-1467.

Szabados, L., Kovács, I., Ábrahám, E., Oberschall, A., Nagy, R., Zsigmond, L., Krasovskaja, I., Kerekes, I., Ben-Haim, A., Koncz, C. (2000). *Arabidopsis* genome project in Hungary: Generating T-DNA tagged *Arabidopsis* genes. *Plant Physiol. Biochem.* 38: S98.

Szabados, L., Kovács, I., Oberschall, A., Ábrahám, E., Kerekes, I., Zsigmond, L., Nagy, R., Alvarado, M., Krasovskaja, I., Gál, M., et al. (2002). Distribution of 1000 sequenced T-DNA tags in the *Arabidopsis* genome. *Plant J.* 32: 233-242.

Tzfira, T., Citovsky, V. (2002). Partners-in-infection: Host proteins involved in the transformation of plant cells by *Agrobacterium*. *Trends Cell Biol.* 12: 121-129.

Valvekens, D., van Montagu, M.V, vanLijsebettens, M.V. (1988). *Agrobacterium tumefaciens*-mediated transformation of *Arabidopsis thaliana* root explants by using kanamycin selection. *Proc. Natl. Acad. Sci. USA* 85: 5536-5540.

Van Larebeke, N., Enbler, G., Holsters, M., Van Den Elsacker, S., Zaenen, I., Schilperoort, R.A., Schell, J. (1974). Large plasmid in *Agrobacterium tumefaciens* essential for crown gall-inducing ability. *Nature* 252: 169-170.

Wang, K., Herrera-Estrella, L., van Montagu, M., Zambryski, P. (1984). Right 25-bp terminus sequences of the nopaline T-DNA is essential for and determines the direction of DNA transfer from *Agrobacterium* to the plant genome. *Cell* 38: 455-462.

Weigel, D., Ahn, J.H., Blazquez, M.A., Borevitz, J.O., Christensen, S.K., Fankhauser, C., Ferrandiz, C., Kardailsky, I., Malancharuvil, E.J., Neff, M.M., et al. (2000). Activation tagging in *Arabidopsis*. *Plant Physiol.* 122: 1003-1013.

Young, J.C., Krysan, P.J., Sussman, M.R. (2001). Efficient screening of *Arabidopsis* T-DNA insertion lines using degenerate primers. *Plant Physiol.* 125: 513-518.

Zaenen, I., van Larebeke, N., Teuch, H., van Montagu, M., Schell, J. (1974). Supercoiled circular DNA in crown-gall inducing *Agrobacterium* strains. *J. Mol. Biol.* 86: 109-127.

Zupan, J., Muth, T.R., Draper, O., Zambryski P. (2000). The transfer of DNA from *Agrobacterium tumefaciens* into plants: A feast of fundamental insights. *Plant J.* 23: 11-28.

Chapter 3

Novel Reverse Genetics Tools in Plant Functional Genomics

Xin Li
Yuelin Zhang

INTRODUCTION

With the complete sequence of the *Arabidopsis* genome known (Arabidopsis Genome Initiative, 2000), understanding the functions of all *Arabidopsis* genes has become the focus of plant biology (Somerville and Dangl, 2000) (see Chapter 10). Reverse genetics will play an essential role in systematically analyzing *Arabidopsis* genes with unknown functions. Meanwhile, sequence information is becoming increasingly available for other plant species. For instance, the complete sequence of rice chromosomes 1 and 4 plus two draft sequences of the rice genome have been reported and are available to the public (Sasaki et al., 2002; Feng et al., 2002; Goff et al., 2002; Yu et al., 2002). There are also vast collections of expressed sequence tags (ESTs) for most economically important plant species in the public database. The availability of the sequence information makes it possible to carry out reverse genetic analysis, not only on biological processes active in *Arabidopsis,* but also on unique phenomena in other plant species, such as nitrogen fixation in legumes, inflorescence architecture of grasses, and fresh fruit development in tomatoes.

The tools for reverse genetics include a set of methods for inactivating gene functions. In yeast and mice, in vivo gene replacement through homologous recombination has been widely used for probing gene functions. Mutants carrying a variety of mutations resulting in either partial or complete loss of functions can be generated by gene replacement. Although success-

---

We thank Dr. Steven Henikoff for permission to use the TILLING gel picture from his previous publication. We thank Sandra Goritschnig and Kristoffer Palma for critical reading and helpful comments on the chapter.

ful homologous gene replacements have been reported in *Arabidopsis* and rice (Kempin et al., 1997; Terada et al., 2002), this method has not been used for routine analysis of gene functions in higher plants, primarily because the efficiency is very low and a large number of transgenic plants need to be generated for each gene replacement event. On the contrary, in the moss *Physcomitrella,* systematic gene replacement by homologous recombination is now possible (Schaefer and Zryd, 1997) (see also Chapter 9).

To overcome the difficulty of gene replacement in plants, a number of alternative methods have been developed to knock out plant genes. In the past, two approaches have been commonly used. One approach is based on insertional mutagenesis. Insertions in a target gene can be obtained by screening large T-DNA or transposon insertion libraries (see Chapter 2) using PCR or by searching databases of sequence-indexed insertion mutants (McKinney et al., 1995; Krysan et al., 1999; Bensen et al., 1995; Parinov et al., 1999). The other approach is gene silencing. Both sense and antisense suppression have been used to probe functions of plant genes (Baulcombe, 1996; Wang and Waterhouse, 2002). More recently, gene-silencing methods based on double-stranded (ds) RNAs are becoming increasingly popular because of their improved effectiveness in inactivating endogenous genes (Chuang and Meyerowitz, 2000; Smith et al., 2000).

During the past few years, Deleteagene and TILLING (**T**argeting **I**nduced **L**ocal **L**esions **IN** **G**enomes) have been developed for obtaining mutations in target genes (Li et al., 2001; McCallum et al., 2000; Colbert et al., 2001). In Deleteagene, deletion mutants for target genes are generated by fast neutron mutagenesis and identified by PCR screening. In TILLING, plants carrying point mutations are obtained by screening mutant populations induced by chemical mutagens such as ethyl methanesulfonate (EMS). Both Deleteagene and TILLING have been shown to be very effective reverse genetics tools. In this chapter, we will discuss the principles behind Deleteagene and TILLING and their applications in plant functional genomics. In addition, we will discuss gene silencing using chimeric gene constructs (termed combinatorial gene silencing) and its potential application in analyzing genes encoding redundant functions.

DELETEAGENE

Strategies for Obtaining Deletion Knockout Mutants

In the basic Deleteagene method (Figure 3.1A), wild-type seeds are mutagenized by fast neutron and the resulting M1 seeds are planted. M2 seeds are harvested from individual M1 plants and inventoried. A small

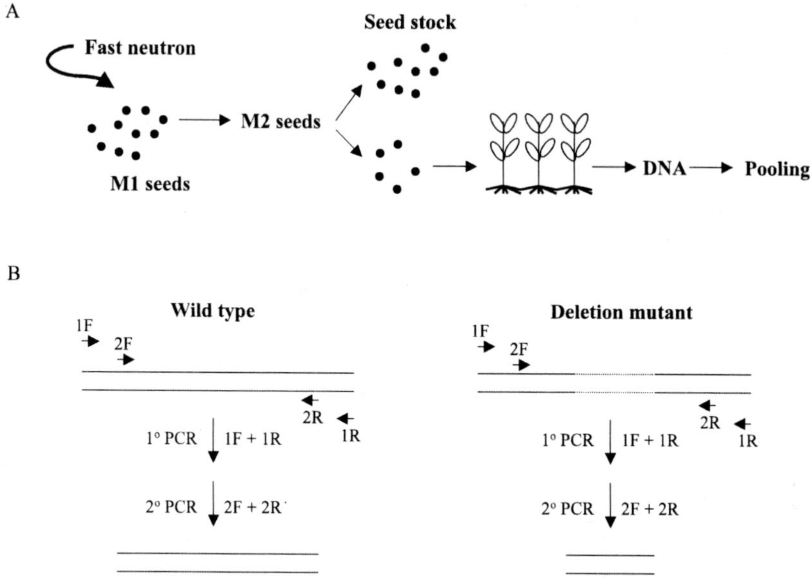

FIGURE 3.1. Scheme for Deleteagene. (A) Flow chart for deletion mutant library construction. (B) Identification of deletion mutants by nested PCR.

number of M2 seeds from each line are replanted to collect tissue for isolating genomic DNA. Once the DNAs are isolated, they are organized into different levels of pools with increasing complexity. To find deletion mutants for a target gene, PCR is first performed on DNA samples of megapools. The PCR product is analyzed by agarose gel electrophoresis to identify DNA fragments amplified from the deletion alleles, which are smaller than the wild-type DNA fragments. Once a deletion mutation is found in a megapool, PCR analysis is subsequently performed on the constituent superpools, pools, and individual lines to identify the single line carrying the deletion.

To identify a deletion mutant, a pair of primers specific to the sequence flanking the target gene and a pair of nested primers are used in PCR screening (Figure 3.1B). Using nested PCR not only increases the detection sensitivity and specificity but also improves the chance of finding small deletions. Under normal PCR conditions, both the wild-type and the mutant DNA will be amplified. To preferentially amplify DNA carrying deletion mutations, the PCR extension time is shortened to suppress amplification of wild-type fragments. Similar strategies were also used to screen for deletion mutants in *Caenorhabditis elegans* (Jansen et al., 1997; Liu et al., 1999).

After the first round of PCR, nested PCR is performed on the 1:50 diluted products of the first-round PCR. The product of the nested PCR is subsequently analyzed by agarose gel electrophoresis. An example of finding a deletion mutant in *AtMyb19* is shown in Figure 3.2. Because deletion mutants can be detected in very large pools, high-throughput screening can be carried out to look for deletion mutants in very large populations consisting of more than 100,000 lines.

Applications of Deletion Knockout Mutants

In *Arabidopsis,* flanking sequences of a large number of T-DNA insertions have been determined (Sessions et al., 2002; Brunaud et al., 2002; http://signal.salk.edu). For the majority of the *Arabidopsis* genes, knockout

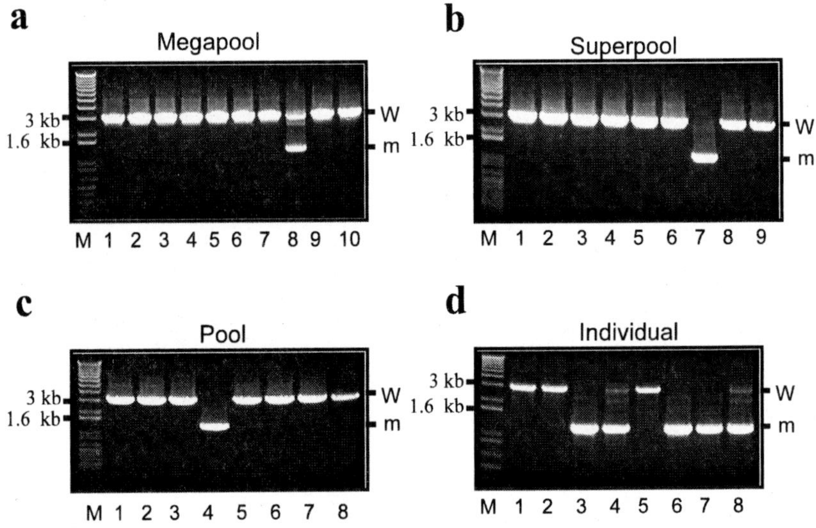

FIGURE 3.2. Identification of a deletion mutant for *AtMyb19.* (A) PCR on DNA from the ten megapools of 2,592 lines is shown (1-10). (B) PCR was performed on DNA from the nine superpools (1-9) of 288 lines that make up megapool number 8. (C) PCR was performed on DNA from the eight pools (1-8) of 36 lines that make up superpool number 7. (D) Eight single plants (1-8) from the line containing the deletion mutant were analyzed by PCR. All PCR was performed using Qiagen HotStarTaq polymerase. Two rounds of PCR were performed on the DNA pools and the nested PCR products were analyzed by agarose gel electrophoresis and stained with ethidium bromide. W, wild type; m, mutant. (*Source:* Li et al., 2001. Reprinted with permission of Blackwell Publishing Ltd.)

mutants can now be obtained simply by searching sequence databases of insertion mutants. However, insertion mutants for a significant percentage of genes are still hard to obtain for various reasons. For example, insertion mutants for small genes are generally difficult to find because the probability of finding an insertion mutant is directly proportional to the size of the target gene (Krysan et al., 1999). In contrast, it is relatively easy to identify deletion mutants for small genes since deletion mutants identified by PCR screening often affect larger regions. Thus, Deleteagene is very useful for reverse genetic analysis of small genes.

Deleteagene is also very useful for obtaining null alleles because almost all deletions are null mutations. Moreover, large deletions can be used to knock out tandem arrays that represent about 8 percent of the *Arabidopsis* genes (Arabidopsis Genome Initiative, 2000). A significant number of genes in tandem arrays probably encode redundant functions. For example, *AHBP-1b (At5g06950)* and *OBF5 (At5g06960)* are two tandemly duplicated genes that encode redundant functions in systemic acquired resistance (Li and Zhang, unpublished data). The *Cf-2* disease resistance locus in tomato also contains two tandemly duplicated genes with redundant functions (Dixon et al., 1996). Simultaneously deleting the highly homologous genes in tandem arrays can greatly simplify the analysis of their functions.

While Deleteagene was initially developed in *Arabidopsis,* it is highly applicable to crop plants. Large numbers of fast neutron-mutagenized seeds can be planted in the field and it will take only one generation to establish a mutant population. The procedure used for screening deletion mutants in *Arabidopsis* can be applied to other plant species directly. For example, we have successfully used similar procedures to screen for deletion mutants in rice. Deleteagene is also useful for plants with large genomes because the number of plants required to give the same mutation coverage are comparable in plants with different genome sizes (Koornneef et al., 1982; Li and Zhang, 2002). One other advantage of Deleteagene is that deletion mutants do not contain foreign DNA. Thus, deletion knockout plants can be directly planted in the field for phenotypic analysis without the need for regulatory approval.

Future Directions of Deletion Mutation Identification

In *Arabidopsis,* deletion mutations identified from various forward genetic screens range from 1 bp to greater than 30 kbp. One of the challenges for Deleteagene is to identify small deletions generated by fast neutron mutagenesis. As the size difference between the wild-type and the mutant

DNA fragment decreases, the difficulty of preferentially amplifying deletion mutant DNA in the presence of a large amount of excessive wild-type DNA increases. In a previous study where 36 deletion mutants were identified by PCR, all deletions were bigger than 0.5 kbp (Li et al., 2001). Deletions smaller than 0.5 kbp have most likely been missed in the PCR screening. While large deletions are easier to find, most of them are not as desirable for genetic analysis because they often affect multiple genes. The current strategy is to screen for medium-sized deletions that are readily detectable by PCR and likely affect only the target gene. To ensure a high probability of identifying such mutants for a target gene, large libraries containing a large number of mutant alleles need to be constructed. Theoretically, TILLING can be used to identify small deletions missed from the PCR screening, but the relatively low abundance of deletion mutations makes it not cost-effective to screen deletion mutations by TILLING. Future development of high-throughput methodology for screening small deletions would allow for identification of useful deletion alleles using smaller populations.

Another challenge is to identify deletions that knock out tandem arrays spanning very large regions. Primers used for PCR screening of a deletion must be located outside the deletion. At the same time, the distance between the primers and the deletion junction must be close enough to allow efficient amplification of the deletion fragment because fragments bigger than 15 kbp are often difficult to amplify using genomic DNA as a template. With no prior knowledge of the size and location of such a large deletion, PCR screening has to be carried out using different primer combinations to find one pair of primers that is located within a short distance outside of the deletion.

While whole genome sequencing can identify all deletions present in a mutant line, it is currently not practical to use sequencing for deletion mutation identification, as it would cost millions of dollars to carry out single-pass sequencing of a small plant genome. However, radical changes in sequencing technology may bring sequencing costs down dramatically. With dozens of research groups in both private and public sectors working on technologies that can potentially cut the costs for quick pass sequencing across the human genome to about $1,000 (Pennisi, 2002), eventually it may become feasible to carry out single-pass sequencing on thousands of fast neutron-generated deletion lines in a plant species with moderate genome size. Unlike PCR screening, sequencing will identify all the deletions, including both small and very large deletions. With each line containing deletions in as many as ten genes, sequencing a few thousand lines would allow the identification of deletions for most genes. Once the flanking sequences for a large number of deletions are identified, deletion mutants for target genes can be obtained simply by searching the mutant DNA sequence database.

TILLING

Basic Concept of TILLING

Ideally, for functional characterization of a target gene, mutants should be obtained with not only complete loss-of-function mutations but also partial loss-of-function and gain-of-function mutations. While knockout plants carrying complete loss-of-function mutations can be routinely obtained by screening insertion or deletion mutant libraries, it remains a big challenge to obtain specific point mutations in a target gene even though those mutations often provide important clues to the functions of the target gene. To obtain mutants with point mutations in plants, TILLING was developed to screen for EMS-induced point mutations (McCallum et al., 2000).

EMS mutagenesis has traditionally been used to establish mutant populations for forward genetic screens because of its high efficiency (Koornneef et al., 1982). The mutations induced by EMS are usually single-base G/C to A/T transitions and are difficult to identify for reverse genetics applications. Using technologies developed for genotyping single nucleotide polymorphisms (SNPs), EMS-induced point mutations can be identified by TILLING.

To construct a mutant library for TILLING, wild-type seeds are first treated with EMS and planted. Seeds from individual M1 plants are collected and planted again to bulk up seeds, and at the same time to obtain tissue samples for DNA extraction. Once the genomic DNAs are isolated from individual mutant lines, they are pooled and arrayed on microtiter plates for screening.

To screen for mutations in a gene, fragments of the target gene are first amplified by gene-specific PCR using the pooled DNAs as templates. The PCR products are subsequently heated and cooled. If a mutation is present in the PCR fragments, cooling of the denatured PCR products allows heteroduplexes to form between the wild-type and mutant fragments. In the original TILLING method, denaturing high-performance liquid chromatography (DHPLC) was used to identify the heteroduplexes, which are detected as chromatographic alterations because DNA samples containing heteroduplexes have reduced retention when resolved on the heated column (McCallum et al., 2000). The current TILLING method is based on CEL I (Oleykowski et al., 1998; Kulinski et al., 2000), a mismatch cleavage enzyme that preferentially cleaves DNA-containing heteroduplexes and was previously shown to be suitable for mutation detection. In this method, PCR products are incubated with CEL I and the cleaved fragments are analyzed by gel electrophoresis to detect mutations (Figure 3.3). Once a mutation is identified in a pool, DNA samples from individual lines are screened to find the plant containing the mutation. The PCR product from the mutant plant is subsequently sequenced to find the exact position of the mutation. For high-throughput screening, the CEL I–based TILLING method has been streamlined to allow automation (Colbert et al., 2001).

Optimized High-Throughput TILLING

Because EMS induces primarily G/C to A/T transitions, potential mutations in a target region can be predicted in advance. Certain mutations, such as nonsense and splice junction mutations, are preferred for knocking out target genes. Missense changes that cause mutations in evolutionarily conserved amino acids are also highly desirable because of their potential detrimental effects on gene function. By targeting regions that are most likely to produce these mutations, the likelihood of recovering deleterious mutations can be improved. To find such regions, a user-friendly Web-based system (http://www.rpoweb.org/coddle), CODDLE (**C**odons **O**ptimized to **D**iscover **D**eleterious **LE**sions), was designed. This program identifies regions with the highest probability of obtaining deleterious mutations and designs primers for amplifying these regions.

For high-throughput analysis of the CEL I–cleaved products, an automated sequencing gel apparatus from LI-COR is used. The two primers used to amplify a target gene are differentially end-labeled with two fluorescent dyes, IR Dye 700 and IR Dye 800. After treatment with CEL I, the PCR product is loaded on a 96-well sequencing gel, separated by denaturing gel electrophoresis, and detected in two separate channels by a LI-COR Global IR^2 gel scanner. Since cleavage of a mismatched DNA fragment by CEL I produces two differentially labeled fragments, it allows simultaneous detection of a mutation on complementary strands; this is very useful for distinguishing the cleaved products from nonspecific amplifications.

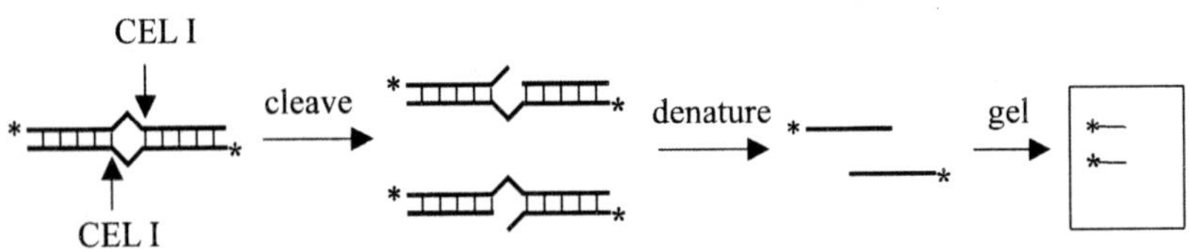

FIGURE 3.3. Point mutation detection using CEL I.

For data analysis, gel image files are automatically converted to compressed JPEG files, which can then be analyzed using Adobe Photoshop. An example of the gel images is shown in Figure 3.4. Due to nonspecific cleavage of the DNA by CEL I, a typical gel image will contain a large number of background bands. New bands resulting from mutations often stand out above the background and can be easily spotted on the two-dimensional gel image. Because of differential labeling of both primers, every mutation will produce two bands in the same lane with one band in each channel. The sum of the two band sizes should be equal to the full-length PCR product.

To provide TILLING service to the *Arabidopsis* community, a high-throughput TILLING facility (ATP, **A**rabidopsis **TILLING P**roject) has been established in the Botany Department at the University of Washington (http://tilling.fhcrc.org:9366/). TILLING customers use CODDLE to identify the best region of a target gene to TILL, choose primers for amplifying

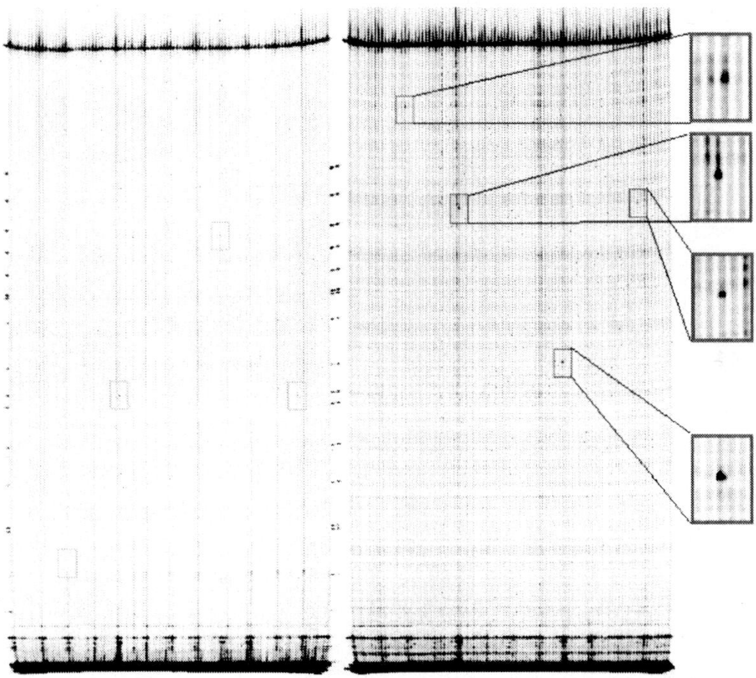

FIGURE 3.4. Example of a TILLING gel used for mutation detection in eightfold pools (see corresponding figure in color photo section). (*Source:* Colbert et al., 2001. Copyrighted by the American Society of Plant Biologists and reprinted with permission.)

this region, and then place an order to TILL the gene. At the same time, customers order the primers from MWG Biotech and request them to be sent to the TILLING facility. ATP will then use these primers to screen for mutation. Initial screening is performed on eightfold pools. For each TILLING request, four gels are run to screen about 3,000 M2 mutant plants (4 gels × 8 plant DNAs × 96 lanes). Once mutant pools are identified, DNAs from the constituent individual M2 plants are screened to find the single plants that harbor the mutations. ATP will then sequence each mutation, release the mutation information to a public database, and send a report on discovered mutations to the user. The user can subsequently order the seeds from the Arabidopsis Biological Resource center.

Since TILLING uses EMS mutagenesis to create random point mutations, it can be applied to virtually any plant species. Since EMS is a very efficient mutagen, only a medium-sized population needs to be screened in order to identify useful mutant alleles. Because EMS mutagenesis induces a broad spectrum of point mutations in a target gene, TILLING not only can provide knockout mutants for genes with unknown functions but also can be used to obtain partial loss-of-function alleles of genes with known functions. The partial loss-of-function mutations are especially useful for studying the functions of essential genes.

Future Directions for Point Mutant Identification

While screening for EMS-induced point mutations using methods for SNP genotyping is a wonderful concept, the throughput of TILLING is in need of improvement. With fewer than ten plants pooled and screened for mutations in each reaction, a large number of PCR reactions need to be performed and analyzed using multiple sequencing gels. Even with the help of robots, this process remains quite laborious. It appears that one of the main factors limiting the detection sensitivity is the background bands derived from nonspecific CEL I cleavages. Future improvement on the specificity of CEL I by protein engineering may reduce the intensity of the background bands and allow higher levels of pooling.

Another aspect of TILLING that waits further improvement is the requirement of sequencing on each mutant to find the exact position of the mutation because analysis of the gel image can provide only the approximate location of a mutation. It is also necessary to carry out sequence analysis on all mutants to identify those with useful mutations, since the majority of the mutations are unlikely to affect the functions of the target gene. At this time, sequencing analysis of point mutations in a large number of plants still requires significant effort.

To fully realize the potential of point mutations in reverse genetic analysis, more advanced methods need to be developed to screen for point mutations induced by chemical mutagenesis. Improvement on the screening method would allow genome-wide mutation discovery and also may make it possible to find rare gain-of-function mutations by screening very large mutant populations. Currently, a large number of methods exist for SNP scoring, but all these methods aim at assaying individuals rather than pools of a large number of samples (Syvanen, 2001). As a result, none of them can be directly applied to screen for variant alleles in populations consisting of tens of thousands of plants in a cost-effective fashion. CEL I–based TILLING remains one of the most cost-effective methods for interrogating a large number of mutations in a target region; however, future improvements on microarray-based SNP scoring methods may allow them to be used to screen mutations in a large number of genes in parallel.

One of the most promising SNP scoring strategies that can be adapted to screen a large number of mutations in pooled samples is microarray primer extension (Pastinen et al., 2000). For each SNP, two allele-specific primers are designed and immobilized on an array. Extension of these primers using a mixture of natural and fluorescent deoxynucleotides on DNA or RNA templates, followed by fluorescence scanning, allows accurate scoring of the SNPs. Because allele-specific primers for a large number of mutant alleles can be spotted on a single array, mutations in multiple genes can be screened in parallel. By designing allele-specific primers with improved specificity, this method can be used to screen pooled samples with higher complexity for the presence of predicted mutations.

Single-base extension (SBE) tag arrays can also be used for high-throughput identification of point mutations (Hirschhorn et al., 2000; Fan et al., 2000). SBE uses DNA polymerase to extend a primer adjacent to the position of an SNP in the presence of fluorescently labeled ddNTPs. The principle of SBE is illustrated in Figure 3.5. Since primer extension using DNA polymerase is very robust, SBE requires minimal assay optimization and allows parallel genotyping of many SNPs under similar reaction conditions. A large number of primers can be designed and used to score for multiple SNPs in multiplexed reactions. To reveal the identity of the SNPs detected by multiplexed SBE reactions, each primer is marked with a unique 5' tag sequence to allow separation of the primer extension products by hybridizing to a tag array containing complementary oligonucleotides. It has been proven that SBE is very accurate in SNP scoring and can detect mutations in pooled samples. The ability to analyze pooled samples in a highly multiplexed fashion makes it practical to use SBE tag arrays to screen EMS-induced point mutations in plants.

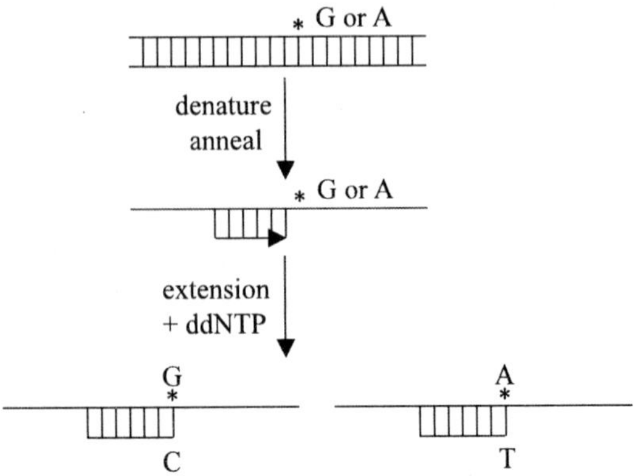

FIGURE 3.5. The principle of single-base extension (SBE). In SBE, a primer is designed to hybridize to its target sequence immediately adjacent to the SNP. In the presence of DNA polymerase and labeled dideoxynucleotides, a primer annealed to the target sequence is extended with a single nucleotide that is complementary to the nucleotide at the SNP. The identity of the SNP is determined by the label of the extended nucleotide.

In theory, microarray-based genotyping methods can be applied to genome-wide mutation discovery. Practically, the number of genes that can be screened in each reaction is still very limited mainly because the templates used in the primer extension reactions must be prepared by PCR amplification to achieve desirable sensitivity and accuracy. Since it remains very difficult to perform multiplex PCR amplification of more than ten DNA fragments reproducibly, efficient generic amplification strategies must be developed to fully realize the power of these high-throughput genotyping methods in mutation identification.

COMBINATORIAL GENE SILENCING

Even though knockout mutants for most *Arabidopsis* genes can now be obtained by different strategies, it is still a great challenge to assign functions to a large number of genes due to the lack of obvious phenotypes in the vast majority of single knockouts (Bouche and Bouchez, 2001). Genetic redundancy is probably one of the reasons that knockout phenotypes are hard to find. When two or more genes encode redundant functions, knockouts for the single genes likely will not show any informative phenotypes that can lead to the functional characterization of these genes.

Analysis of the complete *Arabidopsis* genome sequence revealed extensive structural redundancy. Among the predicted proteins, only 35 percent are unique while 37 percent belong to families with more than five members (Arabidopsis Genome Initiative, 2000). There is clear evidence that some highly similar genes encode partially or completely redundant functions. For example, fruit dehiscence is abolished in the double mutant of two MADS-box genes, *SHP1* and *SHP2,* while no clear phenotype was observed in the single mutants (Liljegren et al., 2000). To assign functions to redundant genes, single mutants can be crossed together to create combined mutants, which can then be used for phenotypic analysis. Alternatively, combinatorial gene silencing using chimeric gene constructs can be used to inactivate multiple genes together. However, the fact that gene silencing rarely results in complete gene inactivation needs to be taken into account in data interpretation.

In *Drosophila,* injecting multiple double-stranded RNAs (dsRNAs) into embryos led to simultaneous silencing of multiple genes (Schmid et al., 2002). In tobacco, chimeric gene constructs were successfully used to inactivate two or three genes involved in lignin biosynthesis (Abbott et al., 2002). In the latter study, a 35S CaMV promoter-driven chimeric transgene with partial sense sequences for multiple lignin biosynthetic enzymes was constructed and transformed into tobacco. All target genes were coordinately suppressed in the transgenic plants.

To improve the effectiveness of silencing multiple genes, dsRNA or intron-containing hairpin RNA constructs (ihpRNA) can be used since they are more effective than conventional sense or antisense constructs (Smith et al., 2000; Wesley et al., 2001). To test the effectiveness of ihpRNA constructs in silencing multiple genes together, we transformed *Arabidopsis* with several chimeric ihpRNA constructs expressing short cDNA fragments from multiple target genes. In most cases, target genes can be simultaneously suppressed in the same transgenic lines (Li and Zhang, unpublished data).

For the construction of chimeric gene constructs, we have been using a PCR-based method to link fragments of multiple genes (Figure 3.6). This method allows fusion of two or more sequences without using restriction enzymes and can be carried out in a high-throughput manner. To express the ihpRNA, we are using a modified pHANNIBAL vector containing the glufosinate resistance (BAR) gene as the selection marker allowing selection of transgenic plants on soil. For high-throughput gene silencing, vectors such as pHELLSGATE that incorporate Gateway technology to facilitate easy cloning of PCR fragments can also be employed (Wesley et al., 2001).

In conventional double-mutant analysis, it takes at least three generations to combine two single mutations. In contrast, it only takes a single transformation step to obtain transgenic plants with multiple genes silenced. Since gene silencing is dominant, T1 plants can be directly assayed for mutant phenotypes. In addition, silencing is more flexible than knock-

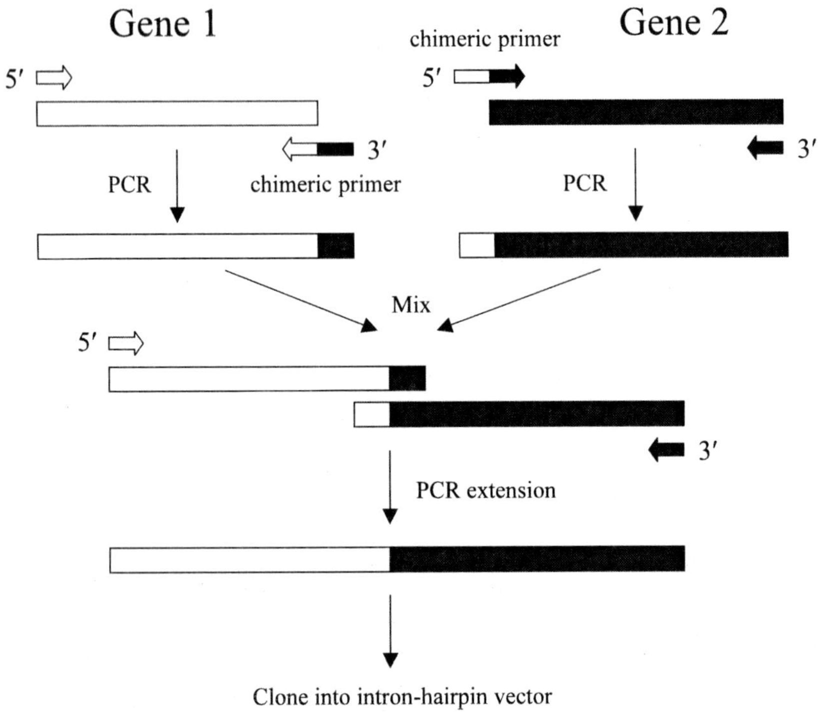

FIGURE 3.6. A general strategy for making chimeras between two genes without the need of restriction digestion. A PCR fragment for gene 1 is first amplified using a normal 5' primer and a chimeric 3' primer. The chimeric 3' primer hybridizes on the 3' side of gene 1 but has sequences at the 5' end that are derived from gene 2. The resulting DNA fragment contains the desired portion of gene 1 with a short stretch of sequence complementary to gene 2. At the same time, a PCR fragment for gene 2 is amplified using a chimeric 5' primer and a normal 3' primer. The chimeric 5' primer hybridizes on the 3' side of gene 2 but has sequences at the 5' end that are derived from gene 1. The resulting DNA fragment contains the desired portion of gene 2 with a short stretch of sequence complementary to gene 1. Mixing of these two fragments and subsequent PCR amplification using the 5' primer of gene 1 and the 3' primer of gene 2 results in a chimeric gene fragment. This fragment is then used to make intron-hairpin constructs.

out analysis because silencing of target genes can be carried out in the appropriate genetic background to simplify phenotypic analysis. Since it is practical to make a large number of silencing constructs and transform them into plants in parallel, combinatorial gene silencing will become a very valuable tool for genome-wide analysis of genes with potential redundant functions.

REFERENCES

Abbott, J.C., Barakate, A., Pincon, G., Legrand, M., Lapierre, C., Mila, I., Schuch, W., Halpin, C. (2002). Simultaneous suppression of multiple genes by single transgenes: Down-regulation of three unrelated lignin biosynthetic genes in tobacco. *Plant Physiol.* 128: 844-853.

Arabidopsis Genome Initiative (2000). Analysis of the genome sequence of the flowering plant *Arabidopsis thaliana*. *Nature* 408: 796-815.

Baulcombe, D.C. (1996). RNA as a target and an initiator of post-transcriptional gene silencing in transgenic plants. *Plant Mol. Biol.* 32: 79-88.

Bensen, R.J., Johal, G.S., Crane, V.C., Tossberg, J.T., Schnable, P.S., Meeley, R.B., Briggs, S.P. (1995). Cloning and characterization of the maize *An1* gene. *Plant Cell* 7: 75-84.

Bouche, N., Bouchez, D. (2001). *Arabidopsis* gene knockout: Phenotypes wanted. *Curr. Opin. Plant Biol.* 4: 111-117.

Brunaud, V., Balzergue, S., Dubreucq, B., Aubourg, S., Samson, F., Chauvin, S., Bechtold, N., Cruaud, C., DeRose, R., Pelletier, G., et al. (2002). T-DNA integration into the *Arabidopsis* genome depends on sequences of pre-insertion sites. *EMBO Rep.* 3: 1152-1157.

Chuang, C.-F., Meyerowitz, E.M. (2000). Specific and heritable genetic interference by double-stranded RNA in *Arabidopsis thaliana*. *Proc. Natl. Acad. Sci. USA* 97: 4985-4990.

Colbert, T., Till, B.J., Tompa, R., Reynolds, S., Steine, M.N., Yeung, A.T., McCallum, C.M., Comai, L., Henikoff, S. (2001). High-throughput screening for induced point mutations. *Plant Physiol.* 126: 480-484.

Dixon, M.S., Jones, D.A., Keddie, J.S., Thomas, C.M., Harrison, K., Jones, J.D. (1996). The tomato *Cf-2* disease resistance locus comprises two functional genes encoding leucine-rich repeat proteins. *Cell* 84: 451-459.

Fan, J.B., Chen, X., Halushka, M.K., Berno, A., Huang, X., Ryder, T., Lipshutz, R.J., Lockhart, D.J., Chakravarti, A. (2000). Parallel genotyping of human SNPs using generic high-density oligonucleotide tag arrays. *Genome Res.* 10: 853-860.

Feng, Q., Zhang, Y., Hao, P., Wang, S., Fu, G., Huang, Y., Li, Y., Zhu, J., Liu, Y., Hu, X., et al. (2002). Sequence and analysis of rice chromosome 4. *Nature* 420: 316-320.

Goff, S.A., Ricke, D., Lan, T.H., Presting, G., Wang, R., Dunn, M., Glazebrook, J., Sessions, A., Oeller, P., Varma, H., et al. (2002). A draft sequence of the rice genome *(Oryza sativa* L. ssp. *japonica)*. *Science* 296: 92-100.

Hirschhorn, J.N., Sklar, P., Lindblad-Toh, K., Lim, Y.M., Ruiz-Gutierrez, M., Bolk, S., Langhorst, B., Schaffner, S., Winchester, E., Lander, E.S. (2000). SBE-TAGS: An array-based method for efficient single-nucleotide polymorphism genotyping. *Proc. Natl. Acad. Sci. USA* 97: 12164-12169.

Jansen, G., Hazendonk, E., Thijssen, K.L., Plasterk, R.H.A. (1997). Reverse genetics by chemical mutagenesis in *Caenorhabditis elegans*. *Nat. Genet.* 17: 119-121.

Kempin, S.A., Liljegren, S.J., Block, L.M., Rounsley, S.D., Yanofsky, M.F., Lam, E. (1997). Targeted disruption in *Arabidopsis*. *Nature* 389: 802-803.

Koornneef, M., Dellaert L.W.M., van der Veen, J.H. (1982). EMS- and radiation-induced mutation frequencies at individual loci in *Arabidopsis thaliana* (L.) Heynh. *Mutat. Res.* 93: 109-123.

Krysan, P.J., Young, J.C., Sussman, M.R. (1999). T-DNA as an insertional mutagen in *Arabidopsis*. *Plant Cell* 11: 2283-2290.

Kulinski, J., Besack, D., Oleykowski, C.A., Godwin, A.K., Yeung, A.T. (2000). CEL I enzymatic mutation detection assay. *Biotechniques* 29: 44-46.

Li, X., Song, Y., Century, K., Straight, S., Ronald, P., Dong, X., Lassner, M., Zhang, Y. (2001). A fast neutron deletion mutagenesis-based reverse genetics system for plants. *Plant J.* 27: 235-242.

Li, X., Zhang, Y. (2002). Reverse genetics by fast neutron mutagenesis in higher plants. *Funct. Integr. Genomics* 2: 254-258.

Liljegren, S.J., Ditta, G.S., Eshed, Y., Savidge, B., Bowman, J.L., Yanofsky, M.F. (2000). *SHATTERPROOF* MADS-box genes control seed dispersal in *Arabidopsis*. *Nature* 404: 766-770.

Liu, L.X., Spoerke, J.M., Mulligan, E.L., Chen, J., Reardon, B., Westlund, B., Sun, L., Abel, K., Armstrong, B., Hardiman, G., et al. (1999). High-throughput isolation of *Caenorhabditis elegans* deletion mutants. *Genome Res.* 9: 859-867.

McCallum, C.M., Comai, L., Greene, E.A., Henikoff, S. (2000). Targeted screening for induced mutations. *Nat. Biotechnol.* 18: 455-457.

McKinney, E.C., Ali, N., Traut, A., Feldmann, K.A., Belostotsky, D.A., McDowell, J.M., Meagher, R.B. (1995). Sequence-based identification of T-DNA insertion mutations in *Arabidopsis*: Actin mutants *act2-1* and *act4-1*. *Plant J.* 8: 613-622.

Oleykowski, C.A., Bronson Mullins, C.R., Godwin, A.K., Yeung, A.T. (1998). Mutation detection using a novel plant endonuclease. *Nucleic Acids Res.* 26: 4597-4602.

Parinov, S., Sevugan, M., Ye, D., Yang, W.-C., Kumaran, M., Sundaresan, V. (1999). Analysis of flanking sequences from *Dissociation* insertion lines: A database for reverse genetics in *Arabidopsis*. *Plant Cell* 11: 2263-2270.

Pastinen, T., Raitio, M., Lindroos, K., Tainola, P., Peltonen, L., Syvanen, A.C. (2000). A system for specific, high-throughput genotyping by allele-specific primer extension on microarrays. *Genome Res.* 10: 1031-1042.

Pennisi, E. (2002). Gene researchers hunt bargains, fixer-uppers. *Science* 298: 735-736.

Sasaki, T., Matsumoto, T., Yamamoto, K., Sakata, K., Baba, T., Katayose, Y., Wu, J., Niimura, Y., Cheng, Z., Nagamura, Y., et al. (2002). The genome sequence and structure of rice chromosome 1. *Nature* 420: 312-316.

Schaefer, D.G., Zryd, J.P. (1997). Efficient gene targeting in the moss *Physcomitrella patens*. *Plant J.* 11: 1195-1206.

Schmid, A., Schindelholz, B., Zinn, K. (2002). Combinatorial RNAi: A method for evaluating the functions of gene families in *Drosophila*. *Trends Neurosci.* 25: 71-74.

Sessions, A., Burke, E., Presting, G., Aux, G., McElver, J., Patton, D., Dietrich, B., Ho, P., Bacwaden, J., Ko, C., et al. (2002). A high-throughput *Arabidopsis* reverse genetics system. *Plant Cell* 14: 2985-2994.

Smith, N.A., Singh, S.P., Wang, M.B., Stoutjesdijk, P.A., Green, A.G., Waterhouse, P.M. (2000). Total silencing by intron-spliced hairpin RNAs. *Nature* 407: 319-320.

Somerville, C., Dangl, J. (2000). Plant biology in 2010. *Science* 290: 2077-2078.

Syvanen, A.C. (2001). Accessing genetic variation: Genotyping single nucleotide polymorphisms. *Nat. Rev. Genet.* 2: 930-942.

Terada, R., Urawa, H., Inagaki, Y., Tsugane, K., Iida, S. (2002). Efficient gene targeting by homologous recombination in rice. *Nat. Biotechnol.* 20: 1030-1034.

Wang, M.B., Waterhouse, P.M. (2002). Application of gene silencing in plants. *Curr. Opin. Plant Biol.* 5: 146-150.

Wesley, S.V., Helliwell, C.A., Smith, N.A., Wang, M.B., Rouse, D.T., Liu, Q., Gooding, P.S., Singh, S.P., Abbott,. D, Stoutjesdijk, P.A., et al. (2001). Construct design for efficient, effective and high-throughput gene silencing in plants. *Plant J.* 27: 581-590.

Yu, J., Hu, S., Wang, J., Wong, G.K., Li, S., Liu, B., Deng, Y., Dai, L., Zhou, Y., Zhang, X., et al. (2002). A draft sequence of the rice genome *(Oryza sativa* L. ssp. *indica). Science* 296: 79-92.

Chapter 4

Transcriptomics in Plants: From Expression to Gene Function

Laurent Zimmerli
Shauna Somerville

INTRODUCTION

The complete sequencing of the genomes of many eukaryotes and prokaryotes has paved the way to a new era in biology. *Arabidopsis thaliana,* a dicotyledonous plant, was the first plant to be completely sequenced (Arabidopsis Genome Initiative, 2000), followed by rice, a monocotyledonous plant (Goff et al., 2002; Yu et al., 2002). The use of whole genome information and high-throughput tools has opened a new field of research called genomics. Among its subdisciplines, metabolomics (the complete set of metabolites expressed in a cell) (Stitt and Fernie, 2003) (see Chapter 5) and proteomics (the complete set of proteins produced in a cell) (Roberts, 2002) (see Chapters 14 and 15) are beginning to be used in the plant science community. However, to date, transcriptomics is the best developed and most widely available of these genomic technologies.

Transcriptomics is based on the idea that a catalog of all the transcripts associated with a specific treatment or developmental state provides a reasonable overview of the underlying biological processes at work. Of course, not all changes in gene expression are reflected in changes in transcript levels, and this has to be considered in any transcriptome-based analysis of a biological process. However, in yeast, differential expression of mRNAs is reflected at the protein level in 60 to 70 percent of cases (Gygi et al., 1999), suggesting that transcriptomic approaches can provide important leads to

L. Z. was supported in part by a Swiss National Fellowship and by funding from the International Research Management (Novartis) Foundation. The Carnegie Institution, the National Science Foundation, and the U.S. Department of Energy also provided research support. Publication number 1629 of the Carnegie Institution of Washington.

genes relevant to the biological process under study. Another outcome of transcriptomics is the identification of potential *cis*-acting regulatory elements based on overrepresentation in the promoters of coexpressed genes. Since promoters function as transducers, integrating information about the environment and internal state of a cell, knowledge of these regulatory sequences will provide insights into the complex network of information influencing gene-expression patterns. Thus, transcriptomics gives access to the circuitry of information, which has generally been inaccessible. In addition, biological roles for unknown or poorly annotated genes may be suggested by their coexpression with genes of known function. Finally, the understanding of the biological consequences of gene-expression patterns, merged to our growing knowledge of the functions of individual genes, will result in a comprehensive, dynamic, molecular image of a plant. This new knowledge, intermixed with other "-omics" disciplines, will surely lead the plant scientific community to *in silico* simulations of plant growth, development, and response to environmental change. Such comprehensive knowledge will allow the engineering of novel plant forms to better serve the needs of an exploding human population with a more productive and environmentally friendly agriculture.

THE TECHNOLOGIES

A number of methods for determining global gene expression have been developed over the past several years (see Table 4.1). The Serial Analysis of Gene Expression (SAGE) method is a first-pass screening that relies on high-throughput sequencing of short fragments of cDNAs (tags) to estimate the relative concentration of mRNA species in a tissue (Vesculescu et al., 1995). Because of its reliance on DNA sequencing, SAGE can select transcripts not yet identified by other high-throughput sequencing projects. However, it is difficult to analyze large numbers of samples, as SAGE is very expensive to perform. In plants, this method has been used to monitor gene expression in rice (Matsumura et al., 1999) and in a tree species (Lorenz and Dean, 2002). In a recent study, SAGE was used to monitor gene expression during the fungal elicitor-induced cell death in rice. Among others, a gene coding for a bax inhibitor, a possible plant cell death regulator (Kawai-Yamada et al., 2001), was identified as down-regulated in the elicitor-treated cells. Interestingly, the authors were able to demonstrate a functional role of this gene in regulating cell death in rice cells (Matsumura et al., 2003).

The transcriptome technology of most widespread use in the plant community is microarrays and thus, we will focus much of the remaining dis-

TABLE 4.1. Examples of current transcriptome technologies.

Expression technology	Review article	Plant example or relevant URL
Macroarrays (filter-based)	Zhao et al., 1995	Hirai et al., 2003
Microarrays—short oligonucleotides	Lipshutz et al., 1999	Puthoff et al., 2003
Microarrays—DNA fragments (cDNAs)	Schena et al., 1995	Ruan et al., 1998
Microarrays—long oligonucleotides	Wang et al., 2003	<http://www.ag.arizona.edu/microarray>
Total Gene Expression Analysis (TOGA)	Sutcliffe et al., 2000	<http://www.dgt.com/wt/disease/platform>
Massive Parallel Signature Sequencing (MPSS)	Brenner et al., 2000	<http://mpss.ucdavis.edu>
Serial Analysis of Gene Expression (SAGE)	Vesculescu et al., 1995	Lorenz and Dean, 2002
GeneCalling	Shimkets et al., 1999	Bruce et al., 2000
Bead-Based Fiber-Optic Array	Steemers et al., 2000	<http://www.illumina.com/tech_plat.htm>
cDNA-AFLP	Bachem et al., 1996	Durrant et al., 2000
Subtractive Hybridization	Diatchenko et al., 1996	Mahalingam et al., 2003

cussion on this technology, with only brief references to other technologies by way of contrast. Furthermore, we will focus our discussion on the use of near full-genome microarrays, which fulfill the promise of the transcriptomics by providing an integrative, comprehensive view of how plants respond to an experimental condition or genetic manipulation. We do not discuss the use of microarrays for single-nucleotide polymorphism analysis. See Botwell and Sambrook (2003) for a recent set of detailed protocols and trouble-shooting guides to producing two-color, cDNA microarrays and macroarrays.

In plants, microarray-based expression profiles have been reported for processes such as seed development (Girke et al., 2000; Ruuska et al., 2002), responses to light (Harmer et al., 2000; Ma et al., 2001; Schaffer et al., 2001), abiotic and biotic stresses (Maleck et al., 2000; Reymond et al., 2000; Kawasaki et al., 2001; Seki et al., 2001; Chen et al., 2002; Kreps et al., 2002; Ramonell et al., 2002; Seki et al., 2002; Vranova et al., 2002; Puthoff et al., 2003; Whitham et al., 2003; Yu and Setter, 2003), hormone treatments (Goda et al., 2002; Sawa et al., 2002), changes in nutrient sources (Negishi et al., 2002; Wang et al., 2000), and analysis of mutants or transgenic plants (Wang et al., 2002; Abe et al., 2003; Glazebrook et al., 2003; Zik and Irish, 2003).

The use of high-density microscopic array elements allows massive parallel data acquisition and analysis. Moreover, the use of solid surface and fluorescent probes usually generates low background noise and good dynamic range, respectively, which allows detection of relatively minor changes in gene expression in weakly expressed genes. For example, Girke and collaborators (2000) estimated that they could detect one to two copies of mRNA per cell with a minimum of detectable fold-change of 1.4 using their cDNA microarrays. Similarly, Lipshutz and collaborators (1995) demonstrated that genes represented at a few copies per cell could be detected using short-oligonucleotide microarrays. From our own experience with two-color, cDNA microarrays, Northern blots based on poly $(A)^+$ RNA samples are roughly tenfold less sensitive than microarrays in detecting weakly expressed genes.

This transcriptome analysis is commonly performed on either one- or two-color microarrays. The one-color format is usually used with microarrays made with short oligonucleotides (Lipshutz et al., 1999). Two-color microarrays are characterized by the hybridization of two different labeled (usually cyanine 3 and cyanine 5) cDNA samples on the same slide. The elements of the microarrays are hybridized with labeled cDNA (two-color microarrays) or cRNA (short-oligonucleotide, one-color microarrays) derived from complex mixtures of either total RNA or poly $(A)^+$ RNA, producing signals that reflect the identity and the relative concentration of each mRNA species in the sample. The slide is than scanned once in the case of the one-color microarrays or twice, once each at two different wavelengths, in the case of the two-color microarrays. Both approaches give equally high-quality gene-expression data.

One-color microarrays use short-oligonucleotide (15-25 nt) array elements, and two-color microarrays use either PCR-amplified cDNAs (100-3,000 bp), gene-specific amplicons (150-500 bp), or, more recently, long oligonucleotides (50-120 nt) as array elements. Short oligos and cDNAs work well for expression analysis, although both types of elements face some limitations. Individual short oligonucleotides sometimes lack the specificity required to report for a single gene. Therefore, it is important to use multiple short oligonucleotides per gene on a microarray. The longer PCR-amplified cDNAs produce strong signals. However, depending on the length and specificity of the amplified cDNAs, cross hybridization to related gene family members is possible (Girke et al., 2000). Furthermore, careful attention must be paid during preparation of the cDNA-based microarray elements to exclude contamination of the PCR products. Long-oligonucleotide microarrays are characterized by strong hybridization signals, good discrimination among closely related genes, and affordability (Kane et al., 2000). However, availability of genomic sequence information is required. A few genes in the genome will have nearly identical sequences even in the 5' and 3' untranslated regions, making it impossible to design oligonucleotides that effectively discriminate between these two genes. In addition, very small genes (e.g., encoding microRNAs) might prove difficult targets for long-oligonucleotide design. For such genes, short-oligonucleotide-based microarrays may be more relevant. Numerous vendors produce high-quality long oligos that can be printed in-house, giving the same flexibility and affordability as with cDNAs elements. For each type of microarray, it is also necessary to include a number of control elements distributed across the microarray that can be used for quality control assessments (e.g., negative controls, positive controls, spiking controls). In addition, for small custom arrays, a number of controls that can be used for data normalization must be included (see discussion in Analysis section).

Substrates for microarray analysis must provide stable attachment of array elements, low background signals, and homogeneous surface chemistry. The most common substrate material used is glass (silicon dioxide). Chemical slide coatings (e.g., silane) allow covalent attachment of array elements to the slides via reactive amine, aldehyde, or epoxide groups. Such coated surfaces allow stable attachment of oligonucleotides, cDNAs, proteins, and other molecules. New slide surfaces continue to be developed and those interested in printing in-house microarrays should survey commercial sources of slides.

Fluorescent cDNA or cRNA probes are prepared from total RNA or mRNA by reverse transcription. For two-color microarrays, control and experimental samples are generally labeled separately with fluorochromes that have nonoverlapping emission spectra, including cyanine, Alexa, and other fluorescent derivatives. Two samples labeled with different fluorochromes can be hybridized to a single microarray to produce data on the relative expression of genes in the two samples. The within-microarray comparisons can be very robust, as technical variation in the hybridization step between samples is minimized. For one-color microarrays, each sample is labeled with the same fluorochrome and hybridized to a separate microarray. Sample-to-sample comparisons are made from data collected from separate microarrays.

The microarray signals are detected by specialized scanners equipped, in some cases, with confocal optics and either photo multiplier tubes or charge-coupled devices as detectors. Detection instruments produce graphical images in standard tagged-image file format (TIFF) with pixel intensity values ranging from 0 to 65,536 (2^{16}). Specialized data extraction software programs are used to determine, at a minimum, the intensity and background values associated with each array element. These data values for each gene are then saved in a tab-delimited format, facilitating the import into software for data mining and modeling.

Microarrays can also be produced for plants for which we do not know the full genome sequence. Although the caveats for data interpretation are more numerous and the need for confirmation of major results by independent methods is stronger, much useful data and interesting leads can be developed from such microarrays. The consensus coding sequences generated from large EST sequencing projects have been used as a basis for designing a short-oligonucleotide Affymetrix microarray for barley (http://www.intl-pag.org/11/abstracts/W07_W50_XI.html), and EST clones have been widely used to produce two-color, cDNA microarrays (e.g., maize [Fernandes et al., 2002]; potato [http://www.tigr.org/tdb/potato/microarray2.shtml], soybean [Thibaud-Nissen et al., 2003]). Although not widely practiced, it is in principle possible to produce microarrays from unsequenced cDNA clones and then to sequence any clones with interesting expression patterns after the data-collection step. However, other methods, such as cDNA-AFLP or subtractive hybridization, may be more appropriate for plants for which few genomic resources are available.

An alternative to the use of microarrays for transcript analysis is filter-based macroarrays in which DNA fragments (e.g., cDNA clones, gene-specific amplicons) are spotted onto nylon or other hybridization membranes using the printing robots employed to array BAC and other similar clones. These macroarrays are probed with radioactively labeled samples and the radioactivity is detected via a phosphorimager. In general, phosphorimagers have a good dynamic range (~3 to 4 orders of magnitude), and this is an advantage for measuring transcript abundance in complex mixtures in which a few transcripts are very highly expressed and a majority of genes are very weakly expressed. As with Northern blots, such membranes can be stripped and reprobed several times. In part because they rely on instrumentation available in many laboratories, macroarrays have found widespread use, especially among those producing small (hundreds to thousands of elements) custom arrays. Many of the same software programs used for microarray analysis have options for the analysis of macroarray data. Macroarrays have been used successfully in such plants as tobacco (Legen et al., 2002), barley (Potokina et al., 2002; Sreenivasulu et al., 2002), lupin (Uhde-Stone et al., 2003), *Medicago truncatula* (Fedorova et al., 2002), rice (Lin et al., 2003), and *Arabidopsis* (Hirai et al., 2003; Richly et al., 2003).

The increasing availability of microarrays and other transcriptomic platforms from commercial sources or academic technology centers will further simplify the adoption of transcriptomics approaches by allowing researchers to focus on data collection rather than on platform production.

EXPERIMENTAL DESIGN

Researchers who want to perform a microarray experiment could either choose an exploratory experiment or be more specific and select a marker-discovery experiment type (Brown and Botstein, 1999). Performing a marker-discovery experiment will lead the researcher to select a small number of genes that are specific for the treatment applied. In this case, the researcher is only interested in genes showing highly specific and drastic changes in expression. Such experiments are considered as a starting point for further investigations, and a limited number of microarrays are generally necessary. In the design of such an experiment, the researcher should be careful to select a time point or treatment that specifically reflects the biological effect studied. When performing an exploratory experiment, the researcher is interested in an integrative view of the plant's response to a particular treatment. In this case, a more detailed and complete profile of expression is desired and it is relevant to analyze a larger number of samples, such as different time points of a biological process or different concentrations of a chemical treatment. When performing complex microarray experiments, involving the comparison of multiple samples, it is often necessary to use a common reference (Type II experiments) (Eisen and Brown, 1999). For two-color microarrays, the common reference could be a mixture of material from each of the samples used during the experiment. This will ideally provide a hybridization signal for each gene represented on the samples and will therefore avoid a zero denominator when calculating the hybridization ratio. Typically, this approach is well suited for the analysis of time-course experiments. Since the expression level of a gene at each time point is measured relative to a fixed reference, this method allows comparison of the behavior of each gene across the entire time course. In all cases, researchers should include the analysis of multiple independent biological replicates to account for the biological variability in their samples and the variability associated with microarray production and use (Lee et al., 2000; Churchill, 2002; Yang and Speed, 2002).

ANALYSIS

Although the practical aspects of microarray experiments have become fairly straightforward, a general agreement of their interpretation has not. Similar to DNA sequence information, large-scale expression data sets represent a challenge for finding new paths to biological meaning, further emphasizing that data analysis and interpretation are vital aspects of the genomic approach.

Methods of Normalization

Aside from quality control checks of the hybridization step, the first step in data analysis is the normalization of data. Normalized data are corrected to rescale and remove systematic biases to compare separate fluorescent channels or experiments. For example, two samples can exhibit differences in intensity values due only to differences in the incorporation of the fluorescent label. In this context, the goal is not to adjust the data to the normal distribution but rather to standardize the two sets of data prior to comparison, in much the same way that the expression level of a housekeeping gene is used to standardize or normalize gene expression in Northern blots. One normalization approach is to select a standard set of housekeeping genes that demonstrate constant expression levels under a variety of conditions and to use their level of expression to calibrate the fluorescent signals of the samples to be compared. This method requires a large set of preliminary data from which to choose a significant number (hundreds) of medium- to highly expressed genes with constant expression in the conditions under study. Moreover, since no gene set has proven to be universally stable across all experimental conditions, this method is better suited for small-scale, specialized arrays used in experiments performed under a limited number of environmental conditions or developmental states. Alternately, artificial transcripts spiked into the RNA samples at known levels can be used for this purpose. With arrays consisting of tens of thousands of elements, one can also make the assumption that most genes (~90 percent or more) will not change in expression under the conditions tested. Thus, the mean, mode, median, or other measure (e.g., median intensity of the upper fiftieth percentile of samples) of fluorescence intensity for each channel can be used to determine a normalization factor (Richmond and Somerville, 2000). This approach has the advantage of being simple to determine and independent of a predetermined set of housekeeping genes, but it is not applicable to small focused microarrays in which a significant portion of the array elements show intensity changes. More complex normalization methods that deal with removal of outlier and/or nonlinear data sets have been described (Carlisle et al., 2000; Finkelstein et al., 2000; Yang et al., 2002). Another option is to center the data. Centering is a process that will adjust the values of your data and will facilitate the comparison of data from independent arrays, such as the microarrays of a time-course experiment. It is generally relevant to center the data to the mean or median of all values of the gene or microarray (Botwell and Sambrook, 2003). Although better methods of detecting and removing biases are constantly under development (Finkelstein et al., 2002), experimental repetition is often the best way to identify aberrant data sets.

Data Mining and Modeling

Microarrays have traditionally represented changes in gene expression as ratios of the intensity values (fold-change) for a gene from two experimental conditions (DeRisi et al., 1997). Initial experiments commonly relied on a single replicate due to the cost of early microarrays. However, it is now widely recognized that replication of the biological materials is essential to minimize the number of false-positive and false-negative data points (Lee et al., 2000; Churchill, 2002; Yang and Speed, 2002). Statistical methods have been developed specifically to identify statistically significant changes in gene expression from microarray experiments (Ideker et al., 2000; Newton et al., 2001; Tusher et al., 2001; Draghici, 2002; Smyth et al., 2003). A straightforward approach is to use the traditional T-test (Devore and Peck, 1997). However, determining the appropriate significance value is complicated by the need to include a multiple-comparison correction factor. The Bonferroni correction is one correction factor that addresses this problem (Bland and Altman, 1995). Another measure of assessing data quality of replicated data is the false-discovery rate factor. To determine this factor, the data set is randomized to provide an empirical measure of the false-discovery rate (i.e., the fraction of the expected false positives among the genes that are called as differentially expressed) (Storey, 2002). For example, the Significance Analysis of Microarrays (SAM) program identifies significant differences in mean expression levels or fold-change values and provides a false-discovery rate for each gene (Tusher et al., 2001). A number of freeware statistically based programs are available, and, increasingly, commercial microarray analysis software programs include statistical tests of significance. See the two Web sites, <http://stat-www.berkeley.edu/users/ terry/zarray/Html/> and <http://www.jax.org/staff/churchill/labsite/index. html>, for more information concerning the statistical analysis of microarrays.

Once a set of genes showing differential expression has been identified, additional analysis tools can aid in deriving biological insight from this gene list. The genes can be grouped by their biological functions to determine if genes associated with one or more biological processes are overrepresented in the data set. Such information may provide insights into biological processes that contribute to a plant's response to the imposed treatment. As an example, transcriptome analysis of early *Arabidopsis* seedling development revealed that 26 cellular pathways were coordinated by light (Ma et al., 2001). Several bioinformatic groups have assigned *Arabidopsis* genes to functional classes; these include The Institute for Genomic Research (TIGR, <http://www.tigr.org>), Munich Information Center for Protein Sequences (MIPS, <http://mips.gsf.de/proj/thal/>; Schoof et al., 2002), and The Arabidopsis Information Resource (TAIR, <http://arabidopsis.org>), working in conjunction with the Gene Ontology project (GO, <http://www.geneontology.org>; Ashburner et al., 2000; Hill et al., 2002). A valuable tool for visualizing such data is to organize the genes within one biological function into a pathway and to overlay this pathway with the expression data. Both genMAPP (Dahlquist et al., 2002; <http://www.genMAPP.org>) and AraCyc (Mueller et al., 2003), a tool within TAIR, provide pathway-viewing tools.

If several treatments or time points are being analyzed using microarray technology, then it is possible to identify genes that are coexpressed or coclustered across all the data sets of the experiment. Reducing the data set from hundreds to thousands of genes with significant differential expression to a small number of clusters of coregulated genes simplifies the problem of drawing biological insights from data sets. A variety of clustering programs exist, including hierarchical clustering, principal component analysis, self-organizing maps, and neuronal networks (Carr et al., 1997; Eisen et al., 1998; Tamayo et al., 1999; Toronen et al., 1999; Evertsz et al., 2000; Zhou et al., 2000; Valafar, 2002). Many of these clustering algorithms have been reported to be successful. However, no one algorithm has emerged as a method of choice and each algorithm has its own advantages and disadvantages (Slonim, 2002). For example, with the hierarchical-clustering method, a researcher can follow an intuitive or subjective approach by first scanning the large-scale clusters and then focusing on the fine-scale clusters to identify which clusters are associated with a biological process (Eisen et al., 1998). However, hierarchical clustering often suffers from lack of robustness, nonuniqueness, and inversion problems (Morgan and Ray, 1995). Therefore, one intuitive way of analyzing the data is to test different clustering algorithms and to choose the one that best groups genes with similar function. Objective cluster-evaluation methods are being developed to assess the robustness of clusters generated by the various clustering programs (Yeung et al., 2001; McShane et al., 2002). Another feature that is important when clustering is to evaluate the stability of the cluster. A stable clustering process should give the same pattern when the experiment is repeated. Kerr and Churchill (2001) applied bootstrapping techniques to evaluate the stability of results from cluster analyses. Often, using the same clustering software, it is also possible to cluster experiments, providing some insight into which experimental treatments produce the most similar gene-expression profiles. This kind of analysis has been used in yeast to suggest a possible mode of action for a drug by comparing the expression profile elicited by the drug to expression profiles of known yeast mutants (Marton et al., 1998).

Placing groups of genes into clusters based on expression patterns can provide clues as to the possible function of uncharacterized genes. This "guilt-by-association" approach relies on the fact that similarity in transcription patterns correlates with similarity in function. Ruan and co-workers (1998) were the first to apply such an approach in plants. This approach has also been used in yeast (DeRisi et al., 1997; Chu et al., 1998), *Drosophila melanogaster* (White et al., 1999), mice (Lee et al., 1999), and humans (Iyer et al., 1999). Such guilt-by-association analysis can be extended to determining the patterns of expression of selected unknown genes in publicly available data sets (Table 4.2). As the number of public microarray data sets increases and structural knowledge of the gene product expands, researchers should be able to postulate gene functions for unknown genes with increasing confidence. Although this method of data analysis is appealing, validating the postulated gene function in mechanistic terms for a totally novel gene is likely to remain challenging.

In the case of plants with fully sequenced genomes, it is possible to look for common *cis*-acting elements in promoters of genes from the same cluster. An underlying assumption of this approach is that genes which are coordinately expressed are likely also to be coordinately regulated. A novel clock-controlled promoter element was discovered in a group of 31 cycling genes that showed a peak in expression near the end of the day (Harmer et al., 2000). Similarly, Maleck and co-workers (2000), in the study of systemic acquired resistance in *Arabidopsis,* identified a cluster containing *PR-1* in which W boxes were highly enriched in the promoters of these genes. A careful analysis of promoters of relevant gene clusters should give a boost to the discovery of transcription factors and DNA-binding proteins. In turn, such analysis could lead to new insights into the networks of gene regulation. Table 4.3 provides a listing of data extraction software, statistical methods for microarrays, and clustering software.

DATA VERIFICATION AND QUALITY CHECKS

A microarray experiment is a multistep process. If optimized protocols and microarray design, high-quality surface chemistry, pure RNA samples, appropriate scanning, careful experimental annotation, and well-designed quantification and data-mining tools are used, then this technology yields highly valuable results. However, microarray data can exhibit undesired variation that may arise from the biological materials or the technical aspects of microarray production, sample labeling and hybridization, or scanning steps. Several people have noted the influence of dye chemistry on the labeling of a small number of genes in two-color microarrays. For this

TABLE 4.2. Transcriptome databases.

Name*	Description	URL
ArrayExpress, European Bioinformatics Institute	Microarray data storage and retrieval; multiple organisms	<http://www.ebi.ac.uk/arrayexpress/>
Gene Expression Omnibus, NCBI	Microarray data storage and retrieval; multiple organisms	<http://www.ncbi.nlm.nih.gov/geo/>
Stanford Microarray Database, Stanford University	Microarray data storage, retrieval, analysis, and visualization; multiple organisms	<http://genomewww5.stanford.edu/MicroArray/SMD/>
Yale Microarray Database, Yale University	Microarray data storage, retrieval, analysis, and visualization; multiple organisms	<http://info.med.yale.edu/microarray/>
NASCArrays, Nottingham Arabidopsis Stock Center	Microarray data retrieval, *Arabidopsis*	<http://ssbdjc2.nottingham.ac.uk/cgi-bin/NARRAYS/experimentbrowse.pl>
The Arabidopsis Information Resource	Microarray data retrieval by element or experiment, *Arabidopsis*	<http://arabidopsis.org>
Complete Arabidopsis Transcriptome MicroArray (CATMA), 11 European laboratories	Microarray element search (GSTs), *Arabidopsis*	<http://www.catma.org/>
Arabidopsis Membrane Protein Library (AMPL), WARD Laboratory	Expression level of membrane proteins, *Arabidopsis* and others	<http://www.cbs.umn.edu/arabidopsis/>
Osmotic Stress Microarray Information Database (OSMID), University of Arizona	Microarray database; specific for plants subjected to abiotic stresses	<http://www.osmid.org/>
Nitrogene networks, New York University	Microarray data retrieval; specific for nitrogen-regulated genes	<http://www.nyu.edu/fas/biology/n2010/>
MaizeGDB, USDA	Maize genetics and genomics database	<http://www.maizegdb.org/>
Rice Microarray Opening Site (RMOS)	Microarray services, data storage and retrieval, rice	<http://cdna01.dna.affrc.go.jp/RMOS/>

*Most of the databases allow anyone to view and download public data. Only ArrrayExpress and GEO allow anyone to deposit data. TAIR (http://arabidopsis.org/info/expression/index.html) provides links to personal Web pages that contain microarray data for individual *Arabidopsis* projects.

TABLE 4.3. DNA array analysis software.

Name	Description	URL
AMAD	Data manager	<http://www.microarrays.org/AMADFaq.html>
Cluster	Data clustering	Eisen Laboratory <http://rana.lbl.gov/EisenSoftware.htm>
TreeView	Data visualization	Eisen Laboratory <http://rana.lbl.gov/EisenSoftware.htm>
XCluster	Data clustering	Stanford University <http://genetics.stanford.edu/~sherlock/cluster.html#history>
TIGR MIDAS	Data extraction	TIGR <http://www.tigr.org/software/#m>
MADAM	Data manager	TIGR <http://www.tigr.org/software/#m>
TIGR MEV	Data analysis	TIGR <http://www.tigr.org/software/#m>
TIGR Spotfinder	Data visualization	TIGR <http://www.tigr.org/software/#m>
J-express	Data analysis and visualization	Molmine <http://www.ii.uib.no/~bjarted/jexpress/install.html>
MAExplorer	Data analysis and visualization	<http://www.lecb.ncifcrf.gov/MAExplorer/>
SAM	Statistical analysis of data	Stanford University <http://www-stat.stanford.edu/~tibs/SAM/>
ArrayViewer	Data analysis	NHGRI <http://www.nhgri.nih.gov/DIR/LCG/15K/HTML/images.html>
AMIADA	Data analysis and visualization	University of Hong Kong <http://aix1.uottawa.ca/~xxia/software/amiada.htm>
Bioconductor	Data extraction and statistical analysis	<http://www.bioconductor.org/>

TABLE 4.3 *(continued)*

Name	Description	URL
GeneX-lite	Data extraction, analysis, and visualization	NCGR <http://www.ncgr.org/genex/>
ArrayVision	Data extraction and visualization	Imaging Research, Inc. <http://www.imagingresearch.com/>
ArrayStat	Statistical analysis of data	Imaging Research, Inc.<http://www.imagingresearch.com/>
BD AtlasImage	Data extraction and visualization	BD Biosciences <http://www.clontech.com/atlas/atlasimage/index.shtml>
ImaGene	Data extraction	BioDiscovery <http://www.biodiscovery.com/imagene.asp>
GeneSight	Data analysis and visualization	BioDiscovery <http://www.biodiscovery.com/genesight.asp>
GeneDirector	Data manager	BioDiscovery <http://www.biodiscovery.com/genedirector.asp>
GD Expressionist	Data extraction, manager, and statistical analysis	GeneData <http://www.genedata.com/products/expressionist/>
GeneSpring	Data analysis	Silicon Genetics <http://www.silicongenetics.com/cgi/SiG.cgi/Products/GeneSpring/index.smf>
IconoClust	Data extraction	Clondiag <http://www.clondiag.com/>
MicroArray Suite	Data extraction and visualization	Scanalytics <http://www.scanalytics.com/product/microarray/index.shtml>
Spotfire	Data analysis and visualization	Spotfire <http://www.spotfire.com/academics/products-fg.htm>
Partek Pro	Data extraction, analysis, and visualization	Partek <http://www.partek.com/html/products/products.html>

reason, it has been suggested that dye-swap experiments be performed when using such microarrays (Kerr et al., 2000). A study designed to better evaluate factors that contribute to technical variation in the production and use of two-color microarrays evaluated four potential sources of variation in two-color arrays: slide (handmade polylysine slides versus commercial slides), clone (8 cDNA clones of varying levels of expression), sector (16 printing pins), and replicate (20 replicate slides per slide type). Although each factor separately did not contribute strongly to variation with the possible exception of sector, interactions among the factors (e.g., replicate × clone) contributed significantly to variation in the data. The major conclusion drawn from this study was that highly standardized and automated fabrication and manipulation of microarrays should significantly improve the quality of the data (Finkelstein and Somerville, unpublished data). In another study of cDNA microarray performance, Yue and collaborators (2001) found that about 10 to 30 percent variation occurs among analogous elements on different microarray slides and the minimum detectable fold-change for differential expression was 1.4. Wang and colleagues (2003) pioneered the evaluation of data quality generated by short-oligonucleotide arrays and focused on issues such as intensity gradients and background problems (Schadt et al., 2000). In a survey of cDNA microarray data from the Arabidopsis Functional Genomics Consortium (AFGC), most large-scale distortions in the intensity values (e.g., gradient effects across the array) could be corrected by normalization methods, such as the Loess normalization method (Yang et al., 2002). However, array element sequence selection and microarray and experimental design are critical factors that, when carefully selected, can improve microarray data quality (Finkelstein et al., 2002). Microarray computational tools are beginning to be designed to evaluate the quality of the data, when replicated data sets are available, and using standardized quality control tests. Furthermore, the move to commercialized and centralized production of microarray platforms combined with greater automation of processing methods will reduce variation from array element to array element and from array to array. Thus, the reliability with which these methods can detect small fold-change differences in weakly expressed genes will be determined more by the biological variation of the samples than by the technology. Together, these improvements will lead to an overall improvement in the standard of quality for microarray data (Brazma et al., 2001; Ball et al., 2002a,b).

Validation of key results by an independent method is necessary, particularly if the number of replicates performed is small. Numerous technologies are available, such as Northern blot, quantitative or competitive RT-PCR, and quantitative ribonuclease protection assay. Quantitative RT-PCR is well suited for confirmations based on samples of small size (Goda et al., 2002;

Puthoff et al., 2003). Quantitative RT-PCR is also the method of choice for studying the expression of highly similar genes from the same family and for evaluating possible cross-hybridization problems.

LIMITATIONS

Since, ideally, a complete genomic DNA sequence is needed to design microarray oligonucleotides or DNA fragment elements covering the entire genome with minimal sequence overlap among genes, only plants such as *Arabidopsis* and rice can be analyzed on this scale at present. *Arabidopsis* arrays consisting of ~22,000 (http://www.affymetrix.com/support/technical/datasheets/arab_datasheet.pdf) to ~27,000 (http://www.catma.org/; http://ag.arizona.edu/microarray/) of ~28,000 predicted genes in the genome (TAIR, http://www.arabidopsis.org/info/agilinks.html) are now available. Although the complete genome sequence is available for *Arabidopsis* and rice, new genes, especially those with no homology to known genes, continue to be added to the genome complement as new EST collections are sequenced or as new technologies for identifying transcribed genes are developed. Often microarrays do not contain elements for chloroplast- or mitochondrial-encoded genes. Thus, even so-called full-genome microarrays are incomplete at this time.

Although the practical aspects of microarray technology are reasonably well established, there are still some limitations to this technology. Since many genes are grouped in gene families, cross hybridization of targets from several closely related genes can confound data interpretation. Girke and collaborators (2000) showed that a cross hybridization could occur among sequences that were greater than 70 percent identical under the conditions of their experiments. In the absence of full knowledge of the genome sequence or when partially characterized DNA fragments (e.g., EST clones) are used, as they sometimes are in two-color microarrays or macroarrays, cross hybridization among closely related gene family members is a distinct possibility. The expression of such genes can be resolved only by further experimentation using other technologies, such as RT-PCR. Oligonucleotide-based probes or genomic amplicons can be designed to distinguish among closely related genes, when full-genome sequence is available, and this is one advantage of using these array elements.

Plants are multicellular organisms, containing numerous organs consisting of different tissues with multiple cell types and the same cell type at different stages of development. Therefore, expression values derived from leaf samples, for example, will be a composite of all the cell types that make up a leaf. For some experiments, it may be appropriate to consider using cell suspension cultures, which ideally should consist of one cell type. Alternatively, laser-capture microdissection (LCM) can be used to sample individual cells. Although this technique has already been used widely in mammals (Emmert-Buck et al., 1996; Simone et al., 1998; Luo et al., 1999), its application to plants has been hampered by the plant cell wall and by ice crystal formation in frozen sections. Recently, Nakazono et al. (2003) optimized fixation methods for plant tissues and used LCM to sample cells from the epidermis and the vascular tissues of maize. An alternative to LCM is to extract cytoplasm from individual cells using micropipettes (Karrer et al., 1995). With amplification, cDNAs prepared from single cells could be used for gene-expression studies as well. Each of these methods has limitations, but when used in conjunction with rigorous confirmation experiments, the single-cell approaches will further refine our understanding of plant responses to environmental or developmental signals.

Another challenge in microarray studies is to distinguish primary from secondary or indirect responses to a treatment. Expression profiles of mutants can be particularly difficult to interpret if the mutant suffers from a number of pleiotropic abnormalities that stress the plants. A careful choice of time points, a detailed time-course spanning the response period (Cho et al., 1998), the use of a regulated promoter to drive expression in a controlled manner (Padidam, 2003), or the use of cycloheximide to inhibit secondary responses (Horvath et al., 1998) have been proposed as methods to identify early or primary responding genes. For example, significant changes in gene expression were observed as early as ten minutes following treatment of young seedlings with chito-oligomers (Ramonell et al., 2002), a much shorter time frame than is commonly used for plant gene-expression studies. Although some of these limitations are challenging, this technology, by its simplicity and affordability, will continue to dramatically enhance our understanding of many aspects of plant biology.

PERSPECTIVE

Microarray techniques have been responsible for the recent accumulation of a broad quantity of biological data. To date, this technology is one of the most-rewarding developments of the postsequencing era. Its capacity for evaluating large-scale gene-expression profiles in rapid and reproducible benchtop experiments is largely responsible for the success. The first microarray article compared gene expression in roots and leaves for 45 *Arabidopsis* genes (Schena et al., 1995). Eight years later, more than 100 organisms are studied with this technology (http://arrayit.com/e-library).

Transcriptomics and other "-omics" are pushing plant biology from hypothesis-driven studies to discovery-based research, which will in turn spur a new phase of hypothesis-driven research (Brown and Botstein, 1999). For example, the use of computer-based analysis of expression profiles will lead to new hypotheses about regulatory networks and interacting genes or gene products controlling a physiological or developmental process. These kinds of results will lead to laboratory tests of the regulatory networks and their components, leading to further refinements of the inferences drawn from computer-based predictions. In turn, new genomics-based projects may be suggested by the laboratory experiments (Figure 4.1). This cycling between discovery-based and hypothesis-based experimentation will surely give rise to a novel golden age for fundamental breakthroughs.

To facilitate such analyses, centralized transcriptome databases are essential. Like GenBank and similar sequence databases, transcriptome databases will stimulate discovery by scientists, who did not generate the primary data, in unexpected ways. The Microarray Gene Expression Data (MGED) working group (http://www.mged.org) was established to facilitate data sharing of microarray expression data. This group promotes two sets of standards. The MIAME (Minimum Information About a Microarray Experiment) standard is a set of recommendations about the information that needs to be deposited along with the expression values to ensure microarray data can be easily interpreted and independently verified (Brazma et al., 2001). The MAGE group "aims to provide a standard for the representation of microarray expression data that would facilitate the exchange of microarray information between different data systems" by developing a data exchange model (text taken from <http://www.mged.org>). Furthermore, this group has recommended that all microarray data be deposited in publicly accessible, MGED-compliant formats, coincident with publication (Brazma et al., 2001). Already, several major journals have adopted this policy. The large sequence database groups have each developed microarray expression databases to meet this need in the scientific community; EBI has established ArrayExpress (Brazma et al., 2002) and NCBI has set up the Gene Expression Omnibus (GEO) (Edgar et al., 2002) (Table 4.2). To date, there are only 24 *Arabidopsis* (21 cDNA microarray and 3 SAGE), 24 maize, 24 soybean, and 3 *Medicago truncatula* data sets available in GEO and no plant data sets available from ArrayExpress. One of the first publicly accessible microarray databases to be established was the Stanford Microarray Database (Table 4.2), which provides access to more than 4,700 two-color or cDNA microarray experiments from a broad range of organisms, including 590 *Arabidopsis* data sets (Gollub et al., 2003). Such centralization of data in a standardized format allows ready access for secondary users and facilitates reanalysis of data. For example,

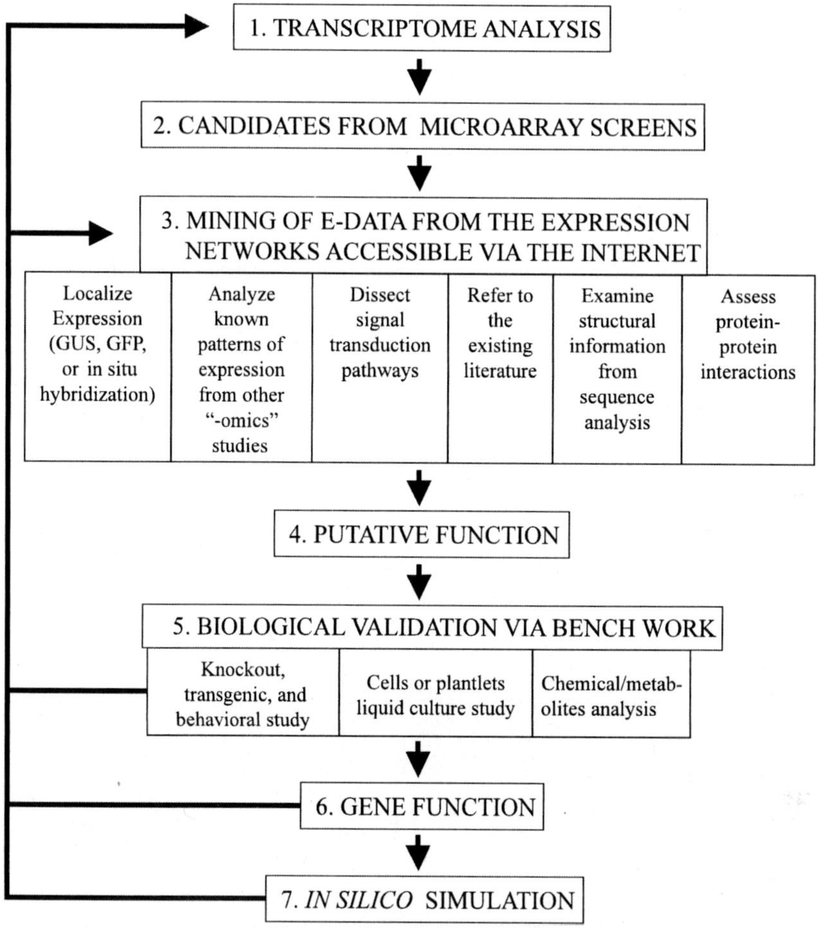

FIGURE 4.1. En route to *in silico* modeling via transcriptomics. (1) The analysis of the transcriptome can follow exploratory or marker-discovery paths. (2) From either approach, candidate genes, which need to be analyzed for their significance in the process studied, are identified. (3) Consulting e-databases will provide complementary information for candidate genes of unknown function. (4) By merging various kinds of e-information, the researcher can hypothesize a gene function. (5) This putative function needs to be tested via bench work. (6) New results will lead to a new assigned gene function. (7) Grouping new knowledge about gene function will permit the development of *in silico* modeling of plant responses to developmental or environmental cues. Additional iterations of this general pattern of experimentation will lead to further refinements of our understanding of the plant.

such databases can be used for *in silico* Northern blots, allowing researchers to assess gene expression of specific genes across a wide range of experimental conditions (Schultz et al., 2002). Such analyses are aided by viewing tools such as "Expression Viewer" and "Spot History" (Figure 4.2). In TAIR, these two tools are available for user-specified *Arabidopsis* genes (http://arabidopsis.org/tools/bulk/microarray/index.html), and the Nottingham Arabidopsis Stock Center provides a "Spot History" viewer for *Arabidopsis* microarray expression data collected as part of the NASCArrays project (http://affy.arabidopsis.info/narrays/spothistory.pl).

A future development that one can imagine is the development of a centralized genomics center(s) that would, among other tasks, regularly release

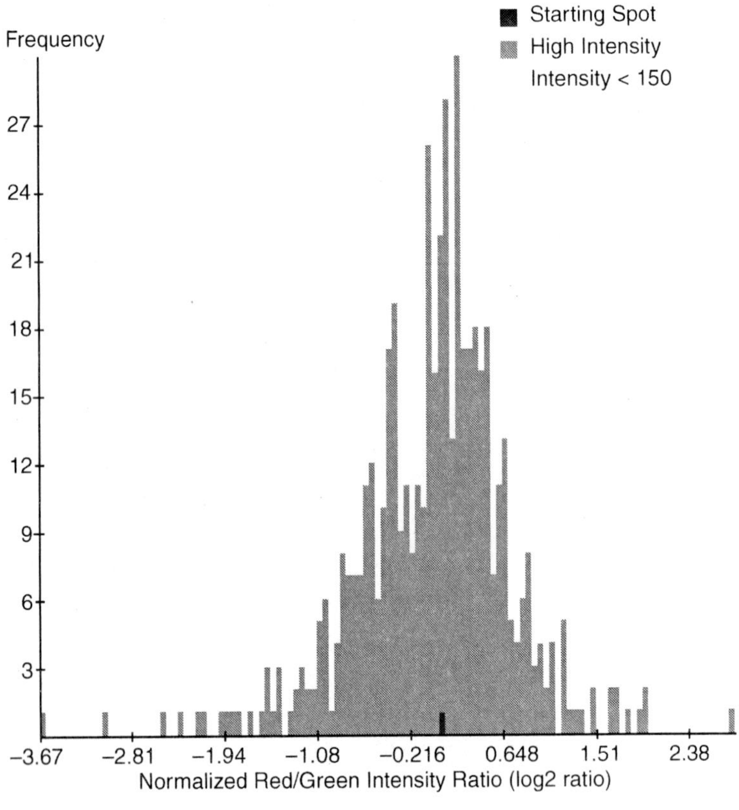

FIGURE 4.2. Spot History for *Arabidopsis* glyceraldehyde-3-phosphate dehydrogenase C subunit (GAPC; Locus: At3G04120; EST:178F2T7) (see corresponding figure in color photo section).

raw or minimally processed gene-expression data for data exploration (Minorsky, 2003). By optimizing and standardizing the various steps of microarray data production, some of the variability associated with this technology could be minimized. If such a centralized facility could perform standardized sets of highly replicated experiments (e.g., surveys of expression patterns in various plant organs or tissues), the need for each lab to perform Northern blots to determine the tissue-expression profiles of specific genes would disappear. One can also imagine the use of large numbers of data points that will provide a "movielike" view of changing gene expression. Such elaborate experiments are currently cost prohibitive for individual labs to perform. This dynamic analysis of gene expression in conjunction with the development of software able to depict variation in gene-expression levels over time would provide researchers an even better evaluation of networks of gene regulation. Such an approach, based on a centralized genomics center, would surely dramatically increase the reliability and accessibility of microarray data. This should lead to an increased use of microarray data by the scientific community and boost the discovery rate of gene function in plants.

This global, interconnected approach would ideally be extended to all genomics technologies, including transcriptomic, proteomic, metabolomic, signal transduction, and protein-protein interactions data, and other related parameters such as gene-expression patterns analyzed by in situ hybridization. As a start in this direction, TAIR (Rhee et al., 2003) was established as a global information access for the *Arabidopsis* scientific community. Similarly, the *Drosophila* community has established a database that combines in situ hybridization studies and microarray-based genomic analyses to create a comprehensive atlas of gene-expression pattern during embryogenesis (Montalta-He and Reichert, 2003). This is an interesting example of what could be added to plant databases to increase visualization of gene expression during specific conditions of development or treatments. Integrated data sets will lead to a profound understanding of gene networks and system rules and will allow the development of *in silico* plants for modeling or "what-if" experiments. Such computer representations have already been tested in the bacterium *Escherichia coli* (Ibarra et al., 2002). Ultimately, through the ability to engineer genomes at the laboratory bench, novel plant forms could be developed that will fit the increasing demands of an exploding human population.

REFERENCES

Abe, H., Urao, T., Ito, T., Seki, M., Shinozaki, K., Yamaguchi-Shinozaki, K. (2003). *Arabidopsis* AtMYC2 (bHLH) and AtMYB2 (MYB) function as transcriptional activators in abscisic acid signaling. *Plant Cell* 15: 63-78.

Arabidopsis Genome Initiative (2000). Analysis of the genome of the flowering plant *Arabidopsis thaliana*. *Nature* 408: 796-815.

Ashburner, M., Ball, C.A., Blake, J.A., Botstein, D., Butler, H., Cherry, J.M., Davis, A.P., Dolinski, K., Dwight, S.S., Eppig, J.T., et al. (2000). Gene ontology: Tool for the unification of biology. The Gene Ontology Consortium. *Nature Genet.* 25: 25-29.

Bachem, C.W., van der Hoeven, R.S., de Bruijn, S.M., Vreugdenhil, D., Zabeau, M., Visser, R.G. (1996). Visualization of differential gene expression using a novel method of RNA fingerprinting based on AFLP: Analysis of gene expression during potato tuber development. *Plant J.* 9: 745-753.

Ball, C.A., Sherlock, G., Parkinson, H., Rocca-Sera, P., Brooksbank, C., Causton, H.C., Cavalieri, D., Gaasterland, T., Hingamp, P., Holstege, F., et al. (2002a). Standards for microarray data. *Science* 298: 539.

Ball, C.A., Sherlock, G., Parkinson, H., Rocca-Sera, P., Brooksbank, C., Causton, H.C., Cavalieri, D., Gaasterland, T., Hingamp, P., Holstege, F., et al. (2002b). The underlying principles of scientific publication. *Bioinformatics* 18: 1409.

Bland, J.M., Altman, D.G. (1995). Multiple significant test: The Bonferroni method. *Brit. Med. J.* 310: 170.

Botwell, D., Sambrook, J. (2003). *DNA Microarrays: A Molecular Cloning Manual.* Cold Spring Harbor, NY: Cold Spring Harbor Laboratory Press.

Brazma, A., Hingamp, P., Quackenbush, J., Sherlock, G., Spellman, P., Stoeckert, C., Aach, J., Ansorge, W., Ball, C.A., Causton, H.C., et al. (2001). Minimum information about a microarray experiment (MIAME): Toward standards for microarray data. *Nat. Genet.* 29: 365-371.

Brazma, A., Parkinson, H., Sarkans, U., Shojatalab, M., Vilo, J., Abeygunawardena, N., Holloway, E., Kapushesky, M., Kemmeren, P., Sansone, S.A. (2002). ArrayExpress: A public repository for microarray gene expression data at the EBI. *Nucleic Acids Res.* 31: 68-71.

Brenner, S., Johnson, M., Bridgham, J., Golda, G., Lloyd, D.H., Johnson, D., Luo, S., McCurdy, S., Foy, M., Ewan, M., et al. (2000). Gene expression analysis by massively parallel signature sequencing (MPSS) on microbead arrays. *Nat. Biotechnol.* 18: 630-634.

Brown, P.O., Botstein, D. (1999). Exploring the new world of the genome with DNA microarrays. *Nat. Genet.* 21(Suppl.): 33-37.

Bruce, W., Folkerts, O., Garnaat, C., Crasta, O., Roth, B., Bowen, B. (2000). Expression profiling of the maize flavonoid pathway genes controlled by estradiol-inducible transcription factors CRC and P. *Plant Cell* 12: 65-80.

Carlisle, A.J., Prabhu, V.V., Elkahloun, A., Hudson, J., Trent, J.M., Linehan, W.M., Williams, E.D., Emmert-Buck, M.R., Liotta, L.A., Munson, P.J., Krizman, D.B.

(2000). Development of a prostate cDNA microarray and statistical gene expression analysis package. *Mol. Carcinog.* 28: 12-22.

Carr, D.B., Somogyi, R., Michaels, G. (1997). Templates for looking at gene expression clustering. *Stat. Comput. Graph. Newslet.* 8: 20-29.

Chen, W., Provart, N.J., Glazebrook, J., Katagiri, F., Chang, H.S., Eulgem, T., Mauch, F., Luan, S., Zou, G., Whitham, S.A., et al. (2002). Expression profile matrix of *Arabidopsis* transcription factor genes suggests their putative functions in response to environmental stresses. *Plant Cell* 14: 559-574.

Cho, R.J., Campbell, M.J., Winzeler, E.A., Steinmetz, L., Conway, A., Wodicka, L., Wolfsberg, T.G., Gabrielian, A.E., Landsman, D., Lockhart, D.J., Davis, R.W. (1998). A genome-wide transcriptional analysis of the mitotic cell cycle. *Mol. Cell* 2: 65-73.

Chu, S., DeRisi, J., Eisen, M., Mulholland, J., Botstein, D., Brown, P.O., Herskowitz, I. (1998). The transcriptional program of sporulation in budding yeast. *Science* 282: 699-705.

Churchill, G.A. (2002). Fundamentals of experimental design for cDNA microarrays. *Nat. Genet.* 32: 490-495.

Dahlquist, K.D., Salomonis, N., Vranizan, K., Lawlor, S.C., Conklin, B.R. (2002). GenMAPP, a new tool for viewing and analyzing microarray data on biological pathways. *Nat. Genet.* 31: 19-20.

DeRisi, J.L., Iyer, V.R., Brown, P.O. (1997). Exploring the metabolic and genetic control of gene expression on a genomic scale. *Science* 278: 680-686.

Devore, J., Peck, R. (1997). *Statistics: The Exploration and Analysis of Data.* Pacific Grove, CA: Duxbury Press.

Diatchenko, L., Lau, Y.F., Campbell, A.P., Chenchik, A., Moqadam, F., Huang, B., Lukyanov, S., Lukyanov, K., Gurskaya, N., Sverdlov, E.D., Siebert, P.D. (1996). Suppression subtractive hybridization: A method for generating differentially regulated or tissue-specific cDNA probes and libraries. *Proc. Natl. Acad. Sci. USA* 93: 6025-6030.

Draghici, S. (2002). Statistical intelligence: Effective analysis of high-density microarray data. *Drug Discov. Today* 7: S55-S63.

Durrant, W.E., Rowland, O., Piedras, P., Hammond-Kosack, K.E., Jones, J.D. (2000). cDNA-AFLP reveals a striking overlap in race-specific resistance and wound response gene expression profiles. *Plant Cell* 12: 963-977.

Edgar, R., Domrachev, M., Lash, A.E. (2002). Gene expression omnibus: NCBI gene expression and hybridization array data repository. *Nucleic Acids Res.* 30: 207-210.

Eisen, M.B., Brown, P.O. (1999). DNA arrays for analysis of gene expression. *Methods Enzymol.* 303: 179-205.

Eisen, M.B., Spellman, P.T., Brown, P.O., Botstein, D. (1998). Cluster analysis and display of genome-wide expression patterns. *Proc. Natl. Acad. Sci. USA* 95: 14863-14868.

Emmert-Buck, M.R., Bonner, R.F., Smith, P.D., Chuaqui, R.F., Zhuang, Z., Goldstein, S.R., Weiss, R.A., Liotta, L.A. (1996). Laser capture microdissection. *Science* 274: 998-1001.

Evertsz, E., Gupta, R., Starink, P., Watson, D. (2000). Technology and applications of gene expression microarrays. In Schena, M. (Ed.), *Microarray Biochip Technology* (pp. 149-166). Natick, MA: Eaton Publishing.

Fedorova, M., van de Mortel, J., Matsumoto, P.A., Cho, J., Town, C.D., Vanden-Bosch, K.A., Gantt, J.S., Vance, C.P. (2002). Genome-wide identification of nodule-specific transcripts in the model legume *Medicago truncatula. Plant Physiol.* 130: 519-537.

Fernandes, J., Brendel, V., Gai, X., Lal, S., Chandler, V.L., Elumalai, R.P., Galbraith, D.W., Pierson, E.A., Walbot, V. (2002). Comparison of RNA expression profiles based on maize expressed sequence tag frequency analysis and microarray hybridization. *Plant Physiol.* 128: 896-910.

Finkelstein, D., Ewing, R., Gollub, J., Sterky, F., Cherry, J.M., Somerville, S. (2002). Microarray data quality analysis: Lessons from the AFGC project. *Plant Mol. Biol.* 48: 119-131.

Finkelstein, D.B., Gollub, J., Ewing, R., Sterky, F., Somerville, S., Cherry, J.M. (2000). Iterative linear regression by sector: Re-normalization of cDNA microarray data and cluster analysis weighted by cross homology. In *Critical Assessment of Microarray Data* (p. 7). Durham, NC: Duke University Medical Center.

Girke, T., Todd, J., Ruuska, S., White, J., Benning, C., Ohlrogge, J. (2000). Microarray analysis of developing *Arabidopsis* seeds. *Plant Physiol.* 124: 1570-1581.

Glazebrook, J., Chen, W., Estes, B., Chang, H.S., Nawrath, C., Metraux, J.P., Zhu, T., Katagiri, F. (2003). Topology of the network integrating salicylate and jasmonate signal transduction derived from global expression phenotyping. *Plant J.* 34: 217-228.

Goda, H., Shimada, Y., Asami, T., Fujioka, S., Yoshida, S. (2002). Microarray analysis of brassinosteroid-regulated genes in *Arabidopsis. Plant Physiol.* 30: 1319-1334.

Goff, S.A., Ricke, D., Lan, T.H., Presting, G., Wang, R., Dunn, M., Glazebrook, J., Sessions, A., Oeller, P., Varma, H., et al. (2002). A draft sequence of the rice genome *(Oryza sativa* L. ssp. *japonica). Science* 296: 92-100.

Gollub, J., Ball, C.A., Binkley, G., Demeter, J., Finkelstein, D.B., Hebert, J.M., Hernandez-Boussard, T., Jin, H., Kaloper, M., Matese, J.C., et al. (2003). The Stanford Microarray Database: Data access and quality assessment tools. *Nucleic Acids Res.* 31: 94-96.

Gygi, S.P., Rochon, Y., Franza, B.R., Aebersold, R. (1999). Correlation between protein and mRNA abundance in yeast. *Mol. Cell. Biol.* 19: 1720-1730.

Harmer, S.L., Hogenesch, J.B., Straume, M., Chang, H.S., Han, B., Zhu, T., Wang, X., Kreps, J.A., Kay, S.A. (2000). Orchestrated transcription of key pathways in *Arabidopsis* by the circadian clock. *Science* 290: 2110-2113.

Hill, D.P., Blake, J.A., Richardson, J.E., Ringwald, M. (2002). Extension and integration of the Gene Ontology (GO): Combining GO vocabularies with external vocabularies. *Genome Res.* 12: 1982-1991.

Hirai, M.Y., Fujiwara, T., Awazuhara, M., Kimura, T., Noji, M., Saito, K. (2003). Global expression profiling of sulfur-starved *Arabidopsis* by DNA macroarray reveals the role of *O*-acetyl-l-serine as a general regulator of gene expression in response to sulfur nutrition. *Plant J.* 33: 651-663.

Horvath, D.M., Huang, D.J., Chua, N.H. (1998). Four classes of salicylate-induced tobacco genes. *Mol. Plant Microbe Interact.* 11: 895-905.

Ibarra, R.U., Edwards, J.S., Palsson, B.O. (2002). *Escherichia coli* K-12 undergoes adaptive evolution to achieve *in silico* predicted optimal growth. *Nature* 420: 186-189.

Ideker, T., Thorsson, V., Siegel, A.F., Hood, L.E. (2000). Testing for differentially expressed genes by maximum-likelihood analysis of microarray data. *J. Comput. Biol.* 7: 805-817.

Iyer, V.R., Eisen, M.B., Ross, D.T., Schuler, G., Moore, T., Lee, J.C., Trent, J.M., Staudt, L.M., Hudson, J. Jr., Boguski, M.S., et al. (1999). The transcriptional program in the response of human fibroblasts to serum. *Science* 283: 83-87.

Kane, M.D., Jatkoe, T.A., Stumpf, C.R., Lu, J., Thomas, J.D., Madore, S.J. (2000). Assessment of the sensitivity and specificity of oligonucleotide (50mer) microarrays. *Nucleic Acids Res.* 28: 4552-4557.

Karrer, E.E., Lincoln, J.E., Hogenhout, S., Bennett, A.B., Bostock, R.M., Martineau, B., Lucas, W.J., Gilchrist, D.G., Alexander, D. (1995). In situ isolation of mRNA from individual plant cells: Creation of cell-specific cDNA libraries. *Proc. Natl. Acad. Sci. USA* 92: 3814-3818.

Kawai-Yamada, M., Jin, L., Yoshinaga, K., Hirata, A., Uchimiya, H. (2001). Mammalian Bax-induced plant cell death can be down-regulated by overexpression of *Arabidopsis* Bax Inhibitor-1 (AtBI-1). *Proc. Natl. Acad. Sci. USA* 98: 12295-12300.

Kawasaki, S., Borchert, C., Deyholos, M., Wang, H., Brazille, S., Kawai, K., Galbraith, D., Bohnert, H.J. (2001). Gene expression profiles during the initial phase of salt stress in rice. *Plant Cell* 13: 889-905.

Kerr, M.K., Churchill, G.A. (2001). Bootstrapping cluster analysis: Assessing the reliability of conclusions from microarray experiments. *Proc. Natl. Acad. Sci. USA* 98: 8961-8965.

Kerr, M.K., Martin, M., Churchill, G.A. (2000). Analysis of variance for gene expression microarray data. *J. Comput. Biol.* 7: 819-837.

Kreps, J.A., Wu, Y., Chang, H.S., Zhu, T., Wang, X., Harper, J.F. (2002). Transcriptome changes for *Arabidopsis* in response to salt, osmotic, and cold stress. *Plant Physiol.* 130: 2129-2141.

Lee, C.K., Klopp, R.G., Weindruch, R., Prolla, T.A. (1999). Gene expression profile of aging and its retardation by caloric restriction. *Science* 285: 1390-1393.

Lee, M.L., Kuo, F.C., Whitmore, G.A., Sklar, J. (2000). Importance of replication in microarray gene expression studies: Statistical methods and evidence from repetitive cDNA hybridizations. *Proc. Natl. Acad. Sci. USA* 97: 9834-9839.

Legen, J., Kemp, S., Krause, K., Profanter, B., Herrmann, R.G., Maier, R.M. (2002). Comparative analysis of plastid transcription profiles of entire plastid chromosomes from tobacco attributed to wild-type and PEP-deficient transcription machineries. *Plant J.* 31: 171-188.

Lin, F., Xu, S.L., Ni, W.M., Chu, Z.Q., Xu, Z.H., Xue, H.W. (2003). Identification of ABA-responsive genes in rice shoots via cDNA macroarray. *Cell Res.* 13: 59-68.

Lipshutz, R.J., Fodor, S.P., Gingeras, T.R., Lockhart, D.J. (1999). High density synthetic oligonucleotide arrays. *Nat. Genet.* 21: 20-24.

Lipshutz, R.J., Morris, D., Chee, M., Hubbell, E., Kozal, M.J., Shah, N., Shen, N., Yang, R., Fodor, S.P. (1995). Using oligonucleotide probe arrays to access genetic diversity. *Biotechniques* 19: 442-447.

Lorenz, W.W., Dean, J.F. (2002). SAGE profiling and demonstration of differential gene expression along the axial developmental gradient of lignifying xylem in loblolly pine *(Pinus taeda). Tree Physiol.* 22: 301-310.

Luo, L., Salunga, R.C., Guo, H., Bittner, A., Joy, K.C., Galindo, J.E., Xiao, H., Rogers, K.E., Wan, J.S., Jackson, M.R., Erlander, M.G. (1999). Gene expression profiles of laser-captured adjacent neuronal subtypes. *Nat. Med.* 5: 117-122.

Ma, L., Li, J., Qu, L., Hager, J., Chen, Z., Zhao, H., Deng, X.W. (2001). Light control of *Arabidopsis* development entails coordinated regulation of genome expression and cellular pathways. *Plant Cell* 13: 2589-2607.

Mahalingam, R., Gomez-Buitrago, A., Eckardt, N., Shah, N., Guevara-Garcia, A., Day, P., Raina, R., Fedoroff, N.V. (2003). Characterizing the stress/defense transcriptome of *Arabidopsis. Genome Biol.* 4: R20.

Maleck, K., Levine, A., Eulgem, T., Morgan, A., Schmid, J., Lawton, K.A., Dangl, J.L., Dietrich, R.A. (2000). The transcriptome of *Arabidopsis thaliana* during systemic acquired resistance. *Nat. Genet.* 26: 403-410.

Marton, M.J., DeRisi, J.L., Bennett, H.A., Iyer, V.R., Meyer, M.R., Roberts, C.J., Stoughton, R., Burchard, J., Slade, D., Dai, H., et al. (1998). Drug target validation and identification of secondary drug target effects using DNA microarrays. *Nat. Med.* 4: 1293-1301.

Matsumura, H., Nirasawa, S., Kiba, A., Urasaki, N., Saitoh, H., Ito, M., Kawai-Yamada, M., Uchimiya, H., Terauchi, R. (2003). Overexpression of Bax inhibitor suppresses the fungal elicitor-induced cell death in rice (*Oryza sativa* L.) cells. *Plant J.* 33: 425-434.

Matsumura, H., Nirasawa, S., Terauchi, R. (1999). Technical advance: Transcript profiling in rice (*Oryza sativa* L.) seedlings using serial analysis of gene expression (SAGE). *Plant J.* 20: 719-726.

McShane, L.M., Radmacher, M.D., Freidlin, B., Yu, R., Li, M.C., Simon, R. (2002). Methods for assessing reproducibility of clustering patterns observed in analyses of microarray data. *Bioinformatics* 18: 1462-1469.

Minorsky, P.V. (2003). Achieving the computable plant: System biology and the future of plant biological research. *Plant Physiol.* 132: 404-409.

Montalta-He, H., Reichert, H. (2003). Impressive expressions: Developing a systematic database of gene-expression patterns in *Drosophila* embryogenesis. *Genome Biol.* 4: 205.

Morgan, B.J.T., Ray, A.P.G. (1995). Non-uniqueness and inversions in cluster analysis. *Appl. Stat.* 44: 117-134.

Mueller, L.A., Zhang, P., Rhee, S.Y. (2003). AraCyc: A biochemical pathway database for *Arabidopsis. Plant Physiol.* 132: 453-460.

Nakazono, M., Qiu, F., Borsuk, L.A., Schnable, P.S. (2003). Laser-capture microdissection, a tool for the global analysis of gene expression in specific plant cell types: Identification of genes expressed differentially in epidermal cells or vascular tissues of maize. *Plant Cell* 15: 583-596.

Negishi, T., Nakanishi, H., Yazaki, J., Kishimoto, N., Fujii, F., Shimbo, K., Yamamoto, K., Sakata, K., Sasaki, T., Kikuchi, S., et al. (2002). cDNA microarray analysis of gene expression during Fe-deficiency stress in barley suggests that polar transport of vesicles is implicated in phytosiderophore secretion in Fe-deficient barley roots. *Plant J.* 30: 83-94.

Newton, M.A., Kendziorski, C.M., Richmond, C.S., Blattner, F.R., Tsui, K.W. (2001). On differential variability of expression ratios: Improving statistical inference about gene expression changes from microarray data. *J. Comput. Biol.* 8: 37-52.

Padidam, M. (2003). Chemically regulated gene expression in plants. *Curr. Opin. Plant Biol.* 6: 169-177.

Potokina, E., Sreenivasulu, N., Altschmied, L., Michalek, W., Graner, A. (2002). Differential gene expression during seed germination in barley (*Hordeum vulgare* L.). *Funct. Integr. Genomics* 2: 28-39.

Puthoff, D.P., Nettleton, D., Rodermel, S.R., Baum, T.J. (2003). *Arabidopsis* gene expression changes during cyst nematode parasitism revealed by statistical analyses of microarray expression profiles. *Plant J.* 33: 911-921.

Ramonell, K.M., Zhang, B., Ewing, R.M., Chen, Y., Dong, X., Stacey, G., Somerville, S. (2002). Microarray analysis of chitin elicitation in *Arabidopsis thaliana*. *Mol. Plant Pathol.* 3: 301-311.

Reymond, P., Weber, H., Damond, M., Farmer, E.E. (2000). Differential gene expression in response to mechanical wounding and insect feeding in *Arabidopsis*. *Plant Cell* 12: 707-720.

Rhee, S.Y., Beavis, W., Berardini, T.Z., Chen, G., Dixon, D., Doyle, A., Garcia-Hernandez, M., Huala, E., Lander, G., Montoya, M., et al. (2003). The Arabidopsis Information Resource (TAIR): A model organism database providing a centralized, curated gateway to *Arabidopsis* biology, research materials and community. *Nucleic Acids Res.* 31: 224-228.

Richly, E., Dietzmann, A., Biehl, A., Kurth, J., Laloi, C., Apel, K., Salamini, F., Leister, D. (2003). Covariations in the nuclear chloroplast transcriptome reveal a regulatory master-switch. *EMBO Rep.* 4: 491-498.

Richmond, T., Somerville, S. (2000). Chasing the dream: Plant EST microarrays. *Curr. Opin. Plant Biol.* 3: 108-116.

Roberts, J.K. (2002). Proteomics and a future generation of plant molecular biologists. *Plant Mol. Biol.* 48: 143-154.

Ruan, Y., Gilmore, J., Conner, T. (1998). Towards *Arabidopsis* genome analysis: Monitoring expression profiles of 1400 genes using cDNA microarrays. *Plant J.* 15: 821-833.

Ruuska, S.A., Girke, T., Benning, C., Ohlrogge, J.B. (2002). Contrapuntal networks of gene expression during *Arabidopsis* seed filling. *Plant Cell* 14: 1191-1206.

Sawa, S., Ohgishi, M., Goda, H., Higuchi, K., Shimada, Y., Yoshida, S., Koshiba, T. (2002). The *HAT2* gene, a member of the HD-Zip gene family, isolated as an auxin inducible gene by DNA microarray screening, affects auxin response in Arabidopsis. *Plant J.* 32: 1011-1022.

Schadt, E.E., Li, C., Su, C., Wong, W.H. (2000). Analyzing high-density oligonucleotide gene expression array data. *J. Cell. Biochem.* 80: 192-202.

Schaffer, R., Landgraf, J., Accerbi, M., Simon, V., Larson, M., Wisman, E. (2001). Microarray analysis of diurnal and circadian-regulated genes in *Arabidopsis. Plant Cell* 13: 113-123.

Schena, M., Shalon, D., Davis, R.W., Brown, P.O. (1995). Quantitative monitoring of gene expression patterns with a complementary DNA microarray. *Science* 270: 467-470.

Schoof, H., Zaccaria, P., Gundlach, H., Lemcke, K., Rudd, S., Kolesov, G., Arnold, R., Mewes, H.W., Mayer, K.F. (2002). MIPS *Arabidopsis thaliana* Database (MAtDB): An integrated biological knowledge resource based on the first complete plant genome. *Nucleic Acids Res.* 30: 91-93.

Schultz, C., Rumsewicz, M.P., Johnson, K.L., Jones, B.J., Gasper, Y.M., Bacic, A. (2002). Using genomics resources to guide research directions. The arabinogalactan protein gene family as a test case. *Plant Physiol.* 129: 1448-1463.

Seki, M., Narusaka, M., Abe, H., Kasuga, M., Yamaguchi-Shinozaki, K., Carninci, P., Hayashizaki, Y., Shinozaki, K. (2001). Monitoring the expression pattern of 1300 *Arabidopsis* genes under drought and cold stresses by using a full-length cDNA microarray. *Plant Cell* 13: 61-72.

Seki, M., Narusaka, M., Ishida, J., Nanjo, T., Fujita, M., Oono, Y., Kamiya, A., Nakajima, M., Enju, A., Sakurai, T., et al. (2002). Monitoring the expression profiles of 7000 *Arabidopsis* genes under drought, cold and high-salinity stresses using a full-length cDNA microarray. *Plant J.* 31: 279-292.

Shimkets, R.A., Lowe, D.G., Tai, J.T., Sehl, P., Jin, H., Yang, R., Predki, P.F., Rothberg, B.E., Murtha, M.T., Roth, M.E., et al. (1999). Gene expression analysis by transcript profiling coupled to a gene database query. *Nat. Biotechnol.* 17: 798-803.

Simone, N.L., Bonner, R.F., Gillespie, J.W., Emmert-Buck, M.R., Liotta, L.A. (1998). Laser-capture microdissection: Opening the microscopic frontier to molecular analysis. *Trends Genet.* 14: 272-276.

Slonim, D.K. (2002). From patterns to pathways: Gene expression data analysis comes of age. *Nat. Genet.* 32(Suppl.): 502-508.

Smyth, G.K., Yang, Y.H., Speed, T. (2003). Statistical issues in cDNA microarray data analysis. *Methods Mol. Biol.* 224: 111-136.

Sreenivasulu, N., Altschmied, L., Panitz, R., Hahnel, U., Michalek, W., Weschke, W., Wobus, U. (2002). Identification of genes specifically expressed in maternal and filial tissues of barley caryopses: A cDNA array analysis. *Mol. Genet. Genom.* 266: 758-767.

Steemers, F.J., Ferguson, J.A., Walt, D.R. (2000). Screening unlabeled DNA targets with randomly ordered fiber-optic gene arrays. *Nat. Biotechnol.* 18: 91-94.

Stitt, M., Fernie, A.R. (2003). From measurements of metabolites to metabolomics: An "on the fly" perspective illustrated by recent studies of carbon-nitrogen interactions. *Curr. Opin. Biotechnol.* 14: 136-144.

Storey, J.D. (2002). A direct approach to false discovery rates. *J. Royal Stat. Soc.* (Series B) 64: 479-498.

Sutcliffe, J.G., Foye, P.E., Erlander, M.G., Hilbush, B.S., Bodzin, L.J., Durham, J.T., Hasel, K.W. (2000). TOGA: An automated parsing technology for analyzing expression of nearly all genes. *Proc. Natl. Acad. Sci. USA* 97: 1976-1981.

Tamayo, P., Slonim, D., Mesirov, J., Zhu, Q., Kitareewan, S., Dmitrovsky, E., Lander, E.S., Golub, T.R. (1999). Interpreting patterns of gene expression with self-organizing maps: Methods and application to hematopoietic differentiation. *Proc. Natl. Acad. Sci. USA* 96: 2907-2912.

Thibaud-Nissen, F., Shealy, R.T., Khanna, A., Vodkin, L.O. (2003). Clustering of microarray data reveals transcript patterns associated with somatic embryogenesis in soybean. *Plant Physiol.* 132: 118-136.

Toronen, P., Kolehmainen, M., Wong, G., Castren, E. (1999). Analysis of gene expression data using self-organizing maps. *FEBS Lett.* 451: 142-146.

Tusher, V.G., Tibshirani, R., Chu, G. (2001). Significance analysis of microarrays applied to the ionizing radiation response. *Proc. Natl. Acad. Sci. USA* 98: 5116-5121.

Uhde-Stone, C., Zinn, K.E., Ramirez-Yanez, M., Li, A., Vance, C.P., Allan, D.L. (2003). Nylon filter arrays reveal differential gene expression in proteoid roots of white lupin in response to phosphorus deficiency. *Plant Physiol.* 131: 1064-1079.

Valafar, F. (2002). Pattern recognition techniques in microarray data analysis. *Ann. N. Y. Acad. Sci.* 980: 41-64.

Velculescu, V.E., Zhang, L., Vogelstein, B., Kinzler, K.W. (1995). Serial analysis of gene expression. *Science* 270: 484-487.

Vranova, E., Atichartpongkul, S., Villarroel, R., Van Montagu, M., Inze, D., Van Camp, W. (2002). Comprehensive analysis of gene expression in *Nicotiana tabacum* leaves acclimated to oxidative stress. *Proc. Natl. Acad. Sci. USA* 99: 10870-10875.

Wang, H., Ma, L., Habashi, J., Li, J., Zhao, H., Deng, X.W. (2002). Analysis of far-red light-regulated genome expression profiles of phytochrome A pathway mutants in *Arabidopsis*. *Plant J.* 32: 723-733.

Wang, H.Y., Malek, R.L., Kwitek, A.E., Greene, A.S., Luu, T.V., Behbahani, B., Frank, B., Quackenbush, J., Lee, N.H. (2003). Assessing unmodified 70-mer oligonucleotide probe performance on glass-slide microarrays. *Genome Biol.* 4: R5.

Wang, R., Guegler, K., LaBrie, S.T., Crawford, N.M. (2000). Genomic analysis of a nutrient response in *Arabidopsis* reveals diverse expression patterns and novel metabolic and potential regulatory genes induced by nitrate. *Plant Cell* 12: 1491-1509.

White, K.P., Rifkin, S.A., Hurban, P., Hogness, D.S. (1999). Microarray analysis of *Drosophila* development during metamorphosis. *Science* 286: 2179-2184.

Whitham, S.A., Quan, S., Chang, H.S., Cooper, B., Estes, B., Zhu, T., Wang, X., Hou, Y.M. (2003). Diverse RNA viruses elicit the expression of common sets of genes in susceptible *Arabidopsis thaliana* plants. *Plant J.* 33: 271-283.

Yang, Y.H., Dudoit, S., Luu, P., Lin, D.M., Peng, V., Ngai, J., Speed, T.P. (2002). Normalization for cDNA microarray data: A robust composite method addressing single and multiple slide systematic variation. *Nucleic Acids Res.* 30: e15.

Yang, Y.H., Speed, T. (2002). Design issues for cDNA microarray experiments. *Nat. Rev. Genet.* 3: 579-588.

Yeung, K.Y., Haynor, D.R., Ruzzo, W.L. (2001). Validating clustering for gene expression data. *Bioinformatics* 17: 309-318.

Yu, J., Hu, S., Wang, J., Wong, G.K., Li, S., Liu, B., Deng, Y., Dai, L., Zhou, Y., Zhang, X., et al. (2002). A draft sequence of the rice genome *(Oryza sativa* L. ssp. *indica). Science* 296: 79-92.

Yu, L.X., Setter, T.L. (2003). Comparative transcriptional profiling of placenta and endosperm in developing maize kernels in response to water deficit. *Plant Physiol.* 131: 568-582.

Yue, H., Eastman, P.S., Wang, B.B., Minor, J., Doctolero, M.H., Nuttall, R.L., Stack, R., Becker, J.W., Montgomery, J.R., Vainer, M., Johnston, R. (2001). An evaluation of the performance of cDNA microarrays for detecting changes in global mRNA expression. *Nucleic Acids Res.* 29: E41-1.

Zhao, N., Hashida, H., Takahashi, N., Misumi, Y., Sakaki, Y. (1995). High-density cDNA filter analysis: A novel approach for large-scale, quantitative analysis of gene expression. *Gene* 156: 207-213.

Zhou, Y.X., Kalocsai, P., Chen, J.Y., Shams, S. (2000). Information processing issues and solutions associated with microarray technology. In Schena, M. (Ed.), *Microarray Biochip Technology* (pp. 167-200). Natick, MA: Eaton Publishing.

Zik, M., Irish, V.F. (2003). Global identification of target genes regulated by *APETALA3* and *PISTILLATA* floral homeotic gene action. *Plant Cell* 15: 207-222.

Chapter 5

Metabolite Profiling in Plants

Richard N. Trethewey

INTRODUCTION

Plants contain a wide diversity of low-molecular-weight chemical constituents. While we now have access to plant genome sequences, we hardly have an idea about the extent of the "metabolome." More than 100,000 secondary metabolites have been identified in plants and this probably represents less than 10 percent of nature's total (Wink, 1988). Estimates of how many metabolites occur in an individual species vary within the 5 to 25,000 range. Whatever the number, current technologies can measure only a small minority of the metabolome and we have a long way to go until we can map the metabolome in any plant species. The subsequent step of understanding the chemical reactions that link the constituents of the metabolome, and assessing the regulatory mechanisms that guide and control metabolic flow within plant cells, is, extrapolating from our current knowledge, a formidable challenge.

We can study only what can be measured. In turn, scientific curiosity drives us to seek to measure as much as we can. As the pace of scientific inquiry has accelerated, catalyzed by profound technological strides in the areas of computation, metabolic engineering, and genomics, there has been an increased focus and effort targeted toward improving methods of metabolite analysis. Technologies are being developed and perfected that provide in a single analysis information on a wide range of chemistries. In this chapter, the current and future applications of plant metabolite profiling will be reviewed.

BACKGROUND AND FUNDAMENTALS OF METABOLITE PROFILING

The Origins of Metabolite Profiling in Plants

Metabolite profiling is not a new concept. The current approaches to metabolite profiling have their origins in the work of a limited number of clinical groups in the late 1960s and early 1970s. One of the first definitions in this field came from Horning and Horning (1971) who, writing in *Clinical Chemistry,* proposed that "[m]etabolic profiles are multi-component GC analyses that define or describe metabolic patterns for a group of metabolically/analytically related metabolites." Prophetically, they argued that "[p]rofiles may prove to be useful in characterizing both normal and pathological states." Niwa (1986) has published an excellent review on the early medical and clinical work in metabolite profiling. The first plant scientist to pick up on this work was Sauter, who began exploratory studies in the late 1970s while working in the area of herbicide development at BASF AG. The methods developed by Sauter rapidly became integrated into the research process at BASF, and an example of their application was published by this group in 1991 (Sauter et al., 1991). This paper describes a gas chromatography (GC)-based procedure for the analysis and quantification of a large number of metabolites from barley seedlings. Sauter et al. extracted frozen plant material in ethanol, homogenized it, and after filtration and evaporation performed a silylation derivatization step. The crude mixture was then subjected directly to gas chromatography (coupled to either a flame ionization detector or a mass spectrometer). The authors were able to resolve between 100 and 200 peaks with a high degree of reproducibility and could determine the chemistries of around 70 of these. Once the method had been established, it was applied to barley plants that had been treated with four different herbicides originating from different chemical classes. Two of the herbicides were known to block the biosynthesis of branched-chain amino acids by inhibition of acetolactate synthase, and the other two were known to act on fatty acid biosynthesis through impairing the activity of acetyl-CoA carboxylase. Comparing the changes in the metabolic profiles following application of these herbicides indicated that they could be classified into two distinct groups with very similar response patterns. Sauter et al. then extended their study to look at several other compounds, including some for which the mode of action was not known. The authors concluded that metabolite profiling is a very powerful tool to aid the chemist working on the design and synthesis of new agrochemicals in the classification of compounds according to their mode of action. They proposed that, through the establishment of a library of responses, "one will learn gradually to interpret the strange language of the profiles."

Definition of the Terms Related to Metabolite Profiling

Parallel to the development of the technologies of metabolite profiling, there has been a bewildering proliferation of the nomenclature associated with this field. At the route of the problem is that some groups have chosen to use the term *metabolomics* and others have opted for *metabonomics* without a consistent definition being generally accepted. A common perception has developed that metabonomics is reserved for application of NMR (nuclear magnetic resonance) methodology (hence the *n* in the middle!), while metabolomics is the preserve of those performing mass spectrometry approaches. However, the Nicholson group, which has pioneered the use of the term *metabonomics*, has offered a more sophisticated definition for this term: the quantitative measurement of the dynamic multiparametric metabolic response of living systems to pathophysiological stimuli or genetic modification (Nicholson et al., 2002; Lindon et at., 2001; Nicholson et al., 1999). This group reserves the term *metabolomics* for the study of isolated systems. In simple terms, if the organism survives the analysis, it is metabonomics, and if the organism has to be killed and extracted, then it is metabolomics. These definitions have recently been contested by Fiehn (2002) and Sumner et al. (2003), who prefer to see metabolomics as a parallel terminology to transcriptomics and proteomics (see Chapters 4, 14, and 15). Indeed, Fiehn (2002) has proposed that metabonomics has been erroneously used to describe what is actually "metabolic fingerprinting." Sumner et al. (2003) have extended this argument by proposing that metabolic fingerprinting should be used for any approach, irrespective of the technology, where pattern-recognition software is used to interpret the results without differentiation of individual metabolites.

A further issue is how wide an analysis should be before the terms *metabolite profiling, metabolomics,* and *metabonomics* apply. In a pure sense, the latter two should be considered only when a considerable fraction of the metabolome is being studied. However, as this is currently technologically unfeasible, then the terms have become commonly applied to technologies where there is a resolution of only several tens of components or fewer. Complicating the picture is also the custom in pharmaceutical research of applying the term *metabolite profiling* in the context of the study of the catabolic degradation products arising from a particular drug. Probably the only route out of this situation would be for the major groups, spanning the competing approaches, to publish a joint statement of nomenclature. In the meantime, considerable caution and discipline is required by all not to complicate the situation further, and when searching literature databases, the range of terminologies associated with this field needs to be taken into account.

METABOLITE-PROFILING TECHNOLOGIES

In contrast to the other levels of genomics investigation (DNA, RNA, or protein), the level of metabolites embraces substances with a wide range of different physical chemical properties. Thus, no one single analytical method or technology is, or will be, capable of measuring the whole metabolome. Currently, many different analytical technologies are deployed in metabolite-profiling studies, and the list is likely to grow rather than contract in the near future. It is beyond the scope of this chapter to perform a thorough assessment of the full range of different technologies, and the reader is referred to more specialist reviews for a detailed appraisal (Sumner et al., 2003; Fiehn, 2002, 2001).

Broadly speaking, the technologies of metabolite profiling can be grouped according to the detection principle into three categories, using the different forms of photon detection, NMR, and mass spectrometry (MS). Examples of the first category include infrared spectroscopy (Oliver et al., 1998; Timmins et al., 1998), liquid chromatography with ultraviolet and photodiode array detection (LC-UV-PDA; Fraser et al., 2000), capillary electrophoresis coupled to ultraviolet absorbance detection (CE-UV; Baggett et al., 2002), and capillary electrophoresis coupled to laser-induced fluorescence detection (CE-LIF; Arlt et al., 2001). Examples of NMR-based metabolite profiling or fingerprinting, as applied to the study of plant biology, will be introduced later in this chapter. The third category of technologies based upon MS detection holds the most long-term promise. All the different technologies offer different balances of speed, ease of use, cost, sensitivity, dynamic range, and selectivity (see Figure 5.1). NMR-based approaches are rapid and selective but are limited in terms of sensitivity and are relatively expensive. The photon-detection methods tend to be very sensitive with a high dynamic range, easy to use, and relatively inexpensive but often lack selectivity. Coupled mass spectrometry methods offer good sensitivity, dynamic range, and selectivity but are less user friendly to operate and slower than many of the other approaches. Coupled mass spectrometry can be relatively inexpensive in terms of a simple benchtop GC-MS system, but high-end mass spectrometers can cost an order of magnitude more. Very few groups have been in a position to choose the technology that is most suitable for the biological question that they are investigating; more often the choice of technology is limited by what is available. This factor needs to be taken into account when assessing the relative merits of the different technologies within example applications and the arguments that are deployed by their protagonists.

As indicated previously, metabolite-profiling methods based on mass spectrometry seem to hold the most promise for the long-term development

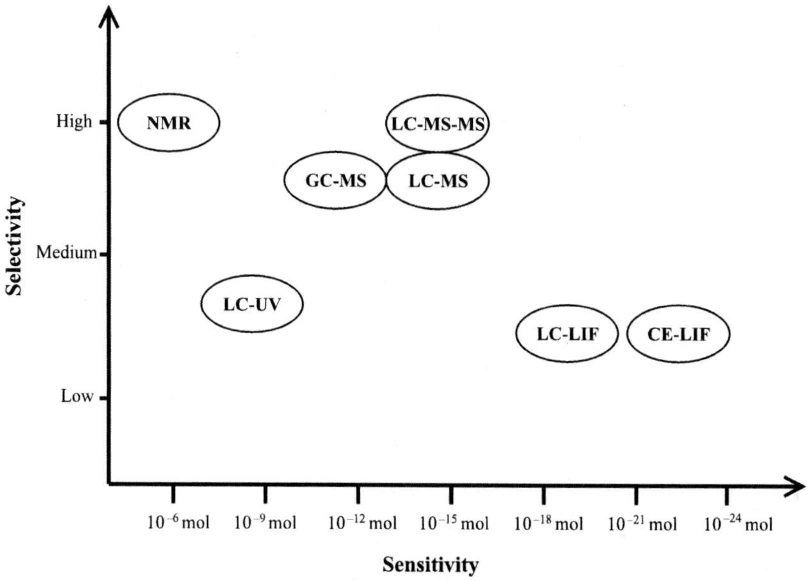

FIGURE 5.1. Performance of different metabolite analysis technologies with respect to their sensitivity and selectivity. NMR has a very high selectivity, but it enables the detection of only highly concentrated metabolites. Laser-induced fluorescence (LIF) provides the highest sensitivity, but this technology is not as selective as mass spectrometry. The best balance of sensitivity and selectivity is currently provided by GC-MS and LC-MS, with time-of-flight mass spectrometers providing the highest sensitivity in scanning mode.

of metabolomics. A bewildering range of different technologies exists, and a biologist is often well advised to seek out an analytical chemist when evaluating the pros and cons of a particular methodology. Gas chromatography-mass spectrometry (GC-MS) is the current workhorse of the field, being relatively inexpensive and offering a high separation efficiency and good sensitivity and dynamic range. However, the key limitation is that GC-MS, by definition, requires that the components to be analyzed are volatile. This can be accomplished by performing derivatization steps within the sample preparation procedure, but these are only effective for a subset of chemical classes. Derivatization also costs both time and money and can introduce increased variability and produce artifacts within the sample. Thus, for wide metabolite-profiling applications, it is advisable to keep the derivatization procedures as simple as possible. Nevertheless, basic benchtop quadrupole GC-MS systems are robust mature technologies with a long history of deployment in the chemical and oil industries and are now affordable for many groups and projects in the biological sciences. As the examples in this chapter indicate, much good basic science can be performed without reaching for expensive, new, cutting-edge technologies. Figure 5.2 shows an expanded section of an example chromatogram of the polar components of *Arabidopsis* leaf material; many hundreds of components can typically be resolved in plant material using this type of approach. Even within the mature GC-MS market there are some promising innovations. First, GC has recently been successfully coupled to time-of-flight mass spectrometry in commercially available products (GC-TOF). This offers the prospect of higher resolution, increased sensitivity, and faster GC (Davis et al., 1999). The first applications of GC-TOF in plant metabolite profiling look promising (Wagner et al., 2003), although it will likely take some time before GC-TOF becomes as robust as the quadrupole technology. Second, the first commercially available GC-GC-TOF systems have recently been marketed. Two-dimensional GC offers the prospect of a detailed separation of analytes that coelute in a one-dimensional chromatography.

Given the intrinsic limitation in the range of substances that can be profiled by GC-MS, with the introduction of the electrospray interface in the early 1990s, liquid chromatography-mass spectrometry (LC-MS) has

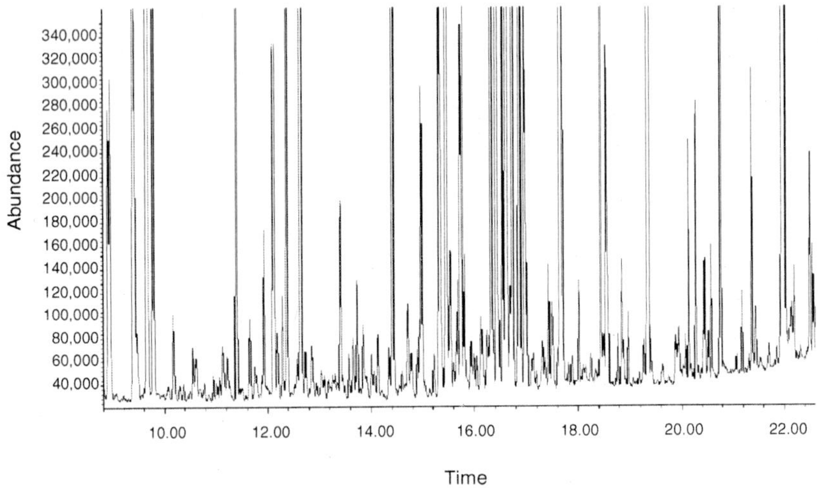

FIGURE 5.2. An expanded section of a GC-MS chromatogram of the polar constituents of an *Arabidopsis* leaf. This representation indicates the density of analytes that can be resolved across a large dynamic range. The analysis time for the whole chromatogram is 40 minutes.

emerged as an exciting and powerful alternative. However, as LC-MS is a much newer technology, it has not yet reached the maturity and robustness of GC-MS. It is also inherently more complicated to maintain and operate, given the higher technical difficulties associated with maintaining stable LC performance and the higher requirements for vacuum systems and ion optics. Nevertheless, the opportunity to be able to analyze nonvolatile and labile species without derivatization has driven a range of groups to explore the application of this technology to metabolite profiling. Recently, hydrophilic interaction LC-MS (HILC-MS) has been used to analyze highly polar *Cucurbita maxima* plant extracts (Tolstikov and Fiehn, 2002), and a reversed phase LC-MS method has been used for the study of saponins in legumes (Huhman and Sumner, 2002) and natural products in ephedra (Schaneberg et al., 2003). Plumb et al. (2002) have also recently published a method for the analysis of rat urine during drug development studies that illustrates the potential of this approach. A further advantage of liquid chromatography is that it can be simultaneously coupled to MS and an UV-PDA detector or even to NMR (Bailey, Cooper, et al., 2000; Bailey, Stanley, et al., 2000).

It is also possible to perform flow-injection mass spectral analyses in the case of metabolic fingerprinting, thus avoiding the time-consuming chromatography step. For example, the application of flow-injection electrospray ionization mass spectrometry (ESI-MS; Castrillo et al., 2003; Goodacre et al., 2003) to the study of salt stress in tomato (Johnson et al., 2003) and the characterization of the composition of olive oils (Goodacre et al., 2002) have been reported. Similarly, direct injection Fourier transform mass spectrometry (FTMS) has been applied to identify metabolites that are associated with the development and ripening of strawberries (Aharoni et al., 2002). FTMS has the advantage of very high resolution and mass accuracy, the latter often allowing the direct computation of elemental compositions. Unfortunately, these approaches cannot discriminate isomers that commonly occur in biological systems. In addition, direct injection ESI of highly complex mixtures is subject to varying ion suppression effects that will limit the potential of this approach to perform quantitative analysis (Annesley, 2003). Given this fact, some of the claims made for direct-injection analysis, in the absence of data to validate the methods, seem overly optimistic. Nevertheless, this is a powerful and comparatively fast approach for metabolic fingerprinting and for screening strategies that seek to ascertain in the first round only whether changes have occurred.

In the rush to apply exciting new technologies to biological applications, it must not be forgotten that a full validation and description of the analytical technology is necessary so that the limits of the approach can be understood and taken into account in the interpretation of the application. In the first instance, data should be available on the performance of the technology in terms of quantification (e.g., sensitivity, dynamic range, linearity, and a characterization of the potential for ion suppression by ionization or detection). Within quantification, two types of approaches can be followed: (1) absolute quantification, whereby the amount of a particular analyte is determined by direct calibration of this substance, or (2) where this is not possible, semiquantification (also known as relative quantification), whereby the amount of an analyte present in a particular sample is expressed as a ratio to that found in a control sample, using internal standards to achieve this normalization. In all quantification work, it is necessary to set limits for detection and quantification below which a signal is simply not to be interpreted. If it is possible to run calibration curves for a standard substance added to example plant material, these limits can be determined easily by standard procedures. Where this is not possible, limits need to be defined by considering the overall behavior of the analyte and the signal-to-noise environment within the chromatographic measurements. Given a rigorous quantification, it is then necessary to assess the overall variability of the analytical process, the recovery of certain key components and chemistries, and the risk and significance of blank contamination. Finally, in a biological application, it is necessary to have an understanding of the biological variability to guide experimental design and facilitate the interpretation of changes. To date, few studies have systematically validated their approach using the full discipline of analytical chemistry, and there is a clear need for standard operating procedures to be developed in the emerging field of metabolomics.

As more complex data sets become available from MS-based metabolite-profiling studies, there is an increasing interest and effort associated with identifying good statistical algorithms for classifying, interpreting, and visualizing the data sets (Fiehn and Weckwerth, 2003; Kell, 2002; Kose et al., 2001; Fell, 2001). Technology development in this area is, in principle, independent of the measurement technology. Given that a typical MS-based metabolite profile will yield significantly in excess of 100 components, an initial focus has been on statistical approaches that reduce the dimensionality of the data. By analogy with expression profiling, the most common approaches that have been followed are unsupervised procedures, such as principle component analysis (PCA) and hierarchical clustering analysis (HCA). Neural networks have recently seen their first applications in this field (Taylor et al., 2002). There also have been the first studies exploring the potential for reconstructing correlation networks from metabolite-profiling data (Fiehn, 2001; Kose et al., 2001). As metabolite-profiling technologies become more widely applied, and data sets are made publicly available on the World Wide Web, it is likely that we will see an exponential growth in the amount of studies that are testing and developing statistical procedures for processing profiling data.

APPLICATIONS OF METABOLITE PROFILING TO PLANTS

Application of Metabolite Profiling to Metabolic-Engineering Studies in Plants

Metabolic-engineering studies in plants have been one of the major drivers for the renewed interest in metabolite profiling. Rapid progress in the ability to clone and transform genes into most plant species of commercial importance has given rise to numerous studies targeting the manipulation of flux in major metabolic pathways, or even the synthesis of novel, high-value substances within plants (Fernie et al., 2002; Galili and Hofgen, 2002; Lessard et al., 2002; Thelen and Ohlrogge, 2002; Siedow and Stitt, 1998; Stitt and Sonnewald, 1995). This work has often been met with mixed success due to our still limited knowledge of metabolic networks and the difficulty of predicting the often wide-reaching metabolic consequences of engineering single steps or pathways. In this section, examples will be covered that demonstrate how metabolite profiling has supported and informed a range of different metabolic-engineering projects.

Metabolic Engineering of Starch Biosynthesis in Potato

There have been a wide and extensive range of approaches taken to the study of the pathway of starch biosynthesis in potato tubers (Fernie et al., 2002). Many of these approaches attempted to demonstrate that there was sufficient understanding of the metabolic system to enable a rational design of tubers that would accumulate more starch in comparison to nonengineered varieties. Despite exhaustive attempts, these approaches met with only limited success; the enzymes and steps that turned out to be important were not necessarily those which were anticipated based upon classical biochemical studies.

Driven by the need in the context of such metabolic engineering work to have a rapid and wide approach to characterizing the metabolism of transgenic tubers, the group of Trethewey set about adapting the method of Sauter et al. (1991) to tuber material. Initial studies with this method revealed exceedingly complex chromatograms with many hundreds of distinct peaks. It was quickly realized that many of these peaks represent multiple isomers of the abundant sugars present in tubers, a consequence of the silylation derivatization method. The method was then modified to introduce a separation of the extracts into polar and nonpolar components and a two-step derivatization procedure that reduced the isomer complexity. A validation of this method for potato tubers was published by Roessner et al. (2000), who showed that it enabled the qualitative and quantitative determination of more than 150 tuber metabolites, including sugars, sugar alcohols, dimeric and trimeric saccharides, amines, amino acids, and organic acids. The variability of the procedure was checked by following the performance of 25 defined metabolites drawn from different chemical classes. The analytical variability was found to be an order of magnitude lower than the biological variability in developing potato tubers. The majority of the metabolites followed were found to have standard deviations in the range of 10 to 25 percent. These figures are exactly comparable with those obtained by other technologies for metabolite measurement (e.g., spectrophotometric assays) and are also similar or superior to those found with other genomics technologies; the average coefficient of variation across an *Arabidopsis* microarray with 2,500 elements was found to be 37 percent (Richmond and Somerville, 2000).

One of the first applications of this wide, nonbiased model was to compare the metabolic composition of soil-grown tubers with those of microtubers cultured in vitro on high-sucrose media. The latter has been a widely used tool for the study of sugar and starch metabolism and was widely believed to be an accurate proxy for field-grown tubers (Veramendi et al., 1999). This conclusion was based upon a limited set of analysis of the enzymes and metabolites directly involved in the sugar-to-starch conversion. However, application of the broader, nonbiased GC-MS protocol revealed many significant differences between these two systems (Roessner et al., 2000). In particular, the in vitro tubers were found to contain higher levels of amino acids and a reduced complement of intermediates related directly to the TCA cycle (organic acids, alanine). Osmoprotectants such as proline, inositol, mannitol, and glycerol were also found to be elevated by many factors. Importantly, the previous studies were confirmed in that the levels of two intermediates of starch biosynthesis, fructose- and glucose-6-phosphate, were found to be similar in the two systems. This case study illustrates the importance when working with metabolic networks of not restricting inquiry to the pathway of interest. Indeed, subsequently a publication from Geigenberger et al. (2000) has demonstrated that developing tubers in field conditions are severely limited by oxygen availability in the starch tissues. It is unlikely that the small in vitro microtubers have a similar limitation, and this would represent a plausible explanation for many of the differences observed in the comparative study of Roessner et al. (2000).

Having demonstrated the potential value of a wide, nonbiased approach when studying starch metabolism, Roessner, Luedemann, et al. (2001) and Roessner et al. (2000) applied the GC-MS profiling method to soil-grown developing potato tubers of lines that overexpress a yeast invertase, specifically in the tuber (Trethewey, Reismeier, et al., 1999; Trethewey et al., 1998; Sonnewald et al., 1997). These lines were generated with the expectation that accelerated cytosolic sucrose metabolism would stimulate flux toward starch biosynthesis; in reality, the reverse occurred, and there was a reduced starch accumulation and an altered partitioning of carbon flux toward glycolysis. By applying the GC-MS protocol, the fundamental conclusions of a time-consuming range of conventional biochemical studies (Trethewey et al., 1998; Trethewey, Reismeier, et al., 1999), particularly the altered-flux partitioning, could be deduced following a single set of analyses taking less than one week from extraction to data analysis (Roessner et al., 2000). In addition, as the GC-MS provides a nonbiased data set of metabolites, further unexpected pleiotropic changes in the metabolism of the transgenic potato tubers were identified. For example, there was a large reduction in the inositol content of the invertase-expressing tubers, and a significant accumulation of shikimate and maltose. A scientist focused on starch pathway engineering would probably have only had the idea to measure maltose. Being open to a wider approach leads to discovery of the unexpected and challenges the metabolic engineer to explain seemingly unrelated correlations. In the long term, though, success in metabolic engineering will depend upon precise, specific, and targeted alterations in flux and there is no alternative but to confront the complexity of the metabolic system.

Metabolite Profiling in Conjunction with Studies of Secondary Metabolism

A few studies already illustrate the power and future potential of metabolite profiling when applied to the study of secondary metabolism in plants. On the whole, these studies have been performed with technologies and methods that are more specialized for one or a limited number of classes of chemical compounds.

One example of a pathway that has been thoroughly researched is the phenylpropanoid pathway (Dixon et al., 2002). In the case of lignin formation, the exact pathway of biosynthesis remains unclear despite considerable effort from a large number of groups (Anterola and Lewis, 2002; Humphreys and Chapple, 2002). Metabolite profiling has recently made a significant contribution to the debate. There are two structurally and functionally distinct enzymes proposed to play a role in lignin biosynthesis: caffeoyl-CoA *O*-methyltransferase and caffeoyl-CoA 3-*O*-methyltransferase. LC-MS-based profiling of both poplar (Meyermans et al., 2000) and alfalfa (Guo et al., 2001) following down-regulation of caffeoyl-CoA 3-*O*-methyltransferase demonstrated an unexpected accumulation of caffeic acid glucoside. However, this compound was not detected following inhibition of caffeic acid *O*-methyltransferase. Given that caffeic acid glucoside probably arises from the glucosylation of free caffeic acid (which in turn can be produced from caffeoyl-CoA by a thioesterase), it has recently been proposed that caffeoyl-CoA is the in vivo substrate of caffeoyl-CoA *O*-methyltransferase, but that caffeic acid is not a substrate in vivo for caffeic acid *O*-methyltransferase. Specific as these results are, they illustrate the difficulty in extrapolating from in vitro results to the in vivo activities and certainly indicate the caution needed when working with genes whose specificity has been estimated from sequence homology comparisons. Metabolite profiling, preferably on a wide and open basis, can play an important role in confirming or rejecting proposed metabolic activities in vivo. The increasing realization that the diversity of plant metabolomes is likely to be a consequence of low enzyme specificity (Schwab, 2003) indicates that metabolite profiling is likely to be needed on a case-by-case basis following overexpression of a particular heterologous enzyme. The exact role of enzymes with low specificity in a particular background will depend greatly on the availability of substrates in this background, thus predictability will always be limited, even in the case where a thorough kinetic characterization of the gene product is available. Recent observations in both strawberry and banana confirm the importance of substrate availability in determining which metabolic reactions are catalyzed by enzymes with low substrate specificity (Aharoni et al., 2000; Wyllie and Fellman, 2000).

Poor enzyme specificity, coupled to diverse metabolic backgrounds, implies that plant biotechnology projects have to consider carefully exactly which gene or allele is selected for metabolic engineering. For example, transgenic petunia plants were recently generated with an overexpressed *S*-linalool synthase gene from *Clarkia breweri*. Although the gene was successfully expressed, no *S*-linalool could be detected. Subsequent LC-MS analysis revealed that the *S*-linalool was probably indeed produced, but then rapidly and completely metabolized by an unknown endogenous activity to a *S*-linalyl glucoside, which was not present in wild-type plants (Lucker et al., 2001).

A further example from the phenylpropanoid pathway also illustrates the concept introduced in the previous paragraphs. Isoflavone phytoestrogens are found primarily in legumes, and given their believed ability to protect against postmenopausal ailments, cardiovascular diseases, and cancers related to hormone dysfunction, there has been considerable interest in engineering their production in nonlegumes. The key enzyme is isoflavone synthase, which in legumes is primarily responsible for the conversion of naringenin to genistein glycoconjugates (Liu et al., 2002; Jung et al., 2000). *Arabidopsis* plants have been engineered that contain both isoflavone synthase and a second enzyme, chalcone isomerase, that catalyzes the step prior to naringenin formation. LC-MS profiling of flavonoid compounds indicated that overexpression of the chalcone isomerase alone led to a general threefold increase in the content of the major glycoconjugates of the flavonols kaempferol and quercitin. In addition, novel conjugates were also detected by the LC-MS profiling. As is so often the case in metabolic engineering, the combined expression of isoflavone synthase and chalcone isomerase did not lead to the desired result of elevated genistein in *Arabidopsis*. However, the expression of isoflavone synthase significantly reduced the levels of flavonols, indicating that novel metabolic reactions are established in the transgenic lines. An even wider metabolite profiling might reveal some clues as to the metabolic fluxes occurring upon heterologous overexpression of these enzymes.

The previous examples, while making the case for a wide metabolic profile, also illustrate that significant progress can be made with dedicated methods. It is anticipated that there will be continued innovation in the development of new focused methodologies. An example of this has been the work of Larson and Graham (2001) in establishing an elegant system for the specific detection of acyl-CoA esters with acyl chain lengths from C4 to C20. The method employs a derivitization step whereby acyl-CoA esters are derived to their fluorescent acyl-etheno-CoA esters and then separated by ion-paired, reversed-phase HPLC and detected fluorometrically. This group is currently applying its technology to the screening of mutant and transgenic plants altered in fatty acid metabolism (Larson et al., 2002). Recently, Ketchum et al. (2003) have reported progress in metabolite profiling of the taxane diterpenoids, and Yamazaki and colleagues have applied metabolite profiling to alkaloids (Yamazaki, Urano, et al., 2003) and anthocyanins (Yamazaki, Nakajima, et al., 2003). Weichert et al. (2000; 2002) have studied oxylipins in cucumber seedlings, and Lange et al. (2001) have used a novel LC-MS method to study isoprenoid biosynthesis in peppermint. Several groups have studied volatile compounds using different metabolite-profiling techniques (Flamini et al., 2003; Brown, 2002; Matich et al., 1996). Recently, Verdonk et al. (2003) coupled solid-phase microextraction techniques to GC-MS analysis for benzenoid, aliphatic aldehyde, and sesquiterpene compounds to study the release of volatile compounds by the white flowers of *Petunia hybrida*. They found that emission of volatiles followed a circadian rhythm that correlated with changes in mRNA levels of key enzymes, indicating in this case that de novo synthesis is tightly coupled to the regulation of gene expression.

Metabolite Profiling in the Study of Polyhydroxyalkanoate Production in Arabidopsis thaliana

Metabolite profiling has also been used in studies related to the engineering of novel compounds in plants. Polyhydroxyalkanoates (PHAs) are a class of biodegradable plastics that were first described in *Bacillus megaterum* early in the twentieth century. It is now known that some 91 different hydroxyalkanoic acids can be constituents of PHA polymers and that PHAs occur in over 90 different genera of bacteria (Steinbüchel and Valentin, 1995). The simplest form of PHA is polyhydroxybutyrate (PHB), and this has served as a test case for a range of academic and commercial groups who are seeking to produce biodegradable plastics from plants. Biodegradable plastics produced from plants would be a renewable resource and, with the right economics, could contribute significantly to supplying the large and rising demand for plastics with a comparatively low impact on the environment.

The production of PHAs in plants requires the introduction of three biosynthetic enzymes (Poirier, 2001). In the case of PHB, starting with a central metabolite, acetyl-CoA, a β-ketothiolase is necessary to produce acetoacetyl-CoA. This substance is then reduced by an NADPH-dependent acetoacetyl-CoA reductase to give rise to 3-hydroxybutyryl-CoA. This is then the substrate for the polymerization reaction catalyzed by polyhydroxybutyrate synthase (Steinbüchel and Schlegel, 1991). The group of Somerville pioneered the transfer of this pathway from the bacteria *Ralstonia eutropha* to the cytosol of *Arabidopsis thaliana,* and the group was the first to demonstrate that PHB could be produced (Poirier et al., 1992). Granules of PHB could be seen in the cytosol, nucleus, and vacuole following microscopic examination, although unfortunately the total level of PHB production was very low. The same group then attempted to produce PHB in the plastid, hoping that this would produce higher levels, simply because there should be higher levels of acetyl-CoA present in the plastids, as this is the site of fatty acid production (Nawrath et al., 1994). After individually transforming each one of the transgenes and assembling them into a pathway via sexual crossing, the group reported PHB levels of up to 1 percent of the fresh weight. Encouragingly, these plants appeared to be normal in terms of growth rates and fertility. Emboldened by this, a second group tested the limits of PHB metabolic engineering by transforming all three genes into *Arabidopsis* in one construct and screening the first-generation transgenic lines for PHB production directly with GC-MS (Bohmert et al., 2000). Lines could be selected with PHB contents up to 4 percent of the fresh weight. However, in this study, a strong negative correlation was found between these higher levels of PHB and plant growth, indicating a limit of viability had been reached.

To investigate these limits further, the PHB-producing lines were subject to exploratory metabolite profiling via a GC-MS method (Bohmert et al., 2000). Contrary to expectations, these lines did not reveal any significant changes in either the composition or the quantity of any of the fatty acids determined. However, significant and widespread changes were found in the polar fraction of the extracts and these often correlated strongly with the PHB levels. Evidence for a reduction in the TCA cycle activity was found, in that decreased levels of isocitrate and fumarate were recorded. This could be linked to the fact that the transgenic lines have a significant new sink for acetyl-CoA. Unexpected was the observation that the levels of common osmoprotectants, such as proline, mannitol, and sugars, were significantly increased. This is strongly indicative of a stress response; however, it is probably not the complete picture, as the osmoprotectant myoinositol and related compounds were also significantly decreased. Although difficult to interpret, these data illustrate that the engineering of metabolic pathways can have significant and wide-reaching consequences. Whether these are a consequence of new sinks being introduced into metabolism or some kind of secondary indication of stress cannot be discerned at this point. Only with a more sophisticated analysis of these and other metabolic-engineering events in a similar region of metabolism will it be possible to start to distinguish cause and effect. However, this example indicates that one should not underestimate the potential scope of changes that might occur following a perturbation of a metabolic network.

Metabolite Profiling with High Spatial Resolution

Sampling often limits the precision with which results can be interpreted in metabolic studies. Given the complexity and diversity of metabolism at the tissue, cellular, and subcellular levels, the sampling strategy needs to be carefully thought through to maximize the chance of obtaining the relevant information; studies with metabolite profiling are no exception to this rule (Kehr, 2001). While for some objectives (e.g., studies on the dietary components of plants) information on the spatial location of metabolites may be of no direct relevance, for other work it is often important to bear in mind that metabolite data represent average values across potentially heterogenous cells and compartments.

Two studies so far have attempted to couple metabolite profiling to higher resolution at the level of sampling. Farré et al. (2001) have adapted a nonaqueous fractionation method for separating vacuole, plastidic, and cytosolic components for application to potato tubers and coupled it to GC-MS-based profiling. Farré et al. (2001) found that the subcellular distribution of sugars, organic acids, amino acids, and sugar alcohols was surprisingly consistent between potato tubers and potato leaves. Meanwhile, Fiehn (2003) has studied vascular exudates from *Cucurbita maxima* using GC-MS. Applying an automated mass spectral deconvolution algorithm, more than 400 components could be separated within the chromatograms. In comparison with leaf tissue, sugars were found to be present at lower concentrations while amino acids were significantly more abundant in the phloem exudates. The study was extended to follow the composition of phloem exudates from four plants over a period of 4.5 days. Fiehn (2003) was surprised to find no evidence of a diurnal rhythm in the concentration of any of the phloem components. However, it was observed that each plant showed a different overall profile and it was estimated that this was caused by variations in some 30 to 40 percent of the analytes. Fiehn (2003) pioneered the interpretation of this complex data set by establishing and visualizing metabolic-correlation networks.

The Application of NMR-Based Profiling to the Study of Plant Metabolism

The emerging field of NMR-based "metabonomics" has been pioneered by the group of Nicholson working at Imperial College in London. As discussed earlier, this group has defined *metabonomics* to be "the quantitative measurement of the dynamic multiparametric metabolic response of living systems to pathophysiological stimuli or genetic modification" (Nicholson et al., 2002; Lindon et al., 2001; Nicholson et al., 1999). Applications have been demonstrated in the areas of natural product chemistry, fingerprinting and classification, and toxicological assessment (Nicholson et al., 2002; Reo, 2002; Nicholson et al., 1999). Excitingly, this group has recently demonstrated how 1-H NMR-based metabonomics can provide a rapid and noninvasive diagnosis of the presence and severity of coronary heart disease (Brindle et al., 2002). This approach is currently being assessed in clinical studies.

The Nicholson group has defined *metabolomics* to be a term restricted for the analysis of isolated systems. Although NMR has a long history of application in questions of plant science (Bligny and Douce, 2001; Ratcliffe and Shachar-Hill, 2001), there have been relatively few publications on the application of high-resolution NMR spectroscopy to plant metabolomics. Noteborn et al. (2000) have proposed the application of offline HPLC-NMR spectroscopy to the evaluation of the composition of transgenic crops. More recently, the Nicholson group has reported a study on the effects of exposure of *Silene cucubalus* cell cultures to cadmium (Bailey et al., 2003). Plant cell extracts were subjected to 1-H NMR spectroscopy and the results subjected to multivariate statistical analysis. PCA revealed a clear discrimination between control and cadmium-dosed groups. From the spectra, it was possible to distinguish that the cadmium treatments led to an increase in malic acid and acetate levels and a decrease in glutamine and branched-chain amino acids. While demonstrating the potential of 1-H NMR spectroscopy in plant metabolomics, there is obviously a considerable way to go before this technology can provide the breadth of information about specific analytes that is provided by MS-based approaches. NMR-based metabolomics is likely to play a role in the rapid screening of biological samples that, at a basic level, allow classification and discrimination (metabolic fingerprinting). Indeed, 1-H NMR coupled to multivariate analysis was recently tested as a metabolic-fingerprinting technology in *Arabidopsis* (Ward et al., 2003). Ward et al. (2003) prepared unfractionated polar extracts from nine different ecotypes. Following PCA, it was found that a nine-component model explained some 99 percent of the variance between the ecotypes, with the first two components accounting for some 83 percent of this variability. The loadings of these two components were dominated by glucose and fumarate. Ward et al. (2003) concluded that an appropriate approach to plant metabolomics is to integrate an NMR primary screen with further investigation and accurate quantification by more targeted (chromatography-linked) methods such as GC-MS. However, these metabolic-fingerprinting approaches are not necessarily as simple as they might first appear; in a careful study of potato, tomato, and tea samples, Defernez and Colquhoun (2003) have indicated some of the pitfalls that may arise, particularly factors that might cause shifts in the NMR spectra. It therefore remains to be demonstrated that high-throughput metabolic fingerprinting of plant extracts with NMR can actually be achieved and the data interpreted with confidence.

The origins of plant metabolomics were in the area of herbicide mode-of-action study (Sauter et al., 1991), and there has recently been some new activity demonstrating the value of NMR in this question (Ott et al., 2003; Aranibar et al., 2001; Hole et al., 2000). Ott et al. (2003), building on the previous study of Aranibar et al. (2001), have shown that it is possible to use 1-H NMR and neural network classification techniques to automatically separate 17 different herbicide mode of actions exerted by 27 herbicidal compounds in corn. Ott et al. (2003) concluded that NMR is sensitive enough to produce fingerprint information that enables a researcher to distinguish between even closely related modes of actions. However, they highlighted the importance of keeping the experimental conditions standardized and well controlled to obtain reproducible results.

The Application of Metabolite Profiling to Functional Genomics

The acquisition of gene sequence information through EST and genome projects has been one of the remarkable scientific successes of the last decade. In plants, several million EST sequences are now available, primarily from important crop species such as soybean or corn. In addition, the complete genome sequence is now available for *Arabidopsis* and the majority of the rice genome has recently been completed (Delseny, 2003; Rice Chromosome 10 Sequencing Consortium, 2003; Schoof and Karlowski, 2003; Bevan, 2002; Hall et al., 2002; Holtorf et al., 2002; Sasaki et al., 2002; Bressan et al., 2001).

However, reading the gene sequence is just the start of the process of achieving scientific understanding. Much more challenging is the process of determining gene function. This needs to occur at both the "simple" level of ascertaining the activity or function of the gene product and at the "complex" level of understanding the role of this gene product in the overall functioning of an organism. The major technologies deployed in functional genomics projects have been different forms of mRNA expression analysis and proteomics (Lin et al., 2003; Phelps et al., 2002; Fiehn et al., 2001; van Wijk, 2001; Celis et al., 2000; Cushmann and Bohnert, 2000; Jacobs et al., 2000; Yates, 2000; Thiellement et al., 1999). Although very powerful, these technologies have the limitation that they do not necessarily determine the function of a gene in the context of the overall phenotype of the organism. For example, expression arrays can provide data on the mRNAs that alter in abundance following an environmental or developmental transition (Phelps et al., 2002; Cushmann and Bohnert, 2000; Richmond and Somerville, 2000). However, changes in steady-state mRNA levels do not necessarily mean that the corresponding gene product actually plays a role within the biological process in question. Further investigation at the metabolic, developmental, or phenotypic level is always required to demonstrate an actual connection. Similarly, proteomic studies that reveal differences in the abundance of a particular protein do not necessarily allow the conclusion that there will be an altered activity within the signaling, metabolic, or structural network under investigation (Trethewey, Krotzky, et al., 1999).

As metabolite profiling is a technology that directly measures indicators of phenotype and protein function (metabolites), it has considerable promise in functional genomics (Fiehn, 2002; Fiehn et al., 2001; Glassbrook and Ryals, 2001; Trethewey, 2001; Trethewey, Krotzky, et al., 1999). One of the first groups to realize the importance of "metabolomics" was that of Oliver, while studying genetic function and redundancy in yeast (Oliver, 1997). This group developed an original approach termed FANCY (**F**unctional **A**nalyses by **C**oresponse in **Y**east) that involved disrupting the activity of genes with unknown function and performing paired comparisons of consequent changes in metabolites linked to the glycolytic pathway (Raamsdonk et al., 2001).

Unusually, the first study that demonstrated the power of wide, nonbiased metabolite profiling in the area of functional genomics was performed by plant researchers. The basic approach described earlier by Roessner et al. (2000) and Roessner, Luedemann, et al. (2001) for potato tubers was in parallel developed and modified for *Arabidopsis* by Fiehn et al. (2000). Using GC-MS analysis, these authors were able to distinguish some 326 distinct compounds in *Arabidopsis* leaves, of which roughly half could be assigned a chemical structure. Fiehn et al. (2000) applied the GC-MS method to the analysis of four different *Arabidopsis* genotypes: the ecotype Columbia (Col-2) and the *dgd1* mutant (reduced digalactosyldiacylglycerol accumulation; Dörmann et al., 1999; Härtel et al., 1998; Dörmann et al., 1995) in this background, as well as the ecotype C24 and the *sdd1-1* mutant in this background (reduced stomatal density; Berger and Altmann, 2000). Given the complexity of the data sets, these authors also explored the use of different data analysis tools and found the most utility in principle component analysis. It was surprising that the "metabolic phenotypes" of the two ecotypes were found to be more divergent, as assessed via principle component analysis, than the mutants were from their respective parental ecotypes. A thorough statistical assessment of the data set revealed 41 significant changes in the *sdd1-1* mutant; two of the most significant changes were in unknown hydrophilic substances. The *dgd1* mutation gave rise to an even more pleiotropic pattern; a total of 153 significant changes were found in comparison to the parental Columbia ecotype (Fiehn et al., 2000).

The study of Fiehn et al. (2000) is not an unambiguous demonstration of the application of metabolite profiling to functional genomics, as it did not lead to the discovery of a novel gene function but rather contributed to the characterization of mutants discovered through other approaches. However, it showed the importance of metabolite profiling in phenotyping, which is in effect an "annotation" of gene function. It is only one further step to envisage a role for metabolite profiling in the actual de novo description of gene function following screening of mutant populations. This expectation has given rise to considerable commercial interest, and metabolite profiling for functional genomics in plants has been followed intensively by such specialist companies as Exelixis, Paradigm Genetics, Phenomenome Discoveries, and Metanomics. Public indications of the success of this approach can also be expected from a steadily increasing number of academic initiatives in this area, such as the Genome Arabidopsis Resource Network (GARNet) in the United Kingdom or the initiative from the Samuel Roberts Noble Foundation that has targeted a model legume, *Medicago truncatula* (May, 2002; Bell et al., 2001).

The Application of Metabolite Profiling to Phenotyping and Breeding

The earlier examples show that metabolite profiling generates a data set on a wide range of metabolites. This information can be used for directed investigation in terms of understanding a particular pathway in the overall context of the metabolic network, or as a way of probing gene function. However, it is clear that metabolic profiles can also be used as a potent tool for classification and phenotyping (Roessner et al., 2002; Roessner, Luedemann, et al., 2001).

Following on from their pioneering application of metabolite profiling to starch pathway engineering, Roessner, Luedemann, et al. (2001) generated data sets on multiple lines from four distinct genotypes, each altered in the starch-sucrose transition through the overexpression of different heterologous enzymes. Given the scale and complexity of the data sets, the authors applied the statistical classification methodology of HCA and PCA to assign clusters to the individual genotypes. First, applying HCA, the authors found that all 18 wild-type samples were clustered into a single distinct group. Likewise, lines expressing a glucokinase in combination with an invertase in the cytosol separated into a distinct group, as did a further set of lines that expressed a sucrose phosphorylase from *Pseudomonas saccharophila* (Trethewey et al., 2001). This latter enzyme cleaves sucrose into glucose-1-phosphate and fructose, thus avoiding the formation of free glucose as occurs through the action of invertase. Of interest was the fact that lines expressing the cytosolic invertase alone (INV) did not cluster into one group. Lines INV-30 and INV-33 clustered closely to the double transgenic lines expressing both the invertase in combination with glucokinase. The line INV-42 was found to cluster close to the wild-type controls. This split behavior reflects the fact that the expression level of the invertase is lowest in the INV-42 line and indicates that such HCA can be quite precise in discriminating different phenotypic behavior based upon metabolite profiles. A similar picture emerged when the authors applied PCA analysis; however, PCA has the advantage that it is possible to distinguish the compounds that contribute the most to the formation of the distinct clusters. Analysis of the data for the four genotypes indicated that the major contributors were maltose, maltitol, homoserine, tryptophane, trehalose, glucose, mannose, glycerate, hexose phosphates, and three unknown peaks (PT00, PT08, and PT16). Again, some of these components would have been analyzed in the context of a classical study of the starch pathway, but many would have escaped attention. The fact that many amino acids were found to have increased in this study, coupled with the observation of the authors that there were no changes in amino acids in the leaves of the potato plants, allowed the authors to speculate that the potato tuber is likely to have an extensive capability for amino acid biosynthesis.

Roessner, Luedemann, et al. (2001) also applied HCA and PCA to metabolite profiles obtained following incubation of potato tuber disks with different concentrations of glucose. In this case, the statistical classification revealed that such disks gave rise to profiles that were a phenocopy of those obtained from transgenic lines that overexpress an invertase in the apoplast. Taken together, this observation is an indication that significant pleiotropic metabolic effects must be the result of the presence of a factor on the plasma membrane that can sense the carbohydrate status of the cell wall space and provoke, through some signaling mechanism, changes in intracellular metabolism (Lalonde et al., 1999; Fernie et al., 2000).

A further study from Roessner, Willmitzer, et al. (2001) extended their analysis to include further genotypes where the endogenous pathway of starch biosynthesis had been down-regulated through antisense repression of the cytosolic or plastidic phosphoglucomutase or ADP glucose pyrophosphorylase (AGPase). In this case, HCA grouped the lines repressed in the adjacent plastidial enzymes (phosphoglucomutase and AGPase) into neighboring and similar clusters. However, PCA analyses offered a higher level of resolution between these two transgenic lines. Determination of the PCA loadings indicated that galactose was the metabolite that contributed most to this separation, but also differences in amino acids and hexose phosphates contributed to the discrimination. Interestingly, in all cases, the endogenous antisense lines formed very distinct clusters from the lines that overexpressed heterologous enzymes, despite all of these lines sharing the overall phenotype of reduced starch accumulation. This illustrates that many routes can lead to the same result and that these results can differ quite profoundly in the consequences for the overall metabolic system. Thus, broad metabolite profiling has a clearly important role to play in the basic phenotyping of transgenic and mutant lines.

Phenotyping via metabolite profiling can also be expected to find a wide application in the characterization of genera and species. An example of the potential of this route has recently been published by Schaneberg et al.

(2003). This group used a simple metabolite-profiling methodology of HPLC coupled to a diode array detector and applied it to the characterization of one of the oldest medicinal herbs of traditional Chinese medicine, *Ephedra sinica,* commonly known as ma huang. Focusing on peaks arising between 52 and 64 min in the chromatograms, this group was able to determine a fingerprint characteristic of the *Ephedra* species. Further, depending on the relative intensity of these peaks, the authors could distinguish between *Ephedra* species originating from Eurasia, North America, or South America. In the case of just two of the peaks, the authors were able to determine the chemical entity. Given that, with a relatively simple analytical procedure, the authors were able to fingerprint the origins of *Ephedra,* a more sophisticated and wider MS-based profiling would seem to offer great potential for achieving a finer discrimination and understanding of the chemical basis of such resolution. Further, such methods have the potential to contribute to routine quality assurance procedures within the herbal and dietary supplement industry.

There would also appear to be considerable potential in the application of metabolite profiling to improve phenotypic characterization during breeding programs. Although this has yet to occur, an example from the model plant *Arabidopsis* illustrates the point. Taylor et al. (2002) set out to determine if GC-MS profiling was able to discriminate between two *Arabidopsis* ecotypes (C24 and Col-0) and two segregating first-generation progeny (F1) populations derived from crosses in which each ecotype was either male or female. The GC-MS profiles could be deconvoluted into some 433 components. Using neural network analysis, it was possible to demonstrate that both the parental lines and the two F1 populations could be distinguished into separate groups. By applying PCA analysis, the authors concluded that malic acid and citrate are the two most important metabolites for distinguishing between the background ecotypes and that glucose and fructose were the metabolites that provided the highest discrimination between the two F1 populations. This example illustrates how wide, nonbiased metabolite profiling may be used to reduce the complexity of a problem in breeding down to certain essential components that perhaps can be followed in a large-scale program by simple targeted analysis.

The Application of Metabolite Profiling to Nutrigenomics

The field of metabolomics has recently started to be actively discussed in connection with advances in the nutritional sciences. As nutrition and medicine are revolutionized by genomics, wide, nonbiased methods for composition analysis are increasingly seen as a powerful tool to enhance the understanding of the relationship between genotype, nutrition, and health (German et al., 2002; van Ommen and Stierum, 2002; Watkins and German, 2002; van der Werf et al., 2001; Watkins et al., 2001). A new term has emerged for this combination of disciplines: *nutrigenomics*.

It is well established that there are clear links between dietary habits and such degenerative diseases as cancer, cardiovascular disease, and diabetes. However, the nature of the exact dietary components that are important and the mechanisms whereby they operate remain largely obscure. While some clear examples of metabolic markers that can be linked to certain disease states and progressions already exist, the potential for the discovery of more via nonbiased metabolite profiling appears to be high. In addition, given that many metabolite-profiling technologies have now been developed to operate robustly at medium to high throughputs, it is conceivable that the careful application of metabolite profiling to the study of particular populations or cohorts of patients, perhaps in combination with intervention studies, will be extremely powerful in understanding the genetic basis of biomarker relevance. This concept is, indeed, directly analogous to the route taken at the DNA level in the field of pharmacogenomics.

To date, only a small number of papers directly address the area of nutrigenomics. All of these have been based around gene-expression analysis of the human subject following particular dietary supplementation (van Ommen and Stierum, 2002). The attraction of metabolomics is that it allows the profiling of metabolic components in both the diet and the human subject. This will certainly open up new levels of knowledge about the nature and importance of uptake and metabolic processes.

The examples of metabolite profiling that have been provided throughout this section indicate the potential of the technology in delivering information on a wide range of the small-molecule constituents of plant matter. However, to date, these studies have largely been on systems chosen by plant scientists for their relevance to questions in plant biology. Until now, only a small number of papers have looked at the composition of crop plants in a similar form to that consumed by humans. Frenzel et al. (2002) have developed a metabolite-profiling procedure based on GC-MS for the analysis of rice grains. Katona et al. (1999) have described a GC-MS profiling procedure for analyzing sugars, sugar alcohols, organic acids, and amino acids in apricots. Burns et al. (2003) have recently reported a study of carotenoid, tocopherol, and chlorophyll metabolite profiles in a selection of ten common dietary fruit and vegetables. While this group used a simple HPLC with diode array detection and its conclusions were largely descriptive, this publication clearly illustrates the potential power of the approach of linking metabolomics to dietary studies.

OUTLOOK FOR METABOLITE PROFILING

The examples and case studies in this chapter illustrate that metabolite profiling is a rapidly emerging approach with a wide range of applications. In the coming years, it can be anticipated that there will be considerable improvements in the basic technologies. Both MS- and NMR-based technologies can be expected to improve in terms of sensitivity and resolution. Improvements in cost and performance will probably foster a wider adaptation of time-of-flight mass spectrometry and FTMS. The need for coupling to a chromatographic separation principle will likely remain, at least for the foreseeable future. However, it can be anticipated that benchtop LC-MS will establish itself as a robust alternative to GC-MS for wide, nonbiased profiling. Newer couplings, such as CE-MS and LC-FT-ICR-MS, can also be expected to grow in importance as more experience is obtained with them.

Parallel to the development of the different technologies, it is important that clear standards for metabolite-profiling work are established and accepted across the community. Initiatives to establish standard operating procedures for plant and technology combinations should be supported. It is a common experience that the true power of an approach is realized after there is standardization and adoption of common operating principles.

Likewise, there is a need for concerted action to establish compound and spectral databases for the small molecules that occur in the plant kingdom. Currently, more than half of the compounds detected with nonbiased, MS-based metabolite profiling are unknown. Bridging this knowledge gap would lead to an exponential progression in the scientific value of metabolite-profiling studies.

With respect to the fields of application of metabolite profiling, there is a very interesting progression occurring. Historically, the first application was in the area of herbicide development, and then after a gap in time, profiling methodologies were extended to the study of plant metabolic pathways and their engineering through plant biotechnology. The recent increase in interest and development effort in this area has been driven by functional genomics. Looking ahead, it can be anticipated that a major future application of metabolomics technologies will be in the area of nutrigenomics and that it will be an important instrument in advancing understanding and applications in the medical and nutritional sciences. Given the power of plant biotechnology to deploy the results from functional genomics programs, the loop may be closed through the breeding or engineering of healthier crop plants. Metabolite profiling is a rare example of a technological approach that has the potential to catalyze the integration of diverse scientific disciplines.

REFERENCES

Aharoni, A., de Vos, R.C., Verhoeven, H.A., Maliepaard, C.A., Kruppa, G., Bino, R., Goodenowe, D.B. (2002). Nontargeted metabolome analysis by use of Fourier Transform Ion Cyclotron Mass Spectrometry. *Omics* 6: 217-234.

Aharoni, A., Keizer, L.C., Bouwmeester, H.J., Sun, Z., Alvarez-Huerta, M., Verhoeven, H.A., Blaas, J., van Houwelingen, A.M., De Vos, R.C., van der Voet, H., et al. (2000). Identification of the SAAT gene involved in strawberry flavor biogenesis by use of DNA microarrays. *Plant Cell* 12: 647-662.

Annesley, T.M. (2003). Ion suppression in mass spectrometry. *Clin. Chem.* 49: 1041-1044.

Anterola, A.M., Lewis, N.G. (2002). Trends in lignin modification: A comprehensive analysis of the effects of genetic manipulations/mutations on lignification and vascular integrity. *Phytochem.* 61: 221-294.

Aranibar, N., Singh, B.K., Stockton, G.W., Ott, K.H. (2001). Automated mode-of-action detection by metabolic profiling. *Biochem. Biophys. Res. Comm.* 286: 150-155.

Arlt, K., Brandt, S., Kehr, J. (2001). Amino acid analysis in five pooled single plant cell samples using capillary electrophoresis coupled to laser-induced fluorescence detection. *J. Chrom. A.* 926: 319-325.

Baggett, B.R., Cooper, J.D., Hogan, E.T. Carper, J., Paiva, N.L., Smith, J.T. (2002). Profiling isoflavonoids found in legume root extracts using capillary electrophoresis. *Electrophoresis* 23: 1642-1651.

Bailey, N.J., Cooper, P., Hadfield, S.T., Lenz, E.M., Lindon, J.C., Nicholson, J.K., Stanley, P.D., Wilson, I.D., Wright, B., Taylor, S.D. (2000). Application of -directly coupled HPLC-NMR-MS/MS to the identification of metabolites of 5-trifluoromethylpyridone (2-hydroxy-5-trifluoromethylpyridine) in hydroponically grown plants. *J. Agric. Food Chem.* 48: 42-46.

Bailey, N.J.C., Oven, M., Holmes, E., Nicholson, J.K., Zenk, M.H. (2003). Metabolomic analysis of the consequences of cadmium exposure in *Silene cucubalus* cell cultures via H-1 NMR spectroscopy and chemometrics. *Phytochem.* 62: 851-858.

Bailey, N.J., Stanley, P.D., Hadfield, S.T., Lindon, J.C., Nicholson, J.K. (2000). Mass spectrometrically detected directly coupled high performance liquid chromatography/nuclear magnetic resonance spectroscopy/mass spectrometry for the identification of xenobiotic metabolites in maize plants. *Rapid Commun. Mass Spectrom.* 14: 679-684.

Bell, C.J., Dixon, R.A., Farmer, A.D., Flores, R., Inman, J., Gonzales, R.A., Harrison, M.J., Paiva, N.L., Scott, A.D., Weller, J.W., May, G.D. (2001). The *Medicago* Genome Initiative: A model legume database. *Nucleic Acids Res.* 29: 114-117.

Berger, D., Altmann, T. (2000). A subtilisin-like serine protease involved in the regulation of stomatal density and distribution in *Arabidopsis thaliana. Genes Dev.* 14: 1119-1131.

Bevan, M. (2002). Plant biology: The first harvest of crop genes. *Nature* 416: 590-591.

Bligny, R., Douce, R. (2001). NMR and plant metabolism. *Curr. Opin. Plant Biol.* 4: 191-196.

Bohmert, K., Balbo, I., Kopka, J., Mittendorf, V., Nawrath, C., Poirier, Y., Tischendorf, G., Trethewey, R.N., Willmitzer, L. (2000). Transgenic *Arabidopsis* plants can accumulate polyhydroxybutyrate to up to 4 percent of their fresh weight. *Planta* 211: 841-845.

Bressan, R.A., Zhang, C., Zhang, H., Hasegawa, P.M., Bohnert, H.J., Zhu, J.K. (2001). Learning from the *Arabidopsis* experience: The next gene search paradigm. *Plant Physiol.* 127: 1354-1360.

Brindle, J.T., Antti, H., Holmes, E., Tranter, G., Nicholson, J.K., Bethell, H.W.L., Clarke, S., Schofield, P.M., McKilligin, E., Mosedale, D.E., Grainger, D.J. (2002). Rapid and noninvasive diagnosis of the presence and severity of coronary heart disease using H-1-NMR-based metabonomics. *Nat. Med.* 8: 1439-1444.

Brown, K. (2002). Plant genetics. Something to sniff at: Unbottling floral scent. *Science* 296: 2327-2329.

Burns, J., Fraser, P.D., Bramley, P.M. (2003). Identification and quantification of carotenoids, tocopherols and chlorophylss in commonly consumed fruits and vegetables. *Phytochem.* 62: 939-947.

Castrillo, J.I., Hayes, A., Mohammed, S., Gaskell, S.J., Oliver, S.G. (2003). An optimized protocol for metabolome analysis in yeast using direct infusion electrospray mass spectrometry. *Phytochem.* 62: 929-937.

Celis, J.E., Kruhoffer, M., Gromova, I., Frederiksen, C., Ostergaard, M., Thykjaer, T., Gromov, P., Yu, J.S., Palsdottir, H., Magnusson, N., Orntoft, T.F. (2000). Gene expression profiling: Monitoring transcription and translation products using DNA microarrays and proteomics. *FEBS Lett.* 480: 2-16.

Cushmann, J.C., Bohnert, H.J. (2000). Genomic approaches to plant stress tolerance. *Curr. Opin. Plant Biol.* 3: 117-124.

Davis, S.C., Makarov, A.A., Hughes, J.D. (1999). Ultrafast gas chromatography using time-of-flight mass spectrometer. *Rapid Comm. Mass Spec.* 13: 237-241.

Defernez, M., Colquhoun, I.J. (2003). Factors affecting the robustness of metabolite fingerprinting using H-1 NMR spectra. *Phytochem.* 62: 1009-1017.

Delseny, M. (2003). Towards an accurate sequence of the rice genome. *Curr. Opin. Plant Biol.* 6: 101-105.

Dixon, R.A., Achnine, L., Kota, P., Liu, C.J., Reddy, M.S.S., Wang, L.J. (2002). The phenylpropanoid pathway and plant defence—A genomics perspective. *Mol. Plant Pathol.* 3: 371-390.

Dörmann, P., Blbo, I., Benning, C. (1999). *Arabidopsis* galactolipid biosynthesis and lipid trafficking mediated by DGD1. *Science* 284: 2181-2184.

Dörmann, P., Hoffmann-Benning, S., Balbo, I., Benning, C. (1995). Isolation and characterization of an *Arabidopsis* mutant deficient in the thylakoid lipid digalactosyl diacylglycerol. *Plant Cell* 7: 1801-1810.

Farré, E.M., Tiessen, A., Roessner, U., Geigenberger, P., Trethewey, R.N., Willmitzer, L. (2001). Analysis of the compartmentation of glycolytic intermediates, nucleotides, sugars, organic acids, amino acids and sugar alcohols in potato tubers using a non-aqueous fractionation method. *Plant Physiol.* 127: 685-700.

Fell, D.A. (2001). Beyond genomics. *Trends Genet.* 17: 680-682.

Fernie, A.R., Riesmeier, J.W., Martiny, A., Sathishkumar, R., Willmitzer, L., Trethewey, R.N. (2000). Consequences of the expression of a bacterial glucokinase, both in combination with and independently of a yeast-derived invertase, in potato tubers. *Aust. J. Plant Physiol.* 27: 827-833.

Fernie, A.R., Willmitzer, L., Trethewey, R.N. (2002). Sucrose to starch: A transition in molecular plant physiology. *Trends Plant Sci.* 7: 35-41.

Fiehn, O. (2001). Combining genomics, metabolome analysis, and biochemical modeling to understand metabolic networks. *Comp. Func. Genomics* 2: 155-168.

Fiehn, O. (2002). Metabolomics: The link between genotypes and phenotypes. *Plant Mol. Biol.* 48: 155-171.

Fiehn, O. (2003). Metabolic networks of *Cucurbita maxima* phloem. *Phytochem.* 62: 875-886.

Fiehn, O., Kloska, S., Altmann, T. (2001). Integrated studies on plant biology using multiparallel techniques. *Curr. Opin. Biotech.* 12: 82-86.

Fiehn, O., Kopka, J., Dörmann, P., Altmann, T., Trethewey, R.N., Willmitzer, L. (2000). Metabolic plant functional genomics. *Nat. Biotechnol.* 18: 1157-1161.

Fiehn, O., Weckwerth, W. (2003). Deciphering metabolic networks. *Eur. J. Biochem.* 270: 579-588.

Flamini, G., Cioni, P.L., Morelli, I. (2003). Differences in the fragrances of pollen, leaves, and floral parts of garland *(Chrysanthemum coronarium)* and composition of the essential oils from flowerheads and leaves. *J. Agric. Food Chem.* 51: 2267-2271.

Fraser, P.D., Pinto, M.E., Holloway, D.E., Bramley, P.M. (2000). Application of high-performance liquid chromatography with photodiode detection to the metabolic profiling of plant isoprenoids. *Plant J.* 24: 551-558.

Frenzel, T., Miller, A., Engel, K.H. (2002). Metabolite profiling: A fractionation method for analysis of major and minor compounds in rice grains. *Cereal Chem.* 79: 215-221.

Galili, G., Hofgen, R. (2002). Metabolic engineering of amino acids and storage proteins in plants. *Metab. Eng.* 4: 3-11.

Geigenberger, P., Fernie, A.R., Gibon, Y., Christ, M., Stitt, M. (2000). Metabolic activity decreases as an adaptive response to low internal oxygen in growing potato tubers. *Biol. Chem.* 381: 723-740.

German, J.B., Roberts, M.A., Fay, L., Watkins, S.M. (2002). Metabolomics and individual metabolic assessment: The next great challenge for nutrition. *J. Nutr.* 132: 2486-2487.

Glassbrook, A., Ryals, J. (2001). A systematic approach to biochemical profiling. *Curr. Opin. Plant Biol.* 4: 186-190.

Goodacre, R., Vaidyanathan, S., Bianchi, G., Kell, D.B. (2002). Metabolic profiling using direct infusion electrospray ionisation mass spectrometry for the characterisation of olive oils. *Analyst* 127: 1457-1462.

Goodacre, R., York, E.V., Heald, J.K., Scott, I.M. (2003). Chemometric discrimination of unfractionated plant extracts analyzed by electrospray mass spectrometry. *Phytochem.* 62: 859-863.

Guo, D., Chen, F., Inoue, K., Blount, J.W., Dixon, R.A. (2001). Downregulation of caffeic acid 3-*O*-methyltransferase and caffeoyl CoA 3-*O*-methyltransferase in transgenic alfalfa: Impacts on lignin structure and implications for the biosynthesis of G and S lignin. *Plant Cell* 13: 73-88.

Hall, A.E., Fiebig, A., Preuss, D. (2002). Beyond the *Arabidopsis* genome: Opportunities for comparative genomics. *Plant Physiol.* 129: 1439-1447.

Härtel, H., Lokstein, H., Dörmann, P., Trethewey, R.N., Benning, C. (1998). Photosynthetic light utilization and xantophyll cycle activity in the galactolipid deficient dgd1 mutant of *Arabidopsis thaliana. Plant Physiol. Biochem.* 36: 407-417.

Hole, S.J., Howe, P.W., Stanley, P.D., Hadfield, S.T. (2000). Pattern recognition analysis of endogenous cell metabolites for high throughput mode of action identification: Removing the postscreening dilemma associated with whole-organism high throughput screening. *J. Biomol. Screen.* 5: 335-342.

Holtorf, H., Guitton, M.C., Reski, R. (2002). Plant functional genomics. *Naturwissenschaften* 89: 235-249.

Horning, E.C., Horning, M.G. (1971). Metabolic profiles: Gas-phase methods for analysis of metabolites. *Clin. Chem.* 17: 802-809.

Huhman, D.V., Sumner, L.W. (2002). Metabolic profiling of saponins in *Medicago sativa* and *Medicago truncatula* using HPLC coupled to an electrospray ion-trap mass spectrometer. *Phytochem.* 59: 347-360.

Humphreys, J.M., Chapple, C. (2002). Rewriting the lignin roadmap. *Curr. Opin. Plant Biol.* 5: 224-229.

Jacobs, D.I., Van Der Heijden, R., Verpoorte, R. (2000). Proteomics in plant biotechnology and secondary metabolism research. *Phytochem. Anal.* 11: 277-287.

Johnson, H.E., Broadhurst, D., Goodacre, R., Smith, A.R. (2003). Metabolic fingerprinting of salt-stressed tomatoes. *Phytochem.* 62: 919-928.

Jung, W., Yu, O., Lau, S.M., O'Keefe, D.P., Odell, J., Fader, G., McGonigle, B. (2000). Identification and expression of isoflavone synthase, the key enzyme for biosynthesis of isoflavones in legumes. *Nat. Biotechnol.* 18: 208-212.

Katona, Z.F., Sass, P., Molnár-Perl, I. (1999). Simultaneous determination of sugars, sugar alcohols, acids and amino acids in apricots by gas chromatography-mass spectrometry. *J. Chrom.* 847: 91-102.

Kehr, J. (2001). High-resolution spatial analysis of plant systems. *Curr. Opin. Plant Biol.* 4: 197-201.

Kell, D.B. (2002). Metabolomics and machine learning: Explanatory analysis of complex metabolome data using genetic programming to produce simple, robust rules. *Mol. Biol. Rep.* 29: 237-241.

Ketchum, R.E.B., Rithner, C.D., Qiu, D., Kim, Y.S., Williams, R.M., Croteau, R.B. (2003). *Taxus* metabolomics: Methyl jasmonate preferentially induces production of taxoids oxygenated at C-13 in *Taxus* x media cell cultures. *Phytochem.* 62: 901-909.

Kose, F., Weckwerth, W., Linke, T., Fiehn, O. (2001). Visualizing plant metabolomic correlation networks using clique-metabolite matrices. *Bioinformatics* 17: 1198-1208.

Lalonde, S., Boles, E., Hellmann, H., Barker, L., Patrick, J.W., Frommer, W.B., Ward, J.M. (1999). The dual function of sugar carriers: Transport and sugar sensing. *Plant Cell* 11: 707-726.

Lange, B.M., Ketchum, R.E.B., Croteau, R.B. (2001). Isoprenoid biosynthesis: Metabolite profiling of peppermint oil gland secretory cells and application to herbicide target analysis. *Plant Physiol.* 127: 305-314.

Larson, T.R., Edgell, T., Byrne, J., Dehesh, K., Graham, I.A. (2002). Acyl CoA profiles of transgenic plants that accumulate medium-chain fatty acids indicate inefficient storage lipid synthesis in developing oilseeds. *Plant J.* 32: 519-527.

Larson, T.R., Graham, I.A. (2001). Technical advance: A novel technique for the sensitive quantification of acyl CoA esters from plant tissues. *Plant J.* 25: 115-125.

Lessard, P.A., Kulaveerasingam, H., York, G.M., Strong, A., Sinskey, A.J. (2002). Manipulating gene expression for the metabolic engineering of plants. *Metab. Eng.* 4: 67-79.

Lin, D., Tabb, D.L., Yates, J.R. (2003). Large-scale protein identification using mass spectrometry. *Biochim. Biophys. Acta* 21: 1-10.

Lindon, J.C., Holmes, E., Nicholson, J.K. (2001). Pattern recognition methods and applications in biomedical magnetic resonance. *Prog. Nucl. Mag. Res. Spec.* 39: 1-40.

Liu, C.J., Blount, J.W., Steele, C.L., Dixon, R.A. (2002). Bottlenecks for metabolic engineering of isoflavone glycoconjugates in *Arabidopsis*. *Proc. Natl. Acad. Sci. USA* 99: 14578-14583.

Lucker, J., Bouwmeester, H.J., Schwab, W., Blaas, J., van der Plas, L.H., Verhoeven, H.A. (2001). Expression of Clarkia *S*-linalool synthase in transgenic petunia plants results in the accumulation of *S*-linalyl-beta-D-glucopyranoside. *Plant J.* 27: 315-324.

Matich, A.J., Rowan, D.D., Banks, N.H. (1996). Solid phase microextraction for quantitative headspace sampling of apple volatiles. *Anal. Chem.* 68: 4114-4118.

May, G.D. (2002). An integrated approach to *Medicago* functional genomics. In Romeo, J.T., Dixon, R.A. (Eds.), *Recent Advances in Phytochemistry,* Volume 36 (pp. 179-195). Oxford, United Kingdom: Pergamon.

Meyermans, H., Morreel, K., Lapierre, C., Pollet, B., De Bruyn, A., Busson, R., Herdewijn, P., Devreese, B., Van Beeumen, J., Marita, J.M., et al. (2000). Modifications in lignin and accumulation of phenolic glucosides in poplar xylem upon down-regulation of caffeoyl-coenzyme A *O*-methyltransferase, an enzyme involved in lignin biosynthesis. *J. Biol. Chem.* 275: 36899-36909.

Nawrath, C., Poirier, Y., Somerville, C. (1994). Targeting of the polyhydroxybutyrate biosynthetic pathway to the plastids of *Arabidopsis thaliana* results in high levels of polymer accumulation. *Proc. Natl. Acad. Sci. USA* 91: 12760-12764.

Nicholson, J.K., Connelly, J., Lindon, J.C., Holmes, E. (2002). Metabonomics: A platform for studying drug toxicity and gene function. *Nat. Rev. Drug Discov.* 1: 153-161.

Nicholson, J.K., Lindon, J.C., Holmes, E. (1999). "Metabonomics": Understanding the metabolic responses of living systems to pathophysiological stimuli via multivariate statistical analysis of biological NMR spectroscopic data. *Xenobiotica* 11: 1181-1189.

Niwa, T. (1986). Metabolic profiling with gas chromatography-mass spectrometry and its application to clinical medicine. *J. Chrom.* 20: 313-345.

Noteborn, H.P., Lommen, A., van der Jagt, R.C., Weseman, J.M. (2000). Chemical fingerprinting for the evaluation of unintended secondary metabolic changes in transgenic food crops. *J. Biotech.* 77: 103-114.

Oliver, S.G. (1997). Yeast as a navigational aid in genome analysis. *Microbiol.* 143: 1483-1487.

Oliver, S.G., Winson, M.K., Kell, D.B., Baganz, F. (1998). Systematic functional analysis of the yeast genome. *Trends Biotechnol.* 16: 373-378.

Ott, K.H., Aranibar, N., Singh, B.J., Stockton, G.W. (2003). Metabonomics classifies pathways affected by bioactive compounds: Artificial neural network classification of NMR spectra of plant extracts. *Phytochem.* 62: 971-985.

Phelps, T.J., Palumbo, A.V., Beliaev, A.S. (2002). Metabolomics and microarrays for improved understanding of phenotypic characteristics controlled by both genomics and environmental constraints. *Curr. Opin. Biotech.* 13: 20-24.

Plumb, R.S., Stumpf, C.L., Gorenstein, M.V., Castro-Perez, J.M., Dear, G.J., Anthony, M., Sweatman, B.C., Connor, S.C., Haselden, J.N. (2002). Metabonomics: The use of electrospray mass spectrometry coupled to reversed-phase liquid chromatography shows potential for the screening of rat urine in drug development. *Rapid Commun. Mass Spectrom.* 16: 1991-1996.

Poirier, Y. (2001). Production of polyesters in transgenic plants. *Adv. Biochem. Eng. Biotechnol.* 71: 209-240.

Poirier, Y., Dennis, D.E., Klomparens, K., Sommerville, C. (1992). Polyhydroxybutyrate, a biodegradable thermoplastic, produced in transgenic plants. *Science* 256: 520-523.

Raamsdonk, L.M., Teusink, B. Broadhurst, D., Zhang, N.S., Hayes, A., Walsh, M.C., Berden, J.A., Brindle, K.M., Kell, D.B., Rowland, J.J., et al. (2001). A functional genomics strategy that uses metabolome data to reveal the phenotype of silent mutations. *Nat. Biotechnol.* 19: 45-50.

Ratcliffe, R.G., Shachar-Hill, Y. (2001). Probing plant metabolism with NMR. *Annu. Rev. Plant Physiol. Plant Mol. Biol.* 52: 499-526.

Reo, N.V. (2002). NMR-based metabolomics. *Drug, Chem. Tox.* 25: 375-382.

Rice Chromosome 10 Sequencing Consortium (2003). In-depth view of structure, activity, and evolution of rice chromosome 10. *Science* 300: 1566-1569.

Richmond, T., Somerville, S. (2000). Chasing the dream: Plant EST microarrays. *Curr. Opin. Plant Biol.* 3: 108-116.

Roessner, U., Luedemann, A., Brust, D., Fiehn, O., Linke, T., Willmitzer, L., Fernie, A.R. (2001). Metabolic profiling allows comprehensive phenotyping of genetically or environmentally modified plant systems. *Plant Cell* 13: 11-29.

Roessner, U., Wagner, C., Kopka, J., Trethewey, R.N., Willmitzer, L. (2000). Simultaneous analysis of metabolites in potato tuber by gas chromatography-mass spectrometry. *Plant J.* 23: 131-142.

Roessner, U., Willmitzer, L., Fernie, A.R. (2001). High-resolution metabolic phenotyping of genetically and environmentally diverse potato tuber systems: Identification of phenocopies. *Plant Physiol.* 127: 749-764.

Roessner, U., Willmitzer, L., Fernie, A.R. (2002). Metabolic profiling and biochemical phenotyping of plant systems. *Plant Cell Rep.* 21: 189-196.

Sasaki, T., Matsumoto, T., Yamamoto, K., Sakata, K., Baba, T., Katayose, Y., Wu, J., Niimura, Y., Cheng, Z., Nagamura, Y., et al. (2002). The genome sequence and structure of rice chromosome 1. *Nature* 420: 312-316.

Sauter, H., Lauer, M., Fritsh, H. (1991). Metabolic profiling of plants: A new diagnostic technique. In Baker, D.R., Fenyes, J.G., Moberg, W.K. (Eds.), *American Chemical Society Symposium Series 443* (pp. 288-299). Washington, DC: American Chemical Society.

Schaneberg, B.T., Crockett, S., Bedir, E., Khan, I.A. (2003). The role of chemical fingerprinting: Application to *Ephedra. Phytochem.* 62: 911-918.

Schoof, H., Karlowski, W.M. (2003). Comparison of rice and *Arabidopsis* annotation. *Curr. Opin. Plant Biol.* 6: 106-112.

Schwab, W. (2003). Metabolome diversity: Too few genes, too many metabolites? *Phytochem.* 62: 837-849.

Siedow, J., Stitt, M. (1998). Plant metabolism: Where are all those pathways leading us? *Curr. Opin. Plant Biol.* 1: 197-200.

Sonnewald, U., Hajiraezaei, M.-R., Kossmann, J., Heyer, A., Trethewey, R.N., Willmitzer, L. (1997). Expression of a yeast invertase in the apoplast of potato tubers increases tuber size. *Nat. Biotechnol.* 15: 794-797.

Steinbüchel, A., Schlegel, H.G. (1991). Physiology and molecular genetics of poly (β-hydroxy-alkanoic acid) synthesis in *Acaligenes eutrophus. Mol. Microbiol.* 5: 535-542.

Steinbüchel, A., Valentin, H.E. (1995). Diversity of bacterial polyhydroxyalkanoic acids. *FEMS Microbiol. Lett.* 128: 219-228.

Stitt, M., Sonnewald, U. (1995). Regulation of metabolism in transgenic plants. *Annu. Rev. Plant Physiol. Plant Mol. Biol.* 46: 341-368.

Sumner, L.W., Mendes, P., Dixon, R.A. (2003). Plant metabolomics: Large-scale phytochemistry in the functional genomics era. *Phytochem.* 62: 817-836.

Taylor, J., King, R.D., Altmann, T., Fiehn, O. (2002). Application of metabolomics to plant genotype discrimination using statistics and machine learning. *Bioinformatics* 18: 241-248.

Thelen, J.J., Ohlrogge, J.B. (2002). Metabolic engineering of fatty acid biosynthesis in plants. *Metab. Eng.* 4: 12-21.

Thiellement, H., Bahrman, N., Damerval, C., Plomion, C., Rossignol, M., Santoni, V., De Vienne, D., Zivy, M. (1999). Proteomics for genetic and physiological studies in plants. *Electrophoresis* 20: 2013-2026.

Timmins, E.M, Quain, D.E., Goodacre, R. (1998). Differentiation of brewing yeast strains by pyrolysis mass spectrometry and Fourier transform infrared spectroscopy. *Yeast* 14: 885-893.

Tolstikov, V.V., Fiehn, O. (2002). Analysis of highly polar compounds of plant origin: Combination of hydrophilic interaction chromatography and electrospray ion trap mass spectrometry. *Anal. Biochem.* 301: 298-307.

Trethewey, R.N. (2001). Gene discovery via metabolic profiling. *Curr. Opin. Biotech.* 12: 135-138.

Trethewey, R.N., Fernie, A.R., Bachmann, A., Fleischer-Notter, H., Geigenberger, P., Willmitzer, L. (2001). Expression of a bacterial sucrose phosphorylase in potato tubers results in a glucose-independent induction of glycolysis. *Plant Cell Environ.* 24: 357-365.

Trethewey, R.N., Geigenberger, P., Riedel, K., Hajirezaei, M.-R., Sonnewald, U., Stitt, M., Riesmeier, J.W., Willmitzer, L. (1998). Combined expression of glucokinase and invertase in potato tubers leads to a dramatic reduction in starch accumulation and a stimulation of glycolysis. *Plant J.* 15: 109-118.

Trethewey, R.N., Krotzky, A.J., Willmitzer, L. (1999). Metabolic profiling: A rosetta stone for genomics. *Curr. Opin. Plant Biol.* 2: 83-85.

Trethewey, R.N., Reismeier, J.W., Willmitzer, L., Stitt, M., Geigenberger, P. (1999). Tuber specific expression of a yeast invertase and a bacterial glucokinase in potato leads to an activation of sucrose phosphate synthase and the creation of a futile cycle. *Planta* 208: 227-238.

van der Werf, M.J., Schuren, F.H.J., Bijlsma, S., Tas, A.C., van Ommen, B. (2001). Nutrigenomics: Application of genomics technologies in nutritional sciences and food technology. *J. Food Science* 66: 772-780.

van Ommen, B., Stierum, R. (2002). Nutrigenomics: Exploiting systems biology in the nutrition and health arena. *Curr. Opin. Biotech.* 13: 517-521.

van Wijk, K.J. (2001). Challenges and prospects of plant proteomics. *Plant Physiol.* 126: 501-508.

Veramendi, J., Willmitzer, L., Trethewey, R.N. (1999). In vitro grown potato microtubers are a suitable system for the study of transgenic lines altered in primary carbohydrate metabolism. *Plant Physiol. Biochem.* 37: 693-697.

Verdonk, J.C., de Vos, C.H.R., Verhoeven, H.A., Haring, M.A., van Tunen, A.J., Schuurink, R.C. (2003). Regulation of floral scent production in petunia revealed by targeted metabolomics. *Phytochem.* 62: 997-1008.

Wagner, C., Sefkow, M., Kopka, J. (2003). Construction and application of a mass spectral and retention time index database generated from plant GC/EI-TOF-MS metabolite profiles. *Phytochem.* 62: 887-900.

Ward, J.L., Harris, C., Lewis, J., Beale, M.H. (2003). Assessment of H-1 NMR spectroscopy and multivariate analysis as a technique for metabolite fingerprinting of *Arabidopsis thaliana. Phytochem.* 62: 949-957.

Watkins, S.M., German, J.B. (2002). Toward the implementation of metabolomic assessments of human health and nutrition. *Curr. Opin. Biotech.* 13: 512-516.

Watkins, S.M., Hammock, B.D., Newman, J.W., German, J.B. (2001). Individual metabolism should guide agriculture toward foods for improved health and nutrition. *Am. J. Clin. Nut.* 74: 283-286.

Weichert, H., Kohlmann, M., Wasternack, C., Feussner, I. (2000). Metabolic profiling of oxylipins upon sorbitol treatment in barley leaves. *Biochem. Soc. Trans.* 28: 861-862.

Weichert, H., Kolbe, A., Kraus, A., Wasternack, C., Feussner, I. (2002). Metabolic profiling of oxylipins in germinating cucumber seedlings lipoxygenase-dependent degradation of triacylglycerols and biosynthesis of volatile aldehydes. *Planta* 215: 612-619.

Wink, M. (1988). Plant breeding: Importance of plant secondary metabolites for protection against pathogens and herbivores. *Theoret. Appl. Genet.* 75: 225-233.

Wyllie, S.G., Fellman, J.K. (2000). Formation of volatile branched chain esters in bananas (*Musa sapientum* L.). *J. Agric. Food Chem.* 48: 3493-3496.

Yamazaki, M., Nakajima, J., Yamanashi, M., Sugiyama, M., Makita, Y., Springob, K., Awazuhara, M., Saito, K. (2003). Metabolomics and differential gene expression in anthocyanin chemo-varietal forms of *Perilla frutescens. Phytochem.* 62: 987-995.

Yamazaki, Y., Urano, A., Sudo, H., Kitajima, M., Takayama, H., Yamazaki, M., Aimi, N., Saito, K. (2003). Metabolite profiling of alkaloids and strictosidine synthase activity in camptothecin producing plants. *Phytochem.* 62: 461-470.

Yates, J.R. (2000) Mass spectrometry: From genomics to proteomics. *Trends Genet.* 16: 5-8.

Chapter 6

Large-Scale Yeast Two-Hybrid Analysis

Barry Causier
John Graham
Brendan Davies

INTRODUCTION

The yeast two-hybrid system has become one of the standard laboratory techniques for the detection and characterization of protein-protein interactions. One of its key advantages is the scale on which interactions can be investigated. On the one hand, it can be used to map individual amino acid residues involved in a specific protein-protein interaction. On the other hand, it can be used to identify novel interactions from complex libraries of expressed proteins. It is the adaptability of the system that has promoted its use in the emerging field of proteomics (the analysis of all proteins expressed in a cell, tissue, or organism). As the various genome and EST sequencing projects continue to generate vast amounts of data, efforts are now concentrating on assigning functions to the many uncharacterized genes and the proteins they encode. Sequence analysis and homology searches will often provide some insight into the function of an unknown factor. However, bioinformatic approaches alone may not provide enough information to place the protein within a biological context. Protein-protein interactions are key to many biochemical processes. Consequently, it may be possible to infer the function of an uncharacterized protein by identifying the proteins with which it interacts. The systematic, large-scale, high-throughput approaches now being taken to build maps of interactions between proteins predicted by genome sequence information have become known as *interactomics*.

In the field of plant biology, the yeast two-hybrid system has been successfully applied to detect interactions with phytochromes, cryptochromes, transcription factors, proteins involved in self-incompatibility mechanisms, the circadian clock, and plant disease resistance (reviewed by Causier and Davies, 2002; Kersten et al., 2002). There also exists a wealth of data generated by the various plant genome-sequencing projects. Taken together with the recent progress made in the development of large-scale yeast two-hybrid screening procedures, this suggests that the time is now ripe for large-scale yeast two-hybrid screens to be applied to organisms such as *Arabidopsis* and rice.

Although a number of biochemical techniques, such as mass spectrometry, surface-plasmon resonance, and immunoprecipitation, lend themselves to interactomics (for reviews, see Auerbach et al., 2002; Uetz and Hughes, 2000), in this chapter we will concern ourselves with only the yeast two-hybrid system. We will begin by briefly describing the basis of the system and then go on to discuss how the system has been developed for large-scale interaction screening, including technologies developed to automate large parts of the screening process. Finally, we will discuss applications of interactomic approaches, how these may be applied in the plant sciences, and how the maps generated may allow evolutionary aspects of protein-protein interactions to be considered.

THE BASIS OF THE YEAST TWO-HYBRID SYSTEM

The yeast two-hybrid system was introduced as a technique in 1989 (Fields and Song, 1989). It is based on research, primarily carried out in Ptashne's laboratory (for review, see Fields and Sternglanz, 1994), which demonstrated that many transcription factors are modular in nature, such that their DNA-binding and transcription activation domains are functionally and physically separable. In their proof-of-principle report, Fields and Song (1989) prepared two expression constructs. The first construct expressed the yeast SNF1 protein as a fusion with the GAL4 transcription factor protein DNA-binding domain (bait protein). The second construct expressed an SNF4-GAL4 activation domain fusion (prey protein). The two constructs were cotransformed into a yeast strain containing a GAL4 responsive *lacZ* reporter gene. A physical interaction between SNF1 and SNF4 resulted in the functional reconstitution of the GAL4 transcription factor and expression of the *lacZ* gene, which was detectable in a colorimetric assay (see Figure 6.1).

The yeast two-hybrid system has now been developed to the point where its use to detect protein-protein interactions has become routine in the laboratory. The system is relatively simple to set up, requires little optimization, and is inexpensive to use. It is also very versatile, since it can be used to test the interaction between two known proteins, to screen expression libraries for new interaction partners, and to identify protein domains involved in a particular interaction. In addition, it has been modified to test for the forma-

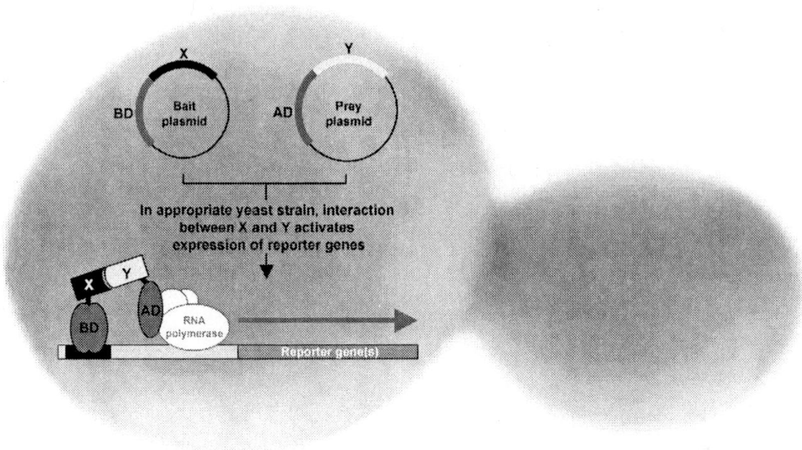

FIGURE 6.1. The yeast two-hybrid system. Two plasmids, bait and prey, are generated. The bait encodes protein X fused in-frame with the DNA-binding domain (BD) of a transcription factor such as GAL4. The prey encodes protein Y fused in-frame with a transcription activation domain (AD). Protein Y may be a specific protein or one encoded by a full-length or random-primed cDNA library. Bait and prey constructs are introduced into a suitable yeast strain either by cotransformation or through the formation of a diploid yeast strain that results from the mating of haploid strains, one containing the bait and the other the prey. Expression of the reporter genes occurs only if proteins X and Y physically interact.

tion of complexes between three or more proteins and to test for protein-nucleic acid interactions (for reviews, see Brent and Finley, 1997; Causier and Davies, 2002).

In terms of protein-protein interactions, two similar yeast two-hybrid systems are in common use. One uses the DNA-binding and transcription activation domains of the GAL4 protein (the GAL4 system; based on the original system of Fields and Song, 1989), while the second utilizes the DNA-binding domain of the bacterial repressor protein LexA in combination with the *Escherichia coli* B42 activation domain (the LexA system; Gyuris et al., 1993). GAL4 and LexA yeast two-hybrid system kits are available from a number of commercial sources (see Causier and Davies, 2002, where the GAL4 and LexA systems are also discussed in more depth), and both systems have been successfully employed in large-scale screens. Other yeast two-hybrid systems have also been developed, such as the SOS recruitment system (Aronheim et al., 1997), the Ras recruitment system (Broder et al., 1998), and the split-ubiquitin system (Stagljar et al., 1998). These systems were introduced to overcome some of the limitations of the GAL4/LexA-based systems (for reviews, see Auerbach et al., 2002; Causier and Davies, 2002). At the time of writing, none of the new variants of the two-hybrid system have been used in large-scale protein-interaction studies. Until their suitability to large-scale screening has been determined, they may, in combination with epitope tagging and coimmunoprecipitation technologies, find a role in confirming protein-protein interactions detected in high-throughput screens.

LARGE-SCALE YEAST TWO-HYBRID STUDIES: A BRIEF HISTORY

A novel aspect of the yeast two-hybrid system, which gives it a significant advantage over many biochemical-based techniques, is the ability to screen expression libraries for new interaction partners. Not only does this allow for large numbers of potential interactors to be tested in a single experiment, but also the cloned cDNAs encoding them are readily available for further analysis.

In the case of interactomics, the adaptation of the yeast two-hybrid system to library screening has been taken one step further. Continued development of the system has meant that it can now be used as a tool to screen for all possible interactions between all proteins expressed in a cell, tissue, or organism (the proteome), systematically and in a partially automated way.

Initial genome-wide screens were restricted to small genomes. For example, Bartel et al. (1996) identified 25 interactions among the 55 proteins encoded by the *E. coli* bacteriophage T7 genome using a yeast two-hybrid approach. The development of large-scale screens of more complex genomes is perhaps best illustrated by the various studies made with yeast proteins. In two complementary reports, a total of 26 yeast proteins involved in pre-mRNA splicing were screened against a random yeast genomic library (encoding fragments of the yeast proteome), resulting in the isolation of a large number (>400) of putative interactions (Fromont-Racine et al., 1997, 2000). Among the interactors were known splicing factors and uncharacterized proteins sharing similarity to splicing factors. The approach validated the technique as a procedure for generating functionally significant maps of the networks of protein-protein interactions that occur in a cell. A similar procedure was used to screen 12 subunits of the yeast RNA polymerase III complex against the yeast proteome (Flores et al., 1999). Interactions between many of the subunits were detected and a new subunit was isolated, suggesting a model for the RNA polymerase III preinitiation complex. The completion of the yeast genome-sequencing project (Goffeau et al., 1996) inevitably led to two ambitious studies to investigate all potential interactions between all proteins predicted from the genome sequence data (Ito et al., 2000, 2001; Uetz et al., 2000). In the first of these projects, two different large-scale approaches were taken. In one, an array of yeast transformants, each expressing a hybrid protein between one of the 6,000 predicted yeast ORFs (open reading frames), and the GAL4 activation domain was produced. Using a mating strategy, the array was screened against yeast transformants expressing 192 bait proteins. Interactions were detected by expression of the *HIS3* reporter gene. Under these conditions, 281 interacting protein pairs were identified. In the second approach, the 6,000 prey transformants were pooled to generate a library that was screened against a library of bait strains expressing the 6,000 ORFs fused to the GAL4 DNA-binding domain. This proteome-against-proteome screen identified 692 interacting protein pairs based on the activation of two reporter genes *(lacZ* and *URA3)*. Taken together, the two approaches identified almost 1,000 interactions involving approximately 1,000 proteins. However, a comparison of the two approaches revealed that the more laborious array-based approach (3.4 interactions per bait) was much more sensitive than the library-versus-library approach (1.8 interactions per bait) (Uetz and Hughes, 2000), perhaps suggesting that the more systematic approach would give a more complete picture of the interactome. In the second project, Ito et al. (2000, 2001) individually cloned the 6,000 yeast ORFs into bait and prey plasmids in yeast strains of opposite mating type, which were divided into pools, each containing 96 clones. Interactions were systematically identified by mating each bait pool against each prey pool, resulting in a total of 3,844 mating reactions, and assaying for the activation of four reporter genes *(ADE2, HIS3, URA3,* and *MEL1)*. Highly significant independent interactions totaled 841 using this procedure. Comparison of the interactions detected in both projects show that they share a surprisingly small number of interactions (141). The reasons for the small overlap are not obvious but may have stemmed from the number of reporter genes used in each study, mutations picked up through PCR amplification of the ORFs that might affect interactions, or because neither screen was saturated (Ito et al., 2001, 2002). Whatever the cause, the results of the two screens suggest that a complete picture of the interactome would require the combination of several independent screening projects (Ito et al., 2001).

Large-scale yeast two-hybrid screens, similar to those described previously, have also been performed to analyze protein-protein interactions in other organisms, including *Drosophila* (Finley and Brent, 1994), the vaccinia virus (McCraith et al., 2000), the hepatitis C virus (Flajolet et al.,

2000), *Caenorhabditis elegans* (Walhout et al., 2000), and *Helicobacter pylori* (Rain et al., 2001). All of these systematic approaches have proved themselves important in the generation of biologically significant protein-protein interaction maps and have proved invaluable in placing previously uncharacterized proteins into a functional context.

Since *Arabidopsis* (Arabidopsis Genome Initiative, 2000) and rice (Goff et al., 2002; Yu et al., 2002) genome sequence data have become available, large-scale yeast two-hybrid interaction studies among plant proteins are now becoming a realistic option. Indeed, such approaches have already been discussed in the recent literature (Kersten et al., 2002) and some progress has been made, using these high-throughput techniques, to identify the network of protein-protein interactions involved in rice defense signal transduction pathways (Fang et al., 2002). More than 130 rice genes involved in the plant defense response, disease resistance, or plant cell signal cascades were expressed as bait fusions. Eight of these were used in a large-scale yeast two-hybrid screen against a rice prey library. Several known interactions were confirmed by this approach and many novel interactions were also detected.

HIGH-THROUGHPUT YEAST TWO-HYBRID METHODOLOGIES

To enable the screening of large numbers of protein-protein interactions, development of new cloning strategies, optimization of the yeast two-hybrid system, and construction of databases have been necessary. As we have described, these technologies have provided the tools with which several large-scale two-hybrid screens have been carried out.

Cloning Technology for High-Throughput Yeast Two-Hybrid Screens

To carry out high-throughput two-hybrid screens, it is first necessary to clone the DNA sequences of interest. Standard cloning strategies dependent on the use of restriction endonucleases and ligase enzymes have been used successfully for large-scale cloning (e.g., see Ito et al., 2000). Recently, however, recombination-based methods that have been developed for DNA cloning are becoming more widespread in large-scale cloning programs.

Homologous Recombination

In the large-scale analysis of yeast protein-protein interactions reported by Uetz and colleagues (2000), the 6,000 yeast ORFs were cloned into the appropriate bait and prey plasmids by homologous recombination in yeast. Each ORF was PCR amplified to include 5' and 3' tails homologous to sequences within the empty bait and prey plasmids. PCR products and a linearized vector were cotransformed into yeast to allow highly efficient recombination to occur between the ORFs and the vector and generation of the appropriate fusion constructs in vivo. Similar procedures have been adopted for other large-scale protein-protein interaction screens (e.g., see Fang et al., 2002). Commercial systems are now available for building yeast two-hybrid libraries by homologous recombination in yeast (see www.clontech.com).

Gateway and Related Cloning Technologies

Gateway technology is an in vitro system that facilitates cloning and subcloning by exploiting the molecular mechanisms of bacteriophage lambda site-specific recombination (Landy, 1989). This system utilizes the DNA sequences (*att*B sites) necessary for site-specific recombination by lambda during the lytic and lysogenic pathways. Gateway can be used to transfer DNA sequences via an intermediary vector (entry vectors) to expression cassettes (destination vectors) in two simple, reversible steps, while maintaining correct orientation and reading frame. The BP reaction is the recombination event used to transfer PCR products containing the site-specific *att*B sites to an appropriate entry vector. A second recombination reaction, the LR reaction, is necessary to transfer the DNA sequence to a destination vector to create an expression cassette. Mixtures of recombination factors specific to either the BP reaction or LR reaction mediate these reactions. Once a DNA sequence is incorporated into the gateway system, it can be transferred to and from the entry and destination vectors using the appropriate reaction mix. Gateway technology was first used in large-scale yeast two-hybrid experiments to study protein interactions involved in *C. elegans* vulval development (Walhout et al., 2000). Gateway cloning systems, including yeast two-hybrid vectors, are now commercially available (see www.invitrogen.com). Other recombination cloning systems, such as the Cre-*lox* site-specific recombination system (Liu et al., 1998), have also been developed for use with the yeast two-hybrid system (see www.clontech.com), but they have yet to find applications in high-throughput protein-protein interaction screens. Recombination cloning procedures can be partially automated and deliver the capacity to provide the high number of accurate cDNA clones required for high-throughput yeast two-hybrid analysis.

Yeast Mating Systems and High-Throughput Yeast Two-Hybrid Screens

Two experimental approaches have been taken to perform large-scale yeast two-hybrid screens: the matrix and the library screening approaches (outlined in Figure 6.2). Both approaches simplify the screening for interactions between large numbers of baits and preys by employing haploid yeast strains of opposite mating type (*MAT*a and *MAT*α). Bait constructs are transformed into yeast of one mating type while prey constructs are transformed into yeast of the other mating type. Once bait strains that autoactivate the yeast reporter genes have been eliminated or modified, the two mating types can be mated to obtain the diploid yeast strain expressing both bait and prey. Optimized interaction-mating protocols have been published and used successfully in large-scale interaction screens using *Arabidopsis* and tobacco proteins (Soellick and Uhrig, 2001).

In the matrix approach (Figure 6.2), full-length ORFs are cloned into two-hybrid vectors, with each ORF being cloned into both bait and prey plasmids. The bait and prey constructs are individually transformed into yeast of opposite mating type. Each bait strain is mated with an ordered and defined array of all prey strains, and diploids expressing interacting proteins are selected based on reporter gene activation. Interactors are identified due to their position on the array. This approach is typically used to identify all possible interactions within a specific family of proteins, between specific families, or between proteins involved in a particular developmental pathway (e.g., see Walhout et al., 2000).

In the second approach, undefined prey libraries, expressing either full-length ORFs or randomly generated cDNAs, are screened with bait strains. This approach has been used in two ways (Figure 6.2). In the first, individual bait strains are screened against the whole prey library. This was the approach taken by Uetz et al. (2000) to study the yeast interactome. The alternative method, used by Ito et al. (2000, 2001), is to produce pools of bait strains and screen each of these against pools of the prey library. In the library screening approach, interactors are identified through DNA sequence analysis of the prey and/or bait constructs isolated from the appropriate diploid yeast cells.

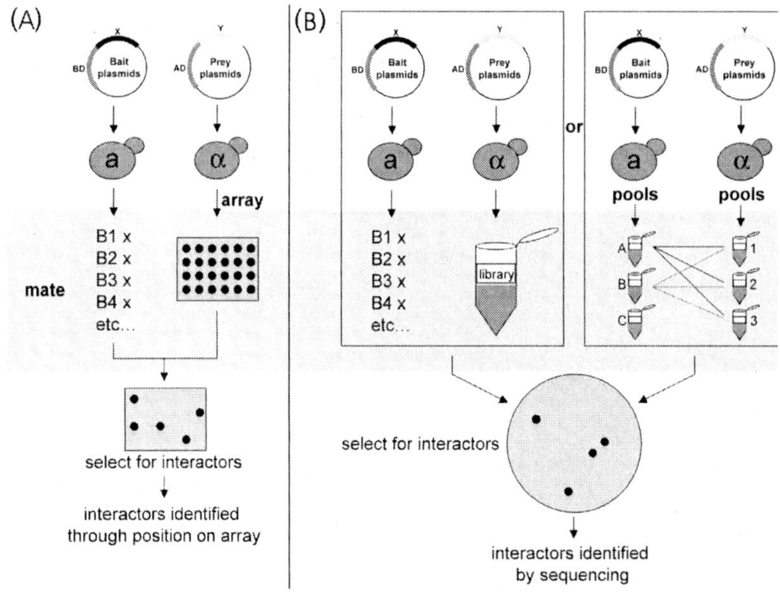

FIGURE 6.2. High-throughput yeast two hybrid screens. (A) *A typical matrix approach:* Yeast strains containing prey constructs are arrayed, usually in microtitre plates, such that the position of each prey is defined. These arrays are separately mated with each individual bait strain (B1, B2, B3, etc.), and diploids expressing interacting bait and prey proteins are selected based on reporter gene activity. Interactors are identified based on their position on the array. (B) *Possible library approaches:* On the left, a prey library (consisting of either full-length ORFs or cDNA fragments) is mated with individual bait strains (B1, B2, B3, etc.). Diploids demonstrating an interaction between bait and prey are selected and the prey identified by DNA sequence analysis. On the right, an alternative library screen is detailed that has been used where the entire proteome is to be analyzed for all potential interactions. The entire set of ORFs to be analyzed are individually cloned into the yeast two-hybrid vectors and transformed into appropriate yeast strains (e.g., baits transformed into a *MAT*a yeast strain and preys into a *MAT*α strain). Bait and prey strains are divided into pools and systematically mated with each other. Diploids expressing interacting bait and prey proteins are selected and the identity of the interacting partners determined through DNA sequence analysis.

Automation of Large-Scale Interaction Screens

Although it is feasible to carry out a large number of matings manually (Walhout and Vidal, 2001), large-scale yeast two-hybrid screens can be performed rapidly and consistently through partial automation of the procedures. Robotics systems have been developed for high-throughput two-hybrid screening that are capable of mating bait and prey strains in high-density arrays on plates. Associated software and visualization systems allow automatic selection of diploids containing potential interactors, as well as the tracking and storage of data during a protein-protein interaction screen.

Suitability of Different Systems and the Problem of False Positives

Although the Gal4 and LexA systems have been successfully used to perform two-hybrid screens, these transcription-based methods are susceptible to a number of problems. Difficulties may be encountered with transcriptional repressors, bait proteins that autoactivate reporter genes, and proteins that do not localize to the yeast cell nucleus (such as integral membrane proteins). In screens where DNA sequences are being cloned individually to assay against each other, it is possible to detect autoactivating clones prior to screening and remove transcription activation domains by creating deletions of the protein. Truncated proteins that have lost the ability to activate transcription can then be included in the screens. Alternative yeast two-hybrid systems have been developed to counter the problems associated with the GAL4/LexA systems. Unfortunately, these new systems are largely untested in large-scale protein interaction studies and some may not be robust enough for large-scale screening without further development (see Causier and Davies, 2002).

False positives are a common problem in two-hybrid screens (Causier and Davies, 2002). The system employed by Fang et al. (2002) to study interactions among rice proteins utilized three positive selection markers *(ADE2, HIS3,* and *lacZ)* and an *URA3*-based negative selection step (to eliminate preys that can activate the reporters in the absence of a bait) to maintain a high level of stringency over the interactions that were detected. It is clear from this study and those of Ito et al. (2000, 2001) previously mentioned that large-scale yeast two-hybrid screens would benefit from the use of several reporter genes to ensure that the minimum number of false positives were isolated, and that only highly significant interactions were identified. A number of the commercially available systems employ at least three reporter genes, which is probably the minimum that should be considered to eliminate false positives (e.g., see details of the ProQuest system at <www.invitrogen.com> and the Matchmaker3 system at <www.clontech.com>).

Databases

The ultimate goal of the high-throughput yeast two-hybrid projects has been to assign a putative function to uncharacterized proteins based on their interaction with characterized proteins. Vast quantities of data generated by these projects have necessitated the construction of databases that can be used to create interaction maps. Such interaction maps provide the potential not only to assign function to previously uncharacterized proteins but also to place the protein within a biological context. However, the use of these bioinformatics algorithms is dependent upon the validity of the data that have been analyzed. Incomplete protein-interaction maps and false positives can give an inaccurate picture of the function of some proteins. To produce a more reliable analysis, it is necessary to integrate and analyze data generated by different screening procedures. Using the Biomolecular Interaction Network Database (BIND, <www.bind.ca>; Bader et al., 2003) to compile data from different sources (including large-scale yeast two-hybrid and mass spectrometry screens), Bader and Hogue (2002) identified 15,143 nonredundant, experimentally determined yeast protein-protein interactions among 4,825 proteins (approximately 75 percent of the yeast genome). Analyses such as this are clearly capable of producing a much more reliable predictive source of information about protein function. As well as assigning function by association, database technologies are also able to make broad comparisons of protein-protein interactions between kingdoms and produce predictive conserved interactions between species.

Internet links to many protein-protein interaction databases are listed at HGMP GenomeWeb (http://www.hgmp.mrc.ac.uk/GenomeWeb/prot-interaction.html), the BIND database (http://www.bind.ca), and at the Database of Interacting Proteins (http://dip.doe-mbi.ucla.edu).

APPLICATIONS OF LARGE-SCALE YEAST TWO-HYBRID SCREENING

In the previous sections, we touched upon some of the applications of large-scale yeast two-hybrid approaches. Bartel et al. (1996), Uetz et al. (2000), and Ito et al. (2000, 2001) have demonstrated that the system can be adapted for proteome-against-proteome screens. As increasingly large amounts of plant genome sequence data become available, there is a pressing need to dissect the function of many of the predicted genes. As such, networks of protein-protein interactions deduced from proteome-versus-proteome screens will inevitably appear in the plant science literature. Before that, however, we are more likely to see reports of small-scale proteome screens, such as that described by Fang et al. (2002). Such studies will screen relatively small numbers of proteins against the proteome so that pathways involved in particular biological processes can be mapped. Another application of high-throughput two-hybrid screening is to search for interactions among functionally related proteins. As an example, we can take one of the interests of our research group: the role of transcription factors in plant development. In 2000, it was predicted that there were >1,500 transcription factors encoded by the *Arabidopsis* genome (Riechmann et al., 2000). Today we believe the number to be in excess of 2,000. To screen all of the potential interactions among this group of proteins would require the analysis of more than 4×10^6 combinations of proteins (i.e., 2,000 baits against 2,000 preys). Only an approach such as the yeast two-hybrid, used in a systematic fashion, can realistically be considered for screens on this scale. Indeed, as part of a consortium, known as REGIA and funded by the European Union, we have already begun to screen for all potential interactions among approximately 1,000 of these transcription factors using the yeast two-hybrid system.

The yeast two-hybrid system has been used extensively to determine protein domains involved in specific protein-protein interactions. In large-scale screens of randomly primed expression libraries, certain fragments of a protein encoded by the library will express interaction domains. By analyzing overlapping prey fragments from the library, a picture of the interaction domain can be generated. As an alternative approach to identifying interaction domains, Deng et al. (2002) used the PFAM protein domain database to analyze the yeast interactome predicted by the large-scale two-hybrid screens of Uetz et al. (2000) and Ito et al. (2000, 2001). Using this approach, they were able to build a picture of the conserved patterns of interactions among protein domains. Furthermore, they were able to utilize their findings to predict protein-protein interactions and found a significant overlap with interactions detected by other methods.

The work of Deng and colleagues (2002), detailed in the previous paragraph, begins to suggest ways in which the data generated by interactome studies may be utilized. We have discussed in earlier sections of this chapter how the function of an uncharacterized protein may be inferred from its interactions with proteins of known function. An extension of this may allow us to make informed predictions about the possible interaction partners of some plant proteins. The protein-protein interaction data generated by a number of large-scale studies are publicly available in a number of databases (see the High-Throughput Yeast Two-Hybrid Methodologies section). It may now be possible to take proteins conserved between species and predict the proteins with which they interact. In this way, complete pathways of potential protein-protein interactions may be generated for a particular species, completely *in silico*.

THE CAVEAT

Large-scale two-hybrid screens designed to investigate all possible protein-protein interactions within the yeast proteome have identified in excess of 5,500 protein-protein interactions between them (Ito et al., 2001; Uetz et al., 2000). However, when we consider that, as a low estimate, there are between 20,000 and 30,000 protein-protein interactions predicted in yeast (Bader and Hogue, 2002; von Merling et al., 2002), it is clear that neither large-scale yeast screen was exhaustive. While improvements in two-hybrid approaches may improve the coverage of such screens, there is an obvious need to incorporate two-hybrid data with that generated using alternative methodologies. Two major studies of the interactions between yeast proteins using mass spectrometry (MS) have recently been published (Gavin et al., 2002; Ho et al., 2002). In each approach, yeast proteins were engineered to contain specific tags. The tagged proteins were expressed in yeast cells so that appropriate protein complexes formed. Based on the tag used, the multiprotein complexes were affinity purified, individual proteins resolved in a denaturing gel, and the protein bands excised and analyzed by MS. In the study by Gavin et al. (2002), 1,440 proteins were identified within 232 distinct complexes. In addition, 91 percent of the complexes contained at least one functionally uncharacterized protein and most complexes shared a protein with a separate complex (Auerbach et al., 2002; Kumar and Snyder, 2002). Ho et al. (2002) generated 725 tagged proteins (approximately 10 percent of predicted yeast proteins) and detected 3,617 interactions, covering 25 percent of the yeast proteome. Both large-scale MS approaches identified significant numbers of protein-protein interactions. However, limitations of MS approaches were also revealed by the two studies. For example, as with the large-scale two-hybrid studies discussed previously, the overlap between the two studies was small (~10 percent of the entire proteins detected) (Ito et al., 2002). Furthermore, a large number of false positives (up to 30 percent in the case of the Gavin et al. [2002] screen) were detected by both groups.

Combined data from all large-scale protein-protein interaction studies indicate that currently only ~15,000 interactions have been identified in yeast (Bader and Hogue, 2002) and, of these, a large proportion (>50 percent) may be false positives (von Merling et al., 2002). Continual development of the technologies used to create interactome maps will be required to reduce the number of false positives isolated. However, to gain a complete, functionally significant picture of the interactome, one would need to integrate the data generated from as many different approaches as possible (Bader and Hogue, 2002; von Merling et al., 2002; Uetz and Hughes, 2000). Protein-protein interaction data from large-scale studies (including yeast two-hybrid, mass spectrometry, protein microarray chips, protein phage display, and other emerging technologies) should be integrated with large-scale expression pattern analysis of the genes encoding the interacting proteins and with the phenotypic analysis of cells containing mutations in these genes (Ito et al., 2002; Legrain et al., 2001; von Merling et al., 2002). Recent studies have demonstrated that proteins which interact and operate in the same biological pathway share similar expression patterns and produce similar phenotypes when mutated. For example, statistical analyses have shown that in the case of *E. coli* bacteriophage T7, yeast, and *C. elegans*, proteins encoded by genes with similar expression profiles are more likely to interact with one another (Ge et al., 2001; Grigoriev, 2001; Walhout et al., 2002). In addition, comparison of the interactions among *C. elegans* germ-line proteins with the phenotypes of nematodes carrying loss-of-function mutations in the genes encoding these proteins demonstrated that essential proteins have a tendency to interact with one another (Walhout et al., 2002). With comprehensive studies such as these, protein-protein interactions can be linked to developmental processes with high confidence.

The various plant genome-sequencing initiatives have propelled plant research into the proteomic era. Exciting times lie ahead as the functions of uncharacterized genes are unraveled and are placed into a biological context. The continued development of versatile techniques such as the yeast two-hybrid system can only accelerate this voyage of discovery.

REFERENCES

Arabidopsis Genome Initiative (2000). Analysis of the genome sequence of the flowering plant *Arabidopsis thaliana*. *Nature* 408: 796-815.

Aronheim, A., Zandi, E., Hennemann, H., Elledge, S.J., Karin, M. (1997). Isolation of an AP-1 repressor by a novel method for detecting protein-protein interactions. *Mol. Cell. Biol.* 17: 3094-3102.

Auerbach, D., Thaminy, S., Hottiger, M.O., Stagljar, I. (2002). The post-genomic era of interactive proteomics: Facts and perspectives. *Proteomics* 2: 611-623.

Bader, G.D., Betel, D., Hogue, C.W.V. (2003). BIND: The biomolecular interaction network database. *Nucleic Acids Res.* 31: 248-250.

Bader, G.D., Hogue, C.W.V. (2002). Analysing yeast protein-protein interaction data obtained from different sources. *Nat. Biotechnol.* 20: 991-997.

Bartel, P.L., Roecklein, J.A., SenGupta, D., Fields, S. (1996). A protein linkage map of *Escherichia coli* bacteriophage T7. *Nat. Genet.* 12: 72-77.

Brent, R., Finley, R.L. (1997). Understanding gene and allele function with two-hybrid methods. *Annu. Rev. Genet.* 31: 663-704.

Broder, Y.C., Katz, S., Aronheim, A. (1998). The Ras recruitment system, a novel approach to the study of protein-protein interactions. *Curr. Biol.* 8: 1121-1124.

Causier, B., Davies, B. (2002). Analysing protein-protein interactions with the yeast two-hybrid system. *Plant Mol. Biol.* 50: 855-870.

Deng, M., Mehta, S., Sun, F., Chen, T. (2002). Inferring domain-domain interactions from protein-protein interactions. *Genome Res.* 12: 1540-1548.

Fang, Y., Macool, D.J., Xue, Z., Heppard, E.P., Hainey, C.F., Tingey, S.V., Miao, G.-H. (2002). Development of a high-throughput yeast two-hybrid screening system to study protein-protein interactions in plants. *Mol. Genet. Genomics* 267: 142-153.

Fields, S., Song, O.-K. (1989). A novel genetic system to detect protein-protein interactions. *Nature* 340: 245-246.

Fields, S., Sternglanz, R. (1994). The yeast two-hybrid system: An assay for protein-protein interactions. *Trends Genet.* 10: 286-292.

Finley, R.L., Brent, R. (1994). Interaction mating reveals binary and ternary connections between *Drosophila* cell cycle regulators. *Proc. Natl. Acad. Sci. USA* 91: 12980-12984.

Flajolet, M., Rotondo, G., Daviet, L., Bergametti, F., Inchauspe, G., Tiollais, P., Transy, C., Legrain, P. (2000). A genomic approach of the hepatitis C virus generates a protein interaction map. *Gene* 242: 369-379.

Flores, A., Briand, J.F., Gadal, O., Andrau, J.C., Rubbi, L., Van Mullem, V., Boschiero, C., Goussot, M., Marck, C., Carles, C., et al. (1999). A protein-protein interaction map of yeast RNA polymerase III. *Proc. Natl. Acad. Sci. USA* 96: 7815-7820.

Fromont-Racine, M., Mayes, A.E., Brunet-Simon, A., Rain, J.C., Colley, A., Dix, I., Decourty, L., Joly, N., Ricard, F., Beggs, J.D., Legrain, P. (2000). Genome-wide protein interaction screens reveal functional networks involving Sm-like proteins. *Yeast* 17: 95-110.

Fromont-Racine, M., Rain, J.C., Legrain, P. (1997). Toward a functional analysis of the yeast genome through exhaustive two-hybrid screens. *Nat. Genet.* 16: 277-282.

Gavin, A.C., Bosche, M., Krause, R., Grandi, P., Marzioch, M., Bauer, A., Schultz, J., Rick, J.M., Michon, A.M., Cruciat, C.M., et al. (2002). Functional organization of the yeast proteome by systematic analysis of protein complexes. *Nature* 415: 141-147.

Ge, H., Liu, Z., Church, G., Vidal, M. (2001). Correlation between transcriptome and interactome mapping data from *Saccharomyces cerevisiae. Nat. Genet.* 29: 482-486.

Goff, S.A., Ricke, D., Lan, T.H., Presting, G., Wang, R., Dunn, M., Glazebrook, J., Sessions, A., Oeller, P., Varma, H., et al. (2002). A draft sequence of the rice genome *(Oryza sativa* L. ssp. *japonica). Science* 296: 92-100.

Goffeau, A., Barrell, B.G., Bussey, H., Davis, R.W., Dujon, B., Feldmann, H., Galibert, F., Hoheisel, J.D., Jacq, C., Johnston, M., et al. (1996). Life with 6000 genes. *Science* 274: 546-567.

Grigoriev, A. (2001). A relationship between gene expression and protein interactions on the proteome scale: Analysis of the bacteriophage T7 and the yeast *Saccharomyces cerevisiae*. *Nucleic Acids Res.* 29: 3513-3519.

Gyuris, J., Golemis, E., Chertov, H., Brent, R. (1993). Cdi1, a human G1 and S phase protein phosphatase that associates with Cdk2. *Cell* 75: 791-803.

Ho, Y., Gruhler, A., Heilbut, A., Bader, G.D., Moore, L., Adams, S.L., Millar, A., Taylor, P., Bennett, K., Boutilier, K., et al. (2002). Systematic identification of protein complexes in *Saccharomyces cerevisiae* by mass spectrometry. *Nature* 415: 180-183.

Ito, T., Chiba, T., Ozawa, R., Yoshida, M., Hattori, M., Sakaki, Y. (2001). A comprehensive two-hybrid analysis to explore the yeast protein interactome. *Proc. Natl. Acad. Sci. USA* 98: 4569-4574.

Ito, T., Ota, K., Kubota, H., Yamaguchi, Y., Chiba, T., Sakuraba, K., Yoshida, M. (2002). Roles for the two-hybrid system in exploration of the yeast interactome. *Mol. Cell. Proteomics* 1: 561-566.

Ito, T., Tashiro, K., Muta, S., Ozawa, R., Chiba, T., Nishizawa, M., Yamamoto, K., Kuhara, S., Sakaki, Y. (2000). Toward a protein-protein interaction map of the budding yeast: A comprehensive system to examine two-hybrid interactions in all possible combinations between the yeast proteins. *Proc. Natl. Acad. Sci. USA* 97: 1143-1147.

Kersten, B., Bürkle, L., Kuhn, E.J., Giavalisco, P., Konthur, Z., Lueking, A., Walter, G., Eickhoff, H., Schneider, U. (2002). Large-scale plant proteomics. *Plant Mol. Biol.* 48: 133-141.

Kumar, A., Snyder, M. (2002). Protein complexes take the bait. *Nature* 415: 123-124.

Landy, A. (1989). Dynamic, structural, and regulatory aspects of lambda-site-specific recombination. *Annu. Rev. Biochem.* 58: 913-949.

Legrain, P., Wojcik, J., Gauthier, J.-M. (2001). Protein-protein interaction maps: A lead towards cellular functions. *Trends Genet.* 17: 346-352.

Liu, Q., Li, M.Z., Leibham, D., Cortez, D., Elledge, S.J. (1998). The univector plasmid-fusion system, a method for rapid construction of recombinant DNA without restriction enzymes. *Curr. Biol.* 8: 1300-1309.

McCraith, S., Hotzam, T., Moss, B., Fields, S. (2000). Genome-wide analysis of vaccinia virus protein-protein interactions. *Proc. Natl. Acad. Sci. USA* 97: 4879-4884.

Rain, J.-C., Selig, L., De Reuse, H., Battaglia, V., Reverdy, C., Simon, S., Lenzen, G., Petel, F., Wojcik, J., Schächter, V., et al. (2001). The protein-protein interaction map of *Helicobacter pylori*. *Nature* 409: 211-215.

Riechmann, J.L., Heard, J., Martin, G., Reuber, L., Jiang, C., Keddie, J., Adam, L., Pineda, O., Ratcliffe, O.J., Samaha, R.R., et al. (2000). *Arabidopsis* transcription factors: Genome-wide comparative analysis among eukaryotes. *Science* 290: 2105-2110.

Soellick, T.-R., Uhrig, J. (2001). Development of an optimized interaction-mating protocol for large-scale yeast two-hybrid analyses. *Genome Biol.* 2: research0052.

Stagljar, I., Korostensky, C., Johnsson, N., Te Heesen, S. (1998). A genetic system based on split-ubiquitin for the analysis of interactions between membrane proteins in vivo. *Proc. Natl. Acad. Sci. USA* 95: 5187-5192.

Uetz, P., Giot, L., Cagney, G., Mansfield, T.A., Judson, R.S., Knight, J.R., Lockshon, D., Narayan, V., Srinivasan, M., Pochart, P., et al. (2000). A comprehensive analysis of protein-protein interactions in *Saccharomyces cerevisiae*. *Nature* 403: 623-627.

Uetz, P., Hughes, R.E. (2000). Systematic and large-scale two-hybrid screens. *Curr. Opin. Microbiol.* 3: 303-308.

von Merling, C., Krause, R., Snel, B., Cornell, M., Oliver, S.G., Fields, S., Bork, P. (2002). Comparative assessment of large-scale data sets from protein-protein interactions. *Nature* 417: 399-403.

Walhout, A.J., Reboul, J., Shtanko, O., Bertin, N., Vaglio, P., Ge, H., Lee, H., Doucette-Stamm, L., Gunsalus, K.C., Schetter, A.J., et al. (2002). Integrating interactome, phenome, and transcriptome mapping data for the *C. elegans* germline. *Curr. Biol.* 12: 1952-1958.

Walhout, A.J., Sordella, R., Lu, X., Hartley, J.L., Temple, G.F., Brasch, M.A., Thierry-Mieg, N., Vidal, M. (2000). Protein interaction mapping in *C. elegans* using proteins involved in vulval development. *Science* 287: 116-122.

Walhout, A.J., Vidal, M. (2001). High-throughput yeast two-hybrid assays for large-scale protein interaction mapping. *Methods* 24: 297-306.

Yu, J., Hu, S., Wang, J., Wong, G.K., Li, S., Liu, B., Deng, Y., Dai, L., Zhou, Y., Zhang, X., et al. (2002). A draft sequence of the rice genome (*Oryza sativa* L. spp. *indica*). *Science* 296: 79-92.

SECTION II: SPECIES EXAMPLES

Chapter 7

Functional Genomics of a Cyanobacterium, *Synechocystis* sp. PCC 6803

Takakazu Kaneko
Shusei Sato
Satoshi Tabata

INTRODUCTION

The genetic systems of cyanobacteria are less complex than that of higher plants, consisting of a nucleus, plastids, and mitochondria. Therefore, cyanobacteria have become model organisms for the study of plant-specific phenomena, including oxygenic photosynthesis. In addition, genetic-engineering technologies have been used successfully in several cyanobacterial species to elucidate gene function. Completion of the identification of the genomic sequence for *Synechocystis* sp. strain PCC 6803 (hereinafter, *Synechocystis*) in 1996 (Kaneko et al., 1996) initiated the strategic study of the function and regulation of cyanobacterial genes. Comprehensive information on the structure of gene components in the genome has accelerated the identification and characterization of genes and their respective biological functions. Such information has led to the development of cutting-edge technologies and approaches in functional genomics, such as systematic gene disruption, transcriptome analysis using array technology, and proteome analysis. As a result, a tremendous amount of information has been generated for individual *Synechocystis* genes involved in a variety of biological processes. In this chapter, the current status of large-scale gene function analyses in *Synechocystis* will be summarized.

A LARGE-SCALE GENE DISRUPTION

Cyanobacteria have long been used as genetic systems for the study of photosynthesis, nitrogen fixation, heterocyst formation, and other biological functions because methods of DNA transformation by electroporation or conjugation are available for several species. *Synechocystis* is one of the most popular species used for this purpose due to its natural transformation ability and photoheterotrophic growth, which facilitates investigation of photosynthetic reactions (Williams, 1988). Two methods of gene disruption, targeted mutagenesis and random mutagenesis, have been implemented in *Synechocystis*.

In targeted mutagenesis, a target gene on a chromosome is replaced by homologous recombination via double crossing-over with a homologous DNA segment containing an antibiotic resistance gene cassette (Figure 7.1A). The DNA construct is readily incorporated into cells and integrated into the chromosome; the resultant mutants can be selected on an agar plate containing antibiotics approximately two weeks after transformation. This method has long been used to investigate gene function in *Synechocystis*. Genome sequencing has led to large-scale and/or systematic gene disruption and characterization of the mutants obtained. Forty-three *Synechocystis* genes may encode sensor histidine kinases of the two-component regulatory system (Mizuno et al., 1996) (see also Chapter 23). Suzuki et al. (2000) performed a systematic disruption of these genes to find novel low-temperature sensors and signal transducers and successfully identified two genes, *hik19 (sll1905)* and *hik33 (sll0698),* as components of the low-temperature signal pathway. Sokolenko et al. (2002) inactivated each of the 62 genes coding for peptidases and showed that disruption of 13 genes resulted in visible phenotypic changes under light, heat, or nutrient stress.

Synechocystis mutant information generated by the gene replacement method has been deposited and cataloged in the CyanoMutants Web database (http://www.kazusa.or.jp/cyano/mutants/) (Nakamura et al., 1999). At the time of this writing, a total of 722 mutants have been registered in this database.

The availability of the complete genome sequence has enabled the generation of mutant libraries in which each mutation is tagged with a specific sequence, such as that of transposons (Tn) (Vidan and Snyder, 2001). An ordered *Synechocystis* cosmid library has been compiled of 107 partially overlapping clones with an average insert length of 40 kbp (Kaneko and Tabata, 1997). Each clone's genomic region has been defined by comparing the end sequences of each clone with the entire genome sequence, and more than 90 percent of the chromosome is covered by the library (CyanoBase at <http://www.kazusa.or.jp/cyano/>). An in vitro Tn7 transposition system was applied to this library to introduce mutations in the *Synechocystis* genes at a genome-wide scale (H. Fukuzawa, personal communication; CyanoBase at <http://www.kazusa.or.jp/cyano/>). The Tn insertion point was determined by utilizing the Tn sequence tag after random insertions, and then each cosmid with the Tn7 insertion was introduced into cells and recombi-

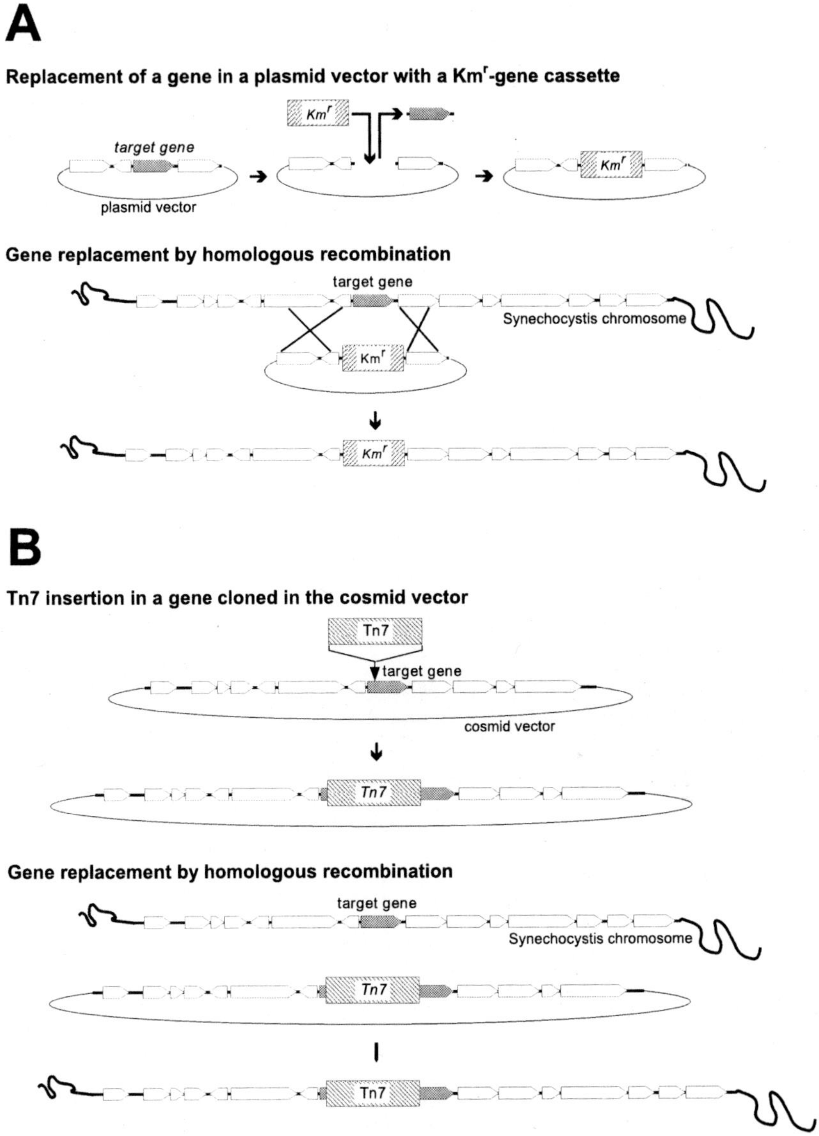

FIGURE 7.1. Outline of the strategy for gene disruption in *Synechocystis*. (A) Generation of mutants by gene replacement. (B) Tn-mediated gene disruption.

nant transformants were selected by antibiotic resistance (Figure 7.1B). This simpler method is superior to targeted mutagenesis in that a larger number of genes can be mutated at a time. With this method, most genes are disrupted by simply increasing the number of insertions, though small ORFs (open reading frames) tend to escape insertions, which may be disadvantageous.

This system has proven to be useful in identifying genes responsible for carbon source uptake. *Synechocystis* possesses a CO_2-concentrating mechanism that involves active CO_2 uptake and HCO_3 transport. An NAD(P)H dehydrogenase complex is essential for CO_2 uptake from air, and two functionally distinct types of CO_2 uptake enzymes not yet fully identified have been reported (Ohkawa, Price, et al., 2000; Ohkawa, Pakrasi, et al., 2000). Shibata et al. (2001) generated a mutant bank by inserting the transposon-bearing cosmid set into the genome of *Synechocystis* cells deficient in one of the two NAD(P)H dehydrogenases (delta-*ndhD3*). From this bank, they isolated mutants unable to grow under air with CO_2, and found Tn insertions in an ORF *slr1302,* which was then designated as *cupB*. This gene was assumed to code for a CO_2 transporter, associated with NAD(P)H dehydrogenase subunits, which binds CO_2 and releases HCO_3^- to the cytoplasm (Shibata et al., 2001). As for the HCO_3^- transport system, an ABC-type HCO_3^- transporter encoded by *cmpABCD* has been identified (Omata et al., 2001). However, inactivation of *cmp* genes had little effect on overall HCO_3^- transport activity, suggesting the presence of one or more HCO_3^- transporters. By using the random Tn insertion method in the *Synechocystis* mutant unable to take up HCO_3^- and grow in low CO_2 at pH 9.0 (delta-*ndhB*/delta-*cmpA*), a novel sodium-bicarbonate transporter (SbtA) gene was identified (Shibata et al., 2002). SbtA was shown to play a central role in HCO_3^- transport in *Synechocystis*.

TRANSCRIPTOME ANALYSIS

Two types of approaches, differential display and DNA microarrays, have been used to monitor cyanobacterial gene expression at the transcriptional level on a genome-wide scale. Contamination of rRNAs in RNA fractions has always been problematic in differential display, and methods in which rRNAs do not impair the detection of mRNAs of protein-encoding genes have been developed.

The differential expression analysis procedure using the customized amplification library (DECAL) is a method in which cosmid clones covering a substantial portion of the genome, except for rRNA gene regions, are fragmented and hybridized with total cellular RNA to isolate probes corresponding only to mRNAs (Alland et al., 1998). The eluted DNA fragments are then amplified by PCR and labeled for screening. Singh and Sherman (2000) performed a global comparison of gene expression between the normal and iron-deficient conditions using DECAL-generated probes and found that, in addition to *isiA* and *idiA,* previously shown to be iron-regulated, *psbA, cpcG,* and *slr0374,* known to be involved in photosynthesis or pigment synthesis, were also regulated by iron.

Bhaya et al. (2000) developed a differential display method utilizing the short palindromic sequence GCGATCGC, designated as the HIP (highly iterated palindrome) element (Robinson et al., 1995). There are 3,160 copies of HIP1 in the *Synechocystis* genome, with an average frequency of one copy per 1,131 bp. More than 50 percent of the protein-encoding genes, but none of the rRNA and tRNA genes, contain HIP. A probe generated by reverse transcription with the HIP1 sequence as a primer and the total RNA as a template successfully identified genes specifically regulated under high- and low-light conditions by differential hybridization.

In microarrays, fluorescently labeled probes can be prepared from non-purified total cellular RNA because large quantities of sequences derived from rRNAs do not seem to hamper the detection of mRNA hybridization signals (see also Chapter 4). A microarray designed for the *Synechocystis* genome is commercially available (TaKaRa Shuzo Co., Japan). This product, CyanoCHIP, contains 3,079 nonredundant DNA segments that cover up to 1 kbp regions from the 3' termini of the protein-encoding genes, except for transposases, in the genome (Hihara et al., 2001). An oligonucleotide set for PCR amplification of the 3,117 genes in the *Synechocystis* genome, available from Dupont Co., can be used to generate a microarray for all the gene components in the genome or for a customized set of genes (Huang, McCluskey, et al., 2002).

Profiles of gene transcription responses to changes in light conditions were intensively studied using the microarray system. Hihara et al. (2001) reported up-regulation of 84 genes and down-regulation of 80 genes during acclimation from low- to high-light intensity using CyanoCHIP. Among the up-regulated genes were genes for CO_2 fixation and protection against photodamage or stress, while the down-regulated genes included genes for photosystems I and II, core and rod components of phycobilisome, and chlorophyll biosynthesis. Gill et al. (2002) also investigated gene transcription during the light-dark transition and identified activation of 783 genes in response to light exposure. UV-light-responsive transcriptional activation in *Synechocystis* cells was also studied by Huang, McCluskey, et al. (2002), who reported that the transcription profile was altered more extensively following UV-B irradiation at 60 $\mu E \cdot m^{-2} \cdot s^{-1}$ for 120 min than following the lower irradiation level of 20 $\mu E \cdot m^{-2} \cdot s^{-1}$. Their results indicate that 146 genes were induced and 159 genes were repressed upon exposure to the higher UV level. Other environmental stresses have also been evaluated in microarray analysis. Salt stress and hyperosmotic stress are thought to have similar physiological effects on cells. Kanesaki et al. (2002) showed that distinct gene sets in addition to common sets of genes are positively or negatively regulated by the respective stresses.

The combination of the microarray technique with mutants deficient in various biological pathways is a powerful approach for locating cascades and gene-regulation networks. Many of the *Synechocystis* genes responsible for the components of photosystem I and phycobilisomes are cold repressible, and genes responsible for transcription and translation are, in general, cold inducible. Suzuki et al. (2001) performed a thorough study to identify the genes downstream of the temperature sensory kinase *hik33 (sll0698)*. By comparing gene expression between the wild type and its mutant using CyanoCHIP, they found that the transcription of *sll1407, sll1404 (exbB), sll1406 (fhuA), sll1408, slr1484,* and *slr1485* was induced by the presence of *hik33* under the low-temperature condition, while transcription of *sll1951 (hlyA)* was repressed. A cAMP signaling pathway was also subjected to microarray analysis, revealing that *sycrp1 (sll1371),* the gene for a cAMP receptor/DNA-binding protein, is in the cAMP signaling pathway (Yoshimura et al., 2000). Yoshimura et al. (2002) showed that 18 genes for the ABC transporter and sensor histidine kinase were under the control of SYCRP1.

PROTEOME ANALYSIS

The first comprehensive analysis of the protein components in cyanobacteria was reported for *Synechocystis* in 1996 by Sazuka and Ohara (1996). They extracted soluble proteins from the whole-cell lysate and separated them on 2D-gels. The protein was extracted from each spot and the amino acid sequence from the N-terminus was partially determined. Combined with the results from refined studies using various cellular fractions, including soluble, insoluble, thylakoid membrane, and secreted proteins, a total of 234 protein spots were identified and matched with their corresponding 143 genes (Sazuka et al., 1999). The multiple protein spots with shared amino acid sequences indicate that posttranslational modification occurs, although such modification has yet to be identified.

Comparison of the N-terminal amino acid sequences with the corresponding genome sequences revealed sequence features around the translation initiation sites (Sazuka et al., 1999). Forty-three out of 143 proteins examined had Met as the first amino acid, for which rare initiation codons such as TTG and CTG were used in addition to the major ATG and GTG codons. The remaining 100 proteins did not have Met at the N-termini, strongly suggesting that the posttranslational cleavage of either Met by methionyl aminopeptidase or a leader peptide by leader peptidase takes place. The Shine-Dalgarno sequences (either GCGG, GGAG, or AGGA) upstream of the actual translation initiation sites were observed for 47.4 percent of the genes examined. All the information on the analyzed proteins and features of the corresponding genes revealed by these studies are presented on the Web database Cyano2Dbase at <http://www2.kazusa.or.jp/cyano/Synechocystis/cyano2D/index.html> (Sazuka et al., 1999).

Recent advances in technologies for cell fractionation, protein manipulation, and sequencing have enabled the analysis of protein components of specific cellular portions at a microscale. Intensive investigation of *Synechocystis* peripheral proteins of thylakoid membranes (Wang et al., 2000), purified photosystem II complex (Kashino et al., 2002), periplasmic proteins (Fulda et al., 2000), plasma membrane proteins (Huang, Parmryd, et al., 2002), light-induced soluble proteins (Choi et al., 2000), and secreted proteins (Sergeyenko and Los, 2000) have been conducted.

ANALYSIS OF PROTEIN INTERACTIONS WITH THE YEAST TWO-HYBRID SYSTEM

Cellular biological activities are based on protein interactions. Therefore, the systematic and comprehensive analysis of protein interactions is expected to provide valuable insight into the complex protein functions involved in these activities. The most established of the various protein interaction investigation methods is the yeast two-hybrid (YTH) screening system that is also suitable for genome-wide screening (see Chapter 6). In this system, a "bait" is expressed by fusion to the DNA-binding domain of a yeast transcription factor and is coexpressed in a yeast cell with a "prey," or a random genomic DNA fragment or a selected cDNA fused to the fragment coding for an activation domain of the transcription factor. The interaction between the bait and prey fusion proteins restores transcriptional activity, resulting in expression of downstream reporter genes. This method has been adopted for large-scale analysis of protein interactions in yeast (Uetz et al., 2000; Ito et al., 2001), *Caenorhabditis elegans* (Walhout et al., 2000), and *Helicobacter pylori* (Rain et al., 2001).

To develop a YTH screening system suitable for systematic analysis of a large number of *Synechocystis* gene products, the original protocol for the *GAL4*-based YTH method was modified as follows (Figure 7.2; Sato et al., manuscript in preparation). First, a recognition site for the restriction en-

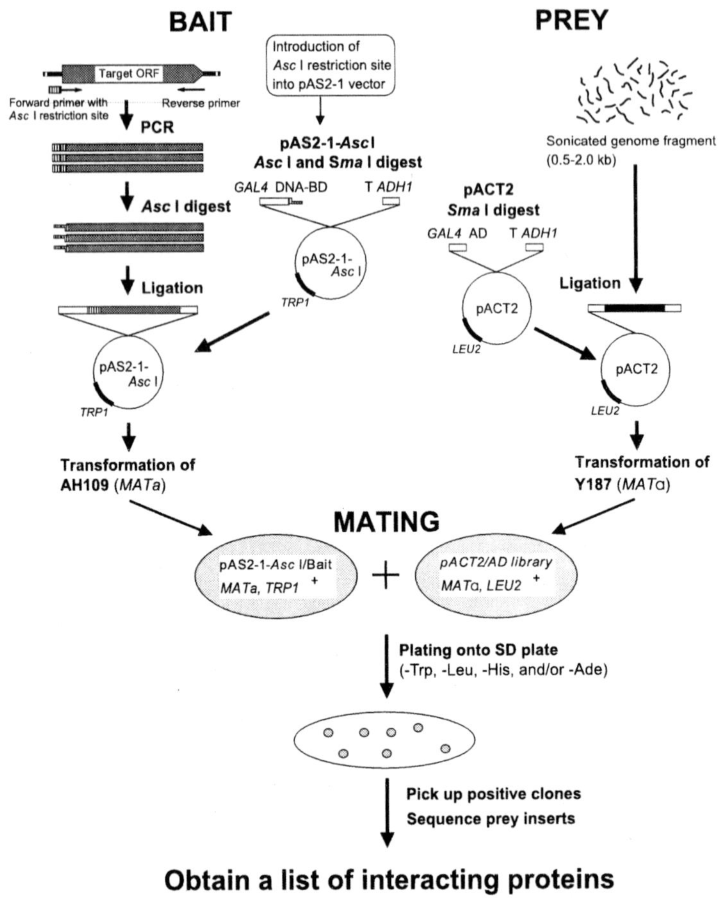

FIGURE 7.2. Outline of the strategy for systematic two-hybrid screening. To construct bait clones, the target ORF was amplified using the ORF specific primer set, with the forward primer containing an in-frame *Asc*I site at the 5' end. The PCR fragment obtained was subjected to *Asc*I digestion and cloned into the *Asc*I and *Sma*I digested pAS2-1-*Asc*I vector. The bait plasmid obtained was introduced into AH109 yeast cells. Prey libraries were generated by sonication of the genomic DNA followed by cloning of a fraction of 500 bp to 2.0 kbp DNA segments into the pACT2 plasmid. A high-complexity library was first obtained in *Escherichia coli* and then introduced into Y187 yeast cells. The two yeast strains containing the bait clone and the prey library, respectively, were mixed and incubated to allow mating. The cells were collected and spread onto agar plates for selection of the *HIS3* and/or the *ADE2* reporter expression. The positive clones were isolated, and prey fragments were individually characterized by sequence analysis followed by comparison with the genomic sequence.

zyme *Asc*I was introduced into the multiple cloning sites of the *GAL4* DNA-binding domain fusion vector (pAS2-1-*Asc*I). This *Asc*I site has enabled cloning of almost all of the *Synechocystis* genes directionally into a vector by simple PCR-based construction, since there are only five *Asc*I sites in the coding regions of the *Synechocystis* genome. Second, prey clones are screened by mating yeast instead of transforming them to reduce the time and cost involved and to allow multiple samples to be handled in parallel (Figure 7.2).

The improved YTH screening system was used to examine the possible protein interactions among two-component signal transducers with certain cognate pairs of sensors and regulators that may be difficult to deduce from the gene organization in the genome. A screening of 80 baits corresponding to genes encoding two-component signal transducers identified a total of 92 independent interactions, including 35 interactions between two-component signal transducers. This study demonstrated that the improved, high-throughput YTH screening system is a powerful method to detect possible protein interactions.

GENOME DATABASES

Genome sequences provide an enormous quantity of valuable information about the structure and function of gene components. To fully utilize such information, well-organized genome databases with user-friendly interfaces are essential. The World Wide Web–based genome database CyanoBase (http://www.kazusa.or.jp/cyano/) was generated when the *Synechocystis* genome sequence was completely identified (Nakamura et al., 1998). CyanoBase provides the graphical images of the circular physical map and the expanded linear gene map of the genome (Figure 7.3). Each box on the linear map is a clickable link to detailed information about the corresponding gene, such as the nucleotide sequence of the gene and the adjacent regions, the translated amino acid sequence, and the results of a similarity search. CyanoBase also provides information about the complete genomes of *Anabaena* sp. PCC 7120 (Kaneko et al., 2001) and *Thermosynechococcus elongatus* BP-1 (Nakamura et al., 2002), as well as the searchable incomplete genome sequences of other cyanobacteria. Newly identified experimental information about gene function can be registered with CyanoBase through a CyanoGenes database (http://www.kazusa.or.jp/cyano/Synechocystis/comments/). CyanoMutants (http://www.kazusa.or.jp/cyano/mutants/), which is part of CyanoGenes, is a repository of *Synechocystis* mutants with known mutations that have been generated utilizing

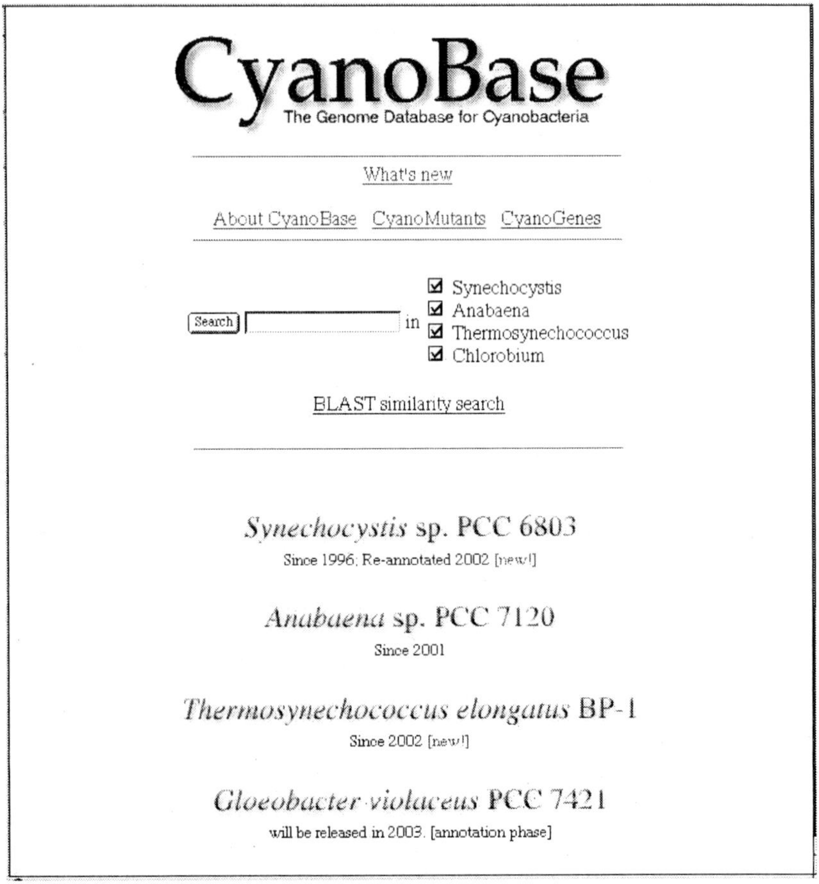

FIGURE 7.3. The first page of the CyanoBase Web site. The first page of CyanoBase provides search engines to quickly retrieve the necessary information from the registered genomes. It also serves as an entrance to the genome information of each organism listed. The URL address of CyanoBase is <http://www.kazusa.or.jp/cyano/>.

the genome sequence (Nakamura et al., 1999). Efforts to collect reannotation information from the community are also documented in the CYORF (Cyanobacteria Gene Annotation Database) database at <http://cyano.genome.ad.jp/>. These genome databases provide and facilitate information exchange on both the structure and function of the *Synechocystis* genes.

CONCLUSIONS AND FUTURE PERSPECTIVES

Since the complete identification of the genome sequence of *Synechocystis* in 1996, the methodologies and strategies for studying gene function and regulation in this organism have changed rapidly. In addition, isolation of homologous genes from other cyanobacteria has been accelerated. The complete genome sequences of seven cyanobacteria with various characteristics (*Anabaena* sp. PCC 7120, *T. elongatus* BP-1, *Prochlorococcus* MED4, *Prochlorococcus* MIT9313, *Prochlorococcus marinus* SS120, *Synechococcus* WH8102, and *Gloeobacter* PCC 7421) have been determined (Kaneko et al., 2001; Nakamura et al., 2002; Hess et al., 2001). Moreover, genome sequencing of *Microcystis aeruginosa, Synechococcus* PCC 6301, *Synechococcus* PCC 7002, *Trichodesmium erythraeum, Nostoc punctiforme* ATCC 29133, *Synechococcus* PCC 7942, *Spirulina/Arthrobacter platensis,* and *Aphanothece halophytical* is currently in progress. The availability of such an enormous amount of sequence data will undoubtedly enable the application of various technologies used in *Synechocystis* to many other cyanobacterial species. However, the simple application of existing methodologies to different species will not provide the most useful data. One of the most promising directions for the future is to intensively analyze the genome information with the aid of computers. Computer predictions potentially confirmed by results from "wet" experiments would be highly beneficial in genetic research. This approach is already feasible with cyanobacteria because of the variety of information and material resources that have been developed.

REFERENCES

Alland, D., Kramnik, I., Weisbrod, T.R., Otsubo, L., Cerny, R., Miller, L.P., Jacobs, W.R., Jr., Bloom, B.R. (1998). Identification of differentially expressed mRNA in prokaryotic organisms by customized amplification libraries (DECAL): The effect of isoniazid on gene expression in *Mycobacterium tuberculosis. Proc. Natl. Acad. Sci. USA* 95: 13227-13232.

Bhaya, D., Vaulot, D., Amin, P., Takahashi, A.W., Grossman, A.R. (2000). Isolation of regulated genes of the cyanobacterium *Synechocystis* sp. strain PCC 6803 by differential display. *J. Bacteriol.* 182: 5692-5699.

Choi, J.S., Kim, D.S., Lee, J., Kim, S.J., Kim, S.I., Kim, Y.H., Hong, J., Yoo, J.S., Suh, K.H., Park, Y.M. (2000). Proteome analysis of light-induced proteins in *Synechocystis* sp. PCC 6803: Identification of proteins separated by 2D-PAGE using N-terminal sequencing and MALDI-TOF MS. *Mol. Cells* 10: 705-711.

Fulda, S., Huang, F., Nilsson, F., Hagemann, M., Norling, B. (2000). Proteomics of *Synechocystis* sp. strain PCC 6803. Identification of periplasmic proteins in cells grown at low and high salt concentrations. *Eur. J. Biochem.* 267: 5900-5907.

Gill, R.T., Katsoulakis, E., Schmitt, W., Taroncher-Oldenburg, G., Misra, J., Stephanopoulos, G. (2002). Genome-wide dynamic transcriptional profiling of the light-to-dark transition in *Synechocystis* sp. strain PCC 6803. *J. Bacteriol.* 184: 3671-3681.

Hess, W.R., Rocap, G., Ting, C.S., Larimer, F., Stilwagen, S., Lamerdin, J., Chisholm, S.W. (2001). The photosynthetic apparatus of *Prochlorococcus:* Insights through comparative genomics. *Photosynthesis Res.* 70: 53-71.

Hihara, Y., Kamei, A., Kanehisa, M., Kaplan, A., Ikeuchi, M. (2001). DNA microarray analysis of cyanobacterial gene expression during acclimation to high light. *Plant Cell* 13: 793-806.

Huang, F., Parmryd, I., Nilsson, F., Persson, A.L., Pakrasi, H.B., Andersson, B., Norling, B. (2002). Proteomics of *Synechocystis* sp. strain PCC 6803: Identification of plasma membrane proteins. *Mol. Cell Proteomics* 1: 956-966.

Huang, L., McCluskey, M.P., Ni, H., LaRossa, R.A. (2002). Global gene expression profiles of the cyanobacterium *Synechocystis* sp. strain PCC 6803 in response to irradiation with UV-B and white light. *J. Bacteriol.* 184: 6845-6858.

Ito, T., Chiba, T., Ozawa, R., Yoshida, M., Hattori, M., Sakaki, Y. (2001). A comprehensive two-hybrid analysis to explore the yeast protein interactome. *Proc. Natl. Acad. Sci. USA* 98: 4569-4574.

Kaneko, T., Nakamura, Y., Wolk, C.P., Kuritz, T., Sasamoto, S., Watanabe, A., Iriguchi, M., Ishikawa, A., Kawashima, K., Kimura, T., et al. (2001). Complete genomic sequence of the filamentous nitrogen-fixing cyanobacterium *Anabaena* sp. strain PCC 7120. *DNA Res.* 8: 205-213.

Kaneko, T., Sato, S., Kotani, H., Tanaka, A., Asamizu, E., Nakamura, Y., Miyajima, N., Hirosawa, M., Sugiura, M., Sasamoto, S., et al. (1996). Sequence analysis of the genome of the unicellular cyanobacterium *Synechocystis* sp. strain PCC 6803. II. Sequence determination of the entire genome and assignment of potential protein-coding regions. *DNA Res.* 3: 109-136.

Kaneko, T., Tabata, S. (1997). Complete genome structure of the unicellular cyanobacterium *Synechocystis* sp. PCC 6803. *Plant Cell Physiol.* 38: 1171-1176.

Kanesaki, Y., Suzuki, I., Allakhverdiev, S.I., Mikami, K., Murata, N. (2002). Salt stress and hyperosmotic stress regulate the expression of different sets of genes in *Synechocystis* sp. PCC 6803. *Biochem. Biophys. Res. Commun.* 290: 339-348.

Kashino, Y., Lauber, W.M., Carroll, J.A., Wang, Q., Whitmarsh, J., Satoh, K., Pakrasi, H.B. (2002). Proteomic analysis of a highly active photosystem II preparation from the cyanobacterium *Synechocystis* sp. PCC 6803 reveals the presence of novel polypeptides. *Biochemistry* 41: 8004-8012.

Mizuno, T., Kaneko, T., Tabata, S. (1996). Compilation of all genes encoding bacterial two-component signal transducers in the genome of the cyanobacterium, *Synechocystis* sp. strain PCC 6803. *DNA Res.* 3: 407-414.

Nakamura, Y., Kaneko, T., Hirosawa, M., Miyajima, N., Tabata, S. (1998). CyanoBase, a www database containing the complete nucleotide sequence of the genome of *Synechocystis* sp. strain PCC6803. *Nucleic Acids Res.* 26: 63-67.

Nakamura, Y., Kaneko, T., Miyajima, N., Tabata, S. (1999). Extension of CyanoBase. CyanoMutants: Repository of mutant information on *Synechocystis* sp. strain PCC6803. *Nucleic Acids Res.* 27: 66-68.

Nakamura, Y., Kaneko, T., Sato, S., Ikeuchi, M., Katoh, H., Sasamoto, S., Watanabe, A., Iriguchi, M., Kawashima, K., Kimura, T., et al. (2002). Complete genome structure of the thermophilic cyanobacterium *Thermosynechococcus elongatus* BP-1. *DNA Res.* 9: 123-130.

Ohkawa, H., Pakrasi, H.B., Ogawa, T. (2000). Two types of functionally distinct NAD(P)H dehydrogenases in *Synechocystis* sp. strain PCC6803. *J. Biol. Chem.* 275: 31630-31634.

Ohkawa, H., Price, G.D., Badger, M.R., Ogawa, T. (2000). Mutation of *ndh* genes leads to inhibition of CO_2 uptake rather than HCO_3^- uptake in *Synechocystis* sp. strain PCC 6803. *J. Bacteriol.* 182: 2591-2596.

Omata, T., Gohta, S., Takahashi, Y., Harano, Y., Maeda, S. (2001). Involvement of a CbbR homolog in low CO_2-induced activation of the bicarbonate transporter operon in cyanobacteria. *J. Bacteriol.* 183: 1891-1898.

Rain, J.C., Selig, L., De Reuse, H., Battaglia, V., Reverdy, C., Simon, S., Lenzen, G., Petel, F., Wojcik, J., Schachter, V., et al. (2001). The protein-protein interaction map of *Helicobacter pylori. Nature* 409: 211-215.

Robinson, N.J., Robinson, P.J., Gupta, A., Bleasby, A.J., Whitton, B.A., Morby, A.P. (1995). Singular over-representation of an octameric palindrome, HIP1, in DNA from many cyanobacteria. *Nucleic Acids Res.* 23: 729-735.

Sazuka, T., Ohara, O. (1996). Sequence features surrounding the translation initiation sites assigned on the genome sequence of *Synechocystis* sp. strain PCC6803 by amino-terminal protein sequencing. *DNA Res.* 3: 225-232.

Sazuka, T., Yamaguchi, M., Ohara, O. (1999). Cyano2Dbase updated: Linkage of 234 protein spots to corresponding genes through N-terminal microsequencing. *Electrophoresis* 20: 2160-2171.

Sergeyenko, T.V., Los, D.A. (2000). Identification of secreted proteins of the cyanobacterium *Synechocystis* sp. strain PCC 6803. *FEMS Microbiol. Lett.* 193: 213-216.

Shibata, M., Katoh, H., Sonoda, M., Ohkawa, H., Shimoyama, M., Fukuzawa, H., Kaplan, A., Ogawa, T. (2002). Genes essential to sodium-dependent bicarbonate transport in cyanobacteria: Function and phylogenetic analysis. *J. Biol. Chem.* 277: 18658-18664.

Shibata, M., Ohkawa, H., Kaneko, T., Fukuzawa, H., Tabata, S., Kaplan, A., Ogawa, T. (2001). Distinct constitutive and low-CO_2-induced CO_2 uptake systems in cyanobacteria: Genes involved and their phylogenetic relationship with homologous genes in other organisms. *Proc. Natl. Acad. Sci. USA* 98: 11789-11794.

Singh, A.K., Sherman, L.A. (2000). Identification of iron-responsive, differential gene expression in the cyanobacterium *Synechocystis* sp. strain PCC 6803 with a customized amplification library. *J. Bacteriol.* 182: 3536-3543.

Sokolenko, A., Pojidaeva, E., Zinchenko, V., Panichkin, V., Glaser, V.M., Herrmann, R.G., Shestakov, S.V. (2002). The gene complement for proteolysis in the cyanobacterium *Synechocystis* sp. PCC 6803 and *Arabidopsis thaliana* chloroplasts. *Curr. Genet.* 41: 291-310.

Suzuki, I., Kanesaki, Y., Mikami, K., Kanehisa, M., Murata, N. (2001). Cold-regulated genes under control of the cold sensor Hik33 in *Synechocystis*. *Mol. Microbiol.* 40: 235-244.

Suzuki, I., Los, D.A., Kanesaki, Y., Mikami, K., Murata, N. (2000). The pathway for perception and transduction of low-temperature signals in *Synechocystis*. *EMBO J.* 19: 1327-1334.

Uetz, P., Giot, L., Cagney, G., Mansfield, T.A., Judson, R.S., Knight, J.R., Lockshon, D., Narayan, V., Srinivasan, M., Pochart, P., et al. (2000). A comprehensive analysis of protein-protein interactions in *Saccharomyces cerevisiae*. *Nature* 403: 623-627.

Vidan, S., Snyder, M. (2001). Large-scale mutagenesis: Yeast genetics in the genome era. *Curr. Opin. Biotechnol.* 12: 28-34.

Walhout, A.J., Sordella, R., Lu, X., Hartley, J.L., Temple, G.F., Brasch, M.A., Thierry-Mieg, N., Vidal, M. (2000). Protein interaction mapping in *C. elegans* using proteins involved in vulval development. *Science* 287: 116-122.

Wang, Y., Sun, J., Chitnis, P.R. (2000). Proteomic study of the peripheral proteins from thylakoid membranes of the cyanobacterium *Synechocystis* sp. PCC 6803. *Electrophoresis* 21: 1746-1754.

Williams, J.G.K. (1988). Construction of specific mutations in photosystem II photosynthetic reaction center by genetic engineering methods in *Synechocystis* 6803. In Packer, L., Glazer, A.N. (Eds.), *Methods in Enzymology*, Volume 167, *Cyanobacteria* (pp. 766-778). San Diego, CA: Academic Press, Inc.

Yoshimura, H., Hisabori, T., Yanagisawa, S., Ohmori, M. (2000). Identification and characterization of a novel cAMP receptor protein in the cyanobacterium *Synechocystis* sp. PCC 6803. *J. Biol. Chem.* 275: 6241-6245.

Yoshimura, H., Yanagisawa, S., Kanehisa, M., Ohmori, M. (2002). Screening for the target gene of cyanobacterial cAMP receptor protein SYCRP1. *Mol. Microbiol.* 43: 843-853.

Chapter 8

Chlamydomonas Genomics: New Approaches for a Classical Model System

Heidi K. Ledford
Rachel M. Dent
Krishna K. Niyogi

INTRODUCTION

Since the late nineteenth century, generations of *Chlamydomonas* researchers have seen waves of scientific trends come and go. As one of the founding genetic systems, *Chlamydomonas* research has made tremendous contributions to the genetic approach to biology. Now, as we advance further into the "-omics" era, *Chlamydomonas* is again at the forefront of research. Currently featuring fully sequenced nuclear, chloroplast, and mitochondrial genomes, *Chlamydomonas* couples genetic and physiological advantages with full genomic information (Table 8.1). These unique advantages promise to make *Chlamydomonas* an exceptional model system for tackling a wide variety of biological questions on a genomic scale (Figure 8.1).

A unicellular, eukaryotic green alga, *Chlamydomonas* offers a number of advantages for plant biologists, cell biologists, phycologists, and ecologists alike (Table 8.1). As a model alga, with close relatives found ubiquitously in soil, freshwater, and even Arctic habitats, studies involving *Chlamydomonas* can shed light on algal ecology and evolution. As a genetically and physiologically tractable unicellular system, *Chlamydomonas* has also con-

We would like to thank Pete Lefebvre and Hank Bestman for communicating results prior to publication. This work was supported by grants to K. K. N. from the National Institutes of Health (R01-GM58799) and the Torrey Mesa Research Institute/Syngenta Research and Technology. H. K. L. was supported by a National Institutes of Health Predoctoral Genetics Training Grant (T32-GM07127).

TABLE 8.1. Features of *Chlamydomonas.*

Lifestyle	**Unicellular, eukaryotic green alga**
Growth rate	Doubling time <10 h
Generation time	Approximately two weeks
Genomic sequencing	Mitochondrial and chloroplast genomes completed. Nuclear genome sequenced with ~10× coverage, assembly in progress.
Mutant screening	Haploid lifestyle simplifies screening for recessive mutations. Microbial lifestyle reduces space required for large-scale screening.
Reverse genetics	Homologous gene disruption in chloroplast and mitochondria. RNAi and TILLING approaches for nuclear reverse genetics.
Genetic engineering	All three genomes transformable. Codon bias and endogenous regulatory sequences should be taken into consideration.
Physical map	BAC library with 8× genomic coverage; 2,500 have been mapped and assembled into overlapping contigs. Sequenced BAC ends to be aligned with nuclear genome sequence assembly.
Genetic map	Nuclear genetic map containing over 250 molecular markers and over 200 phenotypic markers spanning all 17 linkage groups. Any point on the genome within 2 cM of a mapped molecular marker on average.
EST projects	Over 200,000 ESTs deposited in GenBank.
Transcriptional profiling	First-generation microarrays released. Qualitative transcriptome analysis from EST collections.
Full-length cDNA sequencing	In progress.
Databases	Current databases to be consolidated into one integrated database.

tributed to basic cell biology and physiology. And of key significance to the readers of this book, *Chlamydomonas* is a member of the division Chlorophyta, the algal division that shows the closest relationship to vascular plants, making *Chlamydomonas* a useful system for understanding land plants, particularly plant metabolism and photosynthesis.

For the impatient plant biologist, accustomed to waiting during a generation time of a month or more, no doubt one of the biggest attractions of working with *Chlamydomonas* is the rapid generation time and doubling times relative to plants (Figure 8.2). *Chlamydomonas* cultures have a doubling time of less than ten hours, and progeny from crosses can be obtained in less than a month. The haploid lifestyle of *Chlamydomonas* facilitates mutant screening, especially for recessive mutants. *Chlamydomonas* is easy

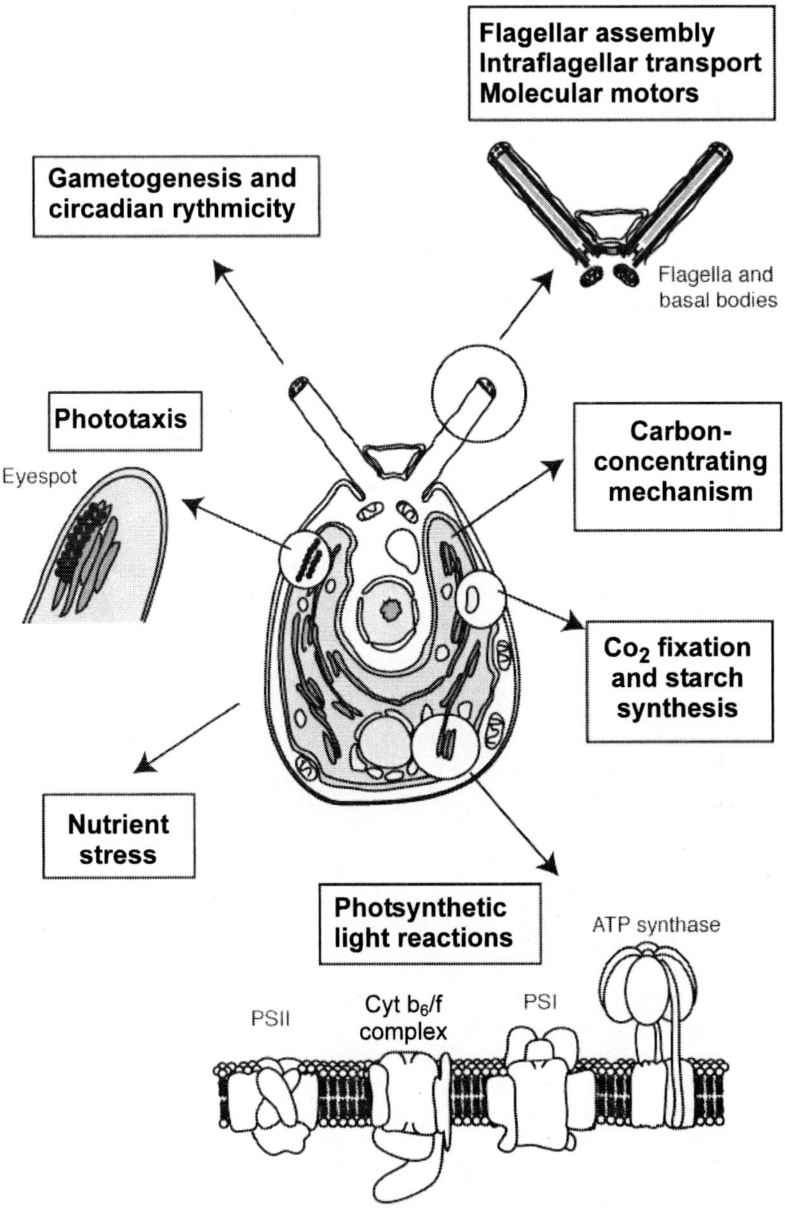

FIGURE 8.1. *Chlamydomonas reinhardtii:* Classical applications and areas of research. (*Source:* Modified from Dent, Han, et al., 2001.)

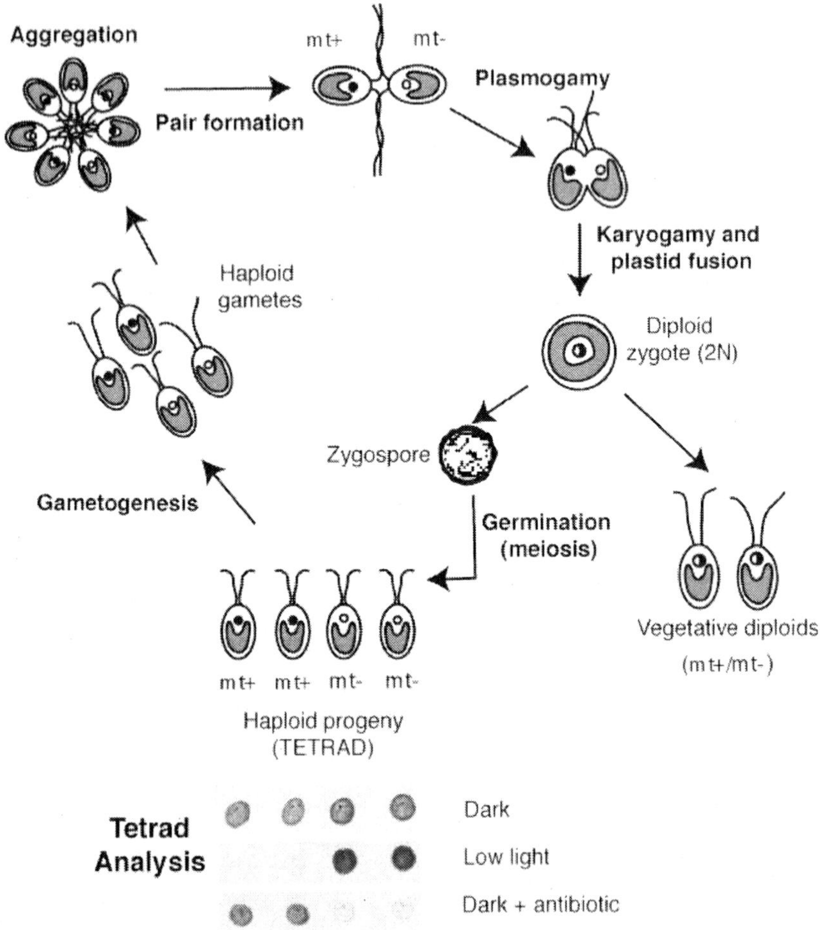

FIGURE 8.2. The life cycle of *Chlamydomonas reinhardtii.* Figure shows tetrad analysis of an insertional mutant, illustrating cosegregation of an antibiotic resistance gene with a low-light-sensitivity phenotype. (*Source:* Modified from Harris, 1989.)

to grow in bulk as homogeneous cultures, so physiological and biochemical characterization is facile compared to working with many higher plants. A unicellular system also greatly simplifies the use of chemical treatments, such as the application of inhibitors, ensuring uniform uptake of the inhibitor by each cell. Thus, chemical treatments are often easily incorporated into genetic screens or physiological measurements. Proteomic characterization may also be facilitated by the ease with which *Chlamydomonas* proteins can be radiolabeled. Furthermore, stress conditions are easily studied because modifying growth conditions is simply a matter of making changes to the growth media, without concerns about heterogeneous uptake or tissue types. This is particularly useful when employing sensitive genomic-scale procedures, such as microarray analysis of transcript abundance (see Chapter 4) or proteomic analysis of protein composition (see Chapters 14 and 15).

In addition to these general advantages, *Chlamydomonas* also features a number of other traits that make it an attractive system for specific biological phenomena (Figure 8.1). For the researcher interested in photosynthesis, *Chlamydomonas* possesses a tremendous advantage over higher plants in that it is able to grow not only photoautotrophically in the light but also heterotrophically in the dark. As discussed in reference to chloroplast genomics, this means that photosynthetic mutants can be easily identified and maintained in the dark. Furthermore, *Chlamydomonas* is a motile alga that is used as an important model for flagellar assembly and function. Again, dispensability is the key, and the ability to isolate nonmotile mutants of *Chlamydomonas* has led to the generation of large collections of flagellar mutants. Because of the ease with which *Chlamydomonas* cultures can be synchronized, circadian rhythm studies are relatively simple to perform. Also, a well-defined sexual cycle (Figure 8.2), involving two haploid mating types (plus and minus) and featuring cell-cell recognition and adhesion, has provided a platform for detailed studies of signal transduction in this alga (Pan et al., 2003).

These are all classical applications of *Chlamydomonas* research, but other interesting avenues are currently developing, such as using *Chlamydomonas* for recombinant antibody production (Mayfield et al., 2003), as a model for testing drug toxicity (Maucourt et al., 2002), and as a metal-sequestering bioremediation system (Rubinelli et al., 2002). The list of research applications is long, but no one topic will be discussed in detail in this chapter. Instead, readers interested in more detail about a specific area of research should sample from the myriad of excellent reviews about everything from starch biosynthesis (Hicks et al., 2001) and photosynthesis (Rochaix, 2002) to intraflagellar transport (Cole, 2003) and flagellar assembly (Dutcher, 2003) to responses to nutrient stress (Grossman, 2000a) and circadian rhythms (Werner, 2002).

This chapter will focus on the characteristics of the three *Chlamydomonas* genomes, as well as the tools and technologies that have contributed to recent advances in functional genomic techniques in this system, referring only to a few illustrative applications of these techniques. To that end, we will introduce each genome by describing salient characteristics of its structure and experimental pliability and then describe the various "-omics" that have been applied to all three genomes, with a brief discussion of future plans for genomic technology in this system. Under the assumption that many readers of this chapter are interested in how *Chlamydomonas* may complement research in plants, we will also highlight the uses and limitations of using *Chlamydomonas* as a model for plant biology. Because the bulk of the work discussed in this chapter has been carried out using *Chlamydomonas reinhardtii*, we will concentrate almost exclusively on this species and refer to it throughout simply as *Chlamydomonas*.

THE CHLOROPLAST GENOME

Comprising approximately 40 percent of the total cell volume, the single, cup-shaped chloroplast of *Chlamydomonas* is one of its most-defining features (Harris, 1989), and it has enticed many scientists to take up *Chlamydomonas* as a model for higher-plant photosynthesis. *Chlamydomonas* offers a number of advantages as a photosynthetic model system, both in its similarities to higher plant systems and in its differences.

Generally speaking, photosynthesis in *Chlamydomonas* bears a strong resemblance to higher plants. Pigment composition, reaction center structure, and electron transport chain components are all highly similar to higher plants, and characterization of photosynthetic mutants in *Chlamydomonas* promises to shed light on higher-plant photosynthesis (Dent, Han, et al., 2001). With all of the benefits that a unicellular system can offer in terms of rapid generation time and uniform uptake of inhibitors, *Chlamydomonas* and cyanobacterial models, such as *Synechocystis,* offer clear advantages. But cyanobacterial species are not always the best model for higher-plant photosynthesis because of differences in pigment compositions, light-harvesting complexes (LHCs), and oxygen-evolving complexes. Moreover, the union of photosynthetic and respiratory electron transport chains at the cytochrome b_6/f complex in cyanobacteria confounds some physiological measurements and ensures that loss-of-function mutants in this complex will be lethal. Also, as a eukaryotic system, *Chlamydomonas* offers insights into interorganellar communication that would not be possible in *Synechocystis*.

Although *Chlamydomonas* shares a number of features with higher plants, many of the differences between these two systems highlight the utility of *Chlamydomonas*. For example, photosynthetic mutants in higher plants often exhibit a reduction in growth rate due to carbon limitation and may not be suitable for detailed phenotypic analysis. In contrast, because *Chlamydomonas* can grow heterotrophically in the dark on acetate-containing media, it is possible to isolate photosynthetic mutants that would be lethal or unhealthy in higher plants. Furthermore, the presence of both light-dependent and -independent chlorophyll biosynthetic pathways means that, unlike angiosperms, *Chlamydomonas* synthesizes pigments and assembles photosynthetic complexes in the dark, making physiological experiments with light-sensitive photosynthetic mutants possible.

Mutants in the light-independent chlorophyll biosynthetic pathway also have been used to study chloroplast development. When grown in the dark, these mutants are "etiolated," lacking chlorophyll and containing only trace amounts of photosynthetic reaction centers. When transferred from the dark to the light, they begin producing chlorophyll and photosynthetic complexes, allowing researchers to characterize the process of import and assembly of the light-harvesting complexes (Hoober et al., 1998).

Chloroplast Genome Sequence

It is fitting that *Chlamydomonas* has become a model for chloroplast genomics, given that studies involving *Chlamydomonas* led to the first discovery of organellar DNA (Ris and Plaut, 1962). Compared to the rather uniform chloroplast genomes of land plants, green algal chloroplast genomes are staggeringly diverse (Maul et al., 2002; Simpson and Stern, 2002). The overall architecture of the *Chlamydomonas* chloroplast genome is similar to higher plants in that it contains an rDNA-containing inverted repeat separating two single-copy regions. Beyond this, the *Chlamydomonas* chloroplast genome is an oddity among chlorophytes, featuring predominant small direct repeats that have so far been found in only *Chlorella* (Maul et al., 2002)

Complete sequencing of the approximately 200 kbp chloroplast genome (Table 8.2) had been confounded for years by the abundance of short dispersed repeats scattered throughout the genome (Maul et al., 2002). The completion of the sequencing effort revealed a number of surprises. Highly reduced, and speckled with fragmented genes and scrambled regulatory units, the chloroplast genome is an evolutionary enigma. The reasons behind this genomic turmoil have not been firmly established, but hypotheses consistently point to the large number of dispersed repeats. This is also unclear, however, as the dispersed repeats could also be the result of gene scrambling and not necessarily the cause.

One surprise to come from the completion of the chloroplast genome sequence was the paucity of new genes discovered. With only 99 genes, the *Chlamydomonas* chloroplast genome is reduced compared to many other green algae and higher plants. As nuclear sequence data have amassed, many of these "lost" genes have been found in the nucleus (Lilly et al.,

TABLE 8.2. Characteristics of the three *Chlamydomonas* genomes.

Characteristic	Chloroplast	Mitochondrion	Nucleus
Genome size	203.4 kbp	15.8 kbp	~100 Mbp
Genome structure	Circular	Linear (single chromosome)	Linear (17 linkage groups)
Percent G+C	34.6	45.2	62
Number of genes	99	13	12,000-16,400 (estimated)
Mode of inheritance	Uniparental from *mt+*	Uniparental from *mt–*	Mendelian
Transformation	Particle bombardment	Particle bombardment	Agitation with glass beads, electroporation
Transgene insertion	Homologous recombination	Homologous recombination	Random insertion
Copies of genome per haploid cell	~80	~50	1
Genome sequencing	Completed	Completed	Assembly currently in progress

2002; Grossman, 2000b). The reasons behind this high frequency of gene transfer to the nucleus are unclear, but some hypotheses implicate the high rate of recombination in the chloroplast. It is possible that the large number of dispersed repeats in the chloroplast genome contributes to a high frequency of recombination events, leading to the excision of genome fragments.

Despite the small number of coding regions, the chloroplast genome size remains relatively large, primarily due to the accumulation of repetitive elements. In fact, >20 percent of the genome consists of repetitive DNAs (Maul et al., 2002). Furthermore, within the *Chlamydomonas* genus there appears to be a loose correlation between the degree of repetitive DNA and the extent of gene rearrangements, leading to speculation that the rearrangements may have arisen from recombination between these repeats (Nedelcu and Lee, 1998b). The short dispersed repeats are mostly randomly dispersed throughout the genome, and the degree of sequence conservation among different groups of short dispersed repeats hints of an as yet undiscovered functional role.

In contrast to land plants, chloroplast gene order is variable among *Chlamydomonas* taxa (Nedelcu and Lee, 1998b). Furthermore, *Chlamydomonas* chloroplast genomes lack the extensive operon structure of their land plant counterparts; the best example is represented by the six *atp* genes that are scattered singly about the *Chlamydomonas* genome but are organized into two operons in other species. This shuffling could be the result of genome restructuring via recombination among short dispersed repeats, or possibly a product of the Wendy transposable element, remnants of which can still be found in the chloroplast genome today (Nedelcu and Lee, 1998b). As more green algal chloroplasts are sequenced as part of the "Deep Green" (http://faculty.washington.edu/mandoli/fuzzy.html) and Joint Genome Institute Comparative Chloroplast Genomics project (http://www.jgi.doe.gov/programs/comparative/cover-page.htm), the evolutionary path of the *Chlamydomonas* chloroplast genome may become more apparent.

Chloroplast Genetics

Chloroplast mutations in *Chlamydomonas* were first identified based on their non-Mendelian segregation (Sager, 1954). Uniparentally inherited from the mating type plus parent, these mutations were used to create the first chloroplast genetic map by calculating the recombination frequency among these non-Mendelian mutants (Mets and Geist, 1983; Gillham, 1974; Sager and Ramanis, 1968). Many different methods were used to create these genetic maps, often with conflicting results. In some cases, it has been hypothesized that recombination frequencies within the chloroplast are affected by alteration of chloroplast architecture in some mutant backgrounds, possibly modifying the mixing frequency among plastomes (Sears, 1998).

There is an extensive collection of photosynthetic mutants, many of which were isolated from screens for acetate-requiring mutants. In fact, nearly every gene in the chloroplast genome has been knocked out, with the exception of those that create a lethal phenotype. The functional role of these essential genes can be addressed by weakening their translation initiation codons (Majeran et al., 2000; Xuemie et al., 1993).

Chloroplast Transformation and Molecular Tools

One of the biggest advantages of *Chlamydomonas* as a model for chloroplast genetics and genomics is the ease with which the chloroplast genome can be transformed via homologous recombination. Although plant chloroplasts also offer the ability to target gene insertion by homologous recombination, the existence of 50 or more chloroplasts per cell, each with multiple copies of the chloroplast genome, poses a serious technical challenge to the isolation of lines that are homoplasmic for a given insertion. With only one chloroplast in *Chlamydomonas*, and roughly 80 copies of the genome per cell, homoplasmicity is easier to achieve. Furthermore, longer generation times and the need for tissue culture regeneration of transformed tissue in higher plants makes them unwieldy, particularly when high-throughput screening is desired.

The first chloroplast transformation in *Chlamydomonas* was performed by biolistically bombarding an acetate-requiring deletion mutant with a chloroplast DNA restriction fragment containing the deleted region (Boynton et al., 1988). Since then, selectable markers for chloroplast transformation have come in two varieties: markers conferring antibiotic or herbicide resistance and markers that restore photosynthetic capacity to a nonphotoautotrophic mutant. Commonly used photosynthetic markers/mutants are *atpB*, *psbA*, and photosystem I (PSI) mutants (Boynton and Gillham, 1993). Mutations that prevent accumulation of PSI often result in high light sensitivity, allowing transformants to be selected based upon their ability to grow in the light.

Chloroplast Engineering

Plant genomics projects are rapidly unveiling gene targets for genetic engineering (Goossens et al., 2003; Somerville and Somerville, 1999). Whether engineering herbicide resistance or metabolite production, researchers are finding that there are many advantages to targeting their transgenes to the chloroplast (Gewolb, 2002; Bock, 2001; Heifetz, 2000). The ability to target a transgene to a specific location via homologous recombination is tantalizing, as is the possibility of introducing multiple genes within a single operon to ensure coregulation. These prokaryotic benefits are coupled with the ability of chloroplasts to properly fold the recombinant proteins, including the proper placement of disulfide bonds. Furthermore, many researchers have found very high levels of transgene expression in the chloroplast (Daniell and Dhingra, 2002; Staub et al., 2000).

Again, the unicellular nature of *Chlamydomonas* makes it a good candidate for metabolic engineering. The simplicity of chloroplast transformation and the abundance of selection strategies available render the chloroplast the most logical site for genetic engineering, including production of foreign proteins. Rapid generation of transgenic strains (a matter of weeks) and relatively cheap growth requirements also make *Chlamydomonas* an attractive system for engineering efforts. This is especially the case for directed evolution, in which microbial-selection methods for screening libraries of recombinant genes can increase the screening capacity approximately 10^2- to 10^6-fold over selection *in planta* (Lassner and Bedbrook, 2001).

Mayfield et al. (2003) have recently engineered *Chlamydomonas* to produce an active antibody against herpes simplex virus. Synthesizing the transgene de novo while accounting for chloroplast codon bias and using endogenous promoters and the 5' untranslated region (UTR), they were able to achieve relatively high levels of protein accumulation. The antibody was properly folded and in no other way posttranslationally modified. This example shows that with careful attention to codon bias and use of endogenous promoters, the *Chlamydomonas* chloroplast can be a useful host for genetic engineering.

THE MITOCHONDRIAL GENOME

Although *Saccharomyces cerevisiae* has provided a constructive model system for mitochondrial studies, questions regarding plant-specific functions—such as interorganellar communication between chloroplasts and mitochondria or the functional significance of the alternative oxidase respiratory pathway—demand a photosynthetic model system. Unfortunately, no higher-plant system for mitochondrial transformation has been established to date, hindering functional studies of the mitochondrial genome.

Chlamydomonas offers a number of advantages as a photosynthetic model system for mitochondrial studies. The capacity for both heterotrophic and photoautotrophic growth has allowed the selection of many *Chlamydomonas* mitochondrial mutants based on an inability to grow on acetate in the dark. The ability to maintain these mutants homoplasmically is also a tremendous advantage, as similar mutants in land plants must be maintained in a heteroplasmic state. Complete with a fully sequenced and transformable mitochondrial genome, *Chlamydomonas* is currently the best photosynthetic organism with which to pursue mitochondrial functional genomics.

Mitochondrial Genome Sequence

The *Chlamydomonas* mitochondrion is very similar to that of higher plants, although again there are a few differences. The basic structure of the respiratory pathway is conserved, consisting of both the classical cyanide-sensitive pathway as well as the plant-specific alternative oxidase pathway (Derzaph and Weger, 1996). There is also a putative rotenone-insensitive NADH dehydrogenase—another feature unique to plant mitochondria—although further work is needed to confirm any functional similarity (Remacle and Matagne, 1998). There are, however, key photorespiratory differences between *Chlamydomonas* and higher plants. The apparent absence of peroxisomes in *Chlamydomonas* has seemingly displaced the peroxisomal steps of photorespiration to the mitochondria, which will likely have a significant effect on protein and metabolite content (Spalding, 1998).

Another difference between *Chlamydomonas* and land plants is the extreme reduction of the mitochondrial genome in *Chlamydomonas* (Table 8.2), which is roughly 10 to 100 times smaller and contains considerably fewer genes than that of most land plants (Nedelcu and Lee, 1998b). Weighing in at a mere 15.8 kbp, the linear genome consists of thirteen intronless genes spread along both strands of DNA and contains the genes encoding respiratory chain proteins and ribosomal RNAs characteristic of mitochondrial genomes in other eukaryotes (Boer et al., 1985). In all, the genome encodes eight proteins, two rRNAs, and three tRNAs (Michaelis et al., 1990; Boer et al., 1985).

Mitochondrial genomes have been sequenced from a number of algae, including *Scenedesmus* (Nedelcu et al., 2000), *Chlorogonium* (Kroymann and Zetsche, 1998), and *Polytomella parva* (Fan and Lee, 2002), among others. Featuring extensive gene rearrangements, algal mitochondrial genomes vary more widely than those of land plants, which show a striking degree of homogeneity (Nedelcu, 1998). The reason for the high degree of variation among algal mitochondrial genomes is as yet unclear, but the presence of short repetitive sequences in *Chlamydomonas* and related taxa may have contributed to gene rearrangements (Nedelcu and Lee, 1998c; Nedelcu, 1997). Similarly, the discovery of a putative remnant of a group II intron in the mitochondrial genome of *Chlamydomonas* suggests that introns may also have contributed to gene rearrangements and deletions in related taxa (Nedelcu and Lee, 1998a).

Surprisingly, rDNA-based phylogenies constructed from nuclear and mitochondrial genomes indicate that while the *Chlamydomonas* nuclear genome groups with the plant clade rather than the ciliate/fungi/animal clade, the mitochondrial genome does not (Gray et al., 1989). At minimum, it can be said that the *Chlamydomonas* mitochondrial genome has taken a distinct evolutionary path, divergent from that of the land plants (Denovan-Wright et al., 1996). As more algal mitochondrial genomes become sequenced, a more precise vision of this path should emerge.

Mitochondrial Genetics

Uniparental inheritance of the mitochondrial genome from the mating type minus parent makes it simple to determine if a given mutation is affecting a mitochondrially encoded gene (Matagne et al., 1989, 1988; Boynton et al., 1987). In fact, many mitochondrial mutants are named *dum* (**d**ark **u**niparental **m**inus inheritance) for this reason. As is the case in the chloroplast, mitochondrial genomes from both parents persist and recombine in vegetative diploids, allowing mutations to be genetically mapped. A genetic map of the mitochondrial genome was constructed by crossing six mitochondrial mutants and calculating the percentage of recombinant progeny resulting from each combination (Remacle et al., 1995). Combining these results with sequence data led to the conclusion that recombination in vegetative zygotes occurs at a frequency of approximately 3.2 percent per kbp.

Screens for respiratory mutants in *Chlamydomonas* have yielded a substantial collection of mitochondrial mutants. So-called "dark dier" mutants have been isolated based on their inability to grow heterotrophically on acetate. Inhibitors of specific respiratory complexes have also been fruitful screening tools for mitochondrial mutants. These mutant screens have yielded mutants reflecting both mutations in the mitochondrial genome and mutations affecting factors encoded by the nucleus that are required for proper mitochondrial function (Lown et al., 2001; Wiseman et al., 1977).

In yeast, it has been possible to isolate ρ^0 "petite" mutants—so named because of their small-colony phenotype—which lack mitochondrial DNA (Chen and Clark-Walker, 2000). Generated via treatment with acriflavin or ethidium bromide, these mutants are able to survive without functional mitochondria if provided a fermentable carbon source. In *Chlamydomonas*, however, the corresponding mitochondrial mutants generated in this fashion undergo only eight or nine mitotic divisions at most and then cease to grow altogether (Alexander et al., 1974). Gillham et al. (1987) confirmed that occurrence of these "minute" mutants in *Chlamydomonas* coincides with loss of mitochondrial DNA. The "minute" phenotype indicates that, unlike yeast, some functions of *Chlamydomonas* mitochondria are essential for survival.

Mitochondrial Molecular Tools

Chlamydomonas is one of only a few organisms for which mitochondrial transformation techniques are available. The ability to transform mitochondria in *Chlamydomonas* was first demonstrated by biolistically transforming a deletion mutant *(dum-1)* using particles coated with mitochondrial DNA (Randolph-Anderson et al., 1993). Homologous recombination was observed in this same study when *dum-1* was transformed with the mitochondrial genome of *Chlamydomonas smithii*, containing a detectable polymorphism.

At present, no mitochondrial markers for transformation exist, although in theory a wild-type copy of a mitochondrial gene could be used to rescue the appropriate respiration-deficient mutant. Another possibility is to use the *MUD2* marker in the *cob* gene, which confers resistance to myxothiazol and has been used previously to assess mitochondrial genome inheritance patterns (Boynton and Gillham, 1996; Bennoun et al., 1991). Bacterial antibiotic resistance markers may also be possible, but little is known about antibiotic sensitivity in *Chlamydomonas* mitochondria. Furthermore, any attempt to engineer a mitochondrial marker will come up against the strong mitochondrial codon bias and lack of information regarding the regulation of gene expression in the mitochondrial genome (Gray and Boer, 1988). Instead, it may be best to cotransform with a chloroplast or nuclear selectable marker and then perform an initial selection for the presence of that marker. Transformants that pass this selection can then be screened for the presence of the mitochondrial transgene. While many of the details remain to be worked out, functional investigation of mitochondria in *Chlamydomonas* has the makings of an underutilized tool that holds a great deal of promise for the future, allowing researchers to investigate topics that are unapproachable in other photosynthetic organisms.

THE NUCLEAR GENOME

The size of the haploid genome of *Chlamydomonas reinhardtii* (Table 8.2) has been estimated at approximately 1×10^8 bp (Harris, 2001). Seventeen nuclear linkage groups are currently recognized. The number of chromosomes has yet to be stated with certainty due to problems with chromosome cytology. Only eight discrete chromosomes are consistently visible by light microscopy in metaphase cells, but electron microscopy of synaptinemal complexes suggests 16 or more chromosomes, a number that is more comparable to the 17 groups identified by genetic analysis (Dutcher et al., 1991).

The nuclear genome is very GC rich. Denaturation studies suggest an overall GC content of 62 percent (Harris, 2001), and this increases to 64.8 percent in coding sequence. This leads to a pronounced codon bias that favors codons with C or G in the third position. This codon bias has led to some difficulty in expressing foreign genes in the nuclear genome of *Chlamydomonas*. In contrast to genomic DNA from higher plants, relatively little methylation of cytosine residues in the CpG dinucleotide was detected in *Chlamydomonas* nuclear DNA (Day et al., 1988; Silfow, 1998).

Reports on the content of repetitive DNA in the nuclear genome vary considerably from none to as much as 30 percent (Silfow, 1998). Initial data from sequencing efforts show that several types of repetitive DNA regions do exist in the genome, and these have been divided into three classes: tandem, inverted, and dispersed repeats (Li et al., 2003). Within these classes, three different types of transposons have been isolated. Class I, or retrotransposons, are represented by the 5.7 kbp *TOC1* element, which was discovered as an insertion in the gene encoding the oxygen-evolving enhancer (OEE1) protein (Day et al., 1988). *Gulliver, Pioneer,* and *TOC2* share features of Class II elements in other systems (Day, 1995; Graham et al., 1995; Ferris, 1989). A family of three elements, *Tcr1* through *Tcr3,* is characterized by inverted terminal repeats and shows some similarity to Class II elements (Wang et al., 1998; Schnell and Lefebvre, 1993). Evidence has also been found for the presence of large tandem repeats, and for tracts of the simple sequence repeat $(CA/GT)_n$ (Silfow, 1998). These microsatellites are of great use as a source of polymorphic DNA markers for genetic mapping.

Gene structure in *Chlamydomonas* is characterized by many small introns, often present at high frequency. It has been reported that, on average, four introns are found per kilobase of coding sequence and the average intron size is 219 bp (Silfow, 1998). A more extensive analysis of intron size will be possible once the genome sequence has been fully annotated. Introns have been found to occur in 5' untranslated regions of several genes (Dame et al., 2002; Sugase et al., 1996; Mitchell and Kang, 1993). Data suggest that intronic sequences play an important role in the regulation of gene expression (Fabry et al., 1995), and they provide a large target for mutations that affect intron splicing. Although the majority of *Chlamydomonas* genes do contain many introns, several intronless genes have also been isolated. These include several core histone genes (Fabry et al., 1995) and the *PsaD* gene (Fischer and Rochaix, 2001). The nucleotide sequences at 5' and 3' splice junctions of *Chlamydomonas* introns generally conform to the eukaryotic consensus, suggesting that the splicing machinery is similar to that of other eukaryotes. *Chlamydomonas* genes are also characterized by long 3' noncoding regions of several hundred base pairs in length.

Although the total gene number and density over the whole nuclear genome cannot be stated until the full annotated sequence is available, the sequence data from BAC sequencing can be used to provide an estimate. In the nearly 524 kbp of BACs that have been sequenced, an upper limit of 86 and a lower limit of 58 genes have been predicted using the three *ab initio* gene-prediction programs, GeneMark, Genscan, and GreenGenie (Li et al., 2003). The upper limit will identify a gene if at least two methods predict it as such, whereas the lower limit requires an EST match or homologue in another organism. Using these data and assuming a genome size of 100 Mbp, the number of genes ranges from 13,084 to 18,273. Using Genscan as a prediction program, the average gene density ranges from 5,154 bp per gene to 7,267 bp per gene, with an average intergenic distance of 745 bp.

Nuclear Genome Sequence

The first draft of the nuclear genome sequence of *Chlamydomonas reinhardtii* was released in January 2003 by the Joint Genome Institute (JGI). This draft of the *Chlamydomonas* genome was generated using a whole-genome shotgun strategy. The Version 1.0 assembly is based on 9X coverage and used more than 2 million sequence fragments. The assembly was constructed using JAZZ, the JGI assembler, which capitalizes on paired-end sequencing reads. The assembly included 2.19 million whole-genome shotgun reads. The version 1.0 assembly contains 4,065 scaffolds, totaling 95.4 Mbp, averaging 23.4 kbp per scaffold. Roughly half of the genome (47.6 Mbp) is contained in 542 scaffolds (http://genome.jgi-psf.org/chlre1/chlre1.info.html).

Initial annotation has been achieved using the programs Genewise and GreenGenie. GreenGenie is a gene finder program that has been trained using *Chlamydomonas* gene sequences (Li et al., 2003). Analysis data for both annotation programs are available on the JGI *Chlamydomonas* genome Web site (http://genome.jgi-psf.org/chlre1/chlre1.home.html), as well as homology analysis against the *Arabidopsis,* yeast, human, and cyanobacterial sequence databases in addition to other *Chlamydomonas* Genbank accessions and EST collections.

Future plans for the sequencing effort involve sequencing of medium-sized insert clones to provide additional long-range linking information and thereby improve assemblies. A high-quality genome assembly became available at the end of 2003, after which a more detailed annotation of the genes will be undertaken by the *Chlamydomonas* community. Complete sequence reconstruction of the *Chlamydomonas* chromosomes will be achieved by linking together sequence scaffolds with the aid of both genetic and molecular maps (Grossman et al., 2003). Selected regions of the genome will likely have to be "finished" by targeted efforts to close gaps between scaffolds and resequencing of regions of low sequence quality. Thus, the ultimate aim of the genome project is to generate a high-quality reference genome that can be used as a foundation for *Chlamydomonas* and algal biology in general (Grossman et al., 2003).

Mapping

The first genetic maps of *Chlamydomonas reinhardtii* were published in 1970 (Levine and Goodenough, 1970) using nonmotile (*pf,* paralyzed flagella), nonphotosynthetic (*ac,* acetate-requiring), antibiotic-resistant, and auxotrophic mutants. Although additional mutant types have now been identified, these groups still represent the major categories of genetic markers. More than 200 genetic loci have now been mapped based on mutation.

In addition to the genetic map, a molecular map containing 264 markers is now available (Kathir et al., 2003). This map was generated using a set of 136 random progeny from a cross of strain 21gr *(mt+)* and the field isolate S1-C5 *(mt–)* (Gross et al., 1988). The S1-C5 strain was used because it exhibits a higher frequency of RFLPs than the interfertile *C. smithii* strain used to generate earlier molecular maps (Ranum et al., 1988). The markers used to generate the molecular map include RFLPs identified in random clones and in specific genes or cDNAs contributed by researchers and sequence tagged site (STS) markers. All molecular markers analyzed thus far have been placed on 17 linkage groups, providing further evidence that this is the definitive number of chromosomes in *Chlamydomonas reinhardtii.* The map covers approximately 1,000 centimorgans (cM). Any position on the *Chlamydomonas* genetic map is, on average, within 2 cM of a mapped molecular marker, which corresponds to about 200 kbp (Kathir et al., 2003). The generation of this map should greatly facilitate map-based cloning of genes identified by mutations.

The genome project has also initiated the development of a physical map that will consist of overlapping BAC clones anchored to the genetic map using the set of 240 markers as probes. Where BACs overlap, these have been used to create contigs. The ultimate aim is to tile the whole genome with anchored BAC clones. To date, 2,500 different BAC clones have been mapped and placed onto overlapping BAC contigs (Grossman et al., 2003). These BAC contigs cover more than 25 percent of the genome. A BAC library is currently available to probe for genes of interest. The library consists of approximately 15,000 clones with an average insert size of 75 kbp and has a seven- to eightfold coverage of the genome (P. Lefebvre, personal communication). To date, BAC contigs from the library have been mapped for linkage groups I-XIII/XIV.

The Toolkit for Nuclear Genome Analysis

Genomic approaches require not only masses of sequencing data, but also the molecular tools necessary for functional studies. Fortunately, the

Chlamydomonas community has generated a number of techniques for elucidation of gene function and is constantly improving upon the selection available. There are tools for everything from testing gene function to using reporter genes or enhancer traps for mutant screens. Furthermore, the process by which these tools were developed illustrates the strategies necessary for expression of exogenous genes in *Chlamydomonas*.

The *Chlamydomonas* nuclear genome can be transformed by agitation with glass beads (Kindle et al., 1989), electroporation (Shimogawara et al., 1998; Brown et al., 1991), or particle bombardment (Sodeinde and Kindle, 1993). Possibly as a consequence of the popular "green yeast" analogy (Goodenough, 1992), many newcomers to *Chlamydomonas* research are unpleasantly surprised to find that, no matter what the transformation method, integration of transgenes in the nuclear genome occurs predominantly by nonhomologous recombination. Of the three transformation methods, the most commonly used is simply vortexing cells in the presence of linearized plasmid DNA and glass beads (Kindle et al., 1989). The high-throughput nature of the glass bead and electroporation methods have facilitated the generation of insertional mutant collections and the cloning of genes by cosmid complementation (Dent, Han, et al., 2001; Funke et al., 1997; Zhang et al., 1994). As is the case in many organisms, transformation of *Chlamydomonas* often results in large deletions around the insertion site, potentially complicating phenotypic interpretation of the resulting transformant.

A wide array of endogenous selectable markers is available. The first to be used were the arginosuccinate lyase *(ARG7)* and the nitrate reductase *(NIT1)* genes, which can be used as selectable markers when transformed into *arg7* and *nit1* mutants, respectively (Smart and Selman, 1991; Debuchy et al., 1989; Fernandez et al., 1989). More recently, a mutant form of the endogenous acetolactate synthase gene has been developed as a dominant selectable marker. When transformed into wild-type *Chlamydomonas,* the mutant form of acetolactate synthase confers resistance to sulfonylurea herbicides (Kovar et al., 2002).

Unfortunately, expression of foreign genes in *Chlamydomonas* has been extremely difficult and often requires modification of the gene to include the proper codon bias and intron structure found in most *Chlamydomonas* genes (Fuhrmann et al., 1999; Lumbreras et al., 1998). In some cases, failure to express a foreign transgene has been linked to extensive methylation of the promoter region, even when the promoter used was endogenous in origin (Blankenship and Kindle, 1992). In other cases, transcriptional silencing is not linked to methylation and may instead have resulted from insertion into heterochromatic chromosomal domains, leading to transcriptional inactivation of the transgene (Cerutti et al., 1997a). Instability of foreign transgenes has also been attributed to rapid transcript degradation (Cerutti et al., 1997b).

In spite of these difficulties, a number of foreign genes have been developed as selectable markers in *Chlamydomonas*. The bacterial phleomycin/zeocin resistance gene, *ble,* is an excellent case study in foreign gene expression in *Chlamydomonas.* Initially, the *ble* gene was used successfully without modification, although it should be noted that its codon bias closely matches that of endogenous genes (Stevens et al., 1996). Fusion of *ble* to 3' and 5' untranslated regions of *RBCS2* enabled expression levels sufficient for selection of transformants based on their resistance to zeocin. This construct was later improved by using smaller 5' and 3' flanking regions, and adding two introns from the endogenous *RBCS2* gene (Lumbreras et al., 1998). In fact, the first intron of the *RBCS2* gene was shown to contain an enhancer element, demonstrating the role of introns in transcriptional modulation of genes in *Chlamydomonas* (Lumbreras et al., 1998).

Promoters for ectopic expression of transgenes have also undergone an extensive evolution over the past few years. The *RBCS2* promoter system was further improved by fusing a region of the *HSP70A* promoter to a reduced *RBCS2* promoter, creating a construct that was able to express the *aadA* gene in *Chlamydomonas* (Schroda et al., 2000). A promising new possibility for constitutive expression is the recently developed *PsaD* promoter system (Fischer and Rochaix, 2001). A *ble*-containing construct with *PsaD* upstream and downstream regions produced considerably more zeocin-resistant transformants than the *ble* construct containing the *RBCS2* promoter. This system is particularly enticing because the endogenous *PsaD* gene is a rare example of a *Chlamydomonas* gene without introns, hinting that it may provide a promoter system which would not require tedious insertion of introns into the transgene.

Inducible promoter systems have not yet been used in *Chlamydomonas,* but a number of endogenous promoters are promising candidates. Examples include copper response elements (Quinn and Merchant, 1995), heat shock-responsive promoters (Schroda et al., 2000), nitrate assimilation promoters (Loppes and Radoux, 2002; Ohresser et al., 1997), and light-responsive promoters (Hahn and Kück, 1999). Also lacking is an efficient promoter-trapping system, although preliminary work has established a system for finding nitrogen starvation-induced promoters (Haring and Beck, 1997)

In addition to selectable markers, a number of reporter genes are available for use in the nuclear genome. The *Chlamydomonas* arylsulfatase gene has been extensively used as a reporter gene, having been applied in a wide range of experiments from mutant screens to promoter analysis (e.g., Loppes and Radoux, 2002; Leisinger et al., 2001; Ohresser et al., 1997). Normally expressed in *Chlamydomonas* under low sulfur conditions, arylsulfatase is a periplasmic protein that, in cell wall–less strains, is secreted into the surrounding media, where it is then accessible to a simple chromogenic assay (Ohresser et al., 1997; Davies et al., 1992). This system has benefited from using an endogenous *Chlamydomonas* gene and does not come with the difficulties often encountered when expressing heterologous genes in *Chlamydomonas*. As is the case with many reporter genes, the relative stability of the arylsulfatase gene product may not reflect the stability of the gene under study, and results must be interpreted with this in mind (Schroda et al., 2000).

Green fluorescent protein, used as a reporter gene in many different systems, has been modified for use in *Chlamydomonas*. Initial attempts to use this reporter were complicated by poor expression, but after accounting for codon bias and inserting introns near the 5' end of the transgene, GFP was successfully modified and used to localize the chlamyopsin protein (COP) (Fuhrmann et al., 1999).

FUNCTIONAL GENOMICS

The data generated by large-scale sequencing projects are invaluable in identifying novel genes and annotating putative gene families and homologues from other organisms. Sequence data do not, however, provide a very precise analysis of gene function. Annotation of putative genes is not always accurate, and functional assignment by similarity comparison can be misleading or incorrect due to complicated evolutionary and structure/function relationships (Phelps et al., 2002). Mutagenesis or targeted down-regulation of genes represents one of the most useful methods for assigning function to a given gene. At the genome level, this can be done in one of two ways. First, genes involved in the control of a particular process can be identified by forward genetics using appropriate mutant screens. Second, the function of a gene whose sequence is known can be assessed by searching for mutations in that gene within a large population of phenotypically uncharacterized mutants, or alternatively by selectively down-regulating the expression of a particular gene. This is termed *reverse genetics*. In *Chlamydomonas,* forward genetics has been used with great success to isolate mutants involved in several processes, including nutrient deficiency responses, chloroplast biogenesis, posttranscriptional gene silencing, and flagellar assembly. Reverse genetics has only recently begun to be used and developed in *Chlamydomonas*. The use of both of these approaches on a genomic scale is still being developed using this organism. The studies in progress and challenges encountered will be covered in the following sections.

Forward Genetics

In recent years, forward genetics in *Chlamydomonas* has generally utilized insertional mutagenesis. Integration of exogenous DNA into the nuclear genome of *Chlamydomonas* occurs predominantly by nonhomologous recombination, resulting in random gene disruption (Tam and Lefebvre, 1993). In most cases, insertional mutagenesis creates null mutations. From a genomic standpoint, it is important to attempt to saturate the genome with mutations when aiming to identify all genes involved in a given process. The number of mutants required to saturate the genome is also influenced by the fact that in *Chlamydomonas* integration of the selectable marker is often accompanied by deletion of part of the adjoining genomic DNA. These deletions can range in size, but can be as large as 50 kbp (Tanaka et al., 1998; Tam and Lefebvre, 1993). Although deletions improve the odds of a particular gene being disrupted by integration of exogenous DNA, the disadvantage is that a number of genes may be deleted at a given site. It is therefore necessary to prove which single gene is responsible for the observed phenotype. The approximate number of transformants *(N)* that need to be screened to identify disruption or deletion of any given gene can be calculated using the formula

$$N = \frac{\ln(1-P)}{\ln(1-f)},$$

where P is the desired probability of finding a mutation of interest and f is the fractional proportion of the genome that is deleted in each cell line (Clarke and Carbon, 1976). Therefore, for the *Chlamydomonas* genome ($\sim 1 \times 10^8$ bp), assuming an average deletion size of 5 kbp, 92,000 lines need to be screened to have a 99 percent chance of isolating a mutation in any one line (Pazour and Witman, 2000). The calculation assumes random distribution of insertion events across the genome. Sufficient studies have not yet been completed to assess if insertion is entirely random, but the problem of site bias can be partially addressed by using a number of vectors and selectable markers when generating a population of mutants. Another factor that needs to be taken into account is the frequency with which the mutations are actually tagged by the transforming DNA. In *Chlamydomonas,* the tagging frequency appears to be approximately 50 to 60 percent for insertional mutagenesis (Moseley et al., 2000; Fleischmann et al., 1999; Niyogi et al., 1997). This compares favorably to the tagging frequency reported for *Arabidopsis* T-DNA lines (McElver et al., 2001).

One problem that has been encountered when using insertional mutagenesis in *Chlamydomonas* is the high frequency of tandem arrays of the selectable marker integrated at a single site. This makes isolation of flanking DNA much more difficult, as techniques such as plasmid rescue or TAIL-PCR that have been used in other systems (Sessions et al., 2002) are not possible. Recent experiments with ligand-mediated suppression PCR (LMS-PCR) (Strauss et al., 2001) or genome walking have solved this problem when isolating flanking DNA for a small number of mutants. It remains to be seen if this technique can be applied to high-throughput studies. In addition, many techniques based upon PCR are technically challenging in *Chlamydomonas* due to the GC-rich nature and high degree of secondary structure of the genome. Thus, isolation of flanking sequences requires considerable optimization for each sample, which is impossible to achieve with a high-throughput approach.

Populations of insertional mutants have been generated to investigate the two major foci of *Chlamydomonas* research, namely eukaryotic flagellar function and photosynthesis. Pazour and Witman (2000) reported the results of a genomic approach, involving both forward and reverse genetics, to isolate mutations affecting the outer dynein arm of the flagella. The outer dynein arm of the *Chlamydomonas* flagellum is composed of three heavy chains (HCs) that are the site of ATP hydrolysis, two intermediate chains (ICs) that are thought to be involved in non-ATP-sensitive microtubule binding and regulation of the dynein complex, eight light chains (LCs) whose function is not well understood, and three proteins in a docking complex (DC) that attaches the dynein motor to the microtubule. Thus, the known number of target genes allowed direct assessment of the percentage success of the approach. From UV-induced mutations that prevent assembly of the outer dynein arm, it is known that most of these mutations lead to a slow, jerky, swimming phenotype (Kamiya, 1988; Mitchell and Rosenbaum, 1985). Using this phenotype, Pazour and Witman (2000) screened 15,000 insertional mutants for swimming defects and identified approximately 400 affected mutant lines. Using these mutants, the authors identified null mutations in one heavy chain gene (βHC), one intermediate chain gene (IC1), four light chain genes (LC2, LC6, LC7, and LC8), and two of the docking complex genes (DC3 and ODA3). Thus, from a population of approximately 15,000 mutants screened for defects in motility and phototaxis, mutations were found in about 50 percent of the genes examined. This is consistent with calculations that assume a random distribution of 5 to 10 kbp deletions.

To date, this is the only report on an insertional mutagenesis project at a genomic level. A study is currently underway to isolate genes involved in photosynthesis and oxidative stress in *Chlamydomonas* (Dent, Kobayashi, et al., 2001). This project aims to generate 90,000 lines and screen these for phenotypes such as photoautotrophic growth, sensitivity to light, and sensitivity to generators of reactive oxygen species. These mutants and the flanking sequences of their insertion sites will be made available to the *Chlamydomonas* community in the same way as for many of the T-DNA lines in *Arabidopsis*.

Reverse Genetics

While the mitochondrial and chloroplast genomes both feature a reasonably high frequency of homologous recombination, the same is unfortunately not the case in the nuclear genome (Kindle et al., 1989). Reverse genetic studies in *Arabidopsis* have extensively utilized PCR screening of T-DNA lines to isolate mutations in genes of interest (Sessions et al., 2002) (see Chapter 2). Several problems have been encountered using this approach in *Chlamydomonas.* First, the technique relies heavily on PCR to amplify flanking regions. As insertional mutagenesis creates only one or two mutations in any one individual, a large number of clones have to be pooled to make this technique viable. As PCR is often challenging in *Chlamydomonas,* amplification of the gene of interest from a large pool is rarely successful. Second, the maintenance of large mutant collections of *Chlamydomonas* is not easy, and storage would require large-scale cryopreservation. Instead, researchers looking for a reverse genetics approach in *Chlamydomonas* have turned to TILLING and RNA interference (RNAi) techniques (see Chapter 3).

TILLING has been successfully developed for use in *Arabidopsis* (McCallum et al., 2000). Mutants populations are generated by techniques, such as EMS treatment and UV-exposure, that result in base pair changes or small deletions. Using these methods, a high number of mutations per clone is possible, thus reducing the number that need to be screened and stored. To screen the population, short regions of a gene of interest are amplified by PCR from pools of DNA from several mutants. After denaturation and slow cooling, the samples are analyzed to screen for heteroduplexes. The screening technique used extensively in *Arabidopsis* research employs an endonuclease that specifically cleaves the heteroduplexes, and the products are analyzed by simple gel electrophoresis (Colbert et al., 2001). A recently described member of the S1 nuclease family, CEL I, has been shown to be particularly suitable for genotyping applications, as it preferentially cleaves mismatches of all types (Yang et al., 2000; Oleykowski et al., 1998) and has now been used to detect heterozygous polymorphisms in DNA pools

(Kulinski et al., 2000). It is likely that this technology will prove most useful in *Chlamydomonas,* as it circumvents many of the problems that have been encountered when using this organism for genomic studies.

RNAi has been useful for analysis of gene function in a wide range of model systems, and it shows tremendous potential in *Chlamydomonas* as well (Cerutti, 2003; Barstead, 2001; Wesley et al., 2001). Targeted gene silencing via RNAi in *Chlamydomonas* was first used by researchers studying phototaxis. For years, biochemical and physiological evidence had been accumulating that chlamyopsin (COP), an opsin-like protein, was a key component of the photoreceptor complex responsible for phototactic behavior. Fuhrmann et al. (2001) were able to address this question functionally by silencing the COP gene. Contrary to expectations, both phototactic and photophobic responses were intact in the silenced strain, putting to rest the notion that COP is necessary for these responses. Meanwhile, Sineshchekov et al. (2002) searched the Kazusa DNA Research Institute EST database and found two sequences with similarity to archaeal-type rhodopsins. By silencing each of the two candidates, Sineshchekov et al. were able to show that each plays a distinct role in phototactic responses. Since then, RNAi has been used successfully to silence a number of *Chlamydomonas* genes, including phototropin (Huang and Beck, 2003) and centrin (Koblenz et al., 2003).

Judging from these examples, RNAi will provide an effective tool for functional evaluation of candidate genes identified in, for example, a transcriptional profiling or proteomics experiment. A tantalizing possibility is the eventual use of RNAi for genomic-scale "mutant" screening. Genome-wide RNAi has now been used in *Caenorhabditis elegans* to screen for genes involved in fat regulation and genome stability (Ashrafi et al., 2003; Kamath et al., 2003; Kamath and Ahringer, 2003; Pothof et al., 2003). This has been facilitated in *C. elegans* by the ease of RNAi induction in this system, which can be achieved at high rates by simply feeding the worms bacteria that have been engineered to produce double-stranded RNA. A different strategy will be needed in *Chlamydomonas,* where a relatively low penetrance of silenced phenotypes makes it necessary to devise a method to select for silenced lines. It may be possible to use a combinatorial approach in which a selective marker is fused to the gene to be tested, allowing for selection of silenced transformants based upon failure to express the selective marker. With these tools in hand, RNAi could potentially be used in a high-throughput, genome-wide screen that would provide a logical complement to the forward genetics approaches already in place.

TRANSCRIPTOMICS

Whether surveying responses to environmental stress or defining developmental pathways, transcriptional-profiling methods offer a powerful approach for capturing a comprehensive snapshot of gene expression under a given set of conditions. The development of transcriptional-profiling methods also complements forward and reverse genetics projects, serving as an important bridge between genotype and phenotype when characterizing mutants (see Chapter 4). There are many different approaches to global profiling of RNA abundance, including expressed sequence tag (EST) sequencing, differential display methods, and microarray analysis. Each of these approaches has been used successfully in *Chlamydomonas* and will be discussed in this chapter.

EST-sequencing projects provide an important foundation for many genomic applications, both by generating large amounts of sequence data and by allowing researchers to compare transcriptional profiles from cDNA libraries of cells grown under specific conditions. The *Chlamydomonas* community has benefited tremendously from two large-scale EST-sequencing projects: one a product of the Kazusa DNA Research Institute (KDRI) (Asamizu et al., 1999) and the other a component of the National Science Foundation (NSF)-sponsored *Chlamydomonas* genomics project (Shrager et al., 2003; Davies and Grossman, 1998). Thanks to these efforts, the current number of *Chlamydomonas* ESTs deposited in GenBank currently exceeds 200,000.

To date, the EST project at KDRI (http://www.kazusa.or.jp/en/plant/chlamy/EST/) has generated a total of 37,990 5' sequence tags from size-selected and normalized cDNA libraries constructed from photoautotrophically grown cells under high and low CO_2 conditions (Table 8.3) (Asamizu et al., 2000, 1999). Although normalization of the cDNA libraries reduces overrepresentation of highly expressed genes, some of these 37,990 sequence tags inevitably represent multiple independent occurrences of the same cDNA within the library. To reduce the amount of redundancy in this collection, ESTs have been further assembled into unique groups. This assembly reduced an initial set of 11,500 sequence tags to 3,400 independent ESTs (Asamizu et al., 1999). KDRI has categorized each of their ESTs into functional groups based upon sequence comparison, and it has provided a particularly valuable service to the community by making EST clones available upon request.

The parallel NSF-funded project sequenced ESTs generated from cells grown under a variety of different conditions (Table 8.3). ESTs from a "core" library comprised of cells grown on acetate in low-light or dark con-

TABLE 8.3. Conditions used for cDNA library construction.

Library	Conditions	Number of clones
Core (Stanford)[a]	Acetate-containing medium, light Acetate-containing medium, dark Minimal medium, light Minimal medium, $+CO_2$	768 (unnormalized) 10,080 (normalized) 12,096 (subtracted) 5' and 3' sequences
Stress I (Stanford)[a]	Nitrogen deprivation Sulfur deprivation Phosphorous deprivation Nitrate-to-ammonium shift Ammonium-to-nitrate shift	12,000 (normalized) 5' and 3' sequences
Stress II (Stanford)[a]	Hydrogen peroxide exposure Cadmium exposure Hydrogen production Osmotic stress (sorbitol) Ammonium-to-nitrate shift	10,752 (normalized) 5' and 3' sequences
Stress III (Stanford)[a]	Iron deprivation Copper deprivation Anaerobiosis High-light stress	Ongoing (normalized) 5' and 3' sequences
Deflagellation (Stanford)[a]	Deflagellated cells (using pH shock)	12,480 (normalized) 5' and 3' sequences
Gamete/Zygotic (Stanford)[a]	Gametogenesis induction Zygote formation	Ongoing (normalized) 5' and 3' sequences
S1D2 (Stanford)[a]	Polymorphic strain used for mapping	124 (normalized) 5' and 3' sequences
Standard (Kazusa)[b]	Minimal medium	11,586 (normalized and subtracted) 5' sequences
CO_2 acclimation (Kazusa)[c]	High CO_2 acclimated Shift from high CO_2 to low CO_2	26,419 (normalized and subtracted) 5' sequences

[a]Shrager et al., 2003.
[b]Asumizu et al., 1999.
[c]Asumizu et al., 2000.

ditions, as well as cells grown photoautotrophically in minimal media under ambient and enriched CO_2 conditions, were sequenced as a baseline of "core" gene expression. Four other libraries were constructed from cells grown under a myriad of stress conditions, ranging from nutrient deficiencies (such as nitrogen, phosphorous, or sulfur limitation) to treatment with sublethal levels of toxic compounds (such as cadmium or hydrogen peroxide). In addition to these stress libraries, libraries generated from deflagellated cells, as well as gametic and zygotic cells at different stages of maturity, have more recently been added to the collection (Shrager et al., 2003). While the analysis of these ESTs to date has focused largely on environmental stress responses, the recent inclusion of developmental libraries will be useful for shedding light on the transcriptional regulation of developmental processes in *Chlamydomonas*.

To maximize the volume of useful sequence information, both 3' and 5' sequences have been generated from these libraries (Shrager et al., 2003). Contig assembly was performed using Phrap, an assembly program designed for assembling shotgun sequencing data that takes sequence quality into account during the assembly (Figure 8.3) (University of Washington Genome Center; http://www.genome.washingtonedu/UWGC). 3' regions were used in two initial rounds of contig assembly, the results of which were then assembled with the 5' sequence set to obtain the longest sequence possible. At best, a contig may represent the full-length sequence of an individual gene, although in practice this is rarely the case. Most of the 8,628 "ACEs" (**A**ssembly of **C**ontiguous **E**STs) contain two contig assemblies—one from the 3' and the other from the 5' end of the cDNA, with sequence between the two assemblies missing. Others contain four or more contigs (Shrager et al., 2003). Contig sequences that did not match a previously identified Volvocales gene were annotated based on similarity to amino acid sequences in GenBank, as inferred from WU-BlastX results. All of this sequence information, including sequence from individual ESTs, contigs, and assembled, annotated ACEs, is available in a searchable public database (http://www.biology.duke.edu/chlamy_genome/search.html). Furthermore, the results have been compiled into a "uni-gene set"—a comprehensive list of every gene identified from these EST assemblies. A final assembly of all *Chlamydomonas* EST collections, including the KDRI collection, is currently underway (Grossman et al., 2003).

Sequence information from EST projects has a number of applications. For example, EST databases have been used to identify additional members of large gene families, such as the LHC proteins of PSII (Elrad et al., 2002; Teramoto et al., 2001). Identification of these genes will allow researchers to perform functional characterization of new gene family members, possibly leading to the delineation of unique functions and evolutionary histories

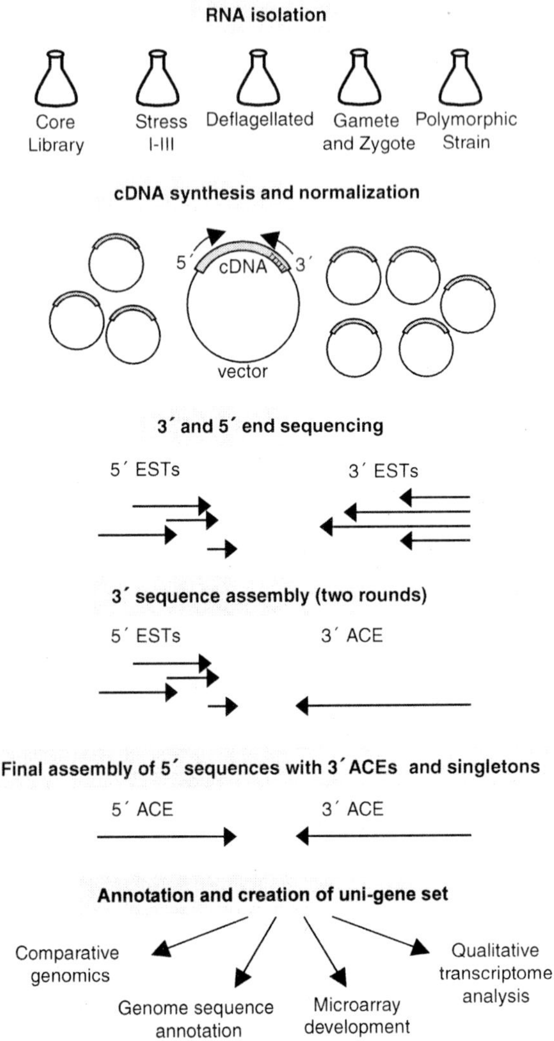

FIGURE 8.3. EST contig assembly. cDNA libraries are cloned into vectors and sequenced from their 5' and 3' ends. These sequences are then assembled, 3' ends first, into contigs of overlapping sequences. Once the 3' end sequences are put through two rounds of assembly, they are assembled with the 5' end sequences. The result is an ACE, which in this diagram consists of two contigs. Although this is the most common result, it is also possible to obtain four or more contigs within a single ACE, and in rare cases, both ends may assemble into one contig. (*Source:* Based on Shrager et al., 2003.)

of individual constituents of this large group of genes. La Fontaine et al. (2002) used a similar method to find genes with similarity to proteins known to be involved in iron or copper metabolism. In *Saccharomyces cerevisiae,* copper is an essential cofactor in high-affinity iron uptake. Searching the *Chlamydomonas* EST database using relevant sequences from fungi, plants, or animals led to the isolation of a number of putative ferritin and copper chaperones. By studying changes in transcript and protein levels of these genes, La Fontaine et al. (2002) found evidence of a unique copper-independent iron-assimilation pathway in *Chlamydomonas.*

Chlamydomonas EST collections have also been mined for transcriptional-profiling data by evaluating the frequency of specific ESTs in libraries generated under different growth conditions. KDRI compared ESTs derived from cells grown in high and low CO_2, deriving a list of transcripts that are present only when cells are grown under one CO_2 concentration or the other (Asamizu et al., 2000). Among the low-CO_2-induced transcripts were a number of carbonic anhydrase genes known to be induced under low-CO_2 conditions, serving as a proof of principle for this approach. Schrager et al. (2003) also used EST collections for transcriptional profiling by comparing the EST composition of the different cDNA libraries generated from cells grown under various stress conditions. This comparison led to the discovery of a group of stress-specific transcripts, including a number of genes predicted to be involved in proteolytic processes.

The growing number of plant EST projects opens the door to new comparative genomics research. These projects help researchers gain access to plant and algal genomes that would simply be too large to sequence at the current level of technology and expense (Mayer and Mewes, 2001). Already, EST projects have begun in a wide range of plants and algae, including *Phaeodactylum tricornutum* (Scala et al., 2002), *Porphyra yezoensis* (Nikaido et al., 2000), *Physcomitrella patens* (Rensing et al., 2002), *Arabidopsis thaliana* (Delseny et al., 1997), and *Triticum aestivum* (Ogihara et al., 2003), among others. As EST-sequencing and -assembly protocols enhance the quality and coverage of EST collections, the ability to compare transcriptomes between species holds increasing promise for gene identification and functional classification in organisms that have large genomes or are otherwise less accessible via traditional genetic approaches (Rensing et al., 2002; Somerville and Somerville, 1999). The presence of an EST project in a genetically tractable algal system means that *Chlamydomonas* ESTs will be useful for functional genomics of other plant systems. Nuclear-sequencing projects are also reaching out to other algal species, such as the centric diatom *Thalassiosira pseudonana,* the coccolithophorid *Emiliania huxleyi,* and the green alga *Ostreococcus* (http://www.ornl.gov/microbial genomes/organisms.html), and annotation in these organisms may also be facilitated by the experimental data cultivated by *Chlamydomonas* researchers over the past century.

Other transcriptional-profiling methods have also been successfully applied in *Chlamydomonas*. Differential display methods have been used to isolate genes induced by conditions ranging from nitrogen starvation and sexual differentiation (Merchan et al., 2001) to cadmium and iron stress (Rubinelli et al., 2002). Im and Grossman (2002) used differential display to investigate transcriptional changes in response to high-light conditions. After identifying a number of transcripts related to carbon metabolism, it was clear that to fully understand high-light responses, it would be necessary to understand the interplay between high-light stress and carbon limitation (Im and Grossman, 2002). Of the nine high-light-inducible gene fragments isolated using differential display, five proved to be regulated by CIA5, a regulatory protein involved in the carbon-concentrating mechanism (Fukuzawa et al., 2001; Xiang et al., 2001).

These differential display methods have since been largely supplanted by the development of *Chlamydomonas* cDNA microarrays. The first *Chlamydomonas* microarrays were fabricated as part of an organellar transcriptional-profiling project. Lilly et al. (2002) thoroughly surveyed changes in organellar gene expression using an array that consisted of the chloroplast genome, eight genes from the mitochondrial genome, and 45 nuclear genes. Cells were subjected to a wide range of environmental stress conditions, including nutrient limitation, UV exposure, high and low temperature, and light intensity. These studies revealed a chloroplast transcriptome that is highly responsive to changes in the environment, although, interestingly, these changes in RNA abundance were rarely reflected as changes in protein content. Furthermore, run-on transcription experiments indicated that much of the increase in chloroplast RNA abundance seen during phosphorous stress is due to changes in RNA stability, rather than transcriptional regulation. These experiments highlight the ability of transcriptional profiling to provide insight not only regarding specific stress responses but also on the basic mechanisms of organellar gene regulation. The results of these studies have been placed in a publicly accessible database, searchable by growth condition (http://bti.cornell.edu/bti2/chlamyweb).

Creation of the uni-gene set based upon EST assembly data has enabled the selection of clones for nuclear cDNA microarrays as well. A first-generation array containing approximately 2,800 clones from the Core, Stress I, and Stress II libraries has been released to the community (Grossman et al., 2003; Im et al., 2003). Microarrays were constructed using 200 to 600 bp of the 3' regions of each clone. Im and colleagues (2003) have repeated their high-light/CO_2 differential display experiments using these first-generation microarrays and have expanded their set of differentially expressed genes over sevenfold.

The second-generation array will feature clones representing the entire unigene set as well as the entire chloroplast and mitochondrial genomes. As part of the NSF-funded *Chlamydomonas* genome project, gene expression changes will be monitored over a time-course of exposure to numerous environmental conditions, such as nutrient deprivation and changes in both light quantity and quality (Grossman et al., 2003). The ultimate aim of this project will be the creation of a functional gene map of the three *Chlamydomonas* genomes by clustering groups of genes with similar expression patterns. The establishment of regulatory networks using this technique is an exciting prospect, and it could contribute to sequence annotation.

The development of microarray platforms for analysis of expression patterns opens up many other doorways as well. An algal microarray system offers the opportunity for researchers using closely related species to try interspecific hybridizations, shown to be a viable means of expression analysis in bacteria (Dong et al., 2001), or intergeneric hybridizations, shown to be effective in the analysis of developing seeds in *Brassica* using *Arabidopsis* microarrays (Jackson et al., 2002; Girke et al., 2000). With this in mind, development of *Chlamydomonas* functional genomics tools has the potential to benefit genomics efforts in other algal species as well.

PROTEOMICS AND METABOLOMICS

Delneri et al. (2001) defined three levels of genomic analysis that can be integrated in a systems-based approach to the biology of an organism, namely, the mRNA level (transcriptome), the protein level (proteome), and the level of low-molecular-weight intermediates or metabolites (metabolome). As described in previous sections, analysis of mRNA can be achieved in a highly comprehensive manner using cDNA microarrays. With respect to true functional genomic analysis, however, mRNA molecules are not strictly functional units within the cell, but simply transmitters of information for protein sequence (Delneri et al., 2001). Thus, transcriptome analysis provides functional information indirectly and by inference. Both proteins and metabolites, on the other hand, have true functional roles within the cell.

Proteomics

Proteomic analysis in *Chlamydomonas* has mainly been developed for the chloroplast compartment. By extrapolation from sequence data, it has been estimated that the *Arabidopsis* chloroplast proteome consists of 87 plastid-encoded proteins and about 2,300 nuclear-encoded proteins (Abdallah et al., 2000). Many of these are related to photosynthesis. The thylakoid membrane system within the chloroplast of algae and higher plants is the subcompartment in which the primary reactions of photosynthesis occur and has therefore been the focus of most of the proteomic analysis in *Chlamydomonas*. The primary reactions of photosynthesis are performed by approximately 100 proteins that are organized into four major multi-subunit protein complexes: PSI, PSII, the ATP-synthase complex, and the cytochrome b_6/f complex. These are supported by an array of additional proteins that are required for the assembly, maintenance, and regulation of the four subcomplexes. Standard proteome analysis combines high-resolution two-dimensional gel electrophoresis (2-DE) with mass spectroscopy (MS) or tandem MS (MS-MS) to identify specific protein spots. The first use of this technique to analyze chloroplast proteins was carried out by Peltier et al. (2000), who analyzed lumenal and peripheral proteins in the *Arabidopsis* chloroplast. The proteomic analysis of intrinsic thylakoid membranes is more technically challenging due to the hydrophobic nature of the proteins involved. In fact, Santoni et al. (2000) reported that trans-membrane proteins could not be identified in any of the proteomic approaches using yeast, tobacco, or *Arabidopsis*. A limited number of animal membrane proteins had been resolved by 2-DE at this date.

Hippler et al. (2001), however, succeeded in developing 2-DE to analyze the protein composition of the thylakoid membrane of *Chlamydomonas*. For the resolution of proteins with several membrane-spanning domains, both the solubilization of proteins and the sample fractionation were shown to be critical. The thylakoid proteins were separated from the pigments and lipids by chloroform/methanol precipitation, and separation of the proteins by 2-DE was achieved by addition of thiourea and the nonionic detergent β-dodecyl maltoside to the dissolving buffer (Pasquali et al., 1997). Using this method, approximately 350 to 400 spots could be resolved on a single gel. The study analyzed the LHC proteins of PSI and PSII in more detail. By using MS-MS and immunological analysis, more than 30 different LHC spots could be identified on the gels. The authors used the techniques to compare the LHCs of wild-type *Chlamydomonas* thylakoids with those of a PSI-deficient *(Δycf4)* and a *crd1* mutant, which is conditionally reduced in PSI and LHCI under copper deficiency. The results showed the presence of most LHCI spots in the *Δycf4* mutant but their absence in the *crd1* mutant. Thus, this initial study demonstrated that the techniques developed allowed for the comparative analysis of membrane protein complexes in wild-type and mutant *Chlamydomonas* cells.

The techniques developed by Hippler et al. (2001) have now been employed in a number of other studies. Most recently, Moseley et al. (2002) utilized the technique to identify specific LHCI proteins that are affected by Fe nutritional status. Thylakoid membrane proteins isolated from Fe-deficient and Fe-replete cells were separated by 2-DE, and members of individual gene families were identified subsequently by immunological methods. Multiple differences were noted between the samples, with some proteins being induced and others significantly diminished. Two polypeptide components of LHCI were shown to decrease in response to Fe starvation, with little or no decrease of the inner antenna polypeptides. The appearance of new anti-LHC cross-reacting polypeptides during Fe starvation may represent alternatively processed forms of previously described LHCI gene products or new, Fe-specific LHCI proteins. Identification of these spots is not reported in this study; however, a more detailed time-course analysis by 2-DE shows that the major novel Fe-deficiency-inducible spot is detectable after just one day of Fe deficiency and is fully present after three days. The changes that are observed in the LHCI and LHCII complexes by proteomic analysis are shown to be highly specific, and it is proposed that they are particular to the Fe-deficiency response. In addition to demonstrating that proteomic analysis is useful for assessing large-scale changes in protein composition in response to nutritional status, the previous study also highlights the use of *Chlamydomonas* as a choice experimental system for uncovering the molecular mechanisms involved in the adaptation of the eukaryotic photosynthetic apparatus to Fe deficiency and other nutritional stresses.

In addition to proteomic analysis of the thylakoid membrane in *Chlamydomonas,* a recent study has also characterized the proteome of the small subunit of the chloroplast ribosome (Yamaguchi et al., 2002) (see also Chapter 14). Plastid gene expression is regulated primarily during translation and involves dynamic interplay between ribosomes, translation factors, and mRNA-protein complexes. Thus, in order to fully understand how mRNAs are translated into functional proteins within the chloroplast, a detailed understanding of the components of translation, including the ribosomes, translation factors, mRNA-binding proteins, and the RNA elements with which they interact, is required. Previous studies have identified 31 small subunit proteins (Randolph-Anderson et al., 1989; Schmidt et al., 1984) and have shown that 14 of these are synthesized in the chloroplast and 17 are synthesized in the nucleus (Schmidt et al., 1983). Using 2-DE followed by liquid chromatography (LC)-MS-MS analysis of peptide fragments, Yamaguchi et al. (2002) identified 20 small subunit ribosomal proteins. These included 19 orthologues of *Escherichia coli* ribosomal proteins (S1-S7, S9, S10, S12-S21) and a homologue of spinach plastid-specific ribosomal protein-3. A novel S1 domain-containing protein, named PSRP-7, was also identified. In addition, the analysis showed that S2, S3, and S5 were larger than their spinach or *E. coli* homologues, with N-terminal extensions occurring in S2 and S5 and an insertion sequence in S3. Evidence suggests that these domains may interact with the S1 protein and PSRP-7 and thus function in aspects of mRNA recognition and translation initiation that are unique to the chloroplast compartment of *Chlamydomonas*.

Unfortunately, mitochondrial proteomics has been limited in *Chlamydomonas* due to the predominance of the chloroplast, which makes isolation of pure mitochondrial fractions free of contaminating thylakoid membranes extremely difficult. This is hardly a surprise, given that chloroplasts occupy greater than 40 percent of the total cell volume, whereas mitochondria comprise only 2 to 4 percent (Harris, 1989). To get around these difficulties, some researchers have resorted to using *Polytomella,* a colorless relative of *Chlamydomonas* (Fan and Lee, 2002; Perez-Martinez et al., 2000; Atteia et al., 1997). Another approach has been to use "yellow in the dark" *Chlamydomonas* mutants that are deficient in the light-independent chlorophyll biosynthetic pathway (Bennoun et al., 1995). When grown in the dark, these "etiolated" cells contain only trace amounts of the PSII reaction centers and LHCs that often contaminate wild-type mitochondrial protein preparations. Both of these approaches are unfortunately unable to investigate interactions between the chloroplastic and mitochondrial proteomes.

Purification of wild-type *Chlamydomonas* mitochondria has been achieved by separating crude extracts on Percoll gradients (Eriksson et al., 1995). Although a thylakoid-free fraction was obtained, the recovery in this fraction was very low, with the majority of the mitochondria continuing to migrate with a thylakoid-contaminated fraction. Despite the low yield, these purified mitochondria were analyzed using blue native-polyacrylamide gel electrophoresis followed by 2-DE (Van Lis et al., 2003). N-terminal sequencing of resolved proteins revealed members of complexes I, III, IV, and V, as well as other proteins localized in the mitochondria. Studies such as these lay the groundwork for future work characterizing differences in the mitochondrial proteome under various physiological conditions and in different mutant strains.

Thylakoid and mitochondrial proteome analysis has in some cases been facilitated by the use of computer predictions of protein targeting (Leister, 2003; Taylor et al., 2003). Although the recent completion of the *Chlamydomonas* nuclear genome makes this approach computationally feasible, there are a number of reasons to proceed with caution. Although chloroplast prediction programs have worked well in *Arabidopsis* and other plants (Leister, 2003), these programs tend to miscall *Chlamydomonas* chloroplast proteins as mitochondrial proteins. As for the mitochondrial proteome, studies from *Arabidopsis* do not bode well for this approach. Analyses of the mitochondrial proteome in *Arabidopsis* have shown that many nucleus-encoded proteins isolated in the mitochondria would not be predicted to localize to that organelle using the standard localization programs available (Millar and Heazlewood, 2003; Millar et al., 2001). This could be due to both limited use of plant mitochondrial proteins during development of these programs, as well as recent evidence that mitochondrial targeting may occur independently of an N-terminal extension.

Metabolomics

Metabolomics has been the term adopted to describe the comprehensive, nonbiased, high-throughput analyses of complex metabolite mixtures present in cell or tissue extracts (Hall et al., 2002) (see also Chapter 5). In organisms such as yeast, where the whole cell contains approximately 600 metabolic intermediates, analysis of the entire metabolome is less complex than analysis of the genome, proteome, or transcriptome (Raamsdonk et al., 2001). In plants, however, this situation is reversed. Approximately 50,000 different compounds have been identified from plants (De Luca and St Pierre, 2000), but it has been predicted that the final figure for the plant kingdom will probably approach 200,000 (Fiehn, 2002). Due to their diversity and ease of analysis, microbes may prove to be the richest overall source of metabolites, although plants will likely be the source of the most complex individual mixtures (Hall et al., 2002). *Chlamydomonas* and other unicellular algae represent the intersection of these two groups and thus possess great potential for the exploration of metabolites and metabolic flux, but technical challenges must inevitably be overcome when attempting to detect and quantify such a diverse range of compounds.

At present, metabolomics represents the branch of functional genomics that has been developed the least in *Chlamydomonas*. There are no detailed reports in the literature of metabolite profiling in this organism. Bestman, Ek, and Vos (personal communication) have, however, utilized the approach to investigate metabolic carbon flux in *Chlamydomonas* during the transition from using CO_2 as the primary carbon source to using acetate via the glyoxylate pathway. After the algal cells were exposed to isotope-labeled acetate, the metabolites were extracted and analyzed using GC-MS. Modeling of the pathways involved indicated that a fivefold greater carbon flux through the glyoxylate pathway could be detected in the dark, as compared to the light, when using acetate as the carbon source. This study shows the advantages of the microbial lifestyle in analyzing the metabolic effects of exogenously applied intermediates and thus pathway redirection.

At this stage, metabolomics remains an underused tool with great promise in *Chlamydomonas*.

TYING IT ALL TOGETHER: DATABASES FOR THE CHLAMYDOMONAS *COMMUNITY*

As referenced throughout this chapter, all genomic and organismal data on *Chlamydomonas* are dispersed among three databases: ChlamyDB (mutant alleles, sequenced genes, genetic loci, strain description, and bibliographical citations), ChlamyEST (sequence data from ESTs, contigs, unigene sets, chloroplast, and mitochondria), and the JGI database (nuclear genome sequence, gene model predictions, and preliminary annotation). A major goal of the *Chlamydomonas* genome project is to develop an integrated database in which all this information is accessible and interconnected (Grossman et al., 2003). ChlamyDB and ChlamyEST will be unified using a version of the Generic Model Organism Database (GMOD) (Stein et al., 2002). The data from ChlamyDB will be incorporated into four modules: genetics, organism, publication, and map modules. A sequence module will be populated by nuclear, chloroplast, and mitochondrial genomic sequences; EST sequences and contigs; complete cDNA sequences; and DNA sequences used to build microarrays. In the future, an expression module will be added to incorporate global gene-expression data derived from microarray analysis. This central *Chlamydomonas* database will be linked to the JGI database, allowing users to enter one database and retrieve data from the other at the same time (Grossman et al., 2003). This system will therefore allow a simple point of access to sequence data from the NSF-sponsored Chlamydomonas Genome Project and the JGI sequencing effort and provide tools for querying both genomic and gene-expression data.

CLOSING THOUGHTS AND FUTURE DIRECTIONS

The genomics revolution has boosted *Chlamydomonas* research to a whole new level, and as the tools described in this chapter are being developed, they are already having a profound impact on the contributions that can be made using this organism. It is exciting to think of what lies in store for the *Chlamydomonas* community, and it is tempting to imagine some of the new roads that *Chlamydomonas* research may travel in the next few years.

For example, there has recently been increasing interest in turning the power of genomic approaches toward ecological problems (Jackson et al., 2002). Given that *Chlamydomonas* strains have been isolated everywhere from arctic landscapes to brackish water to sewage ponds, studies with *Chlamydomonas* could shed light on a wide range of ecological questions. The intimate familiarity with *Chlamydomonas* stress responses in the laboratory from data collected over the past century could be expanded to address ecological questions regarding a diverse range of topics from nutrient cycles to water fouling. Genomics tools such as microarrays and proteomic assays developed for *Chlamydomonas* can be applied to closely related naturally occurring strains isolated from different environments.

Plant biologists have also been taking advantage of accumulating sequence data to facilitate quantitative trait loci (QTL) mapping (Mackay, 2001; Mitchell-Olds, 2001). The extensive collections of natural *Chlamydomonas* isolates from a variety of environments around the world are a rich source of material for the discovery of QTLs. Characteristics such as photosynthetic performance parameters or cell size, for example, could be traced through crosses, eventually narrowing in on the chromosomal region that causes these differences. This approach has been used successfully in *Arabidopsis* (Kliebenstein, Gershenzon, et al., 2001; Kliebenstein, Kroymann, et al., 2001). The large number of progeny required for QTL mapping makes the unicellular nature of *Chlamydomonas* very useful for these projects, and marker generation can be facilitated by the abundant polymorphisms between strains and full genomic sequence data available for this species.

Finally, the union of genomics and evolutionary biologists will no doubt bring some order to what often seems a chaotic assembly of algae within the Chlorophyta. The *Chlamydomonas* chloroplast genome is only one of the many chloroplast genomes comprising the Deep Green project, an ambitious project that aims to reconstruct phylogenies based upon the entire chloroplast sequences of many land plants and green algae. The evolutionary relationship among green algae is still poorly developed, and genome-sequencing projects such as these promise to shed new light on these open questions. As information from the *Chlamydomonas* genome project expands, other algal groups stand to benefit from sequence and annotation data for this alga. With all these possibilities and more on the horizon, without question, the flourishing field of *Chlamydomonas* functional genomics will continue to grow in exciting new directions.

REFERENCES

Abdallah, F., Salamini, F., Leister, D. (2000). A prediction of the size and evolutionary origin of the proteome of chloroplasts of *Arabidopsis*. *Trends Plant Sci.* 5: 141-142.

Alexander, N.J., Gillham, N.W., Boynton, J.E. (1974). The mitochondrial genome of *Chlamydomonas reinhardtii*. Induction of minute colony mutations by acriflavine and their inheritance. *Mol. Gen. Genet.* 130: 275-290.

Asamizu, E., Miura, K., Kucho, K., Inoue, Y., Fukuzawa, H., Ohyama, K., Nakamura, Y., Tabata, S. (2000). Generation of expressed sequence tags from low-CO_2 and high-CO_2 adapted cells of *Chlamydomonas reinhardtii*. *DNA Res.* 7: 305-307.

Asamizu, E., Nakamura, Y., Sato, S., Fukuzawa, H., Tabata, S. (1999). A large scale structural analysis of cDNAs in a unicellular green alga, *Chlamydomonas reinhardtii*. I. Generation of 3433 non-redundant expressed sequence tags. *DNA Res.* 6: 369-373.

Ashrafi, K., Chang, F.Y., Watts, J.L., Fraser, A.G., Kamath, R.S., Ahringer, J., Ruvkun, G. (2003). Genome-wide RNAi analysis of *Caenorhabditis elegans* fat regulatory genes. *Nature* 421: 268-272.

Atteia, A., Dreyfus, G., Gonzalez-Halphen, D. (1997). Characterization of the alpha and beta-subunits of the F_0F_1-ATPase from the alga *Polytomella* spp., a colorless relative of *Chlamydomonas reinhardtii*. *Biochim. Biophys. Acta* 1320: 275-284.

Barstead, R. (2001). Genome-wide RNAi. *Curr. Opin. Chem. Biol.* 5: 63-66.

Bennoun, P., Atteia, A., Pierre, Y., Delsome, M. (1995). Etiolated cells of *Chlamydomonas reinhardtii:* Choice material for characterization of mitochondrial membrane polypeptides. *Proc. Natl. Acad. Sci. U.S.A.* 92: 10202-10206.

Bennoun, P., Delosme, M., Kueck, U. (1991). Mitochondrial genetics of *Chlamydomonas reinhardtii:* Resistance mutation marking the cytochrome b gene. *Genetics* 127: 335-344.

Blankenship, J.E., Kindle, K.L. (1992). Expression of chimeric genes by the light-regulated *cabII-1* promoter in *Chlamydomonas reinhardtii:* A *cabII-1/nit1* gene functions as a dominant selectable marker in a *nit-nit2* strain. *Mol. Cell. Biol.* 12: 5268-5279.

Bock, R. (2001). Transgenic plastids in basic research and plant biotechnology. *J. Mol. Biol.* 312: 425-438.

Boer, P.H., Bonen, L., Lee, R.W., Gray, M.W. (1985). Genes for respiratory chain proteins and ribosomal RNAs are present on a 16-kilobase-pair DNA species from *Chlamydomonas reinhardtii* mitochondria. *Proc. Natl. Acad. Sci. USA* 82: 3340-3344.

Boynton, J.E., Gillham, N.W. (1993). Chloroplast transformation in *Chlamydomonas*. *Methods Enzymol.* 217: 510-536.

Boynton, J.E., Gillham, N.W. (1996). Genetics and transformation of mitochondria in the green alga *Chlamydomonas*. *Methods Enzymol.* 264: 279-296.

Boynton, J.E., Gillham, N.W., Harris, E.H., Hosler, J.P., Johnson, A.M., Jones, A.R., Randolph-Anderson, B.L., Robertson, D., Klein, T.M., Shark, K.B., Sanford, J.C. (1988). Chloroplast transformation in *Chlamydomonas* with high velocity microprojectiles. *Science* 240: 1534-1538.

Boynton, J.E., Harris, E.H., Burkhart, B.D., Lamerson, P.M., Gillham, N.W. (1987). Transmission of mitochondrial and chloroplast genomes in crosses of *Chlamydomonas*. *Proc. Natl. Acad. Sci. USA* 84: 2391-2395.

Brown, L.E., Sprecher, S.L., Keller, L.R. (1991). Introduction of exogenous DNA into *Chlamydomonas reinhardtii* by electroporation. *Mol. Cell. Biol.* 11: 2328-2332.

Cerutti, H. (2003). RNA interference: Traveling in the cell and gaining functions? *Trends Genet.* 19: 39-46.

Cerutti, H., Johnson, A.M., Gillham, N.W., Boynton, J.E. (1997a). Epigenetic silencing of a foreign gene in nuclear transformants of *Chlamydomonas*. *Plant Cell* 9: 925-945.

Cerutti, H., Johnson, A.M., Gillham, N.W., Boynton, J.E. (1997b). A eubacterial gene conferring spectinomycin resistance on *Chlamydomonas reinhardtii:* Integration into the nuclear genome and gene expression. *Genetics* 145: 97-110.

Chen, X.J., Clark-Walker, G.D. (2000). The petite mutation in yeasts: 50 years on. *Int. Rev. Cytol.* 194: 197-238.

Clarke, L., Carbon, J. (1976). A colony bank containing synthetic Col-E-1 hybrid plasmids representative of the entire *Escherichia coli* genome. *Cell* 9: 91-99.

Colbert, T., Till, B.J., Tompa, R., Reynolds, S., Steine, M.N., Yeung, A.T., McCallum, C.M., Comai, L., Henikoff, S. (2001). High-throughput screening for induced point mutations. *Plant Physiol.* 126: 480-484.

Cole, D.G. (2003). The intraflagellar transport machinery of *Chlamydomonas reinhardtii*. *Traffic* 4: 435-442.

Dame, G., Gloeckner, G., Beck, C.F. (2002). Knock-out of a putative transporter results in altered blue-light signalling in *Chlamydomonas*. *Plant J.* 31: 577-587.

Daniell, H., Dhingra, A. (2002). Multigene engineering: Dawn of an exciting new era in biotechnology. *Curr. Opin. Biotechnol.* 13: 136-141.

Davies, J.P., Grossman, A.R. (1998). The use of *Chlamydomonas* (Chlorophyta: Volvocales) as a model algal system for genome studies and the elucidation of photosynthetic processes. *J. Phycol.* 34: 907-917.

Davies, J.P., Weeks, D.P., Grossman, A.R. (1992). Expression of the arylsulfatase gene from the beta 2-tubulin promoter in *Chlamydomonas reinhardtii*. *Nucleic Acids Res.* 20: 2959-2965.

Day, A. (1995). A transposon-like sequence with short terminal inverted repeats in the nuclear genome of *Chlamydomonas reinhardtii*. *Plant Mol. Biol.* 28: 437-442.

Day, A., Schirmer-Rahire, M., Kuchka, M.R., Mayfield, S.P., Rochaix, J.D. (1988). A transposon with an unusual arrangement of long terminal repeats in the green alga *Chlamydomonas reinhardtii*. *EMBO J.* 7: 1917-1927.

De Luca, V., St Pierre, B. (2000). The cell and developmental biology of alkaloid biosynthesis. *Trends Plant Sci.* 5: 168-173.

Debuchy, R., Purton, S., Rochaix, J.D. (1989). The argininosuccinate lyase gene of *Chlamydomonas reinhardtii:* An important tool for nuclear transformation and for correlating the genetic and molecular maps of the *ARG7* locus. *EMBO J.* 8: 2803-2809.

Delneri, D., Brancia, F.L., Oliver, S.G. (2001). Towards a truly integrative biology through the functional genomics of yeast. *Curr. Opin. Biotechnol.* 12: 87-91.

Delseny, M., Cooke, R., Raynal, M., Grellet, F. (1997). The *Arabidopsis thaliana* cDNA sequencing projects. *FEBS Lett.* 405: 129-132.

Denovan-Wright, E.M., Sankoff, D., Spencer D.F., Lee, R.W. (1996). Evolution of fragmented mitochondrial ribosomal RNA genes in *Chlamydomonas. J. Mol. Evol.* 42: 382-391.

Dent, R.M., Han, M., Niyogi, K.K. (2001). Functional genomics of plant photosynthesis in the fast lane using *Chlamydomonas reinhardtii. Trends Plant Sci.* 6: 364-371.

Dent, R.M., Kobayashi, M.C., Niyogi, K.K. (2001). Functional genomics of photosynthesis using *Chlamydomonas reinhardtii. Twelfth International Congress on Photosynthesis:* 1-6.

Derzaph, T., Weger, H.G. (1996). Immunological identification of the alternative oxidase in *Chlamydomonas reinhardtii* (Chlorophyta). *J. Phycol.* 32: 621-623.

Dong, Y., Glasner, J.D., Blattner, F.R., Triplett, E.W. (2001). Genomic interspecies microarray hybridization: Rapid discovery of three thousand genes in the maize endophyte, *Klebsiella pneumoniae* 342, by microarray hybridization with *Escherichia coli* K-12 open reading frames. *Appl. Environ. Microbiol.* 67: 1911-1921.

Dutcher, S.K. (2003). Elucidation of basal body and centriole functions in *Chlamydomonas reinhardtii. Traffic* 4: 443-451.

Dutcher, S.K., Power, J., Galloway, R.E., Porter, M.E. (1991). Reappraisal of the genetic map of *Chlamydomonas reinhardtii. J. Hered.* 82: 295-301.

Elrad, D., Niyogi, K.K., Grossman, A.R. (2002). A major light-harvesting polypeptide of photosystem II functions in thermal dissipation. *Plant Cell* 14: 1801-1816.

Eriksson, M., Gardestrom, P., Samuelsson, G. (1995). Isolation, purification, and characterization of mitochondria from *Chlamydomonas reinhardtii. Plant Physiol.* 107: 479-483.

Fabry, S., Muller, K., Lindauer, A., Park, P.B., Cornelius, T., Schmitt, R. (1995). The organization structure and regulatory elements of *Chlamydomonas* histone genes reveal features linking plant and animal genes. *Curr. Genet.* 28: 333-345.

Fan, J., Lee, R.W. (2002). Mitochondrial genome of the colorless green alga *Polytomella parva:* Two linear DNA molecules with homologous inverted repeat termini. *Mol. Biol. Evol.* 19: 999-1007.

Fernandez, E., Schnell, R., Ranum, L.P., Hussey, S.C., Silflow, C.D., Lefebvre, P.A. (1989). Isolation and characterization of the nitrate reductase structural gene of *Chlamydomonas reinhardtii. Proc. Natl. Acad. Sci. USA* 86: 6449-6453.

Ferris, P.J. (1989). Characterization of a *Chlamydomonas* transposon *Gulliver* resembling those in higher plants. *Genetics* 122: 363-378.

Fiehn, O. (2002). Metabolomics: The link between genotypes and phenotypes. *Plant Mol. Biol.* 48: 155-171.

Fischer, N., Rochaix, J.D. (2001). The flanking regions of *PsaD* drive efficient gene expression in the nucleus of the green alga *Chlamydomonas reinhardtii*. *Mol. Genet. Genomics* 265: 888-894.

Fleischmann, M.M., Ravanel, S., Delosme, R., Olive, J., Zito, F., Wollman, F.-A., Rochaix, J.-D. (1999). Isolation and characterization of photoautotrophic mutants of *Chlamydomonas reinhardtii* deficient in state transition. *J. Biol. Chem.* 274: 30987-30994.

Fuhrmann, M., Oertel, W., Hegemann, P. (1999). A synthetic gene coding for the green fluorescent protein (GFP) is a versatile reporter in *Chlamydomonas reinhardtii*. *Plant J.* 19: 353-361.

Fuhrmann, M., Stahlberg, A., Govorunova, E., Rank, S., Hegemann, P. (2001). The abundant retinal protein of the *Chlamydomonas* eye is not the photoreceptor for phototaxis and photophobic responses. *J. Cell. Sci.* 114: 3857-3863.

Fukuzawa, H., Miura, K., Ishizaki, K., Kucho, K.I., Saito, T., Kohinata, T., Ohyama, K. (2001). *Ccm1*, a regulatory gene controlling the induction of a carbon-concentrating mechanism in *Chlamydomonas reinhardtii* by sensing CO_2 availability. *Proc. Natl. Acad. Sci. U.S.A.* 98: 5347-5352.

Funke, R.P., Kovar, J.L., Weeks, D.P. (1997). Intracellular carbonic anhydrase is essential to photosynthesis in *Chlamydomonas reinhardtii* at atmospheric levels of CO_2: Demonstration via genomic complementation of the high-CO_2-requiring mutant *ca-1*. *Plant Physiol.* 114: 237-244.

Gewolb, J. (2002). Bioengineering: Plant scientists see big potential in tiny plastids. *Science* 295: 258-259.

Gillham, N.W. (1974). Genetic analysis of the chloroplast and mitochondrial genomes. *Annu. Rev. Genet.* 8: 347-391.

Gillham, N.W., Boynton, J.E., Harris, E.H. (1987). Specific elimination of mitochondrial DNA from *Chlamydomonas* by intercalating dyes. *Curr. Genet.* 12: 41-47.

Girke, T., Todd, J., Ruuska, S., White, J., Benning, C., Ohlrogge, J. (2000). Microarray analysis of developing *Arabidopsis* seeds. *Plant Physiol.* 124: 1570-1581.

Goodenough, U.W. (1992). Green yeast. *Cell* 70: 533-538.

Goossens, A., Hakkinen, S.T., Laakso, I., Seppanen-Laakso, T., Biondi, S., De Sutter, V., Lammertyn, F., Nuutila, A.M., Soderlund, H., Zabeau, M., et al.. (2003). A functional genomics approach toward the understanding of secondary metabolism in plant cells. *Proc. Natl. Acad. Sci. USA* 100: 8595-8600.

Graham, J.E., Spanier, J.G., Jarvik, J.W. (1995). Isolation and characterization of *Pioneer1*, a novel *Chlamydomonas* transposable element. *Curr. Genet.* 28: 429-436.

Gray, M.W., Boer, P.H. (1988). Organization and expression of algal *Chlamydomonas reinhardtii* mitochondrial DNA. *Philos. T. Roy. Soc. B* 319: 135-148.

Gray, M.W., Cedergren, R., Abel, Y., Sankoff, D. (1989). On the evolutionary origin of the plant mitochondrion and its genome. *Proc. Natl. Acad. Sci. USA* 86: 2267-2271.

Gross, C.H., Ranum, L.P.W., Lefebvre, P.A. (1988). Extensive restriction fragment length polymorphisms in a new isolate of *Chlamydomonas reinhardtii*. *Curr. Genet.* 13: 503-508.

Grossman, A. (2000a). Acclimation of *Chlamydomonas reinhardtii* to its nutrient environment. *Protist* 151: 201-224.

Grossman, A.R. (2000b). *Chlamydomonas reinhardtii* and photosynthesis: Genetics to genomics. *Curr. Opin. Plant Biol.* 3: 132-137.

Grossman, A.R., Harris, E., Hauser, C., Lefebvre, P., Martinez, D., Rokhsar, D., Shrager, J., Silflow, C., Stern, D., Vallon, O., Zhang, Z. (2003). *Chlamydomonas reinhardtii* at the crossroads of genomics. *Eukaryot. Cell* 2: 1137-1150.

Hahn, D., Kück, U. (1999). Identification of DNA sequences controlling light- and chloroplast-dependent expression of the *lhcb1* gene from *Chlamydomonas reinhardtii*. *Curr. Genet.* 34: 459-466.

Hall, R., Beale, M., Fiehn, O., Hardy, N., Sumner, L., Bino, R. (2002). Plant metabolomics: The missing link in functional genomics strategies. *Plant Cell* 14: 1437-1440.

Haring, M.A., Beck, C.F. (1997). A promoter trap for *Chlamydomonas reinhardtii*: Development of a gene cloning method using 5' RACE-based probes. *Plant J.* 11: 1341-1348.

Harris, E.H. (1989). *The Chlamydomonas Sourcebook: A Comprehensive Guide to Biology and Laboratory Use*. San Diego: Academic Press.

Harris, E.H. (2001). *Chlamydomonas* as a model organism. *Annu. Rev. Plant Phys.* 52: 363-406.

Heifetz, P.B. (2000). Genetic engineering of the chloroplast. *Biochimie* 82: 655-666.

Hicks, G.R., Hironaka, C.M., Dauvillee, D., Funke, R.P., D'Hulst, C., Waffenschmidt, S., Ball, S.G. (2001). When simpler is better. Unicellular green algae for discovering new genes and functions in carbohydrate metabolism. *Plant Physiol.* 127: 1334-1338.

Hippler, M., Klein, J., Fink, A., Allinger, T., Hoerth, P. (2001). Towards functional proteomics of membrane protein complexes: Analysis of thylakoid membranes from *Chlamydomonas reinhardtii*. *Plant J.* 28: 595-606.

Hoober, J.K., Park, H., Wolfe, G.R., Komine, Y., Eggink, L.L. (1998). Assembly of light-harvesting systems. In Rochaix, J.-D., Goldschmidt-Clermont, M., Merchant, S. (Eds.), *The Molecular Biology of Chloroplasts and Mitochondria in Chlamydomonas* (pp. 363-376). Dordrecht: Kluwer Academic Publishers.

Huang, K., Beck, C.F. (2003). Phototropin is the blue-light receptor that controls multiple steps in the sexual life cycle of the green alga *Chlamydomonas reinhardtii*. *Proc. Natl. Acad. Sci. USA* 100: 6269-6274.

Im, C.S., Grossman, A.R. (2002). Identification and regulation of high light-induced genes in *Chlamydomonas reinhardtii*. *Plant J.* 30: 301-313.

Im, C.S., Zhang, Z.D., Shrager, J., Chang, C.W., Grossman, A.R. (2003). Analysis of light and CO_2 regulation in *Chlamydomonas reinhardtii* using genome-wide approaches. *Photosynth. Res.* 75: 111-125.

Jackson, R.B., Linder, C.R., Lynch, M., Purugganan, M., Somerville, S., Thayer, S.S. (2002). Linking molecular insight and ecological research. *Trends Ecol. Evol.* 17: 409-414.

Kamath, R.S., Ahringer, J. (2003). Genome-wide RNAi screening in *Caenorhabditis elegans. Methods* 30: 313-321.

Kamath, R.S., Fraser, A.G., Dong, Y., Poulin, G., Durbin, R., Gotta, M., Kanapin, A., Le Bot, N., Moreno, S., Sohrmann, M., et al. (2003). Systematic functional analysis of the *Caenorhabditis elegans* genome using RNAi. *Nature* 421: 231-237.

Kamiya, R. (1988). Mutations at twelve independent loci result in absence of outer dynein arms in *Chlamydomonas reinhardtii. J. Cell. Biol.* 107: 2253-2258.

Kathir, P., LaVoie, M., Brazelton, W.J., Haas, N.A., Lefebvre, P.A., Silflow, C.D. (2003). Molecular map of the *Chlamydomonas reinhardtii* nuclear genome. *Eukaryot. Cell* 2: 362-379.

Kindle, K.L., Schnell, R.A., Fernandez, E., Lefebvre, P.A. (1989). Stable nuclear transformation of *Chlamydomonas* using the *Chlamydomonas* gene for nitrate reductase. *J. Cell. Biol.* 109: 2589-2601.

Kliebenstein, D.J., Gershenzon, J., Mitchell-Olds, T. (2001). Comparative quantitative trait loci mapping of aliphatic, indolic and benzylic glucosinolate production in *Arabidopsis thaliana* leaves and seeds. *Genetics* 159: 359-370.

Kliebenstein, D.J., Kroymann, J., Brown, P., Figuth, A., Pedersen, D., Gershenzon, J., Mitchell-Olds, T. (2001). Genetic control of natural variation in *Arabidopsis* glucosinolate accumulation. *Plant Physiol.* 126: 811-825.

Koblenz, B., Schoppmeier, J., Grunow, A., Lechtreck, K.F. (2003). Centrin deficiency in *Chlamydomonas* causes defects in basal body replication, segregation and maturation. *J. Cell Sci.* 116: 2635-2646.

Kovar, J.L., Zhang, J., Funke, R.P., Weeks, D.P. (2002). Molecular analysis of the acetolactate synthase gene of *Chlamydomonas reinhardtii* and development of a genetically engineered gene as a dominant selectable marker for genetic transformation. *Plant J.* 29: 109-117.

Kroymann, J., Zetsche, K. (1998). The mitochondrial genome of *Chlorogonium elongatum* inferred from the complete sequence. *J. Mol. Evol.* 47: 431-440.

Kulinski, J., Besack, D., Oleykowski, C.A., Godwin, A.K., Yeung, A.T. (2000). CEL I enzymatic mutation detection assay. *Biotechniques* 29: 44-48.

La Fontaine, S., Quinn, J.M., Nakamoto, S.S., Page, M.D., Gohre, V., Moseley, J.L., Kropat, J., Merchant, S. (2002). Copper-dependent iron assimilation pathway in the model photosynthetic eukaryote *Chlamydomonas reinhardtii. Eukaryot. Cell* 1: 736-757.

Lassner, M., Bedbrook, J. (2001). Directed molecular evolution in plant improvement. *Curr. Opin. Plant Biol.* 4: 152-156.

Leisinger, U., Rufenacht, K., Fischer, B., Pesaro, M., Spengler, A., Zehnder, A.J., Eggen, R.I. (2001). The glutathione peroxidase homologous gene from *Chlamydomonas reinhardtii* is transcriptionally up-regulated by singlet oxygen. *Plant Mol. Biol.* 46: 395-408.

Leister, D. (2003). Chloroplast research in the genomic age. *Trends Genet.* 19: 47-56.

Levine, R.P., Goodenough, U.W. (1970). The genetics of photosynthesis and of the chloroplast in *Chlamydomonas reinhardii*. *Annu. Rev. Genet.* 4: 397-408.

Li, J.B., Lin, S.P., Jia, H.G., Wu, H.M., Roe, B.A., Kulp, D., Stormo, G.D., Dutcher, S.K. (2003). Analysis of *Chlamydomonas reinhardtii* genome structure using large-scale sequencing of regions on linkage groups I and III. *J. Eukaryot. Microbiol.* 50: 145-155.

Lilly, J.W., Maul, J.E., Stern, D.B. (2002). The *Chlamydomonas reinhardtii* organellar genomes respond transcriptionally and posttranscriptionally to abiotic stimuli. *Plant Cell* 14: 2681-2706.

Loppes, R., Radoux, M. (2002). Two short regions of the promoter are essential for activation and repression of the nitrate reductase gene in *Chlamydomonas reinhardtii*. *Mol. Genet. Genomics* 268: 42-48.

Lown, F.J., Watson, A.T., Purton, S. (2001). *Chlamydomonas* nuclear mutants that fail to assemble respiratory or photosynthetic electron transfer complexes. *Biochem. Soc. Trans.* 29: 452-455.

Lumbreras, V., Stevens, D.R., Purton, S. (1998). Efficient foreign gene expression in *Chlamydomonas reinhardtii* mediated by an endogenous intron. *Plant J.* 14: 441-447.

Mackay, T.F. (2001). The genetic architecture of quantitative traits. *Annu. Rev. Genet.* 35: 303-339.

Majeran, W., Wollman, F.A., Vallon, O. (2000). Evidence for a role of ClpP in the degradation of the chloroplast cytochrome b_6f complex. *Plant Cell* 12: 137-150.

Matagne, R.F., Michel-Wolwertz, M.R., Munaut, C., Duyckaerts, C., Sluse, F. (1989). Induction and characterization of mitochondrial DNA mutants in *Chlamydomonas reinhardtii*. *J. Cell. Biol.* 108: 1221-1226.

Matagne, R.F., Rongvaux, D., Loppes, R. (1988). Transmission of mitochondrial DNA in crosses involving diploid gametes homozygous or heterozygous for the mating-type locus in *Chlamydomonas*. *Mol. Gen. Genet.* 214: 257-262.

Maucourt, K., Agarwal, M., Rene, B., Fermandjian, S. (2002). Use of *Chlamydomonas reinhardtii* mutants for anticancer drug screening. *Biochem. Pharmacol.* 64: 1125-1131.

Maul, J.E., Lilly, J.W., Cui, L., dePamphilis, C.W., Miller, W., Harris, E.H., Stern, D.B. (2002). The *Chlamydomonas reinhardtii* plastid chromosome: Islands of genes in a sea of repeats. *Plant Cell* 14: 2659-2679.

Mayer, K., Mewes, H.-W. (2001). How can we deliver the large plant genomes? Strategies and perspectives. *Curr. Opin. Plant Biol.* 5: 173-177.

Mayfield, S.P., Franklin, S.E., Lerner, R.A. (2003). Expression and assembly of a fully active antibody in algae. *Proc. Natl. Acad. Sci. USA* 100: 438-442.

McCallum, C.M., Comai, L., Greene, E.A., Henikoff, S. (2000). Targeting Induced Local Lesions IN Genomes (TILLING) for plant functional genomics. *Plant Physiol.* 123: 439-442.

McElver, J., Tzafrir, I., Aux, G., Rogers, R., Ashby, C., Smith, K., Thomas, C., Schetter, A., Zhou, Q., Cushman M.A., et al. (2001). Insertional mutagenesis of genes required for seed development in *Arabidopsis thaliana*. *Genetics* 159: 1751-1763.

Merchan, F., van den Ende, H., Fernandez, E., Beck, C.F. (2001). Low-expression genes induced by nitrogen starvation and subsequent sexual differentiation in *Chlamydomonas reinhardtii,* isolated by the differential display technique. *Planta* 213: 309-317.

Mets, L.J., Geist, L.J. (1983). Linkage of a known chloroplast gene mutation to the uniparental genome of *Chlamydomonas. Genetics* 105: 559-579.

Michaelis, G., Vahrenholz, C., Pratje, E. (1990). Mitochondrial DNA of *Chlamydomonas reinhardtii:* The gene for apocytochrome *b* and the complete functional map of the 15.8 kb DNA. *Mol. Gen. Genet.* 223: 211-216.

Millar, A.H., Heazlewood, J.L. (2003). Genomic and proteomic analysis of mitochondrial carrier proteins in *Arabidopsis. Plant Physiol.* 131: 443-453.

Millar, A.H., Sweetlove, L.J., Giege, P., Leaver, C.J. (2001). Analysis of the *Arabidopsis* mitochondrial proteome. *Plant Physiol.* 127: 1711-1727.

Mitchell, D.R., Kang, Y. (1993). Reversion analysis of dynein intermediate chain function. *J. Cell. Sci.* 105: 1069-1078.

Mitchell, D.R., Rosenbaum, J.L. (1985). A motile *Chlamydomonas* flagellar mutant that lacks outer dynein arms. *J. Cell. Biol.* 100: 1228-1234.

Mitchell-Olds, T. (2001). *Arabidopsis thaliana* and its wild relatives: A model system for ecology and evolution. *Trends Ecol. Evol.* 16: 693-700.

Moseley, J.L., Allinger, T., Herzog, S., Hoerth, P., Wehinger, E., Merchant, S., Hippler, M. (2002). Adaptation to Fe-deficiency requires remodeling of the photosynthetic apparatus. *EMBO J.* 21: 6709-6720.

Moseley, J., Quinn, J., Eriksson, M., Merchant, S. (2000). The *Crd1* gene encodes a putative di-iron enzyme required for photosystem I accumulation in copper deficiency and hypoxia in *Chlamydomonas reinhardtii. EMBO J.* 19: 2139-2151.

Nedelcu, A.M. (1997). Fragmented and scrambled mitochondrial ribosomal RNA coding regions among green algae: A model for their origin and evolution. *Mol. Biol. Evol.* 14: 506-517.

Nedelcu, A.M. (1998). Contrasting mitochondrial genome organizations and sequence affiliations among green algae: Potential factors, mechanisms, and evolutionary scenarios. *J. Phycol.* 34: 16-28.

Nedelcu, A.M., Lee, R.W. (1998a). A degenerate group II intron in the intronless mitochondrial genome of *Chlamydomonas reinhardtii:* Evolutionary implications. *Mol. Biol. Evol.* 15: 918-922.

Nedelcu, A.M., Lee, R.W. (1998b). Modes and tempos of mitochondrial and chloroplast genome evolution in *Chlamydomonas:* A comparative analysis. In Rochaix, J.-D., Goldschmidt-Clermont, M., Merchant, S. (Eds.), *The Molecular Biology of Chloroplasts and Mitochondria in Chlamydomonas* (pp. 63-91). Dordrecht: Kluwer Academic Press.

Nedelcu, A.M., Lee, R.W. (1998c). Short repetitive sequences in green algal mitochondrial genomes: Potential roles in mitochondrial genome evolution. *Mol. Biol. Evol.* 15: 690-701.

Nedelcu, A.M., Lee, R.W., Lemieux, C., Gray, M.W., Burger, G. (2000). The complete mitochondrial DNA sequence of *Scenedesmus obliquus* reflects an intermediate stage in the evolution of the green algal mitochondrial genome. *Genome Res.* 10: 819-831.

Nikaido, I., Asamizu, E., Nakajima, M., Nakamura, Y., Saga, N., Tabata, S. (2000). Generation of 10,154 expressed sequence tags from a leafy gametophyte of a marine red alga, *Porphyra yezoensis*. *DNA Res.* 7: 223-227.

Niyogi, K.K., Björkman, O., Grossman, A.R. (1997). *Chlamydomonas* xanthophyll cycle mutants identified by video imaging of chlorophyll fluorescence quenching. *Plant Cell* 9: 1369-1380.

Ogihara, Y., Mochida, K., Nemoto, Y., Murai, K., Yamazaki, Y., Shin-i, T., Kohara, Y. (2003). Correlated clustering and virtual display of gene expression patterns in the wheat life cycle by large-scale statistical analyses of expressed sequence tags. *Plant J.* 33: 1001-1011.

Ohresser, M., Matagne, R.F., Loppes, R. (1997). Expression of the arylsulphatase reporter gene under the control of the *nit1* promoter in *Chlamydomonas reinhardtii*. *Curr. Genet.* 31: 264-271.

Oleykowski, C.A., Mullins, C.R.B., Godwin, A.K., Yeung, A.T. (1998). Mutation detection using a novel plant endonuclease. *Nucleic Acids Res.* 26: 4597-4602.

Pan, J., Misamore, M.J., Wang, Q., Snell, W.J. (2003). Protein transport and signal transduction during fertilization in *Chlamydomonas*. *Traffic* 4: 452-459.

Pasquali, C., Fialka, I., Huber, L.A. (1997). Preparative two-dimensional gel electrophoresis of membrane proteins. *Electrophoresis* 18: 2573-2581.

Pazour, G.J., Witman, G.B. (2000). Forward and reverse genetic analysis of microtubule motors in *Chlamydomonas*. *Methods* 22: 285-298.

Peltier, J.B., Friso, G., Kalume, D.E., Roepstorff, P., Nilsson, F., Adamska, I., van Wijk, K.J. (2000). Proteomics of the chloroplast: Systematic identification and targeting analysis of lumenal and peripheral thylakoid proteins. *Plant Cell* 12: 319-341.

Perez-Martinez, X., Vazquez-Acevedo, M., Tolkunova, E., Funes, S., Claros, M.G., Davidson, E., King, M.P., Gonzalez-Halphen, D. (2000). Unusual location of a mitochondrial gene: Subunit III of cytochrome *c* oxidase is encoded in the nucleus of Chlamydomonad algae. *J. Biol. Chem.* 275: 30144-30152.

Phelps, T.J., Palumbo, A.V., Beliaev, A.S. (2002). Metabolomics and microarrays for improved understanding of phenotypic characteristics controlled by both genomics and environmental constraints. *Curr. Opin. Biotechnol.* 13: 20-24.

Pothof, J., van Haaften, G., Thijssen, K., Kamath, R.S., Fraser, A.G., Ahringer, J., Plasterk, R.H., Tijsterman, M. (2003). Identification of genes that protect the *C. elegans* genome against mutations by genome-wide RNAi. *Genes Dev.* 17: 443-448.

Quinn, J.M., Merchant, S. (1995). Two copper-responsive elements associated with the *Chlamydomonas Cyc6* gene function as targets for transcriptional activators. *Plant Cell* 7: 623-628.

Raamsdonk, L.M., Teusink, B., Broadhurst, D., Zhang, N., Hayes, A., Walsh, M.C., Berden, J.A., Brindle, K.M., Kell, D.B., Rowland, J.J., et al. (2001). A functional genomics strategy that uses metabolome data to reveal the phenotype of silent mutations. *Nat. Biotechnol.* 19: 45-50.

Randolph-Anderson, B.L., Boynton, J.E., Gillham, N.W., Harris, E.H., Johnson, A.M., Dorthu, M.P., Matagne, R.F. (1993). Further characterization of the respiratory deficient *dum-1* mutation of *Chlamydomonas reinhardtii* and its use as a recipient for mitochondrial transformation. *Mol. Gen. Genet.* 236: 235-244.

Randolph-Anderson, B.L., Gillham, N.W., Boynton, J.E. (1989). Electrophoretic and immunological comparisons of chloroplast and prokaryotic ribosomal proteins reveal that certain families of large subunit proteins are evolutionarily conserved. *J. Mol. Evol.* 29: 68-88.

Ranum, L.P.W., Thompson, M.D., Schloss, J.A., Lefebvre, P.A., Silflow, C.D. (1988). Mapping flagellar genes in *Chlamydomonas* using restriction fragment length polymorphisms. *Genetics* 120: 109-122.

Remacle, C., Colin, M., Matagne, R.F. (1995). Genetic mapping of mitochondrial markers by recombinational analysis in *Chlamydomonas reinhardtii. Mol. Gen. Genet.* 249: 185-190.

Remacle, C., Matagne, R.F. (1998). Mitochondrial genetics. In Rochaix, J.-D., Goldschmidt-Clermont, M., Merchant, S. (Eds.), Molecular Biology of Chloroplasts and Mitochondria in *Chlamydomonas* (pp. 661-674). Dordrecht: Kluwer Academic Publishers.

Rensing, S.A., Rombauts, S., Van de Peer, Y., Reski, R. (2002). Moss transcriptome and beyond. *Trends Plant Sci.* 7: 535-538.

Ris, H., Plaut, W. (1962). Ultrastructure of DNA-containing areas in the chloroplast of *Chlamydomonas. J. Cell Biol.* 13: 383-391.

Rochaix, J.-D. (2002). *Chlamydomonas,* a model system for studying the assembly and dynamics of photosynthetic complexes. *FEBS Lett.* 529: 34-38.

Rubinelli, P., Siripornadulsil, S., Gao-Rubinelli, F., Sayre, R.T. (2002). Cadmium- and iron-stress-inducible gene expression in the green alga *Chlamydomonas reinhardtii:* Evidence for H43 protein function in iron assimilation. *Planta* 215: 1-13.

Sager, R. (1954). Mendelian and non-Mendelian inheritance of streptomycin resistance in *Chlamydomonas reinhardtii. Proc. Natl. Acad. Sci. USA* 40: 356-363.

Sager, R., Ramanis, Z. (1968). The pattern of segregation of cytoplasmic genes in *Chlamydomonas. Proc. Natl. Acad. Sci. USA* 61: 324-331.

Santoni, V., Molloy, M., Rabilloud, T. (2000). Membrane proteins and proteomics: Un amour impossible? *Electrophoresis* 21: 1054-1070.

Scala, S., Carels, N., Falciatore, A., Chiusano, M.L., Bowler, C. (2002). Genome properties of the diatom *Phaeodactylum tricornutum. Plant Physiol.* 129: 993-1002.

Schmidt, R.J., Myers, A.M., Gillham, N.W., Boynton, J.E. (1984). Immunological similarities between specific chloroplast ribosomal proteins from *Chlamydomonas reindardtii* and ribosomal proteins from *Escherichia coli. Mol. Biol. Evol.* 1: 317-334.

Schmidt, R.J., Richardson, C.B., Gillham, N.W., Boynton, J.E. (1983). Sites of synthesis of chloroplast ribosomal proteins in *Chlamydomonas reinhardtii. J. Cell. Biol.* 96: 1451-1463.

Schnell, R.A., Lefebvre, P.A. (1993). Isolation of the *Chlamydomonas* regulatory gene *Nit2* by transposon tagging. *Genetics* 134: 737-747.

Schroda, M., Blöcker, D., Beck, C.F. (2000). The HSP70A promoter as a tool for the improved expression of transgenes in *Chlamydomonas. Plant J.* 21: 121-131.

Sears, B.B. (1998). Replication, recombination, and repair in the chloroplast genetic system of *Chlamydomonas*. In Rochaix, J.-D., Goldschmidt-Clermont, M., Merchant, S. (Eds.), *The Molecular Biology of Chloroplasts and Mitochondria in Chlamydomonas* (pp. 115-138). Dordrecht: Kluwer Academic Publishers.

Sessions, A., Burke, E., Presting, G., Aux, G., McElver, J., Patton, D., Dietrich, B., Ho, P., Bacwaden, J., Ko, C., et al. (2002). A high-throughput *Arabidopsis* reverse genetics system. *Plant Cell* 14: 2985-2994.

Shimogawara, K., Fujiwara, S., Grossman, A., Usuda, H. (1998). High-efficiency transformation of *Chlamydomonas reinhardtii* by electroporation. *Genetics* 148: 1821-1828.

Shrager, J., Hauser, C., Chang, C.W., Harris, E.H., Davies, J., McDermott, J., Tamse, R., Zhang, Z.D., Grossman, A.R. (2003). *Chlamydomonas reinhardtii* genome project: A guide to the generation and use of the cDNA information. *Plant Physiol.* 131: 401-408.

Silfow, C.D. (1998). Organization of the nuclear genome. In Rochaix, J.-D., Goldschmidt-Clermont M., Merchant, S. (Eds.), *The Molecular Biology of Chloroplasts and Mitochondria in Chlamydomonas* (pp. 25-40). Dordrecht: Kluwer Academic Publishers.

Simpson, C.L., Stern, D.B. (2002). The treasure trove of algal chloroplast genomes. Surprises in architecture and gene content, and their functional implications. *Plant Physiol.* 129: 957-966.

Sineshchekov, O.A., Jung, K.H., Spudich, J.L. (2002). Two rhodopsins mediate phototaxis to low- and high-intensity light in *Chlamydomonas reinhardtii*. *Proc. Natl. Acad. Sci. USA* 99: 8689-8694.

Smart, E.J., Selman, B.R. (1991). Isolation and characterization of a *Chlamydomonas reinhardtii* mutant lacking the gamma-subunit of chloroplast coupling factor 1 (CF1). *Mol. Cell. Biol.* 11: 5053-5058.

Sodeinde, O.A., Kindle, K.L. (1993). Homologous recombination in the nuclear genome of *Chlamydomonas reinhardtii*. *Proc. Natl. Acad. Sci. USA* 90: 9199-9203.

Somerville, C., Somerville, S. (1999). Plant functional genomics. *Science* 285: 380-383.

Spalding, M.H. (1998). CO_2 acquisition: Acclimation to changing carbon availability. In Rochaix, J.-D., Goldschmidt-Clermont, M., Merchant, S. (Eds.), *The Molecular Biology of Chloroplasts and Mitochondria in Chloroplasts* (pp. 529-547). Dordrecht: Kluwer Academic Publishers.

Staub, J.M., Garcia, B., Graves, J., Hajdukiewicz, P.T., Hunter, P., Nehra, N., Paradkar, V., Schlittler, M., Carroll, J.A., Spatola, L., et al. (2000). High-yield production of a human therapeutic protein in tobacco chloroplasts. *Nat. Biotechnol.* 18: 333-338.

Stein, L.D., Mungall, C., Shu, S., Caudy, M., Mangone, M., Day, A., Nickerson, E., Stajich, J.E., Harris, T.W., Arva, A., Lewis, S. (2002). The Generic Genome Browser: A building block for a model organism system database. *Genome Res.* 12: 1599-1610.

Stevens, D.R., Rochaix, J.D., Purton, S. (1996). The bacterial phleomycin resistance gene *ble* as a dominant selectable marker in *Chlamydomonas. Mol. Gen. Genet.* 251: 23-30.

Strauss, C., Mussgnug, J.H., Kruse, O. (2001). Ligation-mediated suppression-PCR as a powerful tool to analyse nuclear gene sequences in the green alga *Chlamydomonas reinhardtii. Photosynth. Res.* 70: 311-320.

Sugase, Y., Hirono, M., Kindle, K.L., Kamiya, R. (1996). Cloning and characterization of the actin-encoding gene of *Chlamydomonas reinhardtii. Gene* 168: 117-121.

Sussman, M.R., Amasino, R., Krysan, P. (2002). The endgame for reverse genetics. Isolation and distribution of a knockout mutant for every gene in *Arabidopsis. Plant Physiol.* 129; available at <http //www plantphysiol org/>.

Tam, L.W., Lefebvre, P.A. (1993). Cloning of flagellar genes in *Chlamydomonas reinhardtii* by DNA insertional mutagenesis. *Genetics* 135: 375-384.

Tanaka, A., Ito, H., Tanaka, R., Tanaka, N.K., Yoshida, K., Okada, K. (1998). Chlorophyll *a* oxygenase *(CAO)* is involved in chlorophyll *b* formation from chlorophyll *a. Proc. Natl. Acad. Sci. USA* 95: 12719-12723.

Taylor, S.W., Fahy, E., Ghosh, S.S. (2003). Global organellar proteomics. *Trends Biotechnol.* 21: 82-88.

Teramoto, H., Ono, T., Minagawa, J. (2001). Identification of *Lhcb* gene family encoding the light-harvesting chlorophyll-a/b proteins of photosystem II in *Chlamydomonas reinhardtii. Plant Cell Physiol.* 42: 849-856.

Van Lis, R., Atteia, A., Mendoza-Hernandez, G., Gonzalez-Halphen, D. (2003). Identification of novel mitochondrial protein components of *Chlamydomonas reinhardtii: A proteomic approach. Plant Physiol.* 132: 318-330.

Wang, S.-C., Schnell, R.A., Lefebvre, P.A. (1998). Isolation and characterization of a new transposable element in *Chlamydomonas reinhardtii. Plant Mol. Biol.* 38: 681-687.

Werner, R. (2002). *Chlamydomonas reinhardtii* as a unicellular model for circadian rhythm analysis. *Chronobiol. Int.* 19: 325-343.

Wesley, S.V., Helliwell, C.A., Smith, N.A., Wang, M.B., Rouse, D.T., Liu, Q., Gooding, P.S., Singh, S.P., Abbott, D., Stoutjesdijk, P.A., et al. (2001). Construct design for efficient, effective and high-throughput gene silencing in plants. *Plant J.* 27: 581-590.

Wiseman, A., Gillham, N.W., Boynton, J.E. (1977). Nuclear mutations affecting mitochondrial structure and function in *Chlamydomonas. J. Cell. Biol.* 73: 56-77.

Xiang, Y., Zhang, J., Weeks, D.P. (2001). The *Cia5* gene controls formation of the carbon concentrating mechanism in *Chlamydomonas reinhardtii. Proc. Natl. Acad. Sci. USA* 98: 5341-5346.

Xuemie, C., Kindle, K., Stern, D. (1993). Initiation codon mutations in the *Chlamydomonas* chloroplast *petD* gene result in temperature-sensitive photosynthetic growth. *EMBO J.* 12: 3627-3635.

Yamaguchi, K., Prieto, S., Beligni, M.V., Haynes, P.A., McDonald, W.H., Yates III, John R., Mayfield, S.P. (2002). Proteomic characterization of the small subunit of *Chlamydomonas reinhardtii* chloroplast ribosome: Identification of a novel S1 domain-containing protein and unusually large orthologs of bacterial S2, S3, and S5. *Plant Cell* 14: 2954-2974.

Yang, B., Wen, X., Kodali, N.S., Oleykowski, C.A., Miller, C.G., Kulinski, J., Besack, D., Yeung, J.A., Kowalski, D., Yeung, A.T. (2000). Purification, cloning, and characterization of the CEL I nuclease. *Biochemistry* 39: 3533-3541.

Zhang, H., Herman, P.L., Weeks, D.P. (1994). Gene isolation through genomic complementation using an indexed library of *Chlamydomonas reinhardtii* DNA. *Plant Mol. Biol.* 24: 663-672.

Chapter 9

Functional Genomics in *Physcomitrella*

Wolfgang Frank
Hauke Holtorf
Ralf Reski

INTRODUCTION

The objective to elucidate the mode of action of all plant genes is an overwhelming task requiring the development of new and the revision of existing technologies. The multifaceted initiatives and strategies to attain this goal created the neologism *plant functional genomics*. As for many other fundamental aspects in plant biology, *Arabidopsis thaliana* has been the precursor for the progression of enabling technologies in the functional genomics research area. The successful completion of the *Arabidopsis* genome-sequencing initiative by the end of the year 2000 (Arabidopsis Genome Initiative, 2000) has been another milestone in understanding the complexity of a plant organism. Recently, the draft genomic sequences of two rice subspecies (Goff et al., 2002; Yu et al., 2002) and high-quality sequences of single rice chromosomes (Feng et al., 2002; Sasaki et al., 2002) have been published that will allow the first comparison of the genome structure of a monocotyledonous and a dicotyledonous species. The rice genomic sequences will also trigger new functional genomics efforts in the major graminaceous crop plants.

The information obtained by the sequencing projects, however, raises the question, What is the most feasible way to assign functions to all the predicted genes? Many independent research efforts have been started following different strategies and utilizing a diverse set of technical tools in *Arabidopsis*. The classical forward genetics approach is based on the phenotypic screening of plants that have been mutagenized by chemical (e.g., ethyl methane sulfonate) or physical (X-ray) treatments. The bottleneck using this strategy is the laborious positional cloning of the mutated gene locus once an interesting phenotype has been identified, even though dense genetic and physical maps in *Arabidopsis* are available (Camilleri et al., 1998; Chang et al., 2001; Lukowitz et al., 2000; Peters et al., 2001). To facilitate a large-scale functional genomics analysis, foreign DNA is introduced into the plant to disrupt genes by way of random integration. The DNA that is used is either a piece of the *Agrobacterium*-derived T-DNA or a transposon (Azpiroz-Leehan and Feldmann, 1997; Bouche and Bouchez, 2001; Krysan et al., 1999; Martienssen, 1998; Parinov and Sundaresan, 2000; Pereira, 2000; Walbot, 2000). The integrated DNA serves as an anchor and allows the identification of its flanking regions by PCR-based methods. The generated disruption lines also enable large reverse genetics screens to assess gene function. Currently, databases are set up that contain sequence information of DNA stretches flanking the insertion sites, enabling the easy identification of a specific plant line carrying a mutation in the gene of interest (Parinov et al., 1999; Sessions et al., 2002; Tissier et al., 1999) (see also Chapter 2). Furthermore, the techniques of gene, promoter, and enhancer trap approaches represent additional tools to gain information about gene function (Campisi et al., 1999; Springer, 2000). Using these reporter gene-based methods, the understanding of the regulation and involvement of certain genes in developmental processes and/or environmental adaptations will be increased. Another T-DNA-based approach uses strong promoter and enhancer elements to create dominant mutations in the plant's genome. This activation tagging method describes a technology that is also amenable to polyploid plant species whose genetic analysis is otherwise limited (Weigel et al., 2000).

Even though pioneering achievements have been obtained by using these technologies, there are some limitations interfering with the comprehensive characterization of all *Arabidopsis* genes. A targeted knockout of a particular gene of interest in a flowering plant is almost impossible because homologous recombination occurs only at marginal frequencies (Mengiste and Paszkowski, 1999; Puchta and Hohn, 1996; Reski, 1998a). In consideration of this feature, the functional analysis of particular genes has been performed either by overexpression or antisense RNA suppression. Additional methods have been developed based on silencing-related phenomena such as cosuppression (Meins, 2000), virus-induced gene silencing (Baulcombe, 1999), double-stranded RNA-mediated interference (Chuang and Meyerowitz, 2000), and chimeraplasty using chimeric RNA/DNA oligonucleotides (Zhu et al., 2000). Another question is, If we unravel the function of all *Arabidopsis* genes, do we understand plant biology? Certainly not, because other plant species have evolved their own gene pool to meet particular demands and these gene pools may be valuable sources to identify new gene functions. Therefore, it is of special interest to launch functional genomics initiatives in species other than *Arabidopsis*. One such species, which has become a versatile alternative to implement plant functional genomics, is the nonvascular, multicellular land plant *Physcomitrella patens,* a representative of the bryophytes (Schaefer, 2001; Schaefer, 2002; Reski, 1998b, 1999). *Physcomitrella* originally had been promoted to study plant differentiation processes facilitated by the haploid status of the moss protonema. Differentiation events in the moss development can be ascribed to single cells enabling the examination of the acquisition of cellular polarity in plants (Bhatla et al., 2002). Besides the use of *Physcomitrella* to investigate basic aspects of plant development, a set of molecular tools was developed to make the moss accessible to state-of-the-art molecular and genetic analyses (Cho et al., 1999; Kranz et al., 2000; Schaefer et al., 1991; Reski et al., 1994, 1998). The use of *Physcomitrella* as a tool for functional genomics circumvents many drawbacks arising from approaches used to study other plant species.

Two major characteristics are the decisive factors why *Physcomitrella* may be advantageous compared to other model species. One feature of the mosses is their heteromorphic *Generationswechsel,* the alternation of two generations that are distinct from each other in terms of nuclear DNA amounts and morphology (Figure 9.1). In contrast to higher plants, the life cycle comprises a predominant haploid gametophytic phase and a short diploid sporophytic phase. The haploid phase provides the most prevalent growth form, which has distinct advantages in the screening for mutant phenotypes. Starting from a germinating haploid spore, the gametophytic phase is initiated by the growth of a filamentous germ tube. The germ tube grows out to form the protonema, which is composed of irregular branched filaments that are subclassified into chloronema and caulonema cells. Chloronema cells contain large numbers of chloroplasts and cross walls that are perpendicular to the growth axis, whereas caulonema cells are characterized by a reduced number of plastids and oblique cross walls with respect to the growth axis. Further development proceeds by the formation of buds that are initially composed of a three-faced apical cell. The bud forms the initial meristem for the development of the erect moss plant and marks the transition of the filamentous juvenile to the leafy adult gametophyte. By further cell division, leaf primordia are formed that give rise to leaves usually consisting of one cell layer. The next developmental phase results in the formation of sex organs. As the sex organs (gametangia) emerge from the adult gametophyte, they are also termed gametophores. Monoecious moss species (e.g., *Physcomitrella*) bear both the male (antheridia) and female (archegonia) sex organs on one plant. Male gametes (spermatozoids) are produced within antheridia, and female gametes (oogonia or egg cells) are produced in archegonia. The fertilization is achieved by the swimming of spermatozoids through a surface water film down to the neck of the archegonia, which normally contains one egg cell. The zygote develops into

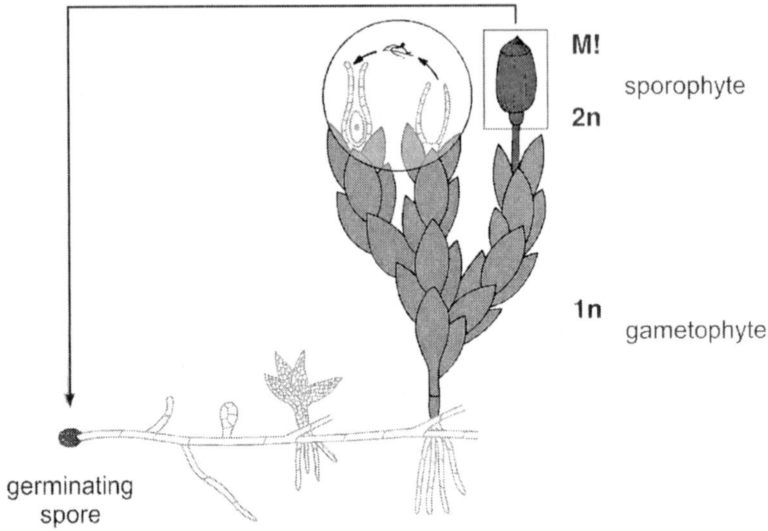

FIGURE 9.1. The heteromorphic *Generationswechsel* of mosses. A haploid spore germinates and grows into protonema existing of chloronema and caulonema cells. Starting with a three-faced apical cell, the bud formation is initiated and gives rise to the leafy adult gametophyte. In monoecious moss species both sex organs (antheridia and archegonia) are present on one and the same plant. Fertilization of the egg cells takes place in the presence of water. From the fertilized egg the sporophyte grows out of the archegonia. The diploid sporophyte is highlighted by the surrounding rectangle. Within the spore capsule the cells undergo meiosis (M!) and new spores are formed.

an embryo that grows out to the diploid sporophyte. The two moss generations are physically connected because the sporophyte grows on top of the gametophyte. Within the spore capsule (sporogonium), the diploid cells undergo meiosis and produce a large number of haploid spores, resulting in the completion of the life cycle. The spores are released from the sporophyte by the disruption of the spore capsule.

The dominant haploid gametophytic phase is a favorable characteristic that facilitates forward genetics screens for altered phenotypes because the disruption of a single gene cannot be counterbalanced by a second allele. Reverse genetics or forward genetics screens of *Arabidopsis* T-DNA or transposon mutagenized populations have identified a large number of gene knockout lines, but only a few of them show informative phenotypes from which conclusions regarding the gene's function can be drawn (Bouché and Bouchez, 2001). Gene disruption in the moss *Physcomitrella* may overcome these difficulties because loss-of-function mutants of a single gene locus will have a stronger impact on the phenotypes in the haploid gametophytic stage. This hypothesis can be confirmed by the fact that gene families in *Physcomitrella* include fewer members compared to *Arabidopsis* (Rensing, Rombauts, Van der Peer, et al., 2002). Thus, problems arising from functional gene redundancy that interfere with large genetic screens may be reduced. A second striking, unique feature among plants is *Physcomitrella*'s ability to integrate transforming DNA at high frequency by way of homologous recombination (Schaefer and Zryd, 1997). The homologous recombination rates in moss are several orders of magnitude higher than in seed plants (Reski, 1999, 1998b), making *Physcomitrella* the plant of choice to study gene function by the precise disruption of single genes. Experiments performed by several independent groups have shown that a targeted disruption of a genomic locus in *Physcomitrella* correlates with a mutant phenotype that reveals the biological function of the disrupted gene (Girke et al., 1998; Girod et al., 1999; Strepp et al., 1998). Imaizumi et al. (2002) have used an elegant approach to study the function of plant cryptochromes. To overcome the problems that arise from gene redundancy, they have generated double disruptants of two redundant cryptochrome genes present in *Physcomitrella.* Besides the approaches utilizing single gene disruptions, new strategies have been developed to exploit the high rate of homologous recombination for large-scale reverse genetic approaches that will complement functional analysis in other plant species (Reski, 1998b; Schaefer, 2001). Large-scale functional genomics approaches have already been initiated in *Physcomitrella* using shuttle mutagenesis of genomic libraries (Nishiyama et al., 2000). The potential of this technique has been demonstrated in yeast, where the mutagenesis of the genome led to the most extensive collection of defined and targeted yeast mutants to date (Ross-MacDonald et al., 1999). The experiments performed in *Physcomitrella* resulted in 5,264 tagged lines, out of which 203 mutants showed altered developmental or morphological phenotypes. The same group successfully initiated a second approach using gene-trap and enhancer-trap systems in *Physcomitrella* (Hiwatashi et al., 2001). Through the high rate of homologous recombination, 235 gene-trap and 1,073 enhancer-trap lines were obtained from 5,637 and 3,726 transgenic lines, respectively. Furthermore, the genomic locus of a gene-trap line displaying rhizoid-specific expression of the *uidA* reporter gene was identified.

In collaboration with our industrial partner, BASF Plant Science GmbH, we have started to use *Physcomitrella* to understand gene function on a genomic scale. The basis of the project was the generation of normalized cDNA libraries derived from different developmental stages and differentially treated tissues of *Physcomitrella*. The libraries have been used for mass sequencing, generating 110,000 EST sequences. Bioinformatic analyses of the sequence data revealed that the *Physcomitrella* transcriptome contains around 25,000 genes, among which 11,000 represent novel plant genes (Rensing, Rombauts, Van der Peer, et al., 2002). The cDNA libraries are used as a source to disrupt the coding regions of the genes via insertional transposon mutagenesis. The transposed DNA fragment encompasses the *nptII* gene (neomycin phosphotransferase), conferring resistance to kanamycin and its derivative G418. The resulting pool of mutagenized cDNAs is used to transform *Physcomitrella* protoplasts to generate a large population of transgenic *Physcomitrella* lines. We expect that the disrupted cDNAs will preferably integrate into the genome by homologous recombination, thereby eliminating the function of the wild-type genomic locus. The aim is to transform *Physcomitrella* with the entire complement of its genes to obtain a highly saturated mutant population. A standardized phenotypic and metabolic screening of the transformants will allow the identification of interesting gene knockouts (Egener et al., 2002). In addition to the large-scale mutagenesis, we also perform targeted knockouts of single genes of interest that are derived from the EST database. In these cases, the gene disruption constructs are not generated by the transposon mutagenesis, but they are obtained by direct cloning of the *nptII* selectable marker cassette into suitable restriction sites present in the cDNA region. Within this chapter, we will describe the joint project with our industrial partner BASF Plant Science GmbH to combine a randomized mutagenesis with the targeted gene disruption to saturate the *Physcomitrella* genome with loss-of-function mutants of any transcribed gene. To realize this aim, we have developed several technical platforms to comply with the requirements to perform high-throughput production of a mutagenized *Physcomitrella* collection. The essential steps that are involved in the large-scale mutagenesis of *Physcomitrella* are summarized in Figure 9.2 and will be discussed in detail in the following sections of this chapter.

MASS PRODUCTION OF STANDARDIZED MOSS MATERIAL

The setup of a plant functional genomics project implies the development of a high-throughput transformation platform. The established protocols for the transformation of *Physcomitrella* are based on a polyethylene glycol-mediated DNA transfer into protoplasts (Reutter et al., 1998; Schaefer et al., 1991). The first protocol on the preparation and regeneration of protoplasts from *Physcomitrella* has been described by Stumm et al. (1975). An improved protocol using the enzyme mixture Driselase for cell wall digestion of *Physcomitrella* protonema has been developed by

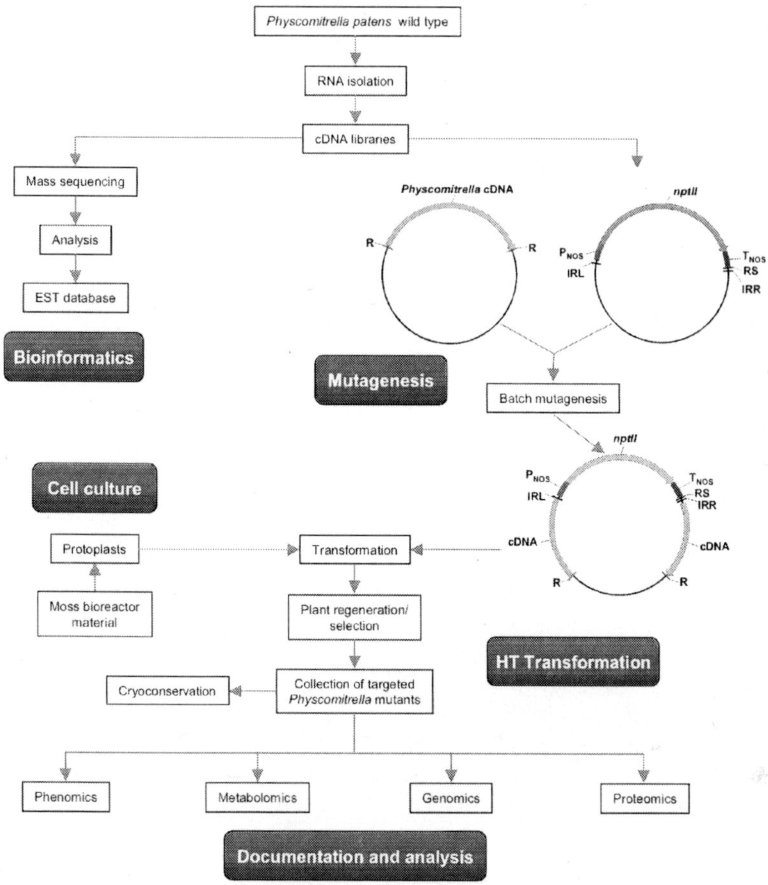

FIGURE 9.2. Schematic overview of the large-scale mutagenesis approach in *Physcomitrella*. The integration of different fields of plant biology, molecular biology, and bioinformatics as well as the development of new technologies were necessary to realize a high-throughput functional genomics project in *Physcomitrella*.

Grimsley et al. (1977). The latter method uses *Physcomitrella* plants grown on a solidified medium in petri dishes supplemented with ammonium tartrate, causing the moss protonema to grow predominantly in the chloronema stage (Ashton and Cove, 1977; Jenkins and Cove, 1983). Protonema tissue from *Physcomitrella* comprising chloronema and young caulonema cells is thought to be the best source for the preparation of protoplasts, whereas tissues from advanced developmental stages are not sufficiently susceptible to cell-wall degrading enzymes (Lal, 1984). Even though these protocols allow the preparation of reasonable amounts of protoplasts from *Physcomitrella,* they are not suited for mass production of moss protoplasts. The cultivation of plants on solidified media is labor intensive, and a high degree of standardized protoplast quality is hard to achieve. The routine analysis of regenerating protoplasts that have undergone transformation with a mutagenized cDNA library requires the growth of moss tissue under controlled environmental conditions. To obtain large quantities of standardized *Physcomitrella* plant material, we have chosen liquid bioreactor cultures. It has been shown by Rother et al. (1994) that *Physcomitrella* is capable of growing in a liquid mineral medium without supplying ammonium tartrate. Under these conditions *Physcomitrella* plants not only grew preferentially as protonema tissue but also developed leafy gametophores that could be detrimental to the production of standardized protoplast material because they would originate from different stages of moss development. The use of bioreactors for the cultivation of *Physcomitrella* has already been reported, but those cultures were grown as batch cultures and, more important, the cell material has not been used for protoplast isolation (Boyd et al, 1988; Cove et al., 1997; Reutter and Reski, 1996).

To obtain large volumes of moss material and to reduce manual labor, semicontinuous bioreactor cultures of *Physcomitrella* have been started. Under semicontinuous conditions, the cultures are maintained for several weeks by daily sampling of plant material and replacement of the sample volume with fresh medium. To increase the transformation capacity, the influence of ammonium tartrate supplementation and pH value on the protoplast yield has been studied (Hohe and Reski, 2002; Hohe et al., 2001). Protonema of *Physcomitrella* was precultured in liquid-modified Knop medium according to Reski and Abel (1985). Using the same amount of precultured *Physcomitrella,* tissue-stirred tank glass bioreactors were inoculated to set up semicontinuous liquid cultures under varying conditions with respect to ammonium tartrate supplementation and pH control. When the dry weight had reached approximately 150 $mg \cdot l^{-1}$, the bioreactors were harvested daily and an equal amount of fresh medium was added, which resulted in an average dilution rate of 0.19 to 0.23 per day. The supplementation of the growth medium with 2.5 mM ammonium tartrate increased the protoplast yield from semicontinuous *Physcomitrella* bioreactor cultures sixfold; this development is ascribed to the inhibition of caulonema development. Under growth conditions without the addition of ammonium tartrate, but with a controlled pH value of 4.5, even higher yields of protoplasts were obtained, although the protonema developed caulonema cells as well. It has been assumed that pH conditions in the medium may influence cell wall composition and a constant pH value of 4.5 may be advantageous for subsequent enzymatic digestion of the cell walls. These results were supported by analyzing the protoplast yield from *Physcomitrella* protonema grown at a constant pH value of 7.5, where the number of isolated protoplasts was severely reduced. Based on these results, a semicontinuous bioreactor culture at pH 4.5 without the addition of ammonium tartrate has been deduced as a standard culture protocol for large-scale suspension cultures of *Physcomitrella* for protoplast isolation and transformation. Such a bioreactor inoculated with moss material is shown in Figure 9.3. The abandonment of ammonium tartrate prevents the severe effects of this supplement on moss development that might also affect transformation and protoplast regeneration. The average harvest of around 930 $ml \cdot d^{-1}$ from a five-liter bioreactor generates protoplasts sufficient for 105 to 114 transformations following standard transformation protocols (Reutter et al., 1998; Schaefer et al., 1991). The semicontinuous growth conditions also guaran-

FIGURE 9.3. Bioreactor culture of *Physcomitrella* (see corresponding figure in color photo section).

tee the uniformity of the starting material and avoid batch-specific variance that may falsify subsequent analysis of the transformants.

PRODUCTION OF A PHYSCOMITRELLA *EST COLLECTION*

The molecular basis for the production of a saturated mutagenized *Physcomitrella* population generated by the combination of transposon mutagenesis and high-throughput transformation is the availability of a comprehensive cDNA collection that nearly covers the entire transcriptome of *Physcomitrella*. Large EST-sequencing projects have been initiated in the past from a wide range of organisms. Why should we start another one using the "primitive" land plant *Physcomitrella?* Complementing the genome-sequencing projects performed in *Arabidopsis* and rice, the resulting data allow scientists to study the evolution of organisms by applying comparative genomics as well as to elucidate the function of previously unknown genes by functional genomics. It has been estimated that bryophytes diverged from ferns and seed plants around 450 million years ago (Theissen et al., 2001). The comparison of the gene pool of these plants will allow the identification of those genes that were conserved during evolution and represent the minimal set of genes found in any plant species as well as those genes that have been expanded or reduced in the different plant lineages. Furthermore, studies have demonstrated that *Physcomitrella* is a valuable source for unknown genes (Machuka et al., 1999; Reski et al., 1998). Within our functional genomics project, we intended to generate different cDNA libraries representing almost the complete transcriptome of *Physcomitrella*. The resulting cDNA libraries will be used, on the one hand, to perform mass sequencing to generate a comprehensive EST database and, on the other hand, for the transposon mutagenesis and transformation of *Physcomitrella* to build a saturated mutagenized *Physcomitrella* population. In addition to our project, other *Physcomitrella* EST-sequencing projects have been initiated, generating sequences that are publicly available (http://www.moss.leeds.ac.uk) and contain around 60,000 EST sequences, a collection that underlines the growing interest in the model species *Physcomitrella*. As a source for the preparation of cDNA libraries from *Physcomitrella,* we have used material collected from all stages of the life cycle and plant material that has undergone treatments with different plant hormones. To reduce redundancy of individual cDNAs, the libraries were subtracted and normalized prior to mass sequencing. A detailed description of the production of the cDNA libraries is given by Rensing, Rombauts, Hohe, et al. (2002). The cDNAs were cloned directionally into a sequencing vector and sequenced using primers, allowing synthesis from the 3' end. Samples that did not yield a sequencing reaction were processed using a primer directed from the 5' end, leading to a total of around 3 percent 5' sequencing reactions. Around 80 percent of the sequences were derived from these libraries, whereas the remaining 20 percent of the sequences were obtained from a protonema library that had neither been normalized nor cloned in a directed manner (Reski et al., 1994). Before mass sequencing, we estimated the number of *Physcomitrella* genes to be around 20,000, compared to the prediction of 23,000 *Arabidopsis* genes at that time. A total number of around 110,000 cDNA clones have been sequenced from the normalized and subtracted libraries, representing more than fivefold the number of estimated genes, to ensure that a high percentage of genes is covered within the EST database. The subsequent clustering to remove redundant sequences (the process of clustering included poly-A removal, quality clipping to remove stretches of the cloning vector, as well as repeat removal) generated 33,581 clusters, comprising 21,219 singletons (built from a single sequence) and 12,362 contigs (built from at least two sequences). A detailed analysis of the clusters to obtain information on the gene content of *Physcomitrella* led to a rate of 0.76 genes per cluster. While allowing a 10 percent margin of error, this would lead to an estimate of 25,500 ± 2,500 expressed protein-coding genes in *Physcomitrella,* which is close to the number of around 26,000 genes predicted for the *Arabidopsis* genome. Even though both species contain a comparable number of genes, the degree of similarity of the genomes is quite low. Only about 50 percent of the *Physcomitrella* expressed protein genes can be matched to an *Arabidopsis* homologue based on an E-value threshold of $1 \cdot 10^{-2}$. Furthermore, the comparison of the *Physcomitrella* EST database with the public sequence databases using different homology search programs revealed that around 11,600 genes did not yield a significant hit, even if protein family or motif searches have been applied. These unique genes with unknown function represent a valuable genetic pool of novel genes, and the elucidation of their function is a major challenge for future investigations.

The analysis of the EST database has produced important information with respect to gene content in *Physcomitrella* and its homology to other species. At least as important as this, the analysis has shown that the cDNA libraries represent a high-quality source to perform the transposon mutagenesis and subsequent transformation of *Physcomitrella* protoplasts. Particularly with regard to a saturated mutagenized *Physcomitrella* population, the low redundancy of cDNA clones within the libraries reduces the number of transformations that are necessary to disrupt every gene in the *Physcomitrella* genome by homologous recombination.

IN VIVO TRANSPOSON MUTAGENESIS OF cDNA LIBRARIES

For the production of a saturated mutant collection, we want to take advantage of the high homologous recombination rate in *Physcomitrella* and the undirected disruption of cDNAs achieved by in vivo transposon mutagenesis. The disruption constructs that are used for protoplast transformation are generated by a transposon-based shuttle mutagenesis system, which has been developed for reliable introduction of the selectable *nptII* selectable marker gene cassette into clones from cDNA libraries (Rak et al., 2001). During this procedure a bacterial transposon harboring the *nos* promoter::*nptII*::*nos* terminator cassette integrates into the cDNA clones at random sites, thereby disrupting the cDNA insert. The use of a bacterial-derived transposon instead of a eukaryotic one guarantees stability of the transposable element in the nuclear genome of transgenic plants. By the use of cDNA-based disruption constructs, we want to accomplish preferential integration of disruption constructs into transcriptionally active loci, preventing a high rate of insertion into noncoding regions. Using this strategy of randomized transposon-mediated mutagenesis of cDNAs together with homology-driven integration of the knockout constructs, we aim to produce about 80,000 independent *Physcomitrella* mutant lines. cDNA was prepared from protonema material that was grown for different time periods in liquid culture. These cDNAs cover those genes expressed during vegetative growth before initiation of further developmental progression. Before cloning of the cDNAs into plasmid vectors, the library was normalized to reduce redundancy (Ko, 1990).

To enrich transcripts that are present during gametophore and sporophyte development, the reverse-transcribed mRNA pools isolated from these tissues were subjected to a subtraction procedure (Egener et al., 2002). The resulting low-redundancy moss cDNA library was cloned into the vector pUCMinIV. This vector represents a 1.7 kpb derivative of the pUC19 plasmid (Yanisch-Perron et al., 1985). Within this vector, most nonessential sequences were removed to minimize potential target sites for transposon integration. A detailed description of the transposon shuttle mutagenesis procedure is given in Figure 9.4. After in vivo mutagenesis, the cells harboring modified pUCMinIV plasmids are selected by growing the cells on the appropriate medium containing ampicillin and streptomycin. The obtained cells are propagated and harvested for plasmid DNA isolation. Batches of the plasmid DNA are subjected to *Sda*I restriction. The *Sda*I recognition site encompasses a nucleotide sequence of 8 bp. Provided that *Sda*I recognition sites are not present in the cDNA sequences, the restriction yields two linear DNA molecules. One of these corresponds to the

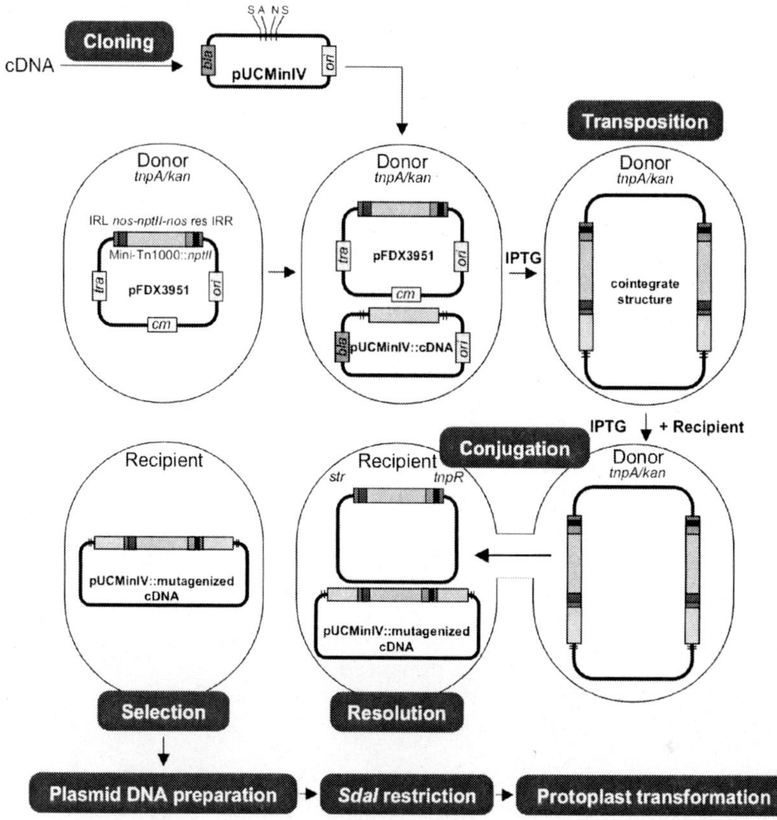

FIGURE 9.4. Transposon shuttle mutagenesis of *Physcomitrella* cDNA libraries. mRNA from protonema, gametophores, and sporophytes was isolated and used for cDNA synthesis. After normalization and subtraction, the low-redundancy cDNA libraries are ligated into the minimized pUCMinIV plasmid. The ligation reactions were transformed into the *Escherichia coli* donor strain R2217 harboring the novel conjugative plasmid pFDX3951. This plasmid carries the transposon mini-Tn1000::*nptII* and retained the chloramphenicol resistance gene *(cm)* and the *tra* function from the original conjugative plasmid R388 (Avila and Cruz, 1988). The mini-Tn1000 transposon represents a minimized derivative of the bacterial transposon Tn1000 that is flanked by the transposon Tn1000 border repeat sequences required for transposition (Sheratt, 1989). The *nptII* selective marker gene within the border sequences is controlled by the *Agrobacterium tumefaciens nos (nopaline synthase)* promoter and terminator elements (Egener et al., 2002). Promoter studies have shown that the *nos* promoter causes gene expression at levels that can be compared to the cauliflower mosaic *35S*

FIGURE 9.4. *(continued)*

promoter (Holtorf et al., 2002). Beside the plasmid pFDX3951, the donor strain R2217 also contains the plasmid pDFX3957 (plasmid not shown) that encodes the bacterial transposase gene *tnpA* and the kanamycin resistance gene *(kan)*. Expression of the *tnpA* gene is mediated by an IPTG-inducible promoter. Upon IPTG, transposition of mini-Tn::*nptII* occurs resulting in the formation of a cointegrate structure between the plasmids pUCMinIV and pFDX3591. For resolution of the cointegrate structure the complex has to be transferred into the recipient *E. coli* strain R1037. This strain harbors the *tnpR* gene, which is driven by an IPTG-responsive promoter and encodes a resolvase and a streptomycin resistance gene *(str)*. Both genes are encoded on the chromosome. The DNA transfer into the recipient cell is mediated by bacterial conjugation of the donor and recipient cell. Resolution of the cointegrate structure is achieved by the IPTG-induced expression of the *tnpR* gene, which results in the release of the pUCMinIV plasmid containing a mutagenized cDNA insert (pUCMinIV::mutagenized cDNA). Recipient cells harboring modified pUCMinIV plasmids are selected by growing the cells on the appropriate medium containing ampicillin and streptomycin. After cell propagation, the plasmid DNA is isolated, digested with the enzyme *SdaI,* and used for transformation of *Physcomitrella* protoplasts. *tnpA:* bacterial transposase gene, *kan:* kanamycin resistance gene, *tra:* transfer region from plasmid R388, *cm:* chloramphenicol resistance gene, *ori:* origin of replication, *bla:* ampicillin resistance gene, IRL: left inverted repeat border sequence, IRR: right inverted repeat border sequence, *nos-nptII-nos:* selectable marker cassette consisting of the *nos* promoter-neomycin resistance gene-*nos* terminator, res: resolution site, *str:* streptomycin resistance gene, *tnpR:* bacterial resolvase gene. Recognition sites of restriction enzymes: S *(SdaI)*, A *(AscI)*, N *(NotI)*. Plasmids are not drawn to scale.

vector backbone of pUCMinIV; the remaining fragment represents the mutagenized cDNA harboring the *nptII* selection marker expression cassette. The mixture of both fragments is subsequently used for the transformation of *Physcomitrella* protoplasts. The elucidated procedure was designed for a randomized mutagenesis of cDNAs without any knowledge about the nucleotide sequence of the concerned cDNA molecules. This enables the large-scale preparation of mutagenized cDNA libraries where the coding regions are disrupted by the insertion of the selection cassette without any preferential site of integration (Miller, 1992). We proved the randomness of transposon insertion within the cDNA region of one representative *Physcomitrella* cDNA clone with an insert size of 808 bp (Egener et al., 2002). This cDNA was subjected to the shuttle mutagenesis procedure. The plasmid DNA of 72 randomly chosen bacterial colonies was sequenced to determine the site of transposon integration (Figure 9.5). Forty-one insertions were found to be in forward orientation; i.e., the *nptII* resistance marker was in the same orientation as the *bla (beta-lactamase)* marker of the plasmid vector; 31 insertions occurred in the opposite direction; and 66

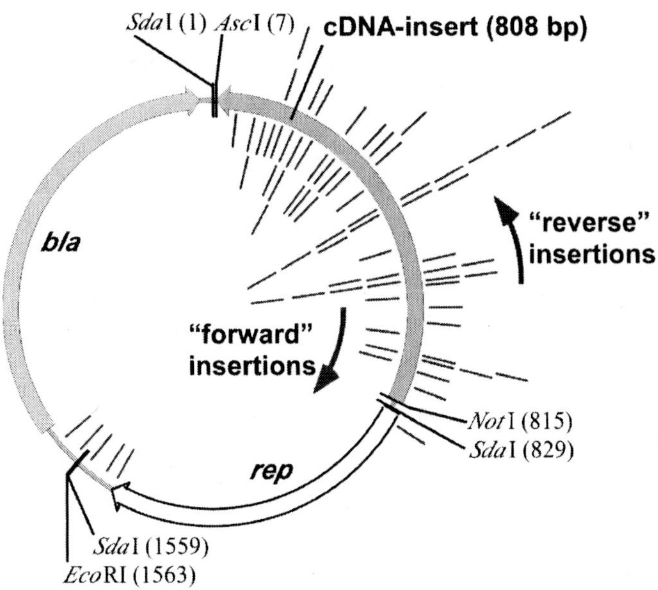

FIGURE 9.5. Random transposon insertion into a *Physcomitrella* cDNA. A representative cDNA clone of 808 bp (ID: S_PP015059353) has been subjected to shuttle mutagenesis. The transposon insertion sites of 72 resulting clones have been mapped by DNA sequencing. Forty-one insertions occurred in the forward orientation; i.e., the *nptII* resistance marker was in the same orientation as the *bla (beta-lactamase)* marker of the plasmid vector (depicted as lines within the circle). Thirty-one insertions took place in the opposite direction (depicted as lines outside the circle). Sixty-six insertions occurred throughout the cDNA; six insertions were found outside of the cDNA. (*Source:* Egener et al., 2002.)

insertions (corresponding to 92 percent) took place within the cDNA region without any apparent preference for the insertion site. In six cases, the transposon integrated into the minimized vector backbone of the pUCMinIV plasmid. Furthermore, in all cases, the transposition of the mini-Tn1000 transposon into the DNA occurred only once per plasmid.

FAST-FORWARD PRODUCTION OF SINGLE KNOCKOUT CONSTRUCTS

Beside the large-scale production of *Physcomitrella* mutant lines using the transposon mutagenized cDNAs, we follow a targeted gene knockout approach with a selection of commercially valuable genes. The usefulness of this approach has recently been shown for a number of *Physcomitrella* genes (Girke et al., 1998; Girod et al., 1999; Imaizumi et al., 2002; Koprivova et al., 2002; Strepp et al., 1998; Zank et al., 2002). Using the developed transformation platform, we intended to generate 100 knockout lines. After searching the EST database for *Physcomitrella* homologues of the identified target genes, the EST clones were retrieved from the clone repository. In contrast to the shuttle mutagenesis approach, the knockout constructs were generated by direct cloning of the *nos-nptII-nos* fragment into the central part of the respective cDNAs. After selection on G418-containing medium, 4,898 putative knockout plants were analyzed by PCR to confirm correct targeting of the knockout construct into the nuclear DNA. For this, PCR primers were designed that flank the predicted genomic insertion site. Using these primers for PCR reactions with genomic DNA templates, we expect deviating sizes of the amplicon in wild-type and knockout plants. An undisrupted genomic locus will yield a short PCR product, whereas a longer product is amplified in the knockout plants. The increase in size is ascribed to the *nos-nptII-nos* cassette that was inserted into the locus. In many cases, the amplification of the longer fragment is not possible. In these cases, a clear loss of the wild-type band is indicative of a successful gene disruption.

Samples of those plants displaying a loss of the wild-type band during the PCR screen were examined in more detail by Southern blot analyses of genomic DNA. In these cases, the disruption of the genomic locus could be confirmed (T. Egener and J. M. Lucht, unpublished results). The analyses of the first 59 candidate genes indicated that the average gene-targeting efficiency was 26 percent. For 55 out of 59 target genes (93 percent), we were able to identify knockout mutants. The highest rate of homologous recombination of one representative target gene was found to be 88 percent. Eight target genes showed only low gene-targeting efficiencies below 5 percent. In four cases, we were not able to generate knockout mutants. One explanation could be that the disruption of these particular genes causes a lethal phenotype. The gene-targeting efficiency was not correlated with the length of the cDNA. The differences in the recombination rate may be attributed to unfavorable exon-intron structures of the respective genomic loci. A secured proof of this hypothesis is currently not possible because genomic sequence data are missing for most of the genes. However, if we use long continuous stretches of genomic DNA for the construction of disruption constructs, we observe an increase of the gene-targeting efficiency. Thus, a one-to-one match of the *nos-nptII-nos* flanking regions and the genomic locus of interest could represent an alternative way to obtain knockout lines for previously unsuccessful approaches using cDNA disruption constructs. Nevertheless, the targeted knockout of single genes in *Physcomitrella* using cDNA-based disruption constructs is a fast and promising tool to address the function of selected genes. Many candidate genes can be disrupted in a short period of time, and mutants can be analyzed efficiently for knockout insertion by a fast PCR screen based on EST sequence information.

HIGH-THROUGHPUT TRANSFORMATION PROTOCOL

The demand to generate a saturated collection of around 80,000 *Physcomitrella* knockout lines requires a standardized transformation protocol to yield a uniform output of mutant lines. The use of normalized and subtracted cDNAs derived from three different tissues for shuttle mutagenesis should result in the disruption of nearly all expressed *Physcomitrella* genes. In contrast to the targeted knockout strategy, the mass production of knockout lines is conducted by transformation of DNA batches into *Physcomitrella* protoplasts. We have adapted a transformation protocol from Strepp et al. (1998) that is based on the polyethylene glycol-mediated DNA transfer into protoplasts. A standard transformation is performed using $3 \cdot 10^5$ protoplasts and 20 µg linearized DNA. After protoplast regeneration, the putative mutant lines have to pass through two times two weeks selection on medium containing 25 $\mu g \cdot ml^{-1}$ G418 interrupted by two weeks release on nonselective medium. Two rounds of selection on G418-containing medium are necessary to eliminate unstable transformants that are caused by the extrachromosomal replication of DNA (Ashton et al., 2000). In addition to the selection of antibiotic-containing medium, the plants are grown on media with different nutritional supplements to identify mutants with metabolic defects. The moss material for large-scale transformation was made available by the setup of *Physcomitrella* liquid bioreactor cultures. Furthermore, to increase the number of transformants, the protocol for protoplast regeneration was improved (Schween, Hohe, et al., 2003). We analyzed the influence of DNA conformation on transformation efficiency. The transformation of circular plasmids yielded only 0.2 percent of stable transformants after two rounds of selection on G418-containing medium. The use of linearized DNA constructs increased the number of stable transformants to 16 percent. Therefore, for routine large-scale transformation of the mutagenized cDNA libraries, we continued to utilize linearized DNA constructs. To optimize the production of plant material, we used three different knockout constructs for transformation and investigated their output of transformants under varying preculture conditions. The moss plants were grown in bioreactor cultures that were either supplemented with ammonium tartrate or pH controlled (pH value 4.5). Both conditions did not differ with respect to protoplast yield. Nevertheless, we observed that transformation efficiency was eightfold higher when using the pH-controlled moss culture compared to the supplemented cultures. The average yield of transformants per transformation was found to be 48 transgenic moss plants. The analyses of the three gene constructs revealed that gene-targeting efficiencies varied between 8 and 48 percent. The highest rates of gene targeting were found among mutants that were generated using pH-controlled plant material. For this, we use pH-controlled plant material for the standard transformation procedure. Currently, we perform 18 independent transformations with a constant production capacity of nine plants per tranformation, yielding about 160 plants per day. The tracking and clear identification of any single transformant is achieved by a standardized bar code system.

The first transformations were carried out using batches of mutagenized cDNAs. In these cases, the number and identity of the clones present in the transformation remained unknown. To alleviate the subsequent molecular analyses of the generated mutant lines, we modified the original protocol. After shuttle mutagenesis, 20 independent clones were pooled for plasmid DNA isolation and used for one transformation. The DNA and the bacterial clones of each pool are stored in a repository. If the transformation of a DNA pool leads to a mutant displaying an interesting phenotype, the corresponding clones can be used for further analyses. Based on the nucleotide sequence of the respective cDNAs, oligonucleotide primers are designed to perform PCR-based screenings. The knowledge about the identity of the individual clones present in one transformation will allow a fast identification of the affected genomic locus causing the mutant phenotype.

MUTANT QUALITY CONTROL

Besides the number of 80,000 mutants to be produced within the project, the establishment of a standardized protocol for mutant quality control is another challenging task. Therefore, to keep up with the speed of the generation of the mutagenized *Physcomitrella* collection, the procedures for the mutant analysis had to be adapted. Two characteristic traits of the generated plants are of special interest. On the one hand, the plants that have undergone selection on G418-containing medium have to be confirmed to be stable transformants containing the *nptII* selectable marker cassette. On the other hand, all plants have to be investigated in terms of their nuclear DNA content because doubling of the chromosomal DNA may have occurred during the transformation procedure by protoplast fusion. To confirm a stable integration of the *nptII* selectable marker cassette, we developed a protocol for a one-step genomic DNA extraction from *Physcomitrella* plants that was subsequently used for a PCR screen utilizing primers derived from the *nptII* gene (Schween et al., 2002). The protocol for the isolation of genomic DNA has to be fast to be able to analyze around 800 to 1,000 plants per week and feasible for small amounts of plant material because some of the generated mutants showed retarded growth during the regeneration process. After protoplast regeneration, putative transformants were subjected to two rounds of selection on medium containing G418. From these plants, 1 to 5 mg were used for genomic DNA extraction in 50 µl of extraction buffer (for a detailed protocol, see Schween et al., 2002). The extraction was performed without mechanical disruption of the plant material, and an incubation step for 15 min at 45°C was found to enhance the extraction of genomic DNA. The PCR conditions were optimized by the addition of polyvinylpyrrolidone, which is known to absorb polyphenols, thereby preventing their interaction with DNA that may result in a poor yield of PCR products. Furthermore, the addition of the linear multivalent polyamine spermidine improved the PCR results from *Physcomitrella* crude extracts of genomic DNA. During the analysis of 15,000 *Physcomitrella* plants, 13.7 percent did not yield a PCR product. These plants were subjected to a third round of selection on antibiotic-containing medium. The majority of the plants survived the additional antibiotic treatment and were considered to be transgenic, whereas the remainder may represent plants harboring extrachromosomal plasmid DNA. Taken together, 98 percent of the generated plants were confirmed to be transgenic. Using this protocol, the analysis of 300 to 400 *Physcomitrella* plants can be achieved per day by a single operator.

To determine the nuclear DNA content, i.e., to distinguish between haploid and diploid mutants, which may have arisen from protoplast fusion during the transformation process, each mutant plant is subjected to flow cytometric (FCM) analysis. FCM analysis of plant material can be performed by using protoplasts or isolated nuclei that are released by chopping up the plant material (Galbraith, 1989; Ulrich and Ulrich, 1991). For routine FCM analysis, 30 mg of fresh plant material was chopped up in the presence of 2.5 ml of 4',6-diamidino-2-phenylindole (DAPI)-containing buffer. Before the measurement, the extracts were filtered through a sieve and kept dark for 15 to 60 minutes. The FCM analyses indicated that the large-scale transformation and regeneration procedure generated only a small number of polyploid mutants. Of 16,000 transformants tested so far, 95 percent were found to be haploid (Egener et al., 2002).

SCREENING FOR MUTANT PHENOTYPES

One of the major aims underlying a large-scale mutagenesis project is the generation of a wide range of mutant phenotypes deviating from the wild type. Therefore, the forward genetics approach using transposon-mediated mutagenesis has to be accompanied by a detailed analysis of each mutant with respect to its physiological, morphological, and metabolic traits. A morphological screening for developmental aberrations and a basic physiological screening to identify mutants lacking essential biosynthesis capacities are performed in-house. To identify mutants that show altered nutritional requirements, each mutant line is subjected to growth on a minimal medium and a supplemented full medium. The minimal medium is modified Knop medium; the full medium contains additional MS microelements (Murashige and Skoog, 1962) and supplements. A detailed composition of the media is summarized in Table 9.1. Regenerating protoplasts were cultivated on full medium to balance nutritional defects in the mutants. The regenerated plants were consecutively subdivided and grown in parallel on full medium and minimal medium to discover mutants affected by metabolic deviations. A fraction of 7.2 percent (1,163 of 16,203) of the plants showed a retarded growth on minimal medium but a normal growth on supplemented full medium. Wild-type control plants grew equally well on both media. The retarded growth on minimal medium suggests a metabolic defect in the concerned mutants that can be rescued by supplying the respective metabolite. We found a retarded growth on minimal medium as well as on full medium in 3.1 percent (508 of 16,203) of the transformants. In these cases, additional supplements in the medium may be required for the complementation of the metabolic defect. One mutant identified during

TABLE 9.1. Composition of plant media used for the metabolic screen of *Physcomitrella* mutant lines.

Medium type	Composition
Minimal medium (modified Knop medium)	250 $mg{\cdot}l^{-1}$ KH_2PO_4, 250 $mg{\cdot}l^{-1}$ $MgSO_4 \times 7\,H_2O$, 250 $mg{\cdot}l^{-1}$ KCl, 1,000 $mg{\cdot}l^{-1}$ $Ca(NO_3)_2 \times 4\,H_2O$, 12.5 $mg{\cdot}l^{-1}$ $FeSO_4 \times 7\,H_2O$, pH 5.8; solid medium supplemented with 1 percent (w/v) agar
Full medium	Minimal medium supplemented with MS microelements (Murashige and Skoog, 1962) and 4 $mg{\cdot}l^{-1}$ myo-inositol, 2.8 $mg{\cdot}l^{-1}$ choline chloride, 1 $mg{\cdot}l^{-1}$ nicotinic acid, 0.5 $mg{\cdot}l^{-1}$ thiamine-HCl, 0.25 $mg{\cdot}l^{-1}$ pyridoxine, 0.01 $mg{\cdot}l^{-1}$ biotin, 0.25 $mg{\cdot}l^{-1}$ p-aminobenzoic acid, 1.9 $mg{\cdot}l^{-1}$ Ca-D-pantothenate, 0.015 $mg{\cdot}l^{-1}$ riboflavine, 6.76 $mg{\cdot}l^{-1}$ adenine, 3.84 $mg{\cdot}l^{-1}$ Na-palmitinic acid, 250 $mg{\cdot}l^{-1}$ peptone, 920 $mg{\cdot}l^{-1}$ ammonium tartrate, and 50 $g{\cdot}l^{-1}$ glucose

the screening for physiological aberrations showed retarded growth on minimal medium but a normal growth rate on the supplemented full medium. To identify the compounds that restore the wild-type phenotype, we supplemented the minimal medium with chemical compounds that are present in the full medium (Schulte et al., 2003). The exogenous application of vitamins in the medium restored the growth rate of the mutant, indicating that the biosynthetic pathway of a particular vitamin is affected (Figure 9.6).

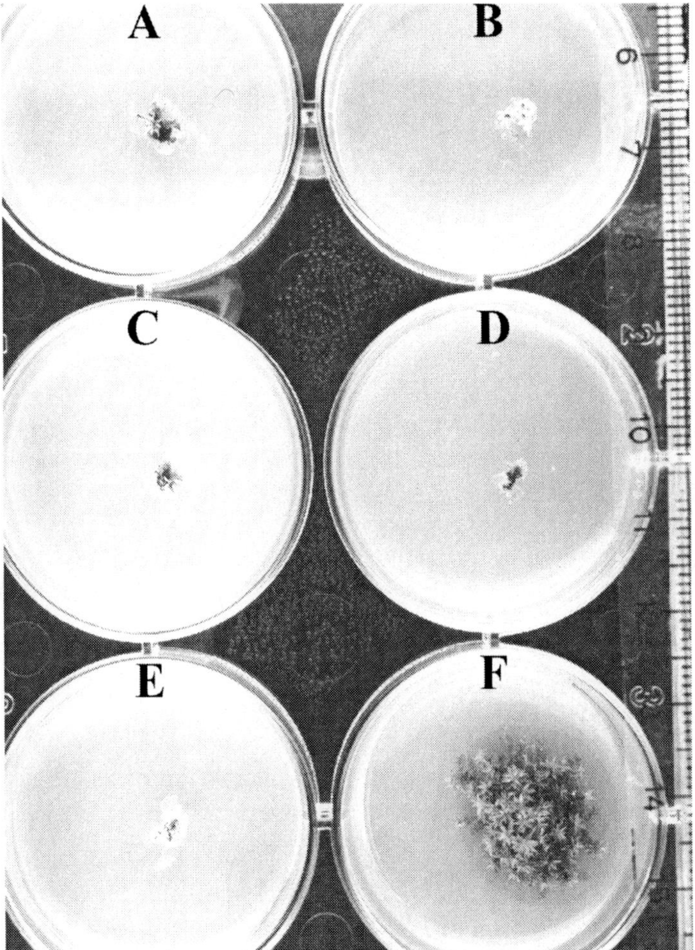

FIGURE 9.6. Vitamin-deficient *Physcomitrella* mutant (see corresponding figure in color photo section).

The screening for mutants with a developmental defect is facilitated by digital imaging and microscopic analysis of each mutant. Every mutant is compared to the wild type with respect to morphological aberrations and growth behavior. After regeneration of protoplasts for 11 weeks on supplemented Knop medium, the plants are transferred for eight weeks onto minimal medium. At this point we start the detailed morphological analysis of the mutants. At this stage, the regular development results in the formation of buds that give rise to leafy gametophores. We observed deviations from normal development in 16.2 percent (2,631 of 16,203) of the *Physcomitrella* mutant collection. The variations include morphological traits such as changed structure and color of the moss plant, deviating coverage of the plants with gametophores, altered shape and uniformity of leaves, and abnormal arrangement of cells within the leaves (Egener et al., 2002). To exclude the idea that the observed morphological deviations were caused by secondary effects based on the growth conditions rather than by the insertion of the mutagenized cDNA libraries into the nuclear DNA, we analyzed control wild-type plants during the same stages of development. We did not observe any morphological alterations in 350 wild-type plants and only less than 1 percent developmental aberrations in 400 plants derived from regenerated protoplasts that have undergone a mock transformation without DNA. A selection of impressive morphological mutants was presented by Egener et al. (2002). The frequency of altered phenotypes in the *Physcomitrella* mutant collection was found to be considerably higher compared to the less than 2 percent reported in *Arabidopsis* knockout plants. One obvious explanation might be the predominant haploid state of *Physcomitrella*. Furthermore, a lower degree of gene redundancy (Rensing, Rombauts, Van der Peer, et al., 2002) may give direction to the high proportion of altered mutant phenotypes.

Besides the phenotypic analyses regarding developmental or physiological aberrations, a third screening of the mutants is performed by the company Metanomics (Berlin, Germany; www.metanomics.de), which has developed new technologies for a comparative analysis of plant metabolic compounds. The aim of this project is to address gene function by the identification of mutant lines with disordered metabolic profiles (Trethewey, 2001; Trethewey et al., 1999) (see Chapter 5). For this, highly automated facilities have been implemented to analyze the metabolome of plants grown under standardized conditions. The metabolic profile of every mutant will be recorded and compared to the wild-type metabolic status. Altered levels of a particular compound might be caused by the disruption of genes that are involved in either the biosynthesis or the breakdown of the respective compound. These analyses will allow us to associate genes with biochemical pathways (Roessner et al., 2002). The use of *Physcomitrella* knockout mutants to study biochemical pathways has been described previously (Girke et al., 1998; Sperling et al., 2003; Zank et al., 2002). In these cases, the altered biochemical composition of the knockout plants allowed us to allocate a clear function to the genes under investigation. We hope to characterize novel gene functions by the metabolic profiling of our mutant collection.

MOLECULAR ANALYSIS OF MUTANTS

The diversity of the identified mutants represents an excellent source to isolate the genes responsible for the observed deviating phenotypes. The isolation of the genes that were affected during the insertion of the disruption constructs into the nuclear DNA is the final challenging task that has to be considered. The high rate of homologous recombination and the use of the *nptII* gene as a selectable marker provide molecular tags required for the identification of the disrupted genomic loci. We have developed several molecular tools that are applied to elucidate the molecular basis of the particular phenotypes. A set of mutants has been analyzed with respect to the number of *nptII* cassettes present in the genome. A fast protocol for the isolation of *Physcomitrella* genomic DNA has been developed to facilitate the investigation of plants by DNA gel blot assays (Schlink and Reski, 2002). Genomic DNA of 154 transformants was digested with *Pvu*II and used for Southern blot experiments. The *nptII* gene harbors one *Pvu*II recognition site and every integration site should give rise to two bands when using a *nptII*-derived hybridization probe. The average copy number of integrated *nptII* marker cassettes was estimated to be about ten (K. Schlink, unpublished results). Further analyses were performed to characterize the number of integration sites of the *nptII* marker. For this, the DNA of 16 independent mutant lines was digested using different enzymes that do not cut within the *nptII* cassette but cut genomic *Physcomitrella* quite frequently (Egener et al., 2002). The pattern of hybridizing bands using a *nptII*-derived probe allows us to discriminate between independent integration sites of *nptII* cassettes into the genome and cointegration of several cassettes at the same site. The majority of the detected bands were found to be larger than 20 kpb, indicating that most of the transposon-mutagenized cDNAs cointegrated into the nuclear DNA in close vicinity. The average number of integration sites harboring at least one *nptII* cassette within the genome was estimated to be three. These results are concordant with previous reports in which integration loci carrying concatemeric copies of the transgene were described (Schaefer and Zryd, 1997; Nishiyama et al., 2000). We were able to confirm these assumptions by sequencing PCR products that were obtained using *nptII*-derived primer pairs. The sequence analysis revealed tandem integrations of independent disruption constructs in close vicinity (T. Egener and H. Holtorf, unpublished results).

Besides the characterization of the number of integration events, we also adapted several techniques to isolate the genomic regions flanking the integration site. Basically, we apply PCR-based methods (Ochman et al., 1993) or follow a plasmid rescue strategy. One of these methods is the inverse PCR (iPCR; Hartl and Ochman, 1994; Triglia, 2000). After restriction, digest, and circularization of genomic DNA, we used inversely oriented primers derived from the *nptII* gene to amplify adjacent sequences. The obtained PCR fragments were small in size, which made it difficult to unambiguously define the gene insertion border.

We have also performed thermal asymmetric interlaced PCR (TAIL-PCR; Liu et al., 1995) that is based on the use of specific nested primers, e.g., derived from the *nptII* gene in combination with degenerated nested primers that should bind to boundary sequences. As we have previously observed concatemeric structures of the disruption constructs at the integration site, we cannot exclude that regions of neighboring *nptII* cassettes will be amplified. Therefore, an accurate sequence analysis of the obtained fragments is necessary. The TAIL-PCR will be successful as long as *nptII* marker gene sequences lie in the vicinity of the disrupted genomic locus. The yielded products can be sequenced directly (Mazars and Theillet, 1996). In addition, the fragments can serve as probes in Southern blotting experiments to study the integrity of the respective genomic locus. Besides the application of PCR-based methods, we follow a plasmid rescue approach that has been successfully applied to clone plant sequences flanking a selection marker gene (Mathur et al., 1998; Nakazawa et al., 2001). We have designed a plasmid vector that allows us to trap boundary sequences of the mini::Tn1000 transposon. The plasmid contains a truncated N-terminal fragment of the *nptII* marker gene under the control of the bacterial *tac* promoter (de Boer et al., 1983). Digestion of mutant genomic DNA with a suitable restriction enzyme cutting within the *nptII* marker gene and cloning of the fragments into the trap vector restores an active *nptII* gene in the plasmid. After transformation of the ligation reactions into *Escherichia coli* cells, the cells are selected on kanamycin-containing medium. Only cells that have taken up a plasmid harboring a restored marker gene are able to survive the antibiotic treatment. The obtained clones are subjected to sequence analysis to identify the genomic integration site. Using this technology, we were able to clone several boundary sequences (T. Egener, unpublished results). Combining PCR-based methods and the plasmid rescue technology, we characterized more than 25 genomic loci from *Physcomitrella* knockout mutants. The sequences of the isolated fragments are compared with public databases and our EST database to obtain additional sequence information on the affected locus. Currently, the application of TAIL-PCR and plasmid rescue are the most promising methods for the efficient isolation of tagged genomic loci in *Physcomitrella*.

The strategy described so far concentrates mainly on the characterization of mutants that have been generated by transformation of DNA batches containing multiple anonymous mutagenized cDNA clones. For the analysis of mutants that have been generated by the transformation of DNA batches, including only 20 independent mutagenized cDNA clones, we follow a different approach. The single cDNA constructs are subjected to sequence analysis to identify the cDNA sequence and the transposon integration site. Based on the sequence information, primers are designed flanking the transposon sequence and used for PCR with genomic DNA isolated from the mutant and control wild-type plants. As described for the analysis of the targeted knockout mutants, the PCR using DNA prepared from wild-type plants will yield a product of predictable size. In the cases of knockout plants harboring the disruption cassette within this locus, the size of the obtained PCR product either is shifted to a larger size or amplification fails because of an unfavorable exon-intron structure. Once a putative integration site has been identified, specific hybridization probes can be deduced for Southern blot analysis to confirm the PCR results. A shift of the hybridizing band in the mutant relative to the wild type is indicative of a knockout insertion at the gene locus under investigation.

CONCLUSION

The moss *Physcomitrella patens* originally attracted many research groups to study processes in plant development. The basic morphological organization of this ancient land plant with only a few cell types led to valuable insights in this field. A new direction of research has been initiated where almost all techniques of modern molecular biology have been successfully applied to *Physcomitrella*. The discovery of the high efficiency of homologous recombination in *Physcomitrella* again led to a new way to study gene function by gene displacement and the generation of targeted knockout plants. The generation of targeted knockout plants has become a routine technology in many laboratories.

Another advantage of the *Physcomitrella* system is based on the predominant haploid phase of its life cycle. Genetic variations are not counterbalanced by second alleles and often result in phenotypic deviations. A major step to implement functional genomics tools in *Physcomitrella* has been made by EST-sequencing projects initiated by independent research groups. The comparison of the *Physcomitrella* transcriptome data with those from *Arabidopsis* and rice has clearly shown that bryophytes and seed plants share a high degree of genetic conservation. Nevertheless, a large pool of genes seems to be specific for bryophytes. Consequently, *Physcomitrella* genes can, on the one hand, be applied for crop plant improvement and, on the other hand, serve as a valuable source to identify genes with novel functions. We have started a large-scale functional genomics approach to isolate those genes which might be of interest with respect to diverse biological questions. The conclusions that can be drawn after the analysis of the first set of about 16,000 mutants are encouraging. In particular, the outcome of mutants showing deviating phenotypes underlines the advantages of the moss system when compared to similar approaches undertaken in other species. Even though we have already cloned a number of candidate genes that have been affected by the mutagenesis approach, further development of the molecular tools will be necessary to speed up the identification of tagged loci. One step in this direction has already been made by using a definite number of clones for the transformation procedure. We expect that molecular analyses of these mutants will allow an easier way to identify the affected genes. Furthermore, the availability of large genomic sequence data would promote the molecular analyses of mutant lines. Until now, the major part of sequence data deposited in the public databases represents EST data. The genome size of *Physcomitrella* is estimated to be around 511 Mbp (Schween, Gorr, et al., 2003). The formation of an international *Physcomitrella* genome initiative will be necessary to face the demanding task of deciphering *Physcomitrella*'s entire genome. In addition to the described strategies to generate *Physcomitrella* knockout lines, the system also provides the opportunity to perform gene replacement studies. Based on homologous recombination, new genes can be introduced into the nuclear DNA at targeted sites, thereby disrupting the respective wild-type gene. Even though this represents an elegant way to investigate structure and function relationships for specific proteins, particular experiments following this approach have not been reported to date. Considering the potential of *Physcomitrella* for plant functional genomics, the system can provide valuable contributions to understanding plant biology.

REFERENCES

Arabidopsis Genome Initiative (2000). Analysis of the genome sequence of the flowering plant *Arabidopsis thaliana*. *Nature* 408: 796-815.

Ashton, N.W., Champagne, C.E.M., Weiler, T., Verkoczy, L.K. (2000). The bryophyte *Physcomitrella patens* replicates extrachromosomal transgenic elements. *New Phytol.* 146: 391-402.

Ashton, N.W., Cove, D.J. (1977). The isolation and preliminary characterisation of auxotrophic and analogue resistant mutants of the moss *Physcomitrella patens*. *Mol. Gen. Genet.* 154: 87-95.

Avila, P., de la Cruz, F. (1988). Physical and genetic map of the IncW plasmid R388. *Plasmid* 20: 155-157.

Azpiroz-Leehan, R., Feldmann, K.A. (1997). T-DNA insertion mutagenesis in *Arabidopsis:* Going back and forth. *Trends Genet.* 13: 152-156.

Baulcombe, D.C. (1999). Fast forward genetics based on virus-induced gene silencing. *Curr. Opin. Plant Biol.* 2: 109-113.

Bhatla, S.C., Kiessling, J., Reski, R. (2002). Observation of polarity induction by cytochemical localization of phenylalkylamine-binding sites in regenerating protoplasts of the moss *Physcomitrella patens*. *Protoplasma* 219: 99-105.

Bouche, N., Bouchez, D. (2001). *Arabidopsis* gene knockout: Phenotypes wanted. *Curr. Opin. Plant Biol.* 4: 111-117.

Boyd, P.J., Hall, J., Cove, D.J. (1988). An airlift fermenter for the culture of the moss *Physcomitrella patens*. In Glime, J.M. (Ed.), *Methods in Bryology* (pp. 41-45). Proc. Bryol. Meth. Workshop Mainz, Nichinan: Journal of the Hattori Botanical Laboratory.

Camilleri, C., Lafleuriel, J., Macadre, C., Varoquaux, F., Parmentier, Y., Picard, G., Carboche, M., Bouchez, D. (1998). A YAC contig map of *Arabidopsis thaliana* chromosome 3. *Plant J.* 14: 633-642.

Campisi, L., Yang, Y., Yi, Y., Heilig, E., Herman, B., Cassista, A.J., Allen, D.E., Xiang, H., Jack, T. (1999). Generation of enhancer trap lines in *Arabidopsis* and characterization of expression patterns in the inflorescence. *Plant J.* 17: 699-707.

Chang, Y.L., Tao, Q., Scheuring, C., Ding, K., Meksem, K., Zhang, H.B. (2001). An integrated map of *Arabidopsis thaliana* for functional analysis of its genome sequence. *Genetics* 159: 1231-1242.

Cho, S.H., Chung, Y.S., Cho, S.K., Rim, Y.W., Shin, J.S. (1999). Particle bombardment mediated transformation and GFP expression in the moss *Physcomitrella patens*. *Mol. Cells* 9: 14-19.

Chuang, C.F., Meyerowitz, E.M. (2000). Specific and heritable genetic interference by double-stranded RNA in *Arabidopsis thaliana*. *Proc. Natl. Acad. Sci. USA* 97: 4985-4990.

Cove, D.J., Knight, C.D., Lamparter, T.L. (1997). Mosses as model systems. *Trends Plant Sci.* 2: 99-105.

de Boer, H.A., Comstock, L.J., Vasser, M. (1983). The *tac* promoter: A functional hybrid derived from the *trp* and *lac* promoters. *Proc. Natl. Acad. Sci. USA* 80: 21-25.

Egener, T., Granado, J., Guitton, M.C., Hohe, A., Holtorf, H., Lucht, J.M., Rensing, S., Schlink, K., Schulte, J., Schween, G., et al. (2002). High frequency of phenotypic deviations in *Physcomitrella patens* plants transformed with a gene-disruption library. *BMC Plant Biology* 2: 6.

Feng, Q., Zhang, Y., Hao, P., Wang, S., Fu, G., Huang, Y., Li, Y., Zhu, J., Liu, Y., Hu, X., et al. (2002). Sequence and analysis of rice chromosome 4. *Nature* 420: 316-320.

Galbraith, D.W. (1989). Analysis of higher plants by flow cytometry and cell sorting. *Int. Rev. Cytol.* 116: 165-228.

Girke, T., Schmidt, H., Zahringer, U., Reski, R., Heinz, E. (1998). Identification of a novel delta 6-acyl group desaturase by targeted gene disruption in *Physcomitrella patens. Plant J.* 15: 39-48.

Girod, P.A., Fu, H., Zryd, J.P., Viestra, R.D. (1999). Multiubiquitin chain binding subunit MCBI (RPN10) of the 26S proteasome is essential for developmental progression in *Physcomitrella patens. Plant Cell* 11: 1457-1472.

Goff, S.A., Ricke, D., Lan, T.H., Presting, G., Wang, R., Dunn, M., Glazebrook, J., Sessions, A., Oeller, P., Varma, H., et al. (2002). A draft sequence of the rice genome *(Oryza sativa* L. ssp. *japonica). Science* 296: 92-100.

Grimsley, N.H., Ashton, N.W., Cove, D.J. (1977). The production of somatic hybrids by protoplast fusion in the moss *Physcomitrella patens. Mol. Gen. Genet.* 154: 97-100.

Hartl, D.L., Ochman, H. (1994). Inverse polymerase chain reaction. *Methods Mol. Biol.* 31: 187-196.

Hiwatashi, Y., Nishiyama, T., Fujita, T., Hasebe, M. (2001). Establishment of gene-trap and enhancer-trap systems in the moss *Physcomitrella patens. Plant J.* 28: 105-116.

Hohe, A., Reski, R. (2002). Optimisation of a bioreactor culture of the moss *Physcomitrella patens* for mass production of protoplasts. *Plant Sci.* 163: 69-74.

Hohe, A., Schween, G., Reski, R. (2001). Establishment of a semicontinuous bioreactor culture of *Physcomitrella patens* for mass production of protoplasts. *Acta Horticult.* 560: 425-428.

Holtorf, H., Hohe, A., Wang, H.L., Jugold, M., Rausch, T., Duwenig, E., Reski, R. (2002). Promoter subfragments of the sugar beet V-type H^+-ATPase subunit c isoform drive the expression of transgenes in the moss *Physcomitrella. Plant Cell Rep.* 21: 341-346.

Imaizumi, T., Kadota, A., Hasebe, M., Wada, M. (2002). Cryptochrome light signals control development to suppress auxin sensitivity in the moss *Physcomitrella patens. Plant Cell* 14: 373-386.

Jenkins, G.I., Cove, D.J. (1983). Light requirements for regeneration of protoplasts of the moss *Physcomitrella patens. Planta* 157: 39-45.

Ko, M.S. (1990). An equalized cDNA library by the reassociation of short double-stranded cDNAs. *Nucl. Acids Res.* 18: 5705-5711.

Koprivova, A., Meyer, A.J., Schween, G., Herschbach, C., Reski, R., Kopriva, S. (2002). Functional knockout of the 5'-phosphosulfate reductase gene in *Physcomitrella patens* revives an old route of sulfate assimilation. *J. Biol. Chem.* 277: 32195-32201.

Kranz, H., Scholz, K., Weisshaar, B. (2000). c-MYB oncogene-like genes encoding three MYB repeats occur in all major plant lineages. *Plant J.* 21: 231-235.

Krysan, P.J., Young, J.C., Sussman, M.R. (1999). T-DNA as an insertional mutagen in *Arabidopsis*. *Plant Cell* 11: 2283-2290.

Lal, M. (1984). The culture of bryophytes including apogamy, apospory, parthenogenesis and protoplasts. In Dyer, A.F., Duckett, J.G. (Eds.), *The Experimental Biology of Bryophytes* (pp. 97-115). London: Academic Press.

Liu, Y.G., Mitsukawa, N., Oosumi, T., Whittier, R.F. (1995). Efficient isolation and mapping of *Arabidopsis thaliana* T-DNA insert junctions by thermal asymmetric interlaced PCR. *Plant J.* 8: 457-463.

Lukowitz, W., Gillmor, C.S., Scheible, W.R. (2000). Positional cloning in *Arabidopsis:* Why it feels good to have a genome initiative working for you. *Plant Physiol.* 123: 795-805.

Machuka, J., Bashiardes, S., Ruben, E., Spooner, K., Cuming, A., Knight, C., Cove, D. (1999). Sequence analysis of expressed sequence tags from an ABA-treated cDNA library identifies stress response genes in the moss *Physcomitrella patens*. *Plant Cell Physiol.* 40: 378-387.

Martienssen, R.A. (1998). Functional genomics: Probing plant gene function and expression with transposons. *Proc. Natl. Acad. Sci. USA* 95: 2021-2026.

Mathur, J., Szabados, L., Schaefer, S., Grunenberg, B., Lossow, A., Jonas-Straube, E., Schell, J., Konzc, C., Koncz-Kalman, Z. (1998). Gene identification with sequenced T-DNA tags generated by transformation of *Arabidopsis* cell suspension. *Plant J.* 13: 707-716.

Mazars, G.R., Theillet, C. (1996). Direct sequencing by thermal asymmetric PCR. *Methods Mol. Biol.* 65: 35-40.

Meins, F. (2000). RNA degradation and models for post-transcriptional gene-silencing. *Plant Mol. Biol.* 43: 261-273.

Mengiste, T., Paszkowski, J. (1999). Prospects for the precise engineering of plant genomes by homologous recombination. *Biol. Chem.* 380: 749-758.

Miller, J.H. (1992). *A Short Course in Bacterial Genetics.* Cold Spring Harbor, NY: Cold Spring Harbor Laboratory Press.

Murashige, T., Skoog, F. (1962). A revised medium for rapid growth and bioassays with tobacco tissue culture. *Physiol. Plant.* 15: 473-497.

Nakazawa, M., Yabe, N., Ichikawa, T., Yamamoto, Y.Y., Yoshizumi, T., Hasunuma, K., Matsui, M. (2001). DFL1, an auxin-responsive GH3 gene homologue, negatively regulates shoot cell elongation and lateral root formation, and positively regulates the light response of hypocotyl length. *Plant J.* 25: 213-221.

Nishiyama, T., Hiwatashi, Y., Sakakibara, I., Kato, M., Hasebe, M. (2000). Tagged mutagensesis and gene-trap in the moss *Physcomitrella patens* by shuttle mutagenesis. *DNA Res.* 7: 9-17.

Ochman, H., Ayala, F.J., Hartl, D.L. (1993). Use of polymerase chain reaction to amplify segments outside the boundaries of known sequences. *Methods Enzymol.* 218: 309-321.

Parinov, S., Sevugan, M., Ye, D., Yang, W.C., Kumaran, M., Sundaresan, V. (1999). Analysis of flanking sequences from dissociation insertion lines: A database for reverse genetics in *Arabidopsis*. *Plant Cell* 12: 319-341.

Parinov, S., Sundaresan, V. (2000). Functional genomics in *Arabidopsis:* Large-scale insertional mutagenesis complements the genome sequencing project. *Curr. Opin. Biotechnol.* 11: 157-161.

Pereira, A. (2000). A transgenic perspective on plant functional genomics. *Transgenic Res.* 9: 245-260.

Peters, J.L., Constandt, H., Neyt, P., Cnops, G., Zethof, J., Zabeau, M., Gerats, T. (2001). A physical amplified fragment-length polymorphism map of *Arabidopsis. Plant Physiol.* 127: 1579-1589.

Puchta, H., Hohn, B. (1996). From centimorgans to base pairs—Homologous recombination in plants. *Trends Plant Sci.* 1: 340-348.

Rak, B., Reski, R., Zimmermann, S., Guitton, M.C., Duwenig, E., Freund, A. (2001). Method for the mutagenesis of nucleotide sequences from plants, algae and fungi. Patent WO 01/38509 A1.

Rensing S.A., Rombauts, S., Hohe, A., Lang, D., Duwenig, E., Rouze, P., Van de Peer, Y., Reski, R. (2002). The transcriptome of the moss *Physcomitrella patens:* Comparative analysis reveals a rich source of new genes. Available at <http://www.plant-biotech.net/Rensing_et_al_trans-criptome2002.pdf>.

Rensing, S.A., Rombauts, S., Van de Peer, Y., Reski, R. (2002). Moss transcriptome and beyond. *Trends Plant Sci.* 7: 535-538.

Reski, R. (1998a). Development, genetics and molecular biology of mosses. *Bot. Acta* 111: 1-15.

Reski, R. (1998b). *Physcomitrella* and *Arabidopsis:* The David and Goliath of reverse genetics. *Trends Plant Sci.* 3: 209-210.

Reski, R. (1999). Molecular genetics of *Physcomitrella. Planta* 208: 301-309.

Reski, R., Abel, W.O. (1985). Induction of budding on chloronemata and caulonemata of the moss, *Physcomitrella patens,* using isopentenyladenine. *Planta* 165: 354-358.

Reski, R., Faust, M., Wang, X.H., Wehe, M., Abel, W.O. (1994). Genome analysis of the moss *Physcomitrella patens* (Hedw.) BSG. *Mol. Gen. Genet.* 244: 352-359.

Reski, R., Reynolds, S., Wehe, M., Kleber-Janke, T., Kruse, S. (1998). Moss *(Physcomitrella patens)* expressed sequence tags include several sequences which are novel for plants. *Bot. Acta* 111: 145-151.

Reutter, K., Atzorn, R., Hadeler, B., Schmülling, T., Reski, R. (1998). Expression of the bacterial ipt gene in *Physcomitrella* rescues mutations in budding and plastid division. *Planta* 206: 196-203.

Reutter, K., Reski, R. (1996). Production of a heterologous protein in bioreactor cultures of fully differentiated moss plants. *Plant Tiss. Cult. Biotech.* 2: 142-147.

Roessner, U., Willmitzer, L., Fernie, A.R. (2002). Metabolic profiling and biochemical phenotyping of plant systems. *Plant Cell Rep.* 21: 189-196.

Ross-Macdonald, P., Coelho, P.S., Roemer, T., Agarwal, S., Kumar, A., Jansen, R., Cheung, K.H., Sheehan, A., Symoniatis, D., Umansky, L., et al. (1999). Large-scale analysis of the yeast genome by transposon tagging and gene disruption. *Nature* 402: 413-418.

Rother, S., Hadeler, B., Orsini, J.M., Abel, W.O., Reski, R. (1994). Fate of a mutant machrochloroplast in somatic hybrids. *J. Plant Physiol.* 143: 72-77.

Sasaki, T., Matsumoto, T., Yamamoto, K., Sakata, K., Baba, T., Katayose, Y., Wu, J., Niimura, Y., Cheng, Z., Nagamura, Y., et al. (2002). The genome sequence and structure of rice chromosome 1. *Nature* 420: 312-316.

Schaefer, D.G. (2001). Gene targeting in *Physcomitrella patens. Curr. Opin. Plant Biol.* 4: 138-141.

Schaefer, D.G. (2002). A new moss genetics: Targeted mutagenesis in *Physcomitrella patens. Annu. Rev. Plant Biol.* 53: 477-501.

Schaefer, D.G., Zryd, J.P. (1997). Efficient gene targeting in the moss *Physcomitrella patens. Plant J.* 11: 1195-1206.

Schaefer, D.G., Zryd, J.P., Knight, C.D., Cove, D.J. (1991). Stable transformation of the moss *Physcomitrella patens. Mol. Gen. Genet.* 226: 418-424.

Schlink, K., Reski, R. (2002). Preparing high-quality DNA from moss *(Physcomitrella patens). Plant Mol. Biol. Rep.* 20: 423a-423f.

Schulte, J., Erxleben, A., Schween, G., Reski, R. (2003). Screening for metabolic mutants among a collection of over 50.000 gene-disrupted *Physcomitrella patens* plants (submitted).

Schween, G., Fleig, S., Reski, R. (2002). High-throughput-PCR screen of 15,000 transgenic *Physcomitrella* plants. *Plant Mol. Biol. Rep.* 20: 43-47.

Schween, G., Gorr, G., Hohe, A., Reski, R. (2003). Unique tissue-specific cell cycle in *Physcomitrella. Plant Biol.* 5: 50-58.

Schween, G., Hohe, A., Koprivova, A., Reski, R. (2003). Effects of nutrients, cell density and culture techniques on protoplast regeneration and early protonema development in *Physcomitrella patens. J. Plant Physiol.* 160: 209-212.

Sessions, A., Burke, E., Presting, G., Aux, G., McElver, J., Patton, D., Dietrich, B., Ho, P., Bacwaden, J., Ko, C., et al. (2002). A high-throughput *Arabidopsis* reverse genetics system. *Plant Cell* 14: 2985-2994.

Sheratt, D.J. (1989). Tn3 and related transposable elements: Site-specific recombination and transposition. In Berg, D.E., Howe, M.M. (Eds.), *Mobile DNA* (pp. 163-184). Washington, DC: American Society for Microbiology.

Sperling, P., Egener, T., Lucht, J.M., Reski, R., Cirpus, P., Heinz, E. (2002). Identification of a Δ5-fatty acid desaturase from *Physcomitrella patens*. In Murata, N., Yamada, M., Nishida, J., Okuyama, H., Sekiya, J., Hajime, W. (Eds.), *Advanced Researches on Plant Lipid* (pp. 113-116). Dordrecht: Kluwer Academic Publishers.

Springer, P.S. (2000). Gene traps: Tools for plant development and genomics. *Plant Cell* 12: 1007-1020.

Strepp, R., Scholz, S., Kruse, S., Speth, V., Reski, R. (1998). Plant nuclear gene knockout reveals a role in plastid division for the homolog of the bacterial cell division protein FtsZ, an ancestral tubulin. *Proc. Natl. Acad. Sci. USA* 95: 4368-4373.

Stumm, I., Meyer, Y., Abel, W.O. (1975). Regeneration of the moss *Physcomitrella patens* (Hedw.) from isolated protoplasts. *Plant Sci. Lett.* 5: 113-118.

Theissen, G., Münster, T., Henschel, K. (2001). Why don't mosses flower? *New Phytol.* 150: 1-8.

Tissier, A.F., Marillonet, S., Klimyuk, V., Patel, K., Torres, M.A., Murphy, G., Jones, J.D. (1999). Multiple independent defective suppressor-mutator transposon insertions in *Arabidopsis:* A tool for functional genomics. *Plant Cell* 11: 1841-1852.

Trethewey, R.N. (2001). Gene discovery by metabolic profiling. *Curr. Opin. Biotechnol.* 12: 135-138.

Trethewey, R.N., Krotzky, A.J., Willmitzer, L. (1999). Metabolic profiling: A rosetta stone for genomics? *Curr. Opin. Plant Biol.* 2: 83-85.

Triglia, T. (2000). Inverse PCR (IPCR) for obtaining promoter sequence. *Methods Mol. Biol.* 130: 79-83.

Ulrich I., Ulrich, W. (1991). High-resolution flow cytometry of nuclear DNA in higher plants. *Protoplasma* 165: 212-215.

Walbot, V. (2000). Saturation mutagenesis using maize transposons. *Curr. Opin. Plant Biol.* 3: 103-107.

Weigel, D., Ahn, J.H., Blazquez, M.A., Borevitz, J.O., Christensen, S.K., Fankhauser, C., Ferrandiz, C., Kardailsky, I., Malancharuvil, E.J., Neff, M.M., et al. (2000). Activation tagging in *Arabidopsis. Plant Physiol.* 122: 1003-1013.

Yanisch-Perron, C., Vieira, J., Messing, J. (1985). Improved M13 phage cloning vectors and host strains: Nucleotide sequences of the M13mp18 and pUC19 vectors. *Gene* 33: 103-119.

Yu, J., Hu, S., Wang, J., Wong, G.K., Li, S., Liu, B., Deng, Y., Dai, L., Zhou, Y., Zhang, X., et al. (2002). A draft of the rice genome *(Oryza sativa* L. ssp. *indica). Science* 296: 79-92.

Zank, T.K., Zahringer, U., Beckmann, C., Pohnert, G., Boland, W., Holtorf, H., Reski, R., Lerchl, J., Heinz, E. (2002). Cloning and functional characterisation of an enzyme involved in the elongation of Δ6-polyunsaturated fatty acids from the moss *Physcomitrella patens. Plant J.* 31: 255-268.

Zhu, T., Mettenburg, K., Peterson, D.J., Tagliani, L., Baszczynski, C.L. (2000). Engineering herbicide-resistant maize using chimeric RNA/DNA oligonucleotides. *Nat. Biotechnol.* 18: 555-558.

Chapter 10

Arabidopsis Functional Genomics Tools: An Overview

Koen Dekker

INTRODUCTION

Most scientists would not consider *Arabidopsis* to be a plant of agronomic importance, except perhaps those who study herbicides. Nevertheless, the weed *Arabidopsis* has invaded the greenhouses of institutions around the world, and for good reasons. The high degree of organization of the *Arabidopsis* community has resulted in a wide array of genomics tools that can be readily accessed, and that help to efficiently test hypotheses concerning gene-phenotype relationships and molecular networks. However, the abundance of tools enables a good plan, but it is no substitute. Even as a provider of one such tool, insertion-tagged lines (Steiner-Lange et al., 2001), I readily admit that progress still depends on the commitment of well-trained, creative scientists. As pointed out by Knight (2003), the tendency to try to create the same range of tools in all sorts of crop plants might turn into a situation where plant biology starts suffering, at least during the time of tool development, from the heavy allocation of resources in technology development. Thus, I set out to explore which tools play or have played a pivotal role in *Arabidopsis*. With about 2,500 publications per year (Somerville and Koornneef, 2002), such a wealth of information about *Arabidopsis* is available that my personal perspective will clearly be far from complete, yet it will illustrate the currently available major tools and some of what is still on the wish list. I conclude with an attempt to stimulate discussion in the *Arabidopsis* community, and to aid with the selection of efficient genomics tools of other plants, by summarizing some "threats" to be faced and some opportunities that could contribute to continued successes at the frontiers of plant sciences.

CURRICULUM VITAE OF A MODEL

Before evaluating the current situation, let us take a brief look at how the current dominant position in the academic world was obtained. "A Fortunate Choice: The History of *Arabidopsis* As a Model Plant" by Somerville and Koornneef (2002) is an excellent overview of how and why *Arabidopsis* became the model plant. Besides covering virtues such as being small, fast growing, and diploid, they highlight the efficient transformability with *Agrobacterium tumefaciens* (Bechtold et al., 1993; Feldmann and Marks, 1987), partially due to the large number of seeds, as a critical technical innovation that enabled the transition into the molecular age. The expanding community was supported by the establishment in 1991 of the U.K. and U.S. stock centers (Table 10.1) that distributed seed stocks of mutants as well as DNA stocks, such as several large insert libraries (BACs, PACs, etc.) or EST (expressed sequence tag) clones. These insert libraries were primar-

TABLE 10.1. Essential *Arabidopsis* community resources on the Web.

Name	URL	Description
TAIR (The *Arabidopsis* Information Resource)	<www.arabidopsis.org>	Provides information about genes, markers, polymorphisms, maps, sequences, clones, DNA and seed stocks, gene families and proteins, publications, and information about *Arabidopsis* researchers
MIPS-MATDB (*Arabidopsis Thaliana* database)	<www.mips.gsf.de/proj/thal/db>	Provides access to the sequence data of the Arabidopsis Genome Initiative that are compiled, analyzed, annotated, and stored at MIPS
TIGR (*Arabidopsis thaliana* database)	<www.tigr.org/tdb/e2k1/ath1/>	Provides access to the sequence data that are compiled, analyzed, annotated, and stored at TIGR
NASC (Nottingham *Arabidopsis* Stock Center)	<www.nasc.nott.ac.uk>	Collects and distributes seeds and information; activities coordinated with those of ABRC
ABRC (*Arabidopsis* Biological Resource Center)	<www.biosci.ohio-state.edu/~plantbio/Facilities/abrc/abrchome.htm>	Collects and distributes seeds, DNA clones, and libraries (According to the distribution agreement, NASC distributes seeds to Europe and ABRC distributes to North America).

ily intended for mapping (Mozo et al., 1999) and sequencing purposes but remain useful in cases where computer prediction of gene structure is unclear or for the construction of vectors with genomic DNA or cDNA, e.g., for complementation, overexpression, or GUS-fusions. These dedicated centers have been very successful in making available material more visible, they have made the rules for access transparent, and they have succeeded in removing the burden of distribution from the scientist who established the material without taking away the credits.

International coordination of efforts finally culminated in the first sequenced genome, 120 Mbp, of a flowering plant in the year 2000 (Arabidopsis Genome Initiative, 2000). An evaluation by Meinke et al. (2003) listed 620 cloned genes with a loss-of-function phenotype, and, currently, an ambitious plan is underway to obtain information about the function of all 26,500 identified genes by 2010 (Somerville, 2000). Since *Arabidopsis* research is firmly embedded in large plant functional genomics programs in the United States (NSF2010), Germany (GABI), France (Genoplante), the United Kingdom (GARNet), and Japan (RIKEN and others), surely the list will expand. The challenge is how to do this efficiently and convincingly.

FUNCTIONAL GENOMICS TOOLS

Some of the more prominent tools for the discovery of gene-function relationships are presented in Figure 10.1. An extensive listing of resources and of ways to use them is given by Weigel and Glazebrook (2002) in their *Arabidopsis* handbook, while for the latest news and materials, a visit to the Web site of The *Arabidopsis* Information Resource (TAIR) is recommended (Garcia-Hernandez et al., 2002; Rhee et al., 2003). In this chapter, for convenience's sake, reverse genetics and forward screens are treated as two separate approaches to study such relationships. However, there are plenty of examples where both have been elegantly combined (Helariutta et al., 2000; Varotto et al., 2000), which then results in a circle, with multiple entry points, where forward screening and reverse genetics are extensions of each other.

Collaborations that have used insertion tagged lines of the Max Planck Institute for Plant Breeding Research (ZIGIA populations) between July 1996 and April 2003 resulted in 15 publications from forward genetics, 20 publications from reverse genetics, and four publications from combined approaches. The time from discovery until publication was on average two years. This demonstrates that both approaches, reverse and forward genetics, can be successfully used and suggests that one might choose the strategy with which one is most comfortable; a good sense for plants and an ele-

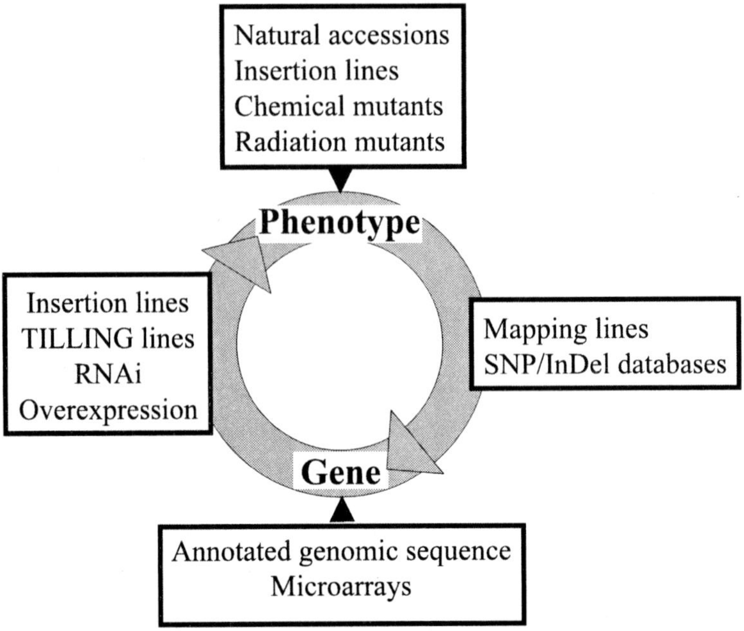

FIGURE 10.1. Overview of well-established functional genomics tools in *Arabidopsis*. Although these tools are used in various ways, only the most common points of application are indicated.

gant screening system favor phenotypic screening, and good molecular biological skills will support reverse genetics.

TOOLS FOR REVERSE GENETICS

One particular strength of *Arabidopsis* is the public availability since the end of 2000 of a very high-quality genome sequence of the Columbia accession (Table 10.2), enabling, e.g., reverse genetics. Hopefully, sufficient funding will be available in the years to come for the MIPS (Schoof et al., 2002) and TIGR teams to maintain this valuable resource and to continue the process of improving gene assignments and annotations in collaboration with the users. Some of the problems still plaguing large-scale functional analysis efforts are undocumented chromosome pseudomolecule changes and gene-naming ambiguities. Although sporadic, the same *Arabidopsis* gene codes are sometimes used by different centers to designate dif-

TABLE 10.2. Some collections providing individual lines for reverse genetics based on TILLING lines or RNAi lines (April 2003).

Collection	Accession	Comment	URL for BLAST search
FHCRC-UW/ TILLING	Col	If no hits are found in the database, a screen can be requested.	<www.tilling.fhcrc.org: 9366/search.html>
AGRIKOLA/ RNAi lines	Col	Project started in 2002. Material is not yet available.	<www.agrikola.org/>

Note: A more extensive list is available from TAIR at <www.arabidopsis.org/links/insertion.html>. Collections of T-DNA lines, including sequence characterized lines (FSTs), are listed in Chapter 2.

ferent genes or diverse gene models, or, conversely, the same sequence or gene model may have different names.

An ideal system would be compatible with the concurrent annotation of alternative splicing variants, of overlapping genes, or of documented diverse gene models covering the same chromosomal region. While the *Arabidopsis* community is bracing itself for the much more challenging functional annotation of the genome, it is a pity that its structural basis is still unstable. Since the genome sequence—together with its structural annotation—is the very basic tool, the main players in that field (TIGR, MIPS, TAIR, and a few others) should continue to work together to adopt, once and for all, a unified structural annotation system that would allow MIPS, TIGR, and other groups involved in large annotation projects to build a coherent and stable genome map.

Still, even a good sequence is not yet a function, and fortunately there exist excellent tools in *Arabidopsis* for gene-function discovery through reverse genetics, such as insertion tagged lines, chemically mutagenized lines, or RNA interference (Table 10.2).

Insertion Tagged Lines

Since the mid-1990s, populations have been available that were mutagenized with a variety of tags, such as the maize transposons Ac/Ds (Sundaresan et al., 1995) or En/Spm (Marsch-Martinez et al., 2002; Wisman et al., 1998), or by T-DNA (Weigel et al., 2000). Each of these systems has its advantages and drawbacks.

The unique feature of transposons is that confirmation of a gene-phenotype relationship is possible by monitoring the reversion to wild type upon transposition of the active transposon out of the original site, which usually results in restoration of the original sequence but sometimes results in a footprint that creates a stable allele.

In general, the Ac/Ds transposon has a stronger tendency to transpose to linked sites than En/Spm, which makes it more difficult to get an even distribution of insertions over the whole genome when starting from a limited number of parental lines. However, this characteristic can be cleverly exploited when the Ac/Ds transposon is added as a payload in a T-DNA insertion. After identification of such an insertion that is close to the region of interest, activation of the transposon to saturate that local region with Ac/Ds insertions is relatively easy, resulting in a small mutant population of high quality for that region ("launchpad" approach; Muskett et al., 2003).

The En/Spm transposon has a much wider range of secondary insertion sites, making it possible to start with five original insertion events and to saturate the genome with insertions through 6 to 12 generations of selfings (Baumann et al., 1998; Figure 10.2). These populations usually contain more than ten insertions per line, making cosegregation analysis to assign a particular insertion with an observed phenotype difficult.

In this context, the SLAT lines (Tissier et al., 1999) deserve special mention because this elegant system combines the strength of T-DNA lines, namely a small number of stable insertions, with the advantages of a transposon, such as reversibility and the possibility to get local saturation. In short, by a number of independent T-DNA transformation events, insertions were introduced where an immobilized transposase and a truncated, but still movable, dSpm element were separated. After one round of transposition of the dSpm element out of the original T-DNA insertion to a random position in the genome, negative selection was applied to remove the T-DNA, including the immobilized transposase, resulting in the stable integration of the dSpm element. Once a candidate for a gene-function relationship has been identified, the confirmation can be performed by reactivating transposition through a cross with a line that contains the transposase. However, this system is labor intensive since it requires both large-scale T-DNA transformation and several selection steps after that.

T-DNA populations, with randomly distributed stable insertions, have been around for more than ten years, but they started to flourish when efficient transformation by dipping (Clough and Bent, 1998) became possible. T-DNA insertion tagging usually results in one to five stable copies with on average fewer than two insertions per line. Additional advantages are that the construct can be modified easily with additional functionalities, such as the cauliflower mosaic virus 35S promoter, or its enhancers, to get domi-

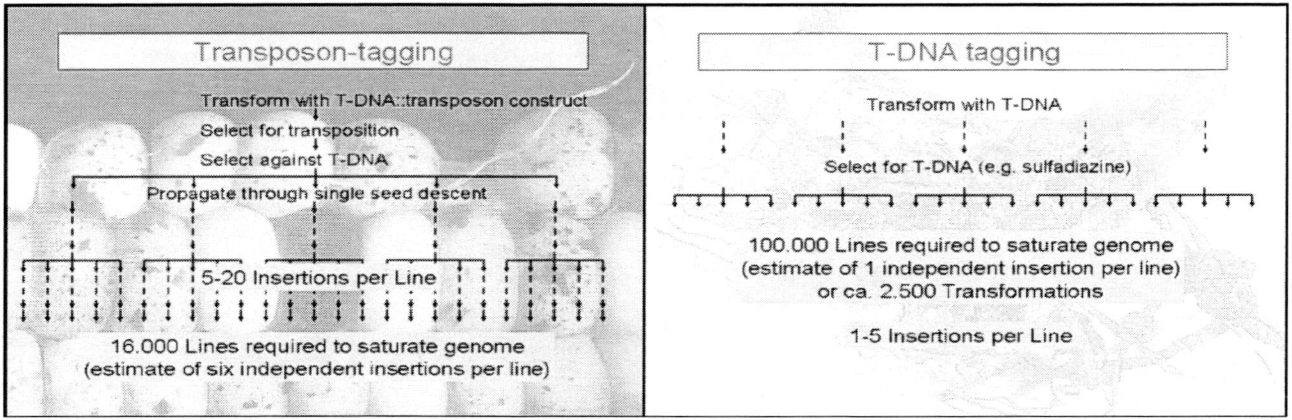

FIGURE 10.2. Different strategies for transposon-tagging and T-DNA tagging. Transposon tagging involves a few transformations and many generations of plants, while T-DNA tagging requires an efficient transformation system with only one or two generations. The background images show spots in corn kernels caused by a somatic reversion characteristic for a transposon, and a root infected by *Agrobacterium*, the microorganism from which the T-DNA vectors were derived.

nant activation tags, or by addition of reporter genes, such as GFP or GUS, to obtain information about promoter or enhancer function. Obviously, such additions require a certain position and orientation of the insertion in the gene of interest; in other cases "normal" knockouts are obtained. Less obvious for novice users of T-DNA lines is the propensity of T-DNA to form complex inserts by inserting in itself (head-to-head, head-to-tail, T-DNA-in-T-DNA, etc.), that truncated borders are not uncommon (problems with PCR primers), and that repaired or aborted integration attempts can generate additional mutations in the genome not linked to the T-DNA insertion (Tax and Vernon, 2001). Thus, the spectrum of phenotypes obtained by T-DNA mutagenesis can be difficult to interpret.

All these types of insertion-tagged populations have been used for reverse genetics (Thorneycroft et al., 2001). Initially, the identification of lines that contained an insertion in a particular gene of interest was achieved by performing PCR reactions with one gene-specific primer and one insert-specific primer on DNA pooled from several lines (Baumann et al., 1998; Sussman et al., 2000), similar to systems that existed in maize (e.g., the Pioneer-Developed Trait Utility System for Corn or transformation; Bensen et al., 1995) or petunia (Maes et al., 1999). This usually requires several rounds of PCR to find an individual or a small group of candidates.

Advances in high-throughput isolation of DNA from plants, in techniques for the amplification of insert flanking sequences, such as linker ligation or TAIL-PCR, and in the development of sequencing in 96-well format have resulted in the availability of large collections of sequence-characterized insertion lines since 2000 (Samson et al., 2002; Sessions et al., 2002). Chapter 2 in this book lists the main collections where such lines are available plus several sites that can be used as a common entry point to such collections. These resources currently supply the community with insertion tagged lines, since the candidates can be identified without experimental work, and the position (exon/intron), as well as the orientation, of the insertions is known from the sequence. About 80 percent of these insertions can be confirmed after retrieval of candidates (Balzergue et al., 2001).

Theoretical calculations predicted that about 180,000 independent insertion lines would result in a 95 percent probability to detect a single specific knockout allele in a 2.1 kbp gene (Krysan et al., 1999), but for practical reasons such a high coverage is unlikely to be reached. T-DNA lines are also subjected to some form of selection for the presence of T-DNA at one stage or another. Thus, during this process, mutations that affect the fitness of the plant in the heterozygous state also are counterselected. Moreover, mutations that are lethal or sterile in the homozygous state become underrepresented when amplification of seed stocks is performed on the bulk level.

Thus, the estimation of the current hit rate of ca. 80 percent could be close to the final saturation level, also since some of the FST projects were completed in 2003.

Another limitation of *Arabidopsis* insertional mutagenesis approaches is that large mutant populations are so laborious to generate that they are built in a few wild-type genetic backgrounds, thereby considerably restricting the study of genetic interactions. The existing insertion mutations are laborious to combine, especially if the targeted loci are linked and genetic screens to identify enhancer or suppressor mutant phenotypes are based on chemical (usually EMS) mutagenesis that requires the subsequent determination of secondary mutation loci via map-based cloning.

Working with Nontagged Populations

Fortunately, additional techniques, such as RNA interference (Cogoni and Macino, 2000) or TILLING (Colbert et al., 2001; Till et al., 2003), complement dissection of a pathway through null mutations in individual homologues (see Chapter 3).

RNA-induced gene silencing has also become popular for the exploration of plant genomes (Waterhouse and Helliwell, 2003). The essence is the introduction of double-stranded RNA (dsRNA) to induce a sequence-specific RNA degradation mechanism that effectively silences a targeted gene. Theoretically, one could down-regulate several members of a gene family with functional redundancy by targeting a region of high homology. Of course, this would be more practical than making, e.g., triple crosses of individual knockout lines. Systems for inducible RNAi can be used for particularly critical genes (Guo et al., 2003). Due to variable levels of remaining transcript in independent transformants, RNAi approaches can result in a spectrum of phenotypes. This can be an added advantage when fishing for genes that were absent from the knockout populations, perhaps for reasons of physiological disadvantage of the null allele. Unfortunately, a hypothesis built solely on an RNAi experiment can be quite difficult to confirm. Off-target gene regulation by RNAi has been revealed in HeLa cells by expression profiling with microarrays (Jackson et al., 2003). Complementation of RNAi lines is not possible, and small amounts of rest activity can be hard to detect with Western analysis (RT-PCR is not suitable), making it difficult to distinguish between rest activity or functional overlap with another pathway or nonrelated genes.

For other approaches, such as populations of gamma-irradiated material and for chemically induced mutations, the detection of nontagged mutations by PCR on DNA pools has been established as well. In the former, primers are selected that are separated by several kbp, and a band is detected only in cases where this distance has been significantly reduced by a deletion (Li et al., 2001).

The latter technique, called TILLING (Colbert et al., 2001), has drawn a lot of attention. The technique uses the enzyme CEL I to detect the heteroduplexes that are present in the PCR product if one of the lines in the pool contains point mutations in the gene of interest. This technique is applicable to all sorts of species, mutants for individual genes are obtained, and confirmation of the gene-phenotype relationship by complementation is possible. The TILLING project of Henikoff (Till et al., 2003) found, on average, ten individual mutated candidates per *Arabidopsis* gene; some of these are subtle replacements and some create stop or nonsense codons that result in null alleles. For practical reasons, people are advised to check for phenotypes with insertion tagged lines when available, since it is unlikely that the more subtle mutations will reveal phenotypes when the knockouts do not. Unfortunately, CEL I has so far not been commercially available, and the technique is technically demanding, but this method is bound to become more widespread.

An often-asked question concerns the efficiency with which mutants do show a phenotype; however, the answer to that is strongly influenced by the preexisting knowledge about the process that the gene could be involved in and by the effort of the scientist to find a phenotype. Our experience with the supply of transposon-tagged lines through reverse genetics (Steiner-Lange et al., 2001) shows that provision of ca. 250 candidate mutants to external groups resulted in 16 publications describing the analysis of gene-phenotype relationships. This would indicate a phenotype in about 6 percent of the cases. However, in one of the first systematic studies using reverse genetics (Meissner et al., 1999), mutants for 73 MYB transcription factors were searched for, which resulted in 26 homozygous mutants being tested in 28 different assays or conditions. This yielded six lines with a phenotype, or, in other words, 23 percent of the 26 mutants studied had a phenotype. Without doubt, this number could be increased by including other extensive assays. The common perception that only a few mutants show a phenotype probably has as much to do with the impracticality of setting up such a battery of assays in one lab, as with true cases of functional redundancy, or even the real absence of function.

The potential for misinterpretation of gene-phenotype relationships was mentioned previously. Some of these problems, caused by secondary mutations, can be compensated for by a few rounds of crosses with the wild type. It is highly recommended that, in the case of a reverse genetics approach, during the first selfing, one sets aside some plants, segregating for the wild-type allele, which should not show any phenotype, and not just focus completely on the heterozygous and homozygous mutant alleles. If the plants that are homozygous for the wild-type allele do not show any altered phenotype, this is an indication that nonlinked secondary site mutations, which are not necessarily tagged, do not interfere. Yet it is considered good practice to perform one of the following experiments: (1) comparison with a second allele, preferably obtained from another population; (2) for transposon tagged lines, looking for reversion back to wild type upon excision of the transposon; (3) complementation, preferably with a construct starting closely after the translation stop of the 5' neighbor up to about the transcription start of the 3' neighbor, or (4) mimicking the phenotype through application of RNAi.

TOOLS FOR FORWARD SCREENING

The insertion-tagged populations that were mentioned previously also have been used extensively for forward screening for phenotypes, since the known insertion can be used as a marker during cloning of the mutated gene. Numerous breakthrough gene-phenotype relationships were discovered by screening for phenotypes of interest and subsequent cloning of the responsible gene. The main prerequisites are sufficient space, a robust screen, and, given that a lot of screens have already been performed, originality in the choice of, e.g., growth conditions, genetic background, or detection method. With 620 gene-phenotype relationships documented (Meinke et al., 2003), early removal of known mutants is important, but not trivial. Frequently, only one aspect of a more pleiotropic phenotype has been highlighted in the description, making queries of databases (e.g., mips.gsf.de/proj/thal/phenotype/index.html) or published lists (Bowman, 1994) error prone, and also access to material for allelic crosses can be problematic. The efficiency of the screen can be improved further by the availability of additional methods, which can be more labor intensive, to remove false positives and, further, to group the hits from the first round into several classes of interest (Page and Grossniklaus, 2002). Also important is the material to be screened, and, fortunately, a wide variety of seed material is available, such as insertion-tagged lines, chemically or radiation, mutagenized material, and panels of diverse *Arabidopsis* accessions.

A factor that has contributed significantly to the success of the several types of populations was their distribution by the stock centers (Table 10.1). These centers offer several sets of insertion-tagged lines, as a bulk or as several pools, that can be screened, and similar material has been created by many institutions that might be willing to share for direct collaborations.

The obvious advantage of using insertion-tagged lines is that cloning is greatly facilitated when the observed phenotype is caused by an insertion-tagged mutation. A further development in this direction could be a set of individual homozygous lines derived from the collections with sequenced tags; i.e., a set of 26,500 lines, each homozygous for a known insertion in one of 26,500 genes theoretically would increase the efficiency of screening a lot. However, the spectrum of gene-phenotype relationships to be found would not change much, since there would still be counterselection against the modulation of critical genes.

Chemical mutagenesis does add another quality due to the presence of, e.g., point mutations or small insertion/deletions (InDels). Another positive factor is that the frequency of the mutations can be influenced, resulting usually in a lower number of lines to be screened than in the case of insertion-tagged lines. Recently, cloning became very feasible, in particular if the accessions Columbia (Col) or Landsberg *erecta* (L*er*) are used. The recombinant inbred lines from a cross of these accessions are available (Alonso-Blanco et al., 1998), and a database with several thousand single nucleotide polymorphisms (SNPs) and small insertions or deletions (InDels) further supports cloning efforts (Table 10.3). With a robust assay it is possible to clone a mutated gene in about 12 months (Jander et al., 2002), although for some positions and depending on the nature of the mutation more time can be required. Besides the availability of other types of polymorphism detection (Cho et al., 1999), sufficient SNP data are available for a lot of other accessions as well (Table 10.3), so this approach offers a lot of flexibility.

Although mutagenesis probably still provides the faster road from phenotypic variation to gene discovery, QTLs can be attractive as well (Nadeau and Frankel, 2000). Studying natural variation offers excellent opportunities for dissecting more complex traits, such as QTLs, in comparison to the single gene-oriented approaches that were mentioned previously (Alonso-Blanco and Koornneef, 2000). Considerable variation for sensitivity to biotic and abiotic stresses and for various developmental or physiological traits has been found in the about 250 available accessions. Such comparisons of different *Arabidopsis* accessions are facilitated by the availability of the multiple-accession SNP databases. By using these modern molecular marker technologies and specific statistical methods, the map position and the effects of these QTLs can be determined. A further boost could come from promising mapping schemes, such as Radiation Hybrids and HAPPY mapping (Waugh et al., 2002), that bridge the resolution gap between meiotic and BAC-based mapping or association mapping (Buckler and Thornsberry, 2002). Another advantage is that the initial number of lines to be investigated for the occurrence of variation is small, so that the decision of whether further pursuit is

TABLE 10.3. Some resources to support forward-screening approaches.

Resource	Comment	Distributor	URL
Arabidopsis accessions and related genera	About 256 *Arabidopsis* accessions exist	NASC	<www.arabidopsis.info/catalogue.html>
		ABRC	<www.arabidopsis.org/servlets/Search?action=new_search&type=ecotype>
		Lehle Seeds	<www.arabidopsis.com/main/cat/seeds/wild types/!wl.html>
T-DNA seed pools	e.g., Feldman lines	NASC	<www.arabidopsis.info/catalogue.html>
		ABRC	<www.arabidopsis.org/abrc/tdna_lines.html>
Chemically mutagenized pools	EMS mutants in Col, L*er,* or Ws	Lehle Seeds	<www.arabidopsis.com/main/cat/seeds/M2/EMS/!2e.html>
Fast Neutron or γ-ray irradiated pools	Irradiated mutants in, e.g. Col, L*er,* or Ws	Lehle Seeds	<www.arabidopsis.com/main/cat/seeds/M1/!m1.html>
Recombinant inbred lines	e.g., Col × L*er,* Col × Nd, Col × Kashmir, L*er* × Cvi, Bay × Sha., Ws × W110F	NASC	<www.arabidopsis.info/catalogue.html>
		ABRC	<www.arabidopsis.org/abrc/ri_mapping.html>
Col × L*er* SNPs and InDels	About 55,000 polymorphisms	Cereon	<www.arabidopsis.org/Cereon/index.html>
SNPs for other accessions	Genome-wide set of SNPs for 13 accessions	Max Planck Institutes	<www.mpiz-koeln.mpg.de/masc>

warranted can be based on a technically demanding assay. A discussion of the rapidly increasing number of SNP detection techniques is outside the scope of this overview, but I expect that with increasing access to these methods such approaches will become more popular. A very interesting innovation is the application of microarrays to determine linkage disequilibrium (Nordborg et al., 2002) or to analyze the molecular processes that could underlie the observed differences between different accessions (Borevitz et al., 2003), while through the same hybridization hints about the responsible alleles can be obtained that can significantly shorten the time required for conclusive identification of involved genes.

TOOLS FOR STUDYING MECHANISMS

After firmly establishing a gene-phenotype relationship, there remains the challenge of clarifying the functional mechanism that underlies a particular phenotype. An important tool for *Arabidopsis* research that deserves attention is the microarray (Table 10.4) (see also Chapter 4). The most obvious use of microarrays is the parallel analysis of the steady-state levels of many transcripts of particular developmental stages, growth conditions, or stresses. This will yield groups of genes from which clustering based on pattern similarity and preexisting knowledge can reveal candidates for reverse genetics approaches to test a certain hypothesis (Harmer et al., 2000). Another application is the study of mutants in order to obtain, e.g., the target genes of transcription factors or other proteins involved in transcription regulation. Here, some alternatives to microarrays for which *Arabidopsis* data are also publicly available should be mentioned: MPSS (Hoth et al., 2002, Meyers et al., 2002), SAGE (Lee and Lee, 2003), or cDNA-AFLP (Brugmans et al., 2002).

Initial arrays were custom-made arrays of the ca. 14,000 selected EST clones from the Ohio State library that represented about 9,000 genes (Maleck et al., 2000). It took quite some time before commercial microarrays became available after the publication of the genomic sequence, as companies were more focused on biomedical sciences, and in the meantime technology and analysis software matured. But, with the latest addition that represents ca. 22,000 genes, of the estimated 26,500 genes in the genome, very complete scans are possible. Still the cost of the arrays, the challenging analysis of results, and the many parameters that can influence expression patterns in plants unintentionally make it necessary to start these studies after careful experimental design. The availability of the CATMA array (Table 10.4) is expected to make these experiments more affordable, since sets of specific gene sequence tags (GSTs) are exchanged to support in-house preparation. It could also offer an alternative to commercial arrays in the respect that the gene assignment, and thus the probe assignment, was performed independently (Crowe et al., 2003).

A number of groups offer the labeling of RNA, hybridization, and initial analysis as a service. The *Arabidopsis* Functional Genomics Network (AFGN) offered such a service, and currently the GARNet project accepts open applications. Besides getting a good deal on reagents, the experience of the people could be invaluable to sort out artifacts or other problems. It also has advantages for the community, since databases with expression data can be established.

Proposals concerning the systematic description of developmental stages (Boyes et al., 2001) and growth conditions have been crucial in order to cre-

TABLE 10.4. Summary of *Arabidopsis* microarray-related resources.

Resource	Provider	Comment
Full array services	NASC's Affymetrix Facility, England <www.arabidopsis.info/ prototype/>	Provide full service, including Affimetrix arrays, hybridization, and initial analysis
	VIB MicroArray Facility, Belgium <www.microarrays.be>	
	TCAG, Canada <tcag.bioinfo.sickkids.on.ca/ microarray.html>	
	State University of New York, United States <www.osa.sunysb.edu/udmf/>	
Expression viewer or spot history	TAIR/AFGC <arabidopsis.org/tools/bulk/ microarray/index.html>	Provides for a particular gene a visual display of expression in up to 576 array experiments
	NASCArrays <ssbdjc2.nottingham.ac.uk/cgi -bin/NARRAYS/spothistory.pl>	Frequency distribution for a particular gene of how often a certain expression level was observed
	CAGE <www.psb.rug.ac.be/CAGE/ objectives.htm>	Planned! Compendium of Arabidopsis Gene Expression using ca. 2,000 hybridization to CATMA arrays
Data downloads	TAIR <arabidopsis.org/info/ expression/>	Provides under "Expression Data" several sites with down-loadable data sets
Common *Arabidopsis* arrays	Affimetrix 22K array <www.affymetrix.com/ products/arrays/>	Contains short oligomer sets for ca. 22,000 genes
	Agilent Oligo Microarray <www.chem.agilent.com>	Contains 60-mers for ca. 15,000 genes
	Qiagen Operon Array-ready. For printed slides look at <www.ag.arizona.edu/ microarray/info.med.yale.edu/ wmkeck/dna_arrays.htm>	Based on 70-mers for 26,000 genes
	CATMA (Complete Arabidopsis Transcriptome MicroArray) <www.catma.org>	Contains 150 to 500 bp genomic fragments amplified by PCR representing ca. 21,000 genes

ate synergies for the potential users. But although public data from expression studies are available (Table 10.4), information about the expression profile of a particular gene can still be hard to obtain. Initial microarrays covered only a part of the genome, and the description of the experiment was not always informative. The proposals concerning the Minimum Information About a Microarray Experiment to be included in data submissions, the so-called MIAME standards (Brazma et al., 2001), in particular have improved the latter aspect.

The initiative to prepare RNA systematically from the Columbia ecotype at many well-defined stages and conditions and to hybridize these samples with one type of array is to be welcomed. This so-called Compendium of Arabidopsis Gene Expression (Table 10.4) would greatly assist in the search for phenotypes in the case of knockout analysis of lesser-known genes by suggesting appropriate conditions or by clarifying whether potentially redundant genes indeed do share a similar expression pattern, or which genes might be involved in the same complex.

The importance of protein complexes is well established. Thus, finding the interacting partners for a certain protein has a high priority if insight into the molecular mechanism is desired. If some of the interacting partners have been studied before, their identification is very informative if little is known about the gene under study. Usually, potential interactions are identified by yeast two-hybrid screening (Immink and Angenent, 2002). The required material for do-it-yourself efforts (e.g., Clontechs Matchmaker or Invitrogen's Gateway-based systems) can be obtained, but to the best of my knowledge, open applications to a central service facility are not possible due to lack of funding, although facilities have been established that could handle approximately 100 genes per week with about 50 percent true interactions discovered (J. Uhrig, personal communication). Yet, I would expect that also in this case the experience of a dedicated facility could be invaluable to sort out artifacts or other problems, besides offering economy of scale and automation possibilities. Although the same argument can be applied to other approaches, such as proteomics, metabolic profiling, or single-cell transcription profiles (Kehr, 2001), I expect that two-hybrid screening in particular is often on the to-do list but is not always successfully completed due to practical obstacles or time constraints.

WEAKNESSES

It is certain that the wide range of tools, as described previously, and their availability currently sets apart *Arabidopsis* from other flowering plants. This and an active *Arabidopsis* community are the strengths on which further successes can be built. However, *Arabidopsis* does also have weaknesses.

A negative trade-off of its small size is the effort required for crosses and the difficulty of isolating specific tissue, especially when large quantities are wanted. This can be a problem for biochemical experiments, for the preparation of cDNA libraries, or for specific RNA preparations for transcriptional-profiling studies using microarrays. Interesting developments in this respect are the sampling of single cells (Kehr, 2001), isolation of plant cells by laser-capture microdissection (Asano et al., 2002), or the combination of fluorescent protein expression with cell sorting by flow cytometry (Galbraith, 2003). However, these are not yet routine techniques, and one sometimes wishes that everything would be just a little bigger.

Further limitations come from a number of traits that can be studied only indirectly, and where other plants might be preferred, e.g., apomixis, fruit or beet development, interaction with rhizobia, C3/C4 metabolism, interaction with certain pests, etc. In general, it has become clear that *Arabidopsis* is a good model for plant science, but much less clear is how far this can be extended toward agronomical business.

Integration has been proposed as the key to applied plant genomics (Osterlund and Paterson, 2002). A comparison of *Arabidopsis* to other flowering plants by Barnes (2002) suggests that this will prove simplest with related plants, such as mustard or rape. Comparative sequence analyses have revealed microcolinearity with, e.g., tomato (Ku et al., 2000; Rossberg et al., 2001) or rice (Mayer et al., 2001), but in general there are many exceptions (Bennetzen, 2000). About 75 percent of genes in dicot species (Ku et al., 2000) and 40 percent in monocots (Mayer et al., 2001) are estimated to be in regions with useful colinearity with *Arabidopsis*. About 65 percent of the *Arabidopsis* protein set could be matched by a maize EST in a study of Brendel et al. (2002). With more sequences from other plants becoming available, educated guesses about functional homology can be based on BLAST data, comparison of intron/exon structure, similarity of neighboring genes, or, if available, expression data. Yet the answer of whether a particular *Arabidopsis* gene with an interesting phenotype has one or more functional homologues in other plants still requires commitment to substantial experimental work at a point where applicability is unclear.

THREATS TO CONTINUED IMPACT

Parts of the threats facing *Arabidopsis* are technical developments that make other species easier to handle. A technique such as TILLING is very suitable for application in all sorts of plants. Doing reverse genetics in other plants extends the range of mechanisms that can be studied to those processes for which *Arabidopsis* might not be the first choice if alternatives are available (e.g., fruit ripening, beet shape, C4 metabolism, apomixis, etc.), and performing studies directly in the plant of interest prevents such problems as translation of results from one plant to another. Best of all, one can try to cross the chemically induced, nontransgenic allele into elite plasm. Currently, there are still problems with the availability of one of the used enzymes (CEL I), and in order to design the primers, one needs sequence information, which requires cloning of the gene and/or its homologues. These are certainly manageable problems, and thus expectations for TILLING are high.

Another potential threat would be a breakthrough in targeted gene replacement, although eagerly awaited by all, which could of course result in a shift toward other species, in particular if some species turn out to be more amenable than others. In general, the application of techniques becomes increasingly complex with increasing genome size, and a good transformation system makes all the difference in a lot of molecular biological experiments. Thus, even under these scenarios, *Arabidopsis* is likely to remain prominent and will also receive stimulating impulses from progress and insights in other plants.

I consider the real threat to be that the contribution of *Arabidopsis* research comes too late or that impact is too indirect to convince skeptics that genomics or molecular biology approaches have a place in crop improvement. At least in some of the plant genomics programs, e.g., GABI and GENOPLANTE, funding for a significant *Arabidopsis* component was obtained by using potential applications as a bait, and certainly working on the basic science and applications side by side helps to obtain critical mass and momentum; yet general strategies for translation from *Arabidopsis* to other plants, and know-how about the efficiency, remain poorly developed. Life in the *Arabidopsis* community has become so comfortable, thanks to the excellent tools available, that stalking a family of genes or hunting for interacting factors or targets of "your favorite gene" is, for scientists, more attractive than moving to another plant in order to reduce-to-practice the originally found proof-of-principle.

I deliberately borrowed some terms from the world of intellectual property since, although patents are normally an incentive to pursue the translation from model to crop, this can be the opposite in cases where an expert on a certain crop or gene has no connection to the people who filed a claim based on results in *Arabidopsis.* Why should a scientist follow up an interesting hint from *Arabidopsis* if he or she cannot fully reap the fruits of the at least equally hard work to develop an application in a crop? Even industry, certainly in Europe, hesitates to invest at the level of *Arabidopsis,* given the uncertainties of efficient implementation of the obtained knowledge, in particular when a transgenic product is less preferred.

A general downturn in interest in plant sciences at this moment would surely be a threat to the existing *Arabidopsis* infrastructure. Only further refinement or expansion of tools is likely to be the right approach. It is in the interest of the *Arabidopsis* community to work on ways to increase the impact of plant molecular biology, e.g., by building bridges to other plants or by integrating the infrastructure from several plants so that switching between plants is enabled.

OPPORTUNITIES

This brings us to the question of where the opportunities are. There is still room for further integration of data on *Arabidopsis* itself. One wish concerns the annotations in the databases, where it would be helpful to have a clear distinction between experimentally verified functionalities and those resulting from gene-prediction algorithms. Another issue is completeness of experimental data. It is not unusual to encounter gene-phenotype descriptions that are highly informative concerning the particular trait or tissue of interest to the authors, but in which it remains unclear if effects outside the primary interest were not investigated or were absent.

There appears to be a high turnover of mutagenized lines in quick-and-dirty studies that have as a pragmatic objective the discovery of one robust, clear phenotype that can be published after focusing research on that line. We all are familiar with the tendency to use resources in a wasteful manner in circumstances of apparent abundance. Although absolutely understandable, it is a pity that the more subtle phenotypes tend to be dropped even though small, local changes without pleiotropic or secondary effects can be very informative in conjunction with data on other genes, in addition to the fact that for potential applications the absence of side effects is often just as important as the distinctiveness of the phenotype. The next technological frontier could be further development and proliferation of techniques for microdissection (Asano et al., 2002) and microanalysis (Kehr, 2001; Galbraith, 2003), *in planta* detection of protein interactions (Subramaniam et al., 2001), and simple *in planta* transient expression systems (Yang et al., 2000). The pace and efficiency of gene function discovery could also increase through assays based on cell cultures or protoplasts to test rapidly a large number of hypotheses that can be confirmed subsequently *in planta* once established in a simpler artificial system (Asai et al., 2002).

Strategically, it would be good to have easier access to some of the other models that could function as a first evaluation tool to test the general validity of observed phenomena and at the same time function as a jumping point into other genera. As a first line one can think of tomato (Emmanuel and Levy, 2002), *Medicago,* tobacco, or another plant with publicly available sequences, and rice (Goff et al., 2002; Yu et al., 2002). After some time, it would become easier to place observations in the context of related mechanisms of other plants, or even to start making reliable predictions about the outcome of experiments in other plants. Initiatives in this direction, such as SPUTNIK, a database platform for comparative plant genomics (Rudd et al., 2003), or the Plant Genome Database (www.plantgdb.org), have started, and with an increasing number of publications describing successful gene swapping between plants (Agius et al., 2003; Zhang and Blumwald, 2001), *Arabidopsis* is unlikely to disappear from research greenhouses for some time to come.

REFERENCES

Agius, F., Gonzalez-Lamothe, R., Caballero, J.L., Munoz-Blanco, J., Botella, M.A., Valpuesta, V. (2003). Engineering increased vitamin C levels in plants by overexpression of a D-galacturonic acid reductase. *Nat. Biotechnol.* 21: 177-181.

Alonso-Blanco, C., Koornneef, M. (2000). Naturally occurring variation in *Arabidopsis:* An underexploited resource for plant genetics. *Trends Plant Sci.* 5: 22-29.

Alonso-Blanco, C., Peeters, A.J., Koornneef, M., Lister, C., Dean, C., van den Bosch, N., Pot, J., Kuiper, M.T. (1998). Development of an AFLP based linkage map of Ler, Col and Cvi *Arabidopsis thaliana* ecotypes and construction of a Ler/Cvi recombinant inbred line population. *Plant J.* 14: 259-271.

Arabidopsis Genome Initiative (2000). Analysis of the genome sequence of the flowering plant *Arabidopsis thaliana. Nature* 408: 796-815.

Asai, T., Tena, G., Plotnikova, J., Willmann, M.R., Chiu, W.L., Gomez-Gomez, L., Boller, T., Ausubel, F.M., Sheen, J. (2002). MAP kinase signalling cascade in *Arabidopsis* innate immunity. *Nature* 415: 977-983.

Asano, T., Masumura, T., Kusano, H., Kikuchi, S., Kurita, A., Shimada, H., Kadowaki, K. (2002). Construction of a specialized cDNA library from plant cells isolated by laser capture microdissection: Toward comprehensive analysis of the genes expressed in the rice phloem. *Plant J.* 32: 401-408.

Balzergue, S., Dubreucq, B., Chauvin, S., Le-Clainche, I., Le Boulaire, F., de Rose, R., Samson, F., Biaudet, V., Lecharny, A., Cruaud, C., et al. (2001). Improved PCR-walking for large-scale isolation of plant T-DNA borders. *Biotechniques* 30: 496-498, 502, 504.

Barnes, S. (2002). Comparing *Arabidopsis* to other flowering plants. *Curr. Opin. Plant Biol.* 5: 128-134.

Baumann, E., Lewald, J., Saedler, H., Schulz, B., Wisman, E. (1998). Successful PCR-based reverse genetic screens using an En-1-mutagenised *Arabidopsis thaliana* population generated via single-seed descent. *Theoret. Appl. Genet.* 97: 729-734.

Bechtold, N., Ellis, J., Pelletier, G. (1993). In planta *Agrobacterium*-mediated gene transfer by infiltration of adult *Arabidopsis thaliana* plants. *C.R. Acad. Sci. Ser. III* 316: 1194-1199.

Bennetzen, J.L. (2000). Comparative sequence analysis of plant nuclear genomes: Microcolinearity and its many exceptions. *Plant Cell* 12: 1021-1029.

Bensen, R.J., Johal, G.S., Crane, V.C., Tossberg, J.T., Schnable, P.S., Meeley, R.B., Briggs, S.P. (1995). Cloning and characterization of the maize *An1* gene. *Plant Cell* 7: 75-84.

Borevitz, J.O., Liang, D., Plouffe, D., Chang, H.S., Zhu, T., Weigel, D., Berry, C.C., Winzeler, E., Chory, J. (2003). Large-scale identification of single-feature polymorphisms in complex genomes. *Genome Res.* 13: 513-523.

Bowman, J.L. (1994). *Arabidopsis: An Atlas of Morphology and Development.* New York: Springer-Verlag.

Boyes, D.C., Zayed, A.M., Ascenzi, R., McCaskill, A.J., Hoffman, N.E., Davis, K.R., Gorlach, J. (2001). Growth stage-based phenotypic analysis of *Arabidopsis:* A model for high throughput functional genomics in plants. *Plant Cell* 13: 1499-1510.

Brazma, A., Hingamp, P., Quackenbush, J., Sherlock, G., Spellman, P., Stoeckert, C., Aach, J., Ansorge, W., Ball, C.A., Causton, H.C., et al. (2001). Minimum information about a microarray experiment (MIAME): Toward standards for microarray data. *Nat. Genet.* 29: 365-371.

Brendel, V., Kurtz, S., Walbot, V. (2002). Comparative genomics of *Arabidopsis* and maize: Prospects and limitations. *Genome Biol.* 3: REVIEWS 1005.1-1005.6.

Brugmans, B., Fernandez del Carmen, A., Bachem, C.W., van Os, H., van Eck, H.J., Visser, R.G. (2002). A novel method for the construction of genome wide transcriptome maps. *Plant J.* 31: 211-222.

Buckler, E.S.T., Thornsberry, J.M. (2002). Plant molecular diversity and applications to genomics. *Curr. Opin. Plant. Biol.* 5: 107-111.

Cho, R.J., Mindrinos, M., Richards, D.R., Sapolsky, R.J., Anderson, M., Drenkard, E., Dewdney, J., Reuber, T.L., Stammers, M., Federspiel, N., et al. (1999). Genome-wide mapping with biallelic markers in *Arabidopsis thaliana. Nat. Genet.* 23: 203-207.

Clough, S.J., Bent, A.F. (1998). Floral dip: A simplified method for *Agrobacterium*-mediated transformation of *Arabidopsis thaliana. Plant J.* 16: 735-743.

Cogoni, C., Macino, G. (2000). Post-transcriptional gene silencing across kingdoms. *Curr. Opin. Genet. Dev.* 10: 638-643.

Colbert, T., Till, B.J., Tompa, R., Reynolds, S., Steine, M.N., Yeung, A.T., McCallum, C.M., Comai, L., Henikoff, S. (2001). High-throughput screening for induced point mutations. *Plant Physiol.* 126: 480-484.

Crowe, M.L., Serizet, C., Thareau, V., Aubourg, S., Rouze, P., Hilson, P., Beynon, J., Weisbeek, P., van Hummelen, P., Reymond, P., et al. (2003). CATMA: A complete *Arabidopsis* GST database. *Nucleic Acids Res.* 31: 156-158.

Emmanuel, E., Levy, A.A. (2002). Tomato mutants as tools for functional genomics. *Curr. Opin. Plant Biol.* 5: 112-117.

Feldmann, K., Marks, M. (1987). *Agrobacterium*-mediated transformation of germinating seeds of *Arabidopsis thaliana:* A non-tissue culture approach. *Mol. Gen. Genet.* 208: 1-9.

Galbraith, D.W. (2003). Global analysis of cell type-specific gene expression. *Comp. Funct. Genom.* 4: 208-215.

Garcia-Hernandez, M., Berardini, T.Z., Chen, G., Crist, D., Doyle, A., Huala, E., Knee, E., Lambrecht, M., Miller, N., Mueller, L.A., et al. (2002). TAIR: A resource for integrated *Arabidopsis* data. *Funct. Integr. Genomics* 2: 239-253.

Goff, S.A., Ricke, D., Lan, T.H., Presting, G., Wang, R., Dunn, M., Glazebrook, J., Sessions, A., Oeller, P., Varma, H., et al. (2002). A draft sequence of the rice genome *(Oryza sativa* L. ssp. *japonica). Science* 296: 92-100.

Guo, H.S., Fei, J.F., Xie, Q., Chua, N.H. (2003). A chemical-regulated inducible RNAi system in plants. *Plant J.* 34: 383-392.

Harmer, S.L., Hogenesch, L.B., Straume, M., Chang, H.S., Han, B., Zhu, T., Wang, X., Kreps, J.A., Kay, S.A. (2000). Orchestrated transcription of key pathways in *Arabidopsis* by the circadian clock. *Science* 290: 2110-2113.

Helariutta, Y., Fukaki, H., Wysocka-Diller, J., Nakajima, K., Jung, J., Sena, G., Hauser, M.T., Benfey, P.N. (2000). The SHORT-ROOT gene controls radial patterning of the *Arabidopsis* root through radial signaling. *Cell* 101: 555-567.

Hoth, S., Morgante, M., Sanchez, J.P., Hanafey, M.K., Tingey, S.V., Chua, N.H. (2002). Genome-wide gene expression profiling in *Arabidopsis thaliana* reveals new targets of abscisic acid and largely impaired gene regulation in the abi1-1 mutant. *J. Cell Sci.* 115: 4891-4900.

Immink, R.G., Angenent, G.C. (2002). Transcription factors do it together: The hows and whys of studying protein-protein interactions. *Trends Plant Sci.* 7: 531-534.

Jackson, A.L., Bartz, S.R., Schelter, J., Kobayashi, S.V., Burchard, J., Mao, M., Li, B., Cavet, G., Linsley, P.S. (2003). Expression profiling reveals off-target gene regulation by RNAi. *Nat. Biotechnol.* 21: 635-637.

Jander, G., Norris, S.R., Rounsley, S.D., Bush, D.F., Levin, I.M., Last, R.L. (2002). *Arabidopsis* map-based cloning in the post-genome era. *Plant Physiol.* 129: 440-450.

Kehr, J. (2001). High resolution spatial analysis of plant systems. *Curr. Opin. Plant Biol.* 4: 197-201.

Knight, J. (2003). Crop improvement: A dying breed. *Nature* 421: 568-570.

Krysan, P.J., Young, J.C., Sussman, M.R. (1999). T-DNA as an insertional mutagen in *Arabidopsis. Plant Cell* 11: 2283-2290.

Ku, H.M., Vision, T., Liu, J., Tanksley, S.D. (2000). Comparing sequenced segments of the tomato and *Arabidopsis* genomes: Large-scale duplication followed by selective gene loss creates a network of synteny. *Proc. Natl. Acad. Sci. USA* 97: 9121-9126.

Lee, J.Y., Lee, D.H. (2003). Use of serial analysis of gene expression technology to reveal changes in gene expression in *Arabidopsis* pollen undergoing cold stress. *Plant Physiol.* 132: 517-529.

Li, X., Song, Y., Century, K., Straight, S., Ronald, P., Dong, X., Lassner, M., Zhang, Y. (2001). A fast neutron deletion mutagenesis-based reverse genetics system for plants. *Plant J.* 27: 235-242.

Maes, T., De Keukeleire, P., Gerats, T. (1999). Plant tagnology. *Trends Plant Sci.* 4: 90-96.

Maleck, K., Levine, A., Eulgem, T., Morgan, A., Schmid, J., Lawton, K.A., Dangl, J.L., Dietrich, R.A. (2000). The transcriptome of *Arabidopsis thaliana* during systemic acquired resistance. *Nat. Genet.* 26: 403-410.

Marsch-Martinez, N., Greco, R., Van Arkel, G., Herrera-Estrella, L., Pereira, A. (2002). Activation tagging using the En-I maize transposon system in *Arabidopsis. Plant Physiol.* 129: 1544-1556.

Mayer, K., Murphy, G., Tarchini, R., Wambutt, R., Volckaert, G., Pohl, T., Dusterhoft, A., Stiekema, W., Entian, K.D., Terryn, N., et al. (2001). Conservation of microstructure between a sequenced region of the genome of rice and multiple segments of the genome of *Arabidopsis thaliana. Genome Res.* 11: 1167-1174.

Meinke, D.W., Meinke, L.K., Showalter, T.C., Schissel, A.M., Mueller, L.A., Tzafrir, I. (2003). A sequence-based map of *Arabidopsis* genes with mutant phenotypes. *Plant Physiol.* 131: 409-418.

Meissner, R.C., Jin, H., Cominelli, E., Denekamp, M., Fuertes, A., Greco, R., Kranz, H.D., Penfield, S., Petroni, K., Urzainqui, A., et al. (1999). Function search in a large transcription factor gene family in *Arabidopsis:* Assessing the potential of reverse genetics to identify insertional mutations in R2R3 MYB genes. *Plant Cell* 11: 1827-1840.

Meyers, B.C., Morgante, M., Michelmore, R.W. (2002). TIR-X and TIR-NBS proteins: Two new families related to disease resistance TIR-NBS-LRR proteins encoded in *Arabidopsis* and other plant genomes. *Plant J.* 32: 77-92.

Mozo, T., Dewar, K., Dunn, P., Ecker, J.R., Fischer, S., Kloska, S., Lehrach, H., Marra, M., Martienssen, R., Meier-Ewert, S., Altmann, T. (1999). A complete BAC-based physical map of the *Arabidopsis thaliana* genome. *Nat. Genet.* 22: 271-275.

Muskett, P.R., Clissold, L., Marocco, A., Springer, P.S., Martienssen, R., Dean, C. (2003). A resource of mapped dissociation launch pads for targeted insertional mutagenesis in the *Arabidopsis* genome. *Plant Physiol.* 132: 506-516.

Nadeau, J.H., Frankel, W.N. (2000). The roads from phenotypic variation to gene discovery: Mutagenesis versus QTLs. *Nat. Genet.* 25: 381-384.

Nordborg, M., Borevitz, J.O., Bergelson, J., Berry, C.C., Chory, J., Hagenblad, J., Kreitman, M., Maloof, J.N., Noyes, T., Oefner, P.J., et al. (2002). The extent of linkage disequilibrium in *Arabidopsis thaliana. Nat. Genet.* 30: 190-193.

Osterlund, M.T., Paterson, A.H. (2002). Applied plant genomics: The secret is integration. *Curr. Opin. Plant Biol.* 5: 141-145.

Page, D.R., Grossniklaus, U. (2002). The art and design of genetic screens: *Arabidopsis thaliana. Nat. Rev. Genet.* 3: 124-136.

Rhee, S.Y., Beavis, W., Berardini, T.Z., Chen, G., Dixon, D., Doyle, A., Garcia-Hernandez, M., Huala, E., Lander, G., Montoya, M., et al. (2003). The Arabidopsis Information Resource (TAIR): A model organism database providing a centralized, curated gateway to *Arabidopsis* biology, research materials and community. *Nucleic Acids Res.* 31: 224-228.

Rossberg, M., Theres, K., Acarkan, A., Herrero, R., Schmitt, T., Schumacher, K., Schmitz, G., Schmidt, R. (2001). Comparative sequence analysis reveals extensive microcolinearity in the lateral suppressor regions of the tomato, *Arabidopsis,* and *Capsella* genomes. *Plant Cell* 13: 979-988.

Rudd, S., Mewes, H.W., Mayer, K.F. (2003). Sputnik: A database platform for comparative plant genomics. *Nucleic Acids Res.* 31: 128-132.

Samson, F., Brunaud, V., Balzergue, S., Dubreucq, B., Lepiniec, L., Pelletier, G., Caboche, M., Lecharny, A. (2002). FLAGdb/FST: A database of mapped flanking insertion sites (FSTs) of *Arabidopsis thaliana* T-DNA transformants. *Nucleic Acids Res.* 30: 94-97.

Schoof, H., Zaccaria, P., Gundlach, H., Lemcke, K., Rudd, S., Kolesov, G., Arnold, R., Mewes, H.W., Mayer, K.F. (2002). MIPS *Arabidopsis thaliana* Database (MAtDB): An integrated biological knowledge resource based on the first complete plant genome. *Nucleic Acids Res.* 30: 91-93.

Sessions, A., Burke, E., Presting, G., Aux, G., McElver, J., Patton, D., Dietrich, B., Ho, P., Bacwaden, J., Ko, C., et al. (2002). A high-throughput *Arabidopsis* reverse genetics system. *Plant Cell* 14: 2985-2994.

Somerville, C. (2000). The twentieth century trajectory of plant biology. *Cell* 100: 13-25.

Somerville, C., Koornneef, M. (2002). A fortunate choice: The history of *Arabidopsis* as a model plant. *Nat. Rev. Genet.* 3: 883-889.

Steiner-Lange, S., Gremse, M., Kuckenberg, M., Nissing, E., Schachtele, D., Spenrath, N., Wolff, M., Saedler, H., Dekker, K. (2001). Efficient identification of *Arabidopsis* knock-out mutants using DNA-arrays of transposon flanking sequences. *Plant Biol.* 3: 391-397.

Subramaniam, R., Desveaux, D., Spickler, C., Michnick, S.W., Brisson, N. (2001). Direct visualization of protein interactions in plant cells. *Nat. Biotechnol.* 19: 769-772.

Sundaresan, V., Springer, P., Volpe, T., Haward, S., Jones, J.D., Dean, C., Ma, H., Martienssen, R. (1995). Patterns of gene action in plant development revealed by enhancer trap and gene trap transposable elements. *Genes Dev.* 9: 1797-1810.

Sussman, M.R., Amasino, R.M., Young, J.C., Krysan, P.J., Austin-Phillips, S. (2000). The *Arabidopsis* knockout facility at the University of Wisconsin-Madison. *Plant Physiol.* 124: 1465-1467.

Tax, F.E., Vernon, D.M. (2001). T-DNA-associated duplication/translocations in *Arabidopsis:* Implications for mutant analysis and functional genomics. *Plant Physiol.* 126: 1527-1538.

Thorneycroft, D., Sherson, S.M., Smith, S.M. (2001). Using gene knockouts to investigate plant metabolism. *J. Exp. Bot.* 52: 1593-1601.

Till, B.J., Reynolds, S.H., Greene, E.A., Codomo, C.A., Enns, L.C., Johnson, J.E., Burtner, C., Odden, A.R., Young, K., Taylor, N.E., et al. (2003). Large-scale discovery of induced point mutations with high-throughput TILLING. *Genome Res.* 13: 524-530.

Tissier, A.F., Marillonnet, S., Klimyuk, V., Patel, K., Torres, M.A., Murphy, G., Jones, J.D. (1999). Multiple independent defective suppressor-mutator transposon insertions in *Arabidopsis:* A tool for functional genomics. *Plant Cell* 11: 1841-1852.

Varotto, C., Pesaresi, P., Meurer, J., Oelmuller, R., Steiner-Lange, S., Salamini, F., Leister, D. (2000). Disruption of the *Arabidopsis* photosystem I gene *psaE1* affects photosynthesis and impairs growth. *Plant J.* 22: 115-124.

Waterhouse, P.M., Helliwell, C.A. (2003). Exploring plant genomes by RNA-induced gene silencing. *Nat. Rev. Genet.* 4: 29-38.

Waugh, R., Dear, P.H., Powell, W., Machray, G.C. (2002). Physical education: New technologies for mapping plant genomes. *Trends Plant Sci.* 7: 521-523.

Weigel, D., Ahn, J.H., Blazquez, M.A., Borevitz, J.O., Christensen, S.K., Fankhauser, C., Ferrandiz, C., Kardailsky, I., Malancharuvil, E.J., Neff, M.M., et al. (2000). Activation tagging in *Arabidopsis. Plant Physiol.* 122: 1003-1013.

Weigel, D., Glazebrook, J. (2002). *Arabidopsis: A Laboratory Manual.* New York: CSHL Press.

Wisman, E., Hartmann, U., Sagasser, M., Baumann, E., Palme, K., Hahlbrock, K., Saedler, H., Weisshaar, B. (1998). Knock-out mutants from an En-1 mutagenized *Arabidopsis thaliana* population generate phenylpropanoid biosynthesis phenotypes. *Proc. Nat. Acad. Sci. USA* 95: 12432-12437.

Yang, Y.N., Li, R.G., Qi, M. (2000). In vivo analysis of plant promoters and transcription factors by agroinfiltration of tobacco leaves. *Plant J.* 22: 543-551.

Yu, J., Hu, S., Wang, J., Wong, G.K., Li, S., Liu, B., Deng, Y., Dai, L., Zhou, Y., Zhang, X., et al. (2002). A draft sequence of the rice genome *(Oryza sativa* L. ssp. *indica). Science* 296: 79-92.

Zhang, H.X., Blumwald, E. (2001). Transgenic salt-tolerant tomato plants accumulate salt in foliage but not in fruit. *Nat. Biotechnol.* 19: 765-768.

Chapter 11

Rice Functional Genomics: From Nucleotide Sequence to Gene Function

Takuji Sasaki

INTRODUCTION

The elucidation of the rice genome sequence has put rice at the forefront of research in functional genomics among cereal crops. After the completion of an accurate rice genome sequence, substantial advances in both forward and reverse genetics are highly anticipated. Characterization of the functions of all of the genes that comprise the rice genome is of utmost importance in agriculture, and, in particular, the identification of genes and genomic regions controlling agronomically important traits such as high yield and disease resistance is a great challenge to the rice research community.

Rice is undoubtedly one of the most, if not even the most, important plant species for humankind, providing a major food source for more than half of the world's population. It is also considered an iconic plant in many of the areas where it is widely cultivated. In southern Asia, where rice cultivation is believed to have originated, the daily life of the local people is deeply intertwined with many myths associated with rice and special events are usually celebrated by cooking delicacies made of waxy rice. Cultivated rice, which consists of *japonica* and *indica* ecotypes of *Oryza sativa,* is thought to have originated from wild rice *O. nivara* (Khush, 1997). During its long history of cultivation and selection, many varieties adapted to local environmental conditions have been generated. As a result, a wide range of alleles for a particular trait exist due to the pressures of artificial selection. This differentiates rice

The author greatly acknowledges Dr. Baltazar A. Antonio of the National Institute of Agrobiological Sciences for valuable suggestions and critical reading of the manuscript.

from the weedy crucifer plant *Arabidopsis thaliana*, the only other plant whose genome sequence has been decoded (see Chapter 10).

Despite the long history of rice cultivation, breeding for agronomic traits depended mainly on observation of phenotypes. However, breeding for some traits that need to be addressed immediately, such as high yield under suboptimal cultivation conditions, is very time-consuming or, in most cases, simply unattainable. Considering the expected increase in the world population in the next 30 years (http://www.un.org/popin/data.html), especially in Asia, Africa, and Latin America where rice is the main staple, improvement of rice production and of its grain quality should be given top priority. In consequence, functional genomics research in rice should focus on a genome-wide approach, combining strategies that utilize sequence variation, mutant phenotypes, and breeding applications.

Rice functional genomics focusing on the identification of genes corresponding to a target trait and on the elucidation of physiological function of these genes is expected to basically contribute to further improvement of the rice plant. Eventually, this will lead to the assignment of some degree of function to all genes in plants. In particular, understanding the function of rice genes is also crucial for understanding the function of orthologous genes in other cereal crops such as maize and wheat. This is because the nucleotide sequences, as well as the order, of genes are similar among these important crops (Bennetzen and Ma, 2003; Devos and Gale, 1997). This information could be useful in both artificial engineering of genes and introgression of new genes among cereal genomes and in applying genomic tools and techniques obtained in rice plants to other cereal species.

High-quality and accurate sequence information of the entire genome is now available for the *japonica* variety *Nipponbare* (Sasaki, Matsumoto, et al., 2002; Feng et al., 2002; The Rice Chromosome 10 Sequencing Consortium, 2003) and, although with a lower quality, also for the *indica* variety *93-11* (Yu et al., 2002). Such an accumulation of sequence information for the two subspecies of rice signifies the start of a new era in rice functional genomics. Recent advancements in forward and reverse genetics to identify gene-function relationships in rice are introduced here with reference to databases and materials for further studies. Strategies using sequence information from other genomes, such as *Arabidopsis*, for the functional analysis of rice genes has, in some cases, been proven to be successful to shortcut both forward and reverse genetics in rice. Moreover, a careful database analysis of information from other species potentially useful for rice research is undoubtedly important, and researchers should commence such efforts prior to starting their experimental work. In this chapter, the current status of rice functional genomics is reviewed to provide new insights for developing novel strategies for further development of plant science and applications to agriculture.

FUNCTIONAL GENOMICS BASED ON FORWARD GENETICS

The sequencing of the entire rice genome and of many rice ESTs (expressed sequence tags) has greatly facilitated genetic analysis from "phenotypes to genes," the so-called forward genetics approach. With the recent availability of high-density genetic maps and physical mapping resources, positional cloning of virtually any gene is now feasible in rice. The strategy of map-based or positional cloning depends on the following factors: (1) an accurately segregating population of the target trait; (2) availability of a large number of DNA markers and feasibility of their application; (3) availability of suitable genomic libraries and/or nucleotide sequence information corresponding to the candidate genomic region; and (4) an efficient transformation system and evaluation system of transgenic plants. The importance of these factors has been reviewed by Tanksley et al. (1995) and by Martin (1998).

The famous garden pea used by Mendel in elucidating the concept and basic rules of heredity should remind us of the importance of the purity of materials used for genetic analyses. One of the reasons why only two of seven traits used by Mendel have been identified in subsequent molecular studies was the lack of basic molecular tools such as DNA markers for common pea.

In the case of rice, almost all mutants were generated in the course of breeding programs. These usually unfavorable traits are generally ignored by breeders so that geneticists do not necessarily benefit from these resources. Fortunately, natural mutants in rice as well as mutants induced by gamma ray and chemical mutagens have been collected (Kinoshita, 1995). Among 463 characterized mutations, 185 were assigned to individual chromosomal regions by linkage analysis using phenotypic markers. Yoshimura (2001) mapped 95 of the 185 mutations to single loci using DNA markers. These phenotypes are called marker genes and are useful for the characterization of the chromosome assignment of unmapped phenotypes by conventional crossing. At present, these mutant phenotypes are the targets for gene identification and the plant resources are maintained under good conditions. Recent tools for insertional mutagenesis by several biological factors described in the following sections are also useful for forward genetics in gene identification by tagging.

DNA Marker Systems

The density of DNA markers, as well as the development of new types of markers, is crucial for fine mapping of target phenotypes. The genome size of rice (430 Mbp) is small compared to other grass species, and the size of intergenic regions is expected to be relatively small. In addition, the annotation of the nucleotide sequence of the rice genome has indicated that most genes in rice are duplicated in a tandem manner (Sasaki, Matsumoto, et al., 2002). This suggests the importance of narrowing down the candidate genomic region to as close as genetic analysis allows. However, it is difficult to identify true correspondence of gene to phenotype even when using an *Agrobacterium*-mediated transformation system that has been successfully established also for rice plants (Hiei et al., 1997). Before the availability of the massive genomic nucleotide sequence, the development of molecular markers, such as the ones based on restriction fragment length polymorphism (RFLP) (Kurata et al., 1994) or simple sequence repeat (SSR), was time-consuming and laborious, though the validity of SSR in polymorphism detection was recognized. McCouch et al. (2002) developed 2,240 new SSR markers using the published genome sequences. At present, accurate genome sequence information of *japonica* and *indica* rice should provide sufficient resources for the systematic development of additional markers. The genome sequence information is primarily useful to detect repeat sequences and other types of polymorphism among rice varieties.

For the detection of polymorphisms among closely related varieties, however, only single nucleotide polymorphism (SNP) analysis is effective. The frequency of polymorphisms detected by SNPs between the *japonica* varieties *Nipponbare* and *Koshihikari* and between *Nipponbare* and *Kitaake* were 0.04 percent and 0.03 percent, respectively (Nasu et al., 2002). On the other hand, SNP frequency between *Nipponbare* and the *indica* variety *Kasalath* was 0.70 percent (Nasu et al., 2002). Although we might not encounter the situation of insertion or deletion of single genes as the course for allelic variation among very closely related varieties, there exist, in fact, differences at the DNA level between these varieties. Association studies of SNPs and phenotypic variations among many closely related varieties will assist in the correlation of sequence information and subtle changes of phenotypes among them.

Another useful polymorphic DNA marker is the miniature inverted-repeat transposable elements (MITEs) of rice (Bureau et al., 1996). Several elements, such as *Tourist, Stowaway,* and *Gaijin,* have been identified and classified into the MITE family (Bureau et al., 1996). There exists sequence similarity within a family, but not between families. MITE sequences are short (100-500 bp) and have conserved terminal inverted repeats. All these characteristics make MITEs suitable DNA markers, on the condition that polymorphisms among MITEs exist among rice varieties. Takagi et al. (2003) identified the frequency of polymorphisms of several rice MITEs as 0.07 to 0.14 percent and 0.48 to 0.73 percent among *japonica* cultivars and between *japonica* and *indica* cultivars, respectively. Komori and Nitta (2003) genetically mapped 20 MITEs as PCR markers. These results indicate the utility of MITEs as DNA markers for fine mapping using the genome sequence information as a resource for MITE-based markers.

The complete rice genome nucleotide sequence of *Nipponbare* with precise positional information will be available by the end of 2004 (http://rgp.dna.affrc.go.jp/IRGSP/). This is primarily useful for the identification of genes in other *japonica* varieties that are supposed to be allelic to *Nipponbare*. However, in the case of genes that have evolved relatively recently, including many disease resistance genes, the corresponding allele may not be found in *Nipponbare* (Leister et al., 1998). In this case, genomic libraries, such as bacterial artificial chromosome (BAC) or cosmid libraries, must be constructed for each target variety.

An efficient system for transformation in rice was established using the Ti plasmid carried by *Agrobacterium tumefaciens* (Zambryski et al., 1980). A drawback of this approach is that infection can be performed only at the callus stage. However, the efficiency of integration of specific DNA fragments by *A. tumefaciens* into the rice genome is higher than the one of the particle bombardment method. Very recently, significant advancement in transformation has been successfully attained employing homologous recombination (Terada et al., 2002). In this case, the rice *Waxy* gene was targeted by the hygromycin-B-resistant gene, *hpt,* with a frequency of about 1 percent (Terada et al., 2002). This has been made possible using active callus material and the diphtheria toxin as a selection marker for the targeting vector. This innovation in transformation technology is recognized as a key step in controlling positional insertion of any gene into the genome and in solving the problems caused by uncontrolled insertion of rearranged genes.

Gene identification by map-based cloning is particularly useful in isolating genes involved in phenotypes of so-called quantitative traits or QTLs. This includes many important traits in breeding that are controlled by several genes. An advanced backcross strategy was first developed by Tanksley to isolate and identify all the QTL genes involved in the fruit weight of tomato (Alpert et al., 1995; Frary et al., 2000). Subsequent studies in rice are focusing on many agronomic traits; however, it takes time and labor to establish a series of chromosomal substitution lines for each combination of *indica* and *japonica* cultivars. In addition, the selection of parent varieties for crossing represents an additional problem because there are many traits to be tackled for gene identification and this requires thorough characterization of the genetic background of the parent varieties. To construct substitution lines in rice, combinations of the *indica* and *japonica* subspecies of *O. sativa,* as well as combinations of *O. sativa* and wild relatives with an AA genome, have been chosen (Yano, 2001; Kurakazu et al., 2001). Yano et al. (2001) tried to identify genes involved in flowering time or heading date that is significant for determination of date of sowing and harvest of rice. Using genetic material derived from a *Nipponbare* × *Kasalath* cross, 14 QTLs have been detected so far (Yano et al., 2001). By selecting nearly isogenic lines for each QTL found among progenies of a backcross of the F1 to *Nipponbare* using DNA markers as indices, these QTLs have been isolated and targeted for gene identification. Factors affecting flowering time are mainly day length, temperature, and nutrition. Among them, day length is the most crucial for rice plants and is most easily and reproducibly controlled under artificial conditions. So far, three genes that are involved in the control of flowering time of rice in response to changes in day length have been identified by map-based cloning (Yano et al., 2000; Takahashi et al., 2001; Kojima et al., 2002).

The importance of gene identification involved in QTLs has been recognized by studying the difference between nucleotide sequences of allelic genes. Among the three identified genes involved in rice flowering time, *Hd1* of *Kasalath* has insertion/deletion of amino acids compared to that of *Nipponbare* (Yano et al., 2000), and *Hd6* of *Nipponbare* has been found to possess a truncated sequence (Takahashi et al., 2001). However, *Hd3a* of both cultivars showed the same amino acid sequence, except for one different residue that looks indifferent in terms of activity of the translated products (Kojima et al., 2002). Although only these three cases have been studied so far, the results clearly indicate the complex nature of genes involved in QTL and suggest the need for further comparison of alleles from many rice varieties, or even wild relatives, to understand the effect of interaction of allelic products. Similar results were also observed in the genus *Lycopersicon* (Nesbitt et al., 2002).

During the process of domestication of rice, humankind has continuously exerted selection pressure on rice that induced favorable change in the rice genome. These variations are not the same as the natural variation observed in *Arabidopsis* (Johanson et al., 2000). However, to obtain more favorable traits for rice cultivation, these variations should be well understood from the point of view of agriculture, as well as from the point of view of evolution. In this sense, positional cloning of relevant loci, followed by comparative analyses of different cultivars, provides an important tool. In Table 11.1, the genes so far identified by map-based cloning (forward genetics) are listed.

TABLE 11.1. Agronomically important traits in rice that have been characterized by forward genetics (F), reverse genetics (R), or cloning by homology (H).

Gene	Phenotype	Characteristics of predicted gene product	Method	Chromosome	Reference
Xa1	resistance to *X. oryzae* pv. *oryzae* (race 1)	NBS-LRR type of plant R-gene	F	4	Yoshimura et al., 1998
Xa21	resistance to *X. oryzae* pv. *oryzae* (race 21)	receptor-LRR type of plant R-gene	F	11	Wang et al., 1998
Pib	resistance to *M. grisea* (race 003)	NBS-LRR type of plant R-gene	F	2	Wang et al., 1999
Pita	resistance to *M. grisea*	NBS-LRR type of plant R-gene	F	12	Bryan et al., 2000
d1	dwarf (Daikoku)	alpha subunit of heterotrimeric GTP-binding protein	F	5	Ashikari et al., 1999; Fujisawa et al., 1999
sd1	semi-dwarf (Dee-geo-woo-gen, IR8)	GA20 oxidase (GA20ox-2)	F	1	Sasaki et al., 2002; Ashikari et al., 2002; Monna et al. 2002; Spielmeyer et al., 2002
Spl7	spotted leaf (lesion-mimic)	heat stress transcription factor	F	5	Yamanouchi et al., 2002
slr1	slender	transcriptional factor of GRAS family	H	3	Ikeda et al., 2001
moc1	abnormal tillering	GRAS family nuclear protein	F	6	Li et al., 2003
d61	dwarf	brassinosteroid receptor kinase	H	1	Yamamura et al., 2000
OsTB1	reduced lateral branching	transcription factor with helix-loop-helix motif	H	3	Takeda et al., 2003
GA3ox2 (d18)	dwarf	GA 3β-hydroxylase (GA3ox2)	H	1	Itoh et al., 2001
GA3ox1	unknown	GA 3β-hydroxylase (GA3ox1)	H	5	Itoh et al., 2001

gid2	dwarf	F-box protein	F	2	Sasaki et al., 2003
RCN1	delay of phase transition	TFL1/CEN homologue	H	11	Nakagawa et al., 2002
RCN2	delay of phase transition	TFL1/CEN homologue	H	2	Nakagawa et al., 2002
brd1	dwarf	brassinosteroid-6-oxidase	H	3	Mori et al., 2002; Hong et al., 2002
Hd1	QTL of flowering time	transcription factor, *CONSTANS* family	F	6	Yano et al., 2001
Hd3a	QTL of flowering time	*FT* family	F	6	Kojima et al., 2002
Hd6	QTL of flowering time	casein kinase *CK2*α family	F	3	Takahashi et al., 2001
OSH15	dwarf, *d6*-type	knotted-type homeobox gene family	R	7	Sato et al., 1999
phyA	insensitivity to far-red light	phytochrome	R	3	Takano et al., 2001
Osaba1	strongly viviparous	zeaxanthin epoxidase	F	n.d.	Agrawal et al., 2001
Ostatc	viviparous	Sec-independent translocase TatC	F	n.d.	Agrawal et al., 2001
OsCHLH	chlorophyll deficiency	largest subunit of magnesium chelatase	F	n.d.	Jung et al., 2003
slg	slender glume	ubiquitin-related modifier	F, R	7	Nakazaki et al., 2003

Note: n.d. = not determined.

FUNCTIONAL GENOMICS BASED ON REVERSE GENETICS

Another powerful strategy for the identification of gene functions is facilitated by biological mutagens, disrupting the gene of interest using a "foreign" piece of DNA, and then determining the phenotypic changes in the resulting mutant. In this sense, both rice and *Arabidopsis* are well-suited species for this strategy. In rice, three main biological mutagens are used for gene tagging or gene disruption, namely, *Ac/Ds* of maize (McClintock, 1956), T-DNA of *Agrobacterium* (Zambryski et al., 1980), and *Tos17* of *O. sativa* (Hirochika, 2001). While *Ac/Ds*- and T-DNA-mediated gene introduction results in genetically modified rice plants, *Tos17* is an endogenous retrotransposon of rice and the modified plants are not classified as being genetically modified. This is a major advantage in the context of the current public concern over growing genetically modified plants in the field that promotes dissemination of pollen. The high costs that may be required in maintaining a closed greenhouse may be a disadvantage of a strategy that involves introduction of foreign genes (T-DNA, transposons). Reverse genetics employing insertional mutagens requires a systematic PCR-based screening to identify insertion-tagged alleles of the genes of interest. The now available accurate genome sequence is a prerequisite to precisely identifying the disrupted site and to providing additional information of allelic variation in disruption, particularly if more than one site of the same gene is disrupted (Ikeda et al., 2001; Takano et al., 2001). This is useful in understanding the crucial portion of the translated gene product by knowing the relationship of the inserted position and the degree of change in phenotype.

Transposons

The maize *Ac/Ds (Activator/Dissociation)* system was first identified by McClintock (1956) and has become a powerful tool for gene isolation. The first gene isolated by transposon tagging was the maize *bronze* gene, a key enzyme of anthocyanin synthesis (Federoff et al., 1984). *Ac/Ds* has also been widely used for gene tagging in heterologous plants, including tobacco, tomato, *Arabidopsis,* petunia, flax, carrot, potato, soybean, and rice. Populations used for reverse genetics based on *Ds* insertions consist of 931 *Arabidopsis* (Parinov et al., 1999) and 1,173 rice (Ito, Motohashi, et al., 2002) gene-disrupted lines, and inserted *Ds* elements were shown to be evenly distributed. In the case of rice, Chin et al. (1999) reported the efficiency of insertion of *Ds* by *Agrobacterium*-mediated transformation with nearly 80 percent of *Ds* elements being excised from the original T-DNA sites when *Ac* cDNA was expressed under control of the CaMV 35S promoter. At present, a large-scale research program on *Ac/Ds* insertional mutagenesis of rice is being undertaken at CSIRO, Australia, to propagate and share the resulting materials worldwide (Upadhayaya et al., 2002). In addition to the use of the nonautonomous *Ds* element, transposition of the autonomous element *Ac* within the rice genome was also attempted. Enoki et al. (1999) have analyzed 559 transgenic rice plants and showed that 18.4 percent had newly transposed *Ac* elements.

T-DNA

T-DNA mutagenesis is another powerful method to study the function of genes (see Chapter 2). T-DNA insertion is a random event and the inserted sequences are stably inherited through multiple generations. T-DNA insertion has been successfully used to generate many insertional mutant populations of *Arabidopsis* since 1991 (Feldman, 1991). A PCR-based screening system of T-DNA insertion lines of *Arabidopsis* was first used to search for mutants of actin (McKinney et al., 1995). At present, DNA pools of T-DNA insertion lines and seeds of the mutants are available from public and private sites. In the case of rice, Jeon et al. (2000) generated 18,358 T-DNA insertion lines with a total of 25,000 tagged genes. They used a vector carrying the GUS reporter gene and detected its expression in leaves, roots, flowers, and seeds, of which a fraction of 1.6 to 2.1 percent of these tested organs were GUS-positive and their expression patterns were organ or tissue specific or ubiquitous in all parts of the plants. It is foreseeable that in rice T-DNA insertion lines can be used for a PCR-based screening system as in *Arabidopsis* and become an effective strategy toward the production of genome-wide mutations in rice (Krysan et al., 1999).

Retrotransposons

Specific to rice is insertional mutagenesis based on the endogenous retrotransposon *Tos17* (Hirochika et al., 1996). The two original copies of endogenous *Tos17* in the *Nipponbare* genome can be transposed under stress conditions such as tissue culture and viral infection. During three to five months tissue culture as a callus, 5 to 30 copies of *Tos17* are generated and transposed, and these transpositions are maintained in the regenerated rice plants. *Tos17* is a retrotransposon of 4,114 bp length and carries identical 138 bp LTRs at its 5' and 3' termini. The transposed *Tos17* is stably maintained in progenies, and the genomic position of its integration site is easily identified by thermal asymmetric interlaced PCR (TAIL-PCR) (Liu and Whittier, 1995) or suppression PCR (Siebert et al., 1995) screening. These characteristics are adequate for PCR screening of extracted and pooled DNAs of each mutant (Hirochika, 1997; Miyao et al., 1998). Since the first report by Sato et al. (1999) on the identification of a mutation of a rice homeobox gene causing dwarfness, several cases of *Tos17* insertion have been reported (Takano et al., 2001; Ito, Hirochika, et al., 2002). Gene tagging by *Tos17* is particularly useful in gene identification in rice-based forward genetics (Agrawal et al., 2001). Examples of gene identification are listed in Table 11.1.

Several points must be considered to ensure success in using reverse genetics strategies. First, the insertions must be evenly distributed to saturate all of the genes within the genome. Assuming that the average size of a rice gene is about 5 kb and the number of transposed *Tos17* in each regenerated rice plant is about eight, Hirochika (2001) estimated that efficient tagging could be performed with 50,000 mutant lines. However, biased preference of *Tos17* to insert into low-copy gene-rich genomic regions (Hirochika, 2001) suggests that saturating the entire genome with only *Tos17* might be difficult. Second, the "purity" of a mutation induced by a particular mutagen should be checked to accurately assign mutant phenotypes to the causing mutations. However, mutants obtained by both *Ds* and *Tos17* insertion systems are not necessarily caused by these biological mutagens. With a certain frequency, such mutations are induced during tissue culture conditions resulting in undefined mutations, the so-called somaclonal variation. This situation is different from that in *Arabidopsis,* which can be transformed very easily with *Agrobacterium* and where the chance of unfavorable genomic variation is reduced. Third, the efficiency of insertional mutagenesis is variety dependent. In the case of *Tos17,* transposition occurs moderately in *Nipponbare,* but no transposition was observed under the same activation condition in another elite *japonica* variety, *Koshihikari,* that carries one original copy. Furthermore, although the existence of *Tos17* is observed in the rice ecotype *indica,* the activity of transposition in some varieties is low. Very recently, the possibility of gene tagging by MITEs was shown in *indica* rice (Jiang et al., 2003). Since about 90 percent of rice production in the world is based on the *indica* variety, it is also desirable to obtain the basic information on gene function in this variety and to study phenotypic and allelic variation in as many rice varieties as possible.

FUNCTIONAL GENOMICS BASED ON ACTIVATION TAGGING

Annotation of the entire genome sequence has indicated that the rice genome not only is richer in genes than *Arabidopsis* but also shows a more prominent degree of redundancy of genes (Sasaki, Matsumoto, et al., 2002). In addition, the rice genome contains many pseudogenes and genes with suboptimal expression. Activation tagging could be an effective method in characterizing these features of the genome and in the identification of novel gene functions. In the case of *Arabidopsis,* a T-DNA vector with multimerized CaMV 35S enhancers has been developed (Weigel et al., 2000; Nakazawa et al., 2003) and more than 30 dominant mutants have been characterized so far (Weigel et al., 2000). It appears that efforts in activation tagging should be increased in accordance with the progress made with loss-of-function methods. If the size of the rice plants is the limiting factor of this method, the system of *Arabidopsis* should be applied temporarily for large-scale screening of orthologues of rice genes, and reconfirmation of the true function of the selected gene must be performed in rice plants.

CONCLUSION

Further development and systematic use of rice functional genomics is a prerequisite to understand—in a global manner—rice gene functions, which in turn is necessary to develop strategies to overcome various stresses, to increase rice production, and to improve the quality of rice as our main staple food. At present, the accurate genome sequence and its annotation are publicly available and systems of both forward and reverse genetics methods are widely applied by the rice research community. However, not all data resulting from functional analyses are fully available to the public. This is partly because of the commercial value of such information, and, therefore, many groups insist on attaching intellectual property rights to newly identified genes. Currently, genomics needs much investment and researchers are required to produce results to facilitate immediate application of new technology in a profit-oriented economy. Even understanding this situation, information on rice gene functions must be much more accessible to researchers to facilitate new scientific discoveries.

Both rice and *Arabidopsis,* which represent the model monocot and dicot plants, should provide the resources for this endeavor. Although they are very different plants and orthologous genes may not necessarily work in the same way (Yano et al., 2000; Kyozuka et al., 1998), combined information may provide the key to unlocking the mystery of the plant world. At the same time, we must be wise enough to share knowledge in rice functional genomics to supply food for all humankind.

REFERENCES

Agrawal, G.K., Yamazaki, M., Kobayashi, M., Hirochika, R., Miyao, A., Hirochika, H. (2001). Screening of the rice viviparous mutants generated by endogenous retrotransposon *Tos17* insertion: Tagging of a zeaxanthin epoxidase gene and a novel *OsTATC* gene. *Plant Physiol.* 125: 1248-1257.

Alpert, K.B., Grandillo, D., Tanksley, S. (1995). *fw2.2:* A major QTL controlling fruit weight is common to both red- and green-fruited tomato species. *Theor. Appl. Genet.* 91: 994-1000.

Ashikari, M., Sasaki, A., Ueguchi-Tanaka, M., Itoh, H., Nishimura, A., Swapan, D., Ishiyama, K., Saito, T., Kobayashi, M., Khush, G.S., et al. (2002). Loss-of-function of rice gibberellin biosynthetic gene, *GA20 oxidase (GA20ox-2),* led to the rice "green revolution." *Breeding Sci.* 52: 143-150.

Ashikari, M., Wu, J., Yano, M., Sasaki, T., Yosimura, A. (1999). Rice gibberellin-insensitive dwarf mutant gene Dwarf 1 encodes the α-subunit of GTP-binding protein. *Proc. Natl. Acad. Sci. USA* 96: 10284-10289.

Bennetzen, J.L., Ma, J. (2003). The genetic colinearity of rice and other cereals on the basis of genomic sequence analysis. *Curr. Opin. Plant Biol.* 6: 128-133.

Bryan, G.T., Wu, K.-S., Farrall, L., Jia, Y., Hershey, H.P., McAdams, S.A., Faulk, K.N., Donaldson, G.K., Tarchini, R., Valent, B. (2000). A single amino acid difference distinguishes resistant and susceptible alleles of the rice blast resistance gene *Pi-ta. Plant Cell* 12: 2033-2045.

Bureau, T.E., Ronald, P.C., Wessler, S.R. (1996). A computer-based systematic survey reveals the predominance of small inverted-repeat elements in wild-type rice genes. *Proc. Natl. Acad. Sci. USA* 93: 8524-8529.

Chin, H.G., Choe, M.S., Lee, S.-H., Park, S.H., Koo, J.C., Kim, N.Y., Lee, J.J., Oh, B.G., et al. (1999). Molecular analysis of rice plants harboring an Ac/Ds transposable element-mediated gene trapping system. *Plant J.* 19: 615-623.

Devos, K.M., Gale, M.D. (1997). Comparative genetics in the grasses. *Plant Mol. Biol.* 35: 3-15.

Enoki, H., Izawa, T., Kawahara, M., Komatsu, M., Koh, S., Kyozuka, J., Shimamoto, K. (1999). *Ac* as a tool for the functional genomics of rice. *Plant J.* 19: 605-613.

Federoff, N.V., Furtek, D.B., Nelson Jr., O.E. (1984). Cloning of the *bronze* locus in maize by a simple and generalizable procedure using the transposable controlling element *Activator (Ac). Proc. Natl. Acad. Sci. USA* 81: 3825-3829.

Feldmann, K.A. (1991). T-DNA insertion mutagenesis in *Arabidopsis:* Mutational spectrum. *Plant J.* 1: 71-83.

Feng, Q., Zhang, Y., Hao, P., Wang, S., Fu, G., Huang, Y., Li, Y., Zhu, J., Liu, Y., Hu, X., et al. (2002). Sequence and analysis of rice chromosome 4. *Nature* 420: 316-320.

Frary, A., Nesbitt, T.C., Grandillo, S., Knapp, E., Cong, B., Liu, J., Meller, J., Elber, R., Alpert, K.B., Tanksley, S.D. (2000). *fw2.2:* A quantitative trait locus key to the evolution of tomato fruit size. *Science* 289: 85-88.

Fujisawa, Y., Kato, T., Ohki, S., Ishikawa, A., Kitano, H., Sasaki, T., Asahi, T., Iwasaki, Y. (1999). Suppression of the heterotrimeric G protein causes abnormal morphology including dwarfism in rice. *Proc. Natl. Acad. Sci. USA* 96: 7575-7580.

Hiei, Y., Komari, T., Kubo, T. (1997). Transformation of rice mediated by *Agrobacterium tumefaciens. Plant Mol. Biol.* 35: 205-218.

Hirochika, H. (1997). Retrotransposons of rice: Their regulation and use for genome analysis. *Plant Mol. Biol.* 35: 231-240.

Hirochika, H. (2001). Contribution of the Tos17 retrotransposon to rice functional genomics. *Curr. Opin. Plant Biol.* 4: 118-122.

Hirochika, H., Sugimoto, K., Otsuki, Y., Tsugawa, H., Kanda, M. (1996). Retrotransposons of rice involved in mutations induced by tissue culture. *Proc. Natl. Acad. Sci. USA* 93: 7783-7788.

Hong, Z., Ueguchi-Tanaka, M., Shimizu-Sato, S., Inukai, Y., Fujioka, S., Shimada, Y., Takatsuto, S., Agematsu, M., Yoshida, S., Watanabe, Y., et al. (2002). Loss-of-function of a rice brassinosteroid biosynthetic enzyme, C-6 oxidase, prevents the organized arrangement and polar elongation of cells in the leaves and stem. *Plant J.* 32: 495-508.

Ikeda, A., Ueguchi-Tanaka, M., Sonoda, Y., Kitano, H., Koshioka, M., Futsuhara, Y., Matsuoka, M., Yamaguchi, J. (2001). Slender rice, a constitutive gibberellin response mutant, is caused by a null mutation of the *SLR1* gene, an ortholog of the height-regulating gene *GAI/RGA/RHT/D8*. *Plant Cell* 13: 999-1010.

Ito, T., Motohashi, R., Kuromoti, T., Mizukado, S., Sakurai, T., Kanahara, H., Seki, M., Shinozaki, K. (2002). A new resource of locally transposed *Dissociation* elements for screening gene-knockout lines in silico on the *Arabidopsis* genome. *Plant Physiol.* 129: 1695-1699.

Ito, Y., Hirochika, H., Kurata, N. (2002). Organ-specific alternative transcripts of KNOX family class 2 homeobox genes of rice. *Gene* 288: 41-47.

Itoh, H., Ueguchi-Tanaka, M., Sentoku, N., Kitano, H., Matsuoka, M., Kobayashi, M. (2001). Cloning and functional analysis of two gibberellin 3β-hydroxylase genes that are differently expressed during the growth of rice. *Proc. Natl. Acad. Sci. USA* 98: 8909-8914.

Jeon, J.-S., Lee, S., Jung, K.-H., Jun, S.-H., Jeong, D.-H., Lee, J., Kim, C., Jang, S., Lee, S., Yang, K., et al. (2000). T-DNA insertional mutagenesis for functional genomics in rice. *Plant J.* 22: 561-570.

Jiang, B., Bao, Z., Zhang, X., Hirochika, H., Eddy, S.R., McCouch, S.R., Wessler, S.R. (2003). An active DNA transposon family in rice. *Nature* 421: 163-167.

Johanson, U., West, J., Lister, C., Michaels, S., Amasino, R., Dean, C. (2000). Molecular analysis of *FRIGIDA,* a major determinant of natural variation in *Arabidopsis* flowering time. *Science* 290: 344-347.

Jung, K.-H., Hur, J., Ryu. C.H., Choi, Y., Chung, Y.-Y., Miyao, A., Hirochika, H., An, G. (2003). Characterization of a rice chlorophyll-deficient mutant using the T-DNA gene-trap system. *Plant Cell Physiol.* 44: 463-472.

Khush, G.S. (1997). Origin, dispersal, cultivation and variation of rice. *Plant Mol. Biol.* 35: 25-34.

Kinoshita, T. (1995). Report of committee on gene symbolization, nomenclature and linkage groups. *Rice Genet. Newsl.* 12: 9-153.

Kojima, S., Takahashi, Y., Kobayashi, Y., Monna, L., Sasaki, T., Araki, T., Yano, M. (2002). *Hd3a,* a rice ortholog of the *Arabidopsis FT* gene, promotes transition to flowering downstream of *Hd1* under short-day condition. *Plant Cell Physiol.* 43: 1096-1105.

Komori, T., Nitta, N. (2003). High frequency of sequence polymorphism in rice MITEs and application to efficient development of PCR-based markers. *Breeding Sci.* 53: 85-92.

Krysan, P.J., Young, J.C., Sussman, M.R. (1999). T-DNA as an insertional mutagen in *Arabidopsis*. *Plant Cell* 11: 2283-2290.

Kurakazu, T., Sobrizal, Ikeda, K., Sanchez, P.L., Doi, K., Angeles, E.R., Khush, G.S., Yoshimura, A. (2001). *Oryza meridionalis* chromosomal segment introgression lines in cultivated rice, *O. sativa* L. *Rice Genet. Newsl.* 18: 81-82.

Kurata, N., Nagamura, Y., Yamamoto, K., Harushima, Y., Sue, N., Wu, J., Antonio, B.A., Shomura, A., Shimizu, T., Lin, S.-Y., et al. (1994). A 300 kilobase interval genetic map of rice including 883 expressed sequences. *Nat. Genet.* 8: 365-372.

Kyozuka, J., Konishi, S., Nemoto, K., Izawa, T., Shimamoto, K. (1998). Down-regulation of *RFL*, the *FLO/LFY* homolog of rice, accompanied with panicle branch initiation. *Proc. Natl. Acad. Sci. USA* 95: 1979-1982.

Leister, D., Kurth, J., Laurie, D.A., Yano, M., Sasaki, T., Devos, K., Graner, A., Schulze-Lefert, P. (1998). Rapid reorganization of resistance gene homologues in cereal genomes. *Proc. Natl. Acad. Sci. USA* 95: 370-375.

Li, X., Qian, Q., Fu, Z., Wang, Y., Xiong, G., Zeng, D., Wang, X., Liu, X., Teng, S., Fujimoto, H., et al. (2003). Control of tillering in rice. *Nature* 422: 618-621.

Liu, Y.G., Whittier, R.F. (1995). Thermal asymmetric interlaced PCR: Automatable amplification and sequencing of insert end fragments from P1 and YAC clones for chromosome walking. *Genomics* 25: 674-681.

Martin, G.B. (1998). Gene discovery for crop improvement. *Curr. Opin. Biotech.* 9: 220-228.

McClintock, B. (1956). Controlling elements and the gene. *Cold Spring Harbor Symp. Quant. Biol.* 21: 197-216.

McCouch, S.R., Teytelman, L., Xu, Y., Lobos, K.B., Clare, K., Walton, M., Fu, B., Maghirang, R., Li, Z., Xing, Y., et al. (2002). Development and mapping of 2240 new SSR markers for rice (*Oryza sativa* L.). *DNA Res.* 9: 199-207.

McKinney, E.C., Ali, N., Traut, A., Feldmann, K.A., Belostotsky, D.A., McDowell, J.M., Meagher, R.B. (1995). Sequence-based identification of T-DNA insertion mutations in *Arabidopsis:* Actin mutants *act2-1* and *act4-1*. *Plant J.* 8: 613-622.

Miyao, A., Yamazaki, M., Hirochika, H. (1998). Systematic screening of mutants of rice by sequencing retrotransposon-insertion sites. *Plant Biotech.* 15: 253-256.

Monna, L., Kitazawa, N., Yoshino, R., Suzuki, J., Matsuda, H., Maehara, Y., Tanji, M., Sato, M., Nasu, S., Minobe, Y. (2002). Positional cloning of rice semidwarfing gene, *sd-1:* Rice "green revolution gene" encodes a mutant enzyme involved in gibberellin synthesis. *DNA Res.* 9: 11-17.

Mori, M., Nomura, T., Ooka, H., Ishizaka, M., Yokota, T., Sugimoto, K., Okabe, K., Kajiwara,H., Satoh. K., Yamamoto, K., et al. (2002). Isolation and characterization of a rice dwarf mutant with a defect in brassinosteroid biosynthesis. *Plant Physiol.* 130: 1152-1161.

Nakagawa, M., Shimamoto, K., Kyozuka, J. (2002). Overexpression of *RCN1* and *RCN2*, rice *TERMINAL FLOWER1/CENTRORADIALIS* homologs, confers delay of phase transition and altered panicle morphology in rice. *Plant J.* 29: 743-750.

Nakazaki, T., Okumoto, Y., Horibata, A., Yamahira, S., Teraishi, M., Nishida, H., Inoue, H., Tanisaka, T. (2003). Mobilization of a transposon in the rice genome. *Nature* 421: 170-172.

Nakazawa, M., Ichikawa, T., Ishikawa, A., Kobayashi, H., Tsuhara, Y., Kawashima, M., Suzuki, K., Muto, S., Matsui, M. (2003). Activation tagging, a novel tool to dissect the function of a gene family. *Plant J.* 34: 741-750.

Nasu, S., Suzuki, J., Ohta, R., Hasegawa, K., Yui, R., Kitazawa, N., Monna, L., Minobe, Y. (2002). Search for and analysis of single nucleotide polymorphisims (SNPs) in rice *(Oryza sativa, Oryza rufipogon)* and establishment of SNP markers. *DNA Res.* 9: 163-171.

Nesbitt, T.C., Tanksley, S.D. (2002). Comparative sequencing in the genus *Lycopersicon:* Implications for the evolution of fruit size in the domestication of cultivated tomatoes. *Genetics* 162: 365-379.

Parinov, S., Sevugan, M., Ye, D., Yang, W.-C., Kumaran, M., Sundaresan, V. (1999). Analysis of flanking sequences from *Dissociation* insertion lines: A database for reverse genetics in *Arabidopsis*. *Plant Cell* 11: 2263-2270.

Rice Chromosome 10 Sequencing Consortium (2003). In-depth view of structure, activity, and evolution of rice chromosome 10. *Science* 300: 1566-1569.

Sasaki, A., Ashikari, M., Ueguchi-Tanaka, M., Itoh, H., Nishimura, A., Swapan, D., Ishiyama, K., Saito, T., Kobayashi, M., Khush, G.S., et al. (2002). A mutant gibberellin-synthesis gene in rice. *Nature* 416: 701-702.

Sasaki, A., Itoh, H., Gomi, K., Ueguchi-Tanaka, M., Ishiyama, K., Kobayashi, M., Jeong, D.-H., An, G., Kitano, H., Ashikari, M., Matsuoka, M. (2003). Accumulation of phosphorylated repressor for gibberellin signaling in an F-box mutant. *Science* 299: 1896-1898.

Sasaki, T., Matsumoto, T., Yamamoto, K., Sakata, K., Baba, T., Katayose, Y., Wu, J., Niimura, Y., Cheng, Z., Nagamura, Y., et al. (2002). The genome sequence and structure of rice chromosome 1. *Nature* 420: 312-316.

Sato, Y., Sentoku, N., Miura, Y., Hirochika, H., Kitano, H., Matsuoka, M. (1999). Loss-of-function mutations in the rice homeobox gene *OSH15* affect the architecture of internodes resulting in dwarf plants. *EMBO J.* 18: 992-1002.

Siebert, P.D., Chenchik, A., Kellogg, D.E., Lukyanov, K.A., Lukyanov. S.A. (1995). An improved PCR method for walking in uncloned genomic DNA. *Nucleic Acids Res.* 23: 1087-1088.

Spielmeyer, W., Ellis, M.H., Chandler, P.M. (2002). Semidwarf *(sd-1)*, "green revolution" rice, contains a defective gibberellin 20-oxidase gene. *Proc. Natl. Acad. Sci. USA* 99: 9043-9048.

Takagi, K., Nagano, H., Kishima, Y., Sano, Y. (2003). MITE-transposon display efficiently detects polymorphisms among the oryza AA-genome species. *Breeding Sci.* 53: 125-132.

Takahashi, Y., Shomura, A., Sasaki, T., Yano, M. (2001). *Hd6*, a rice quantitative trait locus involved in photoperiod sensitivity, encodes the a subunit of protein kinase CK2. *Proc. Natl. Acad. Sci. USA* 98: 7922-7927.

Takano, M., Kanegae, H., Shinomura, T., Miyao, A., Hirochika, H., Furuya, M. (2001). Isolation and characterization of rice phytochrome A mutants. *Plant Cell* 13: 521-534.

Takeda, T., Suwa. Y., Uzuki, M., Kitano, H., Ueguchi-Tanaka, M., Ashikari, M., Matsuoka, M., Ueguchi, C. (2003). The *OsTB1* gene negatively regulates lateral branching in rice. *Plant J.* 33: 513-520.

Tanksley, S.D., Ganal, M.W., Martin, G.B. (1995). Chromosome landing: A paradigm for map-based gene cloning in plants with large genomes. *Trends Genet.* 11: 63-68.

Terada, R., Urawa, H., Inagaki, Y., Tsugane, K., Iida, S. (2002). Efficient gene targeting by homologous recombination in rice. *Nature Biotechnol.* 20: 1030-1034.

Upadhyaya, N.M., Zhou, X.-R., Zhu, Q.-H., Ramm, K., Wu, L., Eamens, A., Sivakumar, R., Kato, T., Yun, D.-W., Santhoshkumar, C., et al. (2002). An *iAc/Ds* gene and enhancer trapping system for insertional mutagenesis in rice. *Funct. Plant Biol.* 29: 547-559.

Wang, G.-L., Ruan, D.-L., Song, W.-Y., Sideris, S., Chen, L., Pi, L.-Y., Zhang, S., Zhang, Z., Fauquet, C., Gaut, B.S., et al. (1998). *XA21D* encodes a receptor-like molecule with a leucine-rich repeat domain that determines race-specific recognition and is subject to adaptive evolution. *Plant Cell* 10: 765-779.

Wang, Z.-X., Yano, M., Yamanouchi, U., Iwamoto, M., Monna, L., Hayasaka, H., Katayose, Y., Sasaki, T. (1999). The *Pib* gene for rice blast resistance belongs to the nucleotide binding and leucine-rich repeat class of plant disease resistance genes. *Plant J.* 19: 55-64.

Weigel, D., Ahn, J.H., Blazquez, M.A., Borevitz, J.O., Christensen, S.K., Fankhauser, C., Ferradiz, C., Kardailsky, I., Malancharuvil, E.J., Neff, M.M., et al. (2000). Activation tagging in *Arabidopsis*. *Plant Physiol.* 122: 1003-1013.

Yamamura, C., Ihara, Y., Wu, X., Noguchi, T., Fujioka, S., Takatsuto, S., Ashikari, M., Kitano, H., Matsuoka, M. (2000). Loss of function of a rice *brassinosteroid insensitive 1* homolog prevents internode elongation and bending of the lamina joint. *Plant Cell* 12: 1591-1605.

Yamanouchi, U., Yano, M., Lin, H., Ashikari, M., Yamada, K. (2002). A rice spotted leaf gene, *Spl7,* encodes a heat stress transcription factor protein. *Proc. Natl. Acad. Sci. USA* 99: 7530-7535.

Yano, M. (2001). Naturally occurring allelic variations as a new resource for functional genomics in rice. In Khush, G.S., Brar, D.S., and Hardy, B. (Eds.), *Rice Genetics IV. Proceedings of the Fourth International Rice Genetics Symposium 2000* (pp. 227-238). New Delhi (India): Science Publishers Inc., and Los Banos (Philippines): International Rice Research Institute.

Yano, M., Katayose, Y., Ashikari, M., Yamanouchi, U., Monna, L., Fuse, T., Baba, T., Yamamoto, K., Umehara, Y., Nagamura, Y., Sasaki, T. (2000). *Hd1,* a major photoperiod sensitivity quantitative trait locus in rice, is closely related to the *Arabidopsis* flowering time gene *CONSTANS*. *Plant Cell* 12: 2473-2484.

Yano, M., Kojima, S., Takahashi, Y., Lin, H., Sasaki, T. (2001). Genetic control of flowering time in rice, a short-day plant. *Plant Physiol.* 127: 1425-1429.

Yoshimura, A. (2001). Utilization of induced and spontaneous mutants for rice genome research. Significance of genetic stocks. *Gamma Field Symposia* 40: 63-70.

Yoshimura, S., Yamanouchi, U., Katayose, Y., Toki, S., Wang, Z.-X., Kono, I., Kurata, N., Yano, M., Iwata, N., Sasaki, T. (1998). Expression of *Xa1,* a bacterial blight-resistance gene in rice, is induced by bacterial inoculation. *Proc. Natl. Acad. Sci. USA* 95: 1663-1668.

Yu, J. Hu, S., Wang, J., Wong, G. K.-S., Li, S., Liu, B., Deng, Y., Dai, L., Zhou, Y., Zhang, X., et al. (2002). A draft sequence of the rice genome *(Oryza sativa* L. spp. *indica). Science* 296: 79-92.

Zambryski, P., Holsters, M., Kruger, K., Depicker, A., Schell, J., van Montagu, M., Goodman, H.M. (1980). Tumor DNA structure in plant cells transformed by *A. tumefaciens*. *Science* 209: 1385-1391.

Chapter 12

Maize Genomics

Joachim Messing

MAIZE AS A CROP

Zea mays, or corn, is probably the most successfully cultivated plant in history. It now leads all crops in the amount harvested worldwide (Table 12.1). Although wheat and rice are grown over a larger area of land than corn, corn produces more grain than other cereals and is therefore more efficient as a natural resource for humankind.

Besides being a staple for a large fraction of the world's population, as are wheat and rice, corn has also gained a broader set of applications. Corn has become the chief ingredient for animal feed and is even used in fish cultivation. Milled corn flour is a raw material for the manufacture of sweeteners used in soft drinks and in processed foods. Other industrial uses of corn include fuel and adhesives, and the list of corn-based products continues to expand (McLaren, 1998). The productivity and versatility of corn has made agriculture the single most important export area for the United States. It has been estimated that U.S. agrobusiness contributes 14 percent of the gross domestic product, 18 percent of employment, and 10 percent of all exports (McLaren, 1998). Because of its economic significance, maize has been a prime organism for genetic, cytogenetic, biochemical, molecular, and now genomic studies.

GENOME SIZE

Maize belongs to the monocotyledonous angiosperms and the family of the Gramineae, which comprises over 10,000 species. This family origi-

I would like to thank two anonymous reviewers and Dr. Benjamin Burr from Brookhaven National Laboratory, Uptown, New York, for their critical reading of the manuscript. The maize genome analysis work was in part supported by National Science Foundation grant #9975618 and DOE grant #DE-FG05-95ER20194.

TABLE 12.1. Worldwide production of the four major crops in 2002.

Crop	Area hectare[a]	Harvest mt tons[b]	Yield hg/ha[c]
Wheat	210.8	568	27
Rice	146	579	40
Maize	138.9	602	43.3
Soybean	79.2	180	22.7

Source: United Nations Food and Agriculture Organization (FAO).
[a]million
[b]million metric tons
[c]hectogram/hectare in thousands

nated some 65 million years ago (mya; Kellogg, 2001). Family members include other important cereal crops besides maize, such as rice, sorghum, wheat, barley, rye, millets, and oats. However, the most important crops economically belong to just three subfamilies: the Panicoideae (sorghum and maize), the Pooideae (barley and wheat), and the Oryzoideae (rice) (Kellogg, 1998, 2001). One of the most striking differences among these species is their genome size (Figure 12.1). From this point of view, one can divide plant genomes into three groups: one group of genomes under 800 Mbp (e.g., rice and sorghum), a middle group of genomes between 0.8 and 3 Gbp (e.g., maize), and a third group of genomes above 3 Gbp (e.g., barley, oat, and wheat). For size comparison, mammalian genomes would be in the middle group, having genomes of about 3 Gbp.

Despite the wide variation in grass genome sizes (Arumuganathan and Earle, 1991) cross-hybridization data indicate that the genes are well conserved (Figure 12.1), while noncoding regions, such as intergenic regions and introns, have diverged relatively rapidly (Weerakoon et al., 1994). This underlying pattern of conservation has allowed the use of heterologous sequences as genetic markers. An interesting outcome of such analyses was the finding that for large chromosome segments the order of genes, as well as their sequences, seems to be conserved across the grasses, at least at the level of genetic maps (Ahn et al., 1993; Gale and Devos, 1998). This finding prompted a proposal to consider the grasses as a single genetic system (Bennetzen and Freeling, 1993). However, if short chromosomal intervals are compared at the DNA sequence level, it appears that collinearity is interrupted by gene insertions and losses. Based on these observations, it has been suggested that alignment of orthologous intervals of closely related grass genomes exhibits a mosaic organization of collinear and noncollinear genes (Song et al., 2002). Furthermore, genomes may have not only ex-

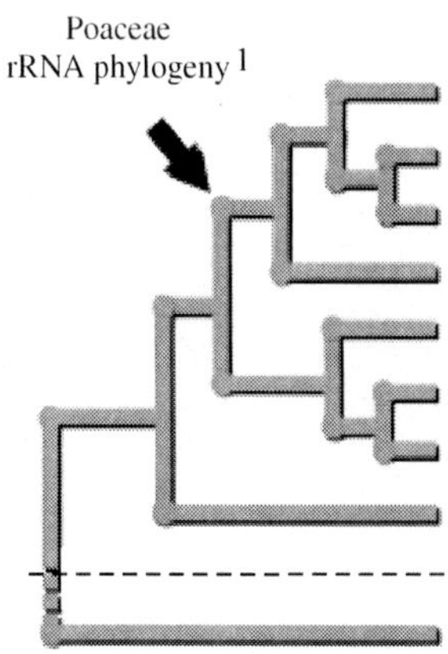

	DNA per haploid genome[2] (Mb)	Cross-hybridization[3] cDNA	Cross-hybridization[3] Genomic DNA
Maize	2716	100%	100%
Sorghum	772	92%	55%
Sugar	3605	92%	42%
Rice	439	58%	23%
Wheat	15966	8.3%	3%
Oat	11315	8.3%	3%
Barley	4873	n/a	16%
Bamboo	NA	42%	23%
Soybean	1115	8.3%	0%

FIGURE 12.1. Comparison of some grass species of economic importance. Eight grass species are presented in their physical relationship. Their phylogeny (1) was taken from Hamby and Zimmer (1988), the genome sizes (2) from Arumuganathan and Earle (1991), and the cross-hybridization data (3) from Weerakoon et al. (1994). Please note that hybridization data stem from a genome-wide analysis and that a subset of genes is highly conserved through all grass species. In this comparison, wheat has the largest genome and rice the smallest, by 36-fold.

panded but also contracted due to gene mobility. In addition, different chromosome numbers in different grass species as well as translocations and inversions of large chromosome fragments relative to ancestral chromosome structures have also led to the interruption of gene collinearity. Still, it was always assumed that gene collinearity is maintained within the same species. However, when orthologous intervals of different maize inbreds were compared, substantial sequence divergence could be observed, including gene insertions or losses (Timmermans et al., 1996; Hu et al., 1998; Fu and Dooner, 2002). A recent study could show that this divergence was absent for the same interval of two subspecies of rice, which maintained complete gene collinearity. This special property of maize inbreds prompted a new hypothesis on heterosis, in particular, since expression studies exhibit the lack of gene dosage, presence of alternate trans-acting factors, and cases of overdominance (Song and Messing, 2003).

Given the conservation of gene sequences as well as the order of genetic markers (termed *synteny*) for chromosomal segments despite the mosaic structures and chromosomal rearrangements that may have taken place during speciation, what accounts for the differences in genome sizes? Polyploidy is one important factor that led to larger genomes in crop plants (Wendel, 2000). Maize, in contrast to rice and sorghum, resulted from the hybridization of two closely related but separately evolved progenitors. As a consequence, many genes in maize exist as duplicate factors (Rhoades, 1951; Helentjaris et al., 1988). This feature of the maize genome was also evident in synteny studies, in which the two subgenomes of maize could be aligned with other grass genomes (Gale and Devos, 1998). As the regulation of the duplicate factors of maize diverged and its two subgenomes underwent massive rearrangements, its genes and chromosomes behaved similar to those of a disomic species. As a result, each of the duplicate factors gave rise to a separate Mendelian trait. However, despite its disomic state, the maize genome size differs from that of one of its closest relatives, sorghum, by a factor of greater than two (770 Mbp versus 2,700 Mbp).

Although it has been proposed that sorghum and maize share the same lineage as one of the two progenitors of maize (Gaut and Doebley, 1997), the source of the additional 43 percent of the maize genomic DNA is still in question. In this regard, it is interesting to note that genome sizes for individual maize lines are also quite variable. Bennett gives an overall range of 2,364 Mbp to about 3,233 Mbp for the size of maize genomes (http://www.agron.missouri. edu/zeadna. html). According to this summary, there seems to be general agreement that the size variation results mainly from the amount of heterochromatin (contained in heterochromatic knobs and B-chromosomes). Older size estimates were usually derived from DNA staining, while more recent estimates have been obtained from chromosome maps built by fingerprinting clones from highly redundant genomic libraries. Although the fingerprint maps still contain many physical gaps, past experience has shown that these gaps mostly represent the heterochromatic regions that are highly variable in size, as described previously. For example, the current estimate for the euchromatic portion of the maize inbred B73 genome is 2,036 Mbp (http://www.genome.arizona.edu/fpc/maize/status.html). If 2 Gbp represents the minimum maize genome size today and the two progenitor species had genomes with the size of sorghum, both constituent genomes expanded by about 12.5 percent after allotetraploidization relative to the sorghum genome. Alternatively, a species with the genome size of sorghum could have combined with a considerably larger genome to form the progenitor to the modern maize genome, or the sorghum genome has lost DNA relative to the maize genome.

GENOME EXPANSION

At this stage, it is still too early to predict whether each subgenome has expanded to the same extent, and what constitutes the extra genomic DNA. Some proportion of the extra DNA could be accounted for by the larger gene family size in maize compared to sorghum, as is the case for storage protein genes (Song et al., 2002). Such new genes usually arise from local duplications—in a 125 Mbp genome, 17 percent of *Arabidopsis thaliana* genes are represented in such locally duplicated gene families (Arabidopsis Genome Initiative, 2000). However, the majority of the additional DNA will likely be noncoding repetitive DNA, as was shown in early analyses of DNA hybridization kinetics (Hake and Walbot, 1980). Most of this repetitive DNA is transposable element families. When a library of sheared genomic DNA was sequenced, and the sequences subjected to BLAST analysis, the majority of the sequences had homology to retrotransposons (Meyers et al., 2001). Because the sample analyzed represented less than 0.1 percent of the genome, one would expect to detect only DNA with more than 700 copies on average using this approach, leaving out coding sequences. Interestingly, nearly a third of the sequences consist of just four retrotransposon families: Huck (10.7 percent), Ji (9.4 percent), Opie (7.1 percent), and Zeon (4.8 percent).

A major difference between DNA transposable elements and retrotransposable elements is that retrotransposable elements cannot excise, whereas DNA transposable elements can move from one position to another position in the genome. Retrotransposons are amplified by a transcript intermediate, which is copied from RNA into DNA; these DNA copies can integrate into multiple sites in the genome, giving rise to clonal events. For example, it has been shown that the *Tos17* retrotransposon can give rise to multiple insertions within the same cell lineage (Hirochika, 2001). Because such a mechanism would increase the DNA content very rapidly, retrotransposition is much more stringently controlled than DNA transposition. Both retrotransposons and DNA transposons are epigenetically silenced to reduce the increase in host mutation rates when transposons are mobile (Bennetzen et al., 1994). The silenced state can be reversed by extreme growth conditions, such as hormone treatment in tissue culture (Hirochika et al., 1996; Peschke et al., 1987) and UV-C and UV-B treatment of pollen (Walbot, 1999). Such activation of transposition has been associated with somaclonal variation (Wessler et al., 1995) and genomic shock (McClintock, 1984), although treatment of haploid pollen with UV-B can also reactivate transposons (Walbot, 1999). Although most insertion events are of somatic origin, many are genetically transmissible because somatic clones within the apical meristems contribute to reproductive structures. In contrast to plants, animals sequester their germline early in development and therefore prevent the transmission of somatic changes (i.e., immune system) to the next generation.

Taken together, these observations indicate that during the evolution of the maize genome, four retrotransposons have been most actively reproduced, and as a result of the increased DNA content, the overall gene density (genes per kbp of DNA) has been reduced. These four retrotransposon families therefore provide a diagnostic signature for the maize genome. This retrotransposon-based signature was evident when maize genomic DNA was hybridized to the chromosomes of the very close relatives of maize, sorghum and coix, in a genome-wide FISH analysis (Takahashi et al., 1999). In the case of sorghum, only the nucleolear organizer, which contains the ribosomal RNA gene cluster, became visible, while coix chromosomes exhibited uniform hybridization. While the maize retrotransposon numbers might have already expanded before maize and coix diverged, this did not occur before the separation of sorghum and maize (16.5 mya). Retrotransposition events studied in the maize *adh1* region occurred between 1 to 6 mya and included all four abundant retrotransposon families (SanMiguel and Bennetzen, 1998). Therefore, I conclude that the degree of activity of specific retrotransposons in different species not only accounts for the range of genome expansion during speciation but has also given different genomes specific molecular signatures.

CLONE MAPS

Maize geneticists were among the earliest to clone plant genes of defined function because of large tissue sizes containing abundant mRNA species and the availability of DNA-transposon tagged mutable alleles (Geraghty et al., 1981; Fedoroff et al., 1984). Despite the ease of gene cloning of highly expressed genes and mutant alleles in maize, the large genome size has presented substantial difficulties in deriving a physical map of the maize chromosomes. The size of a genome is indirectly proportional to the likelihood of isolating genes from a genomic clone library (Clark and Carbon, 1976). To conduct chromosome walking in maize, it was necessary to move from lambda libraries (20 kbp inserts) to cosmid libraries (40 kbp inserts) because of the large genome size (Llaca and Messing, 1998). Still, because of the presence of repetitive sequences even in gene-rich regions, cosmid contigs are not very large and need to be bridged by bacterial artificial chromosome (BAC) clones (Song et al., 2001). Although it has also been possible and useful to construct yeast artificial chromosome

(YAC) libraries of maize with large insert sizes (>100 kbp; Edwards et al., 1992), these libraries have been less useful in chromosome walking than BAC libraries because of clone chimerism and the additional work involved in purification of DNA from yeast rather than bacteria. On the other hand, certain cosmid clones are very unstable because of their high-copy-number origin of replication (Llaca and Messing, 1998). Besides insert size, the lack of insert chimerism, and the ease of DNA purification, the depth of the library and the choice of DNA fragmentation are additional important considerations in the choice of systems for the construction of genomic libraries of medium- and large-size plant genomes. The deeper the library, the more likely overlapping clones can be found that together span the whole genome. A major advance in connecting overlapping clones came from replacing hybridization methods with DNA-fingerprinting methods (Coulson et al., 1986). However, because DNA fingerprinting is based on comparing common restriction patterns, the depth of the library has become very critical for success. While the construction of a clone map for the mouse genome by DNA fingerprinting required a depth of 15X coverage (Gregory et al., 2002), this depth of coverage might not have been sufficient if DNA fragmentation had been done using a single restriction enzyme.

Consequently, maize BAC libraries have now been constructed by fragmenting genomic DNA from a single germplasm source, inbred B73, with three different restriction enzymes: *Hin*dIII, *Eco*RI, and *Mbo*I. Assuming an average insert size of 154 kbp, and a minimal genome size of 2 Gbp, and 450,000 clones, the total depth of coverage in the three libraries would be about 33-fold. When these libraries were examined using different probes, only 1 percent of the clones contained organellar DNA (probably from nuclear DNA containing insertions of organelle DNA fragments), and only 0.01 percent contained lambda DNA, indicating that the libraries are of very high quality (Yim et al., 2002). The composition of the different sublibraries will reflect the bias of the restriction enzymes used in their construction, as illustrated by the ribosomal RNA gene cluster, which has no *Hin*dIII site, one *Eco*RI site, and many *Mbo*I sites. To assess genome coverage, a set of 110,592 *Hin*dIII clones (arrayed on six filters) was used to examine the redundancy of 22 different single-copy genes. These 22 genes were found to be present in the clone set on average 7.4 times each (Yim et al., 2002). Because the *Hin*dIII library has an average insert size of 136 kbp, one would expect 7.5 clones per single-copy gene with a genome size of 2 Gbp, indicating that the size of the genome estimated by DNA fingerprinting is about the same as the size obtained from libraries screened by DNA hybridization.

As of June 6, 2003, 292,039 clones from maize BAC libraries have been DNA fingerprinted (http://www.genome.arizona.edu/fpc/maize/). As the number of fingerprints increases, the number of BAC contigs should decrease, and the sizes of these contigs should increase (Figure 12.2). From a starting point of 85,096 clones, the number of unconnected clones has already dropped from about 11 percent to 4.5 percent and remains at this number despite a doubling in the number of clones fingerprinted. However, manual editing has reduced the number of contigs by 21.5 percent. A total of 178 contigs have already reached a length of 2 to 8.5 Mbp, making them candidates for spanning several cM of genetic distance. The number of anchored contigs has also risen to 1,446; however, the number of contigs is still very high (3,488 fingerprint contigs; FPCs) compared to the rice genome (430 FPCs).

GENETIC MAPS

While the maize physical map is now being assembled for the first time, genetic maps have already been under development for nearly 100 years. Initial maps were based exclusively on genes with visible phenotypes, but with the advent of recombinant DNA, genetic maps now consist primarily of molecular markers. Marker density on genetic maps has risen over the years, mainly because of the construction of recombinant inbred lines (Burr et al., 1988) and immortalized F2 populations (Lee et al., 2002). While mutant phenotypes have been placed on traditional maps built in diverse maize lines, recombinant inbred lines have been used primarily to map DNA sequences.

Many mutant phenotypes that are based on biochemical variation and plant morphology have been described in maize. In most cases, these phenotypes have been mapped in different genetic backgrounds. Therefore, their positions will vary in different maps. A recent public inventory of chromosomal assignments yielded a total of 1,679 mapped genes (Table 12.2) in addition to another 12,315 mapped loci. From both data sets, a total of 3,400 sequences are known and available. This sequence collection will be very valuable in linking the B73 BAC libraries to the genetic map (Cone et al., 2002). Furthermore, loci that have not been cloned should be accessible through the B73 clone map by the genetic linkage to their position. This collection of genes will be an important resource for maize functional genomics.

Because recombination frequencies differ among segregating populations and recombinant inbred lines, the integration of the genetic maps results in only approximate locus coordinates. Despite these inherent difficulties, such integration will be very useful since it attempts to link all available markers to the physical map, which is derived from the inbred B73

Date	Clone	Cov	Marker	Anchor	Contig	Single
7/26/01	83096	4x	2078	—	13150	9361
8/31/01	97949	5x	2078	—	13261	8685
10/24/01	135706	7x	3045	77	12103	7866
12/04/01	156506	8x	3045	77	10580	7422
12/23/01	164701	8x	7760	133	9462	7437
1/23/02	176440	9x	7909	135	8422	7497
2/20/02	191962	9x	7974	135	7315	7730
3/25/02	203765	10x	7978	135	6693	8478
5/3/02	215415	11x	8815	188	6054	9530
6/27/02	232862	12x	10149	405	5318	10786
8/8/02	250065	13x	10191	407	4847	11191
1/6/03	291200	15x	14786	800	4443	13255
6/6//03	292039	15x	15422	1446	3488	14482

Date	Contig Sizes > 200	200:100	100:51	50:26	25:10	9:3
7/26/01	0	0	3	43	1359	8793
8/31/01	0	0	4	127	2238	8385
10/24/01	0	3	80	764	3699	5970
12/04/01	1	19	247	1252	3563	4278
12/23/01	0	25	423	1488	3133	3357
1/23/02	0	62	628	1581	2690	2580
2/20/02	0	165	880	1509	2093	1934
3/25/02	16	248	996	1409	1763	1622
5/3/02	33	360	1004	1237	1442	1323
6/27/02	89	484	982	1050	1081	999
8/8/02	151	572	951	911	805	826
1/6/03	275	712	861	651	574	641
6/6/03	390	528	570	461	400	475

FIGURE 12.2. DNA fingerprinting of maize BAC libraries. The BAC clones of the three maize libraries have been digested with *Hin*dIII, separated by agarose gel electrophoresis, and analyzed by the FPC software (Soderlund et al., 2000). Image analysis of restriction patterns is used to compare all BAC clones for overlaps. The clones are then ordered by their overlaps in stacks of FPCs representing the local *Hin*dIII restriction map of the maize genome. If one of the BAC clones has been anchored to a genetically mapped marker, the entire FPC can be positioned on the genetic map (Cone et al., 2002). The 1,446 FPCs highlighted by a box have been anchored in such a way. Increased coverage of fingerprinted BAC clones marked by the dates decreases the number of FPCs, as indicated by arrows. Contig sizes have been broken down from >200 clones, 200-100 clones, etc. The data can be found at <http://www.genome.arizona.edu/fpc/maize/>.

TABLE 12.2. Genetic maps.

	Sequences	Genes	Additional loci	Total loci
Chromosome 1	539	301	1,803	2,104
Chromosome 2	463	172	1,372	1,544
Chromosome 3	329	182	1,352	1,534
Chromosome 4	374	177	1,300	1,477
Chromosome 5	360	189	1,356	1,545
Chromosome 6	288	143	1,103	1,246
Chromosome 7	258	117	1,025	1,142
Chromosome 8	270	107	1,113	1,220
Chromosome 9	281	132	1,003	1,135
Chromosome 10	238	159	888	1,047
Total	3,400	1,679	12,315	13,994

Source: <http://www.maizegdb.org/>.

line. To accomplish integration, a new map has been constructed based on an **i**ntermated **B**73 and **M**o17 (IBM) population. Both the B73 and Mo17 inbreds are frequently used for commercial breeding of the inbred lines used to produce hybrid seed. This map is expanded threefold compared to maps from segregating populations and has a 15-fold higher resolution than the previous map derived from intermated T × 303 × CO159 recombinant inbred lines (Sharopova et al., 2002; Lee et al., 2002). Common sets of markers are used to recalculate the position of other markers for each interval. The resulting map is therefore referred to as the IBM Neighbors Maps (Table 12.3). Although still under development, the goal is to place a total of 4,960 markers on the integrated map. These integrated markers will constitute important molecular anchors for aligning the genetic map with the B73 clone map (Cone et al., 2002).

GENE TAGGING

As mentioned in the preceding section, gene tagging has been an important tool for maize gene discovery as well as maize functional genomics. Gene tagging in maize is based on the property of DNA transposable elements to insert preferentially into genes (Cowperthwaite et al., 2002, and references therein). Once transposable elements were cloned, their sequences could then be used to identify the genes into which they had in-

TABLE 12.3. IBM Neighbors Maps.

Source	Total
IBM	1,648
SSR T218 × GT119	183
SSR T × 303 × CO159	202
UMC 98	1,386
BNL 2002	1,541
Total IBM neighbors	4,960

Source: <http://www.maizegdp.org/map.php>.

serted. However, there are three major challenges to such a strategy. First, maize has three major transposable element families, Activator/ Dissociation (Ac/Ds), Enhancer-Inhibitor/Suppressor-Mutator (En-I/Spm), and Mutator (Mu), which consist of many copies with internally dissimilar sequences. Second, a mutant phenotype might be closely linked to an insertion site or be caused by another copy of the element. Third, some transposons (e.g., Ac) preferentially move short distances, thereby requiring starting points at many chromosomal locations for full-genome coverage. However, certain properties of transposable elements can be used to favor gene tagging (Zhang and Peterson, 1999). First, active elements cause unstable mutations, which can be visualized as a mosaic phenotype for visible markers, such as pigment formation. These phenotypes can be classified in advance based on their response to known autonomous elements or by their effect on defective elements inserted into reporter genes. Second, active elements are hypomethylated and can be distinguished from inactive elements with restriction enzymes, which is critical in cloning tagged genes because only active elements cause unstable phenotypes. Third, new insertions can be transmitted to give rise to new alleles; multiple alleles are useful in the identification of mutated genes. Fourth, the Ac element exhibits a gene dosage effect that can be used to select for a single active copy in the genome; this can also be observed by segregation behavior given appropriate defective elements in reporter genes. Fifth, Ac and Mu elements have been used to introduce deletions. Sixth, composite elements have been used to cause chromosome breaks, which can create the somatic loss of the distal portion of a specific chromosome arm. Seventh, the Mu element, which is usually employed in a background with multiple elements, confers a high gene-insertion frequency and can therefore be used to increase the number of genes tagged. Eighth, unidirectional PCR methods can be used to isolate junction sequences between transposons and their insertion sites.

Traditionally, transposons have been used to identify specific loci by selecting for the specific associated mutant phenotype, and then isolating the corresponding gene (forward genetics). Since one can create many insertion events in individual seeds, large seed collections can now be screened with any gene sequence to isolate lines mutated in that gene for studying gene function (reverse genetics). Using specific primers, pools of seed DNA can be amplified to identify and recover the seed that might have a Mu element insertion. The Maize Targeted Mutagenesis (MTM) Database (http://mtm.cshl.org/) currently contains 43,776 DNA samples that can be queried for Mu insertions. Another seed collection, also based on the Mu transposon, facilitates a filtration of preexisting insertions versus new insertions by recurring backcrosses. This so-called uniformMu (uMu) collection has two advantages. First, it dilutes the Mu element to ensure a constant mutation rate; if the copy number of Mu gets too high, the mutation rate drops. Second, it ties phenotypes to a single generation of insertions. By using the DNA of the parental generation, new insertions can be enriched by subtraction (http://endosperm.org/). A new feature of this collection is that instead of using microarrays to link insertions to specific transcripts, junction sequences are sequenced to match them with the EST database. Although this collection is rather small, it also focuses on one tissue, endosperm, rather than the entire plant and should be saturated for endosperm mutations. Yet an alternative approach to the ones just described calls for the transformation of maize with a genetically engineered Mu transposable element. The advantage of such a transgenic element is that it can simplify the isolation of tagged sequences. Therefore, such an element has been termed RescueMu, and its application, the Maize Gene Discovery Project (Raizada et al., 2001). The Maize Gene Discovery Project has reported segregating visible phenotypes for ~35,000 families evaluated for seed defects, ~9,000 seedling traits, and 7,000 adult traits (http://gremlin3.zool.iastate.edu/zmdb/RescueMu.html). A disadvantage of such an approach is that the transformed plants contain already native Mu elements so that phenotypes are more likely to arise from native than engineered Mu elements. On the other hand, if phenotypes of interest are mapped, it still would be possible to identify candidate genes by Mu sequences.

GENE-ENRICHMENT PROCEDURES

Transposable elements can also be used to rapidly sequence gene fragments without linking them to a phenotype because they preferentially insert into genes, as described previously. Therefore, it has been proposed that sequencing junction sequences is complementary to sequencing ESTs.

While cDNA sequence representation reflects mainly the expression profile of specific tissues from which they were isolated, junction sequences are unbiased for the expression properties of any gene. Currently, there are 101,796 of such junction sequences in the GSS (Genome Survey Sequences) division in GenBank (http://www.zmdb.iastate.edu/zmdb/GSS/index.html). If one can match a junction sequence to a particular gene, one would also know that there is a gene knockout in the public seed collection, which would be an added benefit over just sequencing cDNAs.

Besides using transposable elements to enrich for gene sequences over noncoding, repetitive DNAs, other biochemical methods can separate the two genomic fractions from each other. As described previously, a large fraction of the noncoding repetitive DNA consists of a few retrotransposon families that are hypermethylated relative to the genes. Therefore, one should be able to use this difference to separate the two fractions (Rabinowicz et al., 1999). There are two approaches to take advantage of the methylation differences (methyl filtration). First, genomic DNA can be digested with a methylation-sensitive restriction enzyme prior to cloning. Second, sheared genomic DNA can be cloned into an *Escherichia coli* strain that possesses a restriction system that destroys methylated plant DNA. In the first case, one can make either BAC libraries or small insert libraries. The BAC libraries have the advantage that they should span larger regions of the genome, while the small insert libraries might provide greater coverage (Yuan et al., 2002). The second approach has the disadvantage that the *E. coli* "filtration" selects for rather small insert libraries. Another approach to separating genes from retrotransposons and noncoding repetitive DNA is based on the reassociation kinetics of DNA sequences (Manning et al., 1975). In this approach, genic DNA is relatively low in copy number and would remain single-stranded during reassociation. This fraction can then be separated from the repetitive, annealed sequences by affinity chromatography (Cot-based libraries). In the GSS division of GenBank, there are currently data sets available for Mu junctions (Stanford, 131,364; Florida, 5,285), for *E. coli*-filtered libraries (TIGR, 251,101; CSH, 135,980), and for Cot-based libraries (TIGR, 259,585). Based on the analysis of a smaller subset of these data, it appears that filtered sequences still contain 20 percent repeat sequences, while the Mu junctions contain only 10 percent repeat sequences (http://pgir.rutgers.edu). One explanation might be that methyl filtration enriches A/T-rich sequences, which would select against exons, which usually are more G/C rich than flanking sequences. Furthermore, genes are also methylated, which would interfere in assembling contigs of gene sequences by this method (Spena et al., 1983; Das and Messing, 1994).

ESTs

The traditional way of determining the gene content of the genome is to sequence transcribed sequences. In this approach, clones from cDNA libraries are subjected to single-pass sequencing from one end to derive short expressed sequence tags (ESTs), which serve as identifiers for genes. In most cases, transcribed sequences present in the cDNA libraries are not normalized, and so EST databases are often biased for highly expressed genes, containing multiple entries of a subset of sequences. Therefore, the number of EST entries does not represent the number of genes of the organism from which it was derived. Overall, maize ranks number 13 in dbEST, with 234,019 entries, and the entire EST collection for the top cereal crops is 1,334,170 (Table 12.4).

To increase the number of different ESTs obtained, cDNA libraries are made from different tissues and times of development to enrich for genes whose expression may be restricted temporally or spatially (Fernandes et al., 2002). The public maize EST collection being generated through the Maize Gene Discovery Project is derived from more than 20 cDNA libraries (http://www.zmdb.iastate.edu/zmdb/EST/). Most of the EST sequences in this collection are derived from the 5' ends of mRNAs so that larger contigs can be derived by assembling sequences derived from the same gene. Since the assembly programs might also align closely related sequences from different gene copies, these contigs are referred to as tentative unique contigs (TUCs), and the ones that do not assemble, as tentative unique singlets (TUSs). The maize EST database currently lists 21,063 TUCs and 19,350 TUSs, which together would represent, at most, 40,413 maize unigenes. Given that rice as a true diploid has, after deducting all transposable element-encoding genes, a predicted gene number of 49,000 (Song et al.,

TABLE 12.4. The top-ranking EST collections (as of July 2003).

Rank	Species	ESTs
6	Wheat	420,070
8	Barley	348,181
9	Soybean	338,122
13	**Maize**	**234,019**
15	Rice	201,794
16	Barrel medic	185,621
17	Thale cress	179,158
20	Tomato	150,228
23	Sorghum	130,106
28	Potato	94,473

2002), the small size of the current maize unigene set could indicate that many genes are, in fact, either silenced or expressed at low levels and/or expressed in only a few cells at specific stages of development. For instance, out of 41 genes of the alpha zein storage protein genes in maize inbred B73, only 21 accumulate mRNA at detectable levels (Song and Messing, 2002, 2003).

MICROARRAYS

To access the potential of maize gene sequences for the analysis of gene function and the regulation of gene expression, efforts are underway to provide DNA chips printed with various collections of DNA sequences (Fernandes et al., 2002). A major effort involves printing cDNA insertions from which ESTs were sequenced in glass slide microarrays. The first arrays were made from ESTs collected from specific tissues; for example, the endosperm-specific microarray contained 8,064 ESTs, comprising ~2,600 different genes. More recently, a unigene set comprising a total of 7,396 different sequences has been assembled with the goal of distributing three slides containing ~25,000 different gene representatives. They PCR-amplified the full-length insert and printed the amplicons onto slides (see <http://www.zmdb.iastate.edu/zmdb/microarray/> for details). Because the DNA samples have not been sequenced completely, hybridization data will include background noise from uncontrolled or unanticipated hybridization and from cross hybridization of sequences that share 85 percent similarity with ~100 bases. One advantage of the currently available cDNA arrays is that they are relatively inexpensive ($75 or $100 per slide, printed with duplicate or triplicate arrays). Once more sequence information is available, as in the case of the *Arabidopsis* genome, microarrays can be fabricated using long oligonucleotides that are gene specific. The more sophisticated arrays are currently much more expensive (minimum of $400 for a single array). This would reduce the background hybridization significantly, except for very closely related genes. However, microarrays of full-length cDNA sets might still be useful for screening genomic DNA fragments for the detection of allelic variants. For instance, the junction sequences of transposon insertion sites could be screened for complementary RNA sequences. Because many of the cDNAs currently available are only partly sequenced, one might not be able to identify matches by aligning EST data with the short junction sequences. Here, the hybridization data could be a useful filter to select gene/cDNA matches.

A more versatile gene chip would contain a complete set of short oligonucleotides designed specifically to hybridize to each predicted gene, as was possible for *Arabidopsis* once the complete genome sequence was available (http://arabidopsis.info/prototype/) (see Chapter 4). There are now at least three different methods for manufacturing this type of gene chip by synthesizing oligonucleotides in situ. One is based on photolithography and requires a specific mask for each array (http://www.affymetrix.com/index.affx). A second system uses a Digital Micromirror Device similar to the Digital Light Processor created by Texas Instruments (http://www.nimblegen.com/), while a third system uses inkjet technology (http://www.chem.agilent.com/).

Interestingly, the question of long versus short oligonucleotides has been an old one, both for gene cloning based on amino acid sequences, and for site-directed mutagenesis. Short oligonucleotides have a high specificity for point mutations, while long oligonucleotides are better for screening complex mixtures of DNA sequences, as was used for the cloning of the human erythropoietin gene (Egrie, 1990). One of the advantages of the NimbleGen and Agilent systems is that they do not require a mask for the array design and might be more flexible in providing different customized platforms at a lower price. Currently, however, there is too little data for an evaluation of the efficacy of different designs in providing gene-specific transcriptome information.

There are many challenges with complex genomes such as maize. In particular, a large number of genes and gene families likely have amplified only recently and therefore are very conserved in sequence. In addition, short insertion or deletion polymorphisms, which occur in maize ESTs approximately every 120 bases, can disrupt hybridization of short oligonucleotides to target RNA or cDNA, preventing transcript detection in different inbred lines and their hybrid crosses. The major technical problem with all in situ oligonucleotide synthesis methods is the efficiency of the reaction and the capping step. Incomplete chains prevent the synthesis of longer oligonucleotides because the mixture of shorter chains will increase with length and the capping efficiency. Therefore, strategies employing long oligonucleotides (60- to 100-mers) might still best be performed by synthesis of oligonucleotides prior to printing onto slides. Methods for quality checking of oligonucleotide length and composition are also advancing rapidly and may permit a higher level of confidence in array specificity in the near future.

GENOMIC SEQUENCES

The major deficiency of sequences from gene-enrichment procedures such as methyl filtration, Cot-based libraries, and EST projects is the loss of positional information. For a small subset (2,036; http://www.agron.missouri.edu/ests.html), positional information has been restored by mapping the EST sequences in segregating populations or recombinant inbred lines. However, genetic mapping is very expensive relative to cloning and sequencing. Furthermore, the positions of some mapped sequences may be ambiguous when a probe detects multiple members of a gene family. For mapping, the longer the available sequence is, the more economical and accurate genetic mapping will be. This is the great value of long fingerprint contigs. Moreover, the ultimate determination of a complete gene set requires a sequence that is collinear with the genetic map of each chromosome.

At this time, available contiguous maize genomic sequences comprise a single BAC (or a fraction of it) or up to three overlapping BAC clones from a few selected regions of the maize genome (Table 12.5). Fortunately, most of these sequences have been obtained from the B73 BAC libraries. Because these libraries are fingerprinted, the clone contigs are also integrated into the physical mapping project. Furthermore, the map position of sequenced BAC clones is also anchoring entire fingerprint contigs to the genetic map.

Although the sample available for analysis is still very small, representing only about 0.25 percent of the euchromatic portion of the maize genome, some potentially interesting features are emerging. This survey does not necessarily provide an accurate estimate of gene numbers but is based on methods of gene predictions currently used to determine genes on completely sequenced chromosomes of other eukaryotic genomes. To gain a larger data set, a number of sequences are included that represent phase 2 sequence data; phase 2 sequences are ordered sequence contigs that still contain gaps, which are closed in phase 3. Rice chromosome 10 was used as a reference sequence since it is representative of the whole rice genome in terms of sequence composition and structure (Rice Chromosome 10 Sequencing Consortium, 2003). Three different data sets are summarized (Table 12.5). Section 1 represents a set that is outside and Section 3 covers the maize/sorghum collinearity project (NSF#9975618). Section 2 represents the maize zein alpha gene family. The zein gene regions were expected to have a higher gene density than average because of tandem arrays of gene family members. The open reading frames (ORFs) were divided into three categories: transposases (t-pons), retrotransposon-related ORFs (retros), and predicted genes.

An automated annotation program was used that first analyzed all clones for hypothetical genes using the gene prediction program FGENESH adjusted to monocot genes (http://www.softberry.com/berry.phtml). The predicted proteins were then subjected to BLAST analysis (http://www.ncbi.

TABLE 12.5. Genomic sequences of maize large insert libraries.

Region	Ref.	# clones	Chr	ORF	retros	t-pons	genes	ORF/kbp	ret/kbp	tp/kbp	g/kbp	Size
Section 1												
Cent1	AC116034	1	—	3	3	0	0	11.36	11.36	n/a	n/a	34,079
Cent2	AC116033	1	—	20	15	0	5	5.00	6.67	n/a	20.00	99,979
Cent3	AC114395	1	—	15	6	2	7	7.60	18.99	56.98	16.28	113,950
BzMcC	AF391808	2	9S	43	18	2	23	5.26	12.56	113.00	9.83	226,001
BzB73	AF448416	1	9S	23	8	2	13	4.62	13.27	53.09	8.17	106,186
Adh1	AF123535	1	1L	30	21	0	9	5.35	7.64	n/a	17.83	160,480
Total		7		134	71	6	57	5.53	10.43	123.45	12.99	740,675
Section 2												
z1A1	AF546186	1	4S	34	18	5	10	4.46	8.43	30.35	15.17	151,728
z1A2	AF546189	1	4S	19	9	0	10	6.11	12.90	n/a	11.61	116,115
z1B1	AF546188	1	7S	39	22	4	13	5.21	9.24	50.84	15.64	203,363
z1B2	AF546190	1	7S	35	20	1	14	5.14	9.00	179.94	12.85	179,942
z1D	AF546187	3	1L	95	65	2	28	5.04	7.36	239.24	17.09	478,484
BSz1C1	AF090447 AF528565	3	4S	84	30	7	47	5.18	14.50	62.15	9.26	435,076
BSz1C2	AF090446	1	4S	6	3	0	3	5.10	10.20	n/a	10.20	30,593
Bz1C1	AC144719 AC144718	2	4S	46	14	1	31	5.73	18.83	263.63	8.50	263,630
Bz1C2	AC144717		4S	3	1	0	2	4.76	14.27	n/a	7.14	14,273
Total		13		361	182	20	158	5.19	10.29	93.66	11.86	1,873,204

Region	Ref.	# clones	Chr	ORF	retros	t-pons	genes	ORF/kbp	ret/kbp	tp/kbp	g/kbp	Size
Section 3												
Rp1-1	AF466931	1	10S	19	6	1	12	5.00	15.85	95.08	7.92	95,078
Rp1-2	AF466932	1	10S	16	8	0	7	6.20	12.39	n/a	14.17	99,156
B1'	AY078063	1	2S	17	7	4	6	6.34	15.41	26.96	17.97	107,840
B1-B73	AF466203 Phase II	2	2S	44	30	0	14	5.66	8.31	n/a	17.80	249,176
R1-B73	AF466202 Phase II	2	10L	41	21	1	19	7.08	13.83	290.35	15.28	290,350
Tb1	AF464738 AY325816	2	1L	37	23	1	13	5.92	9.52	218.96	16.84	218,962
Tb2	AF466646	1	5S	21	8	1	12	6.76	17.74	141.94	11.83	141,939
Tbp1	Phase II	2	5S	34	19	0	15	6.14	10.99	n/a	13.92	208,853
Tbp2	Phase II	2	5S	29	17	2	15	6.77	11.55	98.17	13.09	196,335
C1	Phase II	2	9S	51	26	3	22	6.51	12.78	110.74	15.10	332,217
Pl1	Phase II	2	6L	49	30	1	18	6.45	10.54	316.13	17.56	316,131
Fie/Orp	Phase II	3	4S	66	45	4	17	5.44	7.98	79.03	21.11	358,896
Fie/Orp	Phase II	2	10L	45	19	3	23	6.35	15.04	95.27	12.43	285,802
Total		23		469	259	21	193	6.18	11.20	138.13	15.03	2,900,735
Sum		43		964	512	47	408	5.72	10.77	117.33	13.52	5,514,614
Rice 10				3,647	541	150	2,776	6.46	41.40	149.33	8.07	22,402,313

nlm.nih.gov/BLAST/). Only ORFs of more than 100 amino acids (aa) were counted as genes. In such a program, the numbers of genes and transposons are usually inflated because the program cannot detect a gene that is interrupted by a transposon as one gene. The same is true for nested transposons. For instance, for the z1C1 regions in BSSS53 (B73), manual annotation has resulted in 32 (23) genes, 10 (9) retrotransposons, and 7 (2) transposons. The automated annotation for the same regions predicted 47 (31) genes, 30 (14) retrotransposons, and 7 (1) transposons, exceeding the gene and retrotransposon number by up to 50 percent. However, manual annotation for these data is either already published or still under way. For the comparison to rice chromosome 10, which has not been manually annotated, I have kept the numbers of all clones from the automated annotation.

Based on these data, gene and retrotransposon densities varied between different regions of the maize genome, as expected. Gene density ranged from about 1 gene per 21 kbp to 1 gene per 7 kbp. Even if we assume that the majority of the genome has a gene density closer to the lower number, the potential coding capacity for maize could be as high as 95,000 genes, or even higher. Since rice contains about 49,000 predicted genes (60,000 if one would count in addition transposon-related genes) (see also Chapter 11), an estimate of 95,000 for maize may not be too far off because maize contains two subgenomes, which would double the number of genes. These estimates exclude ORFs from transposable elements because they are all silenced genes. Many of the predicted genes could also represent gene fragments or pseudogenes. As we already know from the analysis of the alpha zein storage protein genes in inbred B73, only 21 out of 41 genes are actually expressed (Song and Messing, 2002, 2003). If, however, in general only half of the genes were expressed, one would expect to find about 45,000 to 50,000 active genes. Therefore, we expect that the calculations of the gene density in maize based on ESTs and the genome sequence will differ significantly. While gene density in maize can reach the average gene density in rice in some regions, the overall retrotransposon density is clearly much higher in maize than it is in rice. It ranges from 1 retrotransposon per 7 kbp to 1 retrotransposon per 19 kbp in maize, compared to an average value of 1 retrotransposon per 41 kbp in rice.

RICE/MAIZE CLONE MAPS

One useful way of determining the set of active genes is by comparative genomics (Messing and Llaca, 1998). As described previously, conserved cDNA clones have been used to developed collinearity maps between maize and rice (Ahn and Tanksley, 1993). However, this initial study involved only 250 gene loci. If maize indeed has 95,000 or more genes, the initial study would represent only 0.0025 percent of the gene collection. Furthermore, chromosome walking in maize indicates that there have been many changes in gene order compared to rice. Orthologous genes have moved to nonorthologous positions in the genome and have amplified to different degrees (Song et al., 2002). The noncollinearity of genes and gene copies could impede the development of a maize/rice clone map. The more recent study, however, indicates that a framework of orthologous markers has been sustained, resulting in a mosaic organization of collinear and noncollinear markers. For instance, in the human/mouse clone map, a total of 51,486 homologous cross-link sequences were found. Interestingly, only 19 percent of the cross-link sequences matched coding sequences (ESTs), while 32 percent matched introns and 49 percent intergenic regions (Gregory et al., 2002). Since the evolutionary distance between human and mouse is about twice the distance between maize and rice (100 versus 50 mya; Burt et al. 1999; Kellogg, 2001) and the number of genes in plants is higher than in humans (Messing, 2001), one would not expect the cross-link sequences to be lower for the two cereal genomes. The human/mouse cross-link sequences were derived mainly from the end sequences of the fingerprinted BAC clones (BAC-end sequences). A database of 453,962 BAC-end sequences of the mouse genome has been used to identify the cross-link sequences with the human genome. Since these BAC clones formed a total of 296 contigs, many of the BAC-end sequences were ordered, giving a complete overview of gene distribution on the various chromosomes. Actually, a human sequence collinear to the human genetic map has been an instrumental tool in reducing the number of mouse contigs; fingerprint contigs were ordered based on collinear markers and then manually edited to merge pairs of overlaps. Therefore, one can envision that rice can play a similar role for maize as human did for mouse for map assembly. Once the two genomes are aligned, one can then ask the question as to which genes have maintained their orthologous position and are expressed in both genomes, which genes have been mobilized and/or amplified, and which are unique to each genome.

REFERENCES

Ahn, S., Tanksley, S. (1993). Comparative linkage maps of the rice and maize genomes. *Proc. Natl. Acad. Sci. USA* 90: 7980-7984.

Arabidopsis Genome Initiative (2000). Analysis of the genome sequence of the flowering plant *Arabidopsis thaliana*. *Nature* 408: 796-815.

Arumuganathan, K., Earle, E. (1991). Nuclear DNA content of some important plant species. *Plant Mol. Biol. Rep.* 9: 208-218.

Bennetzen, J., Freeling, M. (1993). Grasses as a single genetic system: Genome composition, colinearity, and compatibility. *Trends Genet.* 9: 259-261.

Bennetzen, J., Schrick, K., Springer, P., Brown, W., SanMiguel, P. (1994). Active maize genes are unmodified and flanked by diverse classes of modified, highly repetitive DNA. *Genome* 37: 565-576.

Burr, B., Burr, F., Thompson, K., Albertsen, M., Stuber, C. (1988). Gene mapping with recombinant inbreds in maize. *Genetics* 118: 519-526.

Burt, D., Bruley, C., Dunn, I., Jones, C., Ramage, A., Law, A., Morrice, D., Paton, I., Smith, J., Windsor, D., et al. (1999). The dynamics of chromosome evolution in birds and mammals. *Nature* 402: 411-413.

Clark, L., Carbon, J. (1976). A colony bank containing synthetic Col E1 hybrid plasmids representing the entire *E. coli* genome. *Cell* 9: 91-99.

Cone, K., McMullen, M., Bi, I., Davis, G., Yim, Y., Gardiner, J., Polacco, M., Sanchez-Villeda, H., Fang, Z., Schroeder, S., et al. (2002). Genetic, physical, and informatics resources for maize: On the road to an integrated map. *Plant Physiol.* 130: 1598-1605.

Coulson, A., Sulston, J., Brenner, S., Jonathan, K. (1986). Towards a physical map of the genome of the nematode *Caenorhabditis elegans*. *Proc. Natl. Acad. Sci. USA* 83: 7821-7825.

Cowperthwaite, M., Park, W., Xu, Z., Yan, X., Maurais, S., Dooner, H. (2002). Use of the transposon Ac as a gene-searching engine in the maize genome. *Plant Cell* 14: 713-726.

Das, O.P., Messing, J. (1994). Variegated phenotype and developmental methylation changes of a maize allele originating from epimutation. *Genetics* 136: 1121-1141.

Edwards, K., Thompson, H., Edwards, D., deSauzieu, A., Sparks, C., Thompson, J., Greenland, A., Eyers, M., Schuch, W. (1992). Construction and characterization of a yeast artificial chromosome library containing 3 haploid maize genome equivalents. *Plant Mol. Biol.* 19: 299-308.

Egrie, J. (1990). The cloning and production of recombinant human erythropoietin. *Pharmacotherapy* 10: 35-85.

Fedoroff, N.V., Furtek, D.B., Nelson, O.E. (1984). Cloning of the *Bronze* locus in maize by a simple and generalizable procedure using the transposable controlling element *Activator* (Ac). *Proc. Natl. Acad. Sci. USA* 81: 3825-3829.

Fernandes, J., Brendel, V., Gai, X., Lal, S., Chandler, V., Elumalai, R., Galbraith, D., Pierson, E., Walbot, V. (2002). Comparison of RNA expression profiles based on maize expressed sequence tag frequency analysis and micro-array hybridization. *Plant Physiol.* 128: 896-910.

Fu, H., Dooner, H.K. (2002). Intraspecific violation of genetic collinearity and its implication in maize. *Proc. Natl. Acad. Sci. USA* 99: 9573-9578.

Gale, M., Devos, K. (1998). Plant comparative genetics after ten years. *Science* 282: 656-659.

Gaut, B., Doebley, J. (1997). DNA sequence information for the segmental allotetraploid origin of maize. *Proc. Natl. Acad. Sci. USA* 88: 2060-2064.

Geraghty, D., Peifer, M.A., Rubenstein, I., Messing, J. (1981). The primary structure of a plant storage protein: Zein. *Nucleic Acids Res.* 9: 5163-5174.

Gregory, S.G, Sekhom, M., Schein, J., Zhao, S., Osoegawa, K., Scott, C., Evans, R.S., Burridge, P.W., Cox, T.V., Fox, C.A., et al. (2002). A physical map of the mouse genome. *Nature* 418: 743-750.

Hake, S., Walbot, V. (1980). The genome of *Zea mays,* its organization and homology to other grasses. *Chromosoma* 79: 251-270.

Hamby, R.K., Zimmer, E.A. (1988). Ribosomal RNA sequences for inferring phylogeny within the grass family (Poaceae). *Plant Syst. Evol.* 160: 29-37.

Helentjaris, T., Weber, D., Wright, S. (1988). Identification of the genomic locations of duplicate nucleotide sequences in maize by analysis of restriction fragment length polymorphisms. *Genetics* 118: 353-363.

Hirochika, H. (2001). Contribution of the *Tos17* retrotransposon to rice functional genomics. *Curr. Opin. Plant Biol.* 4: 118-122.

Hirochika, H., Sugimoto, K., Otsuki, Y., Tsugawa, H., Kanda, M. (1996). Retrotransposon of rice involved in mutations induced by tissue culture. *Proc. Natl. Acad. Sci. USA* 93: 7783-7788.

Hu, W., Timmermans, M.C.P., Messing, J. (1998). Interchromosomal recombination of *Zea mays. Genetics* 150: 1229-1237.

Kellogg, E. (1998). Relationships of cereal crops and other grasses. *Proc. Natl. Acad. Sci. USA* 95: 2005-2010.

Kellogg, E. (2001). Evolutionary history of the grasses. *Plant Physiol.* 125: 1198-1205.

Lee, M., Sharopova, N., Beavis, W., Grant, D., Katt, M., Blair, D., Hallauer, A. (2002). Expanding the genetic map of maize with the intermated B73 × Mo17 (IBM) population. *Plant Mol. Biol.* 48: 453-461.

Llaca, V., Messing, J. (1998). Amplicons of maize zein genes are conserved within genic but expanded and constricted in intergenic regions. *Plant J.* 15: 211-220.

Manning, J., Schmid, C., Davidson, N. (1975). Interspersion of repetitive and nonrepetitive DNA sequences in the *Drosophila melanogaster* genome. *Cell* 4: 141-155.

McClintock, B. (1984). The significance of responses of the genome to challenge. *Science* 226: 792-801.

McLaren, J. (1998). The success of transgenic crops in the USA. *Pesticide Outlook* 9: 36-41.

Messing, J. (2001). Do plants have more genes than humans? *Trends Plant Sci.* 6: 195-196.

Messing, J., Llaca, V. (1998). Importance of anchor genomes for any plant genome project. *Proc. Natl. Acad. Sci. USA* 95: 2017-2020.

Meyers, B., Tingey, S., Morgante, M. (2001). Abundance, distribution, and transcriptional activity of repetitive elements in the maize genome. *Genome Res.* 11: 1660-1676.

Peschke, V., Phillips, R., Gengenbach, B. (1987). Discovery of transposable element activity among progeny of tissue culture-derived maize plants. *Science* 238: 804-807.

Rabinowicz, P., Schutz, K., Dedhia, N., Yordan, C., Parnell, L., Stein, L., McCombie, W., Martienssen, R. (1999). Differential methylation of genes and retrotransposons facilitates shotgun sequencing of the maize genome. *Nat. Genet.* 23: 305-308.

Raizada, M.N., Nan, G.-L., Walbot, V. (2001). Somatic and germinal mobility of the RescueMu transposon in transgenic maize. *Plant Cell* 13: 1587-1608.

Rhoades, M.M. (1951). Duplicated genes in maize. *Am. Nat.* 85: 105-110.

Rice Chromosome 10 Sequencing Consortium (2003). In-depth view of structure, activity, and evolution of rice chromosome 10. *Science* 300: 1566-1569.

SanMiguel, P., Bennetzen, J. (1998). Evidence that a recent increase in maize genome size was caused by the massive amplification of intergene retrotranspositions of maize. *Ann. Bot.* 81: 27-44.

Sharopova, N., McMullen, M.D., Schultz, L., Schroeder, S., Sanchez-Villeda, H., Gardiner, J., Bergstrom D., Houschins, K., Melia-Hancock, S., Musket, T., et al. (2002). Development and mapping of SSR markers for maize. *Plant Mol. Biol.* 48: 463-481.

Soderlund, C., Humphray, S., Dunham, A., French, L. (2000). Contigs built with fingerprints, markers, and FPC V4.7. *Genome Res.* 10: 1772-1787.

Song, R., Llaca, V., Linton, E., Messing, J. (2001). Sequence, regulation, and evolution of the maize 22-kDa α zein gene family. *Genome Res.* 11: 1817-1825.

Song R., Llaca V., Messing, J. (2002). Mosaic organization of orthologous sequences in grass genomes. *Genome Res.* 12: 1549-1555.

Song, R., Messing, J. (2002). Contiguous genomic DNA sequence comprising the 19-kDa-zein gene family from *Zea mays*. *Plant Physiol.* 130: 1626-1635.

Song, R., Messing, J. (2003). Gene expression of a gene family in maize based on noncollinear haplotypes. *Proc. Natl. Acad. Sci. USA* 100: 9055-9060.

Spena, A., Viotti, A., Pirrotta, V. (1983). Two adjacent genomic zein sequences: Structure, organization, and tissue-specific restriction pattern. *J. Mol. Biol.* 169: 799-811.

Takahashi, C., Marshall, J., Bennet, M., Leitch, I. (1999). Genomic relationships between maize and its wild relatives. *Genome* 42: 12201-12207.

Timmermans, M.C.P, Das, O.P., Messing, J. (1996). Characterization of a meiotic crossover in maize identified by a restriction fragment length polymorphism-based method. *Genetics* 143: 1771-1783.

Walbot, V. (1999). UV-B damage amplified by transposons in maize. *Nature* 397: 398-399.

Weerakoon, L., Weber, D., Schneerman, M. (1994). Estimating genetic relatedness of maize to 23 plant species by RFLP analysis. *Maydica* 38: 231-237.

Wendel, J. (2000). Genome evolution in polyploids. *Plant Mol. Biol.* 42: 225-249.

Wessler, S., Bureau, T., White, S. (1995). LTR-retrotransposons and MITEs: Important players in the evolution of plant genomes. *Curr. Opin. Genet. Dev.* 5: 814-821.

Yim, Y.-S., Davis, G., Duru, N., Musket, T., Linton, E., Messing, J., McMullen, M., Soderlund, C., Polacco, M., Gardiner, J., Coe Jr., E. (2002). Characterization of three maize BAC libraries toward anchoring of the physical map to the genetic map using high density BAC filter hybridization. *Plant Physiol.* 130: 1686-1696.

Yuan, Y., SanMiguel, P., Bennetzen, J. (2002). Methylation-spanning linker libraries link gene-rich regions and identify epigenetic boundaries in *Zea mays*. *Genome Res.* 12: 1345-1349.

Zhang, J., Peterson, T. (1999). Genome rearrangements by nonlinear transposons in maize. *Genetics* 153: 1403-1410.

Chapter 13

Prospects for Functional Genomics in a New Model Grass

Glyn Jenkins
Luis A. J. Mur
Pauline Bablak
Robert Hasterok
John Draper

WHY A NEW MODEL GRASS?

A model organism possesses certain biological features that make it more amenable to scientific investigation compared with other less tractable members of the group it represents. In plant functional genomics programs, the key desirable features that confer model status may include a small diploid genome, rapid life cycle, small stature and undemanding growth requirements, tractable genetics and mapping populations, together with efficient mutagenesis and transformation procedures. These facilitate, for example, the creation of large insert genomic libraries, genome sequencing, the construction of dense genetic and physical maps, and the production of transcriptome, proteome, and metabolome data sets. Models can, therefore, be exploited for gene discovery and the development of bioinformatic tools that are essential to the success of a functional genomics program.

The effectiveness and utility of a model for functional genomics is to a large extent dependent upon the degree of conservation of gene order and repertoire within the group of organisms being compared. Until recently, colinearity was assayed in a number of plant groups, such as the cereals (Gale and Devos, 1998), the Solanaceae (Tanksley et al., 1988; 1992), and the Cruciferae (Kowalski et al., 1994; Lagercrantz, 1998), by comparative genetic mapping using common DNA markers deployed across related species. This approach has shown in the Poaceae, for example, that the genomes of at least ten divergent cereals and a growing number of their wild relatives can be described in terms of only 30 rice linkage blocks (Gale and

Devos, 1998; Devos and Gale, 2000). With the increased availability of DNA sequence data, particularly through the publication of the complete genome sequences of the two model plants *Arabidopsis thaliana* (Arabidopsis Genome Initiative, 2000) and rice (Dickson and Cyranoski, 2001; Sasaki et al., 2002; Feng et al., 2002; Yu et al., 2002; Goff et al. 2002), it is now possible to align and compare the fine structure of orthologous regions from different species. Studies of orthologous segments over 30 kbp in the cereals maize, sorghum, rice, and barley have shown that there is conservation of gene structure, but that microsynteny was disturbed by some small genomic rearrangements (Chen et al., 1997, 1998; Tikhonov et al., 1999; Dubcovsky et al., 2001). A recent, more thorough comparison of larger orthologous sequences in four cereal genomes has shown microcolinearity of the four common genes near the *Vrn1* locus but also has documented sequence divergence and different lineages of genome evolution (Ramakrishna et al., 2002).

The inference from the studies just described is that a model for the temperate cereals and grasses should be chosen on the basis of its phylogenetic proximity to its relatives in order to minimize the risk of breakdown of microsynteny in orthologous chromosome regions. *Arabidopsis thaliana* clearly has utility in comparative genomic studies of near relatives, such as *Brassica* (O'Neill and Bancroft, 2000) and *Capsella* (Rossberg et al., 2001), but is unlikely to serve as a useful "anchor" genome to locate important chromosome regions in cereals and grasses (Bennetzen et al., 1998; Devos et al., 1999; Li and Gill, 2002). Rice is a stronger contender (Havukkala, 1996; Goff, 1999) (Chapter 11), not only because of its international status as a staple food, but also because of its genomic attributes, such as its compact and fully sequenced genome. However, rice belongs to a different subfamily from the temperate cereals and grasses and does not exhibit certain traits that may be important to functional genomics programs, such as resistance to specific types of pathogens, vernalization requirements, and perenniality.

Against this background, it was decided to reexamine the Pooideae in order to identify a grass species that could be developed into a useful model representative of important temperate cereals and forage grasses. The following sections describe the biological attributes of *Brachypodium distachyon,* and the current status and future perspectives of its development as a model for this group of organisms. The final section focuses upon how *B. distachyon* is fulfilling its role as a "niche" model for functional genomics, in terms of understanding the molecular basis of the response of plants to pathogen attack, and its utility in metabolomic research.

UNDERSTANDING THE PLANT

As a result of molecular studies using RFLP and RAPD analysis (Catalan et al., 1995), using nuclear rDNA (ITS) sequences (Hsiao et al., 1995) and phylogenetic analyses of the more variable 3' end of the chloroplast *ndhF* gene (Catalan et al., 1997), the Brachypodieae tribe has been sited in the Poaceae subfamily of the Pooideae occupying the sister position to the pooid "core" clade. The tribe diverged from the more basal lineages together with Diarrheneae, Poeae, Aveneae, Bromeae, and Triticeae (Sharma, 1979) prior to the radiation of the most recently evolved pooids, which include the important cereal crops (Figure 13.1A). It is thought, therefore, that this species is more closely related to the temperate cereals and other monocots than is rice, the species most developed as a model for cereals. Evidence has accumulated to establish the genus *Brachypodium* as a distinctive clade that diverged soon after the division of Pooideae from the Oryzeae, as demonstrated by Shi (1991) and Shi et al. (1993), with reference to length heterogeneity of rDNA repeat units, together with the work of Hsiao et al. (1995) and Catalan et al. (1995, 1997). Interest in this genus has gathered pace following reports of its ancient origin (Kellogg, 1998), as suggested by the variation in chromosome base number, and the inferred relationship to the important cereal crops, as shown by the phylogenetic studies of Sharma (1979).

The collection of *Brachypodium* held at Aberystwyth includes more than 50 accessions harvested in the main from temperate zones, with more than half the species distributed across the Mediterranean and neighboring regions extending to the east. Other representatives of the genus have also been identified in South America and South Africa. This wide distribution of a limited number of species was interpreted as a relic pattern of a very ancient genus (Tateoka, 1968). The diploid accessions of *B. distachyon* were collected mostly from Western Europe (Spain, Italy, and France), with others from Croatia and Turkey. Interestingly, all of the 35 accessions collected from the east of Europe are hexaploid, with only one tetraploid from the island of Formentera, Spain.

The life cycle of *B. distachyon* depends upon its ploidy level, with the tetraploid and hexaploid accessions flowering as little as 30 days after sowing (Figure 13.1B). The diploids require either a vernalization period of at least 35 days (Bablak, 2001) or an equivalent period in short-day growth conditions to induce flowering. In practice, the diploids are vernalized for 42 days, as this increases the number of flowering spikes induced and also synchronizes the flowering cycle, which is completed usually within 80 days. This species has undemanding growth requirements, short generation

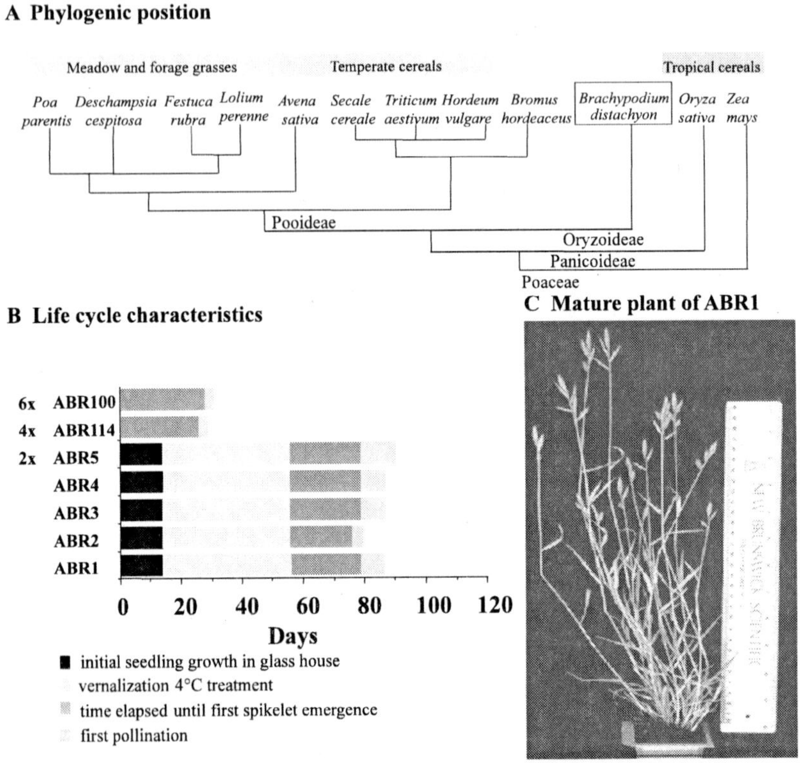

FIGURE 13.1. (A) Phylogeny of *B. distachyon* in relation to important cereals and grasses. (B) Relative duration of parts of the life cycle of seven accessions. (C) Mature plant of diploid genotype ABR1 showing numerous primary, secondary, and tertiary flowering spikes before seed collection.

time, and small stature, which together make it ideal for high-throughput mutant screens or functional genomics programs in which uniform plant development is important. Large numbers of plants can be grown in a limited space without compromising growth, although high-density sowing can cause a reduction in the overall height of the plants and the number of flowering spikes. In practice, modular sowing produces plants with sufficient numbers of leaves and spikelets for experimentation (Figure 13.1C). Mature plants are approximately 30 cm tall with numerous culms supporting three to five primary flowering spikes, each of which carries five to seven spikelets with five to nine fertile florets each. Secondary and tertiary spikes follow these spikes over a period of approximately three to four weeks. As would be expected, the phenotype is plastic depending upon the temperature and light conditions under which the plants are grown.

Construction of a developmental key comparable to that describing *Arabidopsis* (Boyes et al., 2001) is underway, and the main stages and transitions have been identified for the *B. distachyon* diploid accessions ABR1 and ABR5 and the hexaploid ABR100, corresponding wherever possible to equivalents in rice (Councea et al., 2000). The stages relating to flowering are less well defined at the moment, as the primary tiller produces between three and five spikelets over a period of weeks, and the precise start of pollination is difficult to gauge, as this species is autogamous and the process normally begins before the spikelet has emerged fully. Flowering secondary and tertiary tillers are produced throughout the growing period, and each carries more than one spikelet. Pollination begins at or close to the central floret on the rachilla and proceeds in either direction from that point. The end of pollination is easily recognized by the emergence of anthers from each individual floret.

Brachypodium distachyon can be cultured using methods developed for barley and rice (Bablak et al., 1995; Bablak, 2001). Immature zygotic embryos (approximately 14 days after pollination) produce the greatest number of somatic embryos in the shortest period of time and are used routinely to produce embryogenic cultures suitable for transformation by particle bombardment. The most successful result so far was the regeneration of seven transformed plantlets of hexaploid genotype ABR100 per gram of target tissue (Draper et al., 2001). The transformation of diploid accessions is currently being optimized and is currently yielding about three plantlets per gram of bombarded tissue. Once optimized, it is intended to introduce transposable elements into a diploid ecotype and to generate tagged lines.

UNDERSTANDING THE GENOME

The genomes of most higher plants are replete with repetitive DNA sequences, which may either be localized as tandem repeats within heterochromatin or dispersed throughout the chromosome complement. Variation in this repetitive component, together with a high prevalence of polyploidy, accounts for the vast range of nuclear DNA amounts within this group of organisms. Members of the Poaceae clearly contribute to this variation, with 1*C* values ranging from approximately 430 Mbp in rice to about 17,000 Mbp in allohexaploid wheat (Arumuganathan and Earle, 1991). Repetitive DNA and polyploidy conspire to hamper functional genomics programs by reducing the efficiency of isolation, mapping, and characterization of genes. The corollary is that an ideal model for this group of plants should have a compact and economical diploid genome, with demonstrable colinearity of genes with its less tractable, but agronomically more important, relatives. Recent microdensitometric measurements of the 1C DNA contents of two diploid accessions have revealed that *B. distachyon* has the lowest reported value in the Poaceae to date (Draper et al., 2001). Additional analysis by flow cytometry of four diploid accessions confirmed that this species had a genome size that is virtually indistinguishable from that of the model dicot *Arabidopsis thaliana,* i.e., 1*C* ~125 Mbp (Draper et al., 2001). These estimates are consistent with previous observations that *Brachypodium* species contain typically less than 15 percent repetitive DNA (Catalan et al., 1995) and demonstrate the credentials of *B. distachyon* as a model for functional genomics.

Another desirable attribute of a model organism is a karyotype in which the chromosomes are low in number and readily identifiable. These features enable the integration of physical and genetic maps and open up avenues of research into chromosome-specific structure and function. *Brachypodium distachyon* has an exceptionally low basic chromosome number of five, which is probably the archetypal number of the temperature cereals and grasses, given the phylogenetic divergence of this species from an ancestral stock of Pooideae just prior to the radiation of modern core pooids.

The somatic metaphase chromosomes were examined from seven geographically dispersed, diploid accessions of *B. distachyon*. Since there was no gross structural variation between the chromosome complements of these accessions, a consensus karyotype of this species was constructed (Draper et al., 2001). The karyotype is sufficiently asymmetric to permit unequivocal identification of three of the five pairs of chromosomes by light microscopy of preparations stained by acetocarmine alone. Potential confusion between chromosomes 3 and 4 is resolved by fluorescence in situ hybridization (FISH) with ribosomal DNA probes. Figure 13.2A shows a somatic interphase nucleus and chromosome complement of genotype ABR5 probed with 25S and 5S rDNA. The karyogram in Figure 13.2B shows clearly that chromosome 4 bears only a 5S rDNA locus located near its centromere, and chromosome 5 has a 25S locus that occupies a large proportion of the short arm. FISH with the same rDNA probes to the chromosomes of another diploid genotype (ABR1) showed the same distribution of signals.

The structure of the genome of this species was explored further by probing with well-defined chromosomal landmarks. Figure 13.2(A,B) shows that all the telomeres of this species comprise tandem repeats of the typical consensus sequence $(TTTAGGG)_n$, which colocalizes with the telomeric sequence HT100.3 derived from *A. thaliana* (Figure 13.2C). Figure 13.2C also shows that the pericentromeric repeat sequence CCS1, originally iso-

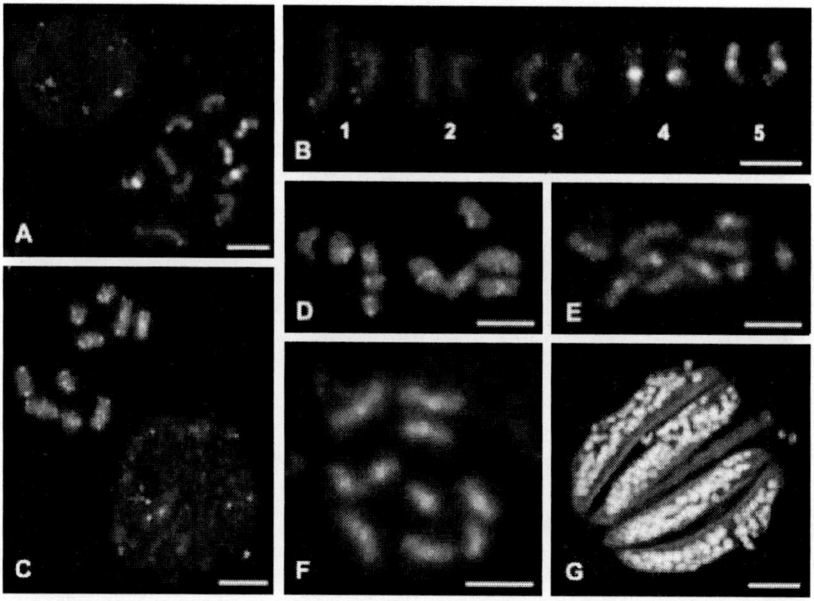

FIGURE 13.2. FISH to somatic interphase and metaphase chromosomes of *B. distachyon* (see corresponding figure in color photo section).

lated from the related species *B. sylvaticum* (Aragon-Alcaide et al., 1996), marks all the centromeres in this species too. A reliable marker of centromeres accurately defines the chromosome arms for mapping purposes and enables studies of the nuclear disposition of this domain at interphase and other stages of the mitotic cell cycle (Figure 13.2A). The pericentromeric repeat CCS1 is a motif within a Ty3-*gypsy* class retrotransposon that appears to have colonized the centromeres of a wide range of species within the Poaceae (Langdon et al., 2000). FISH with conserved regions of the reverse transcriptase (Figure 13.2D) and integrase (Figure 13.2E) genes of this element confirmed that the pericentromeric regions of *B. distachyon* share similar structures with other members of the Poaceae. Retrotransposons may constitute a considerable proportion of the middle repetitive component of the genome, amounting to up to 10 percent of the genome of barley (Suoniemi et al., 1996; Waugh et al., 1997) and more than 60 percent of the genome of maize (SanMiguel et al., 1996; SanMiguel and Bennetzen, 1998; Meyers et al., 2001). Retrotransposons are likely to constitute a sizable proportion of the 12 to 15 percent repetitive DNA in *B. distachyon*. Indeed, somatic chromosomes of ABR1 probed with genomic

DNA from the same genotype reveals that the repetitive DNA in this species is concentrated around the centromeres and maps to the same location as the retrotransposons (Figure 13.2F).

One of the ultimate aims of establishing *B. distachyon* as a new model is to gain access to desirable genetic regions in this species by alignment with syntenic regions in the less tractable relatives. Colinearity may be inferred from the favorable phylogenetic positions of *B. distachyon* and the other temperate cereals and grasses, but such an approach necessitates a clear demonstration that the linear gene order is conserved at the microsyntenic level. Given the current dearth of sequence information and absence of a large insert library in this species, it was decided to test microsynteny in this species by the targeted landing of heterologous BACs onto the genome of this species. This approach requires high-resolution mapping to reconstruct and assign megabase tracts of DNA to particular chromosome arms. Increased resolution is afforded by the use of either meiotic prophase chromosomes, which are substantially longer than their mitotic counterparts, or extended DNA fibers, which have been used to good effect in species such as tomato (Zhong et al., 1999). The immature inflorescence of *B. distachyon* is tiny compared with those at comparable stages in other members of the Poaceae, and it has correspondingly minute anthers (Figure 13.2G). Nevertheless, there is a predictable gradation of development along the spike that permits the retrieval of particular meiotic stages and the construction of an atlas of meiosis comparable to that obtained in *A. thaliana* (Ross et al., 1996). Figure 13.3 shows the meiotic atlas of genotype ABR1, which will serve as a future wild-type reference for describing meiotic mutants in this species as they become available. It should be noted that the chromosomes at metaphase I not only form higher numbers of chiasmata than would be predicted from their size but also do not restrict chiasmata to particular chromosome locations. Both of these features are important in terms of rapid development of accurate and comprehensive genetic and molecular marker maps. Higher-resolution mapping is shown in Figure 13.4A, in which discrete centromeric and telomeric signals can be seen on extended meiotic prophase chromosomes. Further enhancement of resolution is shown in Figure 13.4C, in which extended DNA fibers from ABR1 have been probed with a 2.3 kbp fragment of the 10 kbp 45S rDNA subunit. The 10 kbp periodicity of the signal illustrates the potential for fine mapping of small contiguous DNA regions.

Microscale BAC isolation and DOP-PCR have been developed and used successfully to hybridize rice BACs to somatic chromosomes of the same species (Figure 13.4B). However, FISH with 6 out of 24 rice BACs, each anchored to particular chromosome arms, has so far failed to map specifically and discretely to the chromosomes of *B. distachyon*. Such attempts

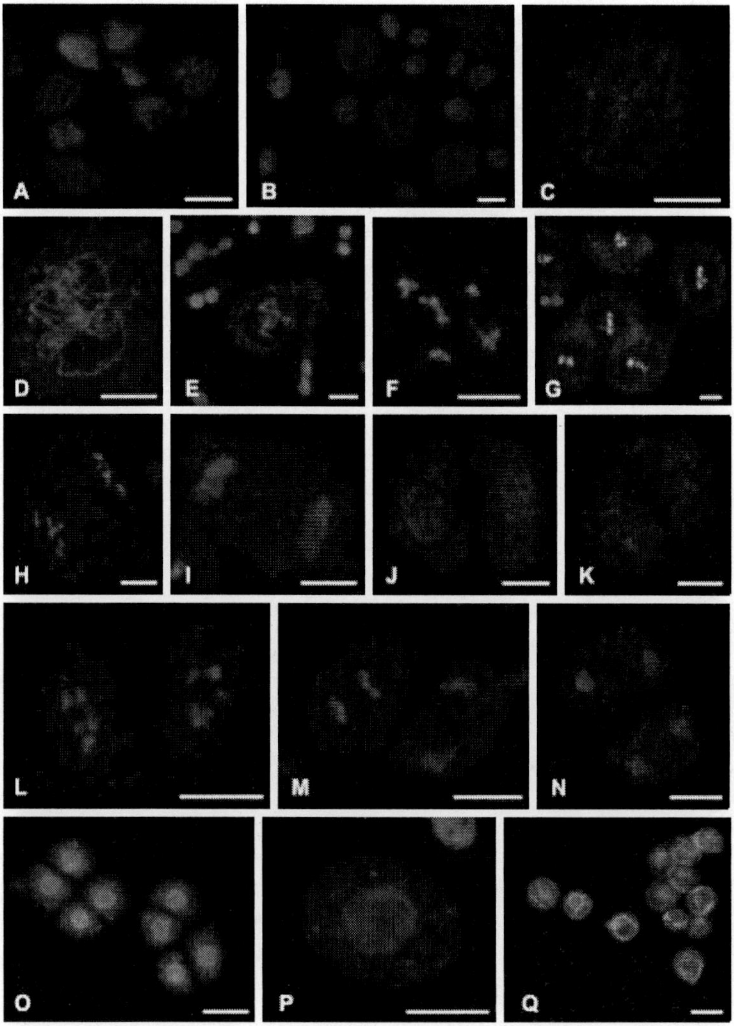

FIGURE 13.3. DAPI-stained atlas of meiosis of diploid *B. distachyon* (see corresponding figure in color photo section).

may be confounded by the small but finite amount of repetitive DNA likely to be present in the large inserts of the rice BACs. Therefore, attention has now been turned to producing a BAC library in *B. distachyon* itself, which could be used, not only for testing synteny in the grasses and cereals, but also for future programs requiring the targeted sequencing of specific re-

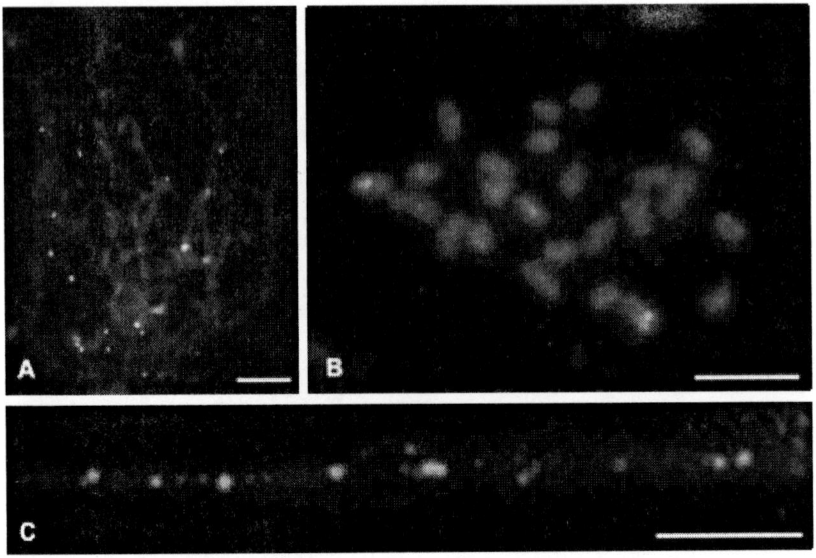

FIGURE 13.4. Chromosomal mapping by FISH (see corresponding figure in color photo section).

gions of the *B. distachyon* genome. The utility of such a comparative genomic approach in *Brachypodium* has been reported recently (Moore, 2003). In this investigation, the syntenic region of the *Ph1* locus was sequenced from a BAC library of *B. sylvaticum*.

Estimation of *C* values and/or cytogenetic analysis in 53 accessions of *B. distachyon* have shown that this species may be diploid, tetraploid, or hexaploid. Current molecular cytogenetic investigations using genomic probes and chromosomal landmarks on *B. distachyon* and its diploid relatives are attempting to determine whether the polyploid series has arisen by spontaneous chromosome doubling of *B. distachyon*, or by hybridization of this species with one or more of its relatives. Preliminary data suggest that the latter is more likely and will be reported elsewhere. However, an interesting and potentially useful feature of the putative allohexaploid is that it forms only bivalents at metaphase I of meiosis (Figure 13.5). In other words, the chromosomes behave in a diploid-like manner analogous to those in allohexaploid wheat. This offers opportunities for modeling a *Ph*-like system in *B. distachyon*, with a view to isolating genes responsible for controlling the way in which chromosomes associate and recombine.

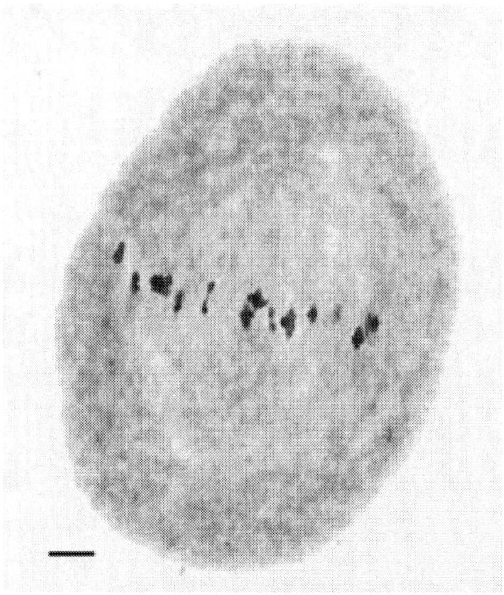

FIGURE 13.5. Metaphase I of meiosis in hexaploid *B. distachyon* showing the formation of 15 regular bivalents. Scale bar represents 5 μm.

TOWARD FUNCTIONAL GENOMICS OF PATHOGENIC INTERACTIONS

In applying functional genomic approaches to plant-pathogen interactions, the suitability of the challenger as much as the host needs to be taken into account. Understanding plant-pathogen interactions in *Arabidopsis thaliana* has been aided by the tractability of bacterial pathogens comprising the *Pseudomonas syringae* and *Xanthomonas campestris* families as much as the host (Mauch-Mani and Slusarenko, 1993). Thus, the discovery of bacterial genes conferring virulence and avirulence, such as type III secretion mechanisms encoded by *hrp* genes through which proteins are delivered into plant cells (reviewed by Mudgett and Staskawicz, 1998), has been matched in *Arabidopsis,* by the cloning of resistance *(R)* genes (Holub, 2001) and downstream plant defense effectors (Glazebrook, 2001). New insights are being made now through large-scale genomic approaches, particularly focusing on the roles of key defense-associated chemical signals (Ramonell and Somerville, 2002). Microarrays have been used to examine gene expression following treatment with salicylic acid (Maleck et al., 2000), jasmonic acid, ethylene (Schenk et al., 2000), H_2O_2 (Desikan et al., 2001), and nitric oxide (Huang et al., 2002) and to reveal coregulated genes—"regulons"—activated following pathogen attack (Schenk et al., 2000, Mahalingam et al., 2003), insect grazing (Reymond et al., 2000), or wounding (Sasaki et al., 2000).

Several pathogenic interactions with cereals have been well-characterized, for example, *Blumeria graminis* with barley and wheat (Schulze-Lefert and Vogel, 2000) or *Ustilago maydis* on maize (Grandel et al., 2000). Of particular note, due to the genetic tractability of both host and pathogen, is the interaction of rice *(Oryza sativa)* with the ascomyete fungus *Magnaporthe grisea* (asexual form *Pyricularia grisea*), the causal agent of the devastating disease rice blast. The utility of rice as a model has been highlighted with the recent development of the Gramene database (www.gramene.org;, Ware et al., 2002) that has begun to collate and annotate rice genomic markers, EST resources, and characterized mutants. All data are displayed with reference to comparative genome maps of maize and Pooid crops. Links are available to the National Institute of Agrobiological Sciences (NIAS), Japan, which provides access to 1,265 or 8,979 sample microarrays constructed using full-insert cDNAs or 3' UTR (http://microarray.rice.dna.affrc.go.jp/; Yazaki et al., 2000). NIA also allows BLAST searching of flanking sequences from mutant lines induced by the rice retrotransposon *Tos17* (http://pc7080.abr.affrc.go.jp/~miyao/pub/tos17/). Rice cv. IR64 mutants are available from the International Rice Research Institute (IRRI, http://www.irri.org/genomics/), including, currently, some 32 lines with altered responses to *Magnaporthe grisea*.

Mutants and transgenic lines are relatively easily generated within *M. grisea,* which has led to a detailed understanding of the processes underlying the development of specific infection structures but not *in planta* development (reviewed by Talbot, 1995; Howard and Valent, 1996; Sweigard et al., 1998). With both rice and *M. grisea* genome sequences being close to completion (Buell, et al., 2002; Martin et al., 2002), it is expected that the increasing number of postgenomic analyses (Rauyaree et al., 2001; Kim et al., 2001; Xu and Xue, 2002) will elucidate events occurring in both the fungus and host as infections develop. Further, the rice/*M. grisea* genomic resources are being used to characterize less tractable interactions. Transformed *M. grisea* mutants have been used to confirm the function of genes from widely differing pathogens, including *Alternaria alternata* (Kawamura et al., 1997), *Botrytis cinerea* (Zheng et al., 2000), ergot (*Claviceps purpurea;* Mey et al., 2002), and *Blumeria graminis* (Bindslev et al., 2001). Phytopathogenic fungal ESTs are also being annotated based on *M. grisea* genomic data (Soanes et al., 2002).

The value of such comparative genomic approaches would be augmented if the plant host being used supported a range of cereal fungal infections. This could reveal common or specific features in plant responses to fungal pathogens. Such information could reveal desirable traits for cereal breeding or targets for broad-action fungicides. Unfortunately, rice is not a satisfactory host for powdery mildews *(Blumeria graminis)* and rust (causal agent, *Puccinia* spp.), which are important diseases of temperate cereals.

Much of our current work involves the development of *Brachypodium distachyon* as a variably interacting host for pathogens for both temperate and tropical crops. Screens have indicated that *B. distachyon* accessions can support interactions with *Blumeria graminis* which can elicit single-cell death and papillae-based resistance (Draper et al., 2001) or lead to the development of disease symptoms (Braun, 1995). Investigations using barley and wheat adapted yellow *(Puccinia striformis)* and brown *(Puccinia recondita)* rust strains identified a number of phenotypes, ranging from localized necrotic flecking (Figure 13.6A) to extensive necrosis (Figure 13.6B) within which uridea formation was observed (Figure 13.6C) and was suggestive of partial resistance to the biotrophic fungus. However, certain *B. distachyon* accessions proved to be fully susceptible to crown rust *(P. coronata)*, a major pathogen of forage grasses, resulting in either chlorotic patches (Figure 13.6D) or uridea formation from asymptomatic leaf tissue (Figure 13.6E).

Responses to genetically tractable pathogens have also been assessed. Pressure injection with cultures of *P. striformis* pv. *phaseolicola* elicited necrotic symptoms (Figure 13.6F) that were not observed with an *hrpL* mutant (Figure 13.6G), suggesting the delivery of necrosis-eliciting factors into a very different host from the pathogen's usual host, French bean *(Phaseolus vulgaris)*. More important, accessions of *B. distachyon* have been identified that display various disease symptoms when challenged with *M. grisea* strain Guy-11 (Draper et al., 2001), and microscopic analysis has revealed that *in planta* hyphal development closely resembles that in rice (Figure 13.6I; Routledge et al., in press). Resistance to *M. grisea* is associated with the formation of a limited necrotic fleck—a hypersensitive response (HR; Figure 13.6J). HR is initiated by the generation of H_20_2, as revealed by staining with 3,3-diaminobenzidine (Figure 13.6K), and seems to correlate with the formation of a "granular" cytoplasm, a preliminary to cell death (Figure 13.6L). Limited lesion growth occurs with the resistant responses of ABR5 to Guy-11, with its eventual cessation associated with callose deposition in epidermal and mesophyll cells (Figure 13.6M). Screens with other blast-eliciting strains have identified a form of resistance to strain FAG 1.1.1 associated with single-cell resistance with little callose deposition (Figure 13.6N). Crosses between ABR5 and ABR1 have indicated that a single dominant resistance gene determined resistance to Guy-11 in ABR5,

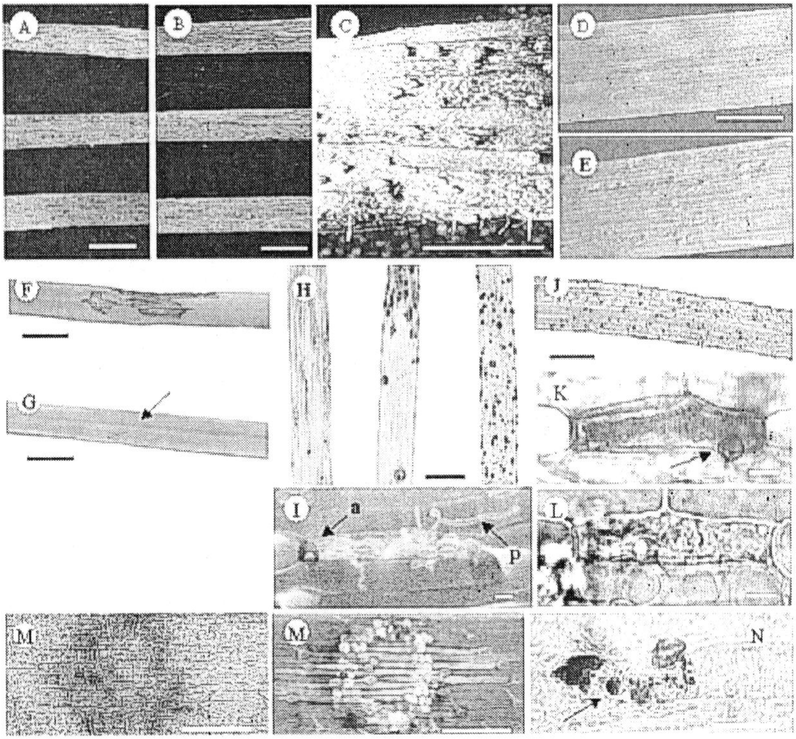

FIGURE 13.6. Variable response to pathogenic challenge by accessions of *B. distachyon*. Challenging *B. distachyon* with *Puccinia striformis* f. sp. *triticae* (wheat yellow stripe rust) strain IP O86053 (ref) leads to (A) brown flecking on accession ABR105 (bar = 1 cm) and (B) brown tissue formation on accession ABR100 (bar = 1 cm) after 14 days. (C) Uridea (pustules) formed within brown tissue in ABR100 seedlings infected with *P. striformis* f. sp. *triticae* (bar = 500 μm). When infecting accession ABR100 with *P. coronata* disease symptoms associated with (D) chlorotic patches with strain 5-02 (bar = 500 μm) or (E) uridea forming from symptomless tissue with stain 7-04 (bar = 500 μm) after 14 days. Pressure injection with a needleless syringe with suspensions of 2×10^8 bacteria per milliliter of wild-type (F) *Pseudomonas syringae* pv. *phaseolicola* and (G) a derived *hrpL* mutant (with the inoculated area arrowed) after two days. (H) Developing disease symptoms forming in *B. distachyon* accession (l-r) ABR7, ABR1, and ABR2 at day five following challenge with *Magnaporthe grisea* Guy-11 (bar = 1 cm). (I) UV-illuminated epifluorescent aniline blue-stained developing Guy-11 infections on ABR1 at 48 h revealed appressorial development (arrowed "a"), epidermal cell filling, and second cell penetration by bulbous secondary hyphae (arrowed "p," bar = 10 μm). (J) Localized lesions elicited by Guy-11 on accession ABR5 at day five following challenge (bar = 1 cm). Lactophenol

FIGURE 13.6 *(continued)*

ABR5 (K) stained with 3,3-diaminobenzidine at 48 h (appressorium arrowed, bar = 10 μm) and (L) cleared only at 72 h (bar = 10 μm). Epifluorescent aniline blue-stained Guy-11-elicited lesions in ABR1 under (M) white-light and (N) UV illumination. (N) Single-cell death in lactophenol-cleared ABR5 at day five following challenge with *Pyricularia grisea* FAG 1.1.1.

but at least two genes controlled the response to FAG-1.1.1 (Routledge et al., in press).

Taking these data together, *B. distachyon* germplasm offers differing disease symptoms as elicited by major pathogens as well as distinctive phenotypes associated with resistance. Work is now under way in our laboratory to develop the required functional genomic expertise and resources to exploit these opportunities. A particular focus is the *B. distachyon/M. grisea* pathosystem for which we aim to gain insight into both plant responses to the pathogen and *in planta* fungal development. Subtractive libraries from *M. grisea*–challenged seedlings have been constructed, and these are serving as the basis for the construction of microrarrays.

The metabolome—the suite of metabolites synthesized by a biological system—has been advanced as the most informative object of functional genomic analysis (Fiehn, 2001) (see Chapter 5). However metabolomic analyses represent perhaps the most challenging due to the exacting necessity for experimental standardization, rigor in sample preparation, careful selection of appropriate mass-spectroscopic technique, and use of suitable programs to process the high dimensionality of data obtained. These difficulties notwithstanding, based on gas chromatography-mass spectroscopy (GC-MS) analysis of 326 polar and lipophillic compounds in wild-type and mutant *Arabidopsis,* sample data could be clustered according to their origin (Fiehn et al., 2000). This approach has been successfully extended to studies involving potato, examining tubers from cultivars differing in sucrose metabolism, analyzing whole plants subjected to environmental stress, and comparing wild-type versus transgenic lines (Roessner et al., 2000, 2001). Though no metabolomic analyses of a plant-microbe interaction have been published, these are currently well under way in our laboratory. Initially, these have focused on estimating the reproducibility of fingerprints of *M. grisea*–challenged material derived by Fourier transform infrared (FT-IR) spectroscopy. Validated cluster analysis grouped differing types and stages of interaction, indicating that there was minimal sample-to-sample variation, thus demonstrating that the *M. grisea/B. distachyon* pathosystem is highly appropriate for functional genomic analysis, including metabolomic profiling, which demands a high degree of reproducibility

(Luis Mur, Royston Goodacre, et al., in process). Currently, data derived from GC-MS and electrospray-ionization mass spectroscopy are being used to identify the chemical differences that are the major sources of variation in *M. grisea* interactions. In the future, we wish also to profile metabolomic changes associated with grass development at stages defined in the aforementioned key, concentrating on flowering and grain formation in particular (Draper et al., 2001).

Justification for a new model organism must be based as much upon need as utility of that species as a scientific tool. We suggest that *B. distachyon* is a highly appropriate bridge species, both in genomic analysis and between different pathosystems in the Poaceae, as detailed both here and in our recent publication (Draper et al., 2001). In addition, as rice, and indeed other cereal models, cannot be used to examine such traits as silaging, sward ecology, or perenniality, we expect *B. distachyon* to emerge as an important model for such niche areas. Clearly, significant obstacles remain to achieving this aim, in particular, optimizing the transformation frequency, the development of tagged mutants, and, ultimately, EST-sequencing programs. These are all either short- or long-term targets for the group. In addition, this work needs to be accompanied with the implementation of a bioinformatic infrastructure. Thus, our Web site will act as a conduit through which it will be possible to access *B. distachyon* germplasm and order microarrays following the establishment of a suitable collaborative agreement. The Web site will also allow the results from microarrays and metabolomic fingerprinting/profiling to be viewed by interested parties. As our BAC landing project allows syntenic regions to be assigned for *B. distachyon,* we fully expect that our rudimentary bioinformatic resource will be absorbed into a more comprehensive structure, such as Gramene.

REFERENCES

Arabidopsis Genome Initiative (2000). Analysis of the genome sequence of the flowering plant *Arabidopsis thaliana*. *Nature* 408: 796-815.

Aragon-Alcaide L., Miller T., Schwarzacher T., Reader S., Moore G. (1996). A cereal centromeric sequence. *Chromosoma* 10: 261-268.

Arumuganathan, K., Earle, E.D. (1991). Nuclear DNA content of some important plant species. *Plant Mol. Biol. Rep.* 9: 208-218.

Bablak, P. (2001). Cultivation, tissue culture and somatic embryogenesis in *Brachypodium distachyon* (L.) P. Beauv. MPhil thesis, University of Leicester.

Bablak, P., Draper, J., Davey, M.R., Lynch, P.T. (1995). Plant regeneration and micropropagation of *Brachypodium distachyon*. *Plant Cell Tiss. Org. Cult.* 42: 97-107.

Bennetzen, J.L., SanMiguel, P., Chen, M., Tikhonov, A., Francki, M., Avramova, Z. (1998). Grass genomes. *Proc. Natl. Acad. Sci. USA* 95: 1975-1978.

Bindslev, L., Kershaw, M.J., Talbot, N.J., Oliver, R.P. (2001). Complementation of the *Magnaporthe grisea* delta cpkA mutation by the Blumeria graminis PKA-c gene: Functional genetic analysis of an obligate plant pathogen. *Mol. Plant Microb. Interact.* 14: 1368-1375.

Boyes, D.C., Zayed, A.M., Ascenzi, R., McCaskill, N.E., Hoffman, A.J., Davis, K.R., Görlach, J. (2001). Growth stage-based phenotypic analysis of *Arabidopsis:* A model for high throughput functional genomics in plants. *Plant Cell* 13: 1499-1510.

Braun, U. (1995). *The Powdery Mildews (Erysiphales) of Europe*. Jena, Germany: Gustav Fischer Verlag.

Buell, C.R. (2002). Current status of the sequence of the rice genome and prospects for finishing the first monocot genome. *Plant Physiol.* 130: 1585-1586.

Catalan, P., Kellogg, E.A., Olmstead, R.G. (1997). Phylogeny of Poaceae subfamily Pooideae based on chloroplast *ndh*F gene sequences. *Mol. Phylogenet. Evol.* 8: 150-166.

Catalan, P., Shi, Y., Armstrong, L., Draper, J., Stace, C.A. (1995). Molecular phylogeny of the grass genus *Brachypodium* P. Beauv. based on RFLP and RAPD analysis. *Bot. J. Linn. Soc.* 117: 263-280.

Chen, M., SanMiguel, P., Bennetzen, J.L. (1998). Sequence organisation and conservation in *sh2/a1*-homologous regions of sorghum and rice. *Genetics* 148: 435-443.

Chen, M., SanMiguel, P., de Oliveira, A.C., Woo, S.S., Zhang, H., Wing, R.A., Bennetzen, J.L. (1997). Microcolinearity in *sh2*-homologous regions of the maize, rice and sorghum genomes. *Proc. Natl. Acad. Sci. USA* 94: 3431-3435.

Councea, P.A., Keisling, T.C., Mitchel, A.J. (2000). A uniform, objective, and adaptive system for expressing rice development. *Crop Sci.* 40: 436-443.

Desikan, R., Mackerness, S., Hancock, J.T., Neill, S.J. (2001). Regulation of the *Arabidopsis* transcriptome by oxidative stress. *Plant Physiol.* 127: 159-172.

Devos, K.M., Beales, J., Nagamura, Y., Sasaki, T. (1999). *Arabidopsis*-rice: Will colinearity allow gene prediction across the eudicot-monocot divide? *Genome Res.* 9: 825-829.

Devos, K.M., Gale, M.D. (2000). Genome relationships: The grass model in current research. *Plant Cell* 12: 637-646.

Dickson, D., Cyranoski, D. (2001). Commercial sector scores success with whole rice genome. *Nature* 409: 551.

Draper, J., Mur, L.A.J., Jenkins, G., Ghosh-Biswas, G.C., Bablak, P., Hasterok, R., Routledge, A.P.M. (2001). *Brachypodium distachyon:* A new model system for functional genomics in grasses. *Plant Physiol.* 127: 1539-1555.

Dubcovsky, J., Ramakrishna, W., SanMiguel, P., Busso, C.S., Yan, L., Shiloff, B.A., Benneteen, J.L. (2001). Comparative sequence analysis of collinear barley and rice BACs. *Plant Physiol.* 125: 1342-1353.

Feng, Q., Zhang, Y., Hao, P., Wang, S., Fu, G., Huang, Y., Li, Y., Zhu, J., Liu, Y., Hu, X., et al. (2002). Sequence and analysis of rice chromosome 4. *Nature* 420: 316-320.

Fiehn, O. (2001). Combining genomics, metabolome analysis, and biochemical modelling to understand metabolic networks. *Comp. Funct. Gen.* 2: 155-168.

Fiehn, O., Kopka, J., Dormann, P., Altmann, T., Trethewey, R.N., Willmitzer, L. (2000). Metabolite profiling for plant functional genomics. *Nat. Biotechnol.* 18: 1157-1161.

Gale, M.D., Devos, K.M. (1998). Comparative genetics in the grasses. *Proc. Natl. Acad. Sci. USA* 95: 1971-1974.

Glazebrook, J. (2001). Genes controlling expression of defence responses in *Arabidopsis*—2001 status. *Curr. Opin. Plant Biol.* 4: 301-308.

Goff, S.A., Ricke, D., Lan, T.H., Presting, G., Wang, R., Dunn, M., Glazebrook, J., Sessions, A., Oeller, P., Varma, H., et al. (2002). A draft sequence of the rice genome *(Oryza sativa* L. ssp. *japonica). Science* 296: 92-100.

Grandel, A., Romeris, T., Kämper, J. (2000). Regulation of pathogenic development in the corn smut fungus *Ustilago maydis. Mol. Plant Pathol.* 1: 61-66.

Havukkala, I.J. (1996). Cereal genome analysis using rice as a model. *Curr. Opin. Genet. Devel.* 6: 711-714.

Holub, E.B. (2001). The arms race is ancient history in *Arabidopsis,* the wildflower. *Nature Rev. Genet.* 2: 516-527.

Howard, R.J., Valent, B. (1996). Breaking and entering: Host penetration by the fungal rice blast pathogen *Magnaporthe grisea. Annu. Rev. Microbiol.* 50: 491-512.

Hsiao, C., Chatterton, N.J., Asay, K.H., Jensen, K.B. (1995). Phylogenetic relationships of the monogenomic species of the wheat tribe, Triticeae (Poaceae), inferred from nuclear rDNA (internal transcribed spacer) sequences. *Genome* 38: 211-223.

Huang, X., von Rad, U., Durner, J. (2002). Nitric oxide induces transcriptional activation of the nitric oxide-tolerant alternative oxidase in *Arabidopsis* suspension cells. *Planta* 215: 914-923.

Kawamura, C., Moriwaki, J., Kimura, N., Fujita, Y., Fuji, S., Hirano, T., Koizumi, S., Tsuge, T. (1997). The melanin biosynthesis genes of *Alternaria alternata* can restore pathogenicity of the melanin-deficient mutants of *Magnaporthe grisea. Mol. Plant Microbe Interact.* 10: 446-453.

Kellogg, E.A. (1998). Relationships of cereal crops and other grasses. *Proc. Natl. Acad. Sci. USA* 95: 2005-2010.

Kim, S., Ahn, I.P., Lee, Y.H. (2001). Analysis of genes expressed during rice-*Magnaporthe grisea* interactions. *Mol. Plant Microbe Interact.* 14: 1340-1346.

Kowalski, S., Lan, T., Feldmann, K., Paterson, A. (1994). Comparative mapping of *Arabidopsis thaliana* and *Brassica oleracea* chromosomes reveals islands of conserved organisation. *Genetics* 138: 499-510.

Lagercrantz, U. (1998). Comparative mapping between *Arabidopsis thaliana* and *Brassica nigra* indicates that *Brassica* genomes have evolved through extensive genome replication accompanied by chromosome fusions and frequent rearrangements. *Genetics* 150: 1217-1228.

Langdon, T., Seago, C., Mende, M., Leggett, M., Thomas, H., Forster, J.W., Thomas, H., Jones, R.N., Jenkins, G. (2000). Retrotransposon evolution in diverse plant genomes. *Genetics* 156: 313-325.

Li, W., Gill, B.S. (2002). The colinearity of the *Sh2/A1* orthologous region in rice, sorghum and maize is interrupted and accompanied by genome expansion in the Triticeae. *Genetics* 160: 1153-1162.

Mahalingam, R., Gomez-Buitrago, A., Eckardt, N., Shah, N., Guevara-Garcia, A., Day, P., Raina, R., Fedoroff, N.V. (2003). Characterizing the stress/defense transcriptome of *Arabidopsis*. *Genome Biol.* 4: R20.

Maleck, K., Levine, A., Eulgem, T., Morgan, A., Schmid, J., Lawton, K.A., Dangl, J.L., Dietrich, R.A. (2000). The transcriptome of *Arabidopsis thaliana* during systemic acquired resistance. *Nat. Genet.* 26: 403-410.

Martin, S.L, Blackmon, B.P, Rajagopalan, R. Houfek, T.D., Sceeles, R.G., Denn, S.O., Mitchell, T.K., Brown, D.E., Wing, R.A., Dean, R.A. (2002). MagnaportheDB: A federated solution for integrating physical and genetic map data with BAC end derived sequences for the rice blast fungus *Magnaporthe grisea*. *Nucleic Acids Res.* 30: 121-124.

Mauch-Mani, B., Slusarenko, A.J. (1993). *Arabidopsis* as a model host for studying plant-pathogen interactions. *Trends Microbiol.* 1: 265-270.

Mey, G., Oeser, B., Lebrun, M.H., Tudzynski, P. (2002). The biotrophic, non-appressorium-forming grass pathogen *Claviceps purpurea* needs a *Fus3/Pmk1* homologous mitogen-activated protein kinase for colonization of rye ovarian tissue. *Mol. Plant Microbe Interact.* 15: 303-312.

Meyers, B.C., Tingley, S.V., Morgante, M. (2001). Abundance, distribution and transcriptional activity of repetitive elements in the maize genome. *Genome Res.* 11: 1660-1676.

Moore, G. (2003). The application of comparative genomics to wheat. Abstract of the Plant and Animal Genomes XI Conference, San Diego.

Mudgett, M.B., Staskawicz, B.J. (1998). Protein signaling via type III secretion pathways in phytopathogenic bacteria. *Curr. Opin. Microbiol.* 1: 109-114.

O'Neill, C., Bancroft, I. (2000). Comparative physical mapping of segments of the genome of *Brassica oleracea* var. *alboglabra* that are homoeologous to sequenced regions of chromosomes 4 and 5 of *Arabidopsis thaliana*. *Plant J.* 23: 233-243.

Ramakrishna, W., Dubcovsky, J., Park, Y.-J., Busso, C., Emberton, J., SanMiguel, P., Bennetzen, J.L. (2002). Different types and rates of genome evolution detected by comparative sequence analysis of orthologous segments from four cereal genomes. *Genetics* 162: 1389-1400.

Ramonell, K.M., Somerville, S. (2002). The genomics parade of defence responses: To infinity and beyond. *Curr. Opin. Plant Biol.* 5: 291-294.

Rauyaree, P., Choi, W., Fang, E., Blackmon, B., Dean, R. (2001). Genes expressed during early stages of rice infection with the rice blast fungus *Magnaporthe grisea*. *Mol. Plant Path.* 2: 347-354.

Reymond, P., Weber, H., Damond, M., Farmer, E.E. (2000). Differential gene expression in response to mechanical wounding and insect feeding in *Arabidopsis*. *Plant Cell* 12: 707-720.

Roessner, U., Luedemann, A., Brust, D., Fiehn, O., Linke, T., Willmitzer, L., Fernie, A. (2001). Metabolic profiling allows comprehensive phenotyping of genetically or environmentally modified plant systems. *Plant Cell* 13: 11-29.

Roessner, U., Wagner, C., Kopka, J., Trethewey, R.N., Willmitzer, L. (2000). Technical advance: Simultaneous analysis of metabolites in potato tuber by gas chromatography-mass spectrometry. *Plant J.* 23: 131-142.

Ross, K.J., Fransz, P., Jones, G.H. (1996). A light microscopic atlas of meiosis in *Arabidopsis thaliana. Chrom. Res.* 4: 507-516.

Rossberg, M., Theres, K., Acarkan, A., Herrero, R., Schmitt, T., Schumacher, K., Schmitz, G., Schmidt, R. (2001). Comparative sequence analysis reveals extensive microcolinearity in the lateral suppressor regions of the tomato, *Arabidopsis* and *Capsella* genomes. *Plant Cell* 13: 979-988.

Routledge, A.P.M., Shelley, G., Smith, J.V., Talbot, N.J., Draper, J., Mur, L.A.J. (2004). *Magnaporthe grisea* interactions with the model grass *Brachypodium distachyon* closely resemble those with rice *(Oryza sativa). Mol. Plant Pathol.* (in press).

SanMiguel, P., Bennetzen, J.L. (1998). Evidence that a recent increase in maize genome size was caused by the massive amplification of intergene retrotransposon. *Ann. Bot.* 82: 37-44.

SanMiguel, P., Tikhonov, A., Jin, Y.-K., Motchoulskaia, N., Zakharov, D., Melake-Berhan, A., Springer, P.S., Edwards, K.J., Lee, M., Avramova, Z., Bennetzen, J.L. (1996). Nested retrotransposon in the intergenic regions of the maize genome. *Science* 274: 765-768.

Sasaki, T. Matsumoto, T., Yamamoto, K., Sakata, K., Baba, T., Katayose, Y., Wu, J., Niimura, Y., Cheng, Z., Nagamura, Y., et al. (2002). The genome sequence and structure of rice chromosome 1. *Nature* 420: 312-316.

Sasaki, Y., Asamizu, E., Shibata, D., Nakamura, Y., Kaneko, T., Awai, K., Masuda, T., Shimada, H., Takamiya, K., Tabata, S., Ohta, H. (2000). Genome-wide expression-monitoring of jasmonate-responsive genes of *Arabidopsis* using cDNA arrays. *Biochem. Soc. Trans.* 28: 863-864.

Schenk, P.M., Kazan, K., Wilson, I., Anderson, J.P., Richmond, T., Somerville, S.C., Manners, J.M. (2000). Coordinated plant defence responses in *Arabidopsis* revealed by microarray analysis. *Proc. Natl. Acad. Sci. USA* 97: 11655-11660.

Schulze-Lefert, P., Vogel, J. (2000). Closing the ranks to attack by powdery mildew. *Trends Plant Sci.* 5: 343-348.

Sharma, M.L. (1979). Some considerations on the phylogeny and chromosomal evolution in grasses. *Cytologia* 44: 679-685.

Shi, Y. (1991). Molecular study of the genus *Brachypodium (Poaceae).* PhD thesis, University of Leicester.

Shi, Y., Draper, J., Stace, C.A. (1993). Ribosomal DNA variation and its phylogenetic implication in the genus *Brachypodium* (Poaceae). *Plant System. Evol.* 188: 125-138.

Soanes, D.M., Skinner, W., Keon, J., Hargreaves, J., Talbot, N.J. (2002). Genomics of phytopathogenic fungi and the development of bioinformatic resources. *Mol. Plant Microbe Interact.* 15: 421-427.

Suoniemi, A., Anamthawat-Jonsson, K., Arna, T., Schulman, A.H. (1996). Retrotransposon BARE-1 is a major, dispersed component of the barley (*Hordeum vulgare* L.) genome. *Plant Mol. Biol.* 30: 1321-1329.

Sweigard, J.A., Carroll, A.M., Farrall, L., Chumley, F.G., Valent, B. (1998). *Magnaporthe grisea* pathogenicity genes obtained through insertional mutagenesis. *Mol. Plant Microbe Interact.* 11: 404-412.

Talbot, N.J. (1995). Having a blast: Exploring the pathogenicity of *Magnaporthe grisea*. *Trends Microbiol.* 3: 9-16.

Tanksley, S., Bernatzky, R., Lapitan, N., Prince, J. (1988). Conservation of gene repertoire but not gene order in pepper and tomato. *Proc. Natl. Acad. Sci. USA* 85: 6419-6423.

Tanksley, S.D., Ganal, M.W., Prince, J.P., de Vicente, M.C., Bonierbale, M.W., Broun, P., Fulton, T.M., Giovannoni, J.J., Grandillo, S., Martin, G.B., et al. (1992). High density molecular maps of the tomato and potato genomes. *Genetics* 132: 1141-1160.

Tateoka, T. (1968). Phytogeographical notes on the genus *Brachypodium* P. Beauv. (Gramineae). *Boletin Soc. Argent. Bot.* 12: 44-56.

Tikhonov, A.P., SanMiguel, P.J., Nakajima, Y., Gorenstein, N.M., Bennetzen, J.L., Avramova, Z. (1999). Colinearity and its exceptions in orthologous *adh* regions of maize and sorghum. *Proc. Natl. Acad. Sci. USA* 96: 7409-7414.

Ware, D.H., Jaiswal, P., Ni, J., Yap, I.V., Pan, X., Clark, K.Y., Teytelman, L., Schmidt, S.C., Zhao, W., Chang, K., Cartinhour, S., Stein, L.D., McCouch, S.R. (2002). Gramene, a tool for grass genomics. *Plant Physiol.* 130: 1606-1613.

Waugh, R., McLean, K., Flavell, A.J., Pearce, S.R., Kumar, A., Thomas, B.B., Powell, W. (1997). Genetic distribution of Bare-1-like retrotransposable elements in the barley genome revealed by sequence-specific amplification polymorphisms (S-SAP). *Mol. Gen. Genet.* 253: 687-694.

Xu, J.-R., Xue, C. (2002). Time for a blast: Genomics of *Magnaporthe grisea*. *Mol. Plant Path.* 3: 173-176.

Yazaki, J., Kishimoto, N., Nakamura, K., Fujii, F., Shimbo, K., Otsuka, Y., Wu, J., Yamamoto, K., Sakata, K., Sasaki, T., Kikuchi, S. (2000). Embarking on rice functional genomics via cDNA microarray: Use of 3' UTR probes for specific gene expression analysis. *DNA Res.* 31: 367-370.

Yu, J. Hu, S., Wang, J., Wong, G.K., Li, S., Liu, B., Deng, Y., Dai, L., Zhou, Y., Zhang, X., et al. (2002). A draft sequence of the rice genome *(Oryza sativa* L. spp. *indica). Science* 296: 79-92.

Zheng, L., Campbell, M., Murphy, J., Lam, S., Xu, J.R. (2000). The *BMP1* gene is essential for pathogenicity in the gray mold fungus *Botrytis cinerea*. *Mol. Plant Microbe Interact.* 13: 724-732.

Zhong, X.-B., Bodeau, J., Fransz, P.F., Williamson, V.M., van Kammen, A., de Jong, J.H., Zabel, P. (1999). FISH to meiotic pachytene chromosomes of tomato locates the root-knot nematode resistance gene *Mi-1* and the acid phosphatase gene *Aps-1* near the junction of euchromatin and pericentromeric heterochromatin of chromosome arms 6S and 6L respectively. *Theoret. Appl. Genet.* 98: 365-370.

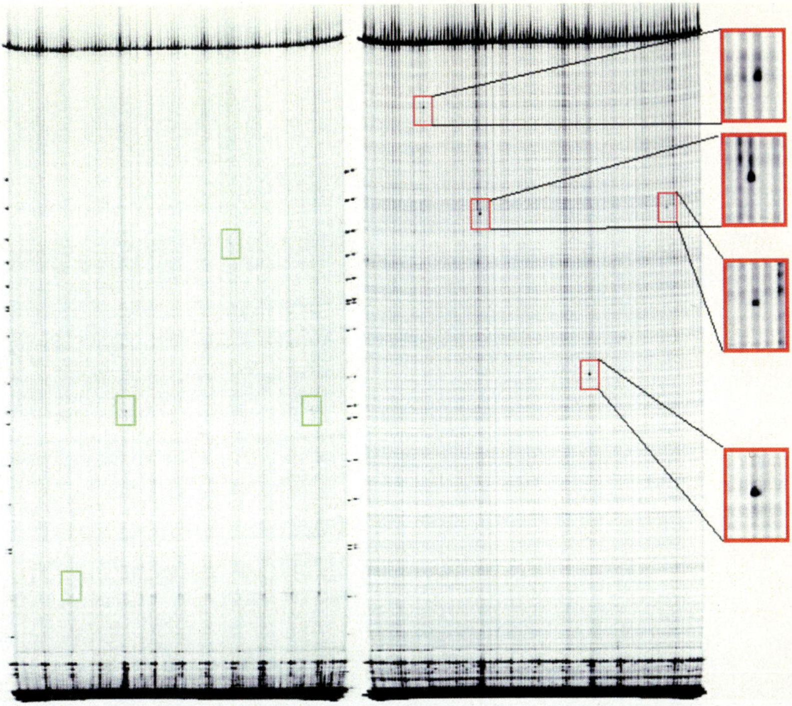

FIGURE 3.4. Example of a TILLING gel used for mutation detection in eightfold pools. The figure shows IR Dye 800 (left) and IR Dye 700 (right) channels of a representative run. Bands corresponding to four of seven mutations detected on this gel are shown boxed, and sections of the IR Dye 700 images are magnified in offsets (far right). Note that these are seen only in one channel but have counterparts in the other channel that add up to the length of the full-sized 1,012 bp product (band at top). Several bands near the bottom of the gel are detected in both channels; these artifactual bands result from random mispriming. (*Source:* Colbert et al., 2001. Copyrighted by the American Society of Plant Physiologists and reprinted with permission.)

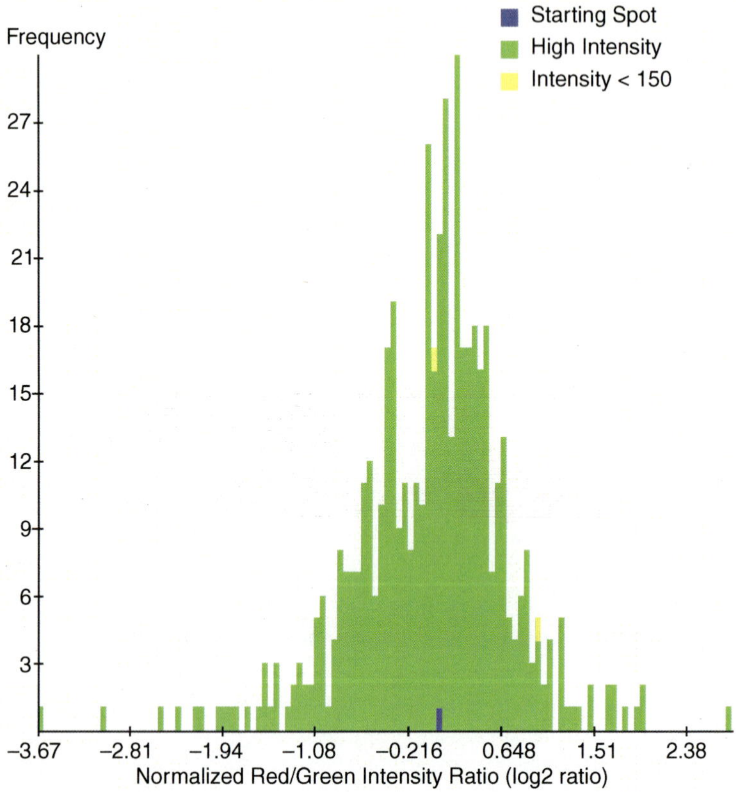

FIGURE 4.2. Spot History for *Arabidopsis* glyceraldehyde-3-phosphate dehydrogenase C subunit (GAPC; Locus: At3G04120; EST:178F2T7). Spot History is a frequency histogram showing the number of experiments (*y*-axis) out of 510 in which the ratio value was as indicated on the *x*-axis. In a majority of experiments, the ratio was at or near 1, suggesting that *GAPC* expression was unchanged in most experimental conditions, as might be expected for a housekeeping gene. To identify experiments in which *GAPC* expression was induced or repressed, it is possible to click on the bar of choice (e.g., the one experiment that gave a ratio value of >2.38), and the database will return details of this experiment and of the *GAPC* response in this experiment. Spot History is one of the many data analysis utilities that can be implemented if microarray data sets are stored in a centralized database. Spot History was developed by Jeremy Gollub and is resident at the Stanford Microarray Database.

FIGURE 9.3. Bioreactor culture of *Physcomitrella*. Protonema tissue for protoplast isolation is grown in liquid culture under highly standardized growth conditions. Each tank glass bioreactor contains five liters of a *Physcomitrella* liquid culture.

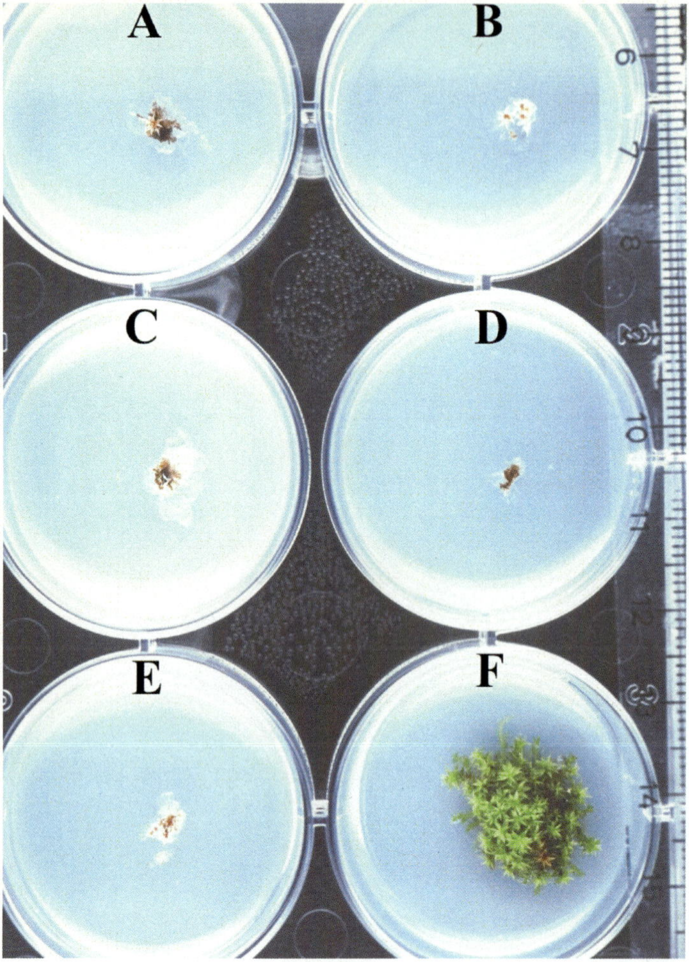

FIGURE 9.6. Vitamin-deficient *Physcomitrella* mutant. One mutant plant that displays retarded growth on minimal medium but normal growth on full medium has been subjected to a medium complementation experiment. Minimal medium was supplemented with particular chemicals present in the full medium to determine the components that are necessary and sufficient to restore the wild-type phenotype. (A) Minimal medium supplemented with compound 1; (B) minimal medium supplemented with compound 2; (C) minimal medium supplemented with compound 3; (D) minimal medium; (E) minimal medium supplemented with compound 4; (F) minimal medium supplemented with vitamins. The additional application of vitamins is sufficient to restore the wild-type phenotype. The scale is given in centimeters.

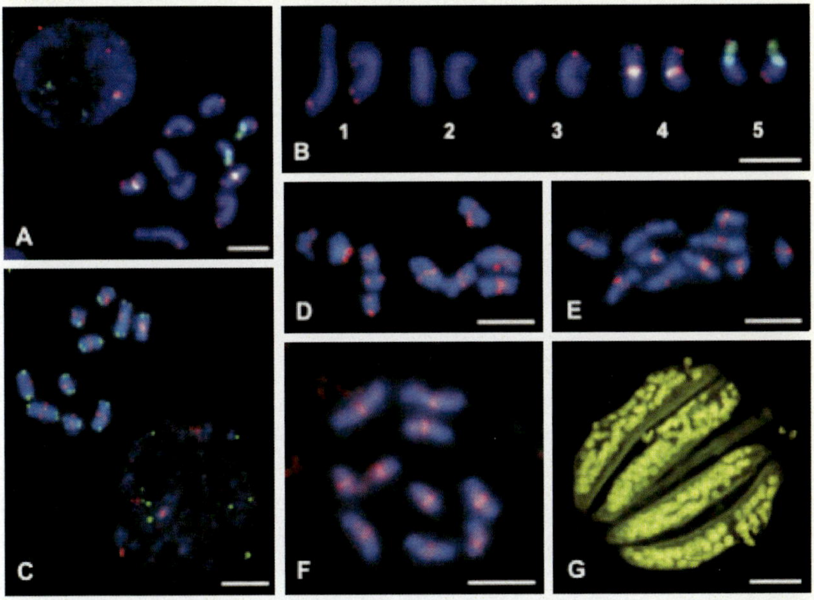

FIGURE 13.2. FISH to somatic interphase and metaphase chromosomes of *Brachypodium distachyon*. (A) The pair of 25S (green) and 5S (yellow) rDNA loci and telomeres (red); (B) the derived karyogram in which all five pairs of chromosomes can be identified unequivocally; (C) highlighted centromeres (red) and telomeres (green); (D and E) the chromosome locations (red) of the *crwydryn* retrotransposon based upon conserved regions of the reverse transcriptase and integrase genes, respectively; (F) a genomic probe (red) that paints pericentromeric regions preferentially; (G) illustration of the minuteness of a mature anther. Scale bars represent 5 μm (A to F) and 50 μm (G).

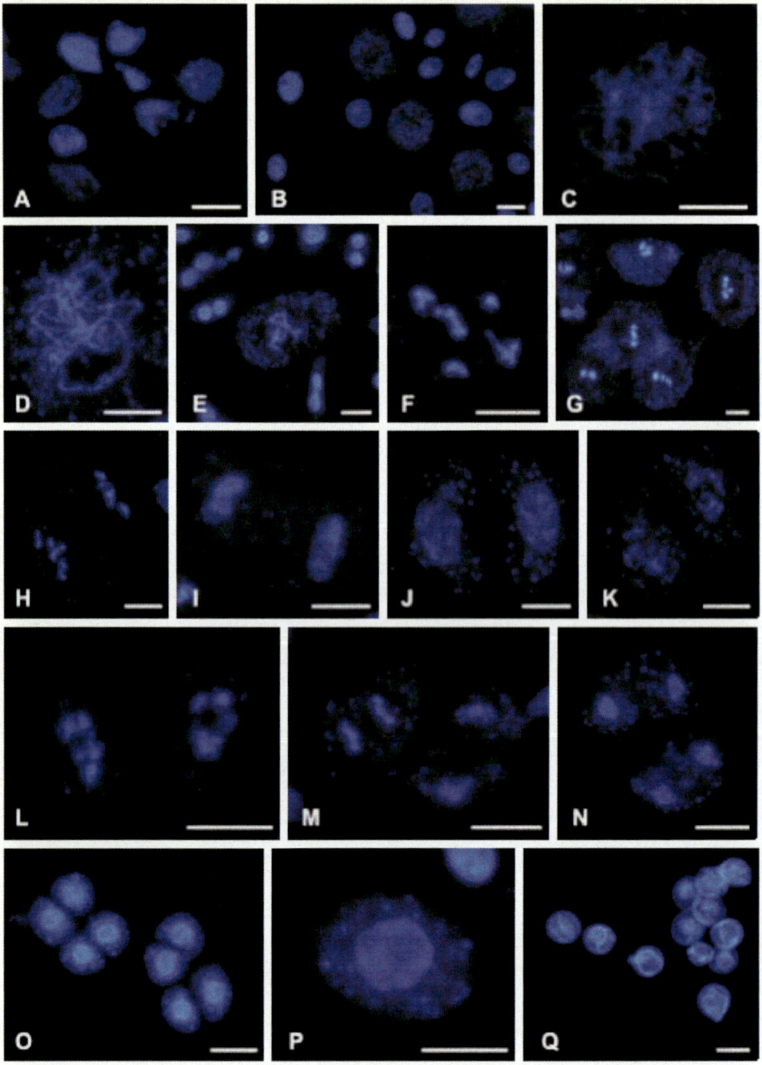

FIGURE 13.3. DAPI-stained atlas of meiosis of diploid *Brachypodium distachyon,* showing (A) premeiotic and nonmeiotic cells; (B) leptotene and nonmeiotic cells; (C) zygotene; (D) pachytene; (E) diplotene and binuclear tapetal cells; (F) diakinesis; (G) metaphase I; (H) late anaphase I; (I) late telophase I; (J) interkinesis; (K) prophase II; (L) metaphase II; (M) anaphase II; (N) telophase II; (O) tetrads; (P) developing microspores; (Q) mature pollen grains. Scale bars represent 10 μm.

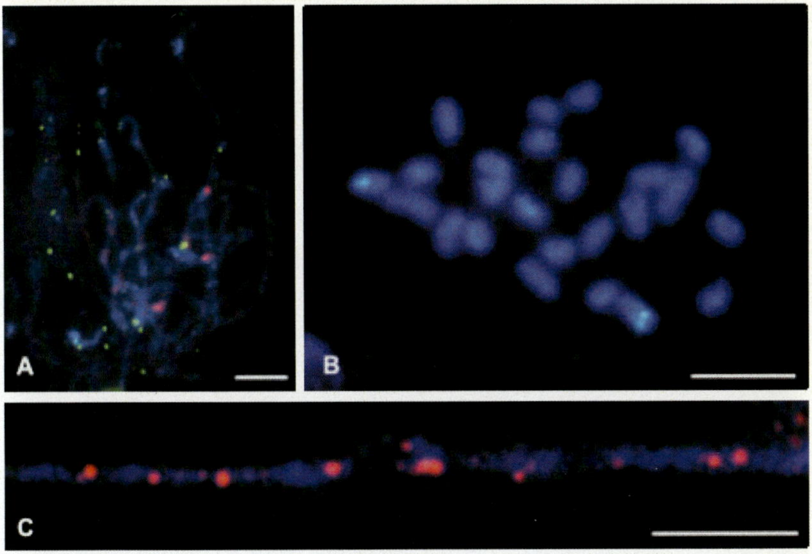

FIGURE 13.4. Chromosomal mapping by FISH, showing (A) Pachytene chromosomes of *Brachypodium distachyon* probed with telomeric (green) and pericentromeric (red) repeat sequences. (B) Somatic chromosomes of rice showing two green signals from a rice BAC hybridized in situ. (C) Extended DNA fiber probed with a *Cla*I fragment of 455 rDNA repeat unit (red) and counterstained with DAPI (blue). Scale bars represent 5 μm.

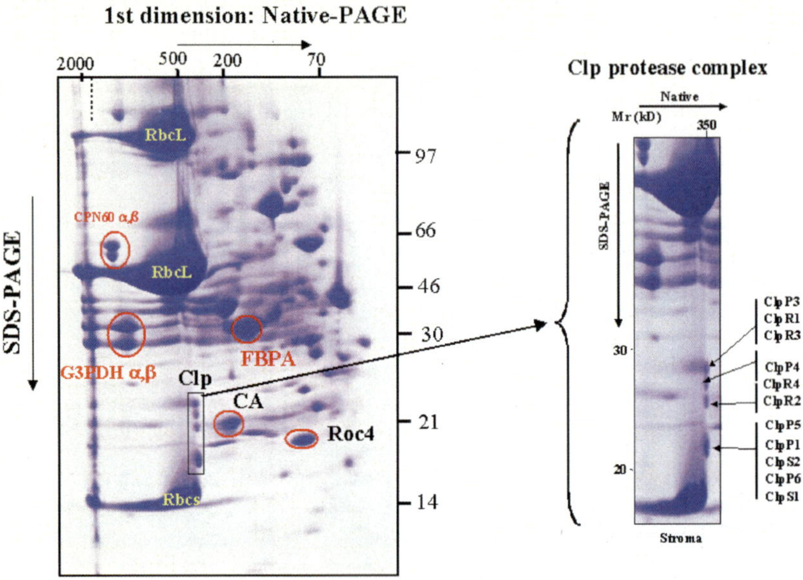

FIGURE 14.2. Protein complexes in the stroma of *Arabidopsis thaliana* chloroplasts. Stromal complexes were first separated on native gels, followed by denaturation and separation by SDS-PAGE. Proteins were stained and identified by MS. The left-hand panel shows an overview of the stromal complexes; the identities of a few protein complexes are indicated. Proteins were identified by automated MS. Abbreviations of proteins: RbcL, Rubisco large subunit; Rbcs, Rubisco, small subunit; FBPA, fructose biphosphatase; Roc4, an isomerase; CPN60, chaperone 60; CA, carbonic anhydrase (Peltier et al., 2004). The right-hand panel (inset) shows a close-up of one such gel in the region of 250-600 kDa. The identities of 11 Clp protease subunits forming the Clp protease core complex are indicated. (*Source:* Adapted from Peltier et al., 2003).

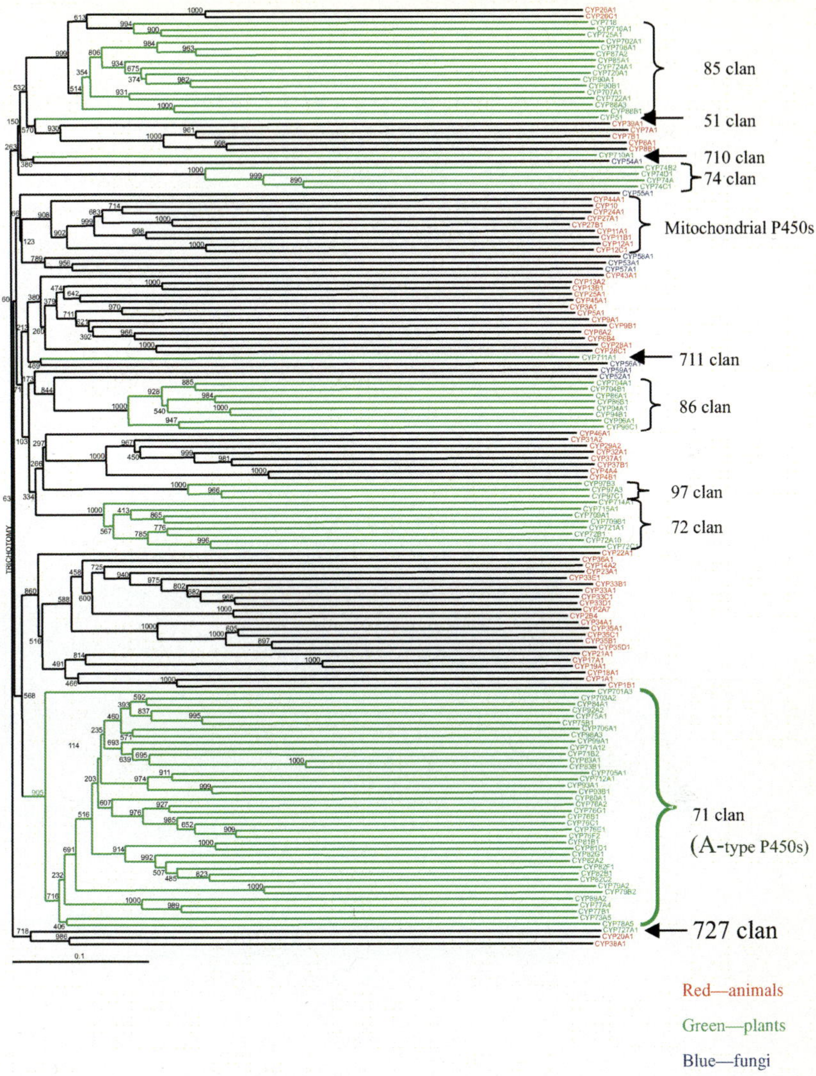

FIGURE 22.1. Phylogenetic tree of cytochrome P450s, illustrating the interrelationships within and among P450s of different kingdoms. The different phyla are color coded (plant: green, animal: red, fungi: blue), and the individual plant clans are shown in brackets. A complete listing of organisms and accession numbers can be found at (http://www.biobase.dk/P450/cyp_allfam_080502_table.pdf). For the animal sequences and non-*Arabidopsis* plant sequences, a single subfamily sequence was chosen at random to represent the individual subfamilies. *Arabidopsis* se-

FIGURE 22.1 *(continued)*

quences took preference over other plant species when both were available. CYP51 is the only family found across phyla and, for simplicity, only a plant sequence is included. The neighbor-join tree was constructed based on a multiple sequence alignment (http://www.biobase.dk/P450/cyp_ allfam_reviewpaper.pdf), generated using a combination of ClustalW and ClustalX. A GONNET protein weight matrix series was used to generate the alignment, and the tree was analyzed using 1,000 bootstrap trials.

SECTION III: ORGANELLES

Chapter 14

Chloroplast Proteomics

Klaas J. van Wijk

THE DYNAMIC NATURE OF THE PLASTID AND ITS PROTEOME

Plastids are essential organelles of prokaryotic origin that are present in every plant cell (except for pollen) and differentiate from meristem-localized proplastids. The plastid number can vary greatly among cell types and also over the lifetime of a given cell. Chloroplasts are the best-known plastid type and contain a thylakoid membrane system that carries the photosynthetic electron transport chain converting light energy into chemical energy. The electron transport chain drives production of ATP and NAD(P)H, which are used for the production of complex carbohydrates in the Calvin cycle and in other anabolic and metabolic pathways. In addition to supporting photosynthesis, chloroplasts also synthesize hormones, fatty acids and lipids, amino acids, vitamins (B_1, K_1, and E), nucleotides, and secondary metabolites, such as alkaloids and isoprenoids, and are required for nitrogen and sulfur assimilation. In C4 plants, such as maize, chloroplasts differentiate into bundle sheath (BS) and mesophyll (M) plastids, each of which has a specialized function in photosynthesis. M plastids have stacked thylakoids, and their primary functions in photosynthesis are "collecting" light energy and generation of ATP and NAD(P)H, as well as fixation of CO_2 as a C4 intermediate, whereas the prime function in photosynthesis of BS plastids is the Calvin cycle. Not much is known concerning how these two plastid types are different in other functions (Sheen, 1999).

Nonphotosynthetic plastids exist in nongreen tissues in plants. They perform many metabolic and anabolic biosynthetic functions, but they derive

I would like to thank all members in my laboratory, as well as numerous colleagues for all stimulating and educating discussions on the topics of proteomics, mass spectrometry, and plastid biology. The National Science Foundation (MCB #0090942 and PGRP#0211935), the U.S. Department of Agriculture (#00186), and NYSTAR are acknowledged for supporting the proteomics activities in my laboratory.

their energy by import of precursors, such as hexosephosphates and ATP (Neuhaus and Emes, 2000). For instance, amyloplasts in root caps are involved in gravity sensing, leucoplasts in trichomes are involved in the biosynthesis and storage of lipids and monoterpenes, and chromoplasts in fruits and flowers synthesize and store carotenoids. Plastids are thus responsible for the synthesis and storage of key molecules required for the basic architecture and functions of plant cells.

Proteins carry out many of the biological functions of plastids. It can therefore be argued that the systematic characterization of the proteome complement of chloroplasts and other plastid types will provide insight into these biological functions. Such a systematic proteome analysis is summarized under the term *proteomics*. This not only concerns the identification of proteins but also addresses posttranslational modifications, protein-protein interactions, and differential protein expression.

With the availability of the complete genomes of *Arabidopsis thaliana* (http://www.arabidopsis.org/) and *Oryza sativa* (rice) (http://www.gramene.org/), the partial genome of *Medicago trunculata* (white clover) (http://www.comparative-legumes.org/), and the genome of the unicellular green algae *Chlamydomonas reinhardtii* (http://genome.jgi-psf.org/chlre1/chlre1.home.html), plant and plastid proteomics has clearly become a very powerful tool. The potential of proteomics for plastid biology, and plant biology in general, has been reviewed previously (Roberts, 2002; Thiellement et al., 2002; van Wijk, 2000, 2001a). For recent general reviews on proteomics, see, e.g., Lin et al. (2003) and Patterson and Aebersold (2003).

THE SCOPE OF THIS CHAPTER

Proteomics is becoming a "mainstream" tool for studying a wide selection of organisms and is now applied in a range of areas, varying from medical applications (e.g., identification of disease marker proteins), basic studies on responses to abiotic and biotic stresses, as well as systematic studies of protein interaction networks. Many of the tools and approaches (e.g., protein separation and identification techniques) used in proteomics are identical, whether applied in studies of plants or animals, either at the level of whole-cell lysates or at the level of purified organelles or membranes (see also Chapter 15).

However, studying the proteome of organelles such as plastids and mitochondria, rather than total-cell lysates, has a number of specific advantages. These include a strongly reduced proteome complexity, potentially leading to a better dynamic resolution, and the possibility to predict the organellar proteins based on targeting information in the protein sequence. Thus, studying organellar proteomes rather than a total-cell lysate is likely to give more in-depth insight into the target proteome.

From a biochemical point of view, the chloroplast can be divided into several compartments, with each compartment having its own specific subset of proteins, or subproteome (Figure 14.1A). To characterize the chloroplast proteome fully, either experimentally or by prediction, it is useful (and probably essential) to subdivide the chloroplast proteome into such subproteomes, as outlined in van Wijk (2000). Only then can one devise optimal experimental strategies to identify and characterize most proteins,

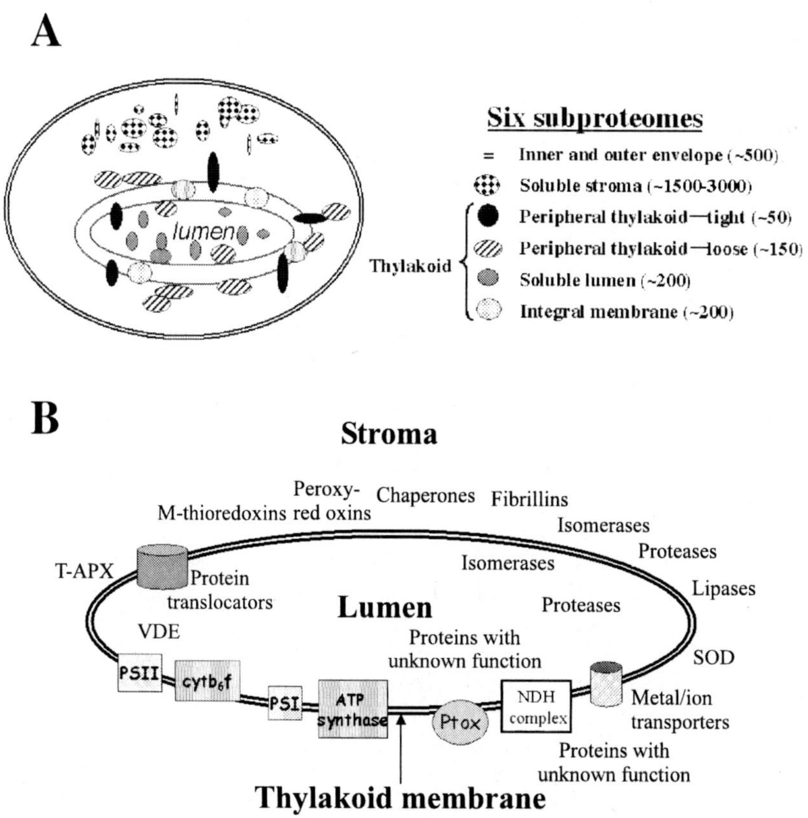

FIGURE 14.1. Overview of the chloroplast proteome. (A) Tentative number of different proteins in the six chloroplast subproteomes: outer and inner envelope, stroma, thylakoid membrane and lumen. (B) Postulated and identified proteins in and around the thylakoid membrane.

including those that are hydrophobic, of low abundance, or transiently expressed.

Indeed, an interesting feedback between prediction studies and experimental proteome analysis is already yielding (candidate) lists of proteins for the chloroplast inner and outer envelope, the stroma, and the thylakoid with its subcompartment, the lumen, as will be discussed. This chapter provides an update of four years of plastid proteomics and an outline of new developments and directions. The chapter is focused on those studies in which large sets of proteins were predicted and/or experimentally identified or characterized; studies that focus on one or a just a few specific plastid proteins will not be discussed here. Thus, this chapter is not about general insights into plastid biology, but rather about the way that proteomics can and has contributed to plastid biology. Efforts using large-scale transcript analysis to address the expression of genes (potentially) encoding for plastid proteins were summarized recently (Leister, 2003). It should be pointed out that transcript and proteome analysis are very complementary. Strong and weak points of both techniques and their complementary nature have been outlined earlier in this context (van Wijk, 2000).

PREDICTION OF THE PLASTID PROTEOME

General Remarks

The vast majority of the plastid proteome is encoded by the nuclear genome. These proteins are generally synthesized as precursor proteins with N-terminal cleavable transit peptides, assigned cTPs. These cTPs are between 30 and 120 amino acids long, are low in acidic residues but high in hydroxylated residues, and are believed to have little structure in aqueous environments—they are "random coils," possibly with a short amphiphilic alpha helix (Emanuelsson and von Heijne, 2001; von Heijne et al., 1989). Several software programs are available that recognize these cTPs, of which TargetP (http://www.cbs.dtu.dk/services/TargetP/) and Predotar (http://www.inra.fr/Internet/Produits/Predotar/) are the most successful (see Table 14.1). Using TargetP, it was earlier estimated that more than 3,000 genes of the *A. thaliana* genome encode for plastid proteins (Emanuelsson et al., 2000). A recent analysis of the latest *A. thaliana* genome annotation from TIGR (version 4.0, April 17, 2003), with 28,581 annotated genes, suggests 4,255 plastid proteins with recognizable cTPs (Emanuelsson, unpublished data). This includes the hits in all five reliability classes (for explanation, see Emanuelsson et al., 2000). By using the rate of false positives and false negatives, this number can in principle be extrapolated to a maximum number.

TABLE 14.1. Predictors cited in this chapter and plastid proteomics databases/data sets.

Database	Web site address
Localization predictors	<http://www.cbs.dtu.dk/services/ChloroP/>
	<http://www.cbs.dtu.dk/services/TargetP/>
	<http://www.cbs.dtu.dk/services/SignalP-2.0/>
	<http://www.inra.fr/Internet/Produits/Predotar/>
Transmembrane predictors	<http://www.cbs.dtu.dk/services/TMHMM/>
	<http://www.enzim.hu/hmmtop/>
Transmembrane protein plant databases (predicted proteins only)	<http://aramemnon.botanik.uni-koeln.de/>
	<http://www.cbs.umn.edu/arabidopsis/>
	<http://www.cbs.umn.edu/rice/>
Chloroplast envelope proteome data set	<http://www.plantbiology.msu.edu/PlastidEnvelope/AtPEMLibrary/FrontPage.htm>
Plastid proteome database	<http://ppdb.tc.cornell.edu/>

A recent phylogenetic analysis of the *Arabidopsis* proteome and potential cyanobacterial ancestors suggested that ~4,500 *Arabidopsis* proteins are derived from bacteria. About 1,300 have N-termini that are being recognized by TargetP as cTPs (Martin et al., 2002). Of course, many additional genes encoding for plastid proteins must have been acquired during the evolution of plants to allow for the compartmentalization of the eukaryotic plant cell, to deal with multicellular structures and with photosynthetic and nonphotosynthetic functions, and to maintain biosynthetic pathways and respond to abiotic and biotic stresses.

Not all plastid proteins can be predicted by TargetP, Predotar, or other localization predictors. The plastid outer envelope proteins are not synthesized as precursor proteins with a cleavable cTP and thus cannot be easily predicted. In addition, a small fraction of known plastid proteins appear to have no obvious N-terminal transit peptide, and this fraction includes some important regulatory proteins, such as TAK kinases (Snyders and Kohorn, 1999), some of the 14-3-3 proteins (Sehnke et al., 2000), the inner envelope protein ceQORH (Miras et al., 2002), and others. It is not clear what information directs these proteins into plastids. Large-scale experimental plastid proteomics studies are likely to extend this set, possibly allowing the recognition of the targeting information or providing new training sets to improve neural network predictors such as TargetP.

About 100 chloroplast genes encode for plastid proteins; one of these proteins (YCF10) is localized in the inner envelope, about 32 encode for integral thylakoid membrane proteins, and the remaining genes encode for peripheral thylakoid proteins or soluble stromal proteins. Interestingly, the plastid genome does not encode for proteins located in the thylakoid lumen. All these plastid-encoded proteins are expressed in chloroplasts, but only a subset is expressed in nonphotosynthetic plastids.

Predictors and databases cited in this chapter are listed in Table 14.1. The predicted numbers of proteins for the various chloroplast compartments and protein classes are summarized in Table 14.2. A schematic overview with the tentative numbers of predicted proteins for the different chloroplast compartments is shown in Figure 14.1A. Identified and postulated proteins in and around the thylakoid membrane are summarized in Figure 14.1B.

Prediction of the Outer Plastid Envelope Proteome

As mentioned previously, outer envelope plastid proteins generally do not have N-terminal cTPs. It is therefore not quite obvious how to predict this interesting membrane proteome. However, based on its evolutionary relation to the outer membrane of gram-negative bacteria, the outer envelope membrane should contain a large number of β-barrel proteins and possibly no α-helical membrane proteins. In a recent study, Schleiff, Eichacker, et al.

TABLE 14.2. Prediction of chloroplast proteins.

Type of protein	Number	Reference
Total plastid proteome predicted by TargetP (TIGR annotation v 4.0, April, 17, 2003)	4,255 out of 28,581 nuclear gene products	Emanuelsson, unpublished data
Outer envelope membrane proteome	891 candidates	Schleiff, Soll, et al., 2003
Inner envelope membrane proteome	136 candidates 183 candidates	Ferro et al., 2003 Koo and Ohlrogge, 2002
α-Helical chloroplast proteins (sum of inner envelope + thylakoid membrane)	957 (out of 25,500) 562 (out of 25,500) 520 (out of 28,585)	Ferro et al., 2003 Koo and Ohlrogge, 2002 Emanuelsson, unpublished data
Thylakoid lumenal proteome	80 candidates 200-250, of which about 90 with TAT motif	Schubert et al., 2002 Peltier et al., 2002; Emanuelsson, unpublished data
Stromal proteome	4,255–520,000 = ~3535 Total cTP set minus α-helical chloroplast proteins minus lumenal proteome	Emanuelsson, unpublished data

(2003) evaluated the probability for the existence of β-sheet, β-barrel, and hairpin structures among all proteins encoded by the *A. thaliana* genome, using a new β-barrel protein prediction strategy (Wimley, 2002). This set was further filtered, based on isoelectric point, by removal of potential mitochondrial proteins, as predicted by TargetP, and by manual selection. This resulted in a set of 891 potential outer membrane proteins, representing about 4.5 percent of all nuclear gene products. Among those, several were earlier experimentally identified, while others seemed bona fide outer membrane proteins, based on the postulated function of these proteins and the biological role of the outer membrane. This prediction strategy resulted in an interesting candidate pool for outer membrane proteins, which however contains a very significant number of false positives. Although several outer membrane proteins have been identified and characterized over the past decade in detailed biochemical and functional studies, a systematic experimental identification using biological mass spectrometry of the outer envelope membrane proteome of chloroplasts and nonphotosynthetic plastids is needed.

Prediction of the Integral Thylakoid Membrane and Inner Envelope Membrane Proteome

In most organisms it is believed that about 20 to 25 percent of all genes encode for helix bundle membrane proteins (Jones, 1998; Krogh et al., 2001). No β-barrel proteins have been reported for the thylakoid membrane or inner envelope, suggesting that all integral membrane proteins in these membranes are of α-helical nature. Numerous α-helical transmembrane predictors have been developed and are freely available on the Internet. Their performance has been evaluated in Moller et al. (2001), who concluded that the hidden Markov Model TMHMM (v. 2.0) (http://www.cbs.dtu.dk/services/TMHMM/) (Sonnhammer, 1998) is by far the best predictor available (Table 14.1).

A set of nine TM predictors has been used to screen the *A. thaliana* genome for integral membrane proteins and was deposited in a database (http://aramemnon.botanik.uni-koeln.de/). Based on "consensus prediction," 6,688 proteins were selected as the most likely TM proteins, thus representing ~25 percent of all gene products in *A. thaliana* (Schwacke et al., 2003). The *A. thaliana* (http://www.cbs.umn.edu/arabidopsis/) and rice (http://www.cbs.umn.edu/rice/) membrane databases developed by Ward used only HMM-TOP (Tusnady and Simon, 1998), which is a modification of TM-DAS, originally described in Cserzo et al. (1997). HMM-TOP was reported not to be a very reliable predictor (Moller et al., 2001; Schwacke et al., 2003). Nevertheless, these rice and *A. thaliana* databases provide a wealth of information (Table 14.1).

Two recent studies attempted to predict the nuclear-encoded α-helical integral membrane proteins in the inner chloroplast envelope (Ferro et al., 2002; Koo and Ohlrogge, 2002). In the study of Ferro et al. (2002), 997 putative chloroplast integral membrane proteins were identified, using the subcellular localization predictor ChloroP (http://www.cbs.dtu.dk/services/ChloroP/) (the old version of TargetP) and the TM predictor HMM-TOP (Cserzo et al., 2002). Koo and Ohlrogge (2002) used TM-HMM and TargetP and identified 562 putative integral membrane proteins. They removed proteins in which the TM domain overlapped with the cTP domain. Both studies started with a similar gene annotation with ~25,500 annotated ORFs (open reading frames) in the *A. thaliana* genome. The two sets are quite different in size, with the 997 proteins very likely representing an overprediction. Screening the latest annotation of the *A. thaliana* genome (TIGR, version 4.0, 28,581 proteins) using TargetP combined with TM-HMM 2.0, and filtering for overlap with the cTP or lumenal TP (lTP), suggested potentially 520 nuclear-encoded α-helical transmembrane proteins (Emanuelsson, unpublished data) (Table 14.2).

With the exception of those proteins which carry an lTP or an L18 domain (in the case of a subset of LHCs [Tu et al., 2000]), it is unclear how these putative integral membrane proteins are sorted to the thylakoid membrane or inner envelope. Therefore, in both studies, further selection of putative inner envelope membrane proteins was based on putative function, known localization (both to remove thylakoid proteins and to select envelope proteins), as well as pI value.

It was suggested in Ferro et al. (2002) that the known inner envelope proteins have typically a rather basic pI value, whereas known thylakoid membrane proteins are more acidic. Out of the 997 predicted TM-containing cTP proteins, 402 were reported to have a pI higher than 8.8. It should be noted that the basic cTP was not removed prior to this calculation, thus potentially overestimating the pI (see Peltier et al., 2002). Out of this set of 402 proteins, 136 proteins had at least four predicted TMs and more than one predicted TM per 100 amino acid residues (Ferro et al., 2002). Among those 136, about 50 proteins had homology to known transporters or were good candidates for valid inner envelope proteins based on predicted function. As discussed in the following, 25 were identified by mass spectrometry from 1-DE of mixed envelopes.

In the study of Koo and Ohlrogge (2002), 21 known thylakoid membrane proteins were removed from the 562 predicted membrane proteins, resulting in a set of 541 putative inner envelope proteins. Based on sequence similarity to proteins of known function, 183 proteins were further selected; 40

percent of these were classified in the category of "transport facilitation," indicating that this collection is highly enriched for membrane transporters. Analysis of expressed sequence tags collected from public databases suggested that the transcript levels of most candidates were relatively constant, with some exceptions, among different tissues, in contrast to the stromal and thylakoid proteins (Koo and Ohlrogge, 2002).

We have recently collected a set of 98 helix bundle thylakoid proteins encoded by the *A. thaliana* nuclear (66) and plastid (32) genome based on manual screening of the literature, experimental data, and orthologue predictions. This represents the most comprehensive set of thylakoid membrane proteins to date and is used to better define prediction criteria of thylakoid membrane proteins among all integral membrane proteins with a cTP (Sun, Emanuelsson, and van Wijk, 2004). Many functional processes related to biogenesis and maintenance can be postulated that require integral membrane proteins, as was recently discussed in more detail in van Wijk (2001b). Therefore, it can be expected that the thylakoid membrane contains at least another 100 additional integral membrane proteins, which are still unknown, however. An extensive experimental thylakoid proteome analysis was carried out to identify a number of these unknown entities, and so far at least 30, earlier unknown, integral thylakoid membrane proteins were identified (Friso et al., 2004; Ytterberg, Peltier, and van Wijk, unpublished data).

Prediction of the Stromal and Thylakoid Lumen Proteomes

The cTP-containing proteins without obvious predicted TM either are soluble in the stroma, are soluble in the thylakoid lumen, or are peripherally associated with the thylakoid or inner envelope membrane. As far as it is known, all thylakoid lumen proteins have a cleavable, N-terminal signal peptide, assigned lTP. The lTPs have no conserved sequence motif but do have a bias in amino acid content, rather similar to bacterial signal peptides that are used for translocation of proteins from the cytosol to the periplast. lTPs can be divided into the n-, a-, h-, and c-domains. The n (N-terminal)-domain has charged residues, the a-domain is variable in length, the h-domain is hydrophobic, and the c-domain is the C-terminus. In addition, the cleavage site is quite well conserved and is mostly "A × A," but a number of variations can be observed (Emanuelsson and von Heijne, 2001; Peltier et al., 2002). Lumenal proteins translocate through the thylakoid membrane either via the so-called "TAT" pathway or via the "Sec" pathway. The lTPs provide most of this pathway specificity, with possible contributions of the "mature" sequence (reviewed in, e.g., Mori and Cline, 2001; Robinson et al., 2001). TAT lTPs have a twin argine in the n-domain. In a recent paper, we described how we collected 203 lumenal proteins from different plant species (based on experimental data and orthologue prediction), with the goal to extract experimental parameters to allow prediction of lumenal proteins (Peltier et al., 2002). This showed that the twin arginines in the TAT substrates are located between 18 and 32 residues upstream of the cleavage site and that the h-domain is typically less hydrophobic than the Sec lTPs. With these and other experimental parameters in hand, we then attempted to predict the lumenal proteome. It turned out that the TAT substrate pool could be fairly well predicted, but that the Sec substrate pool was much more difficult to predict, easily leading to strong overprediction. Nevertheless, from these predictions, we projected that the lumen contains at least 200 different proteins, with about 80 having a typical twin arginine motif (Table 14.2). In the light of the surprising results for *Bacillus subtilis* in which the twin arginine motif-containing proteins did not show a TAT translocase dependency (Jongbloed et al., 2002), experimentation is needed to determine if the predicted twin arginine proteins are indeed translocated via the TAT pathway.

In an effort to improve the prediction of the lumenal proteome, a specific lTP predictor (LumenP) was developed (Westerlund, von Heijne, and Emanuelsson, 2003) and used to rescreen the latest annotation of the *A. thaliana* genome (with ~28,581 predicted ORFs). Similar numbers were observed as earlier with the SignalP predictor, when combined with our experimental parameters defined by the 203 lumenal orthologues. Based on this new prediction, we expect that the lumenal proteome contains up to ~90 TAT proteins and about 200 Sec proteins (Sun, Emanuelsson, and van Wijk, 2004) (Table 14.2).

Adding up the numbers predicted by a combination of TargetP, TM-HMM, LumenP, and additional filtering suggests that the stroma contains up to 3,300 proteins (Figure 14.1; Table 14.2).

PREDICTION OF PROTEOMES OF NONPHOTOSYNTHETIC PLASTIDS

No specific efforts have been made to predict the proteome of nonphotosynthetic plastids. It appears that there is in fact little theoretical basis to distinguish between proteins in nonphotosynthetic plastids and chloroplast proteins based on their amino acid composition. It has always been assumed that the import machineries in the different plastids types are basically the same. Therefore, it is likely that the cTPs are also indistinguishable, thus preventing prediction of the proteomes for different plastid types.

However, with the completion of the sequencing of the *A. thaliana* genome, it has become evident that one or more paralogues exist for several of the proteins of the import apparatus (Jackson-Constan and Keegstra, 2001). It is conceivable that different import complexes with different specificities can form in different plastid types (Yu and Li, 2001), possibly requiring small differences in cTPs. However, it is more likely that gene expression is regulated at the level of transcription via known and unknown regulatory elements (see, e.g., Richly et al., 2003).

SYSTEMATIC EXPERIMENTAL IDENTIFICATION AND CHARACTERIZATION OF THE PLASTID PROTEOME

Protein Separation and Mass Spectrometry and Its Application in Plastid Proteomics

Mass spectrometry plays a very central role in experimental proteomics, and a powerful set of different types of mass spectrometers has been commercially developed and different approaches have been employed. This rapidly expanding field has been extensively reviewed, in, e.g., Aebersold and Mann (2003) and Rappsilber and Mann (2002). It has become relatively easy to identify proteins rapidly and routinely at the 100 to 250 femtomole level if corresponding genomic or EST sequences are available. The leading proteomics laboratories routinely achieve identification of proteins at the subfemtomole level. An update of plant proteomics and the use of mass spectrometry in plant biology was provided in Roberts (2002) and van Wijk (2001a).

Different strategies are applied to identify proteins, and they depend on the complexity of the protein sample, the availability of complementary genome information, mass spectrometers, and expertise. In protein samples of low complexity (presence of 1 to ~3 proteins), proteins can be rapidly identified by peptide mass fingerprinting, using MALDI-TOF-MS. Proteins present in more complex mixtures (3 to ~20 proteins) are better identified using tandem mass spectrometry (MS/MS), either coupled to a nanobore HPLC in the case of ESI-MS/MS or directly analyzed in the case of MALDI-MS/MS. For very complex mixtures (several hundreds of proteins), online two-dimensional HPLC (e.g., MudPIT) directly coupled to mass spectrometry is the most effective way (Washburn et al., 2001). The latter approach has been applied in a systematic proteomic analysis of leaf, root, and seed tissue of rice *(Oryza sativa)* and allowed the detection and identification of 2,528 unique proteins (Koller et al., 2002). For recent reviews concerning peptide mass fingerprinting by MALDI-TOF-MS and

MS/MS approaches, see Aebersold and Mann (2003) and Rappsilber and Mann (2002). For 2-D chromatography, see papers from Yates and colleagues (Liu et al., 2002; Washburn et al., 2001; Wolters et al., 2001).

In most recent years, Fourier transform ion cyclotron resonance MS (FTIC-MS) has been tested in proteomics applications and shows strength for very rapid and accurate mass measurements of small (less than ~40 kDa) proteins and peptides (Johnson et al., 2002; Smith et al., 2001). This also has led to a new approach, coined "top-down" proteomics. In the "top-down" approach, intact proteins are directly analyzed in the mass spectrometer, rather than identifying proteins by mass analysis of corresponding peptides generated by proteolysis ("bottom-up" approach) (Ge et al., 2002; Meng et al., 2001; Smith et al., 2001). This has the important advantage that the primary protein sequence and its posttranslational modifications, so important for cellular functions, can be studied much more effectively. However, several technical barriers still need to be overcome to make this approach feasible for larger (>40 kDa) proteins and complex eukaryotic proteomes. Moreover, extensive front-end separation is needed to obtain a meaningful dynamic resolution (Meng et al., 2002). Recently, we have applied such a "top-down" approach on a small scale to subfractions of the chloroplast proteome of *A. thaliana*, and we showed that indeed such an approach is also potentially feasible for analysis of plant (sub)proteomes (Zabrouskov et al., 2003).

Despite the enormous progress in biological mass spectrometry, it has become evident that fractionation of proteomes is still absolutely needed for comprehensive identification and characterization of posttranslational modifications. Only with fractionation can sufficient dynamic resolution be achieved to identify proteins of low abundance. One should keep in mind that maximum molar ratios between proteins in biological samples are in the range of 1×10^6 to 1×10^9 and that proteins cannot be amplified, as is possible for DNA or RNA.

To visualize protein-expression patterns and detect differential expression of water-soluble proteins, two-dimensional electrophoresis, using isoelectric focusing (IEF) in the first dimension and SDS-PAGE in the second dimension, can be employed (Rabilloud, 2002). Commercialized immobilized ampholine gradient (often referred to as immobilized pH gradients [IPG strips]) is available. In the IPG strips, proteins are separated under fully denaturing conditions based on their isolectric point (pI), followed by separation in the second dimension on SDS-PAGE gels, with separation based on molecular mass.

2-DE is unsuitable for reproducible separation and recovery of integral membrane proteins due to aggregation and insolubility. Although different detergents and protocols have been generated to improve solubility, the more hydrophobic proteins are still resolved only very poorly, or not at all (Celis and Gromov, 1999; Herbert, 1999; Santoni et al., 2000, 1999). It should be noted that the representation of integral membrane proteins on 2-D gels is strongly correlated to the ratio of the number of TM domains and the size of the protein; as a rule of thumb, proteins with a TM/kDa ratio higher than 0.25 easily escape recovery by 2-DE (Santoni et al., 2000). Importantly, β-barrel proteins from, e.g., the outer membrane of gram-negative bacteria, are not very hydrophobic and are thus suitable for 2-DE (Molloy et al., 2000).

Thus, to separate the hydrophobic proteomes with the aim of detecting proteins of low abundance, other methods need to be employed outside the area of 2-DE with IPG strips. RP-HPLC, as well as gel filtration of intact proteins using different solvent systems, has been used to purify different integral membrane proteins (Lew and London, 1997; Sussman, 1988; Tarr and Crabb, 1983; Whitelegge et al., 2002; Zhang et al., 2001) and to separate abundant members of the thylakoid grana proteome (Gomez et al., 2002). The group of Molloy (Molloy et al., 1999) as well as those of Joyard (Ferro et al., 2003, 2002, 2000; Seigneurin-Berny et al., 1999) and van Wijk (Friso et al., 2004) have explored fractionation with organic solvents, followed by 1-D SDS-PAGE. Organic solvent fractionation proved useful since it enriches for hydrophobic proteins, depending on the polarity of the solvents. Organic solvent extraction, however, also has some disadvantages: (1) extraction is incomplete, (2) proteins do tend to aggregate after removal of lipids, and (3) pellets of precipitated proteins can be hard to fully resuspend.

Identification of the Inner and Outer Envelope Proteome of Chloroplasts

Three recent papers describe efforts to experimentally identify the inner and/or outer envelope proteome. Proteins of highly purified outer envelope membranes of pea chloroplasts were analyzed by ESI-MS/MS (Schleiff, Eichacker, et al., 2003). In addition to several known components, four new proteins of the outer envelope membranes were identified. Joyard and colleagues have analyzed the mixed inner and outer envelope membrane proteome of chloroplast of spinach and *A. thaliana,* using sequential extraction with chloroform/methanol in different ratios, followed by 1-DE and offline MS analysis (Ferro et al., 2002; Seigneurin-Berny et al., 1999). Several known as well as unknown inner and outer membrane proteins were identified. Envelope localization was confirmed for a couple of new proteins by transient expression of GFP fusions. In their latest, more extensive study with mixed *A. thaliana* chloroplast envelope membranes, proteins were extracted with chloroform/methanol, alkaline, or saline treatments (Ferro et al., 2003). LC-MS/MS analyses were then performed on each envelope membrane subfraction, leading to the identification of more than 100 proteins. Transient expression of GFP fusions of two phosphate transporters was used to confirm their envelope localization. Almost one-third of the identified proteins were reported to have no known function, whereas more than 50 percent are very likely to be associated with the chloroplast envelope, based on their postulated function or because they were already known envelope proteins. These proteins were (1) involved in ion and metabolite transport, (2) components of the protein import machinery, and (3) involved in chloroplast lipid metabolism. Some soluble proteins, such as proteases and proteins involved in carbon metabolism or in responses to oxidative stress, were associated with envelope membranes and likely represent contaminations.

The Hydrophobic Thylakoid Proteome

Whitelegge and colleagues published several papers in which they explored and developed methodologies to identify hydrophobic proteins of the thylakoid membrane and other biological systems (Gomez et al., 2002; le Coutre et al., 2000). In their studies, different types of chromatography were used to separate intact proteins, and the intact masses were then determined by ESI-MS. Identification of the corresponding proteins was carried out based on these intact mass measurements; this worked well for the abundant members of the thylakoid membrane proteome, since these are well-characterized proteins (e.g., N-terminus is known) and many proteins are encoded by the plastid genome. A particular effort was placed on the identification of proteins in photosystem II-enriched thylakoid membranes from pea and spinach (so-called BBY particles) (Gomez et al., 2002). Electrospray ionization mass spectrometry coupled with reverse-phase chromatography was used to separate and catalog the detectable proteins. Around 90 intact mass tags were detected, corresponding to approximately 40 gene products with different posttranslational modifications. Provisional identification of 30 of these gene products was proposed, based upon coincidence of measured mass with that calculated from genomic sequence and prior identification of the N-terminus.

In a recent study (Friso et al., 2004), we analyzed the hydrophobic integral thylakoid membrane proteome of *A. thaliana* chloroplasts using organic solvent fractionation of salt-stripped thylakoid membranes. The original protocols from Ferro et al. (2000) were adapted to remove the very abundant hydrophobic chromophores (chlorophylls and carotenoids) and to improve extraction efficiency. The different fractions were further separated by RP-HPLC or 1-D PAGE, and online nano-CapLC tandem MS was used to further improve dynamic resolution. Indeed, in addition to the nearly complete set of hydrophobic members of the thylakoid-bound photosynthetic machinery, several low abundant proteins were identified, such as components of the Sec translocon and PGR5 involved in cyclic electron flow (Munekage et al., 2002). The data will become available through a public database (http://ppdb.tc.cornell.edu/).

The Lumenal and Peripheral Thylakoid Proteomes

Comprehensive analyses of protein fractions enriched for lumenal and/or peripheral proteins of the thylakoids from pea, spinach, and *A. thaliana* were carried out by protein separation on two-dimensional electrophoresis gels with IPG strips as the first dimension and SDS-PAGE in the second dimension, followed by MS and Edman degradation sequencing (Friso et al., 2004; Kieselbach et al., 1998; Peltier et al., 2002, 2000; Schubert et al., 2002).

From the analysis of soluble and peripheral proteins in the thylakoids of pea, it was estimated that there are at least 200 to 230 different lumenal and peripheral proteins, after correcting to eliminate possible isoforms and posttranslational modifications (Peltier et al., 2000). Sixty-one proteins were identified; for 33 of these proteins, a clear function or functional domain could be identified, whereas for ten proteins no function could be assigned. For 18 proteins, no expressed sequence tags or full-length genes could be identified in the databases, despite experimental determination of a significant amount of amino acid sequence. Nine previously unidentified proteins with lumenal transit peptides were present along with their full-length genes; seven of these proteins possess the twin arginine motif that is characteristic for substrates of the TAT pathway (Peltier et al., 2000).

In a subsequent study, we focused on a lumenal-enriched proteome of *A. thaliana* in which proteins were separated by two-dimensional electrophoresis and identified by mass spectrometry and Edman sequencing. The identities of 81 proteins were established, and N-termini of many proteins were sequenced to validate localization prediction. Gene annotation of the identified proteins was corrected by experimental data, and an interesting case of alternative splicing was discovered. Expression of a surprising number of paralogues was detected. Expression of five isomerases of different classes suggested strong (un)folding activity in the thylakoid lumen. In addition, a stress defense network against antioxidative stress was identified and included peroxiredoxins, m-type thioredoxins, and a putative lumenal ascorbate peroxidase (Figure 14.1B).

Schroder and colleagues also analyzed the lumenal proteome first with 1-D gels and later with 2-D gels (Kieselbach et al., 1998; Schubert et al., 2002). The purified lumenal proteome was tested by enzymatic measurements and/or N-terminal sequencing, revealing the presence of proteolytic activities, violaxanthin de-epoxidase, polyphenol oxidase, peroxidase, as well as a prolyl *cis/trans*-isomerase (Kieselbach et al., 1998). Using 2-DE from lumen of spinach and *A. thaliana,* they identified additional proteins, leading to a total of 36 lumenal proteins. A large group of peptidyl-prolyl *cis/trans*-isomerases and a family of novel PsbP domain proteins were found (Schubert et al., 2002), similar to results of Peltier et al. (2002).

Thus, all four studies dealing with spinach, pea, and *A. thaliana* gave very similar data sets for the lumenal proteome, showing that proteome analysis of a cellular compartment can be reproducible. Interestingly, no significant differences were observed between the lumenal proteomes of the three tested plant species. We can therefore conclude that prime functions of the lumenal proteome include assistance in the folding and proteolysis of thylakoid proteins, as well numerous unknown functions.

The peripheral thylakoid membrane proteins interact with the membrane either indirectly through protein-protein interactions or directly via electrostatic interactions between the negatively charged polar lipid head groups and positively charged domains on the proteins, via short alpha helices parallel inserted into the lipid bilayer, or possibly via lipid anchors. The latter two interactions are tight and require either detergents or incubation with strong chaotropic agents for their release. In our most recent study (Friso et al., 2004), the peripheral thylakoid proteome was specifically extracted from lumenal-depleted *A. thaliana* thylakoids, using first chaotropic agents, and subsequently Triton X-114 followed by two-phase separation and separation on two-dimensional gels. Several additional proteins were identified, and the affinity of the peripheral proteins for the thylakoid membrane was assessed by comparing the population extracted with chaotropic agents and Triton X-114. These data will also become available via <http://ppdb.tc.cornell.edu/>.

Identification of the Soluble Stromal Proteome of Chloroplasts and Nonphotosynthetic Plastids

The stromal proteomes of chloroplasts and nonphotosynthetic plastids are complex, containing 1,000 to 3,000 proteins. Therefore, a substantial fractionation will be needed to analyze these complex proteomes.

Yates and colleagues described an initial characterization of the wheat amyloplast proteome *(Triticum aestivum)* and identified 171 proteins (Andon et al., 2002). Whole amyloplasts and purified amyloplast membranes were prepared from wheat. Protein extracts were examined by 1-D and 2-D electrophoresis, followed by LC-MS/MS of separated proteins. Using this approach, 108 proteins were identified from whole amyloplasts and 63 proteins from purified amyloplast membranes. The majority of protein identifications were derived from sequences from cereal crops other than wheat, for which relatively few gene sequence data are available (Andon et al., 2002).

No other studies have so far been published that aimed at providing an exhaustive overview of these stromal proteomes, but an extensive proteome analysis of chloroplasts and nonphotosynthetic plastids from roots and petals and their protein interactions is in progress (Peltier, Friso, Cai, and van Wijk, unpublished data).

POSTTRANSLATIONAL MODIFICATIONS OF THE CHLOROPLAST PROTEOME

Stable or transient posttranslational modifications can help to anchor proteins to membrane (in the case of lipid moieties), regulate activity or protein interactions (in the case of, e.g., phosphorylation), stabilize proteins (e.g., in the case of glycosylation, N-terminal formylation), or target proteins for degradation (e.g., in the case of ubiquination). Thus, investigation of posttranslational modifications is important, but often challenging, requiring a significant amount of protein material, mass spectrometry skills, and time. As mentioned earlier, FTIC-MS is a powerful technique to address posttranslational modifications, since many peptide and other bonds can be broken in one MS experiment (especially when using different dissociation techniques), and since such instruments have supreme mass accuracy (Ge et al., 2002; Kelleher et al., 1999). Nevertheless, also relatively cheaper and more accessible instruments (e.g., ESI-Q-TOF-type instruments and ion traps) are suitable. Now we will briefly review the systematic analyses of posttranslational modifications of chloroplast proteins.

In a recent paper concerning chloroplast envelopes, several posttranslational modifications were identified and the literature on posttranslational modifications of chloroplast proteins was reviewed briefly (Ferro et al., 2003). The authors identified several N-acetylated envelope proteins, as was earlier observed similarly for thylakoid proteins (Gomez et al., 2002; Michel et al., 1991, 1988). Interestingly, this N-terminal modification can help to identify the "true" N-terminus (typically difficult in "bottom-up" proteomics using mass spectrometry of tryptic digests) and can help to verify predicted cTP cleavages sites. However, one should note that many nuclear-encoded chloroplast proteins undergo a one or two amino acid N-terminal processing step after removal of the cTP by the stromal processing peptidase (Emanuelsson et al., 1999; Richter and Lamppa, 1998, 1999). Thus, a one to two amino acid mismatch between predicted cTP cleavage site and N-terminus of the mature protein is expected, since the programs are trained to recognize the initial cleavage site.

A physiologically important N-terminal methionine excision by peptide deformylases (PDFs) seems to occur for many of the chloroplast-encoded proteins of *A. thaliana* and *C. reinhardtii* (Giglione and Meinnel, 2001; Giglione et al., 2003). PDFs have been discovered recently in eukaryotic genomes, and it appears that N-terminal methionine excision occurs in all compartments where protein synthesis takes place (reviewed in Giglione and Meinnel, 2001). Chloroplasts of both *A. thaliana* and *C. reinhardtii* contain two PFD (1A and 1B) proteins. Disruption of *PDF1B* in *A. thaliana* led to an albino phenotype, and an extreme sensitivity to the PDF-specific inhibitor actinonin. In contrast, a knockout line for *PDF1A* exhibited no apparent phenotype. Pulse-chase experiments revealed that PDF inhibition leads to destabilization of a crucial subset of chloroplast-encoded photosystem II components in *C. reinhardtii*. The same proteins were destabilized in *pdf1b*. Site-directed substitutions altering the excision site of the most sensitive target, subunit D2, resulted in similar effects. Thus, plastid N-terminal methionine excision is a critical mechanism, specifically influencing the life span of photosystem II polypeptides. However, the mechanisms of proteolysis responsible for the destabilization are unknown.

Vener and colleagues (2001) used so-called "parent ion scanning" to characterize phosphorylated thylakoid proteins. From the analysis of tryptic peptides released from the surface of *Arabidopsis* thylakoids, phosphoproteins were identified by MALDI-TOF-MS and ESI-MS/MS using a triple quadrapole instrument. This showed that the D1, D2, and CP43 proteins of the photosystem II core are phosphorylated at their N-terminal threonines (Thr), the peripheral PsbH protein is phosphorylated at Thr-2, and the mature light-harvesting polypeptides LCHII are phosphorylated at Thr-3. In addition, a doubly phosphorylated form of PsbH modified at both Thr-2 and Thr-4 was detected. By comparing the levels of phospho- and nonphosphopeptides, the in vivo phosphorylation states of these proteins were analyzed under different physiological conditions. None of these thylakoid proteins were completely phosphorylated in the steady state conditions of continuous light, or completely dephosphorylated after a long dark adaptation. However, rapid reversible hyperphosphorylation of PsbH at Thr-4 in response to growth in light/dark transitions and a pronounced specific dephosphorylation of the D1, D2, and CP43 proteins during heat shock were detected.

Analysis of isolated photosystem II complexes using chromatography and intact mass measurements allowed detection and resolution of a minor population of D1 that was apparently palmitoylated (Gomez et al., 2002), similarly as described earlier (Mattoo and Edelman, 1987). Based upon observed +80 Da adducts, D1, D2, CP43, two Lhcbs, and PsbH were confirmed to be phosphorylated, as observed by Vener et al. (2001) and in numerous phosphorylation studies, e.g., Pursiheimo et al. (1998), and a new phosphoprotein was proposed to be the product of *psbT*. The appearance of a second +80 Da adduct on PsbH provides direct evidence for a second phosphorylation site on PsbH, as was described by Vener et al. (2001). Adducts of 32 Da, presumably arising from oxidative modification during illumination, were associated with more highly phosphorylated forms of PsbH, implying a relationship between the two phenomena (Gomez et al., 2002). In conclusion, numerous posttranslational modifications have been detected in the chloroplast by proteomics tools, clearly setting the stage for further functional studies.

PLASTID-LOCALIZED PROTEIN COMPLEXES AND PROTEIN-PROTEIN INTERACTIONS

There must be a very large number of stable and transient interactions between the proteins in the chloroplast. Many contacts between different proteins are transient, with the proteins CPN60 and HSP70 as the best-known examples. Despite the transient nature of the interactions, CPN60 and HSP70 have been shown to interact with several plastid proteins (Hubbs and Roy, 1993; Madueno et al., 1993; Tsugeki and Nishimura, 1993). Many of the plastid-localized biosynthetic pathways are linked to one another, with intermediates from one being used in other pathways. In other cases, biosynthetic pathways branch into two different pathways (for instance, in the case of heme and chlorophyll biosynthesis at the level of protoporphyrin IX), and in many cases it is not known how this branching is regulated. It has been postulated that formation of multisubunit complexes, as well as suborganellar localization of enzymes (e.g., thylakoid membrane versus envelope membrane), controls the flux into the different pathways. Identifying these protein interactions is a daunting task, but necessary to truly understand protein function and plastid metabolic pathways.

Systematic analysis of protein complexes and protein interaction will provide important insights into plastid function, biogenesis, and metabolic regulation. Large epitope tagging projects have already been very successful for yeast (Gavin et al., 2002; Ho et al., 2002). Here, we briefly provide an overview of known protein complexes and report on efforts to obtain a more systematic overview.

The four thylakoid complexes involved in photosynthesis (PSI, PSII, ATP synthase, cyt b_6/f) contain multiple gene products and have been well characterized using a variety of biochemical techniques. This has been discussed in numerous reviews and books (Ort and Yocum, 1996; Wollman et al., 1999) and will not be discussed further here. The thylakoid-bound respiratory NDH complex has been isolated as a 550 kDa complex, with at least three of the *ndh* gene products (Burrows et al., 1998; Zhang et al., 2001), but still remains elusive due to its low expression levels.

Membrane protein complexes in the chloroplast envelope have also been studied intensively over the past decade or so, with much attention focused on the protein import machinery (Jackson-Constan et al., 2001; Jarvis and Soll, 2002; Kessler and Schnell, 2002). Nevertheless, no systematic analysis of envelope membrane complexes has been carried out, but it should be of great interest to elucidate the function of the envelope in protein import, metabolism, and exchange of metabolites.

Given the complexity of the stromal proteome of chloroplasts and nonphotosynthetic plastids, relatively few complexes have been characterized. They include the 30S and 50S ribosomal subunits in *C. reinhardtii* (Yamaguchi et al., 2002) and high plant chloroplasts (Yamaguchi and Subramanian, 2000, 2003; Yamaguchi et al., 2000) and the 550 kDa Rubisco holoenzyme. Complexes in amino acid and starch biosynthesis pathways include a 352 kDa homodimeric arginase complex and an α-glucan phosphorylase, forming a homodimer of 208 kDa. Poorly characterized complexes with known function include a multisubunit form of acetyl-CoA carboxylase (Reverdatto et al., 1999) and the RNA polymerases (NEP and PEP) (Hajdukiewicz et al., 1997; Pfannschmidt and Link, 1994). Another interesting, but uncharacterized, complex is a 300 kDa soluble stromal complex involved in mRNA processing, stability, or translation, with a protein named Mbb1 as the only known component. Mbb1 is a member of a large family of proteins with tetratricopeptide-like repeats (Vaistij et al., 2000). Very recently, a new ribonucleotide particle of 1,700 kDa was identified in the chloroplast of green algae and is likely to be the plastid counterpart of eukaryotic snRNPs involved in nuclear splicing (Rivier et al., 2001). The protein composition of the particle is not yet known.

Heteromeric Clp protease core complexes of ~350 kDa and ~200 kDa dimers of ClpC chaperones were identified in leaf chloroplasts and nonphotosynthetic plastids in petals and roots, using native gels and native IEF, followed by tandem mass spectrometry (Peltier et al., 2003, 2001) (Figure

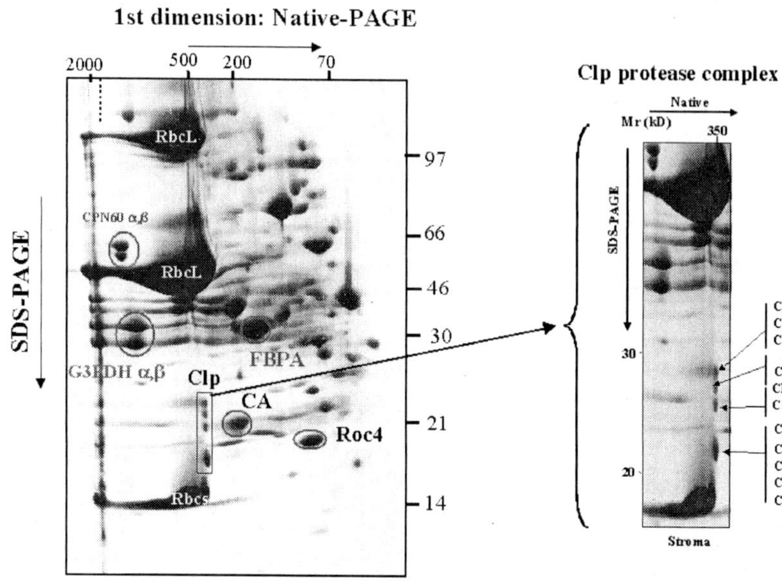

FIGURE 14.2. Protein complexes in the stroma of *A. thaliana* chloroplasts (see corresponding figure in color photo section). (*Source:* Adapted from Peltier et al., 2003.)

14.2). Surprisingly, the Clp core composition was identical in all three plastid types, despite the remarkable differences in plastid proteome composition. The plastid core consisted of five different serine-type protease Clp proteins (ClpP1,3-6), four nonproteolytic ClpR proteins (ClpR1-4), as well as ClpS1,2 unique to higher plants. These native separation techniques are also used systematically to identify the stromal proteomes and their interactions of nonphotosynthetic and photosynthetic plastids (Peltier et al., 2004) (Figure 14.2).

Thioredoxins are small multifunctional redox-active proteins present in many organisms, and their interaction with other proteins is typically of transient nature. In chloroplasts, three types of thioredoxins (f, m, and x) are known (Balmer et al., 2003; Collin et al., 2003). These thioredoxins play central roles in regulating enzyme activity, but only a few of their targets are known. Balmer and colleagues (2003) identified many putative targets of the chloroplast thioredoxins f and m, using an innovative proteomics approach. They mutated thioredoxins f and m and bound them to affinity columns to trap target proteins of chloroplast stroma. The covalently linked proteins were eluted with DTT, separated on gels, and identified by mass spectrometry. This approach led to the identification of 15 potential targets that function in ten chloroplast processes not known to be thioredoxin linked. Included are proteins that seem to function in plastid-to-nucleus signaling and in a previously unrecognized type of oxidative regulation. Approximately two-thirds of these targets contained conserved cysteines. They also identified 11 previously unknown and nine confirmed target proteins that are members of pathways known to be regulated by thioredoxin. In contrast to results with individual enzyme assays, specificity for thioredoxin f or m was not observed in the affinity chromatography (Balmer et al., 2003).

CONCLUSIONS AND FUTURE PERSPECTIVES

Experimental proteomics, involving mass spectrometry combined with different separation techniques, has begun to identify the main and minor components of chloroplasts, as well as to establish some of the protein-protein interactions. Prediction of chloroplast subproteomes has proven to be a viable and complementary approach to obtain candidate pools, thus helping to accelerate understanding of biological processes. The immediate challenges for plastid proteomics over the next one to three years include (1) the cataloging of the proteomes of chloroplasts and the different nonphotosynthetic plastid types, (2) the systematic characterization of protein interactions and complexes, and (3) the improvement of plastid proteome predictors using experimental data sets for training.

Proteomics has the potential to characterize dynamic changes in plastid proteome composition during development, abiotic and biotic stresses, as well as define phenotypes of mutants; I expect this will be a major focus for the next decade in conjunction with more sensitive transcript analysis. This is termed *comparative proteomics,* and several new technologies and approaches (e.g., isotope-labeling techniques and dual dye systems) are emerging to make comparisons more sensitive and less cumbersome (Lill, 2003; Patton et al., 2002). Thus, even though significant progress has been made in plastid proteomics, an enormous challenge lies ahead to unravel plastid biogenesis and differentiation, metabolism, and communication between the plastid and the rest of the plant cell.

REFERENCES

Aebersold, R., Mann, M. (2003). Mass spectrometry-based proteomics. *Nature* 422: 198-207.

Andon, N.L., Hollingworth, S., Koller, A., Greenland, A.J., Yates, J.R., III, Haynes, P.A. (2002). Proteomic characterization of wheat amyloplasts using identification of proteins by tandem mass spectrometry. *Proteomics* 2: 1156-1168.

Balmer, Y., Koller, A., del Val, G., Manieri, W., Schurmann, P., Buchanan, B.B. (2003). Proteomics gives insight into the regulatory function of chloroplast thioredoxins. *Proc. Natl. Acad. Sci. USA* 100: 370-375.

Burrows, P.A., Sazanov, L.A., Svab, Z., Maliga, P., Nixon, P.J. (1998). Identification of a functional respiratory complex in chloroplasts through analysis of tobacco mutants containing disrupted plastid *ndh* genes. *EMBO J.* 17: 868-876.

Celis, J.E., Gromov, P. (1999). 2D protein electrophoresis: Can it be perfected? *Curr. Opin. Biotechnol.* 10: 16-21.

Collin, V., Issakidis-Bourguet, E., Marchand, C., Hirasawa, M., Lancelin, J.M., Knaff, D.B., Miginiac-Maslow, M. (2003). The *Arabidopsis* plastidial thioredoxins: New functions and new insights into specificity. *J. Biol. Chem.* 278: 23747-23752.

Cserzo, M., Eisenhaber, F., Eisenhaber, B., Simon, I. (2002). On filtering false positive transmembrane protein predictions. *Protein Eng.* 15: 745-752.

Cserzo, M., Wallin, E., Simon, I., von Heijne, G., Elofsson, A. (1997). Prediction of transmembrane alpha-helices in prokaryotic membrane proteins: The dense alignment surface method. *Protein Eng.* 10: 673-676.

Emanuelsson, O., Nielsen, H., Brunak, S., von Heijne, G. (2000). Predicting subcellular localization of proteins based on their N-terminal amino acid sequence. *J. Mol. Biol.* 300: 1005-1016.

Emanuelsson, O., Nielsen, H., von Heijne, G. (1999). ChloroP, a neural network-based method for predicting chloroplast transit peptides and their cleavage sites. *Protein Sci.* 8: 978-984.

Emanuelsson, O., von Heijne, G. (2001). Prediction of organellar targeting signals. *Biochim. Biophys. Acta* 1541: 114-119.

Ferro, M., Salvi, D., Brugiere, S., Miras, S., Kowalski, S., Louwagie, M., Garin, J., Joyard, J., Rolland, N. (2003). Proteomics of the chloroplast envelope membranes from *Arabidopsis thaliana. Mol. Cell Proteomics* 2: 325-345.

Ferro, M., Salvi, D., Riviere-Rolland, H., Vermat, T., Seigneurin-Berny, D., Grunwald, D., Garin, J., Joyard, J., Rolland, N. (2002). Integral membrane proteins of the chloroplast envelope: Identification and subcellular localization of new transporters. *Proc. Natl. Acad. Sci. USA* 99: 11487-11492.

Ferro, M., Seigneurin-Berny, D., Rolland, N., Chapel, A., Salvi, D., Garin, J., Joyard, J. (2000). Organic solvent extraction as a versatile procedure to identify hydrophobic chloroplast membrane proteins. *Electrophoresis* 21: 3517-3526.

Friso, G., Giacomelli, L., Ytterberg, A.J., Peltier, J.B., Rudella, A., Sun, Q., van Wijk, K.J. (2004). In-depth analysis of the thylakoid membrane proteome of

Arabidopsis thaliana chloroplasts: New proteins, new functions, and a plastid proteome database. *Plant Cell* 16: 478-499.

Gavin, A.C., Bosche, M., Krause, R., Grandi, P., Marzioch, M., Bauer, A., Schultz, J., Rick, J.M., Michon, A.M., Cruciat, C.M., et al. (2002). Functional organization of the yeast proteome by systematic analysis of protein complexes. *Nature* 415: 141-147.

Ge, Y., Lawhorn, B.G., ElNaggar, M., Strauss, E., Park, J.H., Begley, T.P., McLafferty, F.W. (2002). Top down characterization of larger proteins (45 kDa) by electron capture dissociation mass spectrometry. *J. Am. Chem. Soc.* 124: 672-678.

Giglione, C., Meinnel, T. (2001). Organellar peptide deformylases: Universality of the N-terminal methionine cleavage mechanism. *Trends Plant Sci.* 6: 566-572.

Giglione, C., Vallon, O., Meinnel, T. (2003). Control of protein life-span by N-terminal methionine excision. *EMBO J.* 22: 13-23.

Gomez, S.M., Nishio, J.N., Faull, K.F., Whitelegge, J.P. (2002). The chloroplast grana proteome defined by intact mass measurements from liquid chromatography mass spectrometry. *Mol. Cell Proteomics* 1: 46-59.

Hajdukiewicz, P.T., Allison, L.A., Maliga, P. (1997). The two RNA polymerases encoded by the nuclear and the plastid compartments transcribe distinct groups of genes in tobacco plastids. *EMBO J.* 16: 4041-4048.

Herbert, B. (1999). Advances in protein solubilisation for two-dimensional electrophoresis. *Electrophoresis* 20: 660-663.

Ho, Y., Gruhler, A., Heilbut, A., Bader, G.D., Moore, L., Adams, S.L., Millar, A., Taylor, P., Bennett, K., Boutilier, K., et al. (2002). Systematic identification of protein complexes in *Saccharomyces cerevisiae* by mass spectrometry. *Nature* 415: 180-183.

Hubbs, A.E., Roy, H. (1993). Assembly of in vitro synthesized large subunits into ribulose-bisphosphate carboxylase/oxygenase: Formation and discharge of an L8-like species. *J. Biol. Chem.* 268: 13519-13525.

Jackson-Constan, D., Akita, M., Keegstra, K. (2001). Molecular chaperones involved in chloroplast protein import. *Biochim. Biophys. Acta* 1541: 102-113.

Jackson-Constan, D., Keegstra, K. (2001). Arabidopsis genes encoding components of the chloroplastic protein import apparatus. *Plant Physiol.* 125: 1567-1576.

Jarvis, P., Soll, J. (2002). Toc, Tic, and chloroplast protein import. *Biochim. Biophys. Acta* 1590: 177-189.

Johnson, J.R., Meng, F., Forbes, A.J., Cargile, B.J., Kelleher, N.L. (2002). Fourier-transform mass spectrometry for automated fragmentation and identification of 5-20 kDa proteins in mixtures. *Electrophoresis* 23: 3217-3223.

Jones, D.T. (1998). Do transmembrane protein superfolds exist? *FEBS Lett.* 423: 281-285.

Jongbloed, J.D., Antelmann, H., Hecker, M., Nijland, R., Bron, S., Airaksinen, U., Pries, F., Quax, W.J., van Dijl, J.M., Braun, P.G. (2002). Selective contribution of the twin-arginine translocation pathway to protein secretion in *Bacillus subtilis*. *J. Biol. Chem.* 277: 44068-44078.

Kelleher, N.L., Zubarev, R.A., Bush, K., Furie, B., Furie, B.C., McLafferty, F.W., Walsh, C.T. (1999). Localization of labile posttranslational modifications by electron capture dissociation: The case of gamma-carboxyglutamic acid. *Anal. Chem.* 71: 4250-4253.

Kessler, F., Schnell, D.J. (2002). A GTPase gate for protein import into chloroplasts. *Nat. Struct. Biol.* 9: 81-83.

Kieselbach, T., Hagman, A., Andersson, B., Schröder, W.P. (1998). The thylakoid lumen of chloroplasts: Isolation and characterization. *J. Biol. Chem.* 273: 6710-6716.

Koller, A., Washburn, M.P., Lange, B.M., Andon, N.L., Deciu, C., Haynes, P.A., Hays, L., Schieltz, D., Ulaszek, R., Wei, J., et al. (2002). Proteomic survey of metabolic pathways in rice. *Proc. Natl. Acad. Sci. USA* 99: 11969-11974.

Koo, A.J., Ohlrogge, J.B. (2002). The predicted candidates of *Arabidopsis* plastid inner envelope membrane proteins and their expression profiles. *Plant Physiol.* 130: 823-836.

Krogh, A., Larsson, B., von Heijne, G., Sonnhammer, E.L. (2001). Predicting transmembrane protein topology with a hidden Markov model: Application to complete genomes. *J. Mol. Biol.* 305: 567-580.

le Coutre, J., Whitelegge, J.P., Gross, A., Turk, E., Wright, E.M., Kaback, H.R., Faull, K.F. (2000). Proteomics on full-length membrane proteins using mass spectrometry. *Biochemistry* 39: 4237-4242.

Leister, D. (2003). Chloroplast research in the genomic age. *Trends Genet.* 19: 47-56.

Lew, S., London, E. (1997). Simple procedure for reversed-phase high-performance liquid chromatographic purification of long hydrophobic peptides that form transmembrane helices. *Anal. Biochem.* 251: 113-116.

Lill, J. (2003). Proteomic tools for quantitation by mass spectrometry. *Mass Spectrom. Rev.* 22: 182-194.

Lin, D., Tabb, D.L., Yates, J.R. (2003). Large-scale protein identification using mass spectrometry. *Biochim. Biophys. Acta* 1646: 1-10.

Liu, H., Lin, D., Yates, J.R. III. (2002). Multidimensional separations for protein/peptide analysis in the post-genomic era. *Biotechniques* 32: 898, 900, 902, passim.

Madueno, F., Napier, J.A., Gray, J.C. (1993). Newly imported Rieske iron-sulfur protein associates with both Cpn60 and Hsp70 in the chloroplast stroma. *Plant Cell* 5: 1865-1876.

Martin, W., Rujan, T., Richly, E., Hansen, A., Cornelsen, S., Lins, T., Leister, D., Stoebe, B., Hasegawa, M., Penny, D. (2002). Evolutionary analysis of *Arabidopsis,* cyanobacterial, and chloroplast genomes reveals plastid phylogeny and thousands of cyanobacterial genes in the nucleus. *Proc. Natl. Acad. Sci. USA* 99: 11996-11997.

Mattoo, A.K., Edelman, M. (1987). Intramembrane translocation and posttranslational palmitoylation of the chloroplast 32-kDa herbicide-binding protein. *Proc. Natl. Acad. Sci. USA* 84: 1497-1501.

Meng, F., Cargile, B.J., Miller, L.M., Forbes, A.J., Johnson, J.R., Kelleher, N.L. (2001). Informatics and multiplexing of intact protein identification in bacteria and the archaea. *Nat. Biotechnol.* 19: 952-957.

Meng, F., Cargile, B.J., Patrie, S.M., Johnson, J.R., McLoughlin, S.M., Kelleher, N.L. (2002). Processing complex mixtures of intact proteins for direct analysis by mass spectrometry. *Anal. Chem.* 74: 2923-2929.

Michel, H., Griffin, P.R., Shabanowitz, J., Hunt, D.F., Bennett, J. (1991). Tandem mass spectrometry identifies sites of three post-translational modifications of spinach light-harvesting chlorophyll protein II: Proteolytic cleavage, acetylation, and phosphorylation. *J. Biol. Chem.* 266: 17584-17591.

Michel, H., Hunt, D.F., Shabanowitz, J., Bennett, J. (1988). Tandem mass spectrometry reveals that three photosystem II proteins of spinach chloroplasts contain *N*-acetyl-*O*-phosphothreonine at their NH_2 termini. *J. Biol. Chem.* 263: 1123-1130.

Miras, S., Salvi, D., Ferro, M., Grunwald, D., Garin, J., Joyard, J., Rolland, N. (2002). Non-canonical transit peptide for import into the chloroplast. *J. Biol. Chem.* 277: 47770-47778.

Moller, S., Croning, M.D., Apweiler, R. (2001). Evaluation of methods for the prediction of membrane spanning regions. *Bioinformatics* 17: 646-653.

Molloy, M.P., Herbert, B.R., Slade, M.B., Rabilloud, T., Nouwens, A.S., Williams, K.L., Gooley, A.A. (2000). Proteomic analysis of the *Escherichia coli* outer membrane. *Eur. J. Biochem.* 267: 2871-2881.

Molloy, M.P., Herbert, B.R., Williams, K.L., Gooley, A.A. (1999). Extraction of *Escherichia coli* proteins with organic solvents prior to two-dimensional electrophoresis. *Electrophoresis* 20: 701-704.

Mori, H., Cline, K. (2001). Post-translational protein translocation into thylakoids by the Sec and DeltapH-dependent pathways. *Biochim. Biophys. Acta* 1541: 80-90.

Munekage, Y., Hojo, M., Meurer, J., Endo, T., Tasaka, M., Shikanai, T. (2002). PGR5 is involved in cyclic electron flow around photosystem I and is essential for photoprotection in *Arabidopsis*. *Cell* 110: 361-371.

Neuhaus, H.E., Emes, M.J. (2000). Nonphotosynthetic metabolism in plastids. *Ann. Rev. Plant Physiol. Plant Mol. Biol.* 51: 111-140.

Ort, D.R., Yocum, C.F. (1996). *Oxygenic Photosynthesis: The Light Reactions.* Dordrecht, The Netherlands: Kluwer Academic Publishers.

Patterson, S.D., Aebersold, R.H. (2003). Proteomics: The first decade and beyond. *Nat. Genet.* 33(Suppl): 311-323.

Patton, W.F., Schulenberg, B., Steinberg, T.H. (2002). Two-dimensional gel electrophoresis: Better than a poke in the ICAT? *Curr. Opin. Biotechnol.* 13: 321-328.

Peltier, J.B., Emanuelsson, O., Kalume, D.E., Ytterberg, J., Friso, G., Rudella, A., Liberles, D.A., Soderberg, L., Roepstorff, P., von Heijne, G., van Wijk, K.J. (2002). Central functions of the lumenal and peripheral thylakoid proteome of *Arabidopsis* determined by experimentation and genome-wide prediction. *Plant Cell* 14: 211-236.

Peltier, J.B., Friso, G., Kalume, D.E., Roepstorff, P., Nilsson, F., Adamska, I., van Wijk, K.J. (2000). Proteomics of the chloroplast: Systematic identification and targeting analysis of lumenal and peripheral thylakoid proteins. *Plant Cell* 12: 319-342.

Peltier, J.B., Ripoll, D.R., Friso, G., Rudella, A., Cai, Y., Ytterberg, J., Giacomelli, L., Pillardy, J., van Wijk, K.J. (2004). Clp protease complexes from photo-synthetic and non-photosynthetic plastids and mitochondria of plants, their predicted three-dimensional structures, and functional implications. *J. Biol. Chem.* 279: 4768-4781.

Peltier, J.B., Ytterberg, J., Liberles, D.A., Roepstorff, P., van Wijk, K.J. (2001). Identification of a 350-kDa ClpP protease complex with 10 different Clp isoforms in chloroplasts of *Arabidopsis thaliana. J. Biol. Chem.* 276: 16318-16327.

Pfannschmidt, T., Link, G. (1994). Separation of two classes of plastid DNA-dependent RNA polymerases that are differentially expressed in mustard (*Sinapis alba* L.) seedlings. *Plant Mol. Biol.* 25: 69-81.

Pursiheimo, S., Rintamaki, E., Baena-Gonzalez, E., Aro, E.M. (1998). Thylakoid protein phosphorylation in evolutionally divergent species with oxygenic photosynthesis. *FEBS Lett.* 423: 178-182.

Rabilloud, T. (2002). Two-dimensional gel electrophoresis in proteomics: Old, old fashioned, but it still climbs up the mountains. *Proteomics* 2: 3-10.

Rappsilber, J., Mann, M. (2002). Is mass spectrometry ready for proteome-wide protein expression analysis? *Genome Biol.* 3: COMMENT2008.

Reverdatto, S., Beilinson, V., Nielsen, N.C. (1999). A multisubunit acetyl coenzyme A carboxylase from soybean. *Plant Physiol.* 119: 961-978.

Richly, E., Dietzmann, A., Biehl, A., Kurth, J., Laloi, C., Apel, K., Salamini, F., Leister, D. (2003). Covariations in the nuclear chloroplast transcriptome reveal a regulatory master-switch. *EMBO Rep.* 4: 491-498.

Richter, S., Lamppa, G.K. (1998). A chloroplast processing enzyme functions as the general stromal processing peptidase. *Proc. Natl. Acad. Sci. USA* 95: 7463-7468.

Richter, S., Lamppa, G.K. (1999). Stromal processing peptidase binds transit peptides and initiates their ATP-dependent turnover in chloroplasts. *J. Cell Biol.* 147: 33-44.

Rivier, C., Goldschmidt-Clermont, M., Rochaix, J.D. (2001). Identification of an RNA-protein complex involved in chloroplast group II intron trans-splicing in *Chlamydomonas reinhardtii. EMBO J.* 20: 1765-1773.

Roberts, J.K. (2002). Proteomics and a future generation of plant molecular biologists. *Plant Mol. Biol.* 48: 143-154.

Robinson, C., Thompson, S.J., Woolhead, C. (2001). Multiple pathways used for the targeting of thylakoid proteins in chloroplasts. *Traffic* 2: 245-251.

Santoni, V., Molloy, M., Rabilloud, T. (2000). Membrane proteins and proteomics: Un amour impossible? *Electrophoresis* 21: 1054-1070.

Santoni, V., Rabilloud, T., Doumas, P., Rouquie, D., Mansion, M., Kieffer, S., Garin, J., Rossignol, M. (1999). Towards the recovery of hydrophobic proteins on two-dimensional electrophoresis gels. *Electrophoresis* 20: 705-711.

Schleiff, E., Eichacker, L.A., Eckart, K., Becker, T., Mirus, O., Stahl, T., Soll, J. (2003). Prediction of the plant beta-barrel proteome: A case study of the chloroplast outer envelope. *Protein Sci.* 12: 748-759.

Schleiff, E., Soll, J., Kuchler, M., Kuhlbrandt, W., Harrer, R. (2003). Characterization of the translocon of the outer envelope of chloroplasts. *J. Cell. Biol.* 160: 541-551.

Schubert, M., Petersson, U.A., Haas, B.J., Funk, C., Schroder, W.P., Kieselbach, T. (2002). Proteome map of the chloroplast lumen of *Arabidopsis thaliana*. *J. Biol. Chem.* 277: 8354-8365.

Schwacke, R., Schneider, A., van der Graaff, E., Fischer, K., Catoni, E., Desimone, M., Frommer, W.B., Flugge, U.I., Kunze, R. (2003). ARAMEMNON, a novel database for *Arabidopsis* integral membrane proteins. *Plant Physiol.* 131: 16-26.

Sehnke, P.C., Henry, R., Cline, K., Ferl, R.J. (2000). Interaction of a plant 14-3-3 protein with the signal peptide of a thylakoid-targeted chloroplast precursor protein and the presence of 14-3-3 isoforms in the chloroplast stroma. *Plant Physiol.* 122: 235-242.

Seigneurin-Berny, D., Rolland, N., Garin, J., Joyard, J. (1999). Technical advance: Differential extraction of hydrophobic proteins from chloroplast envelope membranes: A subcellular-specific proteomic approach to identify rare intrinsic membrane proteins. *Plant J.* 19: 217-228.

Sheen, J. (1999). C-4 gene expression. *Ann. Rev. Plant Physiol. Plant Mol. Biol.* 50: 187-217.

Smith, R.D., Pasa-Tolic, L., Lipton, M.S., Jensen, P.K., Anderson, G.A., Shen, Y., Conrads, T.P., Udseth, H.R., Harkewicz, R., Belov, M.E., et al. (2001). Rapid quantitative measurements of proteomes by Fourier transform ion cyclotron resonance mass spectrometry. *Electrophoresis* 22: 1652-1668.

Snyders, S., Kohorn, B.D. (1999). TAKs, thylakoid membrane protein kinases associated with energy transduction. *J. Biol. Chem.* 274: 9137-9140.

Sonnhammer, E.L.L., von Heijne, G., and Krogh A. (1998). A hidden Markov model for predicting transmembrane helices in proteins sequences. In Glasgow, J., Littlejohn, T., Major, F., Lathrop, R., Sankoff, D. and Sensen, C. (Eds.), *Sixth International Conference on Intelligent Systems for Molecular Biology* (pp. 175-182). Menlo Park, CA: AAAI/MIT Press.

Sun, Q., Emanuelsson, O., van Wijk, K.J. (2004). Analysis of curated and predicted plastid subproteomes of *Arabidopsis:* Subcellular compartmentalization leads to distinctive proteome properties *Plant Physiol.* 135: 723-724.

Sussman, M.R. (1988). Purification of integral plasma membrane proteins by reverse-phase high performance liquid chromatography. *Anal. Biochem.* 169: 395-399.

Tarr, G.E., Crabb, J.W. (1983). Reverse-phase high-performance liquid chromatography of hydrophobic proteins and fragments thereof. *Anal. Biochem.* 131: 99-107.

Thiellement, H., Zivy, M., Plomion, C. (2002). Combining proteomic and genetic studies in plants. *J. Chromatogr. B Analyt. Technol. Biomed. Life Sci.* 782: 137-149.

Tsugeki, R., Nishimura, M. (1993). Interaction of homologues of Hsp70 and Cpn60 with ferredoxin-NADP+ reductase upon its import into chloroplasts. *FEBS Lett.* 320: 198-202.

Tu, C.J., Peterson, E.C., Henry, R., Hoffman, N.E. (2000). The L18 domain of light-harvesting chlorophyll proteins binds to cpSRP43. *J. Biol. Chem.* 275: 13187-13190.

Tusnady, G.E., Simon, I. (1998). Principles governing amino acid composition of integral membrane proteins: Application to topology prediction. *J. Mol. Biol.* 283: 489-506.

Vaistij, F.E., Goldschmidt-Clermont, M., Wostrikoff, K., Rochaix, J.D. (2000). Stability determinants in the chloroplast *psbB/T/H* mRNAs of *Chlamydomonas reinhardtii. Plant J.* 21: 469-482.

van Wijk, K.J. (2000). Proteomics of the chloroplast: Experimentation and prediction. *Trends Plant Sci.* 5: 420-425.

van Wijk, K.J. (2001a). Challenges and prospects of plant proteomics. *Plant Physiol.* 126: 501-508.

van Wijk, K.J. (2001b). Proteins involved in biogenesis of the thylakoid membrane. In Andersson, B., Aro, E.-M. (Eds.), *Regulatory Aspects of Photosynthesis: Advances in Photosynthesis,* Volume 8 (pp. 153-175). Dordrecht, The Netherlands: Kluwer Academic Press.

Vener, A.V., Harms, A., Sussman, M.R., Vierstra, R.D. (2001). Mass spectrometric resolution of reversible protein phosphorylation in photosynthetic membranes of *Arabidopsis thaliana. J. Biol. Chem.* 276: 6959-6966.

von Heijne, G., Steppuhn, J., Hermann, S.G. (1989). Domain structure of mitochondrial and chloroplast targeting peptides. *Eur. J. Biochemistry* 80: 535-545.

Washburn, M.P., Wolters, D., Yates, J.R., III. (2001). Large-scale analysis of the yeast proteome by multidimensional protein identification technology. *Nat. Biotechnol.* 19: 242-247.

Westerlund, I., von Heijne, G., Emanuelsson, O. (2003). Lumen P—A neural network predictor for protein localization in the thylakoid lumen. *Protein Sci.* 12: 2360-2366.

Whitelegge, J.P., Zhang, H., Aguilera, R., Taylor, R.M., Cramer, W.A. (2002). Full subunit coverage liquid chromatography electrospray ionization mass spectrometry (LCMS+) of an oligomeric membrane protein: Cytochrome b_6f complex from spinach and the cyanobacterium *Mastigocladus laminosus. Mol. Cell Proteomics* 1: 816-827.

Wimley, W.C. (2002). Toward genomic identifiaction of beta-barrel membrane proteins: Composition and architecture of known structures. *Protein Science* 11: 301-312.

Wollman, F.A., Minai, L., Nechushtai, R. (1999). The biogenesis and assembly of photosynthetic proteins in thylakoid membranes 1. *Biochim. Biophys. Acta* 1411: 21-85.

Wolters, D.A., Washburn, M.P., Yates, J.R., III. (2001). An automated multidimensional protein identification technology for shotgun proteomics. *Anal. Chem.* 73: 5683-5690.

Yamaguchi, K., Prieto, S., Beligni, M.V., Haynes, P.A., McDonald, W.H., Yates, J.R., III, Mayfield, S.P. (2002). Proteomic characterization of the small subunit of *Chlamydomonas reinhardtii* chloroplast ribosome: Identification of a novel S1 domain-containing protein and unusually large orthologs of bacterial S2, S3, and S5. *Plant Cell* 14: 2957-2974.

Yamaguchi, K., Subramanian, A.R. (2000). The plastid ribosomal proteins: Identification of all the proteins in the 50S subunit of an organelle ribosome (chloroplast). *J. Biol. Chem.* 275: 28466-28482.

Yamaguchi, K., Subramanian, A.R. (2003). Proteomic identification of all plastid-specific ribosomal proteins in higher plant chloroplast 30S ribosomal subunit. *Eur. J. Biochem.* 270: 190-205.

Yamaguchi, K., von Knoblauch, K., Subramanian, A.R. (2000). The plastid ribosomal proteins: Identification of all the proteins in the 30S subunit of an organelle ribosome (chloroplast). *J. Biol. Chem.* 275: 28455-28465.

Yu, T.S., Li, H. (2001). Chloroplast protein translocon components atToc159 and atToc33 are not essential for chloroplast biogenesis in guard cells and root cells. *Plant Physiol.* 127: 90-96.

Zabrouskov, V., Giacomelli, L., van Wijk, K.J., McLafferty, F.W. (2003). A new approach for plant proteomics: Characterization of chloroplast proteins of *Arabidopsis thaliana* by top-down mass spectrometry. *Mol. Cell Proteomics* 2: 1253-1260.

Zhang, H., Whitelegge, J.P., Cramer, W.A. (2001). Ferredoxin: $NADP^+$ oxidoreductase is a subunit of the chloroplast cytochrome b_6f complex. *J. Biol. Chem.* 276: 38159-38165.

Chapter 15

Plant Mitochondrial Proteomics

A. Harvey Millar
Joshua L. Heazlewood

DEFINING THE PLANT MITOCHONDRIAL PROTEOME

The compartmentalization of eukaryotic cells remains a major focus of biological research as we seek to understand the separation of specific operations of cells by membrane systems; the complexity of intracellular trafficking; and the communication among multiple, independent, yet integrated genetic systems. The sequencing of whole plant genomes has provided a great mass of data on the fundamental building blocks of plant cells in the form of genes and predicted gene products (Arabidopsis Genome Initiative, 2000; Goff et al., 2002; Yu et al., 2002). Now a massive challenge lies in attributing these genes to both functions and locations in the working cell. Analysis of transcript abundances and gene expression patterns is helping to uncover various aspects of function through comparative and associative genomics, but it is only proteins themselves that have actual locations in cells of functional significance in pathways and networks. Analysis of proteins also opens the world of posttranslational modification for exploration that may prove to be a very significant contributor to what defines cell-to-cell differences.

Compartment-based proteomics utilizes a wealth of experience already gained over many years of organellar isolation and analysis by the scientific community (Jung et al., 2000). It also provides complex sets of proteins that are not so large as to be unmanageable to the academic, noncorporate researcher. Hopefully, it will provide proteomes that can be more easily explained, as the diversity of biological action within an organelle is likely to be more contained and interactions of enzymes and proteins into pathways and networks more easily studied than whole-cell analyses. Finally, the genomes of chloroplasts and mitochondria contain only a fraction of the proteins required for their operation (Abdallah et al., 2000; Gray et al., 1999) (see also Chapter 14). Most organellar proteins are nuclear encoded and, once translated, are transported into the organelle. Many, but not all, of these proteins were originally organellar encoded and have subsequently been transferred to the nucleus over time, since the endosymbiotic capture of their progenitors (Gray et al., 1999). Therefore, unraveling the nature and the origin of organellar proteomes provides a glimpse into the history of the evolution of eukaryotic cells.

Mitochondrial functions are integrated into the life and working of the eukaryotic cells in which they are housed. Thus, different eukaryotic lineages contain mitochondria with somewhat different functions, although all maintain a core role in respiratory carbon oxidation and ATP synthesis. Mitochondria from plants contain extra functions that are directly or indirectly related to photosynthesis and the metabolic plasticity of sedentary organisms. These specific mitochondrial functions range from photorespiration pathway components (Douce et al., 2001); prosthetic groups and cofactor biosynthesis for biotin, folate, lipolate, and vitamin C (Rébeillé et al., 1997); branched respiratory chains (Rasmusson et al., 1999; Vanlerberghe and McIntosh, 1997); variations in protein import and processing machinery (Braun et al., 1992; Braun and Schmitz, 1999; Duby and Boutry, 2002; Lister et al., 2003; Whelan, 1999); and differences in their genetic systems that include a large genome size and gene expression involving posttranscriptional editing by trans-splicing events (Brennicke et al., 1999; Gray et al., 1999; Wissinger et al., 1991).

Soon after the introduction of two-dimensional protein separations based on isoelectric focusing and SDS polyacrylamide gel electrophoresis (SDS-PAGE) in the 1970s, Booz and Travis (1980), closely followed by Diano et al. (1982), published the first two-dimension gels of plant mitochondrial proteomes. During the following twenty years, these techniques were applied to plant mitochondrial research in a wide variety of species, including the following:

1. The mapping of total mitochondrial proteins from different plant species and plant tissues from potato, pea, wheat, and maize (Barent and Elthon, 1992; Colas des Francs-Small et al., 1992, 1993; Humphrey-Smith et al., 1992; Remy et al., 1987; Rios et al., 1991; Rötig and Chauveau, 1987)
2. The mapping of proteins from submitochondrial compartments:
   - Inner membrane proteins (Jänsch et al., 1996)
   - Matrix proteins (Jänsch et al., 1996; Maffey et al., 1997; Pinel et al., 1986)
   - Intermembrane space (Lister et al., 2002; Sweetlove et al., 2001)
   - Outer membrane proteins (Jänsch et al., 1996; Werhahn et al., 2001)

3. The mapping of mitochondrial-encoded proteins (Boutry et al., 1983; Brangeon et al., 2000; De Paepe et al., 1993; Hack and Leaver, 1983; Maffey et al., 1997)
4. The identification of differentially expressed mitochondrial proteins:
   - In mutant and wild-type plants (Gutierres et al., 1997; Jenner et al., 2001; Karpova and Newton, 1999; Mihr et al., 2001; Witt et al., 1997; Zhihai and Zhiyi, 1991)
   - During environmental stress (Banzet et al., 1998; Hourton-Cabassa et al., 1998; Lund et al., 1998; Lund et al., 2001; Pratt et al., 1995, 1994; Wood et al., 1998)

Many of these studies were hampered by the lack, or very high cost, of protein-identification technologies. The advent of mass spectrometry methods and the explosion of their availability over the past five years have transformed our potential to exploit the protein-separating techniques used in many of these studies.

Bioinformatic targeting prediction programs have also given insights into mitochondrial proteomes. The *Arabidopsis* protein set (~28,000) contains 2,900 to 4,500 nuclear-encoded proteins with predicted mitochondrial localizations using bioinformatic tools, while the current TIGR draft rice protein set (~39,000; Yuan et al., 2003) contains 3,000 to 5,000 predicted mitochondrial proteins (Arabidopsis Genome Initiative, 2000; Millar and Heazlewood, unpublished data). More-stringent computer analyses of the genome data will no doubt refine this set to predict significantly fewer proteins in the future. The actual proteomes of mitochondria from these species are most likely composed of at least 500 to 1,000 different protein types, based on gel-based pictorial presentations to date (Heazlewood, Howell, Whelan, and Millar, 2003; Kruft et al., 2001; Millar, Sweetlove, et al., 2001). Approximately 500 to 600 different proteins are believed to be localized in yeast mitochondria, and 700 to 1,300 in mammalian mitochondria, according to current estimates based on experimentation and prediction (Lithgow, 2000; Lopez et al., 2000; Schon, 2001).

In the following pages we attempt to introduce the reader to the recent history of mitochondrial proteomics in plants. We try to provide insights into both the possibilities and the problems associated with the techniques of subcompartment proteomics in plants and the specific results so far obtained in studies of plant mitochondria. We consider that the next few years will greatly advance our understanding of mitochondria in plants and believe that proteomics will play a fundamental role in this journey.

TECHNOLOGY ISSUES

Mitochondrial Isolation

Subcellular proteomics requires robust isolation of subcompartments as free from contaminants as possible. The general principles of isolation, fractionation, and electrophoretic analysis of plant mitochondrial proteins have been reviewed by several laboratories (Douce et al., 1972; Neuburger, 1985; Millar, Liddell, et al., 2001). Breakthroughs in purification over the years included the use of sucrose density gradients (Douce et al., 1972), the later advent of Percoll gradient separations (Neuburger et al., 1982), and the additional use of PVP gradients to yield high-purity mitochondria from green plant tissues (Day et al., 1985). *Arabidopsis* cell culture has dominated plant mitochondrial proteome mapping studies recently. Differential centrifugation and Percoll gradient density separations that yield mitochondria which are largely free of contamination by cytosol, peroxisomes, plastids, and other membranes have been developed from this plant tissue (Kruft et al., 2001; Millar, Sweetlove, et al., 2001; Werhahn et al., 2001). This technique has also been adapted for high-purity isolation of mitochondria from rice seedling coleoptiles (Heazlewood, Howell, Whelan, et al., 2003).

Proteome Display and Protein Identification

The techniques of proteome display and the modern methods of protein identification using mass spectrometry are best considered together. Therefore, we have attempted to introduce the reader to options in an integrated fashion, outlining the options and combinations that are possible, and to explain the drawbacks and advantages of some of these methods in our hands.

Mass Spectrometry Identification Technology

The utilization of mass spectrometry for the analysis of separated proteins has become the primary technique for large-scale protein identification. There are two principal methods used with mass spectrometry for protein identification (Figure 15.1). Both initially rely on the cleavage of isolated samples with a known digestion agent, such as trypsin. After digestion, samples are introduced into the mass spectrometer as peptide ions in the gas phase. The most common method uses the absolute mass of each peptide to produce a peptide mass fingerprint (PMF) of the protein, which is used to search a database of theoretical peptide masses. This method is gen-

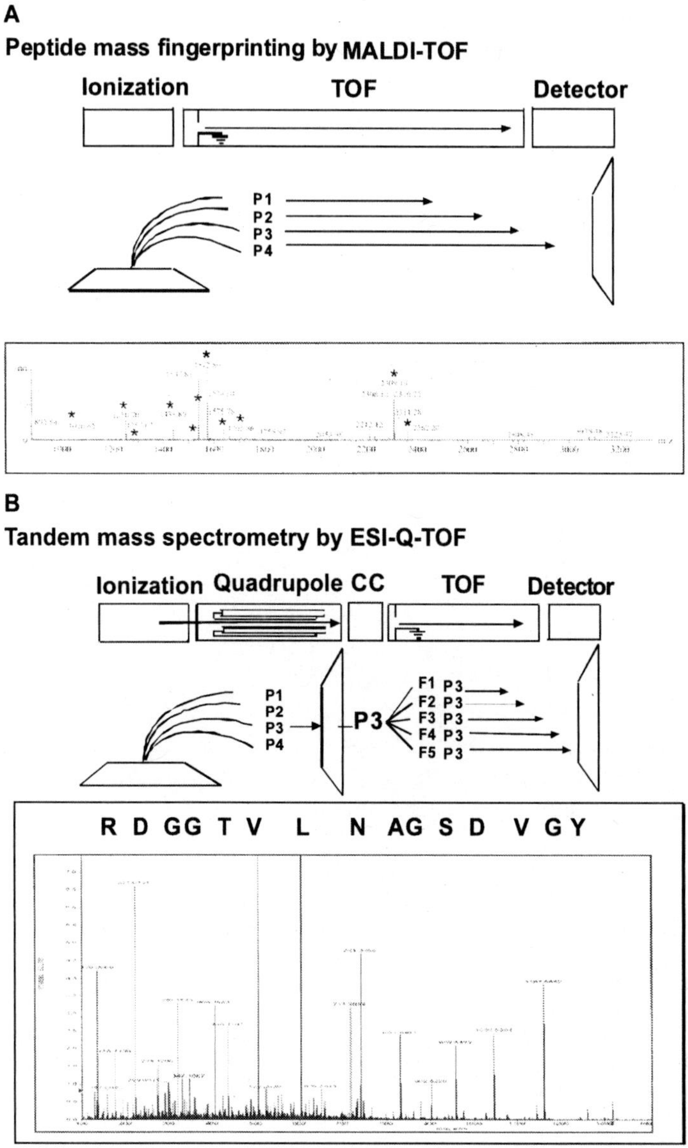

FIGURE 15.1. Mass spectrometry identification of proteins by MALDI-TOF (matrix-assisted laser desorption/ionization time-of-flight)-based peptide mass fingerprinting and ESI-Q-TOF (electrospray ionization quadrupole time-of-flight)-resultant collision-induced dissociation (CID) spectra. (A) Analysis through

peptide mass fingerprinting commonly utilizes a MALDI source for the ionization of samples with the resultant peptides (P1 to P4) accurately measured by the mass detector at the end of the TOF tube. Spectral data from a MALDI-TOF run are shown below the flow diagram, with peptide mass on the *x*-axis expressed as mass/charge (m/z). Ion intensity is shown on the *y*-axis. The ion peaks shown in the chromatogram are used for database searching. *Indicate peptides matched to a protein sample, in this case ATP synthase β subunit from *Arabidopsis*. (B) The production of MS/MS spectra from a digested protein sample requires an initial scan by the TOF mass analyzer only to determine peptide candidates for the collision cell (P1 to P4). The quadrupole is then used to specifically select one ion (P3) to pass through to the collision cell (CC) where peptide fragmentation occurs (F1P3 to F5P3). The resultant fragmented ions are analyzed by the TOF mass detector with MS/MS spectra produced. These MS/MS spectra can be used for protein pattern matching to a database or a de novo sequence can be interpreted for sequence identity matching. Spectrum shown is a peptide from a rice mitochondrial aldehyde dehydrogenase.

erally regarded as a high-throughput technique and can analyze hundreds of samples a day. It is commonly undertaken on a matrix-assisted laser desorption ionization–time-of-flight (MALDI–TOF) instrument (Figure 15.1A). The second method employs a tandem mass spectrometer (MS/MS), typically an electrospray ionization quadrupole time-of-flight (ESI-Q-TOF) or an ESI-ion trap instrument, and is capable of inferring peptide sequence information from individual peptides (Figure 15.1B). This method relies on the fragmentation of the peptide in some sort of collision cell within the mass spectrometer, yielding a fragmentation spectra or CID (collision-induced disassociation) spectra. MS/MS-derived spectra can be interpreted individually to suggest probable amino acid sequences (de novo sequencing). Alternatively, MS/MS spectra can be used to search a database containing theoretical MS/MS spectra (MS/MS pattern matching) to suggest both an amino acid sequence and a probable identity of the protein from which the peptide was derived. The quality of information obtained through MS/MS analysis is superior to data obtained through PMF, but by its very nature, it is more complex and thus has more limitations with respect to throughput in terms of both sample processing and data interpretation time.

IEF/SDS-PAGE: The Classic 2D Gel

Advances in IEF/SDS-PAGE electrophoresis, utilizing immobilized pH gradients (IPGs) for the first IEF dimension, have greatly enhanced the ease and reproducibility of such gels for proteome analysis (Link, 1999). Using this approach, a plant mitochondrial sample can be separated to reveal 400 to 800 protein features, with pI values ranging between 3.5 and 10 and apparent molecular masses between 5 kDa and 150 kDa (Figure 15.2C: Kruft et al., 2001; Millar, Sweetlove, et al., 2001). N-terminal sequencing or MS analysis of protein spots is commonly used to identify the proteins displayed. Alternatively, transfer of 2D separations to nitrocellulose allows probing with antibodies to identify proteins. This classic display technique provides a clear single picture for the researcher; it can be easily compared to other preparations to assess similarities and differences and provides information on two important physical properties of the polypeptide components, notably, their apparent molecular mass and their apparent isoelectric point (pI). Such gels can also be used to assess posttranslational modifications that alter the apparent pI of polypeptides without gross changes in molecular mass. Comparisons of gels are undertaken by software packages that utilize complex algorithms for overlaying and quantifying the protein spot intensities of scanned gels. Current commercial packages include Progenesis and Phoretix (Non-Linear Dynamics, Manchester, United Kingdom), PDQuest (Biorad Laboratories, Hercules, California), and Z3 (Compugen, Tel Aviv, Israel). These display and quantification techniques have been successfully employed to analyze the effects of oxidative stress on the mitochondrial proteome of *Arabidopsis* (Sweetlove et al., 2002). The disadvantage of this presentation technique is the selective loss of specific proteins due to physical properties incompatible with isoelectric focusing. Hydrophobic proteins, basic proteins, and high-molecular-mass proteins are often insoluble during IEF and do not typically appear in the 2D gel presentations (Herbert, 1999; Santoni et al., 1999).

Blue Native/SDS-PAGE: Mapping Native Protein-Protein Interactions

Blue native (BN) PAGE has been used very successfully for separation of mitochondrial electron transport protein complexes (Jänsch et al., 1996; Schägger and von Jagow, 1991). This electrophoresis technique utilizes Coomassie Blue G-250 to bind and provide a net negative charge to protein complexes under nondenaturing conditions and can readily resolve protein complexes in the 50 kDa to 1,000 kDa range. This approach assists in preventing hydrophobic proteins from precipitating during the first dimension. A combination of first-dimension BN-PAGE and second-dimension SDS-PAGE provides evidence of the individual polypeptides incorporated in or associated with mitochondrial complexes. Using this approach, single protein spots or smears usually containing a small number of polypeptide type are arrayed (Figure 15.2B). These presentation techniques are compatible with identification of proteins using immunoreaction techniques, N-termi-

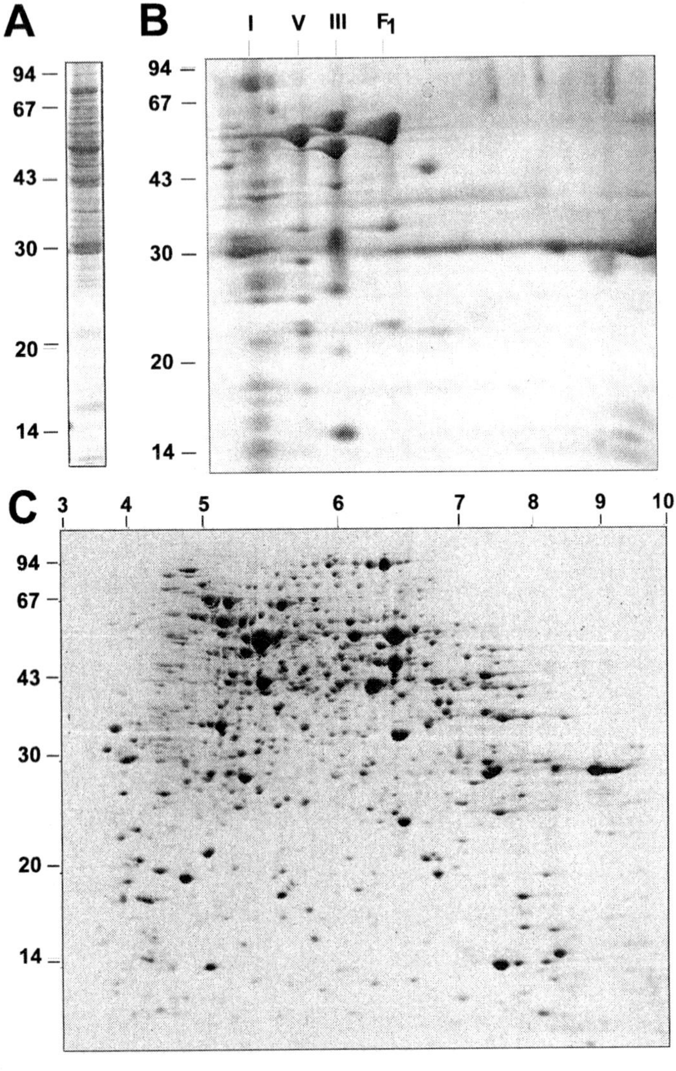

FIGURE 15.2. Gel electrophoresis separation of the plant mitochondrial proteome. Arraying of mitochondrial proteins for mass spectrometric analysis can be undertaken utilizing a variety of techniques. (A) 1D-PAGE, (B) BN-PAGE, and (C) 2D-PAGE have all been successfully used in determining proteins found in the mitochondria.The *y*-axis numbers are apparent molecular masses in kDa; the *x*-axis of C is isoelectric point in pH units. On the BN-PAGE in B the position of complex I (I), ATP synthase (V), complex III (III), and separated F1 of ATP synthase (F1) are shown.

nal sequencing, and MS analysis. Due to the propensity of finding multiple proteins in the same band on these native gels, tandem mass spectrometry and MS/MS data queries are sometimes required to identify conclusively the proteins involved. The limitations in the use of PMF data to resolve adequately the presence of more than two or three proteins in such samples has been previously discussed (Yates, 1998). This native gel approach has been used to identify potato (Jänsch et al., 1996), *Arabidopsis* (Heazlewood, Howell, and Millar, 2003; Heazlewood, Whelan, and Millar, 2003; Kruft et al., 2001; Werhahn and Braun, 2002; Werhahn et al., 2001; Eubel et al., 2003), and rice (Heazlewood, Howell, Whelan, and Millar, 2003) mitochondrial protein complexes.

SDS PAGE–MS/MS: Keeping It Simple

The advent of MS/MS de novo sequencing and MS/MS pattern matching has, for a variety of research objectives, alleviated the need for these aforementioned high-fidelity electrophoresis separation techniques. The mass spectrometer coupled to low-fidelity electrophoresis techniques can be used to decipher complex mixtures. An accumulation of such peptide data and the coverage of such data over the primary sequence of the predicted intact polypeptides are used as evidence for the presence of these proteins in the mixtures. Using this approach, standard 1D SDS-PAGE gels can be used as starting material to identify a significant number of proteins from a band of one apparent molecular mass (Figure 15.2A). This approach avoids many problems of losing hydrophobic, basic, and high-molecular-mass proteins during the protein arraying and still provides valuable information on the apparent molecular mass of the original protein. This approach has been used to identify a series of the hydrophobic carrier proteins of plant mitochondria (Millar and Heazlewood, 2003).

LC-MS/MS: Nongel Proteomics

A further simplification of display can be achieved by utilizing the high-speed, fully automated peptide selection and CID functions of modern mass spectrometers to identify proteins from whole-sample trypsination experiments (Figure 15.3). In such cases, the mass spectrometer is used as the sole tool for prediction and no additional information about the original protein (such as pI or MW) is known. These approaches, in which individual proteins are not separated prior to trypsination, generally provide little information of a quantitative nature about the parent protein compared to other proteins or about the parent protein in different samples. Nonetheless, these

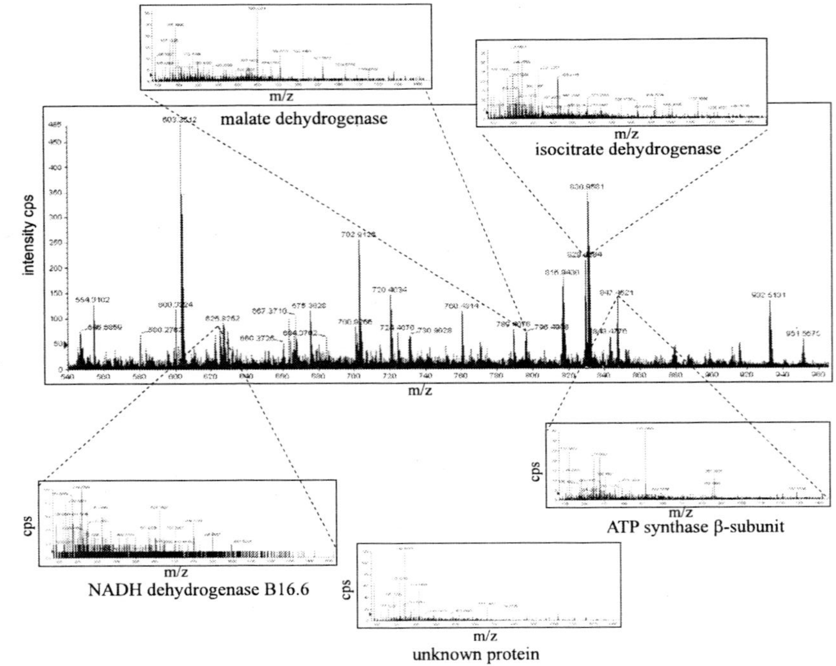

FIGURE 15.3. Automated LC-MS/MS separation and identification of the plant mitochondrial proteome. A trypsinated mitochondrial protein lysate was directly introduced into the mass spectrometer utilizing an online C18 reverse-phase column via a capillary HPLC for peptide separations. The central chromatogram displays ion m/z information at a point midway through the run (a five- to ten-second period). At this particular interval, five ions (peptides) were, in turn, automatically selected for the collision cell by the quadrupole mass spectrometer. Each ion's resultant fragmentation data were analyzed by the TOF mass detector to produce individual spectra for each peptide. These MS/MS-derived spectra were subsequently matched to proteins from a database; proteins matching are indicated under each MS/MS spectrum.

techniques are capable of identifying hundreds of proteins from a single sample in a relatively short period of time with minimal sample handling (Washburn et al., 2001). These techniques are currently being further enhanced through the use of multiple online LC separation steps, increasingly referred to as MudPITs (two-phase and three-phase multidimensional protein-identification technology), which readily allow for the identification of many hundreds of proteins from a single sample (McDonald et al., 2002). The limitations in terms of quantitation are beginning to be overcome by new developments, such as isotope-coded affinity tag (ICAT) technologies, which allow the quantitative comparison of specific cysteine-containing peptides between two complex peptide samples by the use of heavy and light isotope tags by the mass spectrometer (Gygi et al., 1999). This provides new opportunities for quantitative proteomics without gel arraying techniques but as yet has not been applied to plant mitochondrial samples.

Proteomic Choices Dictated by Genomic Data

The choice of which combination of methods to use when undertaking proteomic analyses is significantly dictated by the availability of genomic sequence data. Our analysis of the mitochondrial proteomes from the plant species *Arabidopsis* and rice has incorporated the utilization of both PMF and MS/MS for gel-based protein identifications. The success of positive identifications using PMF data appeared to be directly correlated with sequence availability in the database at the time of searching. Our initial investigations of the rice 2D mitochondrial proteome using PMF coincided with approximately 25 percent of the predicted proteins being available in public databases (~9000, circa mid-2000). This restriction in genomic availability resulted in a positive identification rate of only 30 percent. In contrast, in our work with the *Arabidopsis* 2D mitochondrial proteome, for which the whole genome sequence was publicly available and high-quality protein sets were on hand, our PMF-based positive identification rates were around 60 percent (Millar, Sweetlove, et al., 2001). In both cases, searches were undertaken using MS Fit at Protein Prospector (http://prospector.ucsf.edu/) against species-specific data sets at error tolerances of 50 ppm. These differences clearly reflect the limitations in using PMF data as an identification tool on incomplete or absent genomes. They further illustrate that, even with the complete genomic sequence and a well-annotated protein set to query against, success is only marginally better than 50-50.

A lack of positive species-specific matches using PMF can lead a researcher to use cross-species matches for protein identifications. The reliance on this sort of matching is problematic, given that the basis of the PMF technique is to identify a protein based on its specific theoretical mass profile (Mann et al., 1993). For some highly conserved proteins, a positive match can occur across species, but the number of peptide masses matching the protein will be less and their error tolerances are likely to be larger, leading to matches with much less confidence and greatly increasing the number of false and misleading identifications. Cross-species matching with PMF data has already been widely used to identify proteins of interest, and some of its limitations may be overcome in the future (Liska and Shevchenko, 2003). However, an important point to note is that some sort of match can nearly always be obtained when searching with PMF data, especially when the mass error tolerances are manipulated and large data sets queried. This is best illustrated by the example that a peptide of 1,000 to 2,000 Da will successfully match over 400 proteins at 50 ppm from the *Arabidopsis* protein set. Most current-generation mass spectrometers can produce high-quality PMF spectra that should result in significant matches at mass tolerances of 50 ppm to a protein derived from the same species as the query sample, typically with sequence coverage of more than 30 percent. With only two completed plant genomes currently available, an attempt to undertake a large-scale proteomic exercise through PMF on plant species other than *Arabidopsis* or rice is likely to be extremely challenging. Although the lack of a completed genome can be inhibitive, we have had some level of success matching PMF data to six frame translations of a rice EST database (Heazlewood, Howell, Whelan, and Millar, 2003). This method has also been successfully employed in the identification of arrayed root proteins of the model legume plant *Medicago truncatula,* for which little genomic sequence is currently available (Mathesius et al., 2001). Currently, The Institute for Genomic Research (TIGR) provides publicly available nonredundant EST data sets or gene indices (GI) for 19 commonly researched plant species. These GI sets can be downloaded and incorporated into the end user's mass spectrometric software for data matching and will be a significant help in proteomic studies outside the model plant species.

The utilization of MS/MS data produced on tandem mass spectrometers that incorporate sequence information as well as an accurate peptide mass is far more reliable at producing a positive match. A further advantage of using tandem mass spectrometry for protein identifications is its unmatched ability in analyzing posttranslational modifications. Using MS/MS spectra derived from *Arabidopsis* mitochondrial preparations, we routinely obtain positive identifications for >90 percent of our gel-based samples, while in a small percentage of cases no identifications are made despite the presence of some peptide ions. In some cases multiple proteins are identified from a single protein spot. As with PMF data, the utilization of MS/MS spectra for cross-species matching can also be problematic. Although MS/MS spectra contain peptide sequence information, the matching processes used by most programs (e.g., Mascot, MS Tag, SEQUEST) still rely on an accurate overall peptide mass match for their identifications (Yates, 1998). As a result, successful cross-species matching requires further data interpretation to produce a de novo sequence. A de novo sequence derived from the fragmentation pattern in the MS/MS spectrum can be used to query databases using BLAST (Altschul et al., 1997). In rice, de novo sequencing of MS/MS spectra allows more confident cross-species matching to identify the likely functional class of a protein using near exact BLAST queries of short sequences at NCBI. With the recent availability of the draft genomic sequence from rice, our positive identification rates have improved to approximately 75 percent when using MS/MS matching in combination with de novo cross-species and EST matching.

Based on our comments, the use of PMF data for protein identification in plants may appear to be a less desirable method for proteomics. However, this technique has obvious monetary advantages coupled with the ability to process hundreds of samples a week. In most circles this approach is still regarded as the proteomics workhorse. The clear advantages of MS/MS-based proteomics can potentially be outweighed by the monetary commitment and lack of scalability required in some large-scale mapping exercises. The choice of which method to use is clearly dependent on several factors, but if the desire is routinely to identify proteins from nonmodel species, then there is little current alternative than to employ MS/MS-based tandem mass spectrometry.

Prediction Programs for Generating the Proteome

The accurate sequence-based prediction of the subcellular localization of a protein has been a goal of researchers for a number of years (see also Chapter 14). This is essentially because the biological function of a protein is inexorably linked to its cellular localization. A variety of subcellular prediction programs can be used to discern mitochondrial localization. Most are available through a number of public Web servers, although centralized links to the more popular programs can be found at EXPASY (http://www.expasy.org). These popular programs tend to fall into two groups primarily based on their prediction method. The programs PSORT (Nakai and Horton, 1999), iPSORT (Bannai et al., 2002), and MitoProt II (Claros and Vincens, 1996) utilize predefined experimental rules to determine protein fate, while the programs TargetP (Emanuelsson et al., 2000), Predotar (http://www.inra.fr/predotar/), and SubLoc (Hua and Sun, 2001) use computer-learning techniques to arrive at their predictions. The authors of these prediction programs report relatively high levels of success in their overall prediction rates. The reported sensitivity of these programs to predict mitochondrial localization typically ranges from 80 to 90 percent. An analysis of the mitochondrial prediction numbers for the entire *Arabidopsis* nuclear-encoded protein data set obtained from TIGR is shown in Table 15.1. The total number of proteins predicted to be targeted to the mitochondria by these programs ranges from 3,000 to 5,000. The total number of proteins that any one of these programs predicts to be localized to the mitochondria is over 10,000, more than one-third of the total *Arabidopsis* protein set. Typically, any two programs agree only on a set of 1,500 to 2,000 proteins, and this set changes significantly depending on the two programs chosen. In contrast, the number of proteins estimated to be present in plant mitochondria is thought to be around 1,000, indicating that in *Arabidopsis* these subcellular prediction programs overpredict mitochondrial localization and comprise large nonoverlapping sets of proteins.

Published work mapping mitochondrial proteomes in *Arabidopsis* and rice utilizing mass spectrometry has experimentally identified 130 to 140 proteins from both organisms. The popular bioinformatic predictors are able to determine mitochondrial localizations for 40 to 70 percent of these experimentally validated proteins (Table 15.2). Moreover, a preliminary

TABLE 15.1. Prediction of mitochondrial localization of *Arabidopsis* nuclear-encoded proteins by publicly available prediction programs.

Program	**Mitochondrial prediction**	**Percent of proteome**	**Web site**
Predotar v0.5	4,514	16.5	<http://www.inra.fr/predotar/>
TargetP v1.01	2,955	10.8	<http://www.cbs.dtu.dk/services/TargetP/>
MitoProt II v1.0a4	3,794	13.9	<http://www.mips.biochem.mpg.de/cgi-bin/proj/medgen/mitofilter/>
SubLoc v1.0	3,554	13.0	<http://www.bioinfo.tsinghua.edu.cn/SubLoc/>
iPSORT	4,492	16.5	<http://hypothesiscreator.net/iPSORT/>

Note: Prediction analysis was undertaken on a set of 27,288 *Arabidopsis* nuclear-encoded proteins representing the entire nonredundant proteome. The data set was obtained from the TIGR *Arabidopsis thaliana* database (http://www.tigr.org/tdb/e2k1/ath1/) and encompasses release 3.0, including supplementary sequences. Prediction programs were executed and interpreted as per instructions on their respective Web sites using default settings. A DFM threshold of 0.7 for mitochondrial localization was used with MitoProt II.

TABLE 15.2. A comparison of positive mitochondrial predictions.

Program	*Arabidopsis* (%)	Rice (%)
Predotar 0.5	61	47
TargetP 1.01	56	52
MitoProt II 1.0a4	69	46
SubLoc 1.0	45	38
iPSORT	63	50

Note: The percentage of predicted proteins of the various subcellular localization programs from the experimentally derived nuclear-encoded mitochondrial sets from *Arabidopsis* (133) and rice (130). The protein sets used are outlined in Table 15.3.

mitochondrial data set of over 400 proteins from *Arabidopsis* derived from LC-MS/MS techniques has resulted in positive mitochondrial prediction rates of only 35 to 50 percent (Heazlewood et al., 2004). The discrepancy between these sensitivities and the published ranges for these programs may be accounted for by a lack of experimentally validated plant mitochondrial proteins available in public databases. Most data sets used for program training and rule analysis in these programs are produced from the annotated entries present in the SWISS-PROT database. The number of validated plant-specific mitochondrial entries at SWISS-PROT is usually regarded as not being adequate to be utilized for the training of computer learning programs (Emanuelsson et al., 2000). As a consequence, data sets used for inferring mitochondrial localization in plants incorporate sequences derived from nonplant sources. Interestingly, the sensitivity of mitochondrial predictions from the dicot *Arabidopsis* in Table 15.2 is generally better than that of the monocot rice. This may reflect the differences in annotation quality between the two genomes or a dicot bias in the data sets used to train these programs. A further limitation is that these programs generally utilize only the initial 100 or so N-terminal residues from the query protein for their predictions. Consequently, none of these predictors takes into account the effects of the whole protein in its subcellular destinations, although many reports in the literature indicate the significance of domains outside the initial N-terminus (Adams et al., 2000; Lister et al., 2002; Murcha et al., 2003). A further well-documented complication is the difficulty in differentiating mitochondrial-targeted proteins from those targeted to the chloroplast, which is likely due to similarities in their targeting and import processes (for recent reviews on these and other issues, see Emanuelsson and von Heijne, 2001; Feng, 2001; Peeters and Small, 2001). Thus, a critical limiting factor in the development of reliable mitochondrial prediction is a lack of a large plant-specific data set for interpreting features and to train the predictors. From our analyses of these *Arabidopsis* and rice mitochondrial protein sets, it appears that in general these subcellular prediction programs are capable of identifying many mitochondrial proteins but lack some specificity. With the recent increase in the use of proteomics for the identification of proteins in several organelles from multiple plant species, high-quality data sets should be available in the next few years to further augment and refine the current generation of predictors.

Databases and Data Storage for Proteomics

The biological database came of age during the explosion of genomic-sequencing projects that occurred over the past decade. The development of these biological databases coincided with dramatic technical advances in the areas of computing, scientific automation, and the Internet. Currently available biological databases reflect these genomic-sequencing projects in their structure and content. With the recent increases in availability, sensitivity, and utilization of mass spectrometry by biologists, there is now a requirement for these biological databases to reflect the increases in proteomic-type data that are now being generated. Currently, to the disappointment of many, there is no central store of the primary mass spectrometry data behind the increasing mass of published proteomic reports. These primary data are unavailable for reanalysis by others using newer deciphering software and for the reevaluation of reported matches on newly generated genomic data. The principal method for the presentation and manipulation of proteomic data to date is through 2D maps linked to the protein identified. This is best exemplified by the SWISS–2D-PAGE database (http://www.expasy.org/ch2d/). The effectiveness of these visual databases is largely limited to the cell lines and tissues from the specific organisms displayed, rather than a procedure to accurately identify proteins on your own 2D gel. A recent study comparing specific 2D maps from several plant and bacterial strains found that protein location was not a useful factor in inferring identity (Mathesius et al., 2002). These 2D map databases allow a certain degree of analysis through descriptive searches and cross-linking to multiple species. However, they are limited to visual gel electrophoresis experiments and therefore are not equipped to deal with the newer techniques involving gel-free methods for proteome analysis.

In our view, an essential requirement for effective proteomic databases is the occurrence of the completed genomic sequence of the organism of interest incorporating an advanced annotation of its encoded products. This is essential because it is of real interest to biologists to assign confidently a specific gene to its experimentally identified protein product. This readily enables the direct analysis of the gene in the context of its functional counterpart and is vital for the investigation of multigene families within an organism. Further, it immediately links the researcher to genetic tools, such as gene knockout or knockdown plants, to continue their research in the evolving area of plant functional genomics. In higher plants, the only species that currently fulfills this requirement is *Arabidopsis*. High-quality rice genome annotation was not available at the time of writing but should be by 2005-2006. The *Arabidopsis* Information Resource (TAIR; Huala et al., 2001) has recently expanded to enable an analysis of the *Arabidopsis* proteome through new search routines and has also integrated biochemical pathways and their corresponding gene products that are present in the genome. These sorts of features will be most amenable to the integration of proteomic data into these databases. Similar in scope to the production of a nonredundant transcript set that drives much of the EST databases (Asamizu et al., 2000), the formalization of a nonredundant proteome set will eventually enable large-scale analysis of the functional gene and protein sets within an organism. Interestingly, the data produced through proteomic projects share certain similarities to data currently produced through transcriptome-type projects using microarrays, where information is often experimentally dependent (see Chapter 4). Many advanced routines and queries have been successfully implemented for large-scale analyses in these microarray databases (Gollub et al., 2003). Proteomic databases will thus need to be far more dynamic and flexible to the end user, rather than simply lists of identifications. Features that need to be examined are "on the fly" annotation and analysis procedures with the ability to display and analyze data from multiple experiments and differing techniques simultaneously. More important, systems for tracking experimental procedures utilized in the production of these data need to be cataloged in conjunction with the primary data. These sorts of features will allow the current biological databases to be expanded to more than just proteomic databases; they will enable more advanced analyses that will be required to interpret the vast amount of information currently being produced.

SCIENCE ISSUES

Recent Surveys of the Mitochondrial Proteome

A series of plant mitochondrial proteome mapping studies in plants has so far been presented from *Arabidopsis* (Kruft et al., 2001; Millar, Sweetlove, et al., 2001), pea (Bardel et al., 2002), and rice (Heazlewood, Howell,

Whelan, and Millar, 2003). The *Arabidopsis* papers supplemented with several further identifications (Millar and Heazlewood, 2003; Sweetlove et al., 2002; Werhahn and Braun, 2002; PMF data from Millar and Heazlewood and Eubel and Braun, unpublished) represent a nonredundant set of 135 proteins. The pea proteome is a set of 43 nonredundant proteins and the rice proteome a set of 136 proteins (Tables 15.3 and 15.4). In this case, only data that have led to the matching to a specific gene sequence have been included in the analysis. A comparison between these sets reveals that each contains unknown function proteins that represent about one-quarter of the identifications. Of the remainder, nearly three-quarters of each has been found in at least two species, while only 22 proteins have so far been found in all three proteomes to date. This leaves sets of 34 proteins in *Arabidopsis,* 32 proteins in rice, and 7 proteins in pea for which clear orthologues have not been identified to date in other plant mitochondrial proteomes. This demonstrates the necessity to compare mitochondrial proteomes from different plant species, and most certainly between the major lineages of higher plants, in order to better understand the complex role of these organelles in the plant cell. Looking at the functional classifications of the similar-sized proteome sets in *Arabidopsis* and rice, it is clear that both contain a similar

TABLE 15.3. Number of identified plant mitochondrial proteins in *Arabidopsis,* rice, and pea proteomes.

	Total matched	Total genes
Arabidopsis		
Millar, Sweetlove, et al., 2001	91	80
Kruft et al., 2001	52	41
Sweetlove et al., 2002	37	27
Millar and Heazlewood, 2003	19	19
Werhahn and Braun, 2002	14	14
Heazlewood and Millar (unpublished)	18	18
Eubel and Braun (unpublished)	9	9
Nonredundant ID		135
Rice		
Heazlewood, Howell, Whelan, et al., 2003	221	136
Pea		
Bardel et al., 2002	105	43

Note: The proteins reported have been presented in terms of the number of proteins identified (total matched), and the nonredundant set of genes to which these have been matched (total genes).

TABLE 15.4. Comparisons of identified protein sets in *Arabidopsis*, rice, and pea mitochondrial proteomes.

	Total genes	In all 3	In at least 2	Unique
Unknown function proteins				
Arabidopsis	29			
Rice	35			
Pea	11			
Known function proteins				
Arabidopsis	106	22	72	34
Rice	101	22	69	32
Pea	32	22	25	7

Note: The nonredundant number of unknown function gene products matched (total genes) is presented along with the frequency of identification of products from similar known function genes across the three different plant species.

array of categories (Figure 15.4). Not surprisingly, the lists are dominated by electron transport chain and tricarboxylic acid cycle components that make up about one-third of the identified sets. The rice proteome also contains a series of lower abundance proteins (reflected in the transcription category) and a larger number of hydrophobic proteins (reflected in electron transport chain components), probably due to the more extensive use of LC-MS/MS and blue native-PAGE in this analysis. All three proteomes contain a large series of unknown function proteins that will no doubt form the focus of future attempts to unravel novel mitochondrial functions.

Subproteomes of Mitochondrial Compartments

Although mitochondria are cellular compartments in themselves, they also contain a further set of subcompartments in which different sets of proteins are separated to facilitate biochemical reaction sequences. Using a combination of osmotic shock and differential centrifugation, purified mitochondrial samples can be fractionated into four basic compartments. These comprise the two aqueous compartments, the matrix (MA) and the intermembrane space (IMS), and the two membrane compartments, the inner mitochondrial membrane (IM) and the outer mitochondrial membrane (OM). Isolation of the IM and MA fractions is relatively straightforward, and each together represents nearly 90 percent of total mitochondrial protein. The isolation of the IMS and OM fractions, which usually contain less

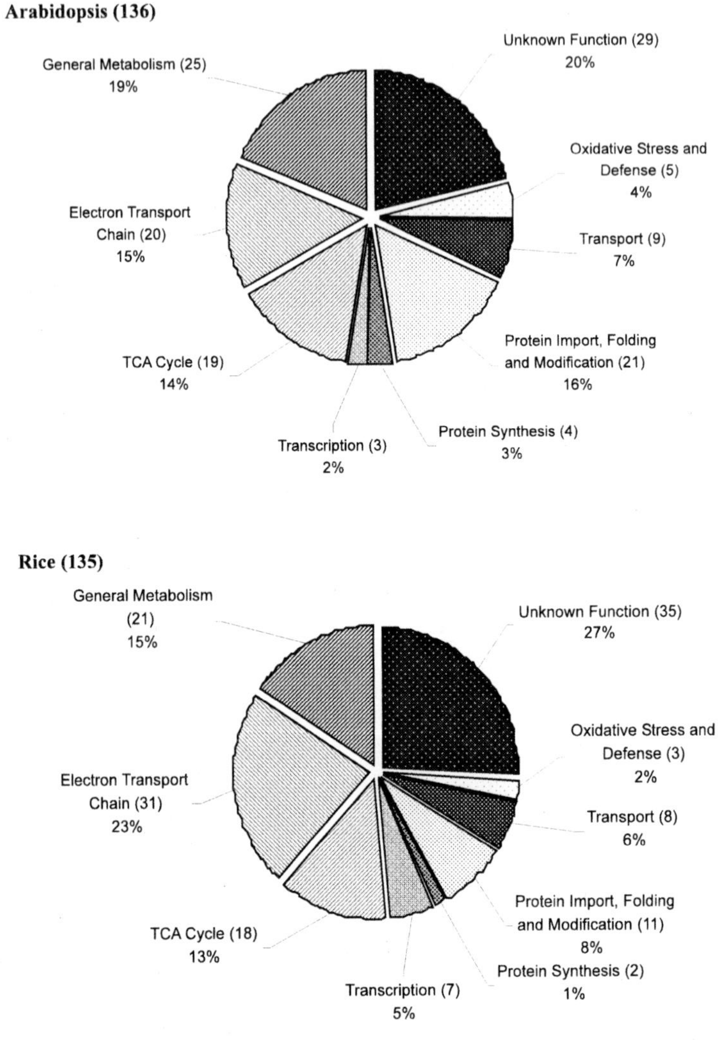

FIGURE 15.4. Distribution of functional categories in the experimentally defined mitochondrial proteomes of *Arabidopsis* and rice. The nonredundant set of 135 *Arabidopsis* proteins (collated from Millar, Sweetlove, et al., 2001; Kruft et al., 2001; Werhahn and Braun, 2002; Sweetlove et al., 2002; Millar and Heazlewood, 2003; unpublished PMF results from Millar and Heazlewood, see <http://www.mitoz.bcs.uwa.edu.au>; Eubel and Braun, unpublished data, available at <http://www.gartenbau.uni-hannover.de/genetik/AMPP>) and the nonredundant set of 136 rice proteins (Heazlewood, Howell, and Millar, 2003) are displayed according to functional category.

than 10 percent of total mitochondrial protein, is a greater challenge. Initial reports by Mannella and co-workers (Mannella, 1985; Mannella and Bonner, 1975) on the plant mitochondrial outer membrane pioneered this work and raised researchers' interests in these compartments. Proteome reports have sought to address the issue of these subcompartments in a number of ways. In one of the first reports on the *Arabidopsis* proteome, the mitochondrial proteins were fractionated on the basis of the degree of membrane association (DMA), yielding sets of soluble proteins from IMS and MA, the peripheral proteins of the two membranes, and the integral membrane proteins (Millar, Sweetlove, et al., 2001). Soluble and/or membrane protein samples have been analyzed by blue native PAGE in potato and *Arabidopsis* (Jänsch et al., 1996; Werhahn and Braun, 2002; Werhahn et al., 2001). The IMS proteome has also been explored in studies localizing a nucleotide diphosphate kinase as a specific IMS protein in *Arabidopsis* and potato mitochondria (Sweetlove et al., 2001), and the functional identification of intermembrane space proteins TIM9 and TIM10 involved in protein import also exploited IMS proteomics (Lister et al., 2002). Identification of a conclusive set of outer membrane proteins is not yet available but would greatly aid us in a molecular understanding of the mitochondrial-cytosol interface and the connection of mitochondria to the cytoskeleton and the endoplasmic reticulum.

Subproteomes of Native Complexes and New Biochemical Pathways

The notion that plant mitochondria contain new enzymes, and even contain novel components in the classic complexes of the electron transport chain, the TCA cycle, and the protein synthesis machinery, has in part driven plant mitochondrial research for many years. Thus, as the number of proteins located in the mitochondrial proteome grows, there is increasing interest in determining the place of these components in physical complexes and in biochemical pathways. For example, all the enzymes required for a GABA shunt of carbon intermediates between 2-oxoglutarate and succinate have now been identified using *Arabidopsis* proteomics. Glutamate dehydrogenase forms glutamate from 2-oxoglutarate to begin the shunt in the mitochondrial matrix (Millar, Sweetlove, et al., 2001), glutamate is decarboxylated in the cytosol (Snedden et al., 1996), a GABA aminotransferase has been identified in the matrix (Sweetlove et al., 2002), and succinic semialdehyde dehydrogenase concludes the shunt in the matrix with the formation of succinate (Millar, Sweetlove, et al., 2001). This pathway allows the bypass of a significant section of the TCA cycle and is often exploited in bacterial systems. The significance of the operation of such a bypass in plant mitochondria still remains to be investigated. In addition there is now evidence for cytochrome P450 enzymes and their electron donors and a suggestion that mitochondria may contain their own nucleotide synthesis pathway (see discussion and references in Kruft et al., 2001, and Millar, Sweetlove, et al., 2001). Our recent studies of the *Arabidopsis* ATP synthase complex (Heazlewood, Whelan, and Millar, 2003) and a comparative study of complex I components in *Arabidopsis* and rice (Heazlewood, Howell, and Millar, 2003) have also revealed the presence of plant-specific and previously unidentified components of these classic protein complexes. The ATP synthase F_0 subcomplex of *Arabidopsis* appears to contain the products of two previously uncharacterized mitochondrial-encoded ORFs, the *orfB* and *orf25* genes. Further, a series of smaller subunits present in yeast and mammals is absent from the plant complex and from sequenced plant genomes (Heazlewood, Whelan, and Millar, 2003). The complex I from higher plants contains a series of conserved components with fungi and animals, while also containing a series of plant-specific proteins of largely unknown function (Heazlewood, Howell, and Millar, 2003).

Subproteomes Based on Protein Domain Characteristics

In the search for a functional assignment of unknown proteins and for discovering new functionalities of known proteins, the identification of protein-binding properties is a valuable tool. Binding-property-based techniques also have the advantage of purification of low-abundance, functionally related sets of proteins from complex mixtures. Phosphoproteomics has become very fashionable due to the availability of phosphoaffinity resins and the critical role of phosphorylation cascades in activity regulation in many biochemical pathways. Several papers have shown that an array of phosphoproteins exists in plant mitochondria (Pical et al., 1993; Sommarin et al., 1990). The published identities of these phosphoproteins have been limited to only three or four proteins (Struglics et al., 1998; Thelen et al., 2000). However, the recent publication by Bykova et al. (2003) using radiolabeled-ATP coupled to 2D gel arraying of potato mitochondrial proteins and mass spectrometry has increased this list by 14, including proteins in the tricarboxylic acid cycle, the electron transport chain, and heat shock/defense proteins. Many proteins in mitochondria also bind divalent metal ligands. Such interactions are able to stabilize the active conformation of proteins, to directly facilitate enzymatic reactions by active site binding and interaction with substrates of products, or to directly participate in the reaction by electron acceptance and donation (Holm et al., 1996). Understanding the range of proteins that undertake such interactions and the biological implications of these interactions is of broad interest to researchers. Using a mobility shift assay and diagonal SDS-PAGE, Herald et al. (2003) have provided a preliminary series of metal-binding proteins in *Arabidopsis* mitochondria that includes Mn-SOD, succinyl CoA ligase beta subunit, Fe-S-centered components of complex I and the b/c_1 complex, along with a series of unknown function proteins. Extension of such studies could now consider NADH-binding proteins using affinity resins and nucleic acid-binding proteins using heparin columns.

Subproteomes of Common Expression Response

Differential proteome analysis is being used to identify changes in expression of the proteome in response to environmental factors and in varying genetic backgrounds. Such studies have the potential not only to identify the proteins involved but also to provide maps of coordinated expression that will ultimately aid attempts to uncover the transcription events and the factors that coordinate the needs of mitochondria with nuclear genome expression. Studies on thermal stress (Lund et al., 1998, 2001; Pratt et al., 1994; Wood et al., 1998) and oxidative stresses (Sweetlove et al., 2002; Taylor et al., 2002) reveal changes in both the level of specific protein expression and the posttranslational modification of mitochondrial protein functions. These studies also largely reveal that the mitochondrial proteome is a robust entity with the number of dramatic changes to protein abundance being relatively small (10-30 proteins) in most cases. Comparison of proteomes between male-sterile and -fertile plant lines have identified a small number of differences, but these are yet to be functionally linked to cause of fertility loss (Gutierres et al., 1997; Mihr et al., 2001; Witt et al., 1997; Zhihai and Zhiyi, 1991). The NCS (nonchromosomal stripe) mutant of maize carrying a deletion of an NADH dehydrogenase complex subunit has been shown to affect expression of other subunits and assembly of the complex (Karpova and Newton, 1999). A mutation in splicing at a specific mitochondrial intron leads to the absence of several mitochondrial-encoded proteins following proteome analysis (Brangeon et al., 2000). Further work looking at gene disruptions in *Arabidopsis* provides a large avenue for future mitochondrial proteome research aimed at determining functional links among mitochondrial proteins.

FUTURE DIRECTIONS

The immediate future for plant mitochondrial proteomics, as with its other plant proteomic counterparts, is to catch up in terms of technology utilization to the rapidly expanding options afforded by recent developments. Thus, immediate musts are the utilization of posttranslation modification identification techniques using mass spectrometry, the exploitation of ICAT and other quantitative nongel techniques, and the more extensive use of multiphase MudPIT detection to further increase sensitivity in mitochondrial protein detection. Mitochondrial proteomics is still very much in a primary discovery phase and this will largely be driven by technology. However, at the same time, a series of biology questions can and should be asked of this data stream to advance our understanding of mitochondria—both in terms of their history and origins and in terms of their regulation today. Much of our current understanding of the proteome is binary presence or absence, but to consider regulation we need more quantitative data. Considering the link between transcript abundances (inferred from microarray data or EST pool sizes) and final proteome abundances can direct attempts to find examples of posttranscriptional regulation of nuclear-encoded mitochondrial gene expression. Linking functional categorization and protein abundances also opens the potential to map biochemical pathways and model fluxes of a whole mitochondrial metabolic network. Linking information on nuclear-encoded mitochondrial proteins to the genes that encoded them forms the platform for identifying common expression patterns to identify common promoters and/or promoter elements. These are the footprints of the master control switches that define the biogenesis of this organelle.

REFERENCES

Abdallah, F., Salamini, F., Leister, D. (2000). A prediction of the size and evolutionary origin of the proteome of chloroplasts of *Arabidopsis*. *Trends Plant Sci.* 5: 141-142.

Adams, K.L., Daley, D.O., Qiu, Y.L., Whelan, J., Palmer, J.D. (2000). Repeated, recent and diverse transfers of a mitochondrial gene to the nucleus in flowering plants. *Nature* 408: 354-357.

Altschul, S.F., Madden, T.L., Schaffer, A.A., Zhang, J., Zhang, Z., Miller, W., Lipman, D.J. (1997). Gapped BLAST and PSI-BLAST: A new generation of protein database search programs. *Nucleic Acids Res.* 25: 3389-3402.

Arabidopsis Genome Initiative (2000). Analysis of the genome sequence of the flowering plant *Arabidopsis thaliana*. *Nature* 408: 796-815.

Asamizu, E., Nakamura, Y., Sato, S., Tabata, S. (2000). A large scale analysis of cDNA in *Arabidopsis thaliana:* Generation of 12,028 non-redundant expressed sequence tags from normalized and size-selected cDNA libraries. *DNA Res.* 7: 175-180.

Bannai, H., Tamada, Y., Maruyama, O., Nakai, K., Miyano, S. (2002). Extensive feature detection of N-terminal protein sorting signals. *Bioinformatics* 18: 298-305.

Banzet, N., Richaud, C., Deveaux, Y., Kazmaier, M., Gagnon, J., Triantaphylidès, C. (1998). Accumulation of small heat shock proteins, including mitochondrial HSP22, induced by oxidative stress and adaptive response in tomato cells. *Plant J.* 13: 519-527.

Bardel, J., Louwagie, M., Jaquinod, M., Jourdain, A., Luche, S., Rabilloud, T., Macherel, D., Garin, J., Bourguignon, J. (2002). A survey of the plant mitochondrial proteome in relation to development. *Proteomics* 2: 880-898.

Barent, R.L., Elthon, T.E. (1992). Two-dimensional gels: An easy method for large quantities of proteins. *Plant Mol. Biol. Rep.* 10: 338-344.

Booz, M.L., Travis, R.L. (1980). Electrophoretic comparison of polypeptides from enriched plasma membrane fractions from developing soybean roots. *Plant Physiol.* 66: 1037-1043.

Boutry, M., Briquet, M., Goffeau, A. (1983). The α-subunit of a plant mitochondrial F_1-ATPase is translated in mitochondria. *J. Biol. Chem.* 258: 8524-8526.

Brangeon, J., Sabar, M., Gutierres, S., Combettes, B., Bove, J., Gendy, C., Chetrit, P., Des Francs-Small, C.C., Pla, M., Vedel, F., De Paepe, R. (2000). Defective splicing of the first nad4 intron is associated with lack of several complex I subunits in the *Nicotiana sylvestris* NMS1 nuclear mutant. *Plant J.* 21: 269-280.

Braun, H.P., Emmermann, M., Kruft, V., Schmitz, U.K. (1992). The general mitochondrial processing peptidase from potato is an integral part of cytochrome c reductase of the respiratory chain. *EMBO J.* 11: 3219-3227.

Braun, H.P., Schmitz, U.K. (1999). The protein import apparatus of plant mitochondria. *Planta* 209: 267-274.

Brennicke, A., Marchfelder, A., Binder, S. (1999). RNA editing. *FEMS Microbiol. Rev.* 23: 297-316.

Bykova, N.V., Egsgaard, H., Møller, I.M. (2003). Identification of 14 new phosphoproteins involved in important plant mitochondrial processes. *FEBS Lett.* 540: 141-146.

Claros, M.G., Vincens, P. (1996). Computational method to predict mitochondrially imported proteins and their targeting sequences. *Eur. J. Biochem.* 241: 779-786.

Colas des Francs-Small, C., Ambard-Bretteville, F., Darpas, A., Sallantin, M., Huet, J.C., Pernollet, J.C., Rémy, R. (1992). Variation of the polypeptide composition of mitochondria isolated from different potato tissues. *Plant Physiol.* 98: 273-278.

Colas des Francs-Small, C., Ambard-Bretteville, F., Small, I.D., Rémy, R. (1993). Identification of a major soluble protein in mitochondria from nonphotosynthetic tissues as NAD-dependant formate dehydrogenase. *Plant Physiol.* 102: 1171-1177.

Day, D.A., Neuburger, M., Douce, R. (1985). Biochemical characterization of chlorophyll-free mitochondria from pea leaves. *Aust. J. Plant Physiol.* 12: 219-228.

De Paepe, R., Forchioni, A., Chétrit, P., Vedel, F. (1993). Specific mitochondrial proteins in pollen: Presence of an additional ATP synthase subunit. *Proc. Natl. Acad. Sci. USA* 90: 5934-5938.

Diano, M. (1982). Electrophoretic comparison of mitochondrial polypeptides from maize lines susceptible and resistant to *Helminthosporium maydis* race T. *Plant Physiol.* 69: 1217-1221.

Douce, R., Bourguignon, J., Neuburger, M., Rebeille, F. (2001). The glycine decarboxylase system: A fascinating complex. *Trends Plant Sci.* 6: 167-176.

Douce, R., Christensen, E.L., Bonner Jr., W.D. (1972). Preparation of intact plant mitochondria. *Biochim. Biophys. Acta* 275: 148-160.

Duby, G., Boutry, M. (2002). Mitochondrial protein import machinery and targeting information. *Plant Sci.* 162: 477-490.

Emanuelsson, O., Nielsen, H., Brunak, S., von Heijne, G. (2000). Predicting subcellular localization of proteins based on their N-terminal amino acid sequence. *J. Mol. Biol.* 300: 1005-1016.

Emanuelsson, O., von Heijne, G. (2001). Prediction of organellar targeting signals. *Biochim. Biophys. Acta* 1541: 114-119.

Eubel, H., Jänsch, L., Braun H.P. (2003). New insights into the respiratory chain of plant mitochondria: Supercomplexes and a unique composition of complex II. *Plant Physiol.* 133: 274-286.

Feng, Z.P. (2001). An overview on predicting the subcellular location of a protein. *In Silico Biol.* 2: 0027, available at <http://www.bioinfo.de/isb/2002/02/0027/>.

Goff, S.A., Ricke, D., Lan, T.H., Presting, G., Wang, R., Dunn, M., Glazebrook, J., Sessions, A., Oeller, P., Varma, H., et al. (2002). A draft sequence of the rice genome *(Oryza sativa* L. ssp. *japonica). Science* 296: 92-100.

Gollub, J., Ball, C.A., Binkley, G., Demeter, J., Finkelstein, D.B., Hebert, J.M., Hernandez-Boussard, T., Jin, H., Kaloper, M., Matese, J.C., et al. (2003). The Stanford Microarray Database: Data access and quality assessment tools. *Nucleic Acids Res.* 31: 94-96.

Gray, M., Burger, G., Lang, F. (1999). Mitochondrial evolution. *Science* 283: 1476-1481.

Gutierres, S., Sabar, M., Lelandais, C., Chetrit, P., Diolez, P., Degand, H., Boutry, M., Vedel, F., de Kouchkovsky, Y., De Paepe, R. (1997). Lack of mitochondrial and nuclear-encoded subunits of complex I and alteration of the respiratory chain in *Nicotiana sylvestris* mitochondrial deletion mutants. *Proc. Natl. Acad. Sci. USA* 94: 3436-3441.

Gygi, S.P., Rist, B., Gerber, S.A., Turecek, F., Gelb, M.H., Aebersold, R. (1999). Quantitative analysis of complex protein mixtures using isotope-coded affinity tags. *Nat. Biotechnol.* 17: 994-999.

Hack, E., Leaver, C. (1983). The alpha-subunit of the maize F_1-ATPase is synthesized in the mitochondrion. *EMBO J.* 2: 1783-1790.

Heazlewood, J.L., Howell, K.A., Millar, A.H. (2003). Mitochondrial complex I from *Arabidopsis* and rice: Orthologs of mammalian and yeast components coupled to plant-specific subunits. *Biochim. Biophys. Acta—Bioenergetics.* 1604: 159-169.

Heazlewood, J.L., Howell, K.A., Whelan, J., Millar, A.H. (2003). Towards the analysis of the rice mitochondrial proteome. *Plant Physiol.* 132: 230-242.

Heazlewood, J.L., Tonti-Filippini, J.S., Gout, A., Day, D.A., Whelan, J., Millar, A.H. (2004). Experimental analysis of the *Arabidopsis* mitochondrial proteome highlights signalling and regulatory components, provides assessment of targeting prediction programs and points to plant specific mitochondrial proteins. *Plant Cell* 16: 241-256.

Heazlewood, J.L., Whelan, J., Millar, A.H. (2003). The products of the mitochondrial ORF25 and ORFB genes are F_O components of the plant F_1F_O ATP synthase. *FEBS Lett.* 540: 201-205.

Herald, V., Heazlewood, J.L., Day, D.A., Millar, A.H. (2003). Proteomic identification of divalent metal cation binding proteins in plant mitochondria. *FEBS Lett.* 537: 96-100.

Herbert, B. (1999). Advances in protein solubilisation for two-dimensional electrophoresis. *Electrophoresis* 20: 660-663.

Holm, R.H., Kennepohl, P., Solomon, E.I. (1996). Structural and functional aspects of metal sites in biology. *Chem. Rev.* 96: 2239-2314.

Hourton-Cabassa, C., Ambard-Bretteville, F., Moreau, F., Davy de Virville, J., Rémy, R., Colas des Francs-Small, C. (1998). Stress induction of mitochondrial formate dehydrogenase in potato leaves. *Plant Physiol.* 116: 627-635.

Hua, S., Sun, Z. (2001). Support vector machine approach for protein subcellular localization prediction. *Bioinformatics* 17: 721-728.

Huala, E., Dickerman, A.W., Garcia-Hernandez, M., Weems, D., Reiser, L., LaFond, F., Hanley, D., Kiphart, D., Zhuang, M., Huang, W., et al. (2001). The Arabidopsis Information Resource (TAIR): A comprehensive database and web-based information retrieval, analysis, and visualization system for a model plant. *Nucleic Acids Res.* 29: 102-105.

Humphrey-Smith, I., Colas des Francs-Small, C., Ambart-Bretteville, F., Rémy, R. (1992). Tissue-specific variation of pea mitochondrial polypeptides detected by computerized image analysis of two-dimensional electrophoresis gels. *Electrophoresis* 13: 168-172.

Jänsch, L., Kruft, V., Schmitz, U.K., Braun, H.P. (1996). New insights into the composition, molecular mass and stoichiometry of the protein complexes of plant mitochondria. *Plant J.* 9: 357-368.

Jenner, H.L., Winning, B.M., Millar, A.H., Tomlinson, K.L., Leaver, C.J., Hill, S.A. (2001). NAD malic enzyme and the control of carbohydrate metabolism in potato tubers. *Plant Physiol.* 126: 1139-1149.

Jung, E., Heller, M., Sanchez, J.C., Hochstrasser, D.F. (2000). Proteomics meets cell biology: The establishment of subcellular proteomes. *Electrophoresis* 21: 3369-3377.

Karpova, O.V., Newton, K.J. (1999). A partially assembled complex I in NAD4-deficient mitochondria of maize. *Plant J.* 17: 511-521.

Kruft, V., Eubel, H., Werhahn, W., Jänsch, L., Braun, H.P. (2001). Proteomic approach to identify novel mitochondrial functions in *Arabidopsis thaliana*. *Plant Physiol.* 127: 1694-1710.

Link, A.J. (1999). *2D Proteome Analysis Protocols*. Totowa, NJ: Humana Press.

Liska, A.J., Shevchenko, A. (2003). Expanding the organismal scope of proteomics: Cross-species protein identification by mass spectrometry and its implications. *Proteomics* 3: 19-28.

Lister, R., Mowday, B., Whelan, J., Millar, A.H. (2002). Zinc dependent intermembrane space proteins stimulate carrier import into plant mitochondria. *Plant J.* 30: 555-566.

Lister, R., Murcha, M.W., Whelan, J. (2003). The mitochondrial protein import machinery of plants (MPIMP) database. *Nucleic Acids Res.* 31: 325-327.

Lithgow, T. (2000). Targeting of proteins to mitochondria. *FEBS Lett.* 476: 22-26.

Lopez, M.F., Kristal, B.S., Chernokalskaya, E., Lazarev, A., Shestopalov, A.I., Bogdanova, A., Robinson, M. (2000). High-throughput profiling of the mitochondrial proteome using affinity fractionation and automation. *Electrophoresis* 21: 3427-3440.

Lund, A.A., Blum, P.H., Bhattramakki, D., Elthon, T.E. (1998). Heat-stress response of maize mitochondria. *Plant Physiol.* 116: 1097-1110.

Lund, A.A., Rhoads, D.M., Lund, A.L., Cerny, R.L., Elthon, T.E. (2001). In vivo modifications of the maize mitochondrial small heat stress protein, HSP22. *J. Biol. Chem.* 276: 29924-29929.

Maffey, L., Degand, H., Boutry, M. (1997). Partial purification of mitochondrial ribosomes from broad bean and identification of proteins encoded by the mitochondrial genome. *Mol. Gen. Genet.* 254: 365-371.

Mann, M., Hojrup, P., Roepstorff, P. (1993). Use of mass spectrometric molecular weight information to identify proteins in sequence databases. *Biol. Mass Spectrom.* 22: 338-345.

Mannella, C.A. (1985). The outer membrane of plant mitochondria. In Douce, R., Day, D.A. (Eds.), *Higher Plant Cell Respiration* (pp. 106-133). Berlin: Springer-Verlag.

Mannella, C.A., Bonner, W.D. (1975). Biochemical characteristics of the outer membranes of plant mitochondria. *Biochim. Biophys. Acta* 413: 213-225.

Mathesius, U., Imin, N., Chen, H., Djordjevic, M.A., Weinman, J.J., Natera, S.H., Morris, A.C., Kerim, T., Paul, S., Menzel, C., et al. (2002). Evaluation of proteome reference maps for cross-species identification of proteins by peptide mass fingerprinting. *Proteomics* 2: 1288-1303.

Mathesius, U., Keijzers, G., Natera, S.H., Weinman, J.J., Djordjevic, M.A., Rolfe, B.G. (2001). Establishment of a root proteome reference map for the model legume *Medicago truncatula* using the expressed sequence tag database for peptide mass fingerprinting. *Proteomics* 1: 1424-1440.

McDonald, W.H., Ohi, R., Miyamoto, D.T., Mitchison, T.J., Yates III, J.R. (2002). Comparison of three directly coupled HPLC MS/MS strategies for identification of proteins from complex mixtures: Single-dimension LC-MS/MS, 2-phase MudPIT, and 3-phase MudPIT. *Int. J. Mass Spectrom.* 219: 245-251.

Mihr, C., Baumgärtner, M., Dieterich, J.H., Schmitz, U.K., Braun, H.P. (2001). Proteomic approach for investigation of cytoplasmic male sterility (CMS) in *Brassica*. *J. Plant Physiol.* 158: 787-794.

Millar, A.H., Heazlewood, J.L. (2003). Genomic and proteomic analysis of mitochondrial carrier proteins in *Arabidopsis thaliana*. *Plant Physiol.* 131: 443-453.

Millar, A.H., Liddell, A., Leaver, C.J. (2001). Isolation and subfractionation of mitochondria from plants. In Pon, L.A., Schon, E.A. (Eds.), *Mitochondria* (pp. 53-74). San Diego, CA: Academic Press.

Millar, A.H., Sweetlove, L.J., Giege, P., Leaver, C.J. (2001). Analysis of the *Arabidopsis* mitochondrial proteome. *Plant Physiol.* 127: 1711-1727.

Murcha, M.W., Lister, R., Ho, A.Y.Y., Whelan, J. (2003). Identification, expression and import of translocase of the inner mitochondrial membrane components 17 and 23 from *Arabidopsis thaliana*. *Plant Physiol.* 131: 1737-1747.

Nakai, K., Horton, P. (1999). PSORT: A program for detecting sorting signals in proteins and predicting their subcellular localization. *Trends Biochem. Sci.* 24: 34-36.

Neuburger, M. (1985). Preparation of plant mitochondria, criteria for assessment of mitochondrial integerity, and purity, survival in vitro. In Douce, R., Day, D.A. (Eds.), *Higher Plant Cell Respiration* (pp. 7-24). Berlin: Springer-Verlag.

Neuburger, M., Journet, E., Bligny, R., Carde, J., Douce, R. (1982). Purification of plant mitochondria by isopycnic centrifugation in density gradients of percoll. *Arch. Biochem. Biophys.* 217: 312-323.

Peeters, N., Small, I. (2001). Dual targeting to mitochondria and chloroplasts. *Biochim. Biophys. Acta* 1541: 54-63.

Pical, C., Fredlund, K.M., Petit, P.X., Sommarin, M., Møller, I.M. (1993). The outer membrane of plant mitochondria contains a calcium-dependent protein kinase and multiple phosphoproteins. *FEBS Lett.* 336: 347-351.

Pinel, C., Douce, R., Mache, R. (1986). A study of mitochondrial ribosomes from the higher plant *Solanum tuberosum* L. *Mol. Biol. Rep.* 11: 93-97.

Pratt, J.R., Wood, C.K., Moore, A.L. (1994). Differential expression of proteins from plant mitochondria subjected to thermal stress. *Biochem. Soc. Trans.* 22: 405S.

Pratt, J.R., Wood, C.K., Moore, A. (1995). Immunochemical characterization of mitochondrial proteins isolated from *Pisum sativum,* subjected to thermal stress and age development. *Biochem. Soc. Trans.* 23: 178S.

Rasmusson, A.G., Svensson, A.S., Knoop, V., Grohmann, L., Brennicke, A. (1999). Homologues of yeast and bacterial rotenone-insensitive NADH dehydrogenases in higher eukaryotes: Two enzymes are present in potato mitochondria. *Plant J.* 20: 79-87.

Rébeillé, F., Macherel, D., Mouillon, J.M., Garin, J., Douce, R. (1997). Folate biosynthesis in higher plants: Purification and molecular cloning of a bifunctional 6-hydroxymethyl-7,8-dihydropterin pyrophosphokinase/7,8-dihydropteroate synthase localized in mitochondria. *EMBO J.* 16: 947-957.

Remy, R., Ambard-Bretteville, F., Colas des Frances, C. (1987). Analysis of two-dimensional gel electrophoresis of the polypeptide composition of pea mitochondria isolated from different tissues. *Electrophoresis* 8: 528-532.

Rios, R., de Buyser, J., Henry, Y., Ambard-Bretteville, F., Remy, R. (1991). Two-dimensional electrophoretic comparison of mitochondrial polypeptides from different wheat (*Triticum aestivum* L.) tissues. *Plant Sci.* 76: 159-166.

Rötig, A., Chauveau, M. (1987). Polypeptide composition of potato tuber mitochondria and submitochondrial fractions analysed by one- and two-dimensional gel electrophoresis. *Plant Physiol. Biochem.* 25: 11-20.

Santoni, V., Rabilloud, T., Doumas, P., Rouquie, D., Mansion, M., Kieffer, S., Garin, J., Rossignol, M. (1999). Towards the recovery of hydrophobic proteins on two-dimensional electrophoresis gels. *Electrophoresis* 20: 705-711.

Schägger, H., von Jagow, G. (1991). Blue native electrophoresis for isolation of membrane protein complexes in enzymatically active form. *Anal. Biochem.* 199: 223-231.

Schon, E.A. (2001). Gene products present in mitochondria of yeast and animal cells. In Pon, L.A., Schon, E.A. (Eds.), *Mitochondria* (pp. 463-482). San Diego, CA: Academic Press.

Snedden, W.A., Koutsia, N., Baum, G., Fromm, H. (1996). Activation of a recombinant petunia glutamate decarboxylase by calcium/calmodulin or by a monoclonal antibody which recognizes the calmodulin binding domain. *J. Biol. Chem.* 271: 4148-4153.

Sommarin, M., Petit, P.X., Møller, I.M. (1990). Endogenous protein phosphorylation in purified plant mitochondria. *Biochim. Biophys. Acta* 1052: 195-203.

Struglics, A., Fredlund, K.M., Møller, I.M., Allen, J.F. (1998). Two subunits of the F_0F_1-ATPase are phosphorylated in the inner mitochondrial membrane. *Biochem. Biophys. Res. Commun.* 243: 664-668.

Sweetlove, L.J., Heazlewood, J.L., Herald, V., Holtzapffel, R., Day, D.A., Leaver, C.J., Millar, A.H. (2002). The impact of oxidative stress on *Arabidopsis* mitochondria. *Plant J.* 32: 891-904.

Sweetlove, L.J., Mowday, B., Hebestreit, H.F., Leaver, C.J., Millar, A.H. (2001). Nucleoside diphosphate kinase III is localized to the inter-membrane space in plant mitochondria. *FEBS Lett.* 16: 272-276.

Taylor, N.L., Day, D.A., Millar, A.H. (2002). Environmental stress causes oxidative damage to plant mitochondria leading to inhibition of glycine decarboxylase. *J. Biol. Chem.* 277: 42663-42668.

Thelen, J.J., Miernyk, J.A., Randall, D.D. (2000). Pyruvate dehydrogenase kinase from *Arabidopsis thaliana:* A protein histidine kinase that phosphorylates serine residues. *Biochem. J.* 349: 195-201.

Vanlerberghe, G.C., McIntosh, L. (1997). Alternative oxidase: From gene to function. *Ann. Rev. Plant Physiol. Plant Mol. Biol.* 48: 703-734.

Washburn, M.P., Wolters, D., Yates III, J.R. (2001). Large scale analysis of the yeast proteome by multidimensional protein identification technology. *Nat. Biotechnol.* 19: 242-247.

Werhahn, W., Braun, H.P. (2002). Biochemical dissection of the mitochondrial proteome from *Arabidopsis thaliana* by three-dimensional gel electrophoresis. *Electrophoresis* 23: 640-646.

Werhahn, W., Niemeyer, A., Jänsch, L., Kruft, V., Schmitz, U.K., Braun, H.P. (2001). Purification and characterization of the preprotein translocase of the

outer mitochondrial membrane from *Arabidopsis thaliana:* Identification of multiple forms of TOM20. *Plant Physiol.* 125: 943-954.
Whelan, J. (1999). Plant mitochondrial protein import: Mechanisms and control. *Aust. J. Plant Physiol.* 26: 725-732.
Wissinger, B., Schuster, W., Brennicke, A. (1991). Trans-splicing in *Oenothera* mitochondria: nad1 mRNAs are edited in exon and trans-splicing group II intron sequences. *Cell* 65: 473-482.
Witt, U., Lührs, R., Buck, F., Lembke, K., Grünegerg-Seiler, M., Abel, W. (1997). Mitochondrial malate dehydrogenase in *Brassica napus:* Altered protein patterns in different nuclear mitochondrial combinations. *Plant Mol. Biol.* 35: 1015-1021.
Wood, C.K., Pratt, J.R., Moore, A.L. (1998). Identification and characterization of cultivar-specific 22-kDa heat shock proteins from mitochondria of *Pisum sativum. Physiol. Plant.* 103: 369-376.
Yates III, J.R. (1998). Database searching using mass spectrometry data. *Electrophoresis* 19: 893-900.
Yu, J., Hu, S., Wang, J., Wong, G.K., Li, S., Liu, B., Deng, Y., Dai, L., Zhou, Y., Zhang, X., et al. (2002). A draft sequence of the rice genome *(Orysa sativa* L. ssp *indica). Science* 296: 79-92.
Yuan, Q., Ouyang, S., Liu, J., Suh, B., Cheung, F., Sultana, R., Lee, D., Quackenbush, J., Buell, C.R. (2003). The TIGR rice genome annotation resource: Annotating the rice genome and creating resources for plant biologists. *Nucleic Acids Res.* 31: 229-233.
Zhihai, S., Zhiyi, L. (1991). Electrophoretic comparative study on polypeptides of mitochondria from cytoplasmic male sterile lines and their maintainers in wheat. *Chin. J. Genet.* 18: 7-13.

SECTION IV: PATHWAYS AND PROCESSES

Chapter 16

Functional Characterization of the Photosynthetic Apparatus in *Arabidopsis thaliana*

Henrik Vibe Scheller
Christina Lunde
Anna Haldrup
Poul Erik Jensen

INTRODUCTION

Photosynthesis is probably the most unique character that distinguishes plants from other multicellular organisms. The photosynthetic apparatus obviously includes the structural proteins involved in capturing light energy and converting it to chemical energy in the form of carbohydrate. In addition, the photosynthetic apparatus includes factors involved in building and maintaining the structural membrane network, the thylakoids, and in assembling and turning over the photosynthetic proteins. In some of these contexts photosynthesis is not clearly separated from other processes in the plant. Nevertheless, it can be estimated that about 10 percent of the genes in a plant are involved in photosynthesis. Many of these genes have origins in cyanobacteria, but it is also clear that plants have recruited a large number of additional proteins (Abdallah et al., 2000; Pesaresi et al., 2001).

Photosynthesis has been extensively studied at all levels for many years by a large scientific community. Traditionally, species such as spinach, pea, and barley have been the most popular experimental plant species for photosynthesis research. Such common crop plants are easy to obtain or to grow in large quantities suitable for classical biochemical studies. Only more recently has *Arabidopsis thaliana* become a common species for photosynthesis research. Biochemically, *Arabidopsis* has some drawbacks due to small size and content of glucosinolates, which form free radicals upon breakdown, but these problems can usually be solved and are outweighed by the advantages offered by molecular genetics. This chapter will focus on studies carried out with *Arabidopsis,* but it should be stressed that both cyanobacteria and green algae remain very important models for photosynthesis research. Especially, *Chlamydomonas reinhardtii* is an excellent eukaryotic model organism with a photosynthetic apparatus that in many respects is similar to that of plants (see Chapter 8). Since it is the intention of this chapter to focus on *Arabidopsis,* the reader should be aware that literature on other species is referenced only when it is particularly relevant for the understanding of plant photosynthesis and no parallel studies with *Arabidopsis* are available. The focus will also be limited to the function of the proteins in the thylakoid membrane that are responsible for energy capture and electron transport. Thus, we will not discuss the stromal processes of carbon dioxide assimilation and also not the chloroplast import and processing apparatus.

OVERVIEW OF ELECTRON TRANSPORT

The light-capturing system in plants comprises a large number of different chlorophyll-containing proteins. Some chlorophyll is contained in the core of the photosystems, photosystem I (PSI) and II (PSII), but the majority of the pigments are contained in special light-harvesting proteins, most of which are associated either with PSII or with PSI. Oxygenic photosynthesis requires the activity of PSII for splitting of water (Figure 16.1). Electrons originating from water oxidation are eventually passed on to plastoquinone (PQ) bound to PSII and released as plastoquinol (PQH_2) in the thylakoid membrane. Reoxidation of PQH_2 is carried out by the cytochrome (Cyt) *b/f* complex, which reduces plastocyanin in the thylakoid lumen. Plastocyanin is reoxidized by photooxidized PSI, which in turn reduces ferredoxin in the stroma. The reduction of $NADP^+$ by ferredoxin is catalyzed by ferredoxin-$NADP^+$ reductase (FNR). During this electron transport from water to $NADP^+$, protons are vectorially transferred from the stroma to the lumen at several points. First, the water oxidation leads to release of H^+ in the lumen. Second, the reduction of PQ is accompanied by uptake of protons from the stroma while the reoxidation by Cyt *b/f* results in release of the protons into the lumen. The Cyt *b/f* complex appears to be capable of Q-cycling, which leads to transport of additional protons during the PQH_2 oxidation process. The electrochemical gradient generated by proton transport into the lumen is utilized by the ATP synthase to generate ATP.

PSII, PSI, Cyt *b/f,* and ATP synthase together with the associated light-harvesting proteins catalyze the main activities of the thylakoid membrane.

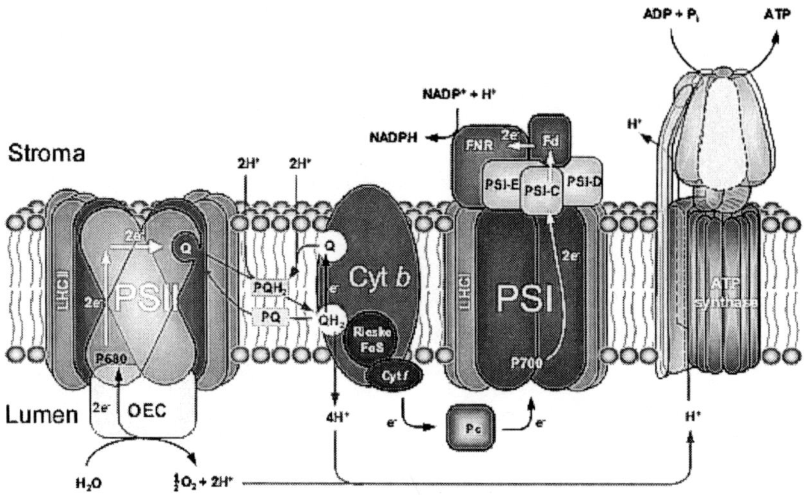

FIGURE 16.1. Schematic illustration of the linear electron transport in the thylakoid membranes. (*Source:* Courtesy of Lars Bredmose.)

However, in addition, several alternative electron transport pathways are responsible for chlororespiration and cyclic electron transport.

PHOTOSYSTEM II

PSII is a pigment-protein complex containing at least 28 subunits in higher plants (Table 16.1). The subunits are encoded by genes designated *psbA, psbB,* etc., but the subunit nomenclature is somewhat confusing, since many of the subunits traditionally have special designations. "PsbG" has turned out not to be a PSII protein at all (Nixon et al., 1989). A further complication is that the composition of PSII differs between prokaryotes and eukaryotes; for example, PsbU and PsbV are present in only prokaryotes, while PsbP and PsbQ are present in only eukaryotes. PsbT has been used for two different subunits, one encoded by the plastid *psbTc* gene and the other by the nuclear *psbTn* gene. The corresponding subunits should be called PsbTc and PsbTn to avoid confusion. Unfortunately, the composition of PSII has not been reviewed recently. An extensive review of PSII composition is outside the scope of this chapter, but a summary of the subunits has been included in Table 16.1. Six different peripheral antenna proteins (Lhcb1 to Lhcb6) also associate with PSII. Lhcb1, Lhcb2, and

TABLE 16.1. Subunit composition of PSII.

Gene	Gene location[a]	Protein	Molecular mass (Da)[b]	Function[c]	Accession number
psbA	C	D1	38,936	Electron transport	ATCG00020
psbB	C	CP47	56,037	Light harvesting	ATCG00680
psbC	C	CP43	51,867	Light harvesting	ATCG00280
psbD	C	D2	39,547	Electron transport	ATCG00270
psbE	C	Cyt b_{559}	9,396	Alpha-subunit of Cyt b559	ATCG00580
psbF	C	Cyt b_{559}	4,395 (MS)	Beta-subunit of Cyt b559	ATCG00570
(psbG)	C	(Ndh subunit K)		Not a PSII subunit!	ATCG00430
psbH	C	PsbH, 10 kDa phospho-protein	7,571		ATCG00710
psbI	C		4,196 (MS)		ATCG00080
psbJ	C		4,117		ATCG00550
psbK	C		4,235		ATCG00070
psbL	C		4,339		ATCG00560
psbM	C		3,783		ATCG00220
psbN	C		4,722		ATCG00700
psbO	N2	OEC 33 kDa protein	26,549/ 26,555	Stabilization of the oxygen-evolving complex	AT5G66570 AT3G50820
psbP	N2	OEC 23 kDa protein	20,201/ 21,180	Stabilization of the oxygen-evolving complex	AT1G06680 AT2G30790
psbQ	N2	OEC 16 kDa protein	16,301/ 16,341	Stabilization of the oxygen-evolving complex	AT4G21280 AT4G05180
psbR	N	10 kDa protein	10,336	Associated with OEC	AT1G79040
psbS	N	PsbS, CP22	21,560	Energy-dependent quenching	AT1G44575

psbT, psbTc	C	PsbT	3,850 (MS)		ATCG00690
psbT, psbTn	N	PsbTn	3,172		AT3G21055
psbU	0	PsbU	11,639		TLL2409
psbV	0	Cyt c550	15,122		TLL1285
psbW	N	PsbW	6,005		AT2G30570
psbX	0	PsbT, 5 kDa protein	5,232		NP_682803
psbY	N	PsbY-1	4,673		AT1G67740
psbY	N	PsbY-2	4,892		AT1G67740
psbZ	0	PsbZ	14,819		NP_485301
lhcb1	N5	LHCII subunit, Lhcb1	24,719/ 24,849/ 24,860	Light harvesting	AT1G29920 AT1G29910 AT1G29930 AT2G34430 AT2G34420
lhcb2	N3	LHCII subunit, Lhcb2	24,889/ 24,891	Light harvesting	AT2G05100 AT2G05070 AT2G27690
lhcb3	N	Lhcb3	24,266	Light harvesting	AT5G54270
lhcb4	N3	CP29	27,298/ 28,146/ 28,154/	Light harvesting	AT5G01530 AT3G08940 AT2G40100
lhcb5	N	CP26	26,043	Light harvesting	AT4G10340
lhcb6	N	CP24	23,093	Light harvesting	AT1G15820

[a]The location of the gene in nuclear (N) or plastid (C) genome in *Arabidopsis* is indicated. When more than one isoform is present in *Arabidopsis,* the number of isoforms is indicated. A "0" indicates that the gene is not present in *Arabidopsis.*

[b]The predicted molecular mass of the mature *Arabidopsis* protein is indicated, except when a mass determined by mass spectrometry (MS) is indicated in the SwissProt database (http://us.expasy.org). For proteins that are not present in *Arabidopsis,* the molecular mass for *Synechococcus elongatus* (PsbX, PsbU, and PsbV) or Anabaena 7120 (PsbZ) is indicated. The predicted masses are for the pure polypeptides and do not include the mass of cofactors and secondary modifications, except for the processing of signal and transit peptides. Cleavage sites were predicted by alignment with homologous sequences with known cleavage sites from other species. For the Lhcb proteins, cleavage sites were from Jansson (1999).

[c]The indicated function is not an extensive list of all the roles for the subunits but merely an indication of the major function.

Lhcb3 form the LHCII trimers, some of which are not associated with only PSII. In contrast, Lhcb4 (CP29), Lhcb5 (CP26), and Lhcb6 (CP24) are genuine PSII subunits with an apparently fixed stoichiometric relationship with the rest of PSII. Most of the PSII subunits are encoded in the chloroplast genome. The chloroplast-encoded subunits include the core D1/D2 heterodimer and the intrinsic Chl *a*-antenna made up of the CP43 and CP47 proteins. Also, the Cyt b_{559} subunits are encoded in the chloroplast. Hence, the nuclear-encoded subunits have accessory roles in the fine-tuning of PSII function and as light-harvesting proteins. The distinction between the chloroplast-encoded and the nuclear-encoded subunits is important for functional genomics. Mutants in chloroplast genes are not easily isolated in plants, even if a screening method is available. Insertional inactivation of plastid genes by biolistic transformation is a powerful method, which has given substantial insight into the function of the chloroplast. However, for some reason this technique is—at least so far—limited to few species, notably *Chlamydomonas* and tobacco. A significant effort has been made to expand the technique to work with *Arabidopsis*, but this has not been successful. Therefore, we have little specific knowledge about the chloroplast-encoded PSII proteins in *Arabidopsis*. With regard to the nuclear-encoded subunits, the situation is very different. Mutagenesis and insertional inactivation as well as RNA interference (RNAi) and related methods work well in *Arabidopsis* (see Chapters 2 and 3) and have been applied to the investigation of PSII.

The PSII Core Subunits

The oxygen-evolving complex of PSII is formed by the PsbO, PsbP, and PsbQ proteins. In *Arabidopsis* there are two *psbO* genes. A mutant with increased chlorophyll fluorescence (high F_0 fluorescence) was isolated after mutagenesis with EMS (Murakami et al., 2002). This mutant turned out to be knocked out in one of the *psbO* genes, resulting in thylakoid membranes with only one of the isoforms. Judging from the immunoblot presented in the paper, the mutant contained about 30 to 50 percent of total PsbO compared to the wild type. Maximum quantum yield of PSII electron transport was only about 0.5 to 0.6 as compared to 0.8 in the wild type, and the oxygen evolution determined in isolated thylakoid membranes was decreased by about 58 percent. The most likely explanation of the results seems to be that only a fraction of the PSII complexes in the mutant contain PsbO and are functional, while the remaining complexes contain inactive PSII without PsbO. The possibility that the two isoforms of PsbO have different functions cannot be excluded, but, on the other hand, having two *psbO* genes is not a conserved trait and appears to be the result of a recent gene duplication in *Arabidopsis* (Murakami et al., 2002).

The most-studied PSII subunit in *Arabidopsis* is PsbS. However, while this subunit can be isolated with PSII, it is not evident that it is present only as a component of PSII. Furthermore, the PsbS subunit has a crucial role in quenching of excess light energy, and, therefore, it is discussed later together with components of the xanthophyll cycle.

The function of the PsbW protein has been studied in *Arabidopsis* using an antisense approach (Shi et al., 2000). The content of the PsbW protein was reduced to below 4 percent of wild-type levels. A destabilization of PSII dimers and possibly a more general destabilization of PSII accompanied this decrease. Hence, the thylakoid membranes in plants lacking PsbW had a substantial decrease in PSII content and a corresponding decrease in PSII activity. In spite of a reduction of PSII activity with about 50 percent, no obvious growth phenotype was observed in the plants. Possibly, a phenotype would have been more apparent had the plants not been grown under conditions for optimal growth but, for example, at high light intensity.

The PsbY-1 and PsbY-2 proteins are a very unusual example of a polyprotein (Thompson et al., 1999). The two proteins have sequence similarity and are both encoded by the same gene, *psbY*. The single preprotein is imported into chloroplasts where it is processed into two integral membrane proteins with the same topology. After removal of the transit peptide, the N-terminal part of the intermediate protein serves as a thylakoid signal peptide for the PsbY-1 protein. However, the C-terminal protein, PsbY-2, is directed to its destination in the thylakoid membrane by an internal peptide in the intermediate protein. This internal peptide serves as a signal peptide and in fact appears to have resulted from duplication of the more conventional signal peptide in the N-terminus.

The role of PsbY-1 and PsbY-2 is not understood. The proteins appear to bind manganese, and they have also been reported to have arginine-metabolizing activity (Gau et al., 1998). However, the significance of these observations for in vivo function of the proteins is not clear.

The Lhcb Proteins

The six different Lhcb proteins (not counting isoforms) have been investigated in *Arabidopsis* by various means. Electron microscopy and single particle analysis has been used to determine the supercomplex structure of *Arabidopsis* PSII (Yakushevska et al., 2001). The structure is clearly dimeric, as has been found for other species. The electron microscopy shows that the outline of *Arabidopsis* PSII is similar to that found in other species. Hence, the Lhcb4, Lhcb5, and Lhcb6 subunits are monomeric and strongly bound to the core dimer. Even though the electron microscopy has a limited resolution, it was possible to locate the Lhcb5 subunit accurately in the complex by comparing PSII complexes from the wild type and a mutant lacking the subunit (see the following material; Yakushevska et al., 2003). Thus, Lhcb5 is located near CP43 and the strongly bound LHCII trimer. The Lhcb1, Lhcb2, and Lhcb3 proteins form LHCII trimers, and the PSII dimer contains two strongly bound LHCII trimers and one or two moderately bound trimers. Loosely bound trimers have been found in other species but were not detected in *Arabidopsis*. However, since the content of LHCII is not markedly different between *Arabidopsis* and other species, the absence of loosely bound trimers is most likely due to dissociation during preparation of PSII, rather than to a fundamental difference in the PSII composition. The composition of LHCII trimers is difficult to determine, but Jackowski et al. (2001) were able to isolate LHCII trimers and to separate them by nondenaturing isoelectric focusing, thereby showing the presence of $(Lhcb1)_2 \cdot Lhcb2$ trimers and Lhcb1·Lhcb2·Lhcb3 trimers. It is not clear if a particular type of trimer corresponds to the strongly, moderately, or loosely bound trimers that are observed by electron microscopy.

Functional analysis of the Lhcb proteins in *Arabidopsis* has been carried out with Lhcb4 (CP29) and Lhcb5 (CP26). The two proteins were down-regulated in plants with an antisense approach (Andersson et al., 2001). Down-regulation was very efficient, also in the case of Lhcb4, in spite of the three different genes encoding the isoforms. Apparently, the three genes are sufficiently similar to allow the simultaneous RNA interference of all three mRNAs. However, the plants lacking Lhcb4 or Lhcb5 were not affected in growth. Pleiotropic effects on other subunits were absent, except for a decrease in Lhcb6 (CP24) in the Lhcb4 antisense plants. Both types of plant showed a slight decrease in PSII quantum yield. On the other hand, the oxygen evolution per leaf area was not much affected and was even somewhat increased in Lhcb5 antisense plants under high light. However, the fact that oxygen evolution was not reduced should be related to the observation that PSII content was increased in the antisense plants. The PSII content was increased by about 30 percent in the Lhcb5 antisense plants and by 15 percent in the Lhcb4 antisense plants grown in high light (in the low-light-grown plants there was no difference from the wild type). Apparently, the ratio between PSI and PSII is very carefully optimized in plants, and mutants with defects in one photosystem have a large ability to compensate for the defect. This was also observed with PSI mutants.

Very recently, a functional study of Lhcb2 was reported (Andersson et al., 2003; Ruban et al., 2003). An antisense construct of one of the three *lhcb2* genes resulted in efficient down-regulation not only of the three Lhcb2 isoforms but also of the five Lhcb1 isoforms. Down-regulation of the eight proteins was shown to take place at the mRNA level. To our knowledge, no other study has proven the simultaneous suppression of so many gene products by this technique. Normally, Lhcb1 and Lhcb2 are part of the LHCII trimers, but in the transformants, trimers with these proteins were of course absent. The stacking of thylakoid membranes is thought to rely on interaction between LHCII proteins in adjacent stacks, and therefore it would have been expected that thylakoid structure was completely disturbed in the transformants. Surprisingly, this was not found; on the contrary, the thylakoid structure appeared normal (Andersson et al., 2003). Furthermore, PSII particles would also have been predicted to be much smaller in the absence of LHCII, but particles of normal size and shape were isolated from the transformants (Ruban et al., 2003). The explanation for these highly unexpected results was found to be the presence of trimers composed of Lhcb5 and Lhcb3. These trimers have not been found in the wild type, but in the transformants they appear to function very much like the normal LHCII trimers. Thus, the antenna size of PSII was only slightly reduced, and maximum quantum yield of PSII electron transport was reduced by only about 2 percent (Andersson et al., 2003). Photosynthetic rates were reduced by about 10 to 15 percent on a leaf area basis, and this was ascribed to a reduced chlorophyll content per leaf area rather than a functional deficiency in the pigment proteins present. A threonine residue in the N-terminus of Lhcb1 and Lhcb2 is reversibly phosphorylated during state transitions, but this threonine residue is not conserved in Lhcb5 or Lhcb3. In agreement with this, the plants lacking Lhcb1 and Lhcb2 did not show any state transitions. Under optimal and controlled growth conditions, the plants grew as well as the wild type. However, under field conditions the plants showed a significantly decreased fitness, with a seed set reduced up to 70 percent.

PSII Assembly and Turnover

Some PSII-deficient mutants have been isolated in *Arabidopsis* using forward genetic screens for high chlorophyll fluorescence, e.g., the *psbO* mutant described earlier (Meurer et al., 1996; Shikanai et al., 1999). However, in many cases, such mutants have turned out not to be affected in one of the structural subunits of PSII, but rather in nuclear factors required for the biogenesis or assembly of the PSII complex. A mutant with high chlorophyll fluorescence designated *hcf136* turned out to be completely and specifically devoid of PSII (Meurer et al., 1998). Apparently, synthesis of the PSII subunits is normal in this mutant, but the Hcf136 protein is somehow required for assembly of PSII. Another mutant, *hcf5*, was also completely devoid of PSII, but many pleiotropic effects were observed (Dinkins et al., 1997). A mutant designated *hcf107* is deficient in PSII and has been shown to be specifically defective in processing of the *psbH* transcript and in the translation of CP47 (Felder et al., 2001).

PSII is well known to be very sensitive to photodamage, which specifically damages the D1 subunit. The damaged D1 is exchanged in a repair cycle. This process has been extensively studied in a number of plants. *Arabidopsis* has not been the major object of these studies, but this species does not appear to differ in any fundamental way from other plants in the D1 photodamage and repair processes. The repair cycle involves proteolytic degradation of damaged D1, and this has recently been shown to involve a homologue of FtsH protease (Bailey et al., 2002). The *var2* mutant lacking the protease is much more sensitive to photoinhibition than the wild type due to the slow repair of damaged PSII.

THE CYTOCHROME **b/f** *COMPLEX*

The Cyt *b/f* complex provides the electronic connection between PSII and PSI. During the plastoquinol-plastocyanin oxido-reductase activity, the complex also pumps protons from the stroma to the lumen via the Q cycle (two protons per electron transferred to plastocyanin). The Cyt *b/f* complex is a dimer in the thylakoid membrane (Cramer et al., 1997) and contains nine subunits per monomer in higher-plant chloroplasts (Table 16.2). The complex contains four redox-active polypeptide subunits: Cyt *f*, Cyt b_6, the Rieske iron-sulfur protein, and FNR (Cramer et al., 1997; Zhang et al., 2001). The small hydrophobic subunits in the complex (PetG, -L, -M, and -N) might be necessary for assembly and/or stability of the complex, as has been demonstrated for PetG and PetN (Berthold et al., 1995; Hager et al., 1999). In addition to the cofactors, heme, and iron-sulfur clusters in the redox-active protein subunits, the Cyt *b/f* complex also contains a chlorophyll and a β-carotene molecule (Huang et al., 1994; Pierre et al., 1995; Zhang et al., 1999).

Unlike the situation for many subunits of PSI and PSII, it seems that most of the subunits of the Cyt *b/f* complex are interdependent, meaning that if one subunit is absent, then the others are not able to assemble or their biosynthesis is suppressed (Wollman et al., 1999, and references within). Thus, an *Arabidopsis* mutant, *hcf152,* which is affected in the processing of the *psbH-petB* part of the chloroplast *psbB-psbT-psbH-petB-petD* transcription unit and therefore lacks the Cyt b_6 protein, is defective in the entire Cyt *b/f* complex (Felder et al., 2001).

TABLE 16.2. Subunit composition of the Cyt *b/f* complex.

Gene	Gene location[a]	Protein	Molecular mass (Da)[b]	Function[c]	Accession number
petA	C	Cyt *f*	35,356	Cyt *f* apoprotein, a high-potential *c*-type cytochrome	ATCG00540
petB	C	Cyt b_6	24,152	Cyt b_6 apoprotein, a *b*-type cytochrome with two *b*-type hemes	ATCG00720
petC	N	Rieske iron-sulfur protein	19,000	Rieske Fe-S apoprotein, a high-potential 2Fe-2S protein	AT4G03280
petD	C	Subunit IV	17,430	Quinone binding at Qp	ATCG00730
petG	C	PetG	4,203		ATCG00600
petH	N4	Ferredoxin: $NADP^+$ oxido-reductase	35,274/ 35,289/ 35,314/ 35,774	FNR apoprotein	AT5G66190 AT1G20020 AT1G30510 AT5G05390
petL	C	PetL	3,400		ATCG00590
petM	N	PetM	4,765		AT2G26500
petN	C	PetN	3,169		ATCG00210

[a]The location of the gene in nuclear (N) or plastid (C) genome in *Arabidopsis* is indicated. When more than one isoform is present in *Arabidopsis,* the number of isoforms is indicated.

[b]The predicted molecular mass of the mature *Arabidopsis* protein is indicated. The predicted masses are for the pure polypeptides and do not include the mass of cofactors and secondary modifications, except for the processing of signal and transit peptides. Cleavage sites were predicted by alignment with homologous sequences with known cleavage sites from other species.

[c]The indicated function is not an extensive list of all the roles for the subunits but merely an indication of the major function.

The chloroplast-encoded subunits of the Cyt *b/f* complex have not been studied by reverse genetic in *Arabidopsis* because of the difficulties with chloroplast transformation in this species. Homoplasmic Δ*petN* tobacco plants displayed a photosynthetically incompetent phenotype that was completely devoid of the Cyt *b/f* complex, whereas both PSI and PSII were present and active (Hager et al., 1999).

The nuclear-encoded Rieske subunit has been studied using *Arabidopsis* mutants. A point mutation changing proline 194 to a leucine in *petC* encoding the Rieske subunit of the Cyt *b/f* complex has been identified as the cause for the *pgr1* (proton gradient regulation) mutation (Munekage et al., 2001). The *pgr1* mutant specifically lacks thermal dissipation of excess absorbed light energy, which was observed as little nonphotochemical quenching of chlorophyll fluorescence. At low light intensity, growth rate and electron transport rate were not affected in *pgr1,* whereas they were significantly restricted at high light intensity, indicating that the lumenal acidification was not sufficient to induce thermal dissipation. The results from this work suggest that thermal dissipation is precisely regulated by lumenal pH in order to maintain maximum photosynthetic activity. That *pgr1* plants were sensitive to changes in light conditions shows that maximum activity of the Cyt *b/f* complex is indispensable for short-term acclimation. Further analysis showed that electron transfer from Cyt *b/f* to PSI is not generally reduced in the *pgr1* mutant, but the pH dependence of this reaction is altered (Jahns et al., 2002).

A T-DNA-tagged *Arabidopsis* mutant defective in biogenesis of the Cyt b_6f complex has been isolated and characterized. This mutant *(hcf164)* shows a high chlorophyll fluorescence phenotype and accumulated less than 10 to 20 percent of the Cyt *b/f* complex (Lennartz et al., 2001). In vivo protein-labeling experiments indicated that the mutation acts posttranslationally by interfering with the assembly of the complex. Hcf164 encodes a thioredoxin-like protein that possesses disulfide reductase activity (Lennartz et al., 2001). The protein is anchored to the thylakoid membrane at its lumenal side. The authors suggest that Hcf164 is involved in the maturation of Cyt *f* probably by catalyzing the reduction of cysteine thiols of the heme-binding motif before covalent binding of the heme group. However, it cannot be ruled out that the protein has other functions during the biogenesis of the Cyt *b/f* complex, such as attachment of the heme group to Cyt b_6 (Lennartz et al., 2001).

Electrons are transferred from the Cyt *b/f* complex by the soluble proteins plastocyanin (Pc) or Cyt c_6. In higher plants, only Pc has been thought to be involved in mediating this electron transport, whereas both Pc and Cyt c_6 have been found in cyanobacteria and algae. The *Arabidopsis* genome contains two genes encoding for plastocyanin *(Pc1* and *Pc2)* that share a high degree of homology at the protein level.

Recently, Gupta et al. (2002) reported the characterization of a Cyt c_6-like protein from *Arabidopsis* (Atc6). Atc6 was shown to be a functional Cyt *c* localized in the thylakoid lumen. A homozygous Atc6 mutant did not display any visible phenotypic changes, suggesting that Atc6 was not essential for growth and development. To examine the function of Atc6 in a plastocyanin-null genetic background, the two *Arabidopsis Pc* genes were silenced using RNA interference (RNAi). The transgenic plants with the *Pc* RNAi construct produced normal amounts of Atc6 but had no detectable Pc and were significantly smaller than wild-type plants. Atc6 mutant plants transformed with the *Pc* RNAi construct contained no detectable Pc or Atc6, and the plants died soon after transfer from agar to soil. Thus, Pc complements the Atc6 mutant, whereas Atc6 only partially replaces Pc. However, analysis of *Arabidopsis* mutants in which both *Pc* genes have been disrupted showed that at least under normal physiological conditions Pc cannot be replaced by Atc6 (Weigel et al., 2003).

PHOTOSYSTEM I

The PSI complex in plants consists of 14 core subunits encoded by the genes *psaA, psaB,* etc. The corresponding subunits are designated PsaA or PSI-A, etc. (Table 16.3). Other nomenclature is rarely used, except that the heterodimer of PSI-A and PSI-B is sometimes known as Chlorophyll *a*-protein 1 or CP1. In addition to the core subunits, the PSI complex contains light-harvesting complexes known as LHCI. LHCI is comprised of four different subunits, Lhca1, Lhca2, Lhca3, and Lhac4, which usually appear to be present in equimolar amounts. In most studies, the stoichiometry between LHCI and PSI corresponds to eight Lhca proteins per P700. In spite of a general agreement about this, it should be noted that careful investigations have shown that the different Lhca proteins can vary independently under different growth conditions (Bailey et al., 2001). Hence, it may be that the composition of PSI-LHCI complexes is somewhat variable. The *Arabidopsis* genome contains genes encoding two proteins, Lhca5 and Lhca6, with similarity to the other Lhca proteins (Jansson, 1999). Lhca5 and Lhca6 are represented by cDNAs, and therefore the proteins are likely to be present in the thylakoid membranes. However, they have not yet been found. Possibly, the two proteins are minor components of PSI and they may be part of PSI only under certain growth conditions. A model of PSI in higher plants is shown in Figure 16.2.

Unlike the situation in PSII, only few of the PSI subunits are encoded in the plastid genome, i.e., subunits PSI-A, -B, -C, -I, and -J. The PSI-A, -B, and -C subunits form the central part of the PSI complex carrying all the electron acceptors. The PSI-A/PSI-B heterodimer is structurally similar to the core of PSII consisting of D1, D2, CP43, and CP47 (Jordan et al., 2001). A number of recent reviews describe the composition and structure of PSI

TABLE 16.3. Subunit composition of PSI.

Gene	Gene location[a]	Protein	Molecular mass (Da)[b]	Function[c]	Accession number
psaA	C	PSI-A	83,230	Electron transport, light harvesting	ATCG00350
psaB	C	PSI-B	82,475	Electron transport, light harvesting	ATCG00340
psaC	C	PSI-C	8,907	Electron transport	ATCG01060
psaD	N2	PSI-D	17,847/ 17,704	Ferredoxin docking	AT4G02770 AT1G03130
psaE	N2	PSI-E	10,399/ 10,476	Ferredoxin docking	AT4G28750 AT2G20260
psaF	N	PSI-F	17,295	Plastocyanin docking, LHCI-730 binding	AT1G31330
psaG	N	PSI-G	10,991	Modulation of PSI activity	AT1G55670
psaH	N2	PSI-H	10,356/ 10,356	State transitions	AT3G16140 AT1G52230
psaI	C	PSI-I	4,134		ATCG00510
psaJ	C	PSI-J	5,009		ATCG00630
psaK	N	PSI-K	8,462	Binding of Lhca2 and Lhca3	AT1G30380
psaL	N	PSI-L	17,970	State transitions	AT4G12800
psaM	(C)	PSI-M	3,424	Not present in *Arabidopsis*	P25903
psaN	N	PSI-N	9,705	Plastocyanin interaction	AT5G64040
psaO	N	PSI-O	10,105	State transitions	AT1G08380
psaX	0	PSI-X		Not present in *Arabidopsis*	P23320
lhca1	N	LHCI-730 subunit, Lhca1	21,544	Light harvesting	AT3G54890
lhca2	N	LHCI-680 subunit, Lhcb2	23,201	Light harvesting	AT3G61470
lhca3	N	LHCI-680 subunit, Lhcb3	24,937	Light harvesting	AT1G61520
lhca4	N	LHCI-730 subunit, Lhca4	22,255	Light harvesting	AT3G47470

lhca5	N	? Lhca5	24,289	AT1G45474
lhca6	N	? Lhca6	24,549	AT1G19150

[a]The location of the gene in nuclear (N) or plastid (C) genome in *Arabidopsis* is indicated. When more than one isoform is present in *Arabidopsis,* the number of isoforms is indicated. A "0" indicates that the gene is not present in *Arabidopsis. psaM* is present in the chloroplast genome of some plants (e.g., gymnosperms) but not in *Arabidopsis* and other angiosperms.

[b]The predicted molecular mass of the mature *Arabidopsis* protein is indicated, except when a mass determined by mass spectrometry (MS) is indicated in the SwissProt database (http://us.expasy.org). For PsaM and PsaX, which are not present in *Arabidopsis,* the molecular masses for *Synechococcus elongatus* PsaM and PsaX are indicated. The predicted masses are for the pure polypeptides and do not include the mass of cofactors and secondary modifications, except for the processing of signal and transit peptides. Cleavage sites were predicted by alignment with homologous sequences with known cleavage sites from other species. For the Lhca proteins, cleavage sites were from Jansson (1999).

[c]The indicated function is not an extensive list of all the roles for the subunits but merely an indication of the major function.

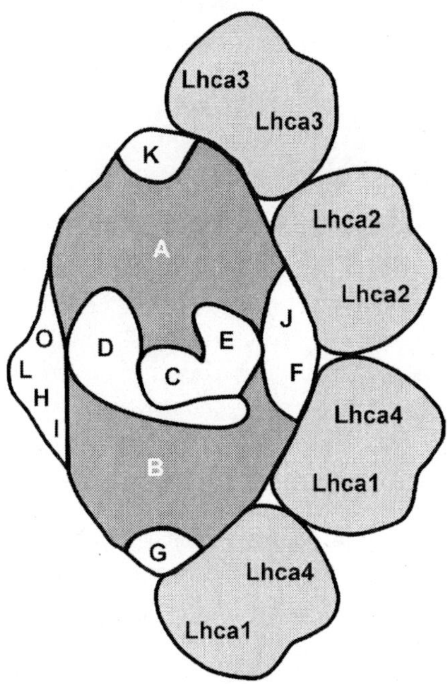

FIGURE 16.2. Schematic model of PSI viewed from the stroma side. The PSI-N subunit is on the lumenal side and not visible in this view.

(Chitnis, 2001; Fromme et al., 2001; Scheller et al., 2001). All the nuclear-encoded subunits of PSI have been investigated in *Arabidopsis* by antisense and cosuppression as well as by using insertional knockout plants.

The Stromal Subunits PSI-D and PSI-E

PSI-D has been known for a long time to interact with ferredoxin and to be necessary for the proper assembly of PSI. *Arabidopsis* plants lacking PSI-D completely have not been reported, but plants with a substantial decrease in the content of PSI-D have been generated using antisense technique (Haldrup et al., 2003). With only about 10 percent PSI-D present compared to the wild type, the plants are tiny and grow very slowly. Analysis of other PSI subunits showed a similar decrease in all the core subunits, while Lhca subunits were reduced to a lower extent. Considering past knowledge about PSI assembly, the most obvious conclusion is that in the absence of PSI-D, the PSI-C protein cannot be properly assembled with PSI and this causes a degradation of the incorrectly assembled complexes. Previous investigations have shown that in *Chlamydomonas* cells with a knockout of the *psaC* gene, the incomplete PSI complexes became unstable and were degraded (Takahashi et al., 1991). Thus, the *Arabidopsis* plants with only 10 percent PSI-D contain only about 10 percent of PSI, but all the PSI present has a normal composition. With this low amount of PSI, the plants can survive but naturally suffer from low photosynthetic rates and highly unbalanced redox conditions.

Two *psaE* genes are present in *Arabidopsis*. An insertional knockout of one of the genes has been reported (Varotto et al., 2000). These plants contain about 40 percent PSI-E compared to the wild type, and a 50 percent decrease in PSI-C and PSI-D was observed. The plants contain normal amounts of most other PSI subunits. The plants with less PSI-E exhibit higher fluorescence, i.e., lower photochemical quenching, indicating a disturbed redox balance of the intersystem chain. Under greenhouse conditions, the plants show a 50 percent decrease in growth rate. The plants contain LHCII in a highly phosphorylated form, presumably due to the more reducing state of the plastoquinone pool and hence more active LHCII kinase. Part of the LHCII was tightly associated with PSI, indicating that the plants to some degree are in a permanent state 2 (Pesaresi et al., 2002). In our laboratory, we have also generated *psaE* sense and antisense plants lacking 50 to 95 percent of the PSI-E protein (Figure 16.3). The most highly down-regulated PSI-E plants were severely affected in growth and produced only a few seeds (Figure 16.3). Consequently, the plants containing very low amounts of PSI-E are found only in the first generation, whereas

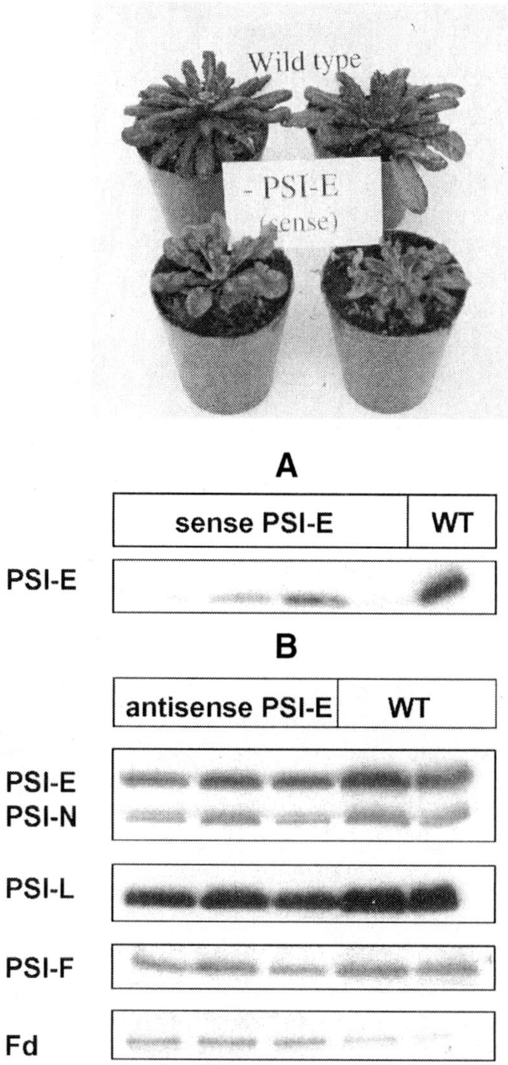

FIGURE 16.3. Wild type and plants with low amounts of PSI-E (5 to 10 percent), as shown by immunoblotting (A). The PSI-E-deficient plants are ten-week-old, first-generation plants obtained by cosuppression. The PSI-E-deficient plants are impaired in growth, exhibit delayed flowering, and show a light green leaf coloration. Cosuppressed plants lacking 40 to 60 percent of PSI-E were analyzed by immunoblotting for the effect on other PSI subunits and ferredoxin (B). The same amount of sample based on chlorophyll was loaded in each lane.

the second generation gave only plants with more than 50 percent PSI-E. The lack of PSI-E resulted in a general reduction of all the tested PSI subunits, whereas the electron acceptor ferredoxin was up-regulated (Figure 16.3).

The Lumenal Subunits, PSI-F and PSI-N

PSI-F has been suggested for a very long time to interact with plastocyanin, and this has been confirmed in a nuclear mutant of *Chlamydomonas* (Farah et al., 1995). *Arabidopsis* plants with very low amounts of PSI-F were generated by an antisense approach (Haldrup et al., 2000). These plants contain down to a few percent of PSI-F compared to the wild type. The plants with low amounts of PSI-F grow extremely slowly and flower only after many months. Recently, a knockout line that appears to have no PSI-F at all was obtained. Surprisingly, these plants grow at least as well as plants with about 5 percent PSI-F. The strong phenotype of *Arabidopsis* plants lacking PSI-F was somewhat surprising, since a growth or fluorescence phenotype was barely seen in *Chlamydomonas* except under high-light conditions (Farah et al., 1995). The functions of PSI-F in plants are several. Electron transport is slowed down significantly in the absence of PSI-F—almost certainly due to the poor interaction with plastocyanin (Haldrup et al., 2000). But PSI-F also appears to be involved in the efficient energy transfer from LCHI-730 (composed of Lhca1 and Lhca4) to P700, as detected by a significant blueshift in the 77 K fluorescence-emission peak from 734 nm in the wild type to 727 nm in the absence of PSI-F. This, together with a low amplitude of P700 absorption changes induced by far-red light excitation, strongly suggests that elimination of PSI-F destabilizes the interaction between the peripheral antenna and the PSI core. Furthermore, the PSI complex becomes more susceptible to photodamage in the absence of PSI-F. PSI lacking PSI-F has a normal subunit composition except for the absence of PSI-N.

PSI-N has also been investigated with a cosuppression approach (Haldrup et al., 1999). The plants lacking any detectable PSI-N were essentially indistinguishable from the wild type in terms of growth and photosynthesis. Careful analysis revealed a 50 percent decrease in the rate of plastocyanin oxidation, but the plants compensate for the lower activity of PSI by increasing the PSI-PSII ratio. Thus, plants lacking PSI-N have about 20 percent higher content of PSI on a chlorophyll basis. Under conditions of sufficient light, this difference does not result in any measurable decrease in photosynthesis. Under light-limited conditions, a 10 percent decrease in quantum yield of oxygen evolution was observed with plants acclimatized to a much higher light intensity. When the antisense-PSI-N plants were grown under very low light they did not show any growth defect. Probably, the plants adjust to any constant condition that they are exposed to, and a selective advantage of the PSI-N subunit may become apparent in only a fluctuating environment.

The PSI-H/PSI-L/PSI-O Cluster

In cyanobacteria, the PSI-L subunit faces the center of the PSI trimers, and in the absence of PSI-L, trimers cannot form (Chitnis and Chitnis, 1993). In plants, PSI trimers are apparently not present. Instead, PSI-L is associated with two other subunits that are not found in cyanobacteria, i.e., PSI-H and PSI-O. The interactions with these two subunits have been shown by cross-linking (Jansson et al., 1996; Zhang and Scheller, 2004). The function of the three subunits has been investigated in *Arabidopsis* plants lacking one of the subunits, generated either by antisense (PSI-L), cosuppression (PSI-H), or RNAi (PSI-O). The first to be reported were plants lacking PSI-H. These plants grow normally, and in the initial investigation, the only effect observed was a 40 percent decrease in the steady state activity of PSI, which was compensated by a 20 percent increase in PSI (Naver et al., 1999). Subsequently, it turned out that the absence of PSI-H has an effect on state transitions that were essentially absent when PSI-H was missing (Lunde et al., 2000). State transitions are a process that serves to balance the activity of PSI and PSII. The mechanism and regulation of state transitions are further described in the section Accessory Electron Transport Systems. The analysis of the transgenic plants showed that in the absence of PSI-H, LHCII became more phosphorylated, probably due to more reduced conditions of plastoquinone and more active LHCII kinase, but the LHCII remained attached to PSII. In wild-type plants, an increase in PSI antenna size of about 30 percent was observed during a state 1–state 2 transition, whereas the plants lacking PSI-H had an unchanged PSI antenna size under all conditions. The implication is that LHCII binds to PSI at or near PSI-H in state 2. Furthermore, the distribution of LHCII is determined by competing binding to PSI and PSII.

Considering the proximity of PSI-L and PSI-O to PSI-H, it is not surprising that plants lacking PSI-L or PSI-O are also affected in state transitions, even though to a lower degree than the plants lacking PSI-H. However, lack of one of these subunits tends to lead to a partial secondary loss of the other subunits in the cluster. Hence, it is not at present possible to clearly distinguish individual roles of the three subunits. The plants lacking PSI-H should give an opportunity to investigate the role of state transitions in the normal life of a plant. This was attempted by growing plants lacking PSI-H under a variety of light conditions as well as under fluctuating light conditions (Lunde et al., 2003). In most cases, no effect on growth or seed set was observed, including conditions of alternating PSI and PSII light. Under conditions of alternating high and low light intensity, a 10 percent lower seed set was observed in plants lacking PSI-H. However, statistically this number was not significantly different from the wild-type seed set due to a high variation among individual plants.

Arabidopsis plants without PSI-L have been used to analyze whether the PSI-H/L/O side is involved in chlorophyll binding. The PSI-L subunit in *S. elongatus* binds three Chl *a* molecules (Jordan et al., 2001). Pigment analysis and comparison of the 5 K absorption spectra of PSI particles from *Arabidopsis* without PSI-L, and therefore also partially deficient in PSI-H and PSI-O, suggest that these subunits together bind approximately five Chl *a* molecules with absorption maxima near 688 and 667 nm (Ihalainen et al., 2002). Thus, if PSI-H binds LHCII in state 2, then the chlorophylls of PSI-H and PSI-L could very well be involved in the connection of the antenna systems of the PSI core complex and LHCII. The relatively long wavelength of at least some of these chlorophylls would then be advantageous for the directed energy flow from LHCII to PSI.

PSI-K and PSI-G

The PSI-K and PSI-G subunits share a limited sequence similarity, suggesting that they have evolved by gene duplication (Scheller et al., 2001). Cyanobacterial PSI contains only one copy designated PSI-K. *Arabidopsis* PSI-K was down-regulated by an antisense approach (Jensen et al., 2000). The plants lacking PSI-K showed normal growth and PSI activity. Careful analysis revealed a decrease in the content of Lhca2 and Lhca3, indicating that PSI-K has a role in proper or efficient binding of these two subunits. Similar results have been reported for an insertional knockout of *psaK* (Varotto et al., 2002). These authors also found that state transitions were decreased in the absence of PSI-K. A minor reduction in state transitions was also seen in the antisense PSI-K plants, but not to the same degree as in the insertional knockout of PSI-K, presumably due to different growth conditions in the two laboratories. Plants lacking PSI-G were generated either by antisense suppression (Jensen et al., 2002) or by insertional knockout (Varotto et al., 2002). In both cases, the plants were found to grow essentially normally. In spite of the similarity between PSI-G and PSI-K, no indication was found that they are functionally related. In the absence of PSI-G, there was no decrease in Lhca proteins and the antenna size of PSI was unaltered. Furthermore, pigment analysis and a comparison of the 5 K absorption spectra of PSI particles with and without PSI-G indicate that the PSI-G subunit binds approximately two redshifted β-carotene molecules (Ihalainen et al. 2002). Surprisingly, plants lacking PSI-G contain less PSI. This was also found in plants lacking PSI-D and PSI-F, but in these cases PSI assembly was clearly affected and the plants suffered from poor growth. In contrast to this, plants lacking PSI-G grow very well. The biggest surprise was that PSI complexes lacking PSI-G had a higher activity, as determined by steady state measurements in vitro. Apparently, the plants compensate for the more active PSI by decreasing the amount of PSI. This response is the opposite of the response found in plants lacking PSI-H, PSI-N, or PSI-K, where a decrease in PSI activity results in increased synthesis and content of PSI. The significance for normal plants of the higher activity in the absence of PSI-G is not clear at present. Our working hypothesis is that the difference in activity between PSI with and without PSI-G indicates a capacity for dynamic regulation of PSI activity under normal conditions. This could be important, for example, under high-light conditions, where decreased activity might increase the content of $P700^+$ and thus allow safe quenching of excess energy.

Excitation energy transfer and trapping dynamics were analyzed in PSI-200 from wild-type and from antisense *Arabidopsis* plants lacking PSI-G or PSI-K (Ihalainen et al., 2002). Two fast phases (5 and 15 ps) could be attributed to the excitation equilibration between bulk chlorophyll and red chlorophyll forms, though the 15 ps phase also contains a contribution from trapping by charge separation. Two other phases (50 and 120 ps) predominantly reflect trapping by charge separation in the PSI reaction center. The trapping of excitation energy that takes place within the 15 ps phase probably arises from photons absorbed in the core antenna, whereas trapping of excitation energy within the 50 ps phase is dominated by photons absorbed by Lhca2 and Lhca3, and trapping within the 120 ps phase is dominated by photons absorbed by Lhca1 and Lhca4. In the PSI particles without PSI-K or PSI-G protein, more excitations are trapped in the 15 ps phase and less in the 50 and 120 ps phases, which is in agreement with these subunits being involved in the interaction between the core and the peripheral antenna proteins.

LHCI

The first of the Lhca proteins to be functionally investigated in *Arabidopsis* was Lhca4 (Zhang et al., 1997). Using antisense suppression, the subunit was down-regulated to low levels and in one line to undetectable levels. The plants did not show any visual phenotype, so obviously Lhca4 is not essential for plants. Lhca4 forms heterodimers with Lhca1, and these heterodimers, known as LHCI-730, appear to be responsible for the long wavelength fluorescence from PSI at 77 K. In the *lhca4* antisense plants, the long wavelength fluorescence was shifted from 728 nm to 722 nm. This shift is very similar to what was observed in plants lacking PSI-F. The blueshift in fluorescence indicates that LHCI-730 is not functional, and this was of course not surprising given the lack of Lhca4. However, it was surprising that only the Lhca4 subunit was missing while Lhca1, Lhca2, and Lhca3 were present in normal amounts. Lhca1 is not supposed to form homodimers (Schmid et al., 1997). It is not clear if Lhca1 in the membranes of the antisense plants was actually connected with PSI. However, it is known from other studies that LHCI that is not attached to PSI complexes can be present in certain mutants, e.g., the *viridis zb-63* mutant of barley (Høyer-Hansen et al., 1988; Nielsen et al., 1996). Under some growth conditions, the Lhca1 subunit was also reduced to very low levels but the 77 K fluorescence spectrum was still shifted to the same extent (Ganeteg et al., 2001). Plants with antisense suppression of either Lhca2 or Lhca3 were also constructed (Ganeteg et al., 2001), and the down-regulated subunits were undetectable. In the Lhca2-repressed plants Lhca3 was almost completely depleted, and Lhca2 decreased to about 30 percent of wild-type levels in Lhca3-repressed plants (Ganeteg et al., 2001). No change was observed in Lhca1 or Lhca4. Surprisingly, the plants lacking Lhca2 and/or Lhca3 showed a shift in the 730 nm fluorescence, which was similar to the shift observed in the Lhca4 antisense plants. Two explanations may explain this unexpected finding: either the 730 nm fluorescence not only is due to LHCI-730 but also has a contribution from LHCI-680, or the absence of LHCI-680 causes a rearrangement of LHCI-730 leading to the blueshift.

Factors required for assembly of PSI have not yet been reported in detail in *Arabidopsis*. However, factors such as *ycf3,* which has been investigated in *Chlamydomonas,* are present also in *Arabidopsis* and therefore likely to be involved in PSI assembly (Naver et al., 2001). Furthermore, a large number of nuclear mutants, which appear to be specifically defective in PSI, have been isolated using forward genetic screens (Meurer et al., 1996; Shikanai et al., 1999). As in the case of PSII mutants, these mutants are apparently not deficient in a structural PSI subunit but in a crucial step in the synthesis or assembly of the complex.

ACCESSORY ELECTRON TRANSPORT SYSTEMS

Under high-light conditions a low pH will be generated in the thylakoid lumen, and this is a prerequisite for the quenching of excess light energy. A high rate of linear electron transport through the two photosystems will lead to a larger number of protons translocated across the thylakoid membrane. However, when CO_2 fixation is low under, e.g., drought and there is a lack of electron acceptors, this mechanism is not effective in forming a ΔpH. One way to overcome this problem is by cyclic electron transport around PSI. In this process electrons are recycled among NADPH, ferredoxin, Cyt *b/f*, plastoquinone, and PSI. Both the significance and the mechanism of cyclic electron transport in C3 plants has, however, remained elusive until the recent finding of a mutant *(pgr5)* that was unable to quench excess excitation energy under high light intensities (Munekage et al., 2002). The *pgr5* mutant lacked a 10 kDa protein, which in the wild type was localized to the thylakoid membrane. The affected protein did not appear to be part of any of the photosystems or the Cyt *b/f* complex. In the absence of PGR5, electrons were trapped in the PSI electron carriers under high light intensities due to acceptor limitations. The reduced level of $P700^+$ in the *pgr5* plant was increased further under O_2-limiting conditions; thus, PGR5 is not involved in any of the alternative O_2-dependent electron sinks. In contrast, the electron donation to plastoquinone via cyclic electron transport was impaired in the *pgr5* plants, and the slow donation could be mimicked in the wild type by the addition of antimycin A. PGR5 therefore appears to be an essential component of the antimycin-sensitive cyclic electron transport and has been suggested to play a role in the efficient electron transfer from ferredoxin to a recently identified Cyt *b/f* and FNR complex (Munekage et al., 2002; Zhang et al., 2001). After exposure to high-light conditions, the *pgr5* plants suffered from severe PSI damage, showing the importance of cyclic electron transport as a photoprotective mechanism.

Cyclic electron transport is one way to contribute to ΔpH over what results from linear electron transport. Biochemical studies in vitro have shown that there are alternative pathways and that only part of the cyclic electron transport is sensitive to antimycin A (Cleland and Bendall, 1992; Scheller, 1996). The significance of other pathways in vivo is difficult to assess, but the PGR5 mutant should allow this question to be answered. A schematic view of the alternative electron pathways is shown in Figure 16.4. The ndh complex may contribute to cyclic electron transport or to proton transport in the dark. The *ndh* genes are present in the *Arabidopsis* chloroplast genome, but the ndh complex has not been studied in *Arabidopsis*. Contribution to the proton gradient may not require a complete cy-

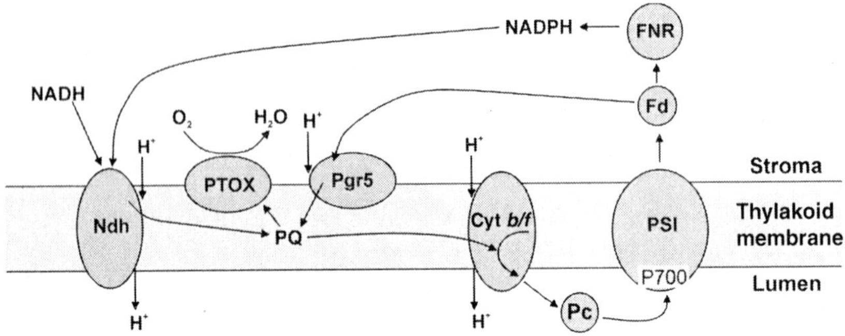

FIGURE 16.4. Schematic model of pathways for cyclic electron transport and chlororespiration. It should be noted that the details of these pathways are poorly understood and some aspects of the model are hypothetical. For example, it is very uncertain at what point(s) protons are translocated across the thylakoid membrane. Also, it has been debated whether the ndh complex can oxidize NADPH or NADH or both.

cle but could also be accomplished by chlororespiration. This has been an issue of discussion for many years, and most convincing studies of chlororespiration were from green algae. However, it has now been found that the thylakoid membranes contain a terminal oxidase (PTOX) that is essential for carotenoid biosynthesis (Carol et al., 1999). This oxidase may also play a minor role in proton transport. The oxidase has not been directly studied in relation to photosynthesis in *Arabidopsis,* but the *Arabidopsis* gene has been expressed in tobacco (Joet et al., 2002). Overexpression of the PTOX oxidase had a substantial effect on plastoquinone reoxidation, and the authors suggest that chlororespiration could be important for regulating cyclic electron transport. Chlororespiration has been reviewed recently (Peltier and Cournac, 2002).

State transitions are a process that aims at restoring an imbalance in the electron flow between the two photosystems. When PSII is overexcited, LHCII is phosphorylated and the mobile antenna dissociates from PSII, migrates through the thylakoid membrane, and binds to PSI. An important question has been whether LHCII binds to PSI or not, and, if so, where the LHCII-binding site is localized on PSI and if PSI plays a regulatory role or is just a passive partner associating by chance with free LHCII. As described previously, the PSI-H-less plants lacked the ability to perform state transition and suggested that the LHCII-binding site is on the PSI-H side of PSI. In contrast, plants having only low levels of PSI-E or PSI-F appear to be locked in state 2 with LHCII permanently bound to PSI. Whether the LHCII bound to PSI in these plants is functional in energy transfer has not been reported. Although there are still open questions about the mechanism of state transitions, a role of phosphorylation of LHCII has been indicated in numerous studies. A hunt for the LHCII kinase has until recently been without much success, but a breakthrough in the search for the LHCII kinase came when a family of proteins, the so-called TAKs (thylakoid-associated kinase), was identified (Snyders and Kohorn, 1999). In a yeast selection system, cDNA clones were screened for kinase activity. A positive clone was identified (WAK1), but this kinase was not involved in the phosphorylation of thylakoid proteins. However, antibodies raised against WAK1 recognized a 55 kDa protein (TAK1) that was found to be integral to the thylakoid membrane. The TAK was immunoprecipitated with Cyt *b/f* complex and LHCII and phosphorylated LHCII in vitro under PQ-reducing conditions (Snyders and Kohorn, 1999, 2001). In addition, antisense *Arabidopsis* plants with low amounts of TAK1 showed a lower level of LHCII phosphorylation and were deficient in state transitions (Snyders and Kohorn, 2001). The kinase epitope of TAK1 was removed by treating intact thylakoids with proteinase K, indicating that the kinase domain is on the stroma side. Two homologues of TAK1 were found by searching the *Arabidopsis* genome (TAK2/3). TAK2 expression was insensitive to the intensity of light, whereas TAK1 and TAK3 under increasing light intensity were down- and up-regulated, respectively. Homology was also found to the TGFβ-I (transforming growth receptor) receptor family (Snyders and Kohorn, 1999). TGFβ-I is involved in a signaling cascade, where activated TGFβ-II recognizes a conserved SGSGSG-motif and phosphorylates TGFβ-I. It has therefore been suggested that TAK3 and TAK2 play a similar role as TGFβ-II in the signaling cascade (Snyders and Kohorn, 1999). In support of this idea, TAK2 was found to bind to the N-terminus of heterologous-expressed TAK1 (Snyders and Kohorn, 2001). Thus, currently, a number of putative kinases might be involved in the phosphorylation of the thylakoid phosphoproteins: (1) the PSII core-associated kinase (Race and Hind, 1996); (2) the four (30, 55, 64, and 84 kDa) kinases, of which the 55 kDa kinase shows homology to a histidine kinase (Weber et al., 1998); and (3) the TAK kinases (Snyders and Kohorn, 1999, 2001).

That TAK homologues are the regulators of state transitions is not completely clear. A *Chlamydomonas* mutant, *stt7,* with inability to perform state transitions and phosphorylate light-harvesting proteins has been described (Fleischmann et al., 1999). The *stt7* gene has been shown to encode a protein kinase, and two homologues of this kinase are encoded in the *Arabidopsis* genome and predicted to be targeted to chloroplasts (Depege et al., 2003)

ELEMENTS INVOLVED IN NONPHOTOCHEMICAL QUENCHING

When plants are exposed to high light and the photochemical pathway is saturated, they can initiate photoprotective mechanims known as nonphotochemical quenching (NPQ). NPQ consists of three mechanisms: qE (energy-dependent quenching), qT (state transitions), and qI (photoinhibition). The energy-dependent quenching (qE) is the largest and fastest contribution and leads to the dissipation of excess absorbed light energy as heat, thereby minimizing the formation of dangerous oxygen radicals (reviewed in Müller et al., 2001).

The qE component of NPQ is closely linked to the xanthophyll cycle. In the dark, violaxanthin is usually the sole xanthophyll cycle pigment and it does not contribute to qE. In the light, zeaxanthin and antheraxanthin are present and these pigments promote qE (Gilmore and Yamamoto, 1993). The xanthophyll cycle consists of the de-epoxidation of violaxanthin first to antheraxanthin and then to zeaxanthin by violaxanthin de-epoxidase (VDE). In low light, antheraxanthin and zeaxanthin are slowly epoxidized to violaxanthin by the enzyme zeaxanthin epoxidase (ZE). VDE is localized to the thylakoid lumen and is activated by a pH below 6.5, with a maximum activity at about pH 5.0 in vitro (Hager, 1969; Hager and Holocher, 1994). It has also been shown in vitro that VDE requires ascorbate as a reductant (Hager, 1969).

To dissect the parameters involved in energy dissipation in vivo, *Arabidopsis* xanthophyll cycle mutants have been isolated by screening for altered nonphotochemical quenching of chlorophyll fluorescence. The *npq1* mutant is lacking VDE activity and is therefore unable to convert violaxanthin to zeaxanthin in excessive light (Niyogi et al., 1998). This mutant exhibits greatly reduced nonphotochemical quenching, demonstrating that violaxanthin de-epoxidation is required for the bulk of rapidly reversible nonphotochemical quenching in *Arabidopsis*. This mutant also shows that zeaxanthin synthesis is necessary for about 70 percent of the total NPQ and 80 percent of the qE. The *npq2* mutants are new alleles of *aba1*, the zeaxanthin epoxidase gene, and accumulate zeaxanthin constitutively. Analysis of the mutants shows that zeaxanthin is not sufficient for qE in the absence of the ΔpH (Niyogi et al. 1998). To identify components in addition to the ΔpH and xanthophylls that are necessary for qE, an *Arabidopsis* mutant, *npq4*, with altered NPQ but normal pigment composition, was isolated (Li et al., 2000). This mutant was defective specifically in the qE component of NPQ, but in contrast to the *npq1* mutants, it exhibited zeaxanthin synthesis in high light that was indistinguishable in extent and kinetics from that of the wild type. The qE phenotype of an *npq4-npq1* double mutant was indistinguishable from that of *npq4,* showing that the *npq4* mutation blocks even the small amount of qE that is detected in *npq1* plants. The *npq4* gene encodes the PsbS protein, which belongs to the light-harvesting complex (LHC) protein superfamily (Kim et al., 1992), and is also known as CP22 (Li et al., 2000). It has been proposed that PsbS binds de-epoxidized xanthophyll and protons, thereby inducing a conformational change that is necessary for qE (Li et al., 2000). A structure-function analysis of PsbS using point mutations generated by classical forward genetics, as well as by site-directed mutagenesis, showed that specific protonable amino acid residues (glutamate E122 and E226) in lumenal loops of PsbS are critical for its function in qE (Li, Phippard, et al., 2002). This indicates that H^+ binding by PsbS is necessary for qE. Resonance Raman spectroscopic experiments suggested that qE-related ΔA535 signal results from a PsbS-dependent 22 nm redshift in the absorbance spectrum of approximately two zeaxanthin molecules per PSII (Ruban et al., 2002). These findings are consistent with the idea that PsbS might form the binding site for the zeaxanthin molecules and H^+ binding both functioning directly in qE.

The small but significant amount of qE that is retained in the *Arabidopsis npq1* mutant demonstrates a component of qE that is not related to operation of the xanthophyll cycle. Analyses of Chl* quenching in lutein-deficient mutants of *Arabidopsis* (Pogson et al., 1998) have suggested the possible involvement of lutein in NPQ. For example, induction of NPQ in the *Arabidopsis lut2* mutant, which lacks lutein, is slower and reaches a lower maximum extent than that in the wild type (Pogson et al., 1998). Construction of a double mutant with *npq1* and *lut2* revealed that the residual qE in the *npq1* single mutant is abolished in the double mutant, which lacks both lutein and the high-light-induced synthesis of antheraxanthin and zeaxanthin (Niyogi et al., 2001). A recent study of *Arabidopsis lut* mutants has shown that lutein is probably not directly involved in qE but is involved in organizing LHCII (Lokstein et al., 2002).

To investigate the effect of low ascorbate levels on NPQ, the induction of NPQ in an ascorbate-deficient mutant of *Arabidopsis, vtc2-2,* was measured (Müller-Moulé et al., 2002). It was found that reduced ascorbate content is a direct substrate limitation for VDE activity.

The biological significance of qE has also been investigated with these mutants. The *npq1* mutant lacking the ability to synthesize zeaxanthin was shown to be more susceptible to low-temperature photodamage of PSII than the wild type (Havaux and Kloppstech, 2001). Somewhat surprisingly, even the *npq4* mutant grows very well under controlled environmental conditions and is essentially indistinguishable from the wild type (Li et al., 2000). However, when grown under controlled fluctuating conditions or under the natural variation in the field, the *npq4* plants showed a substantial reduction in fitness, determined as the number of seeds produced (Kulheim et al., 2002). Li, Müller-Moulé, et al. (2002) recently showed that plants deficient in PsbS were more susceptible to photoinhibition and plants that overexpressed PsbS were more resistant than the wild type. Overexpression of beta-carotene hydroxylase in *Arabidopsis* led to a doubling of the size of the xanthophyll pool, and the plants were more tolerant to photodamage (Davison et al., 2002). These investigations thus confirm the signficance of qE for photoprotection under adverse environmental conditions.

THE ATP SYNTHASE

During electron transport through PSII and PSI, protons are liberated in the thylakoid lumen during water oxidation at PSII and when plastoquinol is oxidized by the Cyt *b/f* complex. The proton gradient that this generated is then used to synthezise ATP via the ATP synthase. The ATP synthase consists of two major regions: an integral membrane portion (CF_0) and an extrinsic stroma exposed portion (CF_1). CF_0 functions as a channel for protons passing through the membrane and consists of at least four different subunits (I, II, III, and IV), with multiple copies (9-12) of subunit III being present (Table 16.4). Subunits I, II, and III are required for CF_1 binding (Lemaire and Wollman, 1989). Subunit IV is not required for the binding of CF_1 to CF_0 but is required for activity (Feng and McCarty, 1990). CF_1 contains the catalytic sites involved in ATP synthesis and consists of five different subunits (α, β, γ, δ, and ε) with an overall stoichiometry of $\alpha_3\beta_3\gamma\delta\varepsilon$. The α- and β-subunits bind ADP and phosphate and catalyze the phosphorylation of ADP into ATP (Bar-Zvi et al., 1983). The δ-subunit links CF_0 to CF_1 and the γ-subunit appears to control proton gating through the enzyme (Engelbrecht and Junge, 1990; Xiao and McCarty, 1989). The ε-subunit blocks catalysis in the dark, preventing breakdown of ATP, and may be involved in proton gating through interaction with the γ-subunit (Richter et al., 1984).

Not many *Arabidopsis* mutants that are affected in thylakoid ATP synthase subunits have been isolated. This is in part due to the fact that only three out of the nine subunits are nuclear encoded and therefore readily accessible for reverse or forward genetics. As with the subunits of the Cyt *b/f* complex, the ATP synthase subunits are interdependent. This has clearly been demonstrated in *Chlamydomonas,* where the whole set of ATP synthase subunits showed a concerted accumulation in vivo, with significantly reduced amounts of CF_1 detected in mutants blocked in the synthesis of a CF_0 subunit (Lemaire and Wollman, 1989; Robertson et al., 1990), which most probably also is the case for the higher-plant ATP synthase.

TABLE 16.4. Subunit composition of the ATP synthase.

	Gene	Gene location[a]	Protein	Molecular mass (Da)[b]	Function[c]	Accession number
CF_1	*atpA*	C	α-subunit	55,327	Catalytic	ATCG00120
	atpB	C	β-subunit	53,932	Catalytic	ATCG00480
	atpC	N2	γ-subunit	35,708/ 36,112	H^+ gating	AT4G04640 AT1G15700
	atpD	N	δ-subunit	20,483	Binding of CF_0 to CF_1, regulation	AT4G09650
	atpE	C	ε-subunit	14,472	ATPase inhibition	ATCG00470
CF_0	*atpF*	C	I	21,056	Binding of CF_0 to CF_1	ATCG00130
	atpG	N	II	16,352	Binding of CF_0 to CF_1	AT4G32260
	atpH	C	III	7,975	Binding of CF_0 to CF_1	ATCG00140
	atpI	C	IV	27,349	H^+ translocation	ATCG00150

[a]The location of the gene in nuclear (N) or plastid (C) genome in *Arabidopsis* is indicated. When more than one isoform is present in *Arabidopsis,* the number of isoforms is indicated.
[b]The predicted molecular mass of the mature *Arabidopsis* protein is indicated. The predicted masses are for the pure polypeptides and do not include the mass of cofactors and secondary modifications, except for the processing of signal and transit peptides. Cleavage sites were predicted by alignment with homologous sequences with known cleavage sites from other species.
[c]The indicated function is not an extensive list of all the roles for the subunits but merely an indication of the major function.

An *Arabidopsis* mutant in the *atpC1* gene has been reported. This *cfq* (coupling factor quick recovery) mutant has an amino acid substitution from glutamic acid (E) to lysine (K) at position 244 in the γ-subunit (Gabrys et al., 1994). Based on a slower decay from peak to steady state transient Chl fluorescence, it is concluded that this *cfq* mutant has decreased ΔpH compared with wild-type plants (Govindjee and Spilotro, 2002).

CONCLUSION

The study of photosynthesis has been greatly aided by the tools of functional genomics that are available in *Arabidopsis*. As described in this chapter, it has been possible to determine the function of many individual proteins in the main pathways for photosynthetic electron transport. *Arabidopsis* mutants have also enabled the identification of accessory electron transport proteins, such as Pgr5, involved in cyclic electron transport. Many researchers have previously searched for these proteins for years, but with little success until the tools of functional genomics became easily available. There are still many aspects of electron transport that can be conveniently studied in *Arabidopsis*. For example, a more-detailed understanding of processes and protein-protein interactions can be elucidated by site-directed mutagenesis studies. The real challenge in the near future, however, is the unraveling of the assembly, turnover, and regulation. Although a few factors have been described, e.g., factors involved in PSII assembly, it is clear that our understanding of these processes is still limited. The disassembly of PSII is beginning to be understood, but much remains to be learned regarding the mechanism of these processes and their regulation. Also, the regulation of photosystem stoichiometry and antenna size remains a mystery. Numerous descriptive studies have been carried out for many years, but they have not led to a good understanding of the underlying regulatory mechanisms. Studies of mutants obtained either by reverse genetics of regulatory factors, such as protein kinases, or by clever forward genetic screens (see, e.g., Walters et al., 2003) hold promise of leading to a much better understanding of the regulatory networks. *Arabidopsis* will remain the most important plant model for these investigations in the foreseeable future. The other important eukaryotic model organism is *Chlamydomonas*. Each model has its advantages and disadvantages: *Chlamydomonas* is a small unicellular organism, and therefore screening of a large number of individual clones is relatively simple, and some studies such as pulse-chase experiments are more easily carried out than in a higher plant. Furthermore, *Chlamydomonas* can be grown nonphotosynthetically; i.e., mutations leading to nonfunctional photosynthesis are not lethal. Finally, efficient genetic transformation of the chloroplast genome is available in *Chlamydomonas*. *Arabidopsis* has the advantage of more efficient methods for obtaining nuclear knockout mutants, either from the large collections available or by RNAi. Finally, *Arabidopsis* has the advantage of being a plant. Thus, even though many regulatory pathways are no doubt very similar, there will also be important differences. In plants the regulation of the photosynthetic apparatus is interfaced with multiple other biochemical and regulatory pathways related to the more complex biology of plants.

REFERENCES

Abdallah, F., Salamini, F., Leister, D. (2000). A prediction of the size and evolutionary origin of the proteome of chloroplasts of *Arabidopsis*. *Trends Plant Sci.* 5: 141-142.

Andersson, J., Walters, R.G., Horton, P., Jansson, S. (2001). Antisense inhibition of the photosynthetic antenna proteins CP29 and CP26: Implications for the mechanism of protective energy dissipation. *Plant Cell* 13: 1193-1204.

Andersson, J., Wentworth, M., Walters, R., Howard, C., Ruban, A.V., Horton, P., Jansson, S. (2003). Absence of the Lchb1 and Lchb2 proteins of the main light-harvesting complex of photosystem II: Effects on photosynthesis, grana stacking, and fitness. *Plant J.* 35: 350-361.

Bailey, S., Thompson, E., Nixon, P.J., Horton, P., Mullineaux, C.W., Robinson, C., Mann, N.H. (2002). A critical role for the Var2 FtsH homologue of *Arabidopsis thaliana* in the photosystem II repair cycle in vivo. *J. Biol. Chem.* 277: 2006-2011.

Bailey, S., Walters, R.G., Jansson, S., Horton, P. (2001). Acclimation of *Arabidopsis thaliana* to the light environment: The existence of separate low light and high light responses. *Planta* 213: 794-801.

Bar-Zvi, D., Tiefert, M.A., Shavit, N. (1983). Interaction of the chloroplast ATP-synthase with the photoreactive nucleotide 3'-*O*-(4-benzoyl)benzoyl adenosine 5'-diphosphate. *FEBS Lett.* 160: 233-238.

Berthold, D.A., Schmidt, C.L., Malkin, R. (1995). The deletion of *petG* in *Chlamydomonas reinhardtii* disrupts the cytochrome *bf* complex. *J. Biol. Chem.* 270: 29293-29298.

Carol, P., Stevenson. D., Bisanz, C., Breitenbach, J., Sandmann, G., Mache, R., Coupland, G., Kuntz, M. (1999). Mutations in the *Arabidopsis* gene *immutans* cause a variegated phenotype by inactivating a chloroplast terminal oxidase associated with phytoene desaturation. *Plant Cell* 11: 57-68.

Chitnis, P.R. (2001). PHOTOSYSTEM I: Function and physiology. *Annu. Rev. Plant Physiol. Plant Mol. Biol.* 52: 593-626.

Chitnis, V.P., Chitnis, P.R. (1993). PsaL subunit is required for the formation of photosystem-I trimers in the cyanobacterium *Synechocystis* SP PCC-6803. *FEBS Lett.* 336: 330-334.

Cleland, R.E., Bendall, D.S. (1992). Photosystem I cyclic electron transport: Measurement of ferredoxin-plastoquinone reductase activity. *Photosynth Res.* 34: 409-418.

Cramer, W.A., Soriano, G.M., Zhang, H., Ponamarev, M.V., Smith, J.L. (1997). The cytochrome b_6f complex: Novel aspects. *Physiol. Plant.* 100: 852-862.

Davison, P.A., Hunter, C.N., Horton, P. (2002). Overexpression of beta-carotene hydroxylase enhances stress tolerance in *Arabidopsis*. *Nature* 418: 203-206.

Depege, N., Bellafiore, S., Rochaix, J.-D. (2003). Rote of chloroplast protein kinase Stt7 in LHCII phosphorylation and state transition in *Chlamydomonas*. *Science* 299: 1572-1575.

Dinkins, R.D., Bandaranayake, H., Baeza, L., Griffiths, A.J.F., Green, B.R. (1997). *hcf5*, a nuclear photosynthetic electron transport mutant of *Arabidopsis thaliana* with a pleiotropic effect on chloroplast gene expression. *Plant Physiol.* 113: 1023-1031.

Engelbrecht, S., Junge, W. (1990). Subunit delta of H^+-ATPases: At the interface between proton flow and ATP synthesis. *Biochim. Biophys. Acta* 1015: 379-390.

Farah, J., Rappaport, F., Choquet, Y., Joliot, P., Rochaix, J.D. (1995). Isolation of a *psaF*-deficient mutant of *Chlamydomonas reinhardtii:* Efficient interaction of plastocyanin with the photosystem I reaction center is mediated by the PsaF subunit. *EMBO J.* 14: 4976-4984.

Felder, S., Meierhoff, K., Sane, A.P., Meurer, J., Driemel, C., Plücken, H., Klaff, P., Stein, B., Bechtold, N., Westhoff, P. (2001). The nucleus-encoded HCF107 gene of *Arabidopsis* provides a link between intercistronic RNA processing and the accumulation of translation-competent *psbH* transcripts in chloroplasts. *Plant Cell* 13: 2127-2141.

Feng, Y., McCarty, R.E. (1990). Subunit interaction within the chloroplast ATP synthase (CF0-CF1) as deduced by specific depletion of CF0 polypeptides. *J. Biol. Chem.* 265: 12481-12485.

Fleischmann, M.M., Ravanel, S., Delosme, R., Olive, J., Zito, F., Wollman, F.A., Rochaix, J.D. (1999). Isolation and characterization of photoautotrophic mutants of *Chlamydomonas reinhardtii* deficient in state transition. *J. Biol. Chem.* 274: 30987-30994.

Fromme, P., Jordan, P., Krauss, N. (2001). Structure of photosystem I. *Biochim. Biophys. Acta* 1507: 5-31.

Gabrys, H., Kramer, D.M., Crofts, A.R., Ort, D.R. (1994). Mutants of chloroplast coupling factor reduction in *Arabidopsis. Plant Physiol.* 104: 769-776.

Ganeteg, U., Strand, A., Gustafsson, P., Jansson, S. (2001). The properties of the chlorophyll *a/b*-binding proteins Lhca2 and Lhca3 studied in vivo using antisense inhibition. *Plant Physiol.* 127: 150-158.

Gau, A.E., Thole, H.H., Sokolenko, A., Altschmied, L., Herrmann, R.G., Pistorius, E.K. (1998). PsbY, a novel manganese-binding, low-molecular-mass protein associated with photosystem II. *Mol. Gen. Genet.* 260: 56-68.

Gilmore, A.M., Yamamoto, H.Y. (1993). Linear models relating xanthophylls and lumen acidity to non-photochemical fluorescence quenching: Evidence that antheraxanthin explains zeaxanthin-independent quenching. *Photosynth. Res.* 35: 67-78.

Govindjee, Spilotro, P. (2002). An *Arabidopsis thaliana* mutant, altered in the γ-subunit of ATP synthase, has a different pattern of intensity-dependent changes in non-photochemical quenching and kinetics of the P-to-S fluorescence decay. *Funct. Plant Biol.* 29: 425-434.

Gupta, R., He, Z., Luan, S. (2002). Functional relationship of cytochrome c_6 and plastocyanin in *Arabidopsis. Nature* 417: 567-571.

Hager, A. (1969). Lichtbedingte pH-Erniedrigung in einem Chloroplasten-Kompartiment als Ursache der enzymatischen Violaxanthin-Zeaxanthin-Umwandlung: Beziehung zur Photophosphorylierung. *Planta* 89: 224-243.

Hager, A., Holocher, K. (1994). Localization of the xanthophyll-cycle enzyme violaxanthin de-epoxidase within the thylakoid lumen and abolition of its mobility by a (light-dependent) pH decrease. *Planta* 192: 581-589.

Hager, M., Biehler, K., Illerhaus, J., Ruf, S., Bock, R. (1999). Targeted inactivation of the smallest plastid genome-encoded open reading frame reveals a novel and essential subunit of the cytochrome b_6f complex. *EMBO J.* 18: 5834-5842.

Haldrup, A., Lunde, C., Scheller, H.V. (2003). *Arabidopsis thaliana* plants lacking the PSI-D subunit of photosystem I suffer severe photoinhibition, have unstable photosystem I complexes and altered redox homeostasis in the chloroplast stroma. *J. Biol. Chem.* 278: 33276-33283.

Haldrup, A., Naver, H., Scheller, H.V. (1999). The interaction between plastocyanin and photosystem I is inefficient in transgenic *Arabidopsis* plants lacking the PSI-N subunit of photosystem I. *Plant J.* 17: 689-698.

Haldrup, A., Simpson, D., Scheller, H.V. (2000). Down-regulation of the PSI-F subunit of photosystem I (PSI) in *Arabidopsis thaliana. J. Biol. Chem.* 275: 31211-31218.

Havaux, M., Kloppstech, K. (2001). The protective functions of carotenoid and flavonoid pigments against excess visible radiation at chilling temperature investigated in *Arabidopsis npq* and *tt* mutants. *Planta* 213: 953-966.

Høyer-Hansen, G., Bassi, R., Hønberg, L.S., Simpson, D.J. (1988). Immunological characterization of chlorophyll *a/b*-binding proteins of barley thylakoids. *Planta* 173: 12-21.

Huang, D., Everly, R.M., Cheng, R.H., Heymann, J.B., Schägger, H., Baker, T.S., Cramer, W.A. (1994). Characterization of the Cyt b_6f complex as a structural and functional dimer. *Biochemistry* 33: 4401-4409.

Ihalainen, J.A., Jensen, P.E., Haldrup, A., van Stokkum, I.H.M., van Grondelle, R., Scheller, H.V., Dekker, J.P. (2002). Pigment organization and energy transfer dynamics in isolated photosystem I (PSI) complexes from *Arabidopsis thaliana* depleted of the PSI-G, PSI-K, PSI-L, or PSI-N subunit. *Biophys. J.* 83: 2190-2201.

Jackowski, G., Kacprzak, K., Jansson, S. (2001). Identification of Lhcb1/Lhcb2/Lhcb3 heterotrimers of the main light-harvesting chlorophyll *a/b*-protein complex of photosystem II (LHCII). *Biochim. Biophys. Acta* 1504: 340-345.

Jahns, P., Graf, M., Munekage, Y., Shikanai, T. (2002). Single point mutation in the Rieske iron-sulfur subunit of cytochrome b_6/f leads to an altered pH dependence of plastoquinol oxidation in *Arabidopsis. FEBS Lett.* 519: 99-102.

Jansson, S. (1999). A guide to the Lhc genes and their relatives in *Arabidopsis. Trends Plant Sci.* 4: 236-240.

Jansson, S., Andersen, B., Scheller, H.V. (1996). Nearest neighbor analysis of higher plant photosytem I holocomplex. *Plant Physiol.* 112: 409-420.

Jensen, P.E., Gilpin, M., Knoetzel, J., Scheller, H.V. (2000). The PSI-K subunit of photosystem I is involved in the interaction between light harvesting complex I-680 and the photosystem I reaction center core. *J. Biol. Chem.* 275: 24701-24708.

Jensen, P.E., Rosgaard, L., Knoetzel, J., Scheller, H.V. (2002). Photosystem I activity is increased in the absence of the PSI-G subunit. *J. Biol. Chem.* 277: 2798-2803.

Joet, T., Genty, B., Josse, E.M., Kuntz, M., Cournac, L., Peltier, G. (2002). Involvement of a plastid terminal oxidase in plastoquinone oxidation as evidenced by expression of the *Arabidopsis thaliana* enzyme in tobacco. *J. Biol. Chem.* 277: 31623-31630.

Jordan, P., Fromme, P., Witt, H.T., Klukas, O., Saenger, W., Krauss, N. (2001). Three-dimensional structure of cyanobacterial photosystem I at 2.5 Å resolution. *Nature* 411: 909-917.

Kim, S., Sandusky, P., Bowlby, N.R., Aebersold, R., Green, B.R., Vlahakis, S., Yocum, C.F., Pichersky, E. (1992). Characterization of a spinach psbS cDNA encoding the 22 kDa protein of photosystem II. *FEBS Lett.* 314: 67-71.

Kulheim, C., Agren, J., Jansson, S. (2002). Rapid regulation of light harvesting and plant fitness in the field. *Science* 297: 91-93.

Li, X.P., Björkman, O., Shih, C., Grossman, A.R., Rosenquist, M., Jansson, S., Niyogi, K.K. (2000). A pigment-binding protein essential for regulation of photosynthetic light harvesting. *Nature* 403: 391-395.

Li, X.P., Müller-Moulé, P., Gilmore, A.M., Niyogi, K.K. (2002). PsbS-dependent enhancement of feedback de-excitation protects photosystem II from photoinhibition. *Proc. Natl. Acad. Sci. USA* 99: 15222-15227.

Li, X.P., Phippard, A., Pasari, J., Niyogi, K.K. (2002). Structure-function analysis of photosystem II subunit S (PsbS) in vivo. *Funct. Plant Biol.* 29: 1131-1139.

Lemaire, C., Wollman, F.A. (1989). The chloroplast ATP synthase in *Chlamydomonas reinhardtii*. II. Biochemical studies on its biogenesis using mutants defective in photophosphorylation. *J. Biol. Chem.* 264: 10235-10242.

Lennartz, K., Plüken, H., Seidler, A., Westhoff, P., Bechtold, N., Meierhof, K. (2001). HCF164 encodes a thioredoxin-like protein involved in the biogenesis of the cytochrome b_6f complex in *Arabidopsis*. *Plant Cell* 13: 2539-2551.

Lokstein, H., Tian, L., Polle, J.E.W., DellaPenna, D. (2002). Photosystem II antenna size and stability. *Biochim. Biophys. Acta* 1553: 309-319.

Lunde, C., Jensen, P.E., Haldrup, A., Knoetzel, J., Scheller, H.V. (2000). The PSI-H subunit of photosystem I is essential for state transition in plants. *Nature* 408: 613-615.

Lunde, C., Jensen, P.E., Rosgaard, L., Haldrup, A., Gilpin, M.J., Scheller, H.V. (2003). Plants impaired in state transitions can to a large degree compensate for their defect. *Plant Cell Physiol.* 44: 44-54.

Meurer, J., Meierhoff, K., Westhoff, P. (1996). Isolation of high-chlorophyll-fluorescence mutants of *Arabidopsis thaliana* and their characterisation by spectroscopy, immunoblotting and Northern hybridisation. *Planta* 198: 385-396.

Meurer, J., Plucken, H., Kowallik, K.V., Westhoff, P. A. (1998). Nuclear-encoded protein of prokaryotic origin is essential for the stability of photosystem II in *Arabidopsis thaliana*. *EMBO J.* 17: 5286-5297.

Müller, P., Li, X.P., Niyogi, K.K. (2001). Non-photochemical quenching: A response to excess light energy. *Plant Physiol.* 125: 1558-1566.

Müller-Moulé, P., Conklin, P.L., Niyogi, K.K. (2002). Ascorbate deficiency can limit violaxanthin de-epoxidase activity in vivo. *Plant Physiol.* 128: 970-977.

Munekage, Y., Hojo, M., Meurer, J., Endo, T., Tasaka, M., Shikanai, T. (2002). PGR5 is involved in cyclic electron flow around photosystem I and is essential for photoprotection in *Arabidopsis*. *Cell* 110: 361-371.

Munekage, Y., Takeda, S., Endo, T., Jahns, P., Hashimoto, T., Shikanai, T. (2001). Cytochrome b_6f mutation specifically affects thermal dissipation of absorbed light energy in *Arabidopsis*. *Plant J.* 28: 351-359.

Murakami, R., Ifuku, K., Takabayashi, A., Shikanai, T., Endo, T., Sato, F. (2002). Characterization of an *Arabidopsis thaliana* mutant with impaired *psbO*, one of two genes encoding extrinsic 33-kDa proteins in photosystem II. *FEBS Lett.* 523: 138-142.

Naver, H., Bourdreau, E., Rochaix, J.D. (2001). Functional studies of Ycf3: Its role in assembly of photosystem I and interactions with some of its subunits. *Plant Cell* 13: 2731-2745.

Naver, H., Haldrup, A., Scheller, H.V. (1999). Cosuppression of photosystem I subunit PSI-H in *Arabidopsis thaliana:* Efficient electron transport and stability of photosystem I is dependent upon the PSI-H subunit. *J. Biol. Chem.* 274: 10784-10789.

Nielsen, V.S., Scheller, H.V., Møller, B.L. (1996). The barley photosystem I mutant *viridis-zb*^{63} contains low amounts of active but unstable photosystem I. *Physiol. Plant* 98: 637-644.

Nixon, P.J., Gounaris, K., Coomber, S.A., Hunter, C.N., Dyer, T.A., Barber, J. (1989). *PsbG* is not a photosystem 2 gene but may be an *ndh* gene. *J. Biol. Chem.* 264: 14129-14135.

Niyogi, K.K., Grossman, A.R., Björkman, O. (1998). *Arabidopsis* mutants define a central role for the xanthophyll cycle in the regulation of photosynthetic energy conversion. *Plant Cell* 10: 1121-1134.

Niyogi, K.K., Shih, C., Chow, W., Soon, P., Barry, J., DellaPenna, D., Björkman, O. (2001). Photoprotection in a zeaxanthin- and lutein-deficient double mutant of *Arabidopsis*. *Photosynth. Res.* 67: 139-145.

Peltier, G., Cournac, L. (2002). Chlororespiration. *Annu. Rev. Plant Biol.* 53: 523-550.

Pesaresi, P., Lunde, C., Jahns, P., Tarantion, D., Meurer, J., Varotto, C., Hirtz, R.D., Soave, C., Scheller, H.V., Salamini, F., Leister, D. (2002). A stable LHCII-PSI aggregate and suppression of photosynthetic state transitions in the *psae1-1* mutant of *Arabidopsis thaliana*. *Planta* 215: 940-948.

Pesaresi, P., Varotto, C., Richly, E., Kurth, J., Salamini, F., Leister, D. (2001). Functional genomics of *Arabidopsis* photosynthesis. *Plant Physiol. Biochem.* 39: 285-294.

Pierre, Y., Breyton, C., Kramer, D., Popot, J.-L. (1995). Purification and characterization of the Cyt b_6f complex from *Chlamydomonas reinhardtii*. *J. Biol. Chem.* 270: 29342-29349.

Pogson, B.J., Niyogi, K.K., Björkman, O., DellaPenna, D. (1998). Altered xanthophylls compositions adversely affect chlorophyll accumulation and nonphotochemical quenching in *Arabidopsis* mutant. *Proc. Natl. Acad. Sci. USA* 95: 13324-13329.

Race, H.L., Hind, G. A. (1996). Protein kinase in the core of photosystem II. *Biochemistry* 35: 13006-13010.

Richter, M.L., Patrie, W.J., McCarty, R.E. (1984). Preparation of the epsilon subunit and epsilon subunit-deficient chloroplast coupling factor 1 in reconstitutively active forms. *J. Biol. Chem.* 259: 7371-7373.

Robertson, D., Boynton, J.E., Gilham, N.W. (1990). Cotranscription of the wild-type chloroplast *atpE* gene encoding the CF_1/CF_0 epsilon subunit with the 3' half of the *rps7* gene in *Chlamydomonas reinhardtii* and characterization of frameshift mutations in *atpE*. *Mol. Gen. Genet.* 221: 155-163.

Ruban, A.V., Pascal, A.A., Robert, B., Horton, P. (2002). Activation of zeaxanthin is an obligatory event in the regulation of photosynthetic light harvesting. *J. Biol. Chem.* 277: 7785-7789.

Ruban, A.V., Wentworth, M., Yakushevska, A.E., Andersson, J., Lee, P.J., Keegstra, W., Dekker, J.P., Boekema, E.J., Jansson, S., Horton, P. (2003). Plants lacking the main light harvesting complex retain PSII macro-organization. *Nature* 421: 648-652.

Scheller, H.V. (1996). Cyclic electron transport in barley follows two independent pathways. *Plant Physiol.* 110: 187-194.

Scheller, H.V., Jensen, P.E., Haldrup, A., Lunde, C., Knoetzel, J. (2001). Role of subunits in eukaryotic photosystem I. *Biochim. Biophys. Acta* 1507: 41-60.

Schmid, V.H.R., Cammarata, K.V., Bruns, B.U., Schmidt, G.W. (1997). In vitro reconstitution of the photosystem I light-harvesting complex LHCI-730: Heterodimerization is required for antenna pigment organization. *Proc. Natl. Acad. Sci. USA* 94: 7667-7672.

Shi, L.X., Lorkovic, Z.J. Oelmuller, R., Schroder, W.P. (2000). The low molecular mass PsbW protein is involved in the stabilization of the dimeric photosystem II complex in *Arabidopsis thaliana*. *J. Biol. Chem.* 275: 37945-37950.

Shikanai, T., Munekage, Y., Shimizu, K., Endo, T., Hashimoto, T. (1999). Identification and characterization of *Arabidopsis* mutants with reduced quenching of chlorophyll fluorescence. *Plant Cell Physiol.* 40: 1134-1142.

Snyder, S., Kohorn, B.D. (1999). TAKs, thylakoids membrane protein kinases associated with energy transduction. *J. Biol. Chem.* 274: 9137-9140.

Snyder, S., Kohorn, B.D. (2001). Disruption of thylakoid-associated kinase 1 leads to alternation of light harvesting in *Arabidopsis*. *J. Biol. Chem.* 276: 32169-32176.

Takahashi, Y. Goldschmidt-Clermont, M. Soen, S.Y., Franzen, L.G., Rochaix, J.D. (1991). Directed chloroplast transformation in *Chlamydomonas reinhardtii:* Insertional inactivation of the *psaC* gene encoding the iron sulfur protein destabilizes photosystem I. *EMBO J.* 10: 2033-2040.

Thompson, S.J., Robinson, C., Mant, A. (1999). Dual signal peptides mediate the signal recognition particle Sec-independent insertion of a thylakoid membrane polyprotein, PsbY. *J. Biol. Chem.* 274: 4059-4066.

Varotto, C., Pesaresi, P., Jahns, P., Lessnick, A., Tizzano, M., Schiavon, F., Salamini, F., Leister, D. (2002). Single and double knockouts of the genes for photosystem I subunits G, K, and H of *Arabidopsis:* Effects on photosystem I composition, photosynthetic electron flow, and state transitions. *Plant Physiol.* 129: 616-624.

Varotto, C., Pesaresi, P., Meurer, J., Oelmuller, R., Steiner-Lange, S., Salamini, F., Leister, D. (2000). Disruption of the *Arabidopsis* photosystem I gene *psaE1* affects photosynthesis and impairs growth. *Plant J.* 22: 115-124.

Walters, R.G., Shephard, F., Rogers, J.J.M., Rolfe, S.A., Horton, P. (2003). Identification of mutants of *Arabidopsis* defective in acclimation of photosynthesis to the light environment. *Plant Physiol.* 131: 472-481.

Weber, P., Fulgosi, H., Sokolenko, A., Karnauchov, I., Andersson, B., Ohad, I., Herrmann, R.G. (1998). Evidence for four thylakoid-located protein kinases. In Gara, G. (Ed.), *Photosynthesis: Mechanisms and Effects,* Volume III (pp. 1883-1886). Dordrecht, The Netherlands: Kluwer Academic Publishers.

Weigel, M., Varotto, C., Pesaresi, P., Finazzi, G., Rappaport, F., Salamini, F., Leister, D. (2003). Plastocyanin is indispensable for photosynthetic electron flow in *Arabidopsis thaliana. J. Biol. Chem.* 278: 31286-31289.

Wollman, F.-A., Minai, L., Nechushtai, R. (1999). The biogenesis and assembly of photosynthetic proteins in thylakoid membranes. *Biochim. Biophys. Acta* 1411: 21-85.

Xiao, J., McCarty, R.E. (1989). Binding of chloroplast coupling factor 1 deficient in the delta subunit to thylakoid membranes. *Biochim. Biophys. Acta* 976: 203-209.

Yakushevska, A.E., Jensen, P.E., Keegstra, W., van Roon, H., Scheller, H.V., Boekema, E.J., Dekker, J.P. (2001). Supermolecular organization of photosystem II and its associated light-harvesting antenna in *Arabidopsis thaliana. Eur. J. Biochem.* 268: 6020-6028.

Yakushevska, A.E., Keegstra, W., Boekema, E.J., Dekker, J.P., Andersson, J., Jansson, S., Ruban, A.V., Horton, P. (2003). The structure of photosystem II in *Arabidopsis:* Localization of the CP26 and CP29 antenna complexes. *Biochemistry* 42: 608-613.

Zhang, H., Goodman, H.M., Jansson, S. (1997). Antisense inhibition of the photosystem I antenna protein Lhca4 in *Arabidopsis thaliana. Plant Physiol.* 115: 1525-1531.

Zhang, H., Huang, D., Cramer, W.A. (1999). Stoichiometrically bound β-carotene in the cytochrome b_6f complex f oxygenic photosynthesis protects against oxygen damage. *J. Biol. Chem.* 274: 1581-1587.

Zhang, H., Whitelegge, J.P., Cramer, W.A. (2001). Ferredoxin: $NADP^+$ oxidoreductase is a subunit of the chloroplast cytochrome b_6f complex. *J. Biol. Chem.* 276: 38159-38165.

Zhang, S., Scheller, H.V. (2004). Light harvesting complex II binds to several small subunits of photosystem I. *J. Biol. Chem.* 279: 3180-3187.

Chapter 17

Functional Genomics of Plant Nitrogen Metabolism

Mathilde Orsel
Karine Boivin
Hélène Roussel
Christelle Thibault
Anne Krapp
Françoise Daniel-Vedele
Christian Meyer

INTRODUCTION

The mechanisms of plant nutrition have always attracted a lot of attention since Aristotle first postulated that plants receive preformed nutrients from their environment (cited in Morot-Gaudry, 2001), and it is only in the past century that it was firmly established that plants can grow with mineral elements alone. Nitrogen is the soil nutrient that plants acquire in greatest abundance. It is thus not surprising that plants have devised strategies to forage and store this limiting element, and also that many developmental processes seem to be regulated by fluctuations in nitrogen availability. Indeed, the development of plant organs does not follow a rigid, predetermined plan but, rather, is mainly controlled by environmental signals, as plants need to continuously adjust their growth pattern to their nutrition conditions. For instance, because they are immobile organisms, they must direct and expand their roots toward nitrogen-rich zones.

Unlike other nutrients, the presence of nitrogen seems to induce the expression of the proteins responsible for its uptake and metabolism, whereas for the other ions it is their absence that is the main stimulatory signal. Nitrate is the preferred and most accessible inorganic nitrogen source for many temperate plants. It appears, indeed, that many plants, unlike fungi and bacteria, prefer nitrate to ammonium as a mineral nitrogen source. After active uptake in the cell, nitrate is reduced to ammonium in two steps: (1) being catalyzed by nitrate reductase (NR) in the cytosol and (2) being

H. R. and part of the experimental work presented here were supported by Génoplante. K. B. was supported by a grant from the Genopole-Evry.

catalyzed by nitrite reductase (NiR), which reduces the product of NR—nitrite—to ammonium in the chloroplasts or plastids. In shoot chloroplasts, ferredoxin serves as reductant for nitrite reduction, whereas in root plastids reducing power is provided by a different ferredoxin protein via ferredoxin-NADP reductase (FNR) and NADPH derived from the oxidative pentose phosphate pathway. In addition, ammonium can also be taken up directly from the environment (soil, atmosphere, etc.) or be generated by a variety of other metabolic pathways, such as photorespiration (which produces the vast majority of this ion in C_3 plants), amino acid metabolism, or nitrogen fixation in legumes. Apart from the primary assimilation through uptake from the soil, nitrogen is also recycled within the cell and within the plants between old or senescing tissues and young growing ones (new leaves, developing seeds). The ammonium ion is subsequently incorporated into organic molecules by glutamine synthetase and glutamate synthase (Figure 17.1; Meyer and Stitt, 2001). In addition to being an essential nutrient, nitrate, even at very low concentrations, is also a signal for developmental processes and regulates other metabolic pathways (Crawford, 1995; Stitt, 1999). For instance, nitrate up-regulates organic acid synthesis needed to counterbalance the alkalinization of cytoplasm resulting from nitrate reduction and to provide the carbon skeleton used for the synthesis of organic nitrogen. Nitrate represses starch accumulation and stimulates leaf elongation and lateral root proliferation (Stitt, 1999). Nitrate can also be a signal through its reduction by the NR enzyme first into nitrite and then into NO, a well-known gaseous signal molecule (Yamasaki and Sakihama, 2000).

For these reasons, nitrogen metabolism has always attracted a lot of attention. Initial molecular studies were devoted to the identification of genes involved in key enzymatic functions of nitrogen metabolism, such as nitrate reduction. Indeed, the NR cDNAs were identified 17 years ago in barley (Cheng et al., 1986), squash (Crawford et al., 1986), and tobacco (Calza et al., 1987) and were among the first plant cDNAs to be cloned. This was achieved through a combination of biochemical and molecular approaches. The protein corresponding to NR activity was first purified and antibodies were then raised and used for the screening of plant cDNA expression libraries. The cloning of the cDNAs allowed the subsequent identification of the genomic sequences corresponding to the NR genes (reviewed in Crawford and Arst, 1993). This breakthrough was closely followed by the cloning of the genes coding for NiR, glutamine synthetase, and glutamate synthase (Morot-Gaudry and Lea, 2001). The role of most of these enzymes is clear, but others, such as glutamate dehydrogenase, have a more mysterious function. The tools provided by the development of the molecular genetics of *Arabidopsis* have helped to better define these obscure parts of nitrogen metabolism. For instance, the study of an *Arabidopsis* mutant

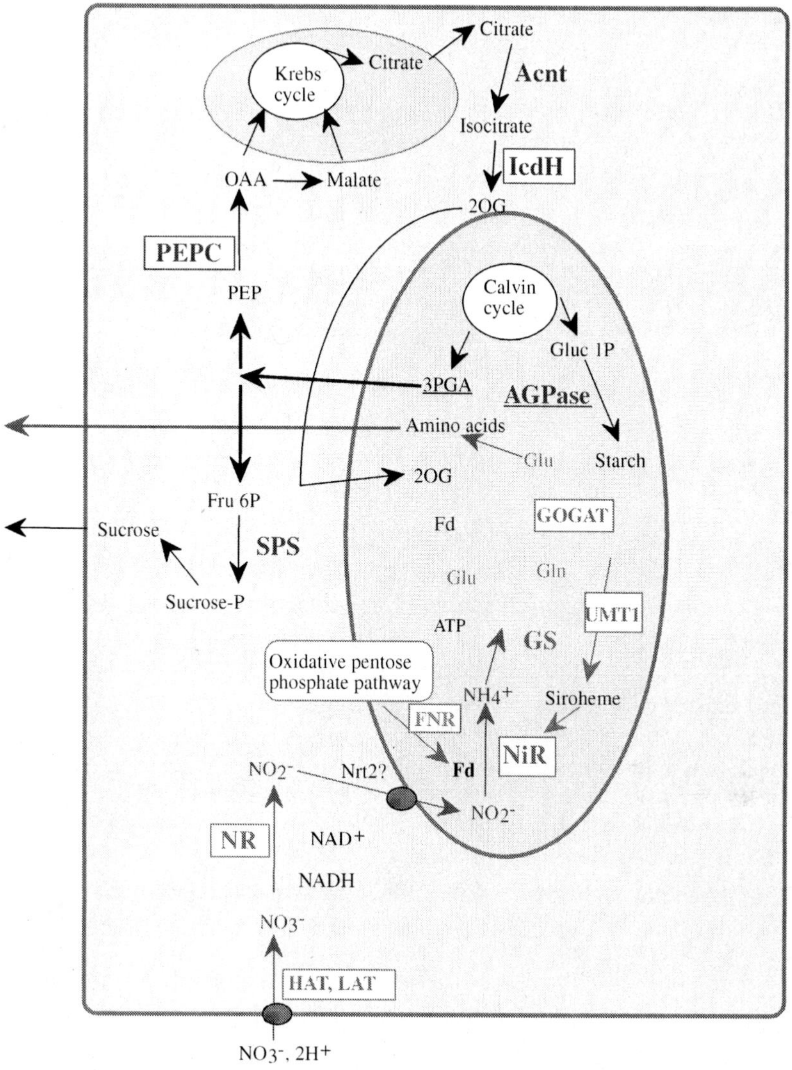

FIGURE 17.1. Relationships between nitrogen assimilation pathway and carbon metabolism. Enzymes that are inducible by nitrate are boxed; those which are repressed are underlined. HAT/LAT, respectively, high- and low-affinity nitrate transporters; SPS, sucrose phosphate synthase; PEPC, phosphoenolpyruvate carboxylase; IcdH, isocitrate dehydrogenase; Acnt, aconitase; AGPase, ADP-glucose pyrophosphorylase; 2OG, 2-oxoglutarate; UMT1, uroporphyrin III methyl transferase; Fd, ferredoxin.

affected in glutamate dehydrogenase showed that in this mutant seedling growth is specifically retarded on media containing inorganic nitrogen in excess and that this enzyme may be involved in ammonium assimilation (Melo-Oliveira et al., 1996). The molecular genetics and physiology of nitrogen metabolism have been revisited recently (Crawford, 1995; Meyer and Stitt, 2001; Campbell, 1999; Coruzzi and Zhou, 2001; Meyer and Stoehr, 2002), and this chapter will focus on two main points: (1) the identification of the function of genes known to be involved in nitrogen metabolism and (2) the identification of novel genes involved in the functioning and regulation of nitrogen metabolism or in nitrogen perception/signaling by analyzing the response of the transcriptome upon variations in nitrogen supply.

FUNCTIONAL GENOMICS OF NITRATE TRANSPORT IN ARABIDOPSIS

The *Arabidopsis* genome sequencing has provided a huge number of predicted gene products, and the next step toward using this information to understand the nitrogen metabolism and its relations to other aspects of plant development involves assigning precise functions to these genes. The analysis of other organisms has allowed some functional predictions that can serve as a basis for precise function determination. For instance, homologues of animal ionotropic glutamate receptors have been found in *Arabidopsis* and have been very recently shown to function as global regulators of carbon and nitrogen metabolism and to control germination through ABA synthesis (Kang and Turano, 2003). Similarly, an orthologue of the yeast TOR (target of rapamycin) gene, a global nitrogen metabolism and translation regulator, has been detected in *Arabidopsis.* Mutants affected in the expression of this gene are blocked at the embryo globular stage (Menand et al., 2002), and we are currently investigating the role of this *Arabidopsis* protein in the control of nitrogen metabolism. Concerning the transport of inorganic nitrogen, a substantial amount of information has now been gathered on the Amt ammonium transporters family (von Wiren et al., 2000). We will more extensively illustrate this functional and comparative approach by our work on the *NRT2* gene family in *Arabidopsis,* which is thought to code for high-affinity nitrate transporters and has been more extensively studied by several research groups.

Nitrate uptake by root cells is a key process for plant nutrition because it is the first step of the nitrate assimilatory pathway (Morot-Gaudry and Lea, 2001). Many physiological investigations have led to the conclusion that plants have evolved two types of nitrate uptake systems to cope with large variations in nitrate concentrations in soils. The low-affinity transport system (LATS) is used preferentially when external nitrate concentration is high (above 1 mM) (Siddiqi et al., 1990). The high-affinity transport system (HATS) is able to take up nitrate at low concentrations (between 1 μM and 1 mM) (Behl et al., 1988). After its entry into the cytoplasm of root cells, nitrate is either reduced or translocated and stored in the vacuole. It is also released into the xylem vessel to reach the aerial organs, in which storage and reduction take place as well. All these nitrate fluxes entail the passage across the plasmalemma or the tonoplasm. The passage through these membranes requires active processes that involve transporters anchored in these structures (Glass et al., 2001).

In higher plants, the molecular basis of root nitrate uptake has been the matter of intensive studies and debates during the past decade (Daniel-Vedele et al., 1998; Forde, 2000). So far, two gene families, *NRT1* and *NRT2,* have been identified on the basis of their deduced amino acid sequences. The NRT1 and NRT2 proteins share the same structural topology, with 12 transmembrane domains, but no primary sequence homology is found between them. The *Arabidopsis chl1* line (Tsay et al., 1993), affected in the *AtNRT1.1* gene, was the first mutant characterized by a defect in nitrate uptake mediated by the LATS system. Further studies demonstrated that the *AtNRT1.1* gene also plays a role in nitrate uptake at low availability (Wang et al., 1999; Liu and Tsay, 2003). The *AtNRT1.1* gene encodes an inducible, dual-affinity nitrate transporter, and switching between the two modes of action is regulated by phosphorylation (Liu and Tsay, 2003).

The first members of the *NRT2* gene family were identified in lower eukaryotes. The *CrnA* gene was identified in the fungus *Aspergillus nidulans* (Unkles et al., 1991) using a screening procedure for chlorate resistance. Two other genes, *CrNRT2.1* and *CrNRT2.2,* were identified in mutants of the green algae *Chlamydomonas reinhardtii* deficient in nitrate and/or nitrite transport (Quesada et al., 1994). The amino acids comparison between the fungus and algae nitrate transporter proteins revealed highly conserved regions and allowed identification of other members of the family on the basis of their sequence identity. Members of the *NRT2* gene family in higher plants were identified by this way in barley, *Nicotiana plumbaginifolia, Glycine max,* and tomato (see Orsel et al., 2002, for a review). So far, in most species, *NRT2* genes are small multigenic families. In *Arabidopsis,* the analysis of the complete genome revealed a *NRT2* gene family with seven members distributed among three chromosomes (Figure 17.2) (Orsel et al., 2002). Sequence comparisons support the idea that these proteins are involved in a high-affinity nitrate transport system. Correlations between regulation of high-affinity nitrate uptake and *AtNRT2.1* gene ex-

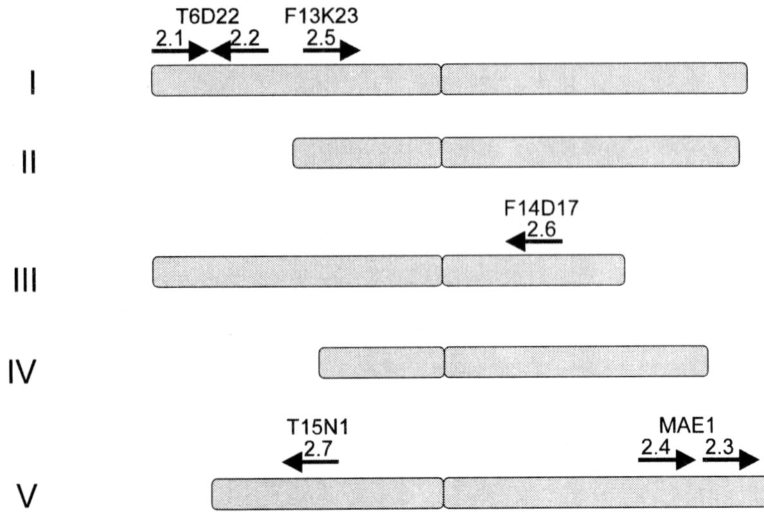

FIGURE 17.2. Localization of the seven *NRT2* genes in *Arabidopsis thaliana*. Numbers above the arrows correspond to the BAC that contains the *NRT2* gene(s).

pression, first identified on the basis of its differential expression under nitrate versus glutamine (Filleur and Daniel-Vedele, 1999), have been extensively studied (Lejay et al., 1999). Until recently, however, no evidence of *NRT2* gene implication in nitrate HATS in higher plants was available.

The first functional evidence supporting the role of *NRT2* genes in the root nitrate uptake has been obtained using an *Arabidopsis* knockout mutant where the T-DNA insertion leads to *AtNRT2.1* and *AtNRT2.2* gene disruptions (Filleur et al., 2001). The initial analysis demonstrated that root nitrate uptake in the *atnrt2.1a* mutant is reduced by 63 percent, providing strong evidence that one or both *AtNRT2.1* and *AtNRT2.2* genes are essential for the inducible component of the HATS. Further investigations have shown that *AtNRT2.1* and *AtNRT2.2* genes play a key role in the regulation of the high-affinity nitrate uptake and in the adaptive responses of the plant to both spatial and temporal changes in nitrogen availability in the environment (Cerezo et al., 2001). At this stage of the investigations, a major challenge arises in understanding the function of each member of the *NRT2* multigene family.

First indications came from comparative analysis of the genes and protein sequences. All members share a high degree of identity, but a complete alignment of the seven *AtNRT2* cDNAs revealed various degrees of diversification in their amino acid sequences (Orsel et al., 2002). The highest percentage of similarity is found between the AtNRT2.3 and AtNRT2.6 proteins (91.5 percent), followed by the AtNRT2.1 and AtNRT2.2 pair (90.4 percent). The lowest score was observed with the AtNRT2.5 and AtNRT2.7 sequences when compared to all other members of the family. A phylogenetic analysis suggests that five of the AtNRT2 proteins are very closely related to the other dicotyledonous NRT2 proteins. Still two others, namely AtNRT2.5 and AtNRT2.7, appear as isolated and individual sequences and might be more closely related to fungi and algae NRT2 nitrate transporters (Figure 17.3) (Orsel et al., 2002). Analyses of the different protein domains confirmed that AtNRT2.5 and AtNRT2.7 are structurally closely related to, respectively, a yeast transporter (YNT1 from *Hansenula polymorpha*) and an algal transporter (CrNRT2.1 from *Chlamydomonas reinhardtii*).

Complementary indications arise from the patterns of the gene expression according to the plant organ and nutritional status (Table 17.1). Six of the seven members are expressed preferentially in the root; only the *AtNRT2.7* gene is expressed mostly in the aerial part (Orsel et al., 2002). The *AtNRT2.4* and *AtNRT2.5* genes exhibit a very different root expression pattern compared to *AtNRT2.1*. They all are induced by low nitrate supply conditions and short-term starvation, but while the *AtNRT2.1* gene is downregulated after a long-term starvation, the *AtNRT2.4* and *AtNRT2.5* genes are strongly and continuously induced. By now, this is the only condition under which the most expressed *AtNRT2* gene in root is not *AtNRT2.1* but *AtNRT2.5*. Nitrate resupply after a long-term starvation strongly induces the *AtNRT2.1* gene but represses the *AtNRT2.4* and *AtNRT2.5* genes (Orsel, unpublished results; Okamoto et al., 2003). Those genes might be important for the uptake and/or redistribution only when nitrate supply is extremely low and leads to a very low nitrate content in the plant. Okamoto et al. (2003) suggested a function of these genes in the nitrate release from the vacuole to the cytoplasm. Alternatively, as the expression is predominant in the root after a long-term starvation, they could be involved in a high-affinity transport system. Cerezo et al. (2001) first suggested the presence of this particular system with a very high affinity for nitrate and which is not encoded by the *AtNRT2.1* and *AtNRT2.2* genes. This hypothesis is reinforced by the analysis of the *atnrt2.1a* mutant: on low-nitrate supply conditions deficiency in the inducible HATS leads to a lower nitrate content in the shoot and to strong up-regulation of *AtNRT2.4* and *AtNRT2.5* genes (Orsel, unpublished results). Furthermore, the mutant is still able to take up nitrate in very low nitrate concentrations (Cerezo et al., 2001).

In contrast, the *AtNRT2.2* gene presents the same expression pattern as *AtNRT2.1,* but always at a very low expression level. *AtNRT2.3* and

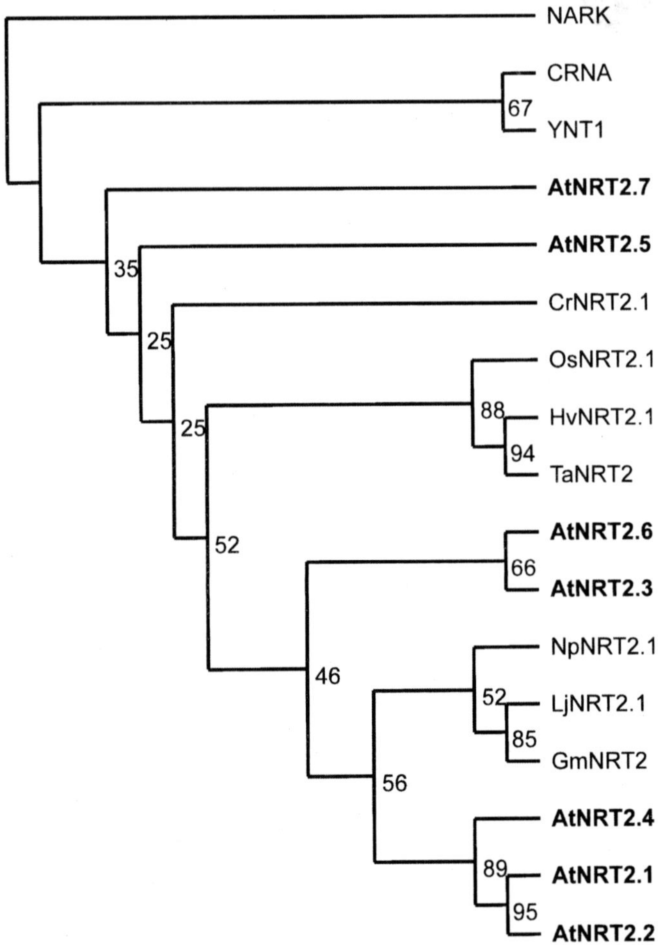

FIGURE 17.3. Unrooted, bootstrapped tree of NRT2 proteins. ClustalW was used to create an alignment of the amino acid sequences that was then bootstrapped (n = 1,000) to create the final tree using the Phylip package. Values indicate the number of times (in percent) that each branch topology was found during bootstrap analysis. GenBank protein ID numbers are AtNRT2.1: AAC64170, AtNRT2.2: AAC35884, AtNRT2.3: BAB10099, AtNRT2.4: BAB10098, AtNRT2.5: AAF78499, AtNRT2.6: CAB89321, AtNRT2.7: CAB87624, *Nicotiana plumbaginifolia* NpNRT2.1: CAA69387, *Lotus japonicus* LjNRT2: CAC35729, soybean GmNRT2: AAC09320, Rice OsNRT2: BAA33382, barley HvNRT2.1: AAC49531, wheat TaNRT2.1: AAK19519, *Chlamydomonas reinhardtii* CrNRT2.1: CAA80925, *Aspergillus nidulans* CRNA: AAA62125, *Hansenula polymorpha* YNT1: CAA93631, *Escherichia coli* NARK: CAA34126.

TABLE 17.1. *AtNRT2* expression profiles in reference to high nitrate supply.

AGI code		Tissue specificity	Low nitrate	Short starvation	Long starvation	Nitrate resupply[b]
At1g08090	*AtNRT2.1*	Root>>Shoot	+ +	++	=	+ +
At1g08100	*AtNRT2.2*	Root>>Shoot	+ +	++	=	+ +
At5g60780	*AtNRT2.3*	Root>Shoot	+			
At5g60770	*AtNRT2.4*	Root>>Shoot	+ + +	+	+ + +	–
At1g12940	*AtNRT2.5*	Root>>Shoot	+ +	+ + +	+ + + + +	– –
At3g45060	*AtNRT2.6*	Root≥Shoot	=			+
At5g14570	*AtNRT2.7*	Root<Shoot	=[a]	=[a]	=[a]	=[a]

\+ Induction
– Repression
= Unchanged
[a] Gene expression in shoots
[b] Gene expression in reference to long-term starvation

AtNRT2.6 are also expressed at low levels in the root part, and it is difficult to analyze their expression pattern even with a sensitive method such as RT-PCR (reverse transcription polymerase chain reaction). Expression of these genes might be tissue specific, and using a promoter-gene reporter system could be more informative.

Subcellular localization of the NRT2 proteins may also lead to the identification of transporters anchored in the tonoplast membrane to allow the passage of nitrate into the vacuole. There is no known peptide signal to the vacuole, but the sequence analysis of the AtNRT2.7 protein reveals a predicted peptide signal to the chloroplast (Schwacke et al., 2003). As there is no nitrate accumulated in the chloroplast, this addressing signal could be interpreted less specifically as a signal to a membrane other than the plasmalemma, such as the tonoplast. The *AtNRT2.7* gene is preferentially expressed in the shoot, and the protein is closely related to the *Chlamydomonas reinhardtii* transporter CrNRT2.1, which is a bispecific nitrate and nitrite porter. This could also indicate that AtNRT2.7 proteins are potentially involved in nitrite transport to the chloroplast.

Physiological characterization of overexpressing plants and mutants affected in the members of the family will help us to further assign a precise function to each *AtNRT2* gene in the nitrate transport and distribution processes.

TRANSCRIPTOME ANALYSIS IN RELATION TO VARIATIONS IN NITROGEN SUPPLY

As already mentioned in the introduction, the absence or presence of nitrogen in the rhizosphere has a profound impact on both plant metabolism and development. Among the possible inorganic N sources, it seems that nitrate has the most influence on plants. Nitrate indeed induces directly and quickly the expression of many genes at the transcriptional level, even at very low concentrations (10-50 μM). The most-studied case of nitrate induction was the one of NR (Cheng et al., 1986; Crawford et al., 1986; Calza et al., 1987), which was among the first examples in plants of an enzyme induced by its own substrate. Conversely, some genes are also known to be repressed by nitrate, such as the one coding for ADP-glucose pyrophosphorylase involved in starch synthesis (Scheible et al., 1997). Furthermore, nitrate influences important developmental processes such as root branching. Interestingly, nitrate can have opposite effects on secondary root proliferation in *Arabidopsis* depending on its concentration; at low concentration, or when applied locally, nitrate enhances lateral root elongation (Zhang et al., 1999), but systemic high nitrate concentrations leads to the decrease of lateral root numbers and of the shoot-root ratio.

All these data strongly suggest that plants have sophisticated regulatory networks responding to nitrate concentration and modulating the expression of many genes. However, little is known about the molecular mechanisms of nitrate signal perception and transduction. Only a few genes have been shown to have a role in these mechanisms (Zhang and Forde, 2000), although calcium, protein kinases, and phosphatases are clearly involved in the transduction of the nitrate signal (Sakakibara et al., 1997; Sueyoshi et al., 1999). Therefore, the identification of the genes whose expression is modulated by the N source has always been a tempting goal, since these genes could be a source of regulatory elements that will contribute to our understanding of nitrogen assimilation. First attempts of a systematic inventory of N-regulated sequences involved the use of 2-D gels to identify nitrate-induced polypeptides in corn seedling roots (Ni and Beevers, 1994) or of the differential display technique (Filleur and Daniel-Vedele, 1999). The latter experiment allowed the cloning of the first representative of the NRT2 nitrate transporter family in *Arabidopsis.* Another approach used subtractive hybridization procedures to screen for mRNAs that are rapidly induced by nitrate in roots (Zhang and Forde, 2000). Several cDNA clones derived from nitrate-inducible mRNAs were identified, and one of these (pANR1) was found to code for a MADS-box transcription factor (Zhang and Forde, 1998). *ANR1* expression is specifically induced by nitrate, and when its expression is down-regulated by antisensing, this leads to an inhibition of lateral root proliferation in nitrate-rich zones.

A systematic microarray analysis of nitrate-responsive genes was subsequently performed in *Arabidopsis* by the group of Nigel Crawford (Wang et al., 2000). This pioneering work was one of the first examples of the use of microarrays to determine global transcriptome changes in plants (Wu et al., 2001). Indeed, based on microarray data, we can now classify genes according to their expression profile, deduct their physiological role, and construct exhaustive lists of genes involved in specific functions such as nitrogen assimilation. In this study, *Arabidopsis* plants were grown in liquid culture with ammonium as the nitrogen source and submitted to either short-term nitrate induction at low concentration (20 min at 250 μM nitrate) or long-term induction at high nitrate concentration (2 h at 5 to 10 mM nitrate). As one would expect, more genes were found to be differentially expressed after 2 h of nitrate treatment than after 20 min. Interestingly, though, more genes were found to be induced by nitrate than repressed. The most-induced gene after short-term nitrate supply encodes uroporphyrin III methyltransferase (UPM1). UPM1 catalyzes an essential and specific step in the synthesis of siroheme, the heme cofactor of NiR, and sulfite reductase (Sakakibara et al., 1996). Interestingly, NiR itself is strongly regulated at the posttranscriptional level by nitrate in *Arabidopsis;* this ion is needed for the NiR protein to appear (Crété et al., 1997). One hypothesis could be that in the absence of nitrate no siroheme is made, which in turn impairs the synthesis of stable and active holo-NiR. In fact, several of the genes that are strongly induced by short-term nitrate treatment are involved in nitrite reduction. This includes NiR itself, ferredoxin, and ferredoxin-NADP oxidoreductase, which supply reducing power to NiR, as well as enzymes of the oxidative pentose phosphate pathway, which produces NADPH in nonphotosynthetic tissues (Wang et al., 2000). Most of these genes were already known to be induced by nitrate, but this study also unveiled novel nitrate-responsive genes, such as a Myb transcription factor, nonsymbiotic hemoglobin (AHB1), or senescence-associated proteins (SAGs). The expression of many of these genes was still affected after 2 h of nitrate induction. In addition, this experiment also revealed mRNAs that are specifically modulated by long-term nitrate supply. This study was very recently extended to the complete *Arabidopsis* transcriptome by using Affymetrix ATH1 chips (Wang et al., 2003).

In another experiment, high-density arrays consisting of 1,280 tomato root cDNAs, whose expression was modulated by mineral nutrition, were used to identify genes regulated by nitrate supply. Probes were obtained from young tomato seedlings starved for nitrate for 48 h and then resupplied with nitrate for 1 to 96 h (Wang et al., 2001). Among these cDNAs, 115 showed increased expression and comprised, as in *Arabidopsis*, genes already known to be regulated by nitrate, as well as 14 new nitrate-induced genes. The latter included genes coding for water channels, ion transporters, a leucine-zipper transcription factor homologous to the TGA1a tobacco transcription factor, a homologue of the auxin response element binding factor ARF1, 14-3-3 proteins, cell wall modifying enzymes, and ribosomal proteins (Wang et al., 2001). The most responsive genes included the ones coding for the elements of the nitrate assimilation pathway (nitrate transporters, NR, NiR, etc.).

Rice genes differentially expressed in a split-root experiment were also identified by suppression subtractive hybridization (SSH) screening (Wang et al., 2002). A split-root experiment was performed between 3 mM nitrate and 3 mM KCl for various periods of time and the obtained SSH library was screened with either forward- or reverse-subtracted probes. This process identified 37 known (with a putative function) and 55 unknown genes that were up-regulated in the nitrate side. Differentially expressed genes with known functions coded for enzymes involved in N, C, or S metabolism, for components of the translation machinery, or for receptor- and SNF1-like kinases. Among these 37 sequences, 11 were also identified in the *Arabidopsis* experiment described earlier, ten in the tomato one, and, interestingly, seven in both. A similar experiment using a split-root method to obtain nitrate-starved and nitrate-induced root tissues was conducted in *Arabidopsis* (Munos, 2002). The differentially expressed genes were identified by SAGE (serial analysis of gene expression; Veculescu et al., 1995), and this is one of the few examples of the use of this method in plants.

In our lab we have started an exhaustive analysis of the response of the *Arabidopsis* transcriptome to variations in nitrate supply. The first step was to design tightly controlled and reproducible growth conditions that would allow us to obtain homogeneous plant material. For this purpose, established *Arabidopsis* plants were transferred to hydroponic cultures for three weeks with sufficient nitrate (10 mM). Plants were then starved for nitrate for ten days and leaves and roots were harvested one, two, four, or ten days after nitrate removal. We first checked that these growth conditions actually resulted in N starvation. Indeed nitrate concentration was found to decrease very rapidly in roots, being almost absent after four days of nitrate starvation, and more slowly in shoots, probably because of its remobilization from roots. After ten days of nitrate starvation, plants were resupplied with either nitrate or ammonium for 6 h. This experimental design thus allows us to follow gene expression during nitrate starvation and also nitrogen supply. Moreover, the concomitant supply of ammonium and nitrate should help in the identification among nitrogen-regulated genes of genes either specifically induced or repressed by nitrate. RNA samples were prepared from each time point and used to obtain probes for microarray hybridization.

A collection of around 10,000 *Arabidopsis* cDNA clones was first built at INRA Evry representing about 6,000 different genes. EST clones have been gathered from different collections: the ABRC EST set (11,946 clones) and the French *Arabidopsis* EST sequencing program. The latter ESTs were issued from different locations: Versailles (1,344 clones), Orsay (312 clones), Gif-sur-Yvette (302 clones), and Perpignan (444 clones). Additional ESTs (25 clones) specifically involved in nitrogen metabolism were also added. Clones whose identity was reconfirmed by sequencing were used to print a first-generation glass microarray containing 10,000 cDNAs in duplicate (20,000 spots per slide). This set of arrayed cDNAs was used for the transcriptome analysis during nitrate starvation and resupply in leaves and roots by labeling first-strand cDNAs with Cy5 and Cy3 fluorescent molecules.

Because of the large amount of data coming from microarray techniques, statistical methods had to be developed and adapted to determine significantly the differences of gene expression among samples as well as to check the quality (reproducibility) of these expression data (Wu et al., 2001). Different types of biases, such as sample variability, which can be either biological or technical (variability in RNA extraction or labeling, array printing, etc.), or signal quantification (quenching, nonlinearity, etc.), require a normalization step. A preliminary analysis of the data obtained from the transcriptome analysis of our nitrate starvation/resupply experiments shows that a reproducible set of genes varies in response to changes in nitrate supply. For example, in the NRT2 nitrate transporter family, the *AtNRT2.1* gene is repressed by starvation and reinduced specifically by nitrate. The increase in expression is much more modest when ammonium is supplied as a nitrogen source. Conversely, the *AtNRT2.5* gene is induced also by nitrate starvation but repressed when nitrogen is resupplied. These results are in accordance with the ones described in the previous paragraph and validate these microarray data. Thus, microarray analysis confirms interesting information on the possible role of these two genes in the management of N resources: AtNRT2.5 could be more involved in the remobilization of stored nitrate, whereas AtNRT2.1 could be dedicated to the uptake of external nitrate.

One of the major drawbacks of EST-based microarrays is that cross hybridization will occur between members of the same gene family if their sequence identity is greater than 75 to 80 percent (Wu et al., 2001). Since a significant number of *Arabidopsis* genes are members of such families, this can seriously limit the use of these arrays. For example, in the *NRT2* gene family, the nucleotide sequences of the *AtNRT2.1* and *AtNRT2.2* genes are more than 87 percent identical. This can be circumvented either by the use of gene-sequence tags (GSTs), as in the CATMA chip (www.catma.org) developed at INRA Evry, or by using oligonucleotide-based chips. We have thus also analyzed transcriptome variations in our experiment by employing a high-density oligonucleotide chip (Affymetrix ATH1 chip). This chip allows the monitoring of the expression level of more than 23,000 different *Arabidopsis* genes (www.affymetrix.com). Affymetrix ATH1 chips were hybridized at the IGBMC facility (Strasbourg, France) with RNA extracted from samples harvested after 24 and 48 h of nitrate starvation as well as after resupply with nitrate or ammonium. Data were analyzed with the Affymetrix Microarray Suite package (MAS 5.0), which provided an accurate measurement of the expression level of each gene and induction-repression ratios between each experiment. In the Affymetrix procedure, only one labeled sample is analyzed per chip (instead of the double labeling of the control and test samples with Cy3 and Cy5 fluorochromes in the case of EST-based microarrays), and the MAS software provides an estimation of the level of expression of each gene (signal), as well as a "detection call" that determines if a given transcript is present or absent in the sample. These signals are then compared, after normalization, between two experiments and a signal $\log_2$ ratio is calculated to identify induced or repressed genes (see <www.affymetrix.com> for details). Leaf and root samples were mixed for the analysis of the response to nitrate starvation. After 24 h of nitrate starvation, the signal of 230 genes (about 1 percent of the *Arabidopsis* genes) showed more than a twofold decrease, whereas after 48 h 187 genes (among which 120 were already repressed after 24 h) showed the same decrease in expression level (Figure 17.4). This implies that 110 genes are repressed only up to 24 h of nitrate starvation and 67 only after this time point. Clearly, this kind of information could bring valuable new insights on how plants respond to nitrate starvation in the short term and how they manage it in the long term, with probably specific gene expression programs. Genes that were induced by nitrate starvation were found in comparable numbers (116 and 220 genes were induced more than twofold after, respectively, 24 and 48 h of nitrate starvation) and included several inorganic and organic ion transporters (data not shown).

When after ten days of starvation the plants were replenished with nitrate or ammonium for 6 h, many genes were also found to be induced or repressed, particularly in roots (Figure 17.5). Interestingly, most of the genes that responded to nitrogen supply were regulated specifically by nitrate and did not respond to ammonium supply. This strongly suggests that nitrate is not only a nutrient, similar to ammonium, but also a signal molecule leading

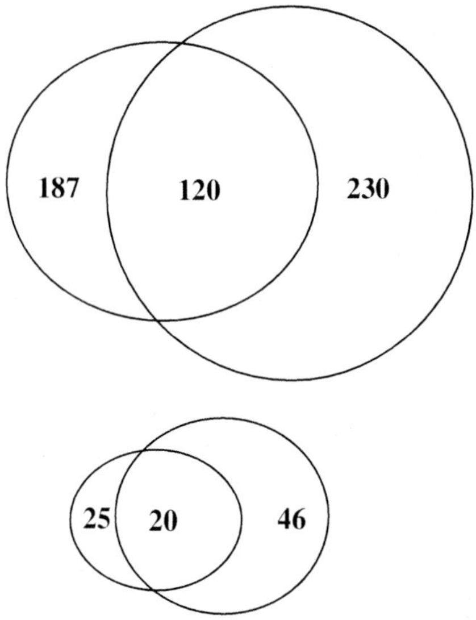

FIGURE 17.4. Number of *Arabidopsis* genes regulated by nitrogen starvation. Circles on the left: number of genes showing more than a twofold decrease after 48 h of nitrate starvation and which are called "present" at T_0 (start of nitrate starvation, top circle) and the subset of genes showing more than a fourfold decrease (bottom circle). Circles on the right: the same as on the left but after 24 h of starvation.

to the appearance of a new expression program in *Arabidopsis.* Interestingly, the genes that have been previously found as being induced by nitrate were also often identified as being up-regulated by nitrate in this experiment (Wang et al., 2000, 2001, 2002).

The efficiency of the Affymetrix ATH1 chip in discriminating between very similar sequences is illustrated by the differences in signal intensity between *AtNRT2.1* and *AtNRT2.2* (signal of about 3,000 and 90 after 48 h of nitrate starvation for, respectively, *AtNRT2.1* and *AtNRT2.2;* Figure 17.6), whereas these two RNAs share more than 87 percent of identity. These levels of expression were also confirmed by quantitative real-time PCR experiments showing that the signal provided by the Affymetrix chip is, in this case at least, a quantitative measurement of the level of gene expression. Microarray experiments have to be considered as preliminary steps of functional analysis of interesting genes, and the validity of the data, is required to be confirmed by complementary techniques such as real-time RT-PCR.

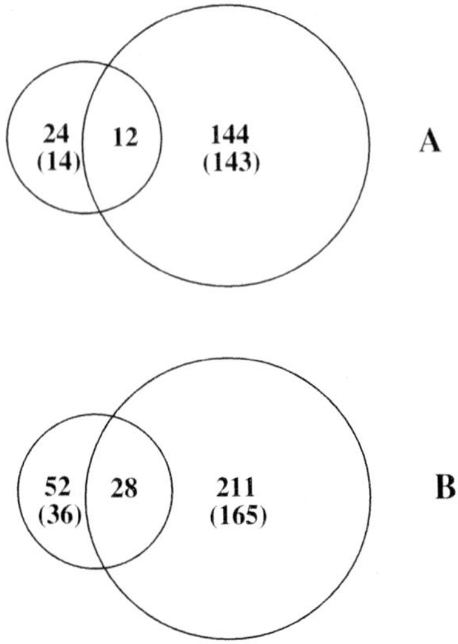

FIGURE 17.5. Number of *Arabidopsis* genes regulated by nitrogen supply. (A) Circles on the left: number of genes in leaves showing more than a fourfold decrease after 6 h of nitrate supply to plants that have been starved for ten days. Numbers of genes specifically repressed by nitrate (no change after ammonium supply for 6 h) are given in parentheses below these numbers. Circles on the right: the same as on the left but in roots. (B) Circles on the left: Number of genes in leaves showing more than a fourfold increase after 6 h of nitrate supply to plants that have been starved for ten days. Numbers of genes specifically induced by nitrate (no change after ammonium supply for 6 h) are given in parentheses below these numbers. Circles on the right: the same as on the left but in roots.

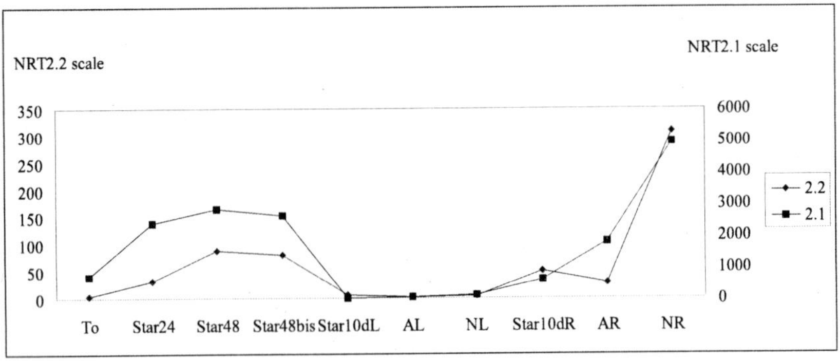

FIGURE 17.6. Expression level of the *AtNRT2.1* and *AtNRT2.2* genes measured by hybridization on Affymetric ATH1 chip. Star, samples from plants starved for 24 or 48 h or ten days; L, leaves; R, roots; A, ammonium induced; N, nitrate induced.

CONCLUSION

Our understanding of the nitrogen metabolism in plants is clearly far from being complete. Thus, it is of importance to first identify and characterize all the genes whose expression is modulated by variations in nitrogen supply. Furthermore, many genes are already known in other (often model) organisms as being involved in the regulation of nitrogen metabolism and in the integration of this metabolic pathway with other developmental processes, such as cell cycling or reproduction. The study of their orthologues in *Arabidopsis* should serve as a starting point to elucidate further this regulation. Indeed, despite extensive studies, little is known in plants about the regulatory pathways involved in the control of these processes. Thus, we can anticipate that the extensive use of the functional genomics data that start to accumulate will open new avenues of research and provide a better knowledge of the role of nitrate as both a nutrient and a signal in plants.

REFERENCES

Behl, R., Tischner, R., Raschke, K. (1988). Induction of a high capacity nitrate uptake mechanism in barley roots prompted by nitrate uptake through a constitutive low capacity mechanism. *Planta* 176: 235-240.

Calza, R., Huttner, E., Vincentz, M., Rouze, P., Galangau, F., Vaucheret, H., Chérel, I., Meyer, C., Kronenberger, J., Caboche, M. (1987). Cloning of DNA fragments complementary to tobacco nitrate reductase mRNA and encoding epitopes common to the nitrate reductase from higher plants. *Mol. Gen. Genet.* 209: 552-562.

Campbell, W.H. (1999). Nitrate reductase structure, function and regulation: Bridging the gap between biochemistry and physiology. *Annu. Rev. Plant Physiol. Plant Mol. Biol.* 50: 277-303.

Cerezo, M., Tillard, P., Filleur, S., Munos, S., Daniel-Vedele, F., Gojon, A. (2001). Major alterations of the regulation of root NO_3^- uptake are associated with the mutation of *Nrt2.1* and *Nrt2.2* genes in *Arabidopsis*. *Plant Physiol.* 127: 262-271.

Cheng, C.L., Dewdney, J., Kleinhofs, A., Goodman, H.M. (1986). Cloning and nitrate induction of nitrate reductase mRNA. *Proc. Natl. Acad. Sci. USA* 83: 6825-6831.

Coruzzi, G.M., Zhou, L. (2001). Carbon and nitrogen sensing and signaling in plants: Emerging "matrix effects." *Curr. Opin. Plant Biol.* 4: 247-253.

Crawford, N.M. (1995). Nitrate: Nutrient and signal for plant growth. *Plant Cell* 7: 859-868.

Crawford, N.M., Arst, H.N.J. (1993). The molecular genetics of nitrate assimilation in fungi and plants. *Annu. Rev. Genet.* 27: 115-146.

Crawford, N.M., Campbell W.H., Davis, R.W. (1986). Nitrate reductase from squash: cDNA cloning and nitrate reduction. *Proc. Natl. Acad. Sci. USA* 83: 8073-8076.

Crété, P., Caboche, M., Meyer, C. (1997). Nitrite reductase expression is regulated at the post-transcriptional level by the nitrogen source in *Nicotiana plumbaginifolia* and *Arabidopsis thaliana. Plant J.* 11: 625-634.

Daniel-Vedele, F., Filleur, S., Caboche, M. (1998). Nitrate transport: A key step in nitrate assimilation. *Curr. Opin. Plant Biol.* 1: 235-239.

Filleur, S., Daniel-Vedele, F. (1999). Expression analysis of a high-affinity nitrate transporter isolated from *Arabidopsis thaliana* by differential display. *Planta* 207: 461-469.

Filleur, S., Dorbe, M.F., Cerezo, M., Orsel, M., Granier, F., Gojon. A., Daniel-Vedele, F. (2001). An *Arabidopsis* T-DNA mutant affected in *Nrt2* genes is impaired in nitrate uptake. *FEBS Lett.* 489: 220-224.

Forde, B.G. (2000). Nitrate transporters in plants: Structure, function and regulation. *Biochim. Biophys. Acta* 1465: 219-235.

Glass, A., Brito, D., Kaiser, B., Kronzucker, H., Okamoto, M., Rawat, S., Siddiqi, M., Silim, S., Vidmar, J., Zhuo, D. (2001). Nitrogen transport in plants, with an emphasis on the regulation of fluxes to match plant demand. *J. Plant Nutr. Soil Sci.* 164: 199-207.

Kang, J., Turano, F.J. (2003). The putative glutamate receptor 1.1 (AtGLR1.1). functions as a regulator of carbon and nitrogen metabolism in *Arabidopsis thaliana. Proc. Natl. Acad. Sci. USA* 100: 6872-6877.

Lejay, L., Tillard, P., Lepetit, M., Olive, F.D., Filleur, S., Daniel-Vedele, F., Gojon, A. (1999). Molecular and functional regulation of two uptake systems by N- and C-status of *Arabidopsis* plants. *Plant J.* 18: 509-519.

Liu, K.H., Huang, C.Y., Tsay, Y.F. (1999). Chl1 is a dual-affinity nitrate transporter of arabidopsis involved in multiple phases of nitrate uptake. *Plant Cell* 11: 865-874.

Liu, K.H., Tsay, Y.F. (2003). Switching between the two action modes of the dual-affinity nitrate transporter CHL1 by phosphorylation. *EMBO J.* 22: 1005-1013.

Melo-Oliveira, R., Oliveira, I.C., Coruzzi, G.M. (1996). *Arabidopsis* mutant analysis and gene regulation define a nonredundant role for glutamate dehydrogenase in nitrogen assimilation. *Proc. Natl. Acad. Sci. USA* 93: 4718-4723.

Menand, B., Desnos, T., Nussaume, L., Berger, F., Bouchez, D., Meyer, C., Robaglia, C. (2002). Expression and disruption of the *Arabidopsis TOR* (target of rapamycin) gene. *Proc. Natl. Acad. Sci. USA* 99: 6422-6427.

Meyer, C., Stitt, M. (2001). Nitrate reduction and signalling. In Morot-Gaudry, J.-F., Lea, P.J. (Eds.), *Plant Nitrogen* (pp. 37-60). Paris, France: Springer-Verlag-INRA Editions.

Meyer, C., Stoehr, C. (2002). Soluble and plasma membrane-bound enzymes involved in nitrate and nitrite metabolism. In Foyer, C.H., Noctor, G. (Eds.), *Photosynthetic Nitrogen Assimilation and Associated Carbon and Respiratory Metabolism* (pp 49-62). Dordrecht, The Netherlands: Kluwer Academic Publisher.

Morot-Gaudry, J.-F. (2001). *Nitrogen Assimilation by Plants.* Enfield, NH: Science Publishers, Inc.

Morot-Gaudry, J.-F., Lea, P.J. (2001). *Plant Nitrogen.* Paris, France: Springer-Verlag-INRA Editions.

Munos, S. (2002). Reprogrammations génétiques induites en réponse à la déficience nutritionnelle en N chez *Arabidopsis thaliana.* Thesis, Université Montpellier II, France.

Ni, M., Beevers, L. (1994). Nitrate-induced polypeptides in membranes from corn seedling roots. *J. Exp. Bot.* 45: 355-365.

Okamoto, M., Vidmar, J.J., Glass, A.D. (2003). Regulation of *NRT1* and *NRT2* gene families of *Arabidopsis thaliana:* Responses to nitrate provision. *Plant Cell Physiol.* 44: 304-317.

Orsel, M, Filleur, S., Fraisier, V., Daniel-Vedele, F. (2002). Nitrate transport in plants: Which gene and which control. *J. Exp. Bot.* 370: 825-834.

Quesada, A., Galvàn, A., Fernàndez, E. (1994). Identification of nitrate transporter genes in *Chlamydomonas reinhardtii. Plant J.* 5: 407-419.

Sakakibara, H., Kobayashi, K., Deji, A., Sugiyama, T. (1997). Partial characterization of the signaling pathway for the nitrate-dependent expression of genes for nitrogen-assimilatory enzymes using detached maize leaves. *Plant Cell Physiol.* 38: 837-843.

Sakakibara, H., Takei, K., Sugiyama, T. (1996). Isolation and characterization of a cDNA that encodes maize uroporphyrinogen III methyltransferase, an enzyme involved in the synthesis of siroheme, which is a prosthetic group of nitrite reductase. *Plant J.*10: 883-892.

Scheible, W.R., Gonzalez-Fontes, A., Lauerer, M., Muller-Rober, B., Caboche, M., Stitt, M. (1997). Nitrate acts as a signal to induce organic acid metabolism and repress starch metabolism in tobacco. *Plant Cell* 9: 783-798.

Schwacke, R., Schneider, A., van der Graaff, E., Fischer, K., Catoni, E., Desimone, M., Frommer, W.B., Flügge, U-I., Kunze, R. (2003). ARAMEMNON, a novel database for *Arabidopsis* integral membrane proteins. *Plant Physiol.* 131: 16-26.

Siddiqi, M.Y., Glass, A.D.M., Ruth, T.J., Rufty, T.W. (1990). Studies of the uptake of nitrate in barley. I. Kinetics of $^{13}NO_3^-$ influx. *Plant Physiol.* 93: 1426-1432.

Stitt, M. (1999). Nitrate regulation of metabolism and growth. *Curr. Opin. Plant Biol.* 2: 178-186.

Sueyoshi, K., Mitsuyama, T., Sugimoto, T., Kleinhofs, A., Warner, R.L., Oji, Y. (1999). Effects of inhibitors for signaling components on the expression of the genes for nitrate reductase and nitrite reductase in excised barley leaves. *Soil Sci. Plant Nutr.* 45: 1015-1019.

Tsay, Y.F., Schroeder, J.I., Feldmann, K.A., Crawford, N.M. (1993). The herbicide sensitivity gene *CHL1* of *Arabidopsis* encodes a nitrate-inducible nitrate transporter. *Cell* 72: 705-713.

Unkles, S.E., Hawker, K.L., Grieve, C., Campbell, E.I., Montague, P., Kinghorn, J.R. (1991). *cnrA* encodes a nitrate transporter in *Aspergillus nidulans. Proc. Natl. Acad. Sci. USA* 88: 204-208.

Velculescu, V.E., Zhang, L., Vogelstein, B., Kinzler, K.W. (1995). Serial analysis of gene expression. *Science* 270: 484-487.

von Wiren, N., Gazzarrini, S., Gojin, A., Frommer, W.B. (2000). The molecular physiology of ammonium uptake and retrieval. *Curr. Opin. Plant Biol.* 3: 254-261.

Wang, R., Liu, D., Crawford, N.M. (1999). The *Arabidopsis* CHL1 protein plays a major role in high-affinity nitrate uptake. *Proc. Natl. Acad. Sci. USA* 95: 15134-15139.

Wang, R., Guegler, K., LaBrie, S.T., Crawford, N.M. (2000). Genomic analysis of a nutrient response in *Arabidopsis* reveals diverse expression patterns and novel metabolic and potential regulatory genes induced by nitrate. *Plant Cell* 12: 1491-1509.

Wang, R., Okamoto, M., Xing, X., Crawford, N.M. (2003). Microarray analysis of the nitrate response in *Arabidopsis* roots and shoots reveals over one thousand rapidly responding genes and new linkages to glucose, trehalose-6-P, iron and sulfate metabolism. *Plant Physiol.* 132: 556-567.

Wang, X., Wu, P., Xia, M., Wu, Z., Chen, Q., Liu, F. (2002). Identification of genes enriched in rice roots of the local nitrate treatment and their expression patterns in split-root treatment. *Gene* 297: 93-102.

Wang, Y.-H., Garvin, D.F., Kochian, L.V. (2001). Nitrate-induced genes in tomato roots: Array analysis reveals novel genes that may play a role in nitrogen nutrition. *Plant Physiol.* 127: 345-359.

Wu, S.-H., Ramonell, K., Gollub, J., Somerville, S. (2001). Plant gene expression profiling with DNA microarrays. *Plant Physiol. Biochem.* 39: 917-926.

Yamasaki, H., Sakihama, Y. (2000). Simultaneous production of nitric oxide and peroxynitrite by plant nitrate reductase: In vitro evidence for the NR-dependent formation of active nitrogen species. *FEBS Lett.* 468: 89-92.

Zhang, H.M., Forde, B.G. (1998). An *Arabidopsis* MADS box gene that controls nutrient-induced changes in root architecture. *Science* 279: 407-409.

Zhang, H.M., Forde, B.G. (2000). Regulation of *Arabidopsis* root development by nitrate availability. *J. Exp. Bot.* 51: 51-59.

Zhang, H.M., Jennings, A., Barlow, P.W., Forde, B.G. (1999). Dual pathways for regulation of root branching by nitrate. *Proc. Natl. Acad. Sci. USA* 96: 6529-6534.

Chapter 18

Functional Genomics of Plant Salinity Tolerance

Hans J. Bohnert

SALINITY TOLERANCE IN CONTEXT

A number of environmental conditions can produce meaningful stresses for plants because they can be life threatening and, when in combination, determine species distribution and the productivity of crops (Boyer, 1982). Drought, high or fluctuating salinity, and low and freezing temperatures account for most production losses, but flooding, high light, ozone, ion deficiency or imbalance, heavy metals, and soil structure are additional challenging factors. Yield reduction due to drought and ion imbalance amounts to approximately two-thirds of all losses in agriculture, much more than the reduction due to pathogens or insect damage combined. Despite considerable breeding efforts, breeding for tolerance to water deficit or salinity has not had much impact. Various strategies have been employed for "stress" tolerance breeding (Flowers and Yeo, 1995), including more recently marker-assisted mapping and QTL (quantitative trait loci) studies (Menz et al., 2002; Quarrie et al., 1997). With the introduction of high-throughput technologies for transcript and protein analysis, and an increasing number of transcript sequences in the form of expressed sequence tags (ESTs) and genome sequences, classical breeding strategies can now be enhanced (Blum, 1988; Bohnert and Bressan, 2001). The value of these new approaches, summarily labeled with the extension "-omics," is yet to be proved in the real world, but, indisputably, the past years of genomics-type

Work in the author's laboratory has been or is supported by the U.S. Department of Agriculture, U.S. Department of Energy, and the U.S. National Science Foundation. I thank Dr. J. C. Cushman, University of Nevada, Reno, for the permission to include unpublished data. Support and help in preparing the manuscript by Mark Fredricksen, Saori Miyazaki, Valeriy Poroyko, Don Shepley, and Debra Townsley is thankfully acknowledged.

studies in plant systems have generated large databases that can be mined to generate a better understanding of plant functions by analyzing genetic diversity.

Equally enhanced by genomics-type work are physiological studies focusing on plant responses to abiotic stresses, although the two lines of research still exist largely in separate arenas. In the search for stress tolerance mechanisms, plant biologists have produced data in a correlative way in many different species. The biochemical and biophysical principles of stress action and plant response were outlined long ago (Levitt, 1980). Subsequently, transcript abundance and stress-induced changes in gene and protein expression have been monitored in many species under different conditions. The focus on a wide array of conditions, lines, and organisms has, at least in part, held up more rapid progress because analyses targeted peripheral specific stress responses rather than the underlying principles. The past decade then has brought drastic changes in our views about what might constitute stress sensitivity, tolerance, or resistance in a species, and most progress was based on results from molecular genetics experiments. Many or most stress responses are based on stress recognition, the interaction of several separate signal transduction pathways leading to the transcriptional activation of downstream biochemical pathways. The individual genes that have since been characterized as encoding relevant endpoint proteins, induced by a stress condition, point to the importance of defense reactions toward radical oxygen compounds, protection of protein and cell structure, and a general restructuring of metabolism. More recently, the focus has shifted to analyzing the underlying mechanisms of stress sensing, the associated signal transduction conduits, and how they connect to the previously known physiological elements of the stress response.

Genomics technologies now provide a course of action that uses an integrated, systems-oriented approach to understanding biological processes (Brazhnik et al., 2002; Kitano, 2002). As of 2004, the almost complete DNA sequence of the *Arabidopsis* genome and a draft version of the rice genome (see Chapter 11) are available, and more than a million plant ESTs and cDNA sequences have been documented, giving rise to an accelerated pace of gene family and pathway analyses, and a beginning of comparative functional genomics (Arabidopsis Genome Initiative, 2000; <http://www.ncbi.nlm.nih.gov/dbEST/index.html>; e.g., Goff et al., 2002; Yu et al., 2002; Bennetzen, 2002; Buell, 2002; Huala et al., 2001; Shimamoto and Kyozuka, 2002). These resources are complemented by an increasing number of mutant lines, predominantly in the major model species, *Arabidopsis thaliana* and *Oryza sativa,* but increasingly also in other species, in which individual genes have been tagged by transposon or T-DNA insertions (e.g., Krysan et al., 1999; Bressan et al., 2001; see also <http://signal.salk.edu/cgi-

bin/tdnaexpress>). Experiments with these mutants, guided by the data and results from high-throughput transcript, protein, and metabolite analyses (Chen et al., 2002; Fowler and Thomashow, 2002; Trethewey, 2001) (see Chapters 4, 5, 14, and 15), have become possible for "proof of concept" experiments.

Interest in the topic of salinity stress genomics includes the scientific justification, satisfying our curiosity about how tolerance or resistance evolved and what its mechanisms are. Another imperative and rationale for further research is that salinization inevitably accompanies long-term irrigation of the soil. What is happening today in growing areas with elaborate irrigation schemes is reminiscent of events that led to the decline of ancient civilizations, with the countries of the former fertile crescent being the most lucid example. The buildup of sodium chloride and other salts in the soil has now affected up to one-half of the area under irrigation (Flowers and Yeo, 1995). Salinization has become a major problem in arid areas of Australia, the Middle East, northern Africa, the west coast of South America, the West and Midwest of the United States, parts of the Indian subcontinent, and central Asia (http://www.undp.org/popln/fao/water.html). Genomics-based strategies that are applied to gain a better understanding of the constituent multiple components of plant salinity stress responses include (1) profiles of transcripts derived from cDNA libraries from plants that experienced a stress episode, (2) the evaluation of microarray hybridizations that compare stress and control conditions, and (3) the analysis of mutants with either reduced or enhanced salinity stress tolerance phenotypes.

PROBLEMS PLANTS FACE IN HIGH SALINITY

Plants are permanently exposed to their physical environment, which includes changes in seasonal and diurnal cycles, and extreme and stressful deviations from the climatic mean. For continued vegetative growth and the development of reproductive organs under stress, plants must, above all, obtain water. To satisfy this essential requirement, each of the many adaptive mechanisms must be subordinate to this goal. When stomata are closed due to water deficit (drought) or high salinity (unfavorable osmotic potential) to limit water loss, a series of events engage to adjust photosynthesis, carbon fixation, and carbohydrate transport, initiating processes to maintain the integrity of the photosynthetic and carbon fixation apparatus (Long et al., 1994; Horton et al., 1996). In high or fluctuating salinity, the main problems are twofold. The first is decreased water potential of the surrounding medium that must be compensated for to sustain water influx, and the second is the necessity either to exclude excess or to partition extra ions into metabolically inert compartments or structures. Associated with these is the need to distinguish the essential macronutrient, K^+, from sodium to maintain potassium homeostasis in high-salinity soils (Niu et al., 1995). Lack of water and high sodium concentrations prolong the growing cycle, increase vulnerability to pathogens, and ultimately result in decreased competitiveness or yield.

High salinity causes both hyperionic and hyperosmotic effects (Niu et al., 1995). Most commonly, the stress is caused by high Na^+ and Cl^- concentrations in the soil solution. Altered water status most likely brings about initial growth reduction; however, the precise contribution of subsequent processes to inhibition of cell division and expansion and acceleration of cell death has not been well elucidated (Munns et al., 2000; Munns, 2002). Membrane disorganization, reactive oxygen species, metabolic toxicity, inhibition of photosynthesis, and attenuated nutrient acquisition are factors that initiate more catastrophic events. Movement of salt into roots and to shoots is a product of the transpirational flux required to maintain the water status of the plant (Flowers and Yeo, 1992). Unregulated, transpiration can result in toxic levels of ion accumulation in the aerial parts of the plant. An immediate response to salinity that mitigates ion flux to the shoot is stomatal closure. However, because of water potential differences between atmosphere and leaf cells, and the need for carbon fixation, this is an untenable long-term strategy.

The search for mechanisms by physiological approaches has generated much data and some correlations, which have in part been tested by transgenic alterations and further physiological studies (Bartels and Nelson, 1994; Yancey, 1994; Jain and Selvaraj, 1997; Hare et al., 1997; Bohnert and Shen, 1999; Hasegawa et al., 2000). Figure 18.1 outlines in general terms classes of processes (e.g., proton pumping), proteins (e.g., chaperones, ion transporters), and metabolites (e.g., glycine betaine, polyols) that are associated with plant salinity responses. The primary physiological mechanisms on which most analyses agree include an increase in proton pumping, compatible solute synthesis and accumulation, partitioning of sodium (either to the vacuole or to the outside of the cell), increased (hydroxyl) radical scavenging, and control over homeostasis of essential nutrient ions, water, and nitrogen and carbon allocation. Several of these putative mechanisms have been analyzed by the overexpression of genes (Apse et al., 1999; Hayashi et al., 1997; McKersie et al., 1996; Van Camp et al., 1996; Tarczynski et al., 1993; Zhifang and Loescher, 2003; Shi et al., 2003). A crucial role for maintenance of potassium homeostasis in the presence of external high sodium concentrations has frequently been stressed (Niu et al., 1995; Zhu et al., 1998; Amtmann and Sanders, 1999; Rus et al., 2001; Serrano and Rodriguez-Navarro, 2001). The ways these mechanisms are induced require altered or increased activity of many other pathways, includ-

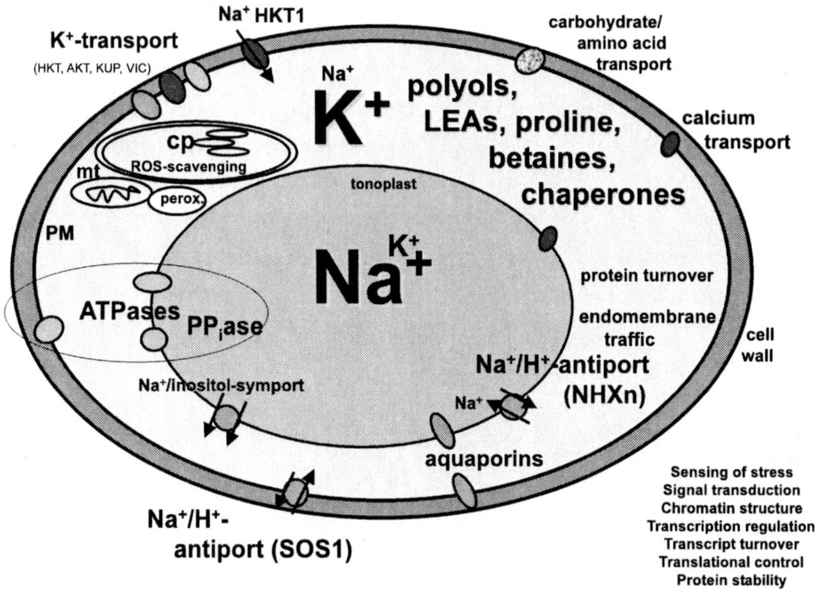

FIGURE 18.1. Schematic view of a plant cell indicating established downstream effectors, metabolites, ions, and functions of salinity stress tolerance. The scheme includes physiological, biochemical, and cell structural categories that have been associated with enhanced tolerance. Additional important aspects are the sensing of stress, transduction of signals, chromatin modification, transcription regulation, transcript maturation and turnover, translational control at the ribosome, or the compartment-targeting or destination levels, and, finally, the control of protein stability and degradation (modified after Hasegawa et al., 2000). Schematically indicated are the cell wall, plasma membrane, cytosol and vacuole (note: several different vacuolar compartments exist), and cell organelles: cp, chloroplast; mt, mitochondrium; perox., peroxisome. The major proton ATPases in the plasma membrane and tonoplast are shown. Their regulation is crucial for the ionic stress response (e.g., Zhu, 2002). The figure includes several transporters (calcium, amino acids/sugars, potassium/sodium) important in the ionic stress response and points to the importance of protein modification/turnover and altered endomembrane traffic. In a coordinated fashion, the partitioning of potassium and sodium in different compartments (or export to the extracellular space) and the osmotic adjustment provided by various accumulating metabolites support homeostasis. These physiological and biochemical parameters represent the "downstream" end result of other processes, identified in the figure, that precede the physiological response. These processes are essential for sensing and appropriately responding to the stress in order to alter cellular and organismal metabolism.

ing protein turnover and endomembrane traffic, protein stability and turnover, and alterations of the transcription and translation machinery. Up until now, the transgenic studies have produced results that are often not easily explained, typically in the form of miniscule improvement under controlled conditions that have not been replicated in fields. Taken together, physiological and transgenic studies have uncovered mechanisms, while more recently, some elements of the sensing and signaling paths were uncovered (see Serrano, 1996; Hasegawa et al., 2000; Zhu et al., 1998; Zhu, 2001, 2002, for reviews). Can we now learn more from a genomics-oriented approach?

THE GENOMICS TOOL KIT APPLIED TO SALINITY STRESS RESEARCH

In the past, the dichotomy in lifestyle that exists between glycophytic and halophytic species let researchers focus on the significant physiological and behavioral differences in the stress response in evolutionarily unrelated species, or on generating transgenic models by transferring a recognized halophytic trait into a glycophyte (Hasegawa et al., 2000; Cushman and Bohnert, 2000). The advance of genomics has reduced the appeal and necessity of this approach in two ways, but genomics-based approaches have equally generated new ways, by transgenically targeting either signal transduction pathways or overexpressing transcription factor genes to manipulate stress-dependent pathways and initiate new tolerance engineering designs (e.g., Fowler and Thomashow, 2002; Kreps et al., 2002; Seki et al., 2002). These reports use a combination of characterized mutants and expression profiles that include a large percentage of the transcriptome to determine groups of stress-regulated transcripts that are under the control of distinct signaling and transcription control circuits. Seemingly diminishing the importance of using halophytes for salinity stress research are the following considerations. Genome sequences and large EST collections from many species now indicate that genes, which had before been characterized in halophytes, seem to be present in all plants, and, second, it has become obvious that significant genetic plasticity exists in reproductively isolated, yet interfertile, populations of a single species, and in closely related species within families. What made the comparison and analysis by wide crosses difficult to interpret between, for example, *Lycopersicon esculentum* and *L. pennellii,* which are distinguished by the abiotic stress tolerance levels, has become possible by applying genomics tools. As another example, the *Arabidopsis* model provides materials for abiotic stress research in the comparison of ecotypes and closely related species in the Brassicaceae family. One of these *Arabidopsis* relatives, *Thellungiella halophila,* is adapted to extreme salinity levels while it provides the advantages of transformation and tagging very similar to *Arabidopsis,* and high sequence identity of its genes to those in *Arabidopsis thaliana* (Bressan et al., 2001). While the choice of a model system to study salinity stress in the past was dictated by physiological or ecological considerations, the genomics tool kit has changed the playing field. Considering the number and type of enlightening mutant phenotypes that have resulted from the tagging of specifically designed *Arabidopsis* salinity stress reporter lines, a much more instructive result can be expected from tagged mutants of *Thellungiella* with its much wider tolerance spectrum.

To study plant salinity stress tolerance reactions, transcript sequences and powerful new tools have become available in recent years. Foremost, these are mutant collections and transcript microarrays and increasingly enhanced by tools monitoring protein-protein interactions, protein location within the cell, and bioinformatics to integrate the data. Salinity stress-related protein and metabolite analyses on a genomic scale are currently less developed and less practiced, but both areas will gain more importance in the near future. Some analyses however, have already addressed the involvement of classes of proteins, such as the *S*-glutathionylated proteins during oxidative stress (Lind et al., 2002), or proteins changing short term under cold treatment (Tafforeau et al., 2002). Catalogs of proteins in tissues or subcellular compartments represent another way to connect proteome dynamics with gene-expression changes (Mathesius et al., 2001; Peltier et al., 2002). The tagged lines for *Arabidopsis thaliana* (http://signal.salk.edu/cgi-bin/tdnaexpress) that cover most of the genome have quickly become an essential element for advancing genomics. Similar collections of mutants are also emerging for tomato, *Medicago truncatula* and *Lotus japonicus* (e.g., Suzuki et al., 2002), and the moss *Physcomitrella patens* (Holttorf et al., 2002). Likewise, mutant lines based on transposon insertion are becoming available for cereal crops, most notably rice and corn (http://www.intlpag.org/10/abstracts/PAGX_P336.html; www.zmdb.iastate.edu/zmdb/NSFgrant.pdf). The *Arabidopsis* tagged lines, and many of the insertion lines, provide location within the genome but require additional analysis to determine a phenotype. Alternative approaches are those that generate lines in which, for example, a stress-affected promoter is linked to a reporter, followed by insertion mutagenesis of the reporter line, which provides immediate recognition of the desired phenotype (Ishitani et al., 1997).

In addition, valuable information has been provided by stress-modified transcript profiles in the form of ESTs that are obtained from cDNA libraries that used "stressed" plant tissues as the starting material (Table 18.1). The EST profiles from tissues stressed before harvest are fundamentally

TABLE 18.1. Plant cDNA libraries derived from "stressed" cells or tissues.

Library	Organism	#ESTs	Stress
Various tissues and times, NaCl (libraries AA, AB, AC)	*Arabidopsis thaliana*	1,982	Salt
AD library (Col-0 *gl1*) library enriched for salt-induced transcripts; 10- to 14-day seedlings; 4 h 160 mM NaCl stress	*Arabidopsis thaliana*	151	Salt
(Col-0 *gl1*) subtracted library enriched for salt-induced transcripts; 10- to 14-day seedlings 4 h 160 mM NaCl stress	*Arabidopsis thaliana*	571	Salt
Acute ozone forward-subtracted library	*Arabidopsis thaliana*	171	Ozone
Acute ozone reverse-subtracted library	*Arabidopsis thaliana*	27	Ozone
Chronic ozone forward-subtracted library	*Arabidopsis thaliana*	129	Ozone
NaCl-treated *Arabidopsis* subtraction library	*Arabidopsis thaliana*	184	Salt
OICL, FMB, TAMU	*Atriplex canescens*	4	Ozone
Leaf cDNA library	*Avicennia marina*	1,005	Salt
Sugar beet germination cDNA library	*Beta vulgaris*	1,228	Salt and mannitol
Sugar beet stress germination cDNA library (subtracted)	*Beta vulgaris*	97	Salt and mannitol
Mannitol stress-inducible cDNA library	*Brassica juncea*	3	Mannitol
Short roots subtractive library	*Brassica napus*	72	Drought
Leaf treated with 0.5 M NaCl	*Bruguiera gymnorhiza*	75	Salt
Strain CC-1690, stress condition I, normalized, lambda Zap II	*Chlamydomonas reinhardtii*	7,803	Nitrogen
Strain CC-1690, stress II (normalized), lambda Zap II	*Chlamydomonas reinhardtii*	23,888	Nitrogen
5 percent to 0.04 percent CO_2	*Chlamydomonas reinhardtii*	12,949	CO_2
Chlamydomonas Uni-Zap XR	*Chlamydomonas reinhardtii*	85	CO_2
HS-5 lambda ZAP II	*Chlamydomonas* sp.	28	Stress
W80 lambda ZAP II	*Chlamydomonas* sp.	82	Stress
Whole cell NaCl up-shift	*Dunaliella salina*	2,844	Salt
Gc01 AAFC ECORC cold-stressed	*Glycine clandestina*	686	Cold
Gc02 AAFC ECORC cold-stressed SSH	*Glycine clandestina*	247	Cold

Gm-c1065	*Glycine max*	8,098	Cold
Gm-c1066	*Glycine max*	3,088	Salt
Gm-c1068	*Glycine max*	3,398	Drought
Cotton drought-tolerant genotype, leaf	*Gossypium hirsutum*	21	Drought
Sunflower drought-stressed leaves	*Helianthus annuus*	5	Drought
Sunflower seedlings	*Helianthus annuus*	12	Salt
Seedling shoot EST library HVcDNA0001 (cold stress)	*Hordeum vulgare*	4,460	Cold
Seedling shoot EST library HVcDNA0002 (dehydration)	*Hordeum vulgare*	4,905	Drought
Young plants, drought stressed (root and shoot libraries)	*Hordeum vulgare*	1,930	Drought
Hordeum vulgare treated with 150 mM NaCl	*Hordeum vulgare*	21	Salt
IGF barley EBro04 library	*Hordeum vulgare*	250	Salt
IGF barley EBro08 library	*Hordeum vulgare*	250	Drought
ITEC ISC barley leaf library	*Hordeum vulgare*	957	Cold
Salt-stressed tomato roots	*Lycopersicon esculentum*	20	Salt
Tomato mixed elicitor, BTI	*Lycopersicon esculentum*	10,014	Variety
Medicago plants, drought	*Medicago truncatula*	8,416	Drought
Medicago nitrogen-deficient, MtBA treatment	*Medicago truncatula*	7,939	Nitrogen
Lambda Uni-Zap XR expression library, 30 h NaCl treatment	*Mesembryanthemum crystallinum*	2,698	Salt
Lambda Uni-Zap XR expression library, 48 h NaCl treatment	*Mesembryanthemum crystallinum*	3,604	Salt
Various tissues, various times of NaCl stress (libraries ME, MG, MH, MI, MJ, MK, ML, MM, MN, MP)	*Mesembryanthemum crystallinum*	2,745	Salt
Drought stress (leaf)	*Oryza sativa*	1,377	Drought
Drought stress (root)	*Oryza sativa*	200	Drought
IRRI drought stress panicle library	*Oryza sativa*	9,659	Drought
Various tissues, various times of NaCl stress (libraries OA, OD, OE, OF, OG)	*Oryza sativa*	1,701	Salt
Subtracted cDNA library of JA/BTH-treated rice leaf	*Oryza sativa*	34	JA
Poplar developing reaction wood first-strand cDNA	*Populus alba*	21	Mechanical bending

TABLE 18.1 *(continued)*

Library	Organism	#ESTs	Stress
Aluminum-stressed root tip cDNA library	*Secale cereale*	1,199	Aluminum
Whole plant, ZAP cDNA library	*Suaeda maritima*	1,000	Salt
Whole plant, ZAP cDNA library	*Thellungiella salsuginea*	1,064	Salt
Whole plant, phagemid cDNA library; 400 mM NaCl	*Thellungiella halophila*	675	Salt
Chinese spring wheat drought-stressed leaf library	*Triticum aestivum*	641	Drought
Wheat ABA-treated embryo cDNA library	*Triticum aestivum*	2,173	ABA
Wheat cold-stressed seedling cDNA library	*Triticum aestivum*	939	Cold
Wheat drought-stressed leaf cDNA library	*Triticum aestivum*	905	Drought
Wheat salt-stressed crown cDNA library	*Triticum aestivum*	943	Salt
Wheat salt-stressed sheath cDNA library	*Triticum aestivum*	900	Salt
B73; 496-stressed shoot cDNA library, Wang/Bohnert	*Zea mays*	1,306	Salt
B73; 603-stressed root cDNA library, Wang/Bohnert lab	*Zea mays*	2,023	Salt
ISUM5-RN	*Zea mays*	1,514	ABA
Zm03 AAFC ECORC cold-stressed seedlings	*Zea mays*	953	Cold
Zm04 AAFC ECORC cold-stressed seedlings	*Zea mays*	756	Cold

Note: The total number of ESTs deposited in GenBank for various plant species listed (June 2003) exceeds 3 million. Databases were searched for such terms as *stress, salinity,* or *drought,* and the list might not be complete.

different, consisting of many more functionally unknown sequences that are often not annotated ("novel") and include a large percentage of known stress-induced transcripts from the EST profiles from tissues of plants that are maintained under optimal conditions (e.g., Öztürk et al., 2002; Wang et al., 2003). Also, high-throughput analysis of transcript abundance and changes in abundance is provided by transcripts printed on microarray slides in the form of ESTs and oligonucleotides, and oligonucleotide chips on which are synthesized short oligonucleotide sequences representing large numbers of genes or the entire genome, together with improved statistical tools for making sense out of the data (Brazhnik et al., 2002; Klok et al., 2002; Stoeckert et al., 2002). Approaching the next level of functional analysis, technologies for protein microarrays are emerging (Zhu and Snyder, 2003). Finally, integration of the phenotypic description of mutants affected in their salinity stress response, expression profiles, and the dynamic changes of transcript expression has recently produced a leap forward in our understanding of the constraints imposed by salinity stress on normal plant functions.

MUTANTS AND MUTANT PHENOTYPES IN SALINITY STRESS TOLERANCE

Mutations generated by treatment with chemical or physical mutagens, by the random insertion of the T-DNA of *Agrobacterium tumefaciens,* or by the induced movement of transposons are available in many species (Feldmann, 1991). EMS- or radiation-induced mutants in salinity stress responses have been isolated based on phenotypic criteria of reduced or increased tolerance to high salinity (see Zhu, 2002), but high-throughput global mutagenesis of a plant species had to wait until both models and suitable technologies had been developed. Owing to its well-known advantages of small size, short generation time, and compact genome, *Arabidopsis thaliana* has become the premier plant version of the yeast, *Drosophila melanogaster,* or *C. elegans* model systems. A model more closely suited to understanding the agronomically important cereal crops has now been added by the publication of the draft rice genome sequence and transposon insertion lines in rice and corn (Brutnell, 2002; Kohli et al., 2001). In *Arabidopsis,* insertional mutagenesis to saturation of the genome has been reached several times over, and several specifically designed mutant populations have been established for reverse genetic screens to find insertions that affect salinity, drought, cold, and/or ABA-dependent processes (Ishitani et al., 1997; Maggio et al., 2002). They involve lines that contain stress-inducible promoters driving the expression of a reporter gene, typically the luciferase coding region, into which T-DNA is randomly inserted. Tagging vectors that lead to either gene disruption and/or activation have been described (Krysan et al., 1999; Weigel et al., 2000). Detection of altered expression of the reporter when seedlings are salt stressed is a main advantage of such lines, which, when compared to insertional tagging of a wild-type line, report stress-relevant phenotypic changes that are easily measured ("camera mutants"). Several hundred mutants have been detected in this fashion (http://www.stress-genomics.org/). A second screen has been developed specifically for salinity stress-affected phenotypes, termed the *root-bending assay* (Wu et al., 1996; Zhu et al., 2002). For this screen, seedlings are deposited on agar containing variable amounts of NaCl in an inverted position, roots facing up. Seedlings affected by high salinity or osmotic stress that are unable to reorient root growth in the gravitational field are isolated from those that show normal root bending. What are the characteristic features of these mutants, or what, if anything, can we learn from the approach?

Table 18.2 presents more than 100 lines from a "camera mutant" screen (http://www.stress-genomics.org). All lines were originally identified as plants with altered expression of luciferase controlled by the stress-induc-

TABLE 18.2. T-DNA tagged mutants identified in an RD29A::luciferase background by altered luciferase expression.

Functional categories	No.	Chromosomal insertion/example/remark
Coding regions termed "unknown," "hypothetical," or "putative"	34	At4g21700; At2g07776; etc.
Between genes in different categories	34	Between At4g18110 (hypothetical protein) and At4g18120 (meiosis protein mei2-related), in putative promoter region of divergently transcribed genes
Signaling (and "possibly" signaling)	14	At1g76360; putative protein kinase
Defense	9	At3g50980; dehydrin-like protein
Transcription	9	At3g49000; RNA polymerase III subunit-like protein
Metabolism	5	At4g39800; *myo*-inositol-1-phosphate synthase
Translation	5	At4g39280; phenylalanyl-tRNA synthetase-like protein
Transposition	4	At2g12950; putative TNP2-like transposase
Transport	2	At5g27360; putative sugar transporter
Protein destination	2	At2g34980; phosphatidylinositol-glycan synthase
Energy	2	At1g65930; isocitrate dehydrogenase
Cell structure	2	At3g27000; actin-related protein
Cell cycle	1	in front of At1g15570 (= cyclin A2;3) in poorly annotated region

Note: Several of the 123 tagged genes (see <www.stress-genomics.org>) included here have been isolated more than once. Several may be categorized in multiple functions, as, for example, HUA2 in the agamous family in signaling and transcription or, similarly, an insertion in the homeodomain protein AHDP (anthocyaninless-2).

ible promoter of the RD29A gene that had been described and analyzed by Yamaguchi-Shinozaki et al. (1992) and Yamaguchi-Shinozaki and Shinozaki (1993, 1994). Subsequent analyses by TAIL-PCR with nested primers for the border of the inserted T-DNA and DNA sequencing of the PCR product provided the insertion site into or close to coding regions. It is obvious that these insertions identify multiple categories of genes affected by the putative knockout (in only a few cases for the genes in Table 18.2 has the phenotype been confirmed by RT-PCR). The insertion in a large number of mutants is connected to coding regions with unknown functions and also to regions of the genome that are either between two coding regions (functionally known and unknown) or in a poorly annotated region of the genome (Table 18.2). Next in frequency are mutant lines with insertions in the categories "signaling," "defense-related functions," and "transcription." Less frequent are lines that show insertions in the categories "metabolism," "translation," and "transposition," and fewer still in the categories "transport," "energy provision," "cell structure," and "cell cycle."

Mutants from this screen identify categories for which an involvement in salinity stress signaling pathways and stress-adapted transcriptional changes can be hypothesized. The disturbed coding regions are homologues of functionally similar genes, such as protein kinases or dehydrin-type coding regions, for which involvement in stress defense is based on prior knowledge. Others, in fact many of the putative mutants, are less intuitively interpreted. As a working hypothesis, it may be stated that the RD29A promoter, and possibly other control elements associated with a particular stress condition, responds to many deviations from cellular homeostasis by altered expression. Allowing this line of thought, a more complex view of the network of signaling interactions leading to stress responses will emerge when we integrate other high-throughput platforms for the analysis of the stress response. Among those are cell- and tissue-specific expression profiles and the dynamic changes that can be monitored by microarray-based expression studies.

The result of a large screen of both chemically and tagging induced lines are the mutants identified by the root-bending assay on sodium-containing media. They are more closely related to salinity stress signaling and response pathways. In this way, the genes of the *SOS* pathway (salt overly sensitive) have been found and functionally extensively characterized by the group of Jian-Kang Zhu (see Zhu, 2002, for a review). A set of genes, *SOS3, SOS2,* and *SOS1,* describes a pathway consisting of a calcium sensor (SOS3) that interacts with a kinase (SOS2) (Liu et al., 2000) in the PKS subfamily of protein kinases, which affects expression and activity of a sodium/proton antiporter in the plasma membrane (SOS1) (Liu and Zhu, 1998; Halfter et al., 2000; Gong, Gong, et al., 2002; Gong, Zhang, et al.,

2002; Guo et al., 2001; Qiu et al., 2002; Quintero et al., 2002; Shi et al., 2000) (Figure 18.2). While the analysis of this pathway had been started before large-scale genomics analyses were initiated, the new tools have contributed to the stupendous progress in recent years. The *SOS-3-2-1* series of genes outlines the first pathway that is strictly in the conduction of ionic stresses, as opposed to osmotic stress pathways (Zhu, 2002). As much as we know, many open questions remain about, for example, the sensing and signal inducers upstream of SOS3; the extent to which the pathway is modulated by different secondary messengers and hormonal cross talk; the number and nature of components, such as transcription factors, downstream of SOS2; and the role of this pathway in negative regulation of responses that are inappropriate under stress conditions. The screens have generated other

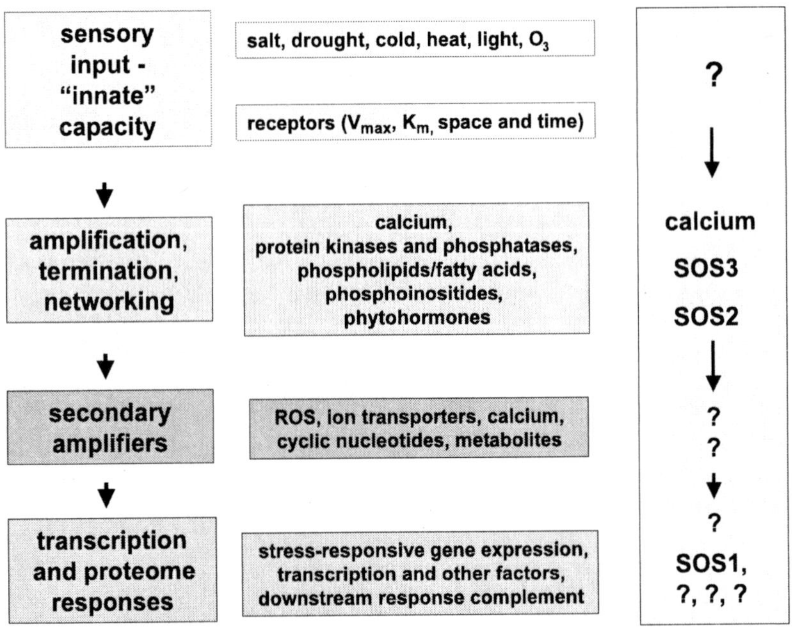

FIGURE 18.2. The SOS ionic stress response pathway in relation to signaling response pathways. The SOS pathway (for salt overly sensitive) has been found to be specific for the ionic stress response at least in *Arabidopsis* (see Zhu, 2002, for a review). The scheme places known elements of the SOS pathway, leading to the expression and/or activity modification of the plasma membrane Na^+/H^+-antiporter SOS1, into a general scheme of input, amplification, multiplication, and response, including known examples for components of other signaling pathways. The scheme is hypothetical and the SOS pathway could be connected to other stress response pathways solely through calcium signaling.

mutants in the SOS category, and camera screens generated mutants in categories termed *los* and *hos* (low/high expression under osmotic stress), where the loss or gain of tolerance may be further qualified by being *los/hos* for different treatments (cold, ABA, drought, salinity) or for all treatments (Zhu, 2001, 2002). One example is *SOS4,* encoding a pyridoxal kinase that is involved in the biosynthesis of pyridoxal-5-phosphate (vitamin B_6), and two transcripts exist due to alternative splicing (Shi, Xiang, et al., 2002). Causative basis for the *sos4* phenotype might be that pyridoxal-5-phosphate acts as a ligand for ion channels, based on results with animal systems, and thus could be involved as an essential component in regulating ion homeostasis.

In the *hos* category, another principle is highlighted that affects salinity stress responses (Koiwa et al., 2002). Two genes, *AtCPL1* and *AtCPL3,* encode phosphatases that affect the phosphorylation state of the C-terminal domain of RNA polymerase II. The CPLs represent negative regulators of stress-responsive gene expression and modulate growth. Yet another approach is presented by Rus et al. (2001). They searched for suppressing mutants in a mutant background characterized by hypersensitivity to high salt. The *sos3-1* mutant (EMS generated) was tagged and screened for suppression of sensitivity, resulting in several lines, most of which showed insertions in *HKT1*. This gene had previously been described as a high-affinity potassium transporter for which later experiments indicated significant influx of sodium under high-salinity stress conditions in *Arabidopsis* (Uozumi et al., 2000). It appears, however, that HKT1 could also be involved in potassium sensing that is affected by high external sodium concentrations.

The high-throughput screening of tagged *Arabidopsis* lines that include a stress-reporting gene cassette provides a general paradigm applicable to many other strategems. In the salinity stress research field, combined with an in-depth biochemical and/or physiological follow-up analysis, the method has provided in a short period insights that could not easily have been gained otherwise.

ESTs AND cDNAs FROM SALT-STRESSED PLANTS

Among the plant ESTs that are publicly available, relatively few cDNA libraries have been generated from tissues that had been stressed in some way. Table 18.1 includes cDNAs associated with different stresses for the 20 most intensively analyzed plant species (>2.1 million ESTs). Libraries generated for stressed cells/tissues resulted in approximately 150,000 ESTs, and ~30,000 of these are ESTs from salt-stressed plants or cells.

A juxtaposition of categories comparing ESTs from control and salt-stressed tissues indicated several yardsticks that are with high probability a general phenomenon. First, the number and percentage of ESTs for functionally unknown and even novel transcripts is significantly higher in the libraries from stressed plants (Bohnert et al., 2001). In a number of cases, coding regions termed hypothetical have been detected in such EST collections, which allowed their recategorization as "expressed, unknown" coding regions. A second noticeable difference between EST collections of control and stress conditions is the number of transcripts that fall into the categories defense, radical scavenging, redox control, and functionally unknown ESTs with a possible function within or associated with membranes, the latter based on putative domain structures. A number of transcripts for those and related functions have been isolated before, but the numbers that have been added are unexpectedly large. In related functionally unknown categories are also ABA-, jasmonate-, salt-, drought-, or ethylene-inducible transcripts that have been described before. The transcript profiles from salt-stressed plants have added more complexity by revealing additional isoforms of coding regions in these categories.

Table 18.3 exemplifies a comparative genomics approach to sorting out the complexity of transcripts that are associated with salinity stress responses. This approach is based on an analysis of the yeast transcriptome response to high salinity in a macroarray probed with radioactive target RNAs (Yale and Bohnert, 2001). Similar categories, and in many cases the same regulated genes in salt-stressed yeast cells, have been reported by Posas et al. (2000) and Rep et al. (2000). When strongly up-regulated yeast transcripts are compared with ESTs derived from cDNA libraries of stressed rice, corn, and ice plant tissues, the number of homologous transcripts that are detected in the collection of cDNAs which amounted to fewer than 3,000 transcripts for each of the higher-plant species proved to be surprisingly large. The list of homologous transcripts includes unknown and hypothetical coding regions as well as transcripts for proteins involved in chaperone action, signaling, transcription/translation, and metabolic pathways that, at least for some of the proteins and transcripts, have been reported for higher plants in the past as salinity stress responsive. Such mining of genes for evolutionary conserved and species-specific responses will gain in importance as database entries increase, functional categorization and annotation improve, and as more dynamic changes in transcript abundance ratios distinguishing control and stressed states are reported.

TABLE 18.3. Plant ESTs homologous to salt stress-induced yeast genes and ORFs.

Order	Systematic name	Gene name	Functional annotation	MIPS category	Ice plant	Rice	Corn
1	YPL218W	SAR1	GTP-binding protein of the ARF family	INTTRA	yes	not found	not found
2	YJR009C	TDH2	glyceraldehyde-3-phosphate dehydrogenase 2	MET	yes	yes	yes
3	YAL004W	—	strong similarity to *Achlya klebsiana* glutamate dehydrogenase	U	yes	yes	yes
4	YAL005C	SSA1	heat shock protein of HSP70 family, cytosolic	CDEF	yes	yes	yes
5	YKR014C	YPT52	GTP-binding protein of the rab family	PDEST	yes	yes	not found
6	YBL087C	RPL23A	60S large subunit ribosomal protein L23.e	PS	yes	yes	yes
7	YER103W	SSA4	heat shock protein of HSP70 family, cytosolic	CDEF	yes	yes	yes
8	YGL166W	CUP2	copper-dependent transcription factor	TRANSC	yes	not found	yes
9	YJL190C	RPS24A	ribosomal protein S15a.e.c10	PS	yes	yes	not found
10	YHR049C-A	—	questionable ORF	U	yes	yes	yes
11	YPL079W	RPL21B	ribosomal protein L21	PS	yes	yes	yes
12	YEL059W	—	hypothetical protein	U	yes	yes	yes
13	YIL052C	RPL34B	ribosomal protein L34.e	PS	yes	not found	not found
14	YNL192W	CHS1	chitin synthase I	MET	yes	yes	not found
15	YER130C	—	similarity to MSN2P and weak similarity to MSN4P	U	yes	yes	not found
16	YGR137W	—	questionable ORF	U	yes	yes	not found
17	YMR110C	—	similarity to aldehyde dehydrogenase	U	yes	not found	yes
18	YDR055W	PST1	strong similarity to SPS2 protein	U	yes	not found	yes
19	YLR167W	UBI3	ubiquitin/40S small subunit ribosomal protein	PS	yes	yes	yes
20	YER141W	COX15	cytochrome oxidase assembly factor	MET	yes	not found	not found
21	YPR035W	GLN1	glutamate-ammonia ligase	MET	yes	yes	not found
22	YJR121W	ATP2	F_1F_0-ATPase complex, F1 beta subunit	IONHOM	yes	yes	not found
23	YDR064W	RPS13C	ribosomal protein	PS	yes	yes	not found

TABLE 18.3 *(continued)*

Order	Systematic name	Gene name	Functional annotation	MIPS category	Ice plant	Rice	Corn
24	YDR461W	MFA1	mating pheromone a-factor 1	CGRO	yes	yes	not found
25	YGL165C	—	questionable ORF	U	yes	yes	not found
26	YHR137W	ARO9	aromatic amino acid aminotransferase II	MET	yes	not found	yes
27	YMR186W	HSC82	heat shock protein	CDEF	yes	yes	yes
28	YLR029C	RPL15A	RPL15A 60S large subunit ribosomal protein L15.e.c12	PS	yes	yes	not found
29	YKL036C	—	questionable ORF	U	yes	not found	yes
30	YLR333C	RPS31	ribosomal protein S25.e.c12	PS	yes	yes	yes
31	YPL090C	RPS6A	ribosomal protein S6.e	PS	yes	yes	not found
32	YOL087C	—	similarity to *Saccharomyces pombe* hypothetical protein	U	yes	yes	yes
33	YLR293C	GSP1	GTP-binding protein of the ras superfamily	TRANSC	yes	yes	not found
34	YAL003W	EFB1	translation elongation factor eEF1beta	TL	yes	yes	yes
35	YGL248W	PDE1	low-affinity 3',5'-cyclic-nucleotide phosphodiesterase	MET	not found	yes	not found
36	YGR085C	RPL11B	ribosomal protein	PS	not found	yes	yes
37	YLL026W	HSP104	heat shock protein	CDEF	not found	yes	yes
38	YML001W	YPT7	GTP-binding protein of the RAB family	INTTRA	yes	yes	not found
39	YPL081W	RPS9A	ribosomal protein S9.e.A	PS	not found	yes	yes
40	YEL054C	RPL15B	60S large subunit ribosomal protein L12.e	PS	yes	yes	yes
41	YGL147C	RPL9A	ribosomal protein L9.e	PS	not found	yes	yes
42	YKL100C	—	similarity to *Caenorhabditis elegans* hypothetical protein	U	yes	yes	yes
43	YHR203C	RPS7A	RPS4B ribosomal protein S4.e.c8	PS	yes	yes	yes
44	YDR154C	—	questionable ORF	U	yes	yes	yes
45	YDR155C	CPH1	cyclophilin (peptidyl-prolyl isomerase)	PDEST	yes	yes	yes
46	YDR025W	RPS18A	ribosomal protein S11.e	PS	yes	yes	yes
47	YDR304C	CYP5	peptidyl-prolyl *cis-trans* isomerase D precursor (cyclophilin D) of the ER	MET	yes	yes	not found

48	YOL040C	RPS21	RPS15, 40S small subunit ribosomal protein	PS	not found	yes	not found
49	YMR119W-A	—	questionable ORF	U	not found	yes	yes
50	YER067W	—	strong similarity to hypothetical protein YIL057c	U	yes	yes	yes
51	YJL053W	PEP8	vacuolar protein sorting/targeting protein	PDEST	yes	yes	yes
52	YJL138C	TIF2	translation initiation factor eIF4A	PS	yes	yes	not found
53	YEL024W	RIP1	ubiquinol-cytochrome-c reductase iron-sulfur protein precursor	ENERGY	yes	not found	not found
54	YBL027W	RPL19A	60S large subunit ribosomal protein L19.e	PS	yes	yes	not found
55	YGR086C	—	strong similarity to hypothetical protein YPL004c	U	yes	yes	yes

Note: Induced yeast genes and ORFs (Yale and Bohnert, 2001) are compared to ESTs from ice plant, rice, and corn that were obtained from cDNA libraries from salt-stressed plants (www.stress-genomics.org). A complete table providing accession numbers is deposited at this Web site. Abbreviations of categories are taken from the yeast MIPS category assignment list.

SALINITY-INDUCED CHANGES OF THE TRANSCRIPTOME

Changes in transcript profiles on a global scale have become possible through microarray analysis of ESTs or oligonucleotides printed to or synthesized on solid supports, a technology that undergoes rapid technology development and sophistication of statistical and bioinformatics tools (e.g., Schena et al., 1995; Lockhart et al., 1996; Stoeckert et al., 2002). Expression profiling with a focus on salinity or abiotic stresses in general began with small experimental arrays, has progressed to larger arrays, and presently includes whole-genome arrays for *Arabidopsis* and rice. Conducted at present are salinity, cold, or drought shock experiments (as opposed to gradual imposition of the stress) that promise to unravel the "response capacity" specific for a species or line (Kawasaki et al., 2001; Chen et al., 2002; Fowler and Thomashow, 2002; Kreps et al., 2002; Öztürk et al., 2002; Seki et al., 2002; Wang et al., 2003). These data provide valuable information but may be of little predictive value for the actual transcriptome status in plants growing under natural, systemic stress conditions (Blum, 1988; Öztürk et al., 2002; Talamé et al., 2003; Bohnert et al., unpublished).

Focusing on salinity stress, published reports utilized small array compilations of up to 2,000 ESTs (Kawasaki et al., 2001; Öztürk et al., 2002), and more recently 5,000 to 8,000 DNA elements (Chen et al., 2002; Kreps et al., 2002; Seki et al., 2002; Wang et al., 2003). The larger microarray slides and chips now available for *Arabidopsis* and rice cover the annotated genome as a set of oligonucleotides for all/most putative coding regions. These "oligo-arrays" include approximately 26,000 DNA elements *(Arabidopsis)* and approximately 55,000 DNA elements (rice) and are based on genome annotation, although many elements included target predicted coding regions whose expression has not been detected (<http://ag.arizona.edu/microarray/workshop2003.html>; David Galbraith, personal communication). Their use in several laboratories for salinity-dependent transcriptome changes promises to significantly advance the database in the coming year.

Chen et al. (2002) assembled a matrix of 402 distinct (including putative) transcription factors (TFs), which were probed by different biotic and abiotic stressors, for organ specificity, and during development. The results indicated regulation by high salinity of individual transcription factors in subsets of these factors in families such as putative zinc-finger TFs, bZIP/HD-ZIPs, and AP2-domain factors. TFs specific to an abiotic stress as well as sets that respond to more than one stress, hormonal, abiotic, and/or biotic, were found. Included in the latter category were GBF3, Athb-12, AtERF1, and AtWRKY6, which have also been characterized in targeted experiments and in other microarray experiments.

Kreps et al. (2002) followed a different strategy for defining the abiotic stress transcriptome in *Arabidopsis thaliana,* focusing on cold, osmotic, and salt stresses with an 8,100-gene GeneChip array that included DNA probes for different functional categories. Root and leaf RNA from plants stressed for 3 h and 27 h, respectively, were used—an attempt to monitor changes over time as the plants adjusted to 100 mM NaCl. The analysis recorded large root/leaf differences, and a limited overlap between drought and salinity responses, as less than 5 percent at 3 h, and even fewer elements after 27 h of stress were common to both. Also in *Arabidopsis,* Seki et al. (2002) reported larger sets of transcripts that overlap between salinity, drought, and to some degree cold, although cold treatment caused fewer changes in the expression of shared regulated genes than either drought or salt (Fowler and Thomashow, 2002). Öztürk et al. (2002) reported minimal overlap between drought and salinity responses in a small array of barley *(Hordeum vulgare)* ESTs, which is probably due to the fact that the ESTs were nearly exclusively obtained from barley cDNA libraries in which the plants had been drought shocked and severely stressed for 6 and 10 h. Specific for salinity stress, Kreps et al. (2002) listed 22 genes, which represented 5 percent of all (440) genes that responded to high salinity. In their majority they were related to oxidative stress (or signaling) responses, such as glutathione reductase and cytochrome P450 coding regions. A small number of coding regions in this screen could be assigned to transcription, membrane transport, and phosphoregulation.

One result of shock stress is the observation of a burst of changes early during the stress. Figure 18.3 outlines this result. When comparing the stress response of the salinity stress response in corn (*Zea mays;* ~5,000 unigenes) and rice (*Oryza sativa* var. Pokkali and IR29; ~1,800 ESTs) in time-course experiments, an immediate or early induction of many transcripts is obvious, while over time fewer coding regions are either up- or down-regulated (Kawasaki et al., 2002; Wang et al., 2003). Both species are glycophytes, which seem to behave different from two halophytes that were subjected to

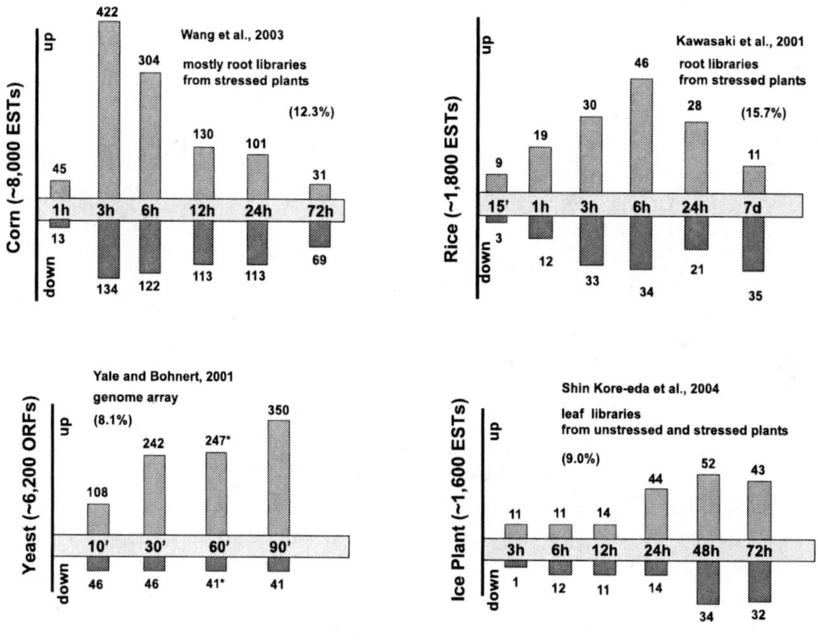

FIGURE 18.3. Comparing salinity shock responses between halophytic and glycophytic species. Numbers of transcripts that are up- and down-regulated (> twofold) between two halophytes (yeast and ice plant) and two glycophytes (corn and rice) in time-course experiments. The data are taken from Wang et al. (2003) for corn, Kawasaki et al. (2001) for rice, Yale and Bohnert (2001) for yeast and for ice plant Shin Kore-eda et al. (2001). Numbers next to bars indicate the number of up- or down-regulated ESTs. Percentage numbers refer to the percent of regulated transcripts. The time scale is included in the *x*-axis.

a similar time course. One of the halophytic species, *Saccharomyces cerevisiae* (on a much shorter time course than the plants), showed a continuous increase of regulated transcripts over the course of the stress (Figure 18.3; Yale and Bohnert, 2001), and this was similar in experiments with yeast reported by others (Posas et al., 2000; Rep et al., 2000). These arrays included DNA elements for the entire yeast genome. Similarly, the halophytic *Mesembryanthemum crystallinum*, for which a small array composed of ~1,200 ESTs was probed (J. C. Cushman, personal communication), responded to a stronger salinity shock (400 mM NaCl, as compared to 150 mM NaCl for rice and corn) by the induction of coding regions that resembled that of yeast and not that of the two other plant species (Figure 18.3). An explanation could be that abiotic stress in glycophytes at sublethal levels leads to the induction of a large number of signaling and response pathways that over time are reduced to the regulation of those coding regions which represent an appropriate set of the defense reactions that are needed. The response by halophytic species in contrast may be guided by an earlier recognition of the nature of the stress or by constitutive presence of the response pathway.

In a pilot experiment, we compared a moderately salt-tolerant rice line (Pokkali) with a salt-sensitive line (IR29) (Kawasaki et al., 2001). While seedlings of Pokkali survive 150 mM NaCl, IR29 plants die. The time-course experiment indicated the fundamentally different behavior of the two lines (Figure 18.4). Pokkali showed a rapid response (15 min) of several signal transduction components that were up-regulated, but IR29 showed a slower response or failed to induce the same transcripts. In the course of the experiment, Pokkali regulated different categories of transcripts in succession, such that four phases could be distinguished (Figure 18.4). Within hours, functions related to restructuring of the proteome were followed by functions that indicated changes in hormonal status, changes in metabolism, and, later, functions that indicated defense reactions and the adaptation of the plants to resume growth (Kawasaki et al., 2001).

Research targeting salinity stress response profiling of transcripts is only beginning—the data we now have corroborate many of the facts that had been worked out in the pregenomics area and thus point to the next sets of experiments. Foremost, these include three focal areas. First, future experiments must record a detailed description of the genotype that is used, as well as include the developmental and health status of the experimental material in great detail for the data to be helpful to others. An *in silico* transcript profile of the widely used model species will be helpful only if a consensus of growth conditions is included. Connected to this topic is the decision about how to stress the plants; shock and gradual imposition of stress generate very different results. An exploratory experiment comparing

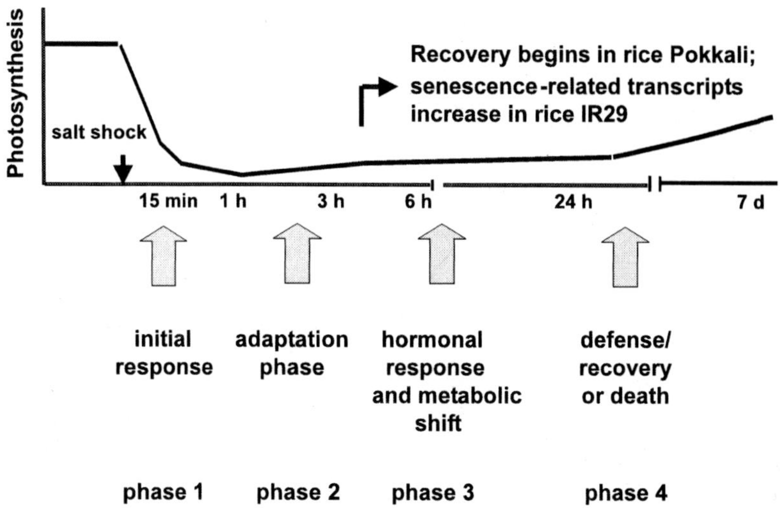

FIGURE 18.4. The path to recovery from salt shock in *Oryza sativa* (var. Pokkali). Transcript profiles and changes in transcript amount during a time-course after the addition of 150 mM NaCl to young rice plants (var. Pokkali [PK] and var. IR29) (Kawasaki et al., 2001). The *y*-axis shows the decline of CO_2-fixation activity and stomatal closure and their gradual recovery over time for rice Pokkali. In rice IR29, the decline is steeper and the plants never recover (Kawasaki et al., 2001). Variety Pokkali is moderately salinity stress-tolerant, while variety IR29 is an extremely salt-sensitive line. The analysis of up- and down-regulated transcripts (~1,800 ESTs arrayed, most of which derived from salt-stressed roots and shoots of young plants) indicated different functional categories, in addition to the typical ~50 percent of functionally unknown transcripts, regulated at different time points. The arrows identify time points at which different categories of regulated transcripts prevail (see Kawasaki et al., 2001). These can be labeled as (1) initial response in less than 1 h, (2) adaptation phase (1-3 h), (3) hormonal response and metabolic shift (~3 to ~12 h), and (4) defense and recovery phase, at which homeostasis is regained while a number of previously low-expression transcripts are increased. Salient differences between PK, which showed signs of recovery and regaining homeostasis after ~6 h of stress, and IR29, which had died after ~24 h of stress, are the following: PK responds faster than IR29 by inducing components of signaling pathways (phase 1), and whatever induction is seen in IR29 is less pronounced. Phase 2 is characterized by a number of up-regulated transcripts in the categories of protein synthesis, protein modification, and protein targeting and turnover. The number of induced transcripts in the hormonal regulation categories (phase 3; e.g., ABA-induced transcripts) is similar in PK and IR29, and transcripts for several metabolic pathways change. After ~24 h, and more pronounced after seven days (phase 4), up-regulated transcripts in PK are found in the categories of cell wall synthesis, oxygen radical scavenging (or possibly peroxide-based signaling), and cellular ion/water homeostasis (e.g., aquaporins).

expression profiles from drought-shocked and slowly desiccated barley plants indicated only marginal overlap, or, rather, what the shock experiments revealed after 6 or 10 h approached the changes in expression of plants, depending on pot size and soil type, that had not been watered for 10 to 12 days (Öztürk et al., 2002; Talamé et al., 2003). A second suggestion is about the use of time-course experiments, which should include as many points as possible. The profiles and categories of regulated transcripts in our analyses changed quite rapidly. Including more time points would certainly reduce the danger of overinterpretation because whatever transcription factor is up-regulated after 3 h of stress may be dependent on a pathway that is induced after 15 min of stress. Third, additional information about overlapping and counteracting regulatory elements can be expected by the inclusion of mutants in known signaling and/or response pathways in transcript profiling.

SALINITY-DEPENDENT SIGNAL TRANSDUCTION

The pathways leading to salinity stress responses are complex and influenced by tissue context, developmental stage, and general stress status. Much of our knowledge comes from the pioneering work in *Arabidopsis* on drought, salinity, and low temperature response reactions (Shinozaki and Yamaguchi-Shinozaki, 1997). The root and leaf reactions to salinity, for example, include different categories of transcripts that change, and the temporal behavior of change is different (Kawasaki et al., 2001; Öztürk et al., 2002). A higher number of ion transporters are affected in roots, and, also, the categories of protein turnover and translation are more strongly induced in roots than in leaves (Kawasaki et al., 2001; H. J. Bohnert et al., unpublished data). Chen et al. (2002) relate a similar finding, reporting that the number of genes encoding (putative) transcription factors that responded to abiotic and biotic stressors was significantly higher in roots compared to the leaf/shoot/flower tissues.

Only partially known are the signaling pathways that lead to ionic stress responses in the SOS category of signaling pathways, but the involvement of changes in (local) calcium concentrations in waves and spikes in the stress responses has been documented in many studies and discussed in detail (Bressan et al., 1998; Knight et al., 1997; Pardo et al., 1998; Schumaker and Sze, 1987; Sheen, 1996). The involvement of ABA, with confirmatory hints from expression profiling, is equally clear (Hobo et al., 1999; Hong et al., 1997; Leung and Giraudat, 1998; Uno et al., 2000, Shinozaki and Yamaguchi-Shinozaki, 1997). Additional pathways obviously exist, connected to salicylic acid-dependent (Mikolajczyk et al., 2000) and possibly multiple phosphoinositide-dependent pathways that act in connection with MAPK cascades (Hirayama et al., 1995; Jonak et al., 1996; Mikami et al., 1998; Drobak and Watkins, 2000; Frank et al., 2000; Munnik et al., 2000; Katagiri et al., 2001; Munnik and Meijer, 2001; Kovtun et al., 2000; Takahashi et al., 2001). Urao et al. (1999) identified a transmembrane hybrid-type histidine kinase in *Arabidopsis* that functions as an osmosensor. The complexity of the network of control pathways has up to now defied connecting or separating the different components unequivocally—expression profiling of mutants should provide the resolution that is presently missing.

SALINITY STRESS ENGINEERING

Apart from satisfying scientific curiosity, what is the value that can be associated with knowledge about plant reactions, leading to tolerance of or sensitivity to high salinity? The major incentive is the agronomical value that crops might acquire if they could be made tolerant to high salinity because most crops are glycophytes, whose metabolism and growth are affected at low concentrations of sodium in the soil, well below 100 mM, that would not pose problems for halophytes. It seems possible to engineer tolerance at this level or at even higher NaCl concentrations, yet productivity might well be compromised, as has been shown in some experiments that used a controlled environment to show the difference between wild type and transformant (Tarczynski et al., 1993; Gaxiola et al., 1994; Quintero et al., 1996; Van Camp et al., 1996; McKersie et al., 1996; Romero et al., 1997; Roxas et al., 1997; Apse et al., 1999; Jaglo-Ottosen et al., 1998; Kasuga et al., 1999; Winicov and Bastola, 1999; Zhifang and Loescher, 2003). It can be expected that future engineering work will include results from expression profiling, for example, in the choice of promoters with enhanced expression strength that mimic the expression kinetics of the resident promoters for the targeted signaling, transcription, or downstream response pathways.

FUTURE DEVELOPMENTS

The present focus on genomics-type plant biology has been ushered in by the generation of the *Arabidopsis* sequencing project initiated near the end of the 1980s, mirrored by the coincidental start of bacterial, yeast, and animal, human in particular, genome-sequencing projects. Genome sequences have become essential requisites for anchoring ESTs and expression profiles. The recognition of syntenic relationships among species is increasingly exploited for comparative genomics analyses. Similar high-throughput methods have emerged for proteins and metabolites, with improvements in tools and technologies continuously changing. Sequences and dynamic expression profiles are only a starting point; the number of predicted reading frames is continuously increasing as bioinformatics tools improve, as small reading frames that have been dismissed in the past are found to be real through EST detection and transcript profiling, and as small RNAs are now being added as important components of gene-expression regulation (Brown et al., 2003). The number of splicing variants, leading to different protein sequences from one gene, can be expected to increase as well. Thus, it would not be surprising if significantly more than 26,000 functions were encoded in the *Arabidopsis* genome, and possibly more than twice that number in, for example, rice and corn.

These numbers are exhilarating and frightening at the same time, considering that the function of fewer than 20 percent of the genes in the *Arabidopsis* genome is known to some degree. Thus, the immediate future will be characterized by high-throughput localization studies for all or most of the salinity stress-related transcripts and proteins, using, for example, cell ablation techniques combined with microarray analysis and cellular and subcellular painting of transcripts and proteins by (partially) automated in situ and real-time in vivo fluorescence detection and localization methods. Also, the analysis of systematic gene disruptions for all encoded proteins, and the analysis of chemically induced small alterations, e.g., by EMS mutagenesis, in a number of "important" proteins will be done, as well as global protein interaction screens. Eventually, we will have a virtual representation of all transcripts, proteins, and major metabolites during the life of a few model species, from seed to seed under optimal conditions and when challenged by biotic and abiotic stresses.

REFERENCES

Amtmann, A., Sanders, D. (1999). Mechanisms of Na^+ uptake by plant cells. *Adv. Botan. Res.* 29: 75-112.

Apse, M.P., Aharon, G.S., Snedden, W.A., Blumwald, E. (1999). Salt tolerance conferred by overexpression of a vacuolar Na^+/H^+-antiport in *Arabidopsis. Science* 285: 1256-1258.

Arabidopsis Genome Initiative (2000). Analysis of the genome sequence of the flowering plant *Arabidopsis thaliana. Nature* 408: 796-815.

Bartels, D., Nelson, D. (1994). Approaches to improve stress tolerance using molecular genetics. *Plant Cell Environ.* 17: 659-667.

Bennetzen, J. (2002). The rice genome: Opening the door to comparative plant biology. *Science* 296: 60-63.

Blum, A. (1988). *Plant Breeding for Stress Environments.* Boca Raton, FL: CRC Press.

Bohnert, H.J., Ayoubi, P., Borchert, C., Bressan, R.A., Burnap, R.L., Cushman, J.C., Cushman, M.A., Deyholos, M., Fischer, R., Galbraith, D.W., et al. (2001). A genomics approach towards salt stress tolerance. *Plant Physiol. Biochem.* 39: 295-311.

Bohnert, H.J., Bressan, R.A. (2001). Abiotic stresses, plant reactions, and approaches towards improving stress tolerance. In Noessberger, J. (Ed.), *Crop Science: Progress and Prospects* (pp. 81-100). Wallingsford, UK: CABI International.

Bohnert, H.J., Shen, B. (1999). Transformation and compatible solutes. *Scientia Hortic.* 78: 237-260.

Boyer, J.S. (1982). Plant productivity and environment. *Science* 218: 443-448.

Brown, J.W., Echeverria, M., Qu, L.H. (2003). Plant snoRNAs: Functional evolution and new modes of gene expression. *Trends Plant Sci.* 8: 42-49.

Brazhnik, P., de la Fuente, A., Mendes, P. (2002). Gene networks: How to put the function in genomics. *Trends Biotech.* 20: 467-472.

Bressan, R.A., Hasegawa, P.M., Pardo, J.M. (1998). Plants use calcium to resolve salt stress. *Trends Plant Sci.* 3: 411-412.

Bressan, R.A., Zhang, C., Zhang, H., Hasegawa, P.M., Bohnert, H.J., Zhu, J.K. (2001). Learning from the *Arabidopsis* experience: The next gene search paradigm. *Plant Physiol.* 127: 1354-1360.

Brutnell, T.P. (2002). Transposon tagging in maize. *Funct. Integrat. Genomics* 2: 4-12.

Buell, C.R. (2002). Obtaining the sequence of the rice genome and lessons learned along the way. *Trends Plant Sci.* 7: 538-542.

Chen, W., Provart, N.J., Glazebrook, J., Katagiri, F., Chang, H.-S., Eulgem, T., Mauch, F., Luan, S., Zou, G., Whitham, S.A., et al. (2002). Expression profile matrix of *Arabidopsis* transcription factor genes suggests their putative functions in response to environmental stresses. *Plant Cell* 14: 559-574.

Cushman, J.C., Bohnert, H.J. (2000). Genomic approaches to plant stress tolerance. *Curr. Opin. Plant Biol.* 3: 117-124.

Drobak, B.K., Watkins, P.A. (2000). Inositol (1,4,5) trisphosphate production in plant cells: An early response to salinity and hyperosmotic stress. *FEBS Lett.* 481: 240-244.

Feldman, K.A. (1991). T-DNA insertion mutagenesis in *Arabidopsis:* Mutational spectrum. *Plant J.* 1: 71-82.

Flowers, T.J., Yeo, A.R. (1992). *Solute Transport in Plants.* Glasgow, Scotland: Blackie.

Flowers, T.J., Yeo, A.R. (1995). Breeding for salinity tolerance in crop plants: Where next? *Aust. J. Plant Physiol.* 22: 875-884.

Fowler, S., Thomashow, M.F. (2002). *Arabidopsis* transcriptome profiling indicates that multiple regulatory pathways are activated during cold acclimation in addition to the CBF cold response pathway. *Plant Cell* 14: 1675-1690.

Frank, W., Munnik, T., Kerkmann, K., Salamini, F., Bartels, D. (2000). Water deficit triggers phospholipase D activity in the resurrection plant *Craterostigma plantagineum. Plant Cell* 12: 111-123.

Gaxiola, R.A., Rao, R., Sherman, A., Grisafi, P. Alper, S., Fink, G.R. (1994). The *Arabidopsis thaliana* proton transporters AtNhx1 and Avp1 can function in cation detoxification in yeast. *Proc. Natl. Acad. Sci. USA* 96: 1480-1485.

Goff, S.A., Ricke, D., Lan, T.H., Presting, G., Wang, R., Dunn, M., Glazebrook, J., Sessions, A., Oeller, P., Varma, H., et al. (2002). A draft sequence of the rice genome *(Oryza sativa* L. ssp. *japonica). Science* 296: 92-100.

Gong, D., Gong, Z., Guo, Y., Chen, X., Zhu, J.K. (2002). Biochemical and functional characterization of PKS11, a novel *Arabidopsis* protein kinase. *J. Biol. Chem.* 277: 28340-28350.

Gong, D., Zhang, C., Chen, X., Gong, Z., Zhu, J.K. (2002). Constitutive activation and transgenic evaluation of the function of an *Arabidopsis* PKS protein kinase. *J. Biol. Chem.* 277: 42088-42096.

Guo, Y., Halfter, U., Ishitani, M., Zhu, J.K. (2001). Molecular characterization of functional domains in the protein kinase SOS2 that is required for plant salt tolerance. *Plant Cell* 13: 1383-1400.

Halfter, U., Ishitani, M., Zhu, J.K. (2000). The *Arabidopsis* SOS2 protein kinase physically interacts with and is activated by the calcium-binding protein SOS3. *Proc. Natl. Acad. Sci. USA* 97: 3730-3734.

Hare, P.D., Cress, W.A., van Staden, J. (1997). The involvement of cytokinins in plant responses to environmental stress. *Plant Growth Regulat.* 23: 79-103.

Hasegawa, P.M., Bressan, R.A., Zhu, J.K., Bohnert, H.J. (2000). Plant cellular and molecular responses to high salinity. *Annu. Rev. Plant Physiol. Plant Mol. Biol.* 51: 463-499.

Hayashi, H., Alia, Mustardy, L., Deshnium, P., Ida, M., Murata, N. (1997). Transformation of *Arabidopsis thaliana* with the coda gene for choline oxidase: Accumulation of glycinebetaine and enhanced tolerance to salt and cold stress. *Plant J.* 12: 133-142.

Hirayama, T., Ohto, C., Mizoguchi, T., Shinozaki, K. (1995). A gene encoding a phosphatidylinositol-specific phospholipase C is induced by dehydration and salt stress in *Arabidopsis thaliana. Proc. Natl. Acad. Sci. USA* 92: 3903-3907.

Hobo, T., Kowyama, Y., Hattori, T. (1999). A bZIP factor, TRAB1, interacts with VP1 and mediates abscisic acid-induced transcription. *Proc. Natl. Acad. Sci. USA* 96: 15348-15353.

Holtorf, H., Guitton, M.C., Reski, R. (2002). Plant functional genomics. *Naturwissenschaften* 89: 235-249.

Hong, S.W., Jon, J.H., Kwak, J.M., Nam, H.G. (1997). Identification of a receptor-like protein kinase gene rapidly induced by abscisic acid, dehydration, high salt, and cold treatments in *Arabidopsis thaliana. Plant Physiol.* 113: 1203-1212.

Horton, P., Ruban, A.V., Walters, R.G. (1996). Regulation of light-harvesting in green plants. *Annu Rev. Plant Physiol. Plant Mol. Biol.* 47: 655-684.

Huala, E., Dickerman, A.W., Garcia-Hernandez, M., Weems, D., Reiser, L., LaFond, F., Hanley, D., Kiphart, D., Zhuang, M., Huang, W., et al. (2001). The Arabidopsis Information Resource (TAIR): A comprehensive database and web-based information retrieval, analysis, and visualization system for a model plant. *Nucleic Acids Res.* 29: 102-105.

Ishitani, M., Xiong, L., Stevenson, B., Zhu, J.K. (1997). Genetic analysis of osmotic and cold stress signal transduction in *Arabidopsis thaliana:* Interactions and convergence of abscisic acid-dependent and abscisic acid-independent pathways. *Plant Cell* 9: 1935-1949.

Jaglo-Ottosen, K.R., Gilmour, S.J., Zarka, D.G., Schabenberger, O., Thomashow, M.F. (1998). *Arabidopsis* CBF1 overexpression induces COR genes and enhances freezing tolerance. *Science* 280: 104-106.

Jain, R.K., Selvaraj, G. (1997). Molecular genetic improvement of salt tolerance in plants. *Biotech. Annu. Rev.* 3: 245-267.

Jonak, C., Kieger, I.S., Lighterink, W., Barker, P.J., Huskisson, N.S., Hirt, H. (1996). Stress signaling in plants: A mitogen-activated protein kinase pathway is activated by cold drought. *Proc. Natl. Acad. Sci. USA* 93: 11274-11279.

Kasuga, M., Liu, Q., Miura, S., Yamaguchi-Shinozaki, K., Shinozaki, K. (1999). Improving plant drought, salt, and freezing tolerance by gene transfer of a single stress-inducible transcription factor. *Nat. Biotech.* 17: 287-291.

Katagiri, T., Takahashi, S., Shinozaki, K. (2001). Involvement of a novel *Arabidopsis* phospholipase D, AtPLD delta, in dehydration-inducible accumulation of phosphatidic acid in stress signaling. *Plant J.* 26: 595-605.

Kawasaki, S., Borchert, C., Deyholos, M., Wang, H., Brazille S., Kawai, K., Galbraith, D., Bohnert, H.J. (2001). Gene expression profiles during the initial phase of salt stress in rice. *Plant Cell* 13: 889-906.

Kitano, H. (2002). Systems biology: A brief overview. *Science* 295: 1662-1664.

Klok, E.J., Wilson, I.W., Wilson, D., Chapman, S.C., Ewing, R.M., Somerville, S.C., Peacock, W.J., Dolferus, R., Dennis, E.S. (2002). Expression profile analysis of the low-oxygen response in *Arabidopsis* root cultures. *Plant Cell* 14: 2481-2494.

Knight, H., Trewavas, A.J., Knight, M.R. (1997). Calcium signaling in *Arabidopsis thaliana* responding to drought and salinity. *Plant J.* 12: 1067-1078.

Kohli, A., Xiong, J., Greco, R., Christou, P., Pereira, A. (2001). Tagged transcriptome display (TTD) in *indica* rice using Ac transposition. *Mol. Genet. Genomics* 266: 1-11.

Koiwa, H., Barb, A.W., Xiong, L., Li, F., McCully, M.G., Lee, B.-L., Sokolchik, I., Zhu, J., Gong, Z., Reddy, M., et al. (2002). C-terminal domain phosphatase-like family members (AtCPLs) differentially regulate *Arabidopsis thaliana* abiotic stress signaling, growth, and development. *Proc. Natl. Acad. Sci. USA* 99: 10893-10898.

Kovtun, Y., Chiu, W.L., Tena, G., Sheen, J. (2000). Functional analysis of oxidative stress-activated mitogen-activated protein kinase cascade in plants. *Proc. Natl. Acad. Sci. USA* 97: 2940-2945.

Kreps, J.A., Wu, Y., Chang, H.S., Zhu, T., Wang, X., Harper, J.F. (2002). Transcriptome changes for *Arabidopsis* in response to salt, osmotic, and cold stress. *Plant Physiol.* 130: 2129-2141.

Krysan, P.J., Young, J.C., Sussman, M.R. (1999). T-DNA as an insertional mutagen in *Arabidopsis. Plant Cell* 11: 2283-2290.

Levitt, J. (1980). *Responses of Plants to Abiotic Stress,* Second Edition. New York: Academic Press, Inc.

Leung, J., Giraudat, J. (1998). Abscisic acid signal transduction. *Annu. Rev. Plant Physiol. Plant Mol. Biol.* 49: 199-222.

Lind, C., Gerdes, R., Hamnell, Y., Schuppe-Koistinen, I., von Lowenheilm, H.B., Holmgren, A., Cotgreave, I.A. (2002). Identification of *S*-glutathionylated cellular proteins during oxidative stress and constitutive metabolism by affinity purification and proteomics analysis. *Arch. Biochem. Biophys.* 406: 229-240.

Liu, J., Ishitani, M., Halfter, U., Kim, C.S., Zhu, J.K. (2000). The *Arabidopsis thaliana* SOS2 gene encodes a protein kinase that is required for salt tolerance. *Proc. Natl. Acad. Sci. USA* 97: 3735-3740.

Liu, J., Zhu, J.K. (1998). A calcium sensor homolog required for plant salt tolerance. *Science* 280: 1943-1945.

Lockhart, D.J., Dong, H., Byrne, M.C., Follettie, M.T., Gallo, M.V., Chee, M.S., Mittmann, M., Wang, C., Kobayashi, M., Horton, H., Brown, E.L. (1996). Expression monitoring by hybridization to high-density oligonucleotide arrays. *Nat. Biotechnol.* 14: 1675-1680.

Long, S., Humphries, S., Falkowski, P.G. (1994). Photoinhibition of photosynthesis in nature. *Annu. Rev. Plant Physiol. Plant Mol. Biol.* 45: 633-663.

Maggio, A., Miyazaki, S., Veronese, P., Fujita, T., Ibeas, J.I., Damsz, B., Narasimhan, M.L., Hasegawa, P.M., Joly, R.J., Bressan, R.A. (2002). Does proline accumulation play an active role in stress-induced growth reduction? *Plant J.* 31: 699-712.

Mathesius, U., Keijzers, G., Natera, S.H., Weinman, J.J., Djordjevic, M.A., Rolfe, B.G. (2001). Establishment of a root proteome reference map for the model legume *Medicago truncatula* using the expressed sequence tag database for peptide mass fingerprinting. *Proteomics* 11: 1424-1440.

McKersie, B.D., Bowley, S.R., Harjanto, E., Leprince, O. (1996). Water-deficit tolerance and field performance of transgenic alfalfa overexpressing superoxide dimutase. *Plant Physiol.* 111: 1177-1181.

Menz, M.A., Klein, R.R., Mullet, J.E., Obert, J.A., Unruh, N.C., Klein, P.E. (2002). A high-density genetic map of *Sorghum bicolor* (L.) Moench based on 2926 AFLP, RFLP and SSR markers. *Plant Mol. Biol.* 48: 483-499.

Mikami, K., Katagiri, T., Luchi, S., Yamaguchi-Shinozaki, K., Shinozaki, K. (1998). A gene encoding phosphatidylinositol 4-phosphate 5-kinase is induced by water stress and abscisic acid in *Arabidopsis thaliana*. *Plant J.* 15: 563-568.

Mikolajczyk, M., Olubunmi, S.A., Muszynska, G., Klessig, D.F., Dobrowolska, G. (2000). Osmotic stress induces rapid activation of a salicylic acid-induced protein kinase and a homolog of protein kinase ASK1 in tobacco cells. *Plant Cell* 12: 165-178.

Munnik, T., Meijer, H.J.G. (2001). Osmotic stress activates distinct lipid and MAPK signaling pathways in plants. *FEBS Lett.* 498: 172-178.

Munnik, T., Meijer, J.H.J., ter Riet, B., Frank, W., Bartels, D., Musgrave, A. (2000). Hyperosmotic stress stimulates phospholipase D activity and elevates the levels of phosphatidic acid and diacylglycerol pyrophosphate. *Plant J.* 22: 147-154.

Munns, R. (2002). Comparative physiology of salt and water stress. *Plant Cell Environ.* 25: 239-250.

Munns, R., Passioura, J.B., Guo, J., Chazen, O., Cramer, G.R. (2000). Water relations and leaf expansion: Importance of time scale. *J. Exp. Bot.* 51: 1495-1504.

Niu, X., Bressan, R.A., Hasegawa, P.M., Pardo, J.M. (1995). Ion homeostasis in NaCl environments. *Plant Physiol.* 109: 735-742.

Öztürk, N.Z., Talame, V., Deyholos, M., Michalowski, C.B., Gozokirmisu, N., Galbraith, D.W., Tuberosa, R., Bohnert, H.J. (2002). Monitoring large-scale changes in transcript abundance in drought- and salt-stressed barley. *Plant Mol. Biol.* 48: 551-573.

Pardo, J.J., Reddy, M.P., Yang, S., Maggio, A., Huh, G.-H., Matsumoto, T., Coca, M.A., Paino-D'Urzo, M., Koiwa, H., Yun, D.-J., et al. (1998). Stress signaling through Ca^{2+}/calmodulin-dependent protein phosphatase calcineurin mediates salt adaptation in plants. *Proc. Natl. Acad. Sci. USA* 95: 9681-9686.

Peltier, J.B., Emanuelsson, O., Kalume, D.E., Ytterberg, J., Friso, G., Rudella, A., Liberles, D.A., Soderberg, L., Roepstorff, P., von Heijne, G., van Wijk, K.J. (2002). Central functions of the lumenal and peripheral thylakoid proteome of *Arabidopsis* determined by experimentation and genome-wide prediction. *Plant Cell* 14: 211-236.

Posas, F., Chambers, J.R., Heyman, J.A., Hoeffler, J.P., de Nada, E., Arino, J. (2000). The transcriptional response of yeast to salt stress. *J. Biol. Chem.* 275: 17249-17255.

Quarrie, S.A., Laurie, D.A., Zhu, J., Lebreton, C., Semikhodskii, A., Steed, A., Witsenboer, H., Calestani, C. (1997). QTL analysis to study the association between leaf size and abscisic acid accumulation in droughted rice leaves and comparisons across cereals. *Plant Mol. Biol.* 35: 155-165.

Qiu, Q.S., Guo, Y., Dietrich, M.A., Schumaker, K.S., Zhu, J.K. (2002). Regulation of SOS1, a plasma membrane Na^+/H^+ exchanger in *Arabidopsis thaliana,* by SOS2 and SOS3. *Proc. Natl. Acad. Sci. USA* 99: 8436-8441.

Quintero, F.J., Garciadeblas, B., Rodriguez-Navarro, A. (1996). The SAL1 gene of *Arabidopsis,* encoding an enzyme with 3'(2'), 5'bisphosphate nucleotide and inositol polyphosphate 1-phosphatase activities, increases salt tolerance in yeast. *Plant Cell* 8: 529-537.

Quintero, F.J., Ohta, M., Shi, H., Zhu, J.K., Pardo, J.J. (2002). Reconstitution in yeast of the *Arabidopsis* SOS signaling pathway for Na^+ homeostasis. *Proc. Natl. Acad. Sci. USA* 99: 9061-9066.

Rep, M., Krantz, M., Thevelein, J.M., Hohmann, S. (2000). The transcriptional response of *Saccharomyces cerevisiae* to osmotic shock. *J. Biol. Chem.* 274: 8920-8300.

Romero, C., Belles, J.M., Vaya, J.L., Serrano, R., Culiañez-Macià, F.A. (1997). Expression of the yeast trehalose-6-phosphate synthase gene in transgenic tobacco plants: Pleiotropic phenotypes include drought tolerance. *Planta* 201: 293-297.

Roxas, V.R., Smigh Jr., R.K., Allen, E.R., Allen, R.D. (1997). Overexpression of glutathione *S*-transferase/glutathione peroxidase enhances the growth of transgenic tobacco seedlings during stress. *Nat. Biotechnol.* 15: 988-991.

Rus, A., Yokoi, S., Sharkhuu, A., Reddy, M., Lee, B.H., Matsumoto, T.K., Koiwa, H., Zhu, J.K., Bressan, R.A., Hasegawa, P.M. (2001). AtHKT1 is a salt tolerance determinant that controls Na^+ entry into plant roots. *Proc. Natl. Acad. Sci. USA* 98: 14150-14155.

Schena, M., Shalon, D., Davis, R.W., Brown, O.P. (1995). Quantitative monitoring of gene expression patterns with a complementary DNA microarray. *Science* 270: 467-470.

Schumaker, K.S., Sze, H. (1987). Inositol 14,5-triphosphate releases Ca^{2+} from vacuolar membrane vesicles of oat roots. *J. Biol. Chem.* 262: 3944-3946.

Seki, M., Ishida, J, Narusaka, M., Fujita, M., Nanjo, T., Umezawa, T., Kamiya, A., Nakajima, M., Enju, A., Sakurai, T., et al. (2002). Monitoring the expression pattern of around 7,000 *Arabidopsis* genes under ABA treatments using a full-length cDNA microarray. *Funct. Integrat. Genomics* 2: 282-291.

Serrano, R. (1996). Salt tolerance in plants and microorganisms: Toxicity targets and defense responses. *Int. Rev. Cytol.* 165: 1-51.

Serrano, R., Rodriguez-Navarro, A. (2001). Ion homeostasis during salt stress in plants. *Curr. Opin. Cell Biol.* 4: 399-404.

Sheen, J. (1996). Ca^{2+} dependent protein kinases and stress signal transduction in plants. *Science* 274: 1900-1902.

Shi, H., Ishitani, M., Kim, C., Zhu, J.K. (2000). The *Arabidopsis thaliana* salt tolerance gene *SOS1* encodes a putative Na^+/H^+ antiporter. *Proc. Natl. Acad. Sci. USA* 97: 6896-6901.

Shi, H., Lee, B.H., Wu, S.J., Zhu, J.K. (2003). Overexpression of a plasma membrane Na^+/H^+ antiporter gene improves salt tolerance in *Arabidopsis thaliana. Nat. Biotechnol.* 21: 81-85.

Shi, H., Quintero, F.J., Pardo, J.M., Zhu, J.K. (2002). The putative plasma membrane Na^+/H^+ antiporter SOS1 controls long distance Na^+ transport in plants. *Plant Cell* 14: 465-477.

Shi, H., Xiong, L., Stevenson, B., Lu, T., Zhu, J.K. (2002). The *Arabidopsis* salt overly sensitive 4 mutants uncover a critical role for vitamin B6 in plant salt tolerance. *Plant Cell* 14: 575-588.

Shimamoto, K., Kyozuka, J. (2002). Rice as a model for comparative genomics of plants. *Annu. Rev. Plant Biol.* 53: 399-419.

Shin Kore-eda, S., Cushman, M.A., Akselrod, I., Bufford, D., Frederickson, M., Clark, E., Cushman, J.C. (2004). Transcript profiling of salinity stress responses by large-scale expressed sequence tag analysis in *Mesembryanthemum crystallinum. Gene* (in press).

Shinozaki, K., Yamaguchi-Shinozaki, K. (1997). Gene expression and signal transduction in the water-stress response. *Plant Physiol.* 115: 327-334.

Stoeckert, C.J., Jr., Causton, H.C., Ball, C.A. (2002). Microarray databases: Standards and ontologies. *Nat. Genet.* 32(Suppl.): 469-473.

Suzuki, H., Achnine, L., Xu, R., Matsuda, S.P., Dixon, R.A. (2002). A genomics approach to the early stages of triterpene saponin biosyntesis in *Medicago truncatula. Plant J.* 32: 1033-1048.

Tafforeau, M., Verdus, M.C., Charlionet, R., Cabin-Flaman, A., Ripoll, C. (2002). Two-dimensional electrophoresis investigation of short-term response of flax seedlings to a cold shock. *Electrophoresis* 15: 2534-2540.

Takahashi, S., Katagiri, T., Hirayama, T., Yamaguchi-Shinozaki, K., Shinozaki, K. (2001). Hyperosmotic stress induced a rapid and transient increase in inositol 1,4,5-triphosphate independent of abscisic acid in *Arabidopsis* cell culture. *Plant Cell Physiol.* 42: 214-222.

Talamé, V., Öztürk, N.Z., Michalowski, C.B., Bohnert, H.J., Tuberosa, R. (2002). *Expression profiling of barley under water deficit using cDNA microarrays.* Paper presented at Plant and Animal Genome X, San Diego, CA.

Tarczynski, M.C., Jensen, R.G., Bohnert, H.J. (1993). Stress protection of transgenic tobacco by production of the osmolyte mannitol. *Science* 259: 508-510.

Trethewey, R.N. (2001). Gene discovery via metabolic profiling. *Curr. Opin. Biotech.* 12: 135-138.

Uno, Y., Furihata, T., Abe, H., Yoshida, R., Shinozaki, K., Yamaguchi-Shinozaki, K. (2000). *Arabidopsis* basic leucine zipper transcription factors involved in an abscisic acid-dependent signal transduction pathway under drought and high-salinity conditions. *Proc. Natl. Acad. Sci. USA* 97: 11632-11637.

Uozumi, N., Kim, E.J., Rubio, F., Yamaguchi, T., Muto, S., Tsuboi, A., Bakker, E.P., Nakamura, T., Schroeder, J.I. (2000). The *Arabidopsis* HKT1 gene homolog mediates inward Na^+ currents in *Xenopus laevis* oocytes and Na^+ uptake in *Saccharomyces cerevisiae. Plant Physiol.* 122: 1249-1260.

Urao, T., Yakubov, B., Satoh, R., Yamaguchi-Shinozaki, K., Seki, B., Hirayama, T., Shinozaki, K. (1999). A transmembrane hybrid-type histidine kinase in *Arabidopsis* functions as an osmosensor. *Plant Cell* 11: 1743-1754.

Van Camp, W., Capiau, K., Van Montagu, M., Inze, D., Slooten, L. (1996). Enhancement of oxidative stress tolerance in transgenic tobacco plants overproducing Fe-superoxide dimutase in chloroplasts. *Plant Physiol.* 112: 1703-1714.

Wang, H., Miyazaki, S., Kawai, K., Deyholos, M., Galbraith, D.W., Bohnert, H.J. (2003). Temporal progression of gene expression responses to salt shock in maize roots. *Plant Mol. Biol.* 52: 873-891.

Weigel, D., Ahn, J.H., Blazquez, M.A., Borevitz, J.O., Christensen, S.K., Fankhauser, C., Ferrandiz, C., Kardailsky, I., Malancharuvil, E.J., Neff, M.M., et al. (2000). Activation tagging in *Arabidopsis. Plant Physiol.* 122: 1003-1013.

Winicov, I., Bastola, D.R (1999). Transgenic overexpression of the transcription factor A1fin1 enhances expression of the endogenous *MsPRP2* gene in alfalfa and improves salinity tolerance of the plants. *Plant Physiol.* 120: 473-480.

Wu, S.J., Lei, D., Zhu, J.K. (1996). *SOS1,* a genetic locus essential for salt tolerance and potassium acquisition. *Plant Cell* 8: 617-627.

Yale, J., Bohnert, H.J. (2001). Transcript expression in *S. cerevisiae* at high salinity. *J. Biol. Chem.* 276: 15996-16007.

Yamaguchi-Shinozaki, K., Koizumi, K., Urao, S., Shinozaki, K. (1992). Molecular cloning and characterization of 9 cDNAs for genes that are responsive to desiccation in *Arabidopsis thaliana:* Sequence analysis of one cDNA clone encodes a putative transmembrane channel protein. *Plant Cell Physiol.* 33: 217-224.

Yamaguchi-Shinozaki, K., Shinozaki, K. (1993). Characterization of the expression of a desiccation-responsive *rd29* gene of *Arabidopsis thaliana* and analysis of its promoter in transgenic plants. *Mol. Gen. Genet.* 213: 331-340.

Yamaguchi-Shinozaki, K., Shinozaki, K. (1994). A novel *cis*-acting element in an *Arabidopsis* gene is involved in responsiveness to drought, low-temperature, or high-salt stress. *Plant Cell* 6: 251-264.

Yancey, P.H. (1994). Compatible and counteracting solutes. In Strance, K. (Ed.), *Cellular and Molecular Physiology of Cell Volume Regulation* (pp. 347-361). Boca Raton, FL: CRC Press.

Yu, J., Hu, S., Wang, J., Wong, G.K., Li, S., Liu, B., Deng, Y., Dai, L., Zhou, Y., Zhang, X., et al. (2002). A draft sequence of the rice genome *(Oryza sativa* L. ssp. *indica). Science* 296: 79-92.

Zhifang, G., Loescher, W.H. (2003). Expression of a celery mannose 6-phosphate reductase in *Arabidopsis thaliana* enhances salt tolerance and induces biosynthesis of both mannitol and a glucosyl-mannitol dimer. *Plant Cell Environ.* 26: 275-283.

Zhu, H., Snyder, M. (2003). Protein chip technology. *Curr. Opin. Chem. Biol.* 7: 55-63.

Zhu, J.K. (2001). Plant salt tolerance. *Trends Plant Sci.* 6: 66-71.

Zhu, J.K. (2002). Salt and drought stress signal transduction in plants. *Annu. Rev. Plant Biol.* 53: 247-273.

Zhu, J., Gong, Z., Zhang, C., Song, C.P., Damsz, B., Inan, G., Koiwa, H., Zhu, J.K., Hasegawa, P.M., Bressan, R.A. (2002). OSM1/SYP61: A syntaxin protein in *Arabidopsis* control abscisic acid-mediated and non-abscisic acid-mediated responses to abiotic stress. *Plant Cell* 14: 3009-3028.

Zhu, J.K., Hasegawa P.M., Bressan, R.A. (1997). Molecular aspects of osmotic stress in plants. *Crit. Rev. Plant Sci.* 16: 253-277.

Zhu, J.K., Liu, J., Xiong, L. (1998). Genetic analysis of salt tolerance in *Arabidopsis thaliana:* Evidence of a critical role for potassium nutrition. *Plant Cell* 10: 1181-1192.

Chapter 19

Functional Genomics Applied to Plant Fatty Acid Biosynthesis

Sari A. Ruuska
John B. Ohlrogge

INTRODUCTION

This chapter will review some recent advances in understanding regulation of fatty acid synthesis in plants and, in particular, how functional genomics has been used to gain new insight into this pathway. As will be evident, this is a new and emerging field with the majority of the research related to fatty acid synthesis in seeds. This focus has occurred because oils obtained from seeds represent major economic revenue and unusual fatty acids with commercial interest occur primarily in seed oils. For the purposes of this chapter, we consider that the term *functional genomics* refers to studies involving large numbers of genes.

Fatty Acids: Structure and Functions

Chemically, fatty acids are carboxylic acids that usually contain an even number of carbon atoms in a straight chain, most often 12 to 24 carbons in length. Fatty acids can be saturated or unsaturated. If unsaturated, one or more *cis* double bonds are generally separated by a methylene group between the carbon atoms. Although in vegetative tissues unusual fatty acid structures are rare, seed fatty acids can contain different functional groups, such as epoxy or hydroxy groups, or even ring structures. Figure 19.1 shows a few examples of structures of common and unusual fatty acids; for more details, see Gunstone et al. (1994).

The plant fatty acid synthesis pathway produces the most abundant forms of highly reduced carbon found in nature. These include polar lipids, such as membrane phospholipids and chloroplast galactolipids. In addition, waxes and cutins that coat plant surfaces are mainly long-chain fatty acid derivatives. Seeds produce abundant storage oils and fats, which are eco-

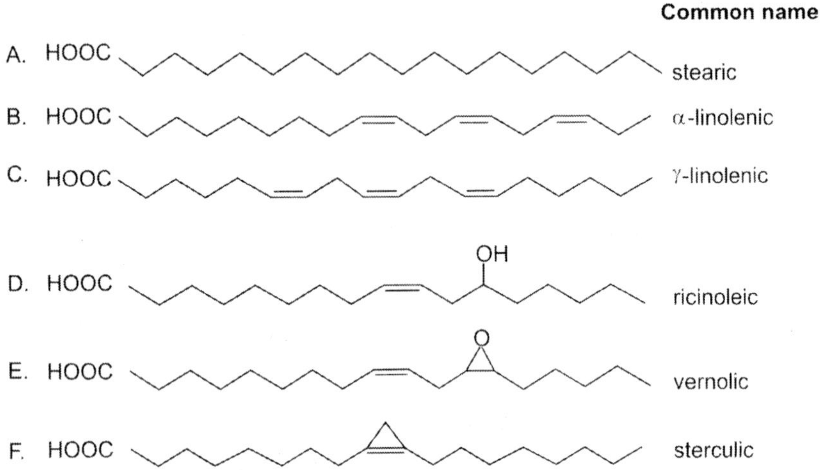

FIGURE 19.1. Structures of selected fatty acids. (A) Saturated fatty acid: stearic acid (18:0). Polyunsaturated fatty acids: (B) α-linolenic acid ($18{:}3^{\Delta 9,12,15}$) and (C) γ-linolenic acid ($18{:}3^{\Delta 6,9,12}$). (D) Hydroxy fatty acid: ricinoleic acid (12-OH-$18{:}1^{\Delta 9}$). (E) Epoxy fatty acid: vernolic acid (12-epoxy-$18{:}1^{\Delta 9}$). (F) Cyclopropane fatty acid: sterculic acid (9,10-methylene-$18{:}1^{\Delta 9}$).

nomically important lipids composed of three fatty acids esterified to a glycerol backbone. These triacylglycerols serve as a carbon and energy reserve in plants as well as in animals. Fatty acid derivatives, such as jasmonic acid and oxylipins, function also in signal transduction. In the next section we will give a short overview of fatty acid synthesis in plants. For more detailed description, see, for example, Somerville et al. (2000), Browse (1997), or Ohlrogge and Browse (1995).

Plant Fatty Acid Synthesis Is Initiated by Acetyl-CoA Carboxylase

Plants are generally considered not capable of bulk, long-distance transport of fatty acids, so each cell must contain the machinery for their synthesis. In plants, fatty acid synthesis takes place almost exclusively inside plastids: the chloroplasts of green cells and leucoplasts and chromoplasts of nongreen tissues. The first committed step in fatty acid synthesis is the formation of malonyl-CoA by carboxylation of acetyl-CoA by acetyl-CoA carboxylase (ACCase; Figure 19.2). There are two different types of ACCase structures in plants. Plastids of dicotylenous and many monocot

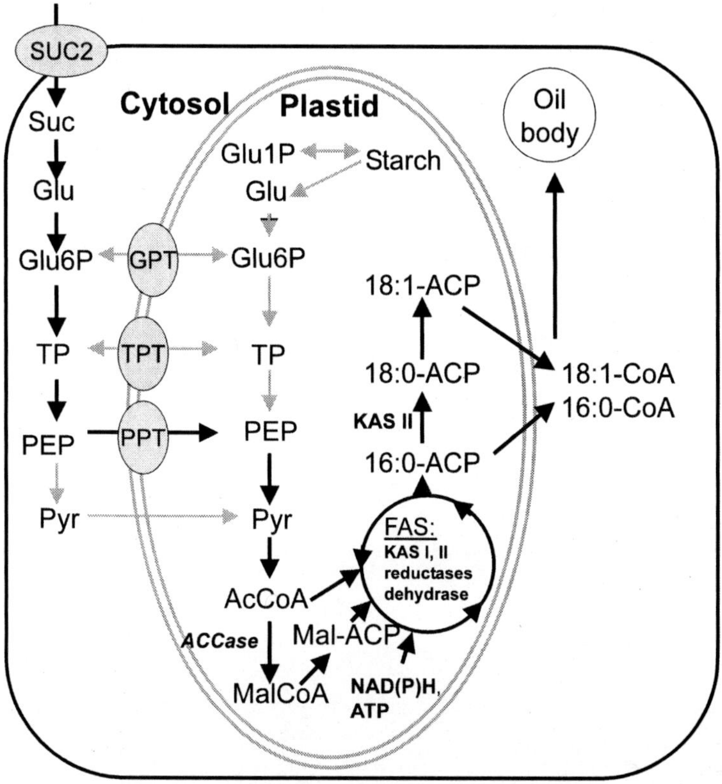

FIGURE 19.2. A simplified scheme of metabolism to produce oil in seeds. Only the major enzymes and transporters involved in conversion of sucrose to oil are shown. The major pathway of carbon flow during oil synthesis, based on EST and microarray analysis, is indicated with black arrows. Abbreviations: Suc, sucrose; Glu, glucose; Glu1P, glucose-1-phosphate; Glu6P, glucose-6-phosphate; TP, triose phosphate; PEP, phosphoenol pyruvate; Pyr, pyruvate; AcCoA, acetyl-CoA; MalCoA, malonyl-CoA; KAS, ketoacyl synthase; FAS, fatty acid synthase; ACP, acyl carrier protein; TAG, triacylglycerol; GPT, glucose-6-phosphate transporter; TPT, triose phosphate transporter; PPT, phosphoenol pyruvate transporter. (*Source:* Adapted from White et al., 2000.)

plants have a multisubunit form that consists of four different polypeptides encoded by separate genes (Sasaki et al., 1995). Biotin carboxylase carrier protein (BCCP), biotin carboxylase (BC), as well as α-carboxyltransferase subunits are encoded by the nuclear genome, whereas the fourth subunit, β-carboxyltransferase, is plastid encoded. In the cytosol, all plants have an

ACCase isozyme that is one large, multifunctional polypeptide encoded by a single gene (Shorrosh et al., 1994). The family Graminae is different in the sense that both the plastid and cytosolic forms of ACCase are of the multifunctional type (Egli et al., 1993). The cytosolic form of ACCase generates malonyl-CoA, which is used for fatty acid chain elongation reactions as well as precursors for secondary metabolites.

Growing Fatty Acids Are Attached to Acyl Carrier Proteins

After the formation of malonyl-CoA, the malonyl group is transferred to an acyl carrier protein (ACP). Through the rest of the fatty acid synthesis pathway, fatty acids are attached to ACPs, which are small-phosphopanteine-group-containing proteins. Multiple ACP isoforms have been found in plants, with distinctive gene-expression patterns (Ohlrogge and Kuo, 1984), and it is possible that these isoforms have a regulatory role in determining the fate of the attached acyl chain (Branen et al., 2001).

Fatty Acid Synthase Complex Elongates the Growing Chain by Two Carbon Atoms

The enzymes that synthesize fatty acids from acetyl- and malonyl-CoA form a loosely associated complex called fatty acid synthase (FAS). The first reaction of FAS is the condensation of the malonyl-ACP with acetyl-CoA by the enzyme 3-ketoacyl-ACP synthase III (KAS III). In this and all subsequent condensations, one carbon is lost as CO_2 and two carbons are attached to the growing chain. The product is then reduced by 3-ketoacyl-ACP reductase, dehydrated by 3-hydroxyacyl-ACP dehydratase, and again reduced by enoyl-ACP reductase to form a new saturated acyl chain. The next cycle is initiated by a condensation of the acyl-ACP with malonyl-ACP by a second condensing enzyme, 3-ketoacyl-ACP synthase I (KAS I), which catalyzes seven elongation steps of the chain up to 16-carbon length. The last condensation step is catalyzed by 3-ketoacyl-ACP synthase II (KAS II) that adds two carbons to the C16 chain, producing C18-ACP. The final products of the plastid FAS are then C16-ACP (palmitoyl-ACP) or C18-ACP (stearoyl-ACP) saturated fatty acids. In the plastids, stearoyl-ACP is usually desaturated to oleoyl-ACP ($18{:}1^{\Delta 9}$) by a soluble stearoyl-ACP desaturase.

Thioesterases Terminate the Fatty Acid Synthesis

The acyl-ACPs formed inside the plastids can be used for two main purposes. First, they can be utilized as acyl donors for synthesis of plastid membrane glycerolipids. Alternatively, the acyl groups can be hydrolyzed from ACPs by thioesterases to produce free fatty acids that leave the plastids. Two main types of acyl-ACP thioesterases occur in plant cells. The most active is the FatA-class, which preferentially cleaves 18:1-ACPs, whereas the other class, FatB, prefers saturated and/or shorter-chain acyl-ACPs, such as 16:0-ACP (Knutzon et al., 1992; Doermann et al., 1995). Plants that produce unusually short-chained fatty acids do this by means of special forms of thioesterases that have a high affinity toward shorter acyl chains. For example, coconut and California bay accumulate 10 and 12 carbon fatty acids in their seed oils. This feature has been successfully transferred to *Brassica* by expressing 12:0 (lauric acid)-ACP thioesterase from the California bay specifically in developing seeds (Voelker et al., 1996). These transgenic plants have been grown commercially to produce lauric acid as an alternative to coconut and palm kernel sources for this fatty acid.

Elongases and Desaturases Modify the Fatty Acid Chains

After hydrolysis by thioesterases, free fatty acids leave the plastid and are attached to CoA by acyl-CoA synthetase. Any further elongation from the C18 chain takes place in the cytosol and is catalyzed by ER-located elongases. Similarly to the plastid 3-ketoacyl-ACP synthases, ER elongases use malonyl-CoA to add two carbon units to the acyl chain, up to C30. Studies with *Arabidopsis* and jojoba suggest that, in plants, elongases belong to a large gene family, possibly because each elongase may only catalyze additions of the C2 unit to a specific chain-length acyl-CoA (Todd et al., 1999).

The introduction of additional double bonds to fatty acids is catalyzed by fatty acid desaturases both inside and outside the plastids. Plants contain a large number of desaturases, which are membrane bound, except the plastid-soluble stearoyl-ACP desaturase discussed previously. The desaturases have been best studied in *Arabidopsis* through mutations, and eight different genes have been well characterized (Wallis and Browse, 2002). Interestingly, many of the enzymes found to be responsible for the production of unusual fatty acids resemble desaturases (Shanklin and Cahoon, 1998), and some of them will be discussed further in this chapter.

Storage Lipids Are Packed into Oil Bodies

One major fate for newly synthesized fatty acids is their allocation to triacylglycerides (TAG), the neutral storage lipids that make up the seed oils. The assembly of TAG is catalyzed by acyltransferase enzymes associated with ER or oil bodies, which sequentially add acyl groups to the backbone of glycerol-3 phosphate. The final step in oil storage is the packing of triacylglyceride molecules into oil bodies, which consist of a TAG matrix surrounded by a phospholipid monolayer embedded with a high concentration of proteins called oleosins.

Why Functional Genomics Has Been Used to Study Plant Fatty Acid Synthesis

Vegetable oils are a major agricultural commodity, with worldwide production of 90 billion pounds, worth nearly $50 billion per year. Their major use is for human consumption, such as frying oil, shortening, and spreads, but there are also other, nonfood uses, such as detergents, paints, and lubricants. The properties of an oil are determined by its fatty acid composition, and so there has been considerable interest in modifying plant fatty acid production from this point of view. For example, it is advantageous to increase the content of oleic acid (18:1) in canola oil, to increase its health benefits, stability, and shelf life (Schierholt and Becker, 2001). On the other hand, it is of interest to produce industrially valuable fatty acids in transgenic crops, for example, to provide materials for biodegradable plastics (Nawrath et al., 1994; Ohlrogge, 1994).

Major aims for molecular genetic improvement of plant oils have been to improve the total oil content of seeds and to increase the content of specific target fatty acids in the oil to improve functional properties and expand uses. In several instances, a functional genomics approach has been successfully used to identify the genes responsible for the production of unusual fatty acids. More recently, the development of genomics technology has opened doors for large-scale approaches aiming to uncover and understand the regulatory factors that govern synthesis and accumulation of fatty acids in tissues. In the following section, we will discuss examples of these approaches.

DIRECTED GENE DISCOVERY: APPROACH

The first genomic approaches in plant fatty acid synthesis were directed toward discovery of genes responsible for production of unusual fatty acids. These projects initially utilized small-scale EST sequencing, but as the technology became more advanced, the sequencing efforts have also expanded. As an example of this approach, the discovery of the enzyme responsible for ricinoleic acid biosynthesis in castor bean is presented.

Isolation of the Gene Responsible for Ricinoleic Acid Biosynthesis

Ricinoleic acid is a substituted oleic acid derivative that has a hydroxyl group on the twelfth carbon (12-hydroxyoleic acid; Figure 19.1D). The commercially important source for ricinoleic acid is the seed oil of castor bean (*Ricinus communis* L.), in which this fatty acid can constitute up to 90 percent of all the fatty acids. The industrial uses for ricinoleate include manufacture of nylon, varnishes, paints, and lubricants. Biochemical studies of ricinoleic acid synthesis suggested that the enzyme responsible for the hydroxylation of oleic acid resembled more common, microsomal fatty acid desaturases. A typical feature of desaturases is that they contain histidine-rich motifs that bind catalytically active di-iron clusters of the enzyme. To find the putative oleic acid hydroxylase, van de Loo et al. (1995) assumed that the enzyme should also contain this motif. In addition, since ricinoleic acid is found only in seeds, the putative hydroxylase gene should be seed specific. The hydroxylase's mRNA level should also be equivalent to other fatty acid desaturases. A cDNA library was prepared from developing castor bean endosperm tissue and was enriched for clones that were expressed at moderate levels in endosperm but not in leaves, which do not contain ricinoleic acid. After partial sequencing of 468 clones of the enriched set, two clones were identified that were similar in sequence to the membrane-bound oleate desaturase ($18{:}1^{\Delta 9}$ desaturase, *FAD2*) from *Arabidopsis*. The full-length cDNA clone was isolated and expressed in tobacco, resulting in the synthesis of ricinoleic acid in transgenic plants, and thus confirming that the correct gene was isolated.

Other Examples

The success of isolating the castor hydroxylase stimulated a number of other similar approaches. For example, Cahoon and co-workers have identified several fatty acid conjugases, which are enzymes that produce polyunsaturated fatty acids with conjugated double bonds (Cahoon et al., 1999, 2001). In conjugated fatty acids such as α-eleostearic acid ($18{:}3^{\Delta 9cis,11cis,13trans}$), α-parinaric acid ($18{:}3^{\Delta 9cis,11trans,13trans,15cis}$), and calendic acid ($18{:}3^{\Delta 8trans,10trans,12cis}$), the double bonds occur without the methylene group in between. Seed oils containing conjugated double bonds are chemically reactive and typically have high rates of oxidation, which makes them suitable as drying agents in paints and varnishes. Developing seeds of *Momordica charantia, Impatiens balsamica,* and *Calendula officinalis,* species whose seed oil contains α-eleostearic, α-parinaric, and calendic acid, respectively, were used for preparation of cDNA libraries. Although the synthesis mechanism of these fatty acids was not known, the search focused on finding variant forms of the common $18{:}1^{\Delta 9}$ desaturase. Altogether more than 6,000 ESTs were sequenced from these libraries; homology searches revealed the existence of clones related to but distinct from *FAD2*. In all three cases, expression of the identified genes in soybean embryos resulted in the accumulation of the desired conjugated fatty acid. Using EST analysis, Cahoon and co-workers also isolated an enzyme that is involved in synthesis of an unusual $18{:}1^{\Delta 12}$ epoxy fatty acid, vernolic acid (Figure 19.1E; Cahoon et al., 2002). Oils containing vernolic acid could potentially be useful in industrial applications, such as the production of nylon, adhesives, or paints. A cytochrome P450 enzyme was identified from developing seeds of *Euphorbia lagascae,* and its expression in transgenic tobacco callus or somatic soybean embryos produced vernolic acid. An EST approach also was used to isolate a gene encoding a Δ6 desaturase from developing borage *(Borago officianalis)* embryos (Nunberg et al., 1996). This gene is responsible for the accumulation of γ-linolenic acid (Figure 19.1C) and octadecatetrenoic acid, which both are important in human nutrition and have a number of medical and industrial applications. Recently, a cyclopropane fatty acid synthase was identified from seeds of *Sterculia foetida,* based on similarity of EST sequences to a bacterial enzyme (Bao et al., 2002). This enzyme is the first step in synthesis of fatty acids containing carbocyclic rings (Figure 19.1F).

The many successful examples of gene discovery through EST sequencing cited previously have confirmed this approach as the method of choice for isolating genes involved in synthesis of plant natural products. Although once considered an expensive approach, major reductions in DNA sequencing costs and dramatic increases in through-put have placed EST projects within reach of most research labs, particularly compared to other more labor-intensive gene isolation methods. Furthermore, in addition to discovery of a single target gene, EST approaches provide a wealth of additional information on sequences and expression levels of other genes that may be relevant to understanding or manipulating plant metabolism.

LARGE-SCALE DNA SEQUENCING AND GENE-EXPRESSION ANALYSIS

As genomic technologies, most notably high-throughput DNA sequencing, have become less expensive and more easily accessible, many large-scale projects have been initiated (see Tables 19.1 and 19.2). The aim of many recent projects is not so much to study or discover a single or a few

TABLE 19.1. Fatty acid-related EST projects.

cDNA library	Enzyme cloned	# of ESTs sequenced	Reference
Castor bean (*Ricinus communis* L.)	Oleic acid hydroxylase	468	van de Loo et al. (1995)
Momordica charantia *Impatiens balsamica*	MomoFadX ImpFadX (conjugases)	3,000	Cahoon et al. (1999)
Calendula officinalis	CoFADX1 CoFADX2 (conjugases)	3,036	Cahoon et al. (2001)
Euphorbia lagascae	CYP726A1 (cytochrome P450) epoxidase	1,006	Cahoon et al. (2002)
Sterculia foetida	Cyclopropane synthase	5,300	Bao et al. (2002)
Arabidopsis thaliana		10,500	White et al. (2000)
Sesamum indicum		3,328	Suh et al. (2003)
Glycine max		245,000; >50,000 from seed-related material	Soybean Genomics Initiative

genes, but to gain understanding of the control of metabolic networks, for example, to understand why certain tissues or plants accumulate more lipids than others do. Another driving force has been the realization that, in many cases, manipulation of a single gene has not provided a desired effect in metabolic-engineering projects. For example, although biochemical evidence suggests that ACCase has a major control over fatty acid synthesis rate, twofold overexpression of *Arabidopsis* ACCase in *Brassica* increased seed oil by only 5 percent, suggesting that other enzymes or factors besides fatty acid synthesis rate became limiting (Roessler et al., 1997). In addition, expression of a California bay laurel 12:0 ACP thioesterase in developing *Brassica* seeds leads to production of oil containing a maximum of 50 percent of lauric acid (12:0), but coexpression of a coconut acyltransferase, which increases the incorporation of 12:0 into triacylglycerides, can further raise the total laurate levels (Knutzon et al., 1999).

TABLE 19.2. Relevant Web sites.

Project	Web site
Sesame developing-seed EST collection	<http://biotech.korea.ac.kr/member/michung/index.html>
Soybean EST database	<http://soybean.ccgb.umn.edu>
Catalog of genes for plant glycerolipid synthesis	<http://www.canr.msu.edu/lgc/>
Arabidopsis lipid gene database	<http://www.plantbiology.msu.edu/lipids/genesurvey/>
Arabidopsis developing-seed microarrays	<http://www.bpp.msu.edu/Seed/SeedArray.htm>
Docosahexaenoic acid (DHA) production in oil crops	<http://www.york.ac.uk/org/cnap/01_research/01c_labB/01c5_func_gen/01c_func_gen.htm>
Sunflower genomics	<http://cgb.indiana.edu/genomics/projects/sunflower/>
HeliantGene: functional genomics of sunflower initiative	<http://bap.ensat.fr/HeliantGene/index.html>

Large-Scale EST Projects

Michigan State University Arabidopsis *Developing-Seed EST Collection*

The first publicly available, large-scale EST set from a plant tissue active in lipid synthesis was the *Arabidopsis* developing-seed EST collection from Michigan State University (MSU) (for details, see White et al., 2000). At that time, the available *Arabidopsis* EST collections did not contain sequences specifically derived from seeds. To initiate the project, a cDNA library was constructed from RNA extracted from developing seeds between 5 and 13 days after flowering, the time of most active storage product synthesis. After subtraction of the library using highly abundant mRNAs (such as seed storage proteins and oleosins), altogether 10,500 clones were sequenced. After classification of the ESTs according to their predicted function, about 40 percent did not have a match in other databases and, thus, at the time represented novel genes expressed possibly only in seeds. This large number of novel sequences reflected the lack of previous sequence information from seed material in general. Recently, the ESTs have been mapped to the *Arabidopsis* genome, and they represent about 3,500 genes.

A detailed bioinformatics analysis of the EST abundances focused on the metabolic pathway of carbon from photosynthates to seed oil. The main conclusions of this analysis can be summarized as follows:

- *The overall frequency of ESTs for FAS genes was higher in the seed EST set than in nonseed* Arabidopsis *ESTs.* Approximately 0.79 percent of the seed ESTs encode plastid proteins of fatty acid synthesis, compared to 0.38 percent of the mixed *Arabidopsis* tissue ESTs (F. Beisson, personal communication). However, there was no clear up-regulation of any one particular gene. This suggests that seeds do not achieve high flux through the fatty acid pathway by increasing the expression of a few individual enzymes, but rather somehow up-regulate the entire pathway at the transcriptional level.
- *ESTs encoding cytosolic glycolytic enzymes were more abundant than ESTs of their plastid counterparts.* Developing oilseeds have both cytosolic and plastid glycolytic pathways, but the participation of the two in fatty acid synthesis has been unclear thus far. These results at the transcriptional level support a major flux for the cytosolic pathway.
- *Phosphoenolpyruvate (PEP) is possibly a main form of carbon that is imported into plastids for fatty acid synthesis.* The analysis uncovered several ESTs of the plastid PEP transporter in seeds, but no putative pyruvate transporter could be identified.

Together, these findings led to a hypothesis that a major route of carbon in developing *Arabidopsis* seeds involves cytosolic glycolysis from sucrose to PEP, and then PEP is imported to plastids via its transporter and finally converted to pyruvate and acetyl-CoA (Figure 19.2). This is further supported by the high abundance of ESTs encoding plastid pyruvate dehydrogenase complex compared to the mitochondrial isoform, as well as by the timing of expression of carbon transporters and cytosolic/plastid isoforms of pyruvate kinase. These aspects will be discussed in the subsection Large-Scale Gene-Expression Analysis.

In addition to providing preliminary information about transcript abundances of >3,000 structural genes involved in seed metabolism, this data set also identified >500 known and putative regulatory factors (transcription factors, protein kinases + phosphatases) that represent new, potential candidates for seed-specific metabolic regulators. A large part of current knowledge about the regulation of gene expression in developing seeds has been generated from analysis of a few transcription factors that function in embryo development, such as *ABI3* (Parcy et al., 1994), *FUS3* (Luerssen et al.,

1998), and *LEC1* (Lotan et al., 1998). Almost no protein phosphatases and kinases have been analyzed in similar detail from seeds, so the data set is likely to contain many novel regulators.

Although the *Arabidopsis* seed EST collection has provided a wealth of information about seed metabolism, it also revealed some limitations to the EST approach. It is obvious that it is hard to identify genes with low expression levels, such as with many transcription factors. Furthermore, even cDNAs of the central fatty acid synthesis enzyme enoyl-ACP reductase were not found in the EST collection, although the mRNA must be present in seed tissue.

Other EST Collections

Two other public EST collections from tissues with high oil synthesis include developing-seed ESTs from soybean and sesame. Soybean *(Glycine max)* is the world's largest source of vegetable oil and is a subject for a large public genomics initiative (Soybean Genomics Initiative, http://soybean.ccgb.umn.edu). As of April 2004, this database consists of about 286,868 EST sequences, originating from over 80 different libraries. The ESTs are organized to a total of 61,127 contigs and singletons. The collection contains more than 53,000 ESTs from different stages of soybean seed and seed coat development, as well as somatic embryos. This very large sequence database will provide a rich resource of information about expression of fatty acid-related and other genes in this crop.

Sesame *(Sesamum indicum)* is an important oilseed crop of Asia that produces seeds with 50 percent of dry weight oil, which is commonly used in oriental cooking due to its distinctive flavor. Sesame oil also contains lignans that act as antioxidants, stabilizing the oil. To identify genes involved in storage product accumulation, as well as lignan synthesis, a total of 3,328 ESTs from 5 to 25-day-old immature sesame seeds were sequenced (Suh et al., 2003). To search for novel or seed-specific sequences, the ESTs were analyzed by BLASTX to the nonredundant protein database of GenBank and also compared against the *Arabidopsis* proteome. The project identified several putative genes or specific isoforms of fatty acid synthesis gene families that may be characteristics to sesame seed (see the following subsection).

Sequence Analysis and Data Mining

The increased amount of sequence information from *Arabidopsis* and other plant species has the capacity to provide new insights into the identity and number of proteins involved in plant lipid metabolism. Similarly, as the number of ESTs available from different species has increased, the opportunities for bioinformatics approaches have multiplied, especially the possibility for performing "digital Northerns" to compare the EST abundances across different libraries.

Unfortunately, the identification of genes in public databases is often incorrect due to the fact that most annotations are done automatically. To improve the annotation of the genes involved in acyl lipid metabolism in *Arabidopsis,* a catalog of 65 proteins involved in fatty acid synthesis was constructed (Mekhedov et al., 2000). This catalog contain genomic, cDNA, and EST sequences for plant enzymes involved in fatty acid and lipid metabolism available in public databases as of August 8, 1999. For each enzyme, sequence entries are listed for several different plant species. Some results of the sequence study are summarized here:

- *More than 160,000 publicly available ESTs were analyzed in terms of their abundance.* Specifically, lipid-related ESTs were compared for *Arabidopsis* and rice. In general, there was a correlation between the number of ESTs for a given reaction in rice and *Arabidopsis*. Therefore, between these evolutionarily diverse species, the relative expression levels of the fatty acid synthesis pathway enzymes seem to have been conserved.
- *Previously undescribed genes for plant glycerolipid synthesis were putatively identified.* For example, palmitate-specific monogalactosyl diacylglycerol desaturase *(FAD5)* was previously identified only through the existence of mutants in *Arabidopsis* (Hugly et al., 1991). The survey of Mekhedov et al. (2000) identified a putative genomic DNA sequence for this gene based on its similarity to known desaturase genes, its chromosome location, and the presence of a transit peptide. Its identity was later confirmed (Mekhedov and Browse, unpublished results). In addition, a candidate gene for a mitochondrial KAS enzyme was found, which has now been confirmed as a component of fatty acid synthesis in plant mitochondria (Yasuno and Wada, 2002).
- *The number of lipid-related gene families and their sizes in* Arabidopsis *were estimated.* About two-thirds of the molecular functions are represented by more than one gene. One consequence of this functional redundancy may be difficulty in observing effects from antisense or gene knockout experiments.

After the completion of the sequencing of the *Arabidopsis* genome, the aforementioned study was extended to include many more proteins and genes involved in *Arabidopsis* fatty acid signaling, transport, and degradation (Beisson et al., 2003). A total of 210 cellular activities were investigated. After a careful analysis of multiple alignments and conserved motif searches, the catalog consists of about 600 genes encoding 644 *Arabidopsis* candidate polypeptides. This catalog also contains data about the total number of *Arabidopsis* ESTs associated with each gene, plus a statistical analysis of their distribution across libraries representing different organs.

A major value of this catalog is in its ability to aid in future experimental work. Only 13 percent of the 644 predicted proteins have been experimentally demonstrated to have lipid-related activity, so the vast majority of the genes represent uncharacterized proteins. Moreover, about one-third of these proteins were previously annotated as "unknown function" or with functions unrelated to acyl lipid metabolism. For each molecular function, the catalog gives a list of most probable candidate genes. For example, there are 11 candidate genes for endoplasmic 2-lysophosphatidate acyltransferase. It is likely that some of these genes are also candidates for lysophosphatidylcholine acyltransferase, a function with no known or putative genes in *Arabidopsis*. In addition, data for EST tissue specificity provide information about enzyme isoforms involved in tissue-specific reactions, such as TAG synthesis in seeds. It can also help in the design of gene knockout strategies and guide the search for knockout phenotypes. This aspect is illustrated in Figure 19.3, where the EST abundances for two isoforms of fatty acid elongases (ketoacyl-CoA synthase, KCS) are compared in different libraries, and the results strongly suggest that one gene is preferentially expressed in seeds and whole siliques whereas the other is expressed in all other tissues except roots.

A recent example of a cross-species study of EST abundances is the comparative analysis of ESTs between sesame and *Arabidopsis* developing seeds (Suh et al., 2003). More than 3,300 sesame seed ESTs and 10,000 *Arabidopsis* seed ESTs were clustered to contigs. After BLASTX analysis of the sesame contigs and singleton ESTs against the *Arabidopsis* proteome, a locus of the *Arabidopsis* protein with the highest similarity was assigned to each sesame EST. Some main findings of the comparison were as follows:

- As expected, ESTs encoding seed storage proteins and oleosins were highly abundant in both *Arabidopsis* and sesame. However, there were differences in the expression levels of different isoforms. For example, two oleosin isoforms, matching to *Arabidopsis* loci *At3g27760* and

At5g40420 were predominant isoforms expressed in *Arabidopsis,* whereas ESTs corresponding to *At4g25140* and *At3g01570* were abundant in sesame.

- Transcripts specific for sesame seed were also identified. For example, although the *Arabidopsis* genome contains predicted multigene families for both metallothionein-like protein and "major latex protein," ESTs associated with these genes were absent from *Arabidopsis* seeds but highly abundant in sesame seeds.
- Relatively high expressions of cytosolic malate dehydrogenase and low expressions of plastidic pyruvate kinase in sesame suggest that a major pathway of carbon for plastid fatty acid synthesis in sesame may involve synthesis of malate in cytosol and its import to plastids, as opposed to PEP import suggested for *Arabidopsis*.

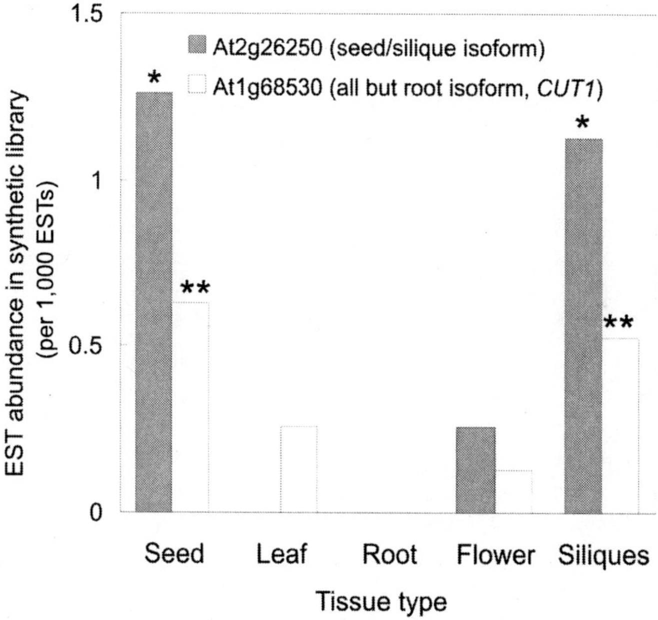

FIGURE 19.3. Differential expression of two fatty acid elongase isoforms in EST libraries originating from different tissues. *Statistically significant differential expression ($p > 0.95$) as compared to leaf, root and flower. **Significantly differential expression ($p > 0.998$) as compared to root. The analysis suggests that for the At1g68530 isoform (also identified as *CUT1*) there is no significantly different expression among seed, leaf, flower, and siliques. (*Source:* Data from Beisson et al., 2003.)

In summary, although EST abundance is an imperfect estimate of gene expression, if proper attention is paid to statistical issues (Audic and Claverie, 1997), these data can lead to useful preliminary insights and hypotheses, which can be further tested by other experiments.

Large-Scale Gene-Expression Analysis

The technology of DNA microarrays has revolutionized the ability of researchers to think broadly about gene-expression research (see Chapter 4). Microarray technology, in brief, involves the printing and immobilization of a set of DNAs on a nylon membrane or a glass slide. Fluorescent or radiolabeled cDNA probes derived from different RNA populations are hybridized on the arrays. The signal intensities for each printed spot are measured, and differential gene expression can be assessed from the different signal intensities. This technology allows for the first time the measurement of thousands of transcripts at a time, and direct comparison of RNA populations from different sources. First pioneered with yeast (Schena et al., 1995), the arrays quickly were adopted by the plant research community and used for studies of, for example, tissue-specific gene expression (Ruan et al, 1998) or plant defense (Arimura et al., 2000). More recently, oligonucleotide-based arrays have been developed, and in spring 2002, arrays with the complete genome of *Arabidopsis* became commercially available (GeneChip Arabidopsis ATH1 Genome Array, see <www.affymetrix.com>). Expression profiling of seeds active in fatty acid synthesis has attracted less attention, but a few attempts are described here.

Arabidopsis *Developing-Seed Microarrays*

Tissue-profiling experiments. The first plant fatty acid-related transcriptional-profiling project was derived from the MSU *Arabidopsis* developing-seed EST project described earlier in this chapter. Clones from this collection were used for the construction of cDNA-based microarrays. The emphasis was on selecting many unknown genes as well as carbohydrate and lipid-related genes. For the first-generation arrays, about 3,000 DNA elements were chosen (summarized in Girke et al., 2000). The set was then complemented with about 80 additional clones from lipid and carbohydrate metabolism genes that were not found from the EST collection. These microarrays were initially used for tissue profiling in *Arabidopsis* to find genes that are expressed highly in seeds by comparing RNA isolated from developing seeds, leaves, and roots. In accordance with the EST study, the microarray experiments showed that most seed-expressed transcripts, including lipid biosynthesis-related genes, were also expressed in other tissues and so could not be classified as seed specific. Of the >100 lipid genes studied, only ten were found to be expressed at least tenfold more in seeds. These genes did not encode FA biosynthetic enzymes, but were more related to FA modification or storage (such as fatty acid elongase or oleosins). However, a subset of >250 highly seed-specific transcripts were identified and will provide candidates for further studies. For details and data, see the Web site <http://www.bpp.msu.edu/Seed/SeedArray.htm>.

Time-course studies. The second-generation *Arabidopsis* seed cDNA microarrays contained ~6,000 DNA elements, which were later mapped to about 3,500 *Arabidopsis* genes. These microarrays were used in time-course experiments to study the temporal patterns of gene expression during *Arabidopsis* seed filling, as well as comparing gene expression between wild-type seeds and seeds from a mutant with a defect in seed storage oil accumulation (*wri1;* Focks and Benning, 1998). Some major observations from this study are listed here; for details, see Ruuska et al. (2002):

- *Genes associated with different storage products had distinct expression patterns (see Figure 19.4):*
  1. Transcripts for many core fatty acid-synthesizing enzymes had a bell-shaped expression pattern and so began to be down-regulated during the active oil synthesis. Examples of this group include ACCase subunits, KAS I, and FAD2 (oleate desaturase).
  2. In contrast, transcripts for some fatty acid-modifying enzymes, such as linoleate desaturase (FAD3) and elongase (FAE1), as well as for oleosins and storage proteins were induced later and remained high throughout the time-course.
  3. Transcripts for starch metabolism enzymes such as amylases were highest in young seeds, which also transiently accumulate starch.
- *Proteins involved in carbon delivery through cytosolic glycolysis via phosphoenolpyruvate to plastid fatty acid synthesis were coordinately expressed.* The expression of many proteins (but not all) involved in sucrose import, metabolism, and cytosolic glycolysis increased toward seed filling or followed the profile of FAS enzymes. In addition, the expression of PEP transporter followed the profile of the core FAS enzymes, as did the plastid pyruvate dehydrogenase, further strengthening the idea that cytosolic carbon delivery for FAS and fatty acid synthesis enzymes are coordinately expressed (Figure 19.2).
- *Several lipid- and carbohydrate metabolism-related genes were down-regulated in* wri1*-mutant seeds with reduced seed lipid content.* Most of these genes belonged to the group with a bell-shaped expres-

sion pattern during seed development and included *BCCP, KAS I, FAD2,* and the gene for the chloroplast pyruvate dehydrogenase E1α subunit. In contrast, genes coding for proteins, such as oleosins and fatty acid elongase, that are involved in lipid synthesis but did not belong to the bell-shaped expression group were not altered in their expression in *wri1*.

Brassica napus *and Maize Embryo FAS Gene Analysis*

Recently, O'Hara et al. (2002) used RNA blots and DNA filter arrays together with EST analysis to study the expression of a group of fatty acid synthesis genes in developing *Brassica napus* embryos. They found that mRNAs for the biotin carboxylase subunit of plastid ACCase, as well as 3-ketoacyl-ACP reductase, enoyl-ACP reductase, and thioesterase all accumulated in a coordinate manner. Their expression profile peaked around three and four weeks after flowering and dramatically decreased afterward. These expression patterns were similar to the reported mRNA and protein levels of enoyl-ACP reductase and stearoyl-ACP desaturase in *Brassica* seeds (Fawcett et al., 1994) and also to many FAS enzymes in *Arabidopsis* seeds (Ruuska et al., 2002). This study again highlighted the fact that increased demand for fatty acid synthesis during embryogenesis was met with the up-regulation of a group of synthesis enzymes. In addition, the transcripts were present in constant molar ratio, such that at all times ACP mRNAs were most abundant, and those of the FAS complex were similar to each other's. The results agreed with earlier EST comparisons and a study of developing *Arabidopsis* siliques, which showed that the accumulation of mRNAs encoding the subunits of the heteromeric ACCase subunits was temporally correlated (Ke et al., 2000).

O'Hara et al. (2002) also conducted a careful comparative study of the relative mRNA levels of the aforementioned proteins in both developing embryos and in leaves. The genes studied were expressed at 3- to 15-fold higher levels in developing embryos, apparently correlating with higher fatty acid synthesis activity in seeds. These results are in apparent disagreement with the *Brassica* seed versus leaf transcript comparison using *Arabidopsis* seed cDNA microarrays (Girke et al., 2000). The reasons for such disagreements may include the fact that, although *Brassica* is genetically closely related to *Arabidopsis* (about 85 percent sequence similarity), averaged fluorescence signal intensities from *Brassica* experiments were about twofold lower than from *Arabidopsis* experiments. This leads to a lower resolution in the array experiments when *Brassica* RNA is used. In addition, it has been shown that array experiments tend to produce smaller gene-expression differentials than RNA blots (Hihara et al., 2001).

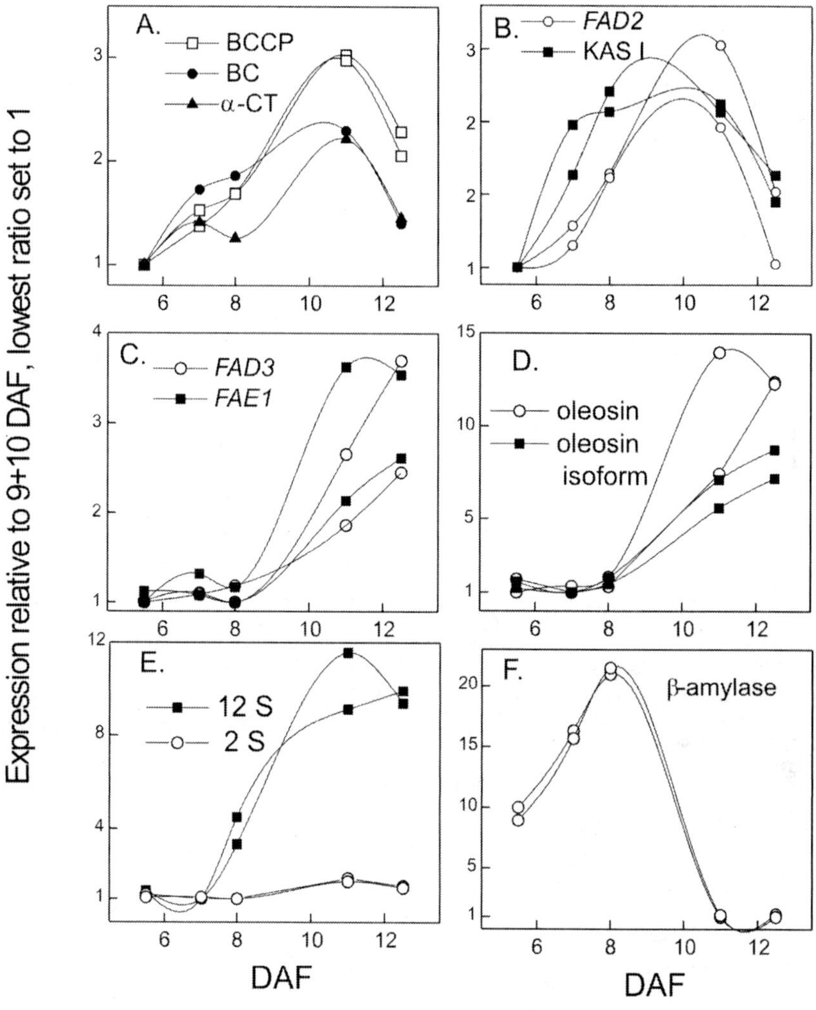

FIGURE 19.4. Genes associated with different storage products have distinctive expression profiles during *Arabidopsis* seed development. Expression of (A) biotin carboxyl carrier protein (BCCP), biotin carboxylase (BC), and α-carboxytransferase (α-CT) subunits of plastidic acetyl-CoA-carboxylase; (B) oleate desaturase *(FAD2); and 3-ketoacyl-ACP-synthase I (KAS I);* (C) *linoleate desaturase (FAD3);* and fatty acid elongase 1 *(FAE1);* (D) 18.5 kDa oleosin and oleosin isoform; (E) 12S and 2S storage proteins; and (F) β-amylase. Expression ratios were rescaled by setting the lowest ratio to 1. See Ruuska et al. (2002) for details. DAF = days after flowering.

Expression of FAS-related genes has also been studied from developing maize embryos (Lee et al., 2002). Transcript profiles were quantitated using glass-based cDNA arrays, nylon membrane arrays, RNA blots, and EST abundance analysis. Similar to results in previous studies, several FAS genes (ACCase, KAS III, ACPs, desaturase) exhibited a coordinated expression pattern, such that their expression peaked several days before the maximal fatty acid accumulation would be expected. Several oleosin isoforms had a different profile, such that their expression was induced at a later stage. The authors also clearly demonstrated that, for a given gene, EST numbers from libraries constructed at different stages of seed development correlated well with microarray results. The mean correlation coefficient between electronic Northern and nylon array expression profiles was 0.86, and between electronic Northern and glass array expression profiles was 0.84. This outcome provided an independent confirmation that ESTs and microarray signals both reflect transcript abundance. Together, these studies on *Arabidopsis, Brassica napus,* and maize all demonstrate that the up-regulation of several fatty acid-synthesizing enzymes during seed development seems to be under similar and coordinated control.

Other Fatty Acid-Related Plant Functional Genomics Projects

Additional functional genomics projects related to plant fatty acid or lipid synthesis follow:

- *Perception and transduction of low-temperature signals to induce desaturation of fatty acids:* Studies with *Synechocystis* have started to identify the components of the signal transduction chain that causes an up-regulation of fatty acid desaturases in response to low temperature. This project has used whole-genome random mutagenesis and disruption of all putative genes for histidine kinases together with screening for the transcriptional activity of the desaturase promoter (Suzuki et al., 2000). A membrane-associated histidine kinase that appeared to be involved in the temperature sensing was identified, and the effect of mutation in this kinase on cold-regulated genes was studied using microarrays (Suzuki et al., 2001).
- *Docosahexaenoic acid (DHA) production in oil crops:* Ian Graham's group from the University of York uses genomics to discover genes in marine algae that could be used to produce docosahexaenoic acid (DHA) in oilseed crops (Tonon et al., 2002). DHA is a long-chain (C22) polyunsaturated fatty acid with important nutritional and pharmaceutical applications, and its current main dietary source is fish.

- *Transcriptional profiling of acetyl-CoA utilization in* Arabidopsis: In addition to fatty acid synthesis, acetyl-CoA in the cytosol is used for synthesis of waxes, isoprenoids, and flavonoids. Eve Wurtele from Iowa State University is conducting transcriptional profiling of acetyl-CoA metabolism in *Arabidopsis*. Her group has recently completed an extensive analysis of day length- and UV stress-related changes in gene expression in leaves of transgenic *Arabidopsis* plants with decreased activity of ATP citrate lyase, the enzyme producing cytosolic acetyl-CoA (http://abstracts.aspb.org/pb2002/public/P66/ 0415. html and http://abstracts.aspb.org/pb2002/public/P66/0920. html).
- *Sunflower genomics:* A project at the University of Indiana (Bloomington) is identifying single nucleotide polymorphisms (SNPs) in the sunflower genome. Together with the EST database, this project will facilitate mapping and cloning candidate genes controlling agronomically important traits (http://www.cgb.indiana.edu/genomics/projects/sunflower/). A large sunflower genomics initiative is also under way in Europe (http://bap.ensat.fr/HeliantGene/Sunfloweramodel.html), which includes cDNA sequencing and construction of macroarrays. A sunflower BAC library has also been constructed and initially has been used to identify genes for putative transmembrane receptors that may have a role in the development of the embryo (Gentzbittel et al., 2002).

SUMMARY AND FUTURE PROSPECTS

So, now that we know a complete plant genome sequence and have the tools to simultaneously measure the expression of thousands of genes, proteins, and metabolites, what do we do with this information? One clear outcome of the genomic revolution has been the shift in focus of many biologists from working on one or a few enzymes/genes to new and broader questions about regulatory networks and systems biology. In addition, it has brought with it the challenge to make sense of experiments that generate a thousandfold or more data than similar experiments carried out a few years ago. In the future, this challenge will be compounded, as integration of transcript profiling with proteomics and metabolomics demands even more sophisticated bioinformatics approaches and thought processes.

Functional genomics is clearly in its infancy and, with regards to the goal of understanding how plants control fatty acid and glycerolipid biosynthesis, only a few studies have been published. So far, no magic bullets or revolutionary new concepts have been described from the published works. Nevertheless, the studies have begun to reveal patterns of gene expression that are at least associated with, and perhaps control, the changes in metabolism and storage products synthesized in plants and, in particular, in developing seeds. The crucial next stages will be to test the clues gleaned from this first generation of functional genomics studies by constructing transgenic plants altered in candidate regulatory genes. Thus, the global-scale data gathering and analysis already accomplished has quickly led back to the essential "gene by gene" tests of the candidates identified so far. Only when candidate genes are shown to modify functions in transgenic plants is it possible to establish which genes truly are control factors over plant biology. Such tests come with all the rate limitations inherent in our ability to produce and analyze transgenic plants, particularly when analysis of seed traits is the desired outcome. Therefore, high-throughput strategies for production and analysis of tests in transgenic plants will become more and more important. An easily transformed cell or tissue culture model system, such as *Arabidopsis* somatic embryo cultures (Ikeda-Iwai et al., 2002), that accurately mimics seed biology would provide a substantial increase in the discovery rate.

REFERENCES

Arimura, G., Tashiro, K., Kuhara, S., Nishioka, T., Ozawa, R., Takabayashi, J. (2000). Gene responses in bean leaves induced by herbivory and by herbivore-induced volatiles. *Biochem. Biophys. Res. Comm.* 277: 305-310.

Audic, S., Claverie, J.M. (1997). The significance of digital expression profiles. *Genome Res.* 25: 3389-3402.

Bao, X., Katz, S., Pollard M., Ohlrogge, J.B. (2002). Carbocyclic fatty acids in plants: Biochemical and molecular genetic characterization of cyclopropane fatty acid synthesis of *Sterculia foetida*. *Proc. Natl. Acad. Sci. USA* 99: 7172-7177.

Beisson, F., Koo, A.J.K., Ruuska, S., Schwender, J., Pollard, M., Thelen, J.J., Paddock, T., Salas, J.J., Savage, L., Milcamps, A., et al. (2003). *Arabidopsis* genes involved in acyl lipid metabolism: A 2003 census of the candidates, a study of the distribution of expressed sequence tags in organs, and a web-based database. *Plant Physiol.* 132: 681-692.

Branen, J.K., Chiou, T.J., Engeseth, N.J. (2001). Overexpression of acyl carrier protein-1 alters fatty acid composition of leaf tissue in *Arabidopsis*. *Plant Physiol.* 127: 222-229.

Browse, J. (1997). Synthesis and storage of fatty acids. In Larkins, B.A. Vasil, I.K. (Eds.), *Cellular and Molecular Biology of Plant Seed Development* (pp. 407-440). Dordrecht, The Netherlands: Kluwer Academic.

Cahoon, E.B., Carlson, T.J., Ripp, K.G., Schweiger, B.J., Cook, G.A., Hall, S.E., Kinney, A.J. (1999). Biosynthetic origin of conjugated double bonds: Production of fatty acid components of high-value drying oils in transgenic soybean embryos. *Proc. Natl. Acad. Sci. USA* 96: 12935-12940.

Cahoon E.B., Ripp, K.G., Hall, S.E., Kinney, A.J. (2001). Formation of conjugated Δ^8, Δ^{10}-double bonds by Δ^{12}-oleic acid desaturase-related enzymes. *J. Biol. Chem.* 276: 2637-2643.

Cahoon, E.B., Ripp, K.G., Hall, S.E., McGonigle, B. (2002). Transgenic production of epoxy fatty acids by expression of a cytochrome P450 enzyme from *Euphorbia lagascae* seed. *Plant Physiol.* 128: 615-624.

Doermann, P., Voelker, T.A., Ohlrogge, J.B. (1995). Cloning and expression in *Escherichia coli* of a novel thioesterase from *Arabidopsis thaliana* specific for long-chain acyl-acyl carrier proteins. *Arch. Biochem. Biophys.* 316: 612-618.

Egli, M.A., Gengenbach, B.G., Gronwald, J.W., Somers, D.A., Wyse, D.L. (1993). Characterisation of maize acetyl-CoA carboxylase. *Plant Physiol.* 101: 499-506.

Fawcett, T., Simon, W.J., Swinhoe, R., Shanklin, J., Nishida, I., Christie, W.W., Slabas, A.R. (1994). Expression of mRNA and steady-state levels of protein isoforms of enoyl-ACP reductase from *Brassica napus*. *Plant Mol. Biol.* 26: 155-163.

Focks, N., Benning, C. (1998). *Wrinkled1:* A novel, low-seed-oil mutant of *Arabidopsis* with a deficiency in the seed-specific regulation of carbohydrate. *Plant Physiol.* 118: 91-101.

Gentzbittel, L., Abbott, A., Galaud, J.P., Georgi, L., Fabre, F., Liboz, T., Alibert, G. (2002). A bacterial artificial chromosome (BAC) library for sunflower, and identification of clones containing genes for putative transmembrane receptors. *Mol. Genet. Genom.* 266: 979-987.

Girke, T., Todd, J., Ruuska, S., White, J., Benning, C., Ohlrogge, J.B. (2000). Microarray analysis of developing *Arabidopsis* seeds. *Plant Physiol.* 124: 1570-1581.

Gunstone, F.D., Harwood, J.L., Padley, F.B. (1994). *The Lipid Handbook,* Second Edition. Boca Raton, FL: CRC Press.

Hihara, Y., Kamei, A., Kanehisa, M., Kaplan, A., Ikeuchi, M. (2001). DNA microarray analysis of cyanobacterial gene expression during acclimation to high light. *Plant Cell* 13: 793-806.

Hugly, S., Kunst, L., Somerville, C. (1991). Linkage relationships of mutants that affect fatty acid composition in *Arabidopsis*. *J. Hered.* 82: 484-488.

Ikeda-Iwai, M., Satoh, S., Kamada, H. (2002). Establishment of a reproducible tissue culture system for the induction of *Arabidopsis* somatic embryos. *J. Exp. Bot.* 53: 1575-1580.

Ke, J., Wen, T.N., Nikolau, B.J., Wurtele, E.S. (2000). Coordinate regulation of the nuclear and plastidic genes coding for the subunits of the heteromeric acetyl-coenzyme A carboxylase. *Plant Physiol.* 122: 1057-1071.

Knutzon, D.S., Bleibaum, J.L., Nelsen, J., Kridl, J.C., Thompson, G.A. (1992). Isolation and characterization of two safflower oleoyl-acyl carrier protein thioesterase cDNA clones. *Plant Physiol.* 100: 1751-1758.

Knutzon, D.S., Hayes, T.R., Wyrick, A., Xiong, H., Maelor, D.H., Voelker, T.A. (1999). Lysophosphatidic acid acyltransferase from coconut endosperm mediates the insertion of laurate at the sn-2 position of triacylglycerols in lauric rapeseed oil and can increase total laurate levels. *Plant Physiol.* 120: 739-746.

Lee, J.M., Williams, M.E., Tingey, S.V., Rafalski, J.A. (2002). DNA array profiling of gene expression changes during maize embryo development. *Funct. Integr. Genomics* 2: 13-27.

Lotan, T., Ohto, M., Yee, K.M., West, M.A.L., Lo, R., Kwong, R.W., Yamagishi, K., Fischer, R.L., Goldberg, R.B., Harada, J.J. (1998). *Arabidopsis* LEAFY COTYLEDON1 is sufficient to induce embryo development in vegetative cells. *Cell* 93: 1195-1205.

Luerssen, K., Kirik, V., Herrmann, P., Misera, S. (1998). *FUSCA3* encodes a protein with a conserved VP1/ABI3-like B3 domain which is of functional importance for the regulation of seed maturation in *Arabidopsis thaliana*. *Plant J.* 15: 755-764.

Mekhedov, S., de Ilárduya, O., Ohlrogge, J. (2000). Toward a functional catalog of the plant genome: A survey of genes for lipid biosynthesis. *Plant Physiol.* 122: 389-401.

Nawrath, C., Poirier, Y., Somerville, C. (1994). Targeting of the polyhydroxybutyrate biosynthetic pathway to the plastids of *Arabidopsis thaliana* results in high levels of polymer accumulation. *Proc. Natl. Acad. Sci. USA* 91: 12760-12764.

Nunberg, A.N., Beremand, P., Thomas, T.L. (1996). Cloning and expression of a delta(6)-desaturase from *Borago officianalis*. *Plant Physiol.* 11(Suppl.): 576.

O'Hara, P., Slabas, A.R., Fawcett, T. (2002). Fatty acid and lipid biosynthetic genes are expressed at constant molar ratios but different absolute levels during embryogenesis. *Plant Physiol.* 129: 310-320.

Ohlrogge, J.B. (1994). Design of new plant products: Engineering of fatty acid metabolism. *Plant Physiol.* 104: 821-826.

Ohlrogge, J.B., Browse, J. (1995). Lipid biosynthesis. *Plant Cell* 7: 957-970.

Ohlrogge, J.B., Kuo, T.M. (1984). Plants have isoforms of acyl carrier protein that are expressed differently in different tissues. *J. Biol. Chem.* 260: 8032-8037.

Parcy, F., Valon, C., Raynal., M. Gaubier-Comella, P., Delseny, M., Giraudat, J. (1994). Regulation of gene expression programs during *Arabidopsis* seed development: Roles of the *ABI3* locus and of endogenous abscisic acid. *Plant Cell* 6: 1567-1582.

Roessler, K., Shintani, D., Savage, L., Boddupalli, S., Ohlrogge, J.B. (1997). Targeting of the *Arabidopsis* homomeric acetyl-coenzyme A carboxylase to plastids of rapeseeds. *Plant Physiol.* 113: 75-81.

Ruan, Y., Gilmore, J., Conner, T. (1998). Towards *Arabidopsis* genome analysis: Monitoring expression profiles of 1400 genes using cDNA microarrays. *Plant J.* 15: 821-833.

Ruuska, S.A, Girke, T., Benning, C., Ohlrogge, J.B. (2002). Contrapuntal networks of gene expression during *Arabidopsis* seed filling. *Plant Cell* 14: 1191-1206.

Sasaki, Y., Konishi, K.T., Nagano, Y. (1995). The compartmentation of acetyl CoA carboxylase in plants. *Plant Physiol.* 108: 445-449.

Schena, M., Shalon, D., Davis, R.W., Brown, P.O. (1995). Quantitative monitoring of gene expression patterns with a complementary DNA microarray. *Science* 270: 467-470.

Schierholt, A., Becker, H.C. (2001). Environmental variability and heritability of high oleic acid content in winter oilseed rape. *Plant Breeding* 120: 63-66.

Shanklin, J., Cahoon, E.B. (1998). Desaturation and related modifications of fatty acids. *Annu. Rev. Plant Physiol. Mol. Biol.* 49: 611-641.

Shorrosh, B.S., Dixon, R.A., Ohlrogge, J.B. (1994). Molecular cloning, characterization and elicitation of acetyl-CoA carboxylase from alfalfa. *Proc. Natl. Acad. Sci. USA* 91: 4323-4327.

Somerville, C., Browse, J., Jaworski, J.G., Ohlrogge, J.B. (2000). Lipids. In Buchanan, B.B., Gruissem, W., Jones, R.L. (Eds.), *Biochemistry and Molecular Biology of Plants* (pp. 456-527). Rockville, MD: American Society of Plant Physiology, The Courier Companies, Inc.

Suh, M.C., Kim, M.J., Hur, C.G., Bae, J.M., Park, Y.I., Chung, C.H., Kang, C.W., Ohlrogge, J.B. (2003). Comparative analysis of expressed sequence tags between *Sesamum indicum* and *Arabidopsis thaliana* developing seeds. *Plant Mol. Biol.* 52: 1107-1123.

Suzuki, I., Kanesaki, Y., Mikami, K., Kanehisa, M., Murata, N. (2001). Cold-regulated genes under control of the cold sensor Hik33 in *Synechocystis. Mol. Microbiol.* 40: 235-244.

Suzuki, I., Los, D.A., Kanesaki, Y., Mikami, K., Murata, N. (2000). The pathway for perception and transduction of low-temperature signals in *Synechocystis. EMBO J.* 19: 1327-1334.

Todd, J., Post-Beittenmiller, D., Jaworski, J.G. (1999). *KCS1* encodes a fatty acid elongase 3-ketoacyl-CoA synthase affecting wax biosynthesis in *Arabidopsis thaliana. Plant J.* 17: 119-130.

Tonon, T., Harvey, D., Larson, T.R., Graham, I.A. (2002). Long chain polyunsaturated fatty acid production and partitioning into triacylglycerols by four microalgae. *Phytochemistry* 61: 15-24.

Van de Loo, F.J., Broun, P., Turner, S., Somerville, C. (1995). An oleate 12-hydroxylase from *Ricinus communis* L. is a fatty acid desaturase homolog. *Proc. Natl. Acad. Sci. USA* 92: 6743-6747.

Voelker, T.A., Hayes, T.R., Cranmer, A.M., Turner, J.C., Davies, H.M. (1996). Genetic engineering of a quantitative trait: Metabolic and genetic parameters influencing the accumulation of laurate in rapeseed. *Plant J.* 9: 229-241.

Wallis, J.G., Browse, J. (2002). Mutants of *Arabidopsis* reveal many roles for membrane lipids. *Prog. Lipid Res.* 41: 254-278.

White, J.A, Todd, J., Newman, T., Focks, N., Girke, T., de Ilárduya, O., Jaworski, J.G., Ohlrogge, J.B., Benning, C. (2000). A new set of *Arabidopsis* expressed sequence tags from developing seeds: The metabolic pathway from carbohydrates to seed oil. *Plant Physiol.* 124: 1582-1594.

Yasuno R., Wada, H. (2002). The biosynthetic pathway for lipoic acid is present in plastids and mitochondria in *Arabidopsis thaliana. FEBS Lett.* 517: 110-114.

Chapter 20

Genetic and Molecular Control of Seed Development in *Arabidopsis*

Loïc Lepiniec
Martine Devic
Frédéric Berger

INTRODUCTION

Seed development is a critical phase in the life cycle of plants, allowing survival and dispersal of the species. In angiosperms, it gives rise to the formation of the embryo, endosperm, and the surrounding seed coat. Through cell division, growth, and differentiation, embryogenesis is the transformation of a single cell into a discrete multicellular organism. Essential developmental processes, that will be reiterated throughout the life of the plant are initiated within the embryo. The development of the embryo has to be coordinated with the formation of endosperm and integuments to produce a functional seed. This implies a tight spatiotemporal regulation of the activity of many proteins involved in the determination of cell identity and pattern formation, or in basal functions such as cell division, elongation, and metabolism. How the plant coordinates these functions during seed development is far from fully understood. However, significant progress has been made in recent years through genetic and, more recently, molecular approaches. The new tools of functional genomics are opening new prospects

---

We wish to thank B. Jones for correcting the manuscript, G. Pelletier for providing the T-DNA collection, and B. Dubreucq for his help with the figures. We thank J. Guilleminot and A. Martinez for their help in screening transgenic lines in the group of M. D., and people in the group of L. L who have contributed during the past years to the systematic screening and study of the *emb* mutants (D. Grain, E. Harscoet, K. Madiona, C. Vasnier, and V. Tanty) and to the FST program (S. Balzergue, B. Dubreucq, and M. Caboche). F. B. was supported by the French Ministry of Research (ACI Développement and Jeune Equipe, and the EMBO Young Investigator Program). The authors have received some financial supports from Genoplante programs (AF1999-020 and AF2001-015).

in terms of system biology toward a better understanding of seed development. In this chapter, we will focus on *Arabidopsis* as a model plant for seed development. The availability of the genomic sequence (Arabidopsis Genome Initiative, 2000; Somerville and Koornneef, 2002) and an increasing array of resources make *Arabidopsis* a useful tool in the study of basic cellular and developmental mechanisms and their genetic controls (Arabidopsis Genome Initiative, 2000; Somerville and Koornneef, 2002). This chapter summarizes recent progress in our understanding of the control of seed development. It is not exhaustive but rather concentrates on the successes of the molecular genetics approach while outlining its limitations.

FROM GENETIC TO MOLECULAR BASES OF SEED DEVELOPMENT

As a first step toward a genetic analysis of seed development, a significant number of laboratories have performed large-scale screens and isolated several mutants affected in this process (Castle et al., 1993; Gasser et al., 1998; Grossniklaus and Schneitz, 1998; Jürgens et al., 1995, 1991; Mayer et al., 1991; McElver et al., 2001; Meinke, 1991, 1995; Meinke and Sussex, 1979; Schneitz et al., 1997; Tzafrir et al., 2003). More than 2,000 embryo mutants have been isolated from these studies during the past few years, and estimates suggest that at least 500 genes are essential for seed development (Meinke et al., 2003). Although the identification of alleles by chromosome walking has been greatly facilitated by the availability of markers and the genomic sequence, it is still time-consuming compared to insertional mutagenesis. Furthermore, homologous recombination remains inefficient in higher plants, preventing large-scale targeted gene disruption. Large collections of T-DNA or transposon insertion mutants have therefore been developed as a first step in a functional genomics approach. More recently, the TILLING approach has provided an efficient way to isolate new allelic mutants from classical mutagenesis (McCallum et al., 2000).

A large number of laboratories are participating in the long-standing effort to search for seed-defective mutants by screening T-DNA insertion lines. For example, the group of David Meinke has screened a collection of *Arabidopsis* transformants produced by Syngenta. This work has been reported in collective publications (McElver et al., 2001; Tzafrir et al., 2003). We have screened the Versailles collection of T-DNA insertion lines (Bechtold et al., 1993). The T-DNA integrates in a random manner in the *Arabidopsis* genome, and, therefore, most genes can be disrupted using this method (Brunaud et al., 2002).

The aim of the screen was to identify genes linked to specific functions involved in seed formation. As a prerequisite for molecular studies, the putatively tagged mutants were classified based on cytological observations of phenotypes. For each mutant line, the stage of developmental arrest was determined by clearing the seeds and observing the embryo morphology. Since in most cases the mutant *(emb)* embryos were not arrested at a single stage of development, the most-advanced stage reached by the *emb* embryos was used for classification. The *emb* lines fell within five major classes: preglobular (14 percent); globular (34 percent); triangular-heart (23 percent); torpedo-cotyledon (27 percent), so-called for the shape of the embryo proper; and *embni* (2 percent) (when the mutant embryo does not resemble any wild-type stage of embryo development). The *albino* class of mutations was excluded from our selection. In Perpignan, a subset of 16,000 independent transgenic *Arabidopsis* T1 plants (from the Versailles collection) was screened for the presence of *emb* mutations by dissection of immature siliques and scoring the frequency of aborted seeds. This type of screening allows the detection of many of the mutations affecting early embryo development. Approximately 10 percent of the plants were selected after a first round of screening. Individually harvested T2 seeds were grown under growth chamber conditions for further screening and characterization. The presence of an *emb* mutation was confirmed for 438 lines, suggesting that about 3 percent of the lines carry mutations in genes essential for early seed development. In Versailles, about 30,000 lines have been screened for the production of defective seeds by direct observation of the mature T3 seeds. This type of screening is biased toward late-stage seed developmental arrest (usually postglobular), for which mature mutant seeds are clearly identifiable. About 250 mutants have been confirmed in the progenies, and only some of them correspond to the lines identified in Perpignan, suggesting that about 0.5 percent of the lines display such a phenotype. Together, our data suggest that about 3 percent of the lines from the Versailles collection carry *emb* mutations. As a comparison, McElver and collaborators found 2.5 percent of *emb* in their collection (McElver et al., 2001).

The linkage between the T-DNA and *emb* mutations has been assessed. Each T1 plant has been studied for the presence and number of T-DNA loci by germination on kanamycin selective medium in a segregating progeny of at least eight T2 plants. The linkage between the mutation and the T-DNA integration is assessed in plants showing a single locus of T-DNA insertion. A kanamycin resistance-to-sensitivity ratio of 2:1 (or lower) is significant for linkage. This bias in segregation ratio of T3 seeds from 3:1 to 2:1 is due to the loss of one phenotypic class, the homozygous *emb* seedlings, which do not exist because of lethality. In the entire *emb* mutant collection, approximately 25 percent of all lines showed linkage with a "functional" T-DNA, that is, harboring a functional *nptII* gene. The functional complementation and characterization of the allelic series for several mutants has confirmed that about one-fifth of the mutations are tagged by a functional T-DNA at the molecular level (Lepiniec and Devic, unpublished results). Preliminary analyses suggest that preglobular and globular classes present higher rates of linkage (25.4 percent and 29.4 percent, respectively), compared to postglobular mutants, i.e., triangular-heart (20.3 percent), torpedo-cotyledon (14 percent), and *embni* (9 percent). These results suggest that mutation-producing embryos arrested at early stages are more likely to be "knockout" mutants giving null phenotypes. It is possible that the classes of mutants with embryos arrested at later stages contain a higher proportion of weaker phenotypes not directly due to T-DNA integration. Our screen has identified around 80 *EMB* genes. In general, our results are similar to those obtained by McElver and collaborators in terms of frequency of mutants in each phenotypic class (McElver et al., 2001).

To systematically characterize the mutations at the molecular level, we have developed efficient protocols to recover genomic flanking sequences (FST) of tagged mutations using PCR-walking (Balzergue et al., 2001; Devic et al., 1997). This technique enables rapid identification of mutants affected in genes of interest and provides a powerful tool for reverse genetics (Samson et al., 2002).

As a result of the development of the large collections of insertional mutants, several new *EMB* genes are described every year through forward and reverse genetics approaches. However, as the number is still low in comparison to the estimation of 500 to 700 *EMB* genes, it is expected that additional genes essential for seed development will continued to be isolated. The systematic work described earlier has already resulted in the identification of the genes listed in the following material. In most cases, however, to validate the identity of candidate genes, alleles still need to be found (by forward and reverse genetics) and/or phenotypical complementation has to be performed. This is, of course, one of the aims of the systematic work. In the meantime, the data collected in the Seedgenes database (http://www.seedgenes.org) are already valuable for suggesting new functions essential for seed development (Tzafrir et al., 2003). In the following sections, we summarize the characterization of genes published in the literature, as their identity has been confirmed (by phenotypic complementation and/or characterization of allelic series) and their phenotype and cellular functions described.

OVULE DEVELOPMENT

Ontogeny of Ovule Development

Development of the *Arabidopsis* ovule has been extensively described (Mansfield and Briarty, 1991; Robinson-Beers et al., 1992; Schneitz et al., 1995) and is summarized in Figure 20.1. Briefly, the ovule primordia develop from the placenta through periclinal divisions within the epidermis or subepidermal layers of the placenta (Jenik and Irish, 2000). Anticlinal divisions in the epidermis result in the formation of ovule primordia (Gasser et al., 1998). They grow initially as radially symmetrical protrusions, but at maturity they display polarity along two axes, the proximal-distal first and the abaxial-adaxial subsequently (Schneitz et al., 1995).

Three domains can be defined along the proximal-distal axis (P-D, also called chalazal-micropylar axis), the funiculus (proximal), the chalaza (central), and the nucellus (distal). The funiculus contains a vascular strand and links the ovule to the placenta. The chalaza gives rise to the two integuments from epidermal origin that will surround the embryo sac. The inner integument (Ii) initiates first, before the outer integument (Oi). At the distal end, the nucellus gives rise to the megaspore mother cell, which forms the embryo sac. At maturity, the embryo sac comprises seven cells (the egg cell that gives rise to the embryo after fertilization, two synergids, the large central cell, and three antipodal cells). Megasporogenesis and gametogenesis have been described in detail elsewhere and will not be reviewed here (Drews et al., 1998; Drews and Yadegari, 2002; Gasser et al., 1998; Grossniklaus and Schneitz, 1998; Yang and Sundaresan, 2000). The second axis of the primordia, referred to as the anterior-posterior axis (A-P, also called abaxial-adaxial, Ab-Ad), is perpendicular to the P-D axis. The integuments grow extensively from the abaxial side and to a limited extent from the adaxial side. This asymmetric growth gives rise to the characteristic shape of the ovule, such that the micropyle is adjacent to the funiculus (Robinson-Beers et al., 1992; Schneitz et al., 1997, 1995). The zone between the chalazal and micropylar regions is called the curving zone or basal body (see Beeckman et al., 2000, and references therein). It is interesting to note that all ovules are oriented in the same direction. This suggests that positional information may come from outside the ovule (Schneitz et al., 1995).

Both the outer and inner integuments consist initially of two cell layers. The two cell layers of the outer integuments (Oi1 and Oi2) consist of large and vacuolated cells in the micropylar and chalazal areas, and smaller cells in the basal body (Beeckman et al., 2000). The inner layer of the inner integument (Ii1, also called endothelium or pigmented layer) generates a third

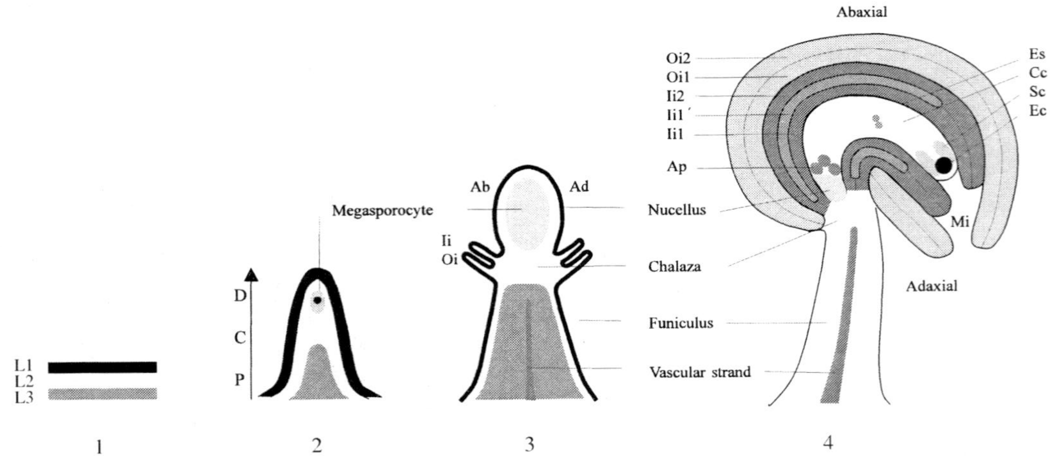

Specification and pattern formation
AP2?, AG ?, ANT, HLL, LUG, STK, SHP1, SHP2

Initiation of morphogenesis
ANT, ATS, BEL1, HLL, INO, NZZ, STK, UCN, WUS

Morphogenesis and differentiation
AP2, BAG, GL2, INO, LAL, LUG, MOG, SIN1, SIN2, SUB, SUP, TSO1, TSL, TT1, TT16, TTG1, TTG2

FIGURE 20.1. Ovule development in *Arabidopsis* and some of the genes involved. (1) The ovule primordia initiate from the placenta through periclinal division within epidermal (L1) and subepidermal (L2, L3) layers. (2) The primordia grow along the proximal-distal axis. At the distal end the nucellus gives rise to the megasporocyte (mother cell) that will form the embryo sac. D, distal; C, central; P, proximal area. (3) The chalaza area gives rise to the two integuments from epidermal origin. Ii, inner integument; Oi, outer integument. The second axis of the primordia (abaxial-adaxial, Ab-Ad) is perpendicular to the P-D axis. (4) At maturity, the embryo sac comprises seven cells, the egg cell, two synergids, the large central cell, and three antipodal cells. Ap, antipodal cells; Cc, central cell; Ec, egg cell; Es, embryo sac; Ms, megasporocyte; Mi, micropyle; Sc, synergid cells. The integuments grow extensively from the abaxial side and to a limited extent from the adaxial side. This asymmetric growth gives rise to the characteristic shape of the ovule, such that the micropyle is adjacent to the funiculus (see references in the text).

layer (Ii1') in the basal body area, toward the outer layer (Ii2) (Beeckman et al., 2000; Schneitz et al., 1995). The cells of the Ii1' layer are morphologically more closely related to those of the Ii2 (large vacuolated cells) than to the cells of the endothelium layer (small and isodiametric resembling meristematic cells). Later integument development has been described in detail elsewhere (Windsor et al., 2000). In mature seed, the outer epidermis (Oi2) is differentiated into mucilage-containing cells and the Oi1 layer is almost completely collapsed. The endothelium (Ii1) is initially pigmented inside the vacuoles and later at the periphery of the cell, before being crushed and incorporated into the "brown pigment layer" together with the Ii1' and Ii2 layers (Leon-Kloosterziel et al., 1994; Western et al., 2000; Windsor et al., 2000). Specific genetic ablation of the endothelium in *Arabidopsis* demonstrated that this cell layer is not absolutely essential for the development of a testa, although it may be important for different aspects of seed development and physiology (Debeaujon et al., 2003). In petunia, it has been suggested that abnormal development of the endothelium may strongly affect endosperm development (Colombo et al., 1997).

Characterization of Genes Involved in the Development of the Ovule

Several mutations that disrupt normal development of the ovule have been identified and are presented in this discussion (i.e., characteristic phenotypes of strong alleles). Some of the genes may respond to positional information and be involved in site-specific initiation of primordia, while others affect less specific cellular processes involved in the growth and differentiation of the primordia.

The *aintegumenta (ant)* ovules failed to develop integuments and an embryo sac, suggesting that ANT is involved in ovule primordia initiation (Gasser et al., 1998; Schneitz, 1999). *ANT* encodes a putative transcription factor of the AP2 family (Elliott et al., 1996; Klucher et al., 1996). It has been suggested that it regulates cell proliferation and growth of different organs by maintaining the meristematic competence of cells during organogenesis (Mizukami and Fischer, 2000). WUSCHEL (WUS) is a homeodomain protein of the nucellus. It acts non-cell-autonomously and is necessary for integument initiation (Mayer et al., 1998, Gross-Hardt et al., 2002). Ectopic expression in the chalazal area (i.e., under the control of the *ANT* promoter) induces the formation of integument-like organs in the basal region of the ovule (Gross-Hardt et al., 2002). No relationship has been demonstrated between WUS and ANT, suggesting that integument formation requires two independent signals from both chalaza and nucellus. The cellular function of WUS would be to coordinate the differentiation of adjacent cell groups. In *huellenlos (hll)* ovules, both integuments are reduced or absent and the cells in the distal regions of the ovule primordia often collapse (Schneitz et al., 1998). HLL has an additive effect to ANT and *hll ant* double-mutant ovules are strongly affected in elongation along the P-D axis (Schneitz, 1999). However, contrary to *ant,* the effects of the *hll* mutation are restricted to the ovule. An *HLL* paralogue has been found with a different pattern of expression, giving an explanation for the specificity of *hll* defects. In addition to their effects on ovule initiation and outgrowth, they function in integument initiation, and it is likely that ANT and HLL are repeatedly required for cell proliferation during ovule development. Interestingly, *HLL* codes for a mitochondrial ribosomal protein, indicating that a metabolic defect can have specific effects on developmental patterning (Meinke and Sussex, 1979).

In *bel1* ovules, no integuments are formed, although an integument-like structure arises in the place of the outer integuments, and the embryo sac is lacking (Modrusan et al., 1994; Robinson-Beers et al., 1992; Schneitz et al., 1997). *BEL1* encodes a homeodomain protein and is specifically expressed in the central region of the primordia before the initiation of the integuments (Reiser et al., 1995). It has been suggested that BEL1 specifies the chalazal area and determines integument identity (Balasubramanian and Schneitz, 2002). In *inner no outer (ino)* ovules, the development of the outer integument is abnormal or does not occur (Baker et al., 1997). This phenotype suggests that the chalazal area may be subdivided into inner and outer integument areas. *INO* encodes for a transcription factor of the YABBY family (Villanueva et al., 1999). Interestingly, ANT is essential for *INO* expression and BEL1 is a negative regulator of *INO*. In *aberrant testa shape (ats)* ovules, the two integuments are initially fused and consist later of three cell layers, including both the endothelium and the outer layer that will form the columella (Leon-Kloosterziel et al., 1994). ATS would also play a role in pattern formation, regulating the expression of *INO* (Balasubramanian and Schneitz, 2002).

Last, it has been shown recently that *nozzle (nzz)* (or *sporocyteless*), which encodes a novel protein, links several aspects of ovule development and coordinates pattern formation along both axes (Balasubramanian and Schneitz, 2002). Indeed, NZZ is essential for normal development of the funiculus, integuments, and nucellus (Schiefthaler et al., 1999). Through its interactions with BEL1 and ANT, it couples P-D pattern formation and growth. In addition, by negatively regulating INO expression in a temporal manner, NZZ confers the onset of the Ab-Ad axis.

The other known genes required for the correct morphogenesis of the integuments probably have more general functions, such as the control of cell division or cell shape. They encode proteins that act later than ANT, HLL, NZZ, or INO, and their function is usually not restricted to the ovule. For instance, in *superman (sup)* ovules, the outer integument grows equally on both adaxial and abaxial sides, suggesting that SUP is necessary for the maintenance of asymmetric growth and Ab-Ad polarity (Gaiser et al., 1995). *SUP,* which encodes a transcription factor of the zinc finger family, was first described as a floral homeotic gene (Gaiser et al., 1995; Sakai et al., 2000). It has been proposed that SUP would restrict the expression of some factors, promoting the growth of different organs, to the cells where SUP is present (Gasser et al., 1998). More recently, it has been shown that SUP attenuates positive *INO* autoregulation and maintains polar development of the outer integument (Meister et al., 2002). The *tousled (tsl)* ovules display an abnormal protruding inner integument (Roe et al., 1993), but additional floral phenotypes suggest that *TSL,* which codes for a putative kinase, regulates cell proliferation in different plant organs (Gasser et al., 1998; Roe et al., 1997). The *leunig (lug)* ovules occasionally display abnormal extension of the inner integument and often lack an embryo sac (Schneitz et al., 1997). *LUG,* which codes a putative transcription factor (Conner and Liu, 2000), is also involved in early flower development (Liu and Meyerowitz, 1995). In *sin1 (short integuments 1)* ovules, the elongation of the integuments is affected and there is an abnormal proliferation of the nucellus (Lang et al., 1994; Robinson-Beers et al., 1992). *SIN1* mutations have pleiotropic effects on flower morphogenesis and embryogenesis. In fact, recent cloning of *SIN1* revealed that it is identical to *CAF (CARPEL FACTORY)* and *SUS1 (SUSPENSOR 1)* (Golden et al., 2002). The gene has been renamed *DICER-LIKE 1 (DCL1)* because the predicted protein displays similarities with DICER, a drosophila multidomain ribonuclease identified by its role in RNA silencing (Schauer et al., 2002). How DCL1 is involved in the control of integument elongation remains to be understood. The analysis of the different phenotypes suggests that a common feature of the *dcl1* mutants is retardation in their ability to progress through different developmental stages (Golden et al., 2002).

Other mutations that have been described act later in the development of the integuments and/or have pleiotropic phenotypes (Schneitz et al., 1997; Schneitz, 1999), but the genes have not been cloned yet. With the exception of the *laelli (lal)* phenotype that seems to be restricted to the ovule, these genes, in general, affect not only the ovule but also whole-plant development. The *blasig (bag)* mutation gives irregular integuments. The *mollig (mog)* mutation gives abnormal enlarged cells in both integuments but also affects other aspects of plant development. The *lal* mutation gives a protruding inner integument, similarly to *sin1,* but its effect is restricted to the ovule. The *unicorn (ucn)* ovule primordia display a small extra protrusion, which may represent a supernumerary integument (Schneitz et al., 1997). UCN may therefore have a negative role, restricting the activity of other genes. Similarly, the *strubbelig (sub)* ovules form several protrusions from the outer integument at the distal tip. Finally, mutations in *TSO1* affect probably both cell division and cell expansion (Hauser et al., 1998; Liu et al., 1997), and SHORT INTEGUMENT 2 (SIN2) is involved in the control of cell division during integument and flower development (Broadhvest et al., 2000).

Ovule Specification

To date, no mutations in *Arabidopsis* specifically and completely block ovule development. This may illustrate a limitation of the classical genetic approach due to functional redundancy. For instance, in petunia it has been demonstrated that two MADS box genes, *Floral Binding Protein 7 (FBP7)* and *FBP11,* play central roles in the determination of ovule identity (Angenent et al., 1995; Colombo et al., 1995) and later during seed development (Colombo et al., 1997). Simultaneous inhibition of these two genes results in a homeotic transformation of ovules into carpeloid structures (Angenent et al., 1995). In addition, overexpression of *FBP11* resulted in ectopic ovule formation on the two outer floral whorl organs (Colombo et al., 1995). These results demonstrate the involvement of both *FBP7* and *FBP11* in ovule initiation. The putative *Arabidopsis* orthologue *AGL11* has been isolated and characterized (Rounsley et al., 1995) and renamed *SEEDSTICK (STK),* since a corresponding mutant fails to detach from the enlarged funiculus (Pinyopich et al., 2003). *STK* is closely related to two other functionally redundant MADS box genes, namely, *SHATTERPROOF 1 (SHP1)* and *SHP2* that are required for fruit dehiscence (Liljegren et al., 2000). Remarkably, *stk shp1 shp2* triple-mutant ovules do not develop and some of them are converted into leaflike or carpeloid structures (Pinyopich et al., 2003). These results suggest that the *STK, SHP1,* and *SHP2* genes share overlapping activities that are required for ovule identity.

In *ant lug* double mutants, carpels are formed but do not produce septum, placenta, and ovule, suggesting that ANT and LUG act redundantly to initiate the carpel margin that is necessary for placenta and ovule production (Liu et al., 2000). Since both genes code for transcription factors, it is possible to search for their target genes. Chevalier et al. (2001) have proposed several candidates, such as *CRABS CLAW (CRC), SPATULATA (SPT), TOUSLED (TSL), PERIANTHA (PAN),* and *CUP-SHAPED COTYLEDON 2 (CUC2).* Indeed, *CRC* is under positive control of LUG and is a member of the YABBY family of transcription factors involved in polarity control of different organs. Double mutants *lug tsl* and *tsl pan* show reduced marginal tissue development. The *cuc1 cuc2* double mutants lack septum formation and show a reduced number of ovules (Chevalier et al., 2001). Interestingly, *CUC1* and *CUC2* code for two homologous NAC-domain proteins that are required for the expression of *SHOOT MERISTEMLES (STM),* the formation of shoot apical meristem (SAM), and organ separation (Ishida et al., 2000; Takada et al., 2001). The function of CUC proteins, therefore, may be to maintain undifferentiated cells during SAM formation and organ separation.

Some *apetala 2 (ap2)* mutations lead to carpeloid outgrowths, suggesting that this gene may be involved in the determination of ovule identity. However, normal ovules are also produced, indicating that AP2 is not essential for ovule development (Gasser et al., 1998; Modrusan et al., 1994; Schneitz, 1999). It has been shown that *AGAMOUS (AG),* which encodes a transcription factor of the MADS family, is not necessary for ovule development but is involved in the specification of ovule identity (Bowman et al., 1991; Western and Haughn, 1999). More recently, it has been demonstrated that the *SHP* genes are responsible for AG-independent ovule development observed in the *ap2 ag* double mutant (Pinyopich et al., 2003). Taken together, these data reveal the partly redundant function of *SHP, STK,* and *AG* in promoting ovule identity (Pinyopich et al., 2003).

Differentiation of the Integuments

Several mutations affecting differentiation of the integuments have been characterized, and some of the corresponding genes have been cloned. Most of the mutations, including *transparent testa glabra 1 (ttg1)* and *ttg2* mutants, affect mucilage, flavonoid, or both metabolisms.

Plants lacking a functional *APETALA 2 (AP2)* gene are affected in the differentiation of both outer layers and are deficient in the development of columellae and the biosynthesis of mucilage (Jofuku et al., 1994). From the available data, it is difficult to conclude whether this deficiency is due to the possible role of AP2 in ovule identity or to a more general cellular function involved in the differentiation of the outer integuments. In addition to their well-characterized roles in trichome and root atrichoblast specification, the *TTG1* and *GLABRA 2 (GL2)* genes that encode a WD40 repeat protein (Walker et al., 1999) and a homeodomain transcription factor (Rerie et al., 1994), respectively, are essential for mucilage extrusion during imbibition (Koornneef, 1981; Koornneef et al., 1982). However, in *ttg1* and *gl2,* columellae are present and some mucilage is produced, suggesting that they are involved in cell morphogenesis rather than in an early step of cell differentiation (Western et al., 2001). Mutants of *TTG2*, which encode a WRKY transcription factor, also lack mucilage in the seed coat (Johnson et al., 2002). Interestingly, TTG2 functions independently of TTG1 and GL2 in mucilage production. Mutations in *MYB61* also reduce both deposition and extrusion of the mucilage (Penfield et al., 2001). The effect of combining the *ttg1-1* and *myb61-1* mutations is additive, suggesting that MYB61 performs a novel role in the development of the testa epidermis. As described previously, in *ats* ovules the two integuments are fused and consist of three cell layers, including both the endothelium and the outer layer that will form the columella. This mutant also accumulates a lower amount of mucilage (Leon-Kloosterziel et al., 1994). In *abscisic acid deficient 1 (aba1)* seeds, a lower amount of mucilage is produced, probably as an indirect consequence of their lack of ABA (Karssen et al., 1983). Recently, mutants for five new loci have been isolated, but the corresponding genes have not been identified (Western et al., 2000). MUCILAGE MODIFIED 4 (MUM4), similar to TTG1 and GL2, is required for complete mucilage synthesis. MUM3 and MUM5 are involved in the regulation of mucilage composition, and MUM1 and MUM2 are necessary for mucilage extrusion (Western et al., 2001).

Some mutations affecting pigmentation (i.e., flavonoid metabolism) in the endothelium have been described, and the corresponding genes have been identified (Debeaujon et al., 2000, 2001; Nesi et al., 2002; Winkel-Shirley, 2001, 2002; Xie et al., 2003). To date, only two of them appear to be involved in cell identity or differentiation processes, namely TT1 (Sagasser et al., 2002) and TT16/ABS (Nesi et al., 2002). *TT1* is expressed in ovules and young seed and encodes a novel type of nuclear zinc finger protein. It is involved in the differentiation of the seed endothelium, although its precise function is still unknown (Sagasser et al., 2002). *TT16* encodes a MADS box protein of the "B-sister" clade, which contains genes of unknown function that are expressed preferentially in female organs (Becker et al., 2002; Nesi et al., 2002). TT16/ABS is necessary for the expression of genes coding enzymes in the endothelium, with the exception of the chalazal-micropylar area. Furthermore, *tt16* endothelial cells (Ii1) are abnormal, resembling cells of the adjacent layers (Ii1' and Ii2), suggesting a role for TT16 in the control of endothelium cell identity (Nesi et al., 2002). However, in both cases, no data show that the deficiency in flavonoid metabolism is the consequence of abnormal cellular differentiation or is due to pleiotropic effects of the mutations. Other mutations specifically affect flavonoid biosynthesis but apparently do not affect endothelium development (Debeaujon et al., 2000; Winkel-Shirley, 2002). In fact, although it has been suggested in transgenic petunia (Jorgensen et al., 2002), there is no indication in *Arabidopsis* that the flavonoid content may influence any aspect of ovule or embryo development. Among the flavonoid biosynthesis genes,

TTG1 and *TTG2* have been already cited for their involvement in mucilage production. The role of TTG2 in flavonoid biosynthesis is unknown, although epistatic relationships have suggested that it acts downstream of TTG1 (Johnson et al., 2002). Finally, there is evidence that TTG1 acts on flavonoid metabolism through its interactions with TT2 and TT8 to control the specific expression of structural genes (Nesi et al., 2000, 2001).

ENDOSPERM DEVELOPMENT

The endosperm originates from the central cell fertilized by one of the male gametes delivered by the pollen tube (Berger, 2003; Olsen, 2001). Unlike embryogenesis, endosperm development is determinate and is limited to the end of the life of the seed at germination. In *Arabidopsis,* as in many other species, the endosperm develops in two phases, consecutively syncytial and cellular (see Figure 20.2). At a specific time, the syncytium is divided into individual cells by a specific cytokinesis termed *cellularization*. The cellular endosperm undergoes further cell divisions and differentiates in a few types of tissues patterned along an anterior-posterior axis. This axis runs from the anterior pole of the ovule (the micropyle at the site of entry of the pollen tube) to the posterior pole (marked in the seed integument by the chalaza adjacent to the maternal vascular tissue).

The Syncytial Phase of Development of Endosperm

Extensive cytological studies on the development of endosperm have been carried out in *Arabidopsis* and in closely related species (Brown, 1999; Mansfield and Briarty, 1990; Schneitz et al., 1995; Schultz and Jensen, 1971; Scott et al., 1998). In these species, the endosperm develops as a syncytium within a few days of fertilization. Endosperm-specific expression of a fluorescent HISTONE2B::YFP fusion, achieved using a GAL4-*UAS* enhancer trap strategy (Haseloff, 1999), enabled measurement of the dynamic parameters for mitotic phases in *Arabidopsis* endosperm. Time-lapse confocal observations showed that successive rounds of mitosis punctuate syncytial endosperm development in *Arabidopsis* and define a series of developmental stages with 2, 4, 8, 12 to 16, 24 to 27, 46 to 48, and 90 to 96 nuclei. Following the initial three synchronous cycles of nuclear division, independent mitotic domains are established and functionally delineated by the expression of the *Arabidopsis* mitotic cyclin B1;1. This suggests that precise control of cyclin accumulation constitutes part of the overall program responsible for syncytial development. Mitotic domains form distinct compartments organized along the anterior-posterior axis,

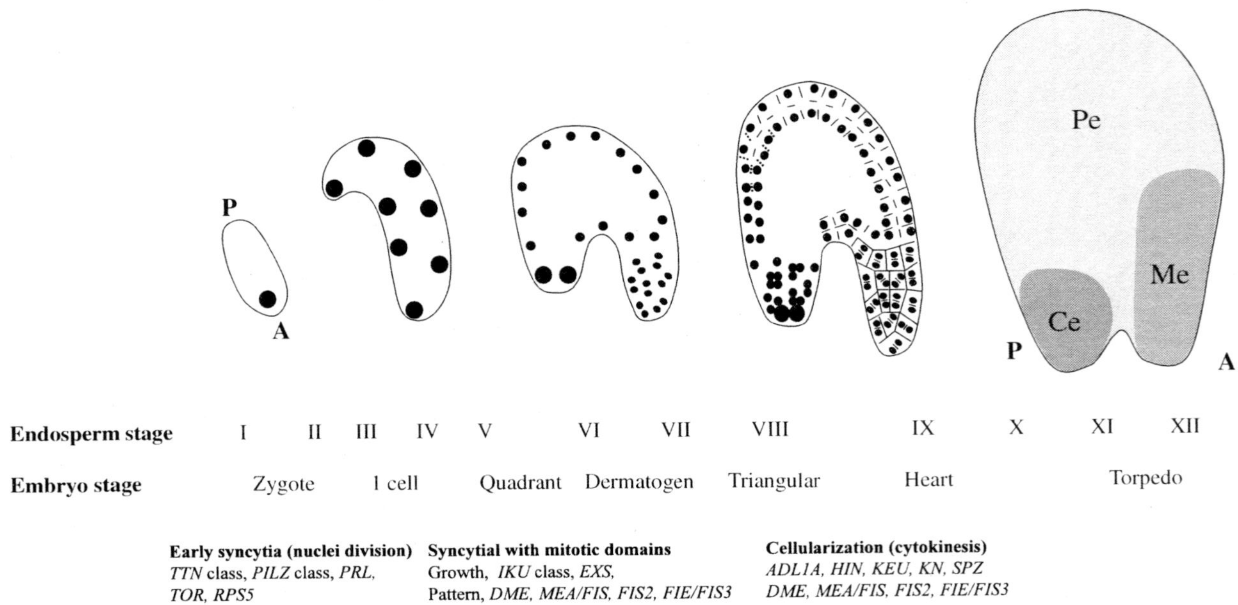

FIGURE 20.2. Endosperm development in *Arabidopsis* and some of the genes involved. Fertilization of the central cell (see Figure 20.1) produces the zygotic endosperm. Endosperm development is divided into two major phases, successively, syncytial and cellular. Stages of development are defined by successions of pseudosynchronous mitoses (roman numerals). The three first mitoses are synchronous and define the early syncytial phase. After this stage, three mitotic domains are defined from the anterior pole (A) to the posterior pole (P). Me, micropylar endosperm; Pe, peripheral endosperm; Ce, chalazal endosperm. The eighth mitotic cycle is followed by cellularization of the syncytial endosperm.

suggesting that precise spatial regulation of expression of cell cycle control genes may play a role in the functional organization of the developing syncytial endosperm. The large multinucleated cell that constitutes the early endosperm is an interesting model for cell cycle studies in a developmental context akin to the *Drosophila* early syncytial embryo (Edgar and Lehner, 1996). The simple organization of the endosperm is very sensitive to dysfunction of essential components of the cell. The class of embryo lethal mutants titan *(ttn)* covers a large range of mutants that show enlarged nuclei in the endosperm (Liu and Meinke, 1998; Tzafrir et al., 2002). Formally belonging to the *titan* class, the mutant for the MCM7 protein PROLIFERA shows a complex variable phenotype with enlarged nuclei in the endosperm (Springer et al., 2000). *PROLIFERA* is expressed in the endosperm and is expected to control the onset of DNA replication.

Control of Seed Growth

Evidence has accumulated showing that the control of endosperm growth during the syncytial phase is an essential factor that determines the ultimate size of the seed. Loss of function of "target of rapamycin" (TOR), which is involved in the control of the G1/S checkpoint, causes an arrest of endosperm syncytial development during the first phase of rapid synchronous division (Menand et al., 2002). This arrest is accompanied by an arrest of growth of the endosperm correlated with a reduction in seed size. Similarly, mutants for the production of the ribosomal protein S5 show a semidominant arrest of division and of growth in the syncytial endosperm, resulting in the production of very small aborted seeds (Weijers et al., 2001). In the latter two cases, mutations also affect embryo development and it is difficult to ascribe the resulting decrease of seed size to the endosperm. The new class of mutants *haiku (iku)* that produce viable seeds of reduced size supports the hypothesis that endosperm growth controls seed size (Garcia et al., 2003). Mutations in *iku* are sporophytic recessive and cause a severe reduction of seed size without affecting seed viability. The mutations in *iku* affect many features of endosperm development and only limited aspects of late embryogenesis. Moreover, the combination of *iku* and a mutation in one of the *pilz* group genes, *HALLIMASCH,* which severely prevents embryo development, does not modify the effect of *iku* on the reduction of seed size. Hence, the reduction of seed size by *iku* is not directly mediated by the embryo, but rather by the endosperm. The earliest phenotypic alteration in the *iku* seed is a premature arrest of growth of the endosperm three days after fertilization. The size of the *iku* seed does not subsequently increase further and thus remains half the size of wild-type seed.

Other mutants with reduced seed size have been reported and can be readily distinguished from *iku* since they are affected in other aspects of the plant growth. These include *extra sporogenous cell (exs)* (Canales et al., 2002) that causes male sterility, *ctr1* that prevents cell elongation and ethylene signal transduction (Kieber et al., 1993), and *aberrant testa shape (ats)* (Leon-Kloosterziel et al., 1994) that has a reduction of the number of layers in seed integuments. Whereas the effects on endosperm development of *ctr1* and of *ats* have not been studied to date, the mutant *exs* has been studied in detail (Canales et al., 2002). While the homozygous mutant plant is male sterile, a quarter of seeds of the heterozygous plant are reduced in size. Size reduction originates from misregulation of the cell cycle with delayed endosperm cellularization and a reduction of the cell size in the embryo. *EXS* encodes a putative receptor kinase and is expressed in the young embryo and endosperm and in the proliferating regions of the shoot.

Cellularization of the Endosperm

The endosperm syncytial stage ends with cellularization, a specific cytokinetic process. Evidence for the accumulation of both KNOLLE (Lauber et al., 1997) and the dynamin-like protein ADL1A (Otegui et al., 2001) in both somatic- and syncytial-type cell plates hints at shared components in the two processes. This is confirmed by genetic evidence, since most mutants defective for cytokinesis appear to be defective for endosperm cellularization. Two exceptions are *keule,* where endosperm cellularization is not affected, and the recently isolated mutant *spätzle,* where cytokinesis is normal although the endosperm does not cellularize (Sorensen et al., 2002). Thus, cellularization probably uses the basic machinery of cytokinesis (microtubules, proteins involved in vesicular transport, and enzymes responsible for cell wall synthesis). Recent time-lapse live observations indicate that cellularization of the *Arabidopsis* endosperm is coupled to the eighth cycle of synchronous mitosis in the peripheral endosperm (Sorensen et al., 2002). Such coupling to a specific cycle of syncytial division is observed in the *Drosophila* embryo (cycle 14). We hypothesize that, akin to cellularization of the syncytial *Drosophila* embryo (Edgar and Lehner, 1996), cellularization of the *Arabidopsis* endosperm depends on the achievement of a critical threshold of the nucleo-cytoplasmic ratio. This is supported by the fact that endosperm cellularizes precociously in *iku* mutants (Garcia et al., 2003). In *iku* endosperm, mitotic activity is not affected while size is reduced during syncytial development. This would cause premature achievement of a threshold nucleo-cytoplasmic ratio and result in the precocious onset of cellularization.

In contrast to cytokinesis, ultrastructural studies have shown a specific organization of the cell plate during cellularization (Otegui and Staehelin, 2000; Otegui et al., 2001). Detailed studies of endosperm cellularization in *Arabidopsis* have confirmed that cell wall formation is mediated by specialized types of phragmoplasts and cell plates (Otegui and Staehelin, 2000). The syncytial-type cell plates go through a series of maturation steps that are similar, but not identical, to those observed in conventional cytokinesis (Otegui et al., 2001). Cellularization establishes one endosperm cell layer and precedes further cell division. In maize and in barley, complex microtubule rearrangements take place prior to endosperm cellularization. During cellularization, so-called adventitious or cytoplasmic phragmoplasts are observed at the junctions of opposing ends of radial microtubules that encircle each nucleus. Similar adventitious phragmoplasts are seen during *Arabidopsis* endosperm cellularization (Brown et al., 1999).

Endosperm Patterning Along the Anterior-Posterior Axis

Endosperm is characterized by two major patterns, an anterior-posterior axis (A-P) and a radial symmetry. In *Arabidopsis* endosperm, three regions are organized along the A-P axis: (1) at the anterior pole close to the embryo, the micropylar endosperm, (2) the peripheral endosperm that represents the central larger portion, and (3) the posterior chalazal endosperm. In *Arabidopsis* the posterior endosperm is the site of endoreduplication after the first three cycles of synchronous divisions (Boisnard-Lorig et al., 2001). Later, during syncytial development, nuclei in the posterior part of the peripheral endosperm assemble into multinucleate structures called nodules. At the end of the syncytial phase and during the early phase of cellular development, the posterior pole becomes the site of recruitment of nuclei and of nodules that migrate into a cytoplasm pool designated as "cyst" (Berger, unpublished data). In *Arabidopsis,* the cytological organization of the posterior endosperm and of the integument suggests similarities with the transfer zone of cereals (Thompson, 2001). In spite of cytological similarities to cereals, the role of this zone in nutrient transport to the endosperm has not been demonstrated in *Arabidopsis*. The posterior endosperm remains transcriptionally active until germination, when it expresses an extensin-like protein in the vicinity of the emerging radicle (Dubreucq et al., 2000).

Central cells from flowering plants contain a relatively large amount of cytoplasm. Thus, asymmetrically distributed, maternally derived gene products may be important in setting up polarized information in the endosperm. Mutations in such genes are predicted to lead to maternal effects under gametophytic controls. Mutations have been isolated where patterning of the posterior zone of *Arabidopsis* endosperm is affected. Endosperm development in mutants that belong to the *fertilization independent seed (fis)* class (Chaudhury et al., 1997; Ohad et al., 1996) is characterized by an expansion of the posterior domain toward the anterior pole. This extension is marked by the presence of nodules and cysts at ectopic positions and by the ectopic expression of the posterior marker linked to the enhancer trap line KS117 (Sorensen et al., 2001). The *FIS* genes encode proteins with homology to "polycomb group" proteins (PcG) (Grossniklaus and Schneitz, 1998; Luo et al., 1999; Ohad et al., 1999). The PcG proteins were initially isolated in *Drosophila,* where they participate in the stabilization of repression of zygotic gene expression to maintain patterns established earlier by maternal controls. *MEDEA/FIS1, FIS2,* and *FIE/FIS3* encode homologues of "enhancer of zeste," the recently characterized PcG su(z)12 (Birve et al., 2001) and "extra sex combs," respectively. MEDEA and FIE, but not FIS2, interact, and it has been hypothesized that the three proteins belong to a larger complex (Luo et al., 2000; Spillane et al., 2000; Yadegari et al., 2000). All *FIS* genes are expressed during the late stages of maturation of the embryo sac (Luo et al., 2000; Vielle-Calzada et al., 1999; Yadegari et al., 2000). The expression of *MEDEA* and *FIS2* is uniform until the middle of syncytial development, when it becomes restricted to the posterior pole and may therefore be linked with maintenance of polarity of the endosperm.

Interactions Between Endosperm and the Other Components of the Seed

In the plant seed, endosperm is at the interface with the embryo and with the maternal integuments. This pivotal position suggests communications among the different genetic compartments of the seed. Maternal effect on endosperm development has been linked to an impaired function of the maternal structures that nourish the seed in barley. Transfer of nutrients from the mother plant to the endosperm that stores reserves involves the specialized placentochalazal tissue of seed integument and the transfer layer in the endosperm (Thompson et al., 2001). Mutants affected in the development of the placentochalazal tissue show defects in seed growth (Cheng, 1996; Felker, 1985; Maitz et al., 2000). Most of these mutants have sporophytic maternal effects on endosperm and embryo development. In mutants *seg1, seg3, seg6,* and *seg7,* the chalazal maternal transmitting tissue becomes necrotic and, as a consequence, the supply of nutrients is presumably cut off and reserve storage in the endosperm is prevented, ultimately causing seed lethality.

In contrast to the sporophytic maternal effect mutant isolated in barley, the maize mutant *miniature 1* is sporophytic recessive, produces small seeds, and shows specific defects in the development of the placentochalazal area of the integument. The activity of the Miniature 1 cell wall invertase 2 is localized to the transfer layer of the endosperm that neighbors the placentochalazal tissue of the integument (Cheng, 1996). The abnormal development of the placentochalazal tissue in *miniature 1* seeds substantiates evidence for communications between the endosperm and the integument that would be involved in the coordination of maternal nutrients supply to the seed. The nature of such communications remains unknown. In *Arabidopsis,* the cytological organization of the posterior endosperm and of the integument suggests similarities with the transfer zone of cereals. Deep cell wall ingrowths from the endosperm are present above a specialized tissue in the integument (Schultz and Jensen, 1971). In spite of cytological similarities to cereals, the role of this zone in nutrient transport to the endosperm has not been demonstrated in *Arabidopsis*.

Similarly to the maize *miniature 1* mutant, the *Arabidopsis iku* mutants demonstrate a feedback from the filial generation to the maternal generation in plant development. The endosperm growth is arrested in *iku* seeds and this appears to cause an arrest of cell elongation in the integument. The endosperm is most likely the source of a signal toward the maternal tissues. Unlike the *miniature 1* mutant, the placentochalazal region in the maternal integument is not affected in *iku,* suggesting that different functions are affected in both classes of mutants. The identification of the *IKU* genes might shed some light on the nature of signals involved in this communication.

Exchanges between the endosperm and the embryo in seeds of dicots have been demonstrated using a physiological approach. The endosperm acts a sink for nutrient unloading from the phloem, which is essential for its storage function, whether directly or indirectly in the embryo cotyledons (Hill et al., 2003). In *Arabidopsis,* there is genetic evidence for trophic contribution from the endosperm to the embryo. The *Arabidopsis* mutants *iku* (Garcia et al., 2003), *titan 3* (Liu and Meinke, 1998), *fis1/medea, fis2* (Chaudhury et al., 1997; Sorensen et al., 2001), *demeter* (Choi et al., 2002), and *spätzle* (Sorensen et al., 2002), affected in endosperm development, produce viable embryos with reduced growth that develop into normal-looking plants. Although *Arabidopsis* endosperm does not store reserves, the reduced embryo growth as a consequence of reduced endosperm size suggests that nutrients are delivered from the endosperm to the embryo.

Epigenetic Control of Endosperm Development

In flowering plants, the megaspore undergoes a series of three synchronous mitoses and cellularizations, leading to a female gametophyte (the embryo sac) reduced to seven cells. The central cell inherits two haploid nuclei that ultimately fuse. The egg cell inherits only one nucleus and is therefore haploid. Since the two male gametes and the two female gametes (the egg cell and the central cell) result from mitosis of a common mother cell, the embryo and the endosperm are genetically identical; however, the program of development of these two organisms is different. By using crosses between *Arabidopsis* parents of different ploidies, it has been shown that the ratio of the number of maternal to paternal genome complements (1m:1p for the embryo and 2m:1p for the endosperm) is critical for normal development of the seed (Scott et al., 1998). Reciprocal crosses between diploid and tetraploid plants produce seeds containing triploid embryos and abnormal endosperm development. When a tetraploid plant is used as the female, the endosperm contains 4m:1p and exhibits a dramatic reduction of size and precocious cellularization. The opposite type of cross creates a paternal excess (2m:2p) with opposite effects in endosperm development. In comparison to the endosperm, the development of the embryo is relatively less affected in imbalanced crosses. Similar results have been obtained in maize using the indeterminate gametophyte line that produces male gametes with abnormal ploidy (Lin, 1984). Such results can be explained by unequal parental gene imprinting. In maize, seeds could be obtained with normal parental balance in the embryo, but not in the endosperm which showed abnormal development (Lin, 1984). This demonstrates that parental imprinting is primarily involved in endosperm development.

Imprinting refers to epigenetic modifications of loci resulting in differential expression of genes depending on the parent from which they are inherited (Messing and Grossniklaus, 1999). Imprinting is a prominent feature of parental-specific expression of genes in mammalians. Parent-specific DNA methylation has been correlated so far with nearly all imprinted mammalian genes. The global role of DNA methylation in *Arabidopsis* has been investigated using plants with hypomethylated DNA. These plants carry a transgene coding for an antisense for the predominant DNA maintenance methyl transferase 1 (lines *MET1* a/s) (Finnegan et al., 1996). Reciprocal crosses between *MET1* a/s and WT plants phenocopy the parental effects of dosage imbalance (Adams et al., 2000; Luo et al., 2000). For example, a wild-type female plant pollinated with a *MET1* a/s plant produces endosperm and seed phenotypes similar to paternal excess imbalance of the 2m:1p ratio. It is thus likely that DNA methylation plays a role in the epigenetic control of endosperm development linked to the parental ratio

2m:1p. It has been shown that at least twelve loci corresponding to genes expressed in the endosperm are imprinted (Vielle-Calzada et al., 2000). It is not clear whether any of these loci are involved in the phenotypic alterations linked to dosage imbalance. Conversely, *HAIKU* genes are likely targets of the epigenetic regulation related to dosage imbalance, since they exhibit strikingly similar phenotype, but this hypothetical link awaits their identification (Garcia et al., 2003).

The first gene from *Arabidopsis thaliana* shown to be imprinted was *MEDEA (MEA)* (Kinoshita et al., 1999; Luo et al., 2000; Vielle-Calzada et al., 1999). Loss-of-function mutations in the *MEA* gene display parent-of-origin effects on seed development. Pollination of wild-type plants with *mea*/+ pollen has no effect on seed development. In contrast, seeds that inherit a maternal *mea* allele show defects in endosperm development regardless of the genotype of the paternal *MEA* allele (Chaudhury et al., 1997; Grossniklaus et al., 1998; Yadegari et al., 2000). Maternal and paternal *MEA* alleles show a characteristic differential expression in endosperm with a delayed onset of the paternal expression, whereas both paternal and maternal alleles are expressed simultaneously in the embryo. This observation suggested that the paternal *MEA* allele is silenced specifically in the endosperm (Kinoshita et al., 1999; Vielle-Calzada et al., 1999). Interactions between plants with demethylated DNA and *fis* mutants have shown that DNA methylation plays an indirect role in paternal imprinting of the *FIS* genes (Luo et al., 2000; Vinkenoog et al., 2000). Recently, *DEMETER* encoding a DNA endoglycosylase has been identified as an essential activator for the expression of *MEA* (Choi et al., 2002). Mutants for *DEMETER* show a gametophytic maternal effect and a phenotype similar to that of *fis* mutants. *DEMETER* encodes a DNA glycosylase and is required for transcriptional activation of *MEDEA*. DEMETER activity appears to generate nicks in the promoter of *MEDEA* that may correspond to cytosine methylated in relation to imprinting. The authors hypothesize that paternal imprinting of *MEDEA* would result from the absence of expression of *DEMETER* in the pollen, whereas the maternal copy is activated prior to fertilization in the embryo sac, its only site of expression. However, this scenario does not explain the expression of *MEDEA* in the embryo where no activity of *DEMETER* has been observed. Alternatively, it could be proposed that the maternal effect of *FIS* mutations originates not only from paternal imprinting but also from the action of the *FIS* genes in the mature female gametophyte via a true maternal effect akin to that defined and observed during animal development in most species. Recently, the two *capulet* mutants with a gametophytic maternal effect on embryo and endosperm development have been reported (Grini et al., 2002). The new class of mutants present a novel phenotype with very early arrest of the endosperm and do not appear to interact in the *FIS* pathway.

EMBRYOGENESIS

Embryo development can be divided conceptually into three main steps. In the first phase the cells divide, start to differentiate, and establish the embryo pattern (Figure 20.3). The second phase is characterized by cell expansion and accumulation of storage compounds. Finally, the embryo desiccates and arrests its development. These different steps have been well described, as have models for pattern formation (Kaplan and Cooke, 1997; Mayer and Jurgens, 1998).

Classification of the emb *Mutations*

It seems that there are two main stages of arrest of the embryos in preglobular mutant seeds, at one cell or eight cells of the embryo proper. These two subclasses of mutants would represent sensitive steps of early embryogenesis: the induction of division in the embryo proper and the formation of the protoderm, the first tissue of the radial pattern. Globular mutants represent the largest class of *emb* mutants and the most sensitive step of seed development. This most sensitive step of embryo development has been described previously in an independent collection of T-DNA transformants (Yadegari, Depaiva, et al., 1994). In this class, 14 percent of the mutants present a phenotype similar to the *suspensor (sus)* (Schwartz et al., 1994) or *raspberry (ras)* (Yadegari, Depaiva, et al., 1994) phenotypes described earlier. The cell divisions are persistent, though at a lower rate than in normal embryos, and the mutant embryo can reach two to three times the size of the normal globular embryos. Frequently, the suspensor undergoes additional divisions and, in some cases, can produce a structure similar to a second globular embryo. The endosperm nuclei are numerous and a network of cell walls is frequently formed. In the second type of globular mutants, the embryos do not undergo additional cell divisions and retain a fairly normal morphology, with the exception of the bloated cell shape.

Mutant embryos arrested at the triangular-heart stage can be separated into the two groups according to their ability to elongate the hypocotyl and the cotyledons. The triangle mutant embryos initiated the globular-heart transition by flattening their upper part. However, they apparently fail to develop cotyledons and to fully elongate the hypocotyl. The heart mutant embryos have completed the acquisition of bilateral symmetry and started to form and elongate the cotyledons but will not develop further. In some

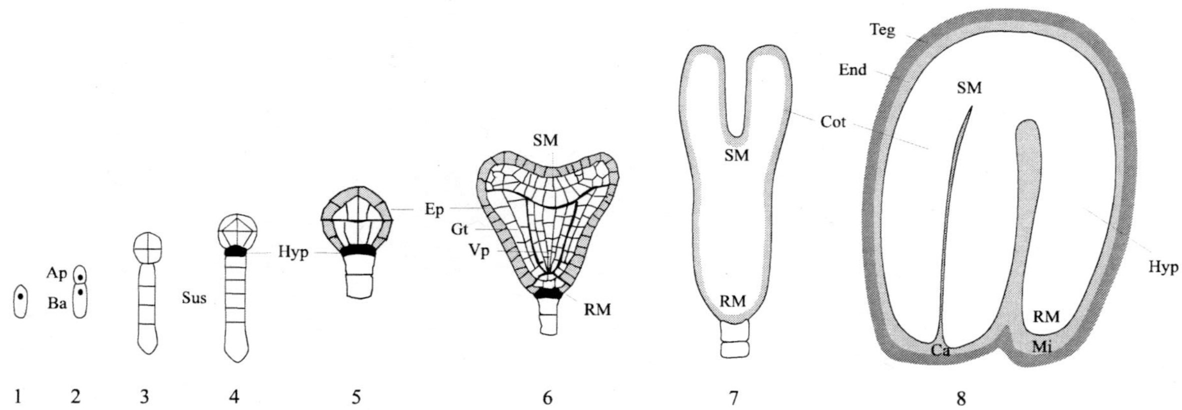

Initiation	Cell division, cytokinesis	Morphogenesis, autotrophy	Cell expansion	Maturation and desiccation
BBM, LEC1, LEC2, PKL, SERK, WUS	*CUL1, KN, KEU, KIS, PFL, OPT3, PILZs, TOR, TTNs, UBP14, VCL1*	*ABP1, ACC1, BDL, BIO1, BIO2, EDD1, EMB506, GLA1, GNOM/EMB30, MP, PIN1, RSY3, SLP, SUS1, SSR16*	*CYT1, FK, GCS1/KNF, HYD1, SMT1, SSE1, TPS1*	*ABI3, ABI4, ABI5, EEL, FUS3, LEC1 LEC2*

FIGURE 20.3. Embryo development in *Arabidopsis* and some of the genes involved. (1) One-cell stage (zygote). (2) Two-cell stage proembryo. The zygote has divided asymmetrically into an apical (Ap) and a basal (Ba) daughter cell. (3) Octant stage. The basal cell gives the hypophysis (Hyp) and the suspensor cells (Sus). (4) Dermatogen stage (16 cells). (5) Globular stage. Ep, epiderm; Gt, ground tissue; Vp, vascular primordium. (6) Heart stage. SM, shoot apical meristem, RM, root meristem. (7) Torpedo stage. Cot, cotyledons. (8) Mature stage. Teg, teguments; End, residual endosperm; Hyp, hypocotyl, Ca, chalazal area; Mi, micropyle area.

cases, cotyledons continue to elongate while the hypocotyl remains short, resulting in the formation of an embryo in a boomerang shape. The suspensor disappears at maturation stage in all heart mutants but not in triangle mutants, suggesting that the initiation of the cotyledons is a prerequisite for the induction of suspensor disappearance.

The main characteristic of torpedo-cotyledon mutants is often a severe defect in pigmentation. Mutant seeds and embryos range from yellow to pale green or purple but rarely reach a chlorophyll content close to the one of wild-type torpedo or cotyledonary stage embryos. In most cases, embryo morphology resembles wild-type torpedo, with often an asymmetric elongation of the cotyledons. In a smaller proportion of mutants, the embryos have developed an apparently normal basal structure (hypocotyl and root meristem) but fail to elongate the cotyledons, and the apical meristem is also more protruding and larger than in wild-type embryos. The suspensor does not persist in torpedo-cotyledon mutants and disappears at the heart-torpedo stage. At the torpedo-cotyledon stage, the pattern of the embryo has already been achieved; therefore, mutations leading to late arrest may be impaired in maturation and/or desiccation processes.

The *embni* group comprises mutants that cannot be classified by reference to a normal development stage. This class probably corresponds to the phenotypic class "other" with 2 percent in McElver et al. (2001). After the asymmetric division of the zygote, which is defined by a large basal cell and a small apical cell, the upper cell corresponding to the embryo proper will continue to divide but without apparent organization. Later during silique development, a suspensor structure can be recognized; however, the upper part is composed of a group of cells ranging from a few cells (two to four) to several dozen, depending on the mutant line. These cells are often rounded and offer few contacts with their neighboring cells. No embryonic structure or organs can be determined by cytologic observations.

General conclusions can be drawn for the observation of *emb* phenotypes. The systematic classification of *emb* mutants reveals several sensitive points in early seed development: the first divisions of the cell of the embryo proper and the endosperm nuclei, the formation of the radial pattern, and the globular-to-heart-stage transition. These stages of development are particularly sensitive to gene disruption. In normal embryo development, the first division of the egg cell, asymmetric in *Arabidopsis,* will define an apical cell, which will produce the embryo proper and a basal cell, which will form the suspensor. In the vast majority of mutants, the first division of the zygote is asymmetrical and the small upper cell and large lower cell will undergo distinct developments. Suspensor- and embryo-like structures can always be distinguished, even though they often present morphological abnormalities. This result suggests that the asymmetry in the egg cell is probably determined prior to fertilization during ovule development. Even in the *gnom* mutant where this division is symmetrical, the fates of the upper and basal cells are not affected (Mayer et al., 1993). In most of the cases, embryonic mutants can be classified by the shape of the embryo (e.g., globular, heart). The stage of arrest or deviation from normal embryo development can be identified precisely when observing the kinetics of *emb* embryo development. The majority of the mutations affect the progression of embryo development at most stages, either by a total block in cell division or by aberrant cell divisions and growth. Globular arrested embryos frequently show a suspensor with additional cells, produced mainly by transversal division, leading sometimes to the formation of a wide suspensor. This phenotype is similar to the *sus* or *ras* phenotypes described earlier. It is hypothesized that the embryo proper, during normal development, sends a signal to the suspensor to prevent its embryogenic growth. This signal cannot be produced in embryos, which fail to proceed to heart stage. The observation that mutant embryos in *heart* and *torpedo* classes, but not in *triangle* class, lose their suspensor at the maturation stage suggests that the initiation of the elongation of cotyledons is also necessary to induce disappearance of the suspensor.

Genes Involved in the Initiation of Embryogenesis

The genetic and molecular mechanisms involved in the initiation of embryogenesis remain elusive. For instance, although mutations that enlarged shoot apical meristem (SAM) may induce the formation of somatic embryo (Mordhorst et al., 1998), somatic embryogenesis may occur from tissues lacking functional SAM (Mordhorst et al., 2002). Furthermore, mutations affecting the formation of apical and basal meristems often present a more spectacular phenotype postembryonically. Thus, these mutations will not be specifically described here, and several reviews on the formation of the meristems and their embryonic origins are available (Weigel and Jürgens, 2002). However, it is to be noted that the ectopic expression of *WUS,* a gene previously identified for its function in shoot and floral meristems (see previous paragraph on ovule development), may be sufficient to induce the formation of viable somatic embryos on vegetative tissues (Zuo et al., 2002). WUS would be involved in maintaining the identity of the embryonic stem cells or, more generally, necessary to maintain stem cells undifferentiated and dividing (Laux et al., 1996). More recently, it has been proposed that WUS would be involved in a general signaling mechanism that is used in different cellular contexts to coordinate the differentiation of adjacent cell groups (Gallois et al., 2002; Gross-Hardt et al., 2002).

There are several other genes the ectopic expression of which may induce or increase embryogenesis. For instance, constitutive overexpression of the *AtSERK (SOMATIC EMBRYOGENESIS RECEPTOR)* gene, which encodes a leucine-rich repeat (LRR) transmembrane receptor-like kinase, is sufficient to increase embryogenic competence in culture (Hecht et al., 2001). Ectopic expression of BABY BOOM (BBM), a transcription factor of the AP2 family, led to spontaneous formation of somatic embryos on seedlings (Boutilier et al., 2002). Ectopic expression of *LEAFY COTYLEDON 1 (LEC1)* induces the expression of embryo-specific genes and the formation of embryo-like structures (Lotan et al., 1998). *LEC1-Like (L1L)* is a closely related homologue, but it may have a different function during embryogenesis due to its different pattern of expression (Kwong et al., 2003). Ectopic expression of *LEC2,* which encodes a transcription factor, also induces the formation of somatic embryos (Stone et al., 2001). The *lec1* and *lec2* mutant phenotypes are different and the double mutant displays synergistic phenotypes (Lotan et al., 1998; Meinke et al., 1994; West et al., 1994). The majority of embryos produced by ectopic expression of *LEC1* failed to develop further (Lotan et al., 1998), in contrast to *LEC2* seedlings that continue to proliferate, producing callus-, cotyledon-, and leaf-like structures (Stone et al., 2001). Therefore, LEC proteins would have partially redundant functions to establish an embryogenic competence.

Last, mutations in *PICKLE (PKL)* induce somatic embryogenesis on roots, and study of the mutants suggested that gibberellic acid signaling may be important in controlling embryogenesis (Ogas et al., 1997). *PKL* encodes a chromatin remodeling factor that may be a repressor of embryo gene-expression programs and regulates the transition from embryonic to vegetative development in *Arabidopsis* (Ogas et al., 1999). Interestingly, PKL may act by repressing *LEC1* in wild-type roots.

Function of EMB Genes in Early Embryogenesis

Fertilization triggers the development of the embryo and endosperm by the initiation of cell division. It is therefore not surprising that the main functions affected in preglobular mutants are related to the regulation and/or the mechanisms of cell division. These mutants are separated into two main groups: (1) mutants with general defects in cell division, mainly represented by the *pilz* (Steinborn et al., 2002) and *titan (ttn)* (Tzafrir et al., 2002) mutants, and (2) mutants affected in cytokinesis (Assaad, 2001; Lukowitz et al., 1996).

The TTN group contains at least nine members and could be divided into subclasses linked to a specific function. Even though most *titan* mutants can be described as preglobular embryo mutants, since their embryos are composed of only one or a few cells (with the exception of *ttn3*), the most striking phenotype is observed in the endosperm. Endosperm nuclei are greatly enlarged and highly polyploid (Liu and Meinke, 1998). The *PILZ* group consists of four members: *CHAMPIGNON (CHO), PORCINO (POR), PFIFFERLING (PFI),* and *HALLIMASH (HAL).* The *pilz* mutant seeds contain embryos that are mushroom shape (hence their name), with one or few cells containing enlarged nuclei. The endosperm does not cellularize and contains a reduced number of large nuclei (Mayer et al., 1999). Since the phenotypes of *pilz* and *ttn* mutants are quite similar, it was expected that these mutants may be defective in similar functions and that some of them were actually allelic. The recent discovery of the identities of *PILZ* and *TITAN (TTN)* genes provided a considerable leap forward in our understanding of the early steps of cell division: the condensation and the segregation of the chromosomes in plant nuclei.

TTN3, TTN7, and *TTN8* encode structural maintenance chromosome (SMC) proteins (Liu et al., 2002). SMC proteins are myosin-like chromosomal ATPases that form two types of complexes. SMC2/SMC4, the condensin complex, is involved in chromosome condensation. SMC1/SMC3, the cohesin complex, plays a role in sister chromatid cohesion and chromosome segregation. In the absence of these complexes, chromosome dynamics and cell division are severely perturbed. The *ttn7* and *ttn8* embryos are arrested after a few cell divisions, whereas the endosperm remains syncytial with very large nuclei. In contrast to the other *titan* and *pilz* mutants, *ttn3* develops a normal embryo and the mutant seed is viable (Liu and Meinke, 1998). Overall organization of the microtubule cytoskeleton appears normal, and, remarkably, endosperm cellularization takes place at the same time as in the wild type. The weak phenotype observed for *ttn3* (subtle phenotype of the embryo that is viable) is explained by partial functional redundancy. Indeed, a second *SMC2* gene *(AtSMC2)* is present in the *Arabidopsis* genome. The *ttn3 Atsmc2* double mutants are expected to produce a stronger phenotype in the embryo.

PILZ genes encode the components of the tubulin folding complex that are conserved in eukaryotes (Steinborn et al., 2002; Tzafrir et al., 2002). In consequence, *pilz* mutants are characterized by an absence of microtubules, are unable to form mitotic spindles and phragmoplast, and are thus defective for cell division. The *pilz* seed contains an embryo that consists of a few enlarged multinucleate cells. The *pilz* endosperm remains syncytial and contains a few enlarged nuclei. Remarkably, chromosome condensation and cell growth do not appear to depend primarily on the presence of microtubules. Another gene, *KIESEL (KIS),* previously described as involved in cell plate formation (Nacry et al., 2000), encodes one of the TFC, namely, TFC A. Mutations in *KIS* cause related but weaker defects than those observed in *pilz* mutants. Particularly, *kis* endosperm cannot be distinguished from wild-type endosperm. The four PILZ proteins probably form a supercomplex of 400 kDa with α- and β-tubulin in order to produce and fold microtubules. TFC A binds to β-tubulin and passes it to TFC D and does not take part in the TFC complex. The weaker phenotype of TFC A cannot be explained by partial redundancy of function, but rather by a newly synthesized tubulin readily utilized to form spindle (Nacry et al., 2000).

Cytokinesis in plants involves the de novo formation of a new cell wall and plasma membrane. Cytokinesis mutations are recognized by specific cellular defects, such as incomplete cell walls and enlarged cells with one or more enlarged nuclei. At least seven genes are involved in *Arabidopsis* (Nacry et al., 2000) with variable embryo phenotype. The defects in *KNOLLE (KN)* and *KEULE (KEU)* genes have been described previously as defects in the radial pattern of the embryos, since *kn* and *keu* embryos are unable to form periclinal division to define the protoderm. On this morphological basis, they can be considered as preglobular mutants. Furthermore, the *kn keu* double mutant has a more severe phenotype and produces embryo development arrested at the one-cell stage (Waizenegger et al., 2000). *KNOLLE* encodes a cytokinesis-specific synthaxin protein that localizes to the cell plate. *KEULE* encodes a SEC1 protein that binds to KNOLLE. Their interaction promotes vesicle fusion necessary for cell plate formation. The function of another protein associated to the cell plate, ADL1A, has been assessed by reverse genetics. Homozygous *adl1a* seeds exhibit a mild phenotype in comparison to *kn* and *keu,* mainly at the cotyledonary stage embryo (Kang et al., 2001). Furthermore, no cytokinesis defects could be observed. Taken together, the phenotypes of *ttn, pilz,* and cytokinesis mutants suggest that cytokinesis is not necessary in the progression from M to G1. Furthermore, since disruption of these genes leads to sporophytic and not gametophytic mutations, different mechanisms in cytokinesis may exist in haploid cells (Nacry et al., 2000).

Very few cell cycle genes essential for embryo development have been identified. In general, their phenotypic effects are not strictly confined to the preglobular development of the embryos. They also display defects in gametophytic and/or globular-heart developments. Disruption of *AtCUL1,* a member of the SCF complex, causes a simultaneous arrest of zygote and central nuclei division (Shen et al., 2002). Even though there are 11 *CULLIN* genes in *Arabidopsis, AtCUL1,* which complements the defects of the yeast *cdc53* mutant strain, has a specialized function in promoting embryo and endosperm cell divisions. Since the SCF ubiquitin ligase complexes target a large variety of proteins and are involved in many cellular processes (including the control of cell division), it is difficult to determine whether one or more proteins that fail to be degraded in *atcul1* are responsible for the block in cell division. Another protein involved in the ubiquitin degradation pathway is also essential for embryo development. Inactivation of the *AtUBP14* gene leads to an arrest of embryo development at the globular stage (Doelling et al., 2001). The mutation equally affects the endosperm, since *ubp14* and *ttn6* are allelic and display abnormal nuclei division. However, the requirement of the embryo for UBP14 is less drastic than for AtCUL1. The second *Arabidopsis* CDC homologue involved, but not essential, in embryo development is the *HOBBIT* gene (Blilou et al., 2002). The phenotypes of *hobbit* embryos and seedlings are mild, probably due to gene duplication, and are predominantly at the level of the root. HOBBIT/CDC27 is required for cell division in the meristems. Finally, *PROLIFERA*, an *Mcm7* homologue, has multiple roles during megagametophyte and embryo development. In addition to its role in DNA replication during the S phase, it plays a role in cytokinesis in the embryo and the endosperm (Holding and Springer, 2002).

Two novel unrelated functions have recently been discovered by the molecular characterization of preglobular mutants. The inactivation of the *Arabidopsis VACUOLELESS 1 (VCL)* blocks vacuole formation in the zygote and leads to alteration of the pattern of cell divisions of the embryo (Rojo et al., 2001). Mutations in the *OPT3* (oligopeptide transporter) gene induce an arrest in embryo development at or before the eighth cell stage (Stacey et al., 2002). These results suggest that establishment of vacuole formation and oligopeptide transport is necessary for normal embryo development.

Globular and Transition Mutants

Diverse functions are involved in this complex process that implies the acquisition of a bilateral symmetry, and the differentiation of the tissues, organs, and meristems. The *emb* mutants affected in this step have been assigned to two distinct classes, globular and transition, since it is easy to discriminate morphologically embryos arrested at the globular stage and embryos that have initiated the transition to heart stage and then aborted. However, the characterization of some of these *EMB* genes suggests that the functions involved in both classes cannot be clearly differentiated. The morphological differences are probably due to the strength of the alleles.

Among the functions that were proposed, it was hypothesized that auxin transport was necessary for the formation of the cotyledons. This idea relied on the experiments of in vivo application of auxin inhibitors on isolated immature embryos (Liu et al., 1993). By means of reverse genetics, the role of the putative auxin receptor ABP1 was assessed (Chen et al., 2001). A homozygous null mutation in *ABP1* leads to embryonic lethality. The mutant embryos are arrested at the globular stage and are unable to perform the transition and acquire a bilateral symmetry because the cells fail to elongate and divide properly. Several mutations that are related to auxin have also been described. In general, these mutations do not lead to the lethality of the embryo but produce seedlings with an abnormal morphology. The structural defects of the seedlings can be traced back to embryo development, where often the primary expression of the mutations can be observed. Mutations in the *MONOPTEROS (MP)* and *BODENLOS (BDL)* genes affect the orientation of the division plane of the embryo, leading to the formation of double-octant embryos. *MP* encodes a ARF5 (auxin response factor) transcription factor (Hardtke and Berleth, 1998) that interacts in yeast two-hybrid assay with BDL/IAA12 (auxin-response gene). It is hypothesized that *BDL* inhibits *MP* action in root meristem initiation (Hamann et al., 2002). In the *hobbit/cdc27* mutant, the mRNA of AXR3/IAA17 accumulates and the expression of DR5::GUS auxin reporter is reduced (Blilou et al., 2002). To be effective, auxin requires a series of transporters such as PIN-FORMED (PIN1), an auxin efflux carrier (Galweiler et al., 1998). The PIN1 protein is distributed randomly in early embryogenesis and shows a polarized localization at the globular stage in wild-type embryos but not in the *gnom* mutant (Steinmann et al., 1999). This observation suggested that the vesicle trafficking of the GNOM/EMB30 (Shevell et al., 1994), an ARF-GEF protein, may establish cell polarity by the establishment of polar auxin transport. The GNOM protein is required for the recycling of auxin transport components and regulates a specific endosomal trafficking pathway (Geldner et al., 2003). For the moment, the full role of auxin in embryogenesis is far from being elucidated, but there are now several lines of evidence highlighting its involvement in pattern formation. Even though auxin has certainly an important role to play in embryo development, it cannot account for the large number of globular mutants.

The globular-to-heart transition may also mark the beginning of the autotrophy of the embryo, while the maternal tissues supported the early development. The main function that stands out is the requirement of active plastids and probably mitochondria. Up to now, four genes encoding proteins that are imported into chloroplast have been shown to play essential roles in seed development. Disruption of these genes produces an arrest of embryo development at the globular stage or later at the heart-torpedo stage with asymmetric and incomplete growth of the cotyledons. These chloroplast functions have to be different from those involved in photosynthesis (at least in the light cycle), since *albino* and *chlorina* embryos are morphologically normal. The *EMBRYO-DEFECTIVE DEVELOPMENT (EDD1)* gene encodes a glycyl-tRNA synthetase, a component of the chloroplast translation machinery (Uwer et al., 1998). Inactivation of the *SCHLEPPERLESS (SLP)* gene causes a delay in embryo development before the heart stage (Apuya et al., 2001). *SLP* encodes the plastid chaperonin-60 alpha subunit, probably the homologue of the bacterial GroEL. The two last genes, *EMB506* and *RASPBERRY 3 (RSY3)*, encode plastid proteins of unknown function. EMB506 possesses five ankyrin repeats involved in protein-protein interaction (Albert et al., 1999). Partial complementation of the *emb506* embryos revealed that the EMB506 protein does not have a housekeeping function in the plastid but is required in the plastid only during embryo and inflorescence development (Despres et al., 2001). The RSY3 protein contains a putative SGG-DS core sequence that could be responsible for an ATP pyrophosphatase activity (Apuya et al., 2002). However, its role in the chloroplast remains to be discovered. The requirement of chloroplast biogenesis in embryo development is also strengthened by the systematic work of McElver and collaborators (McElver et al., 2001) on *emb* mutants. One-quarter of putative *EMB* genes encode proteins that are predicted to be localized in the plastid using TargetP. These data should be confirmed by independent algorithms for the prediction of subcellular targeting. The integrity of mitochondria for proper embryo development is not as well documented yet but is probably equally important. The *SSR16* gene of *Arabidopsis* encodes a mitochondrial ribosomal protein S16 (Tsugeki et al., 1996). Disruption in the *SSR16* gene induces an arrest of embryo development at the globular-heart transition indistinguishable from the phenotype of embryos obtained after inactivation of essential chloroplast proteins.

The *bio1* mutant was the first auxotrophic mutant to be identified by the rescue of the mutant embryos on biotin-supplemented medium (Schneider et al., 1989). The site of synthesis of biotin, an essential vitamin, remains unclear. Some studies suggest that the synthesis occurs primarily in the chloroplast, others in the cytoplasm or mitochondria. *BIO1* and *BIO2* encode two enzymes of this pathway, and their activities are necessary for the development of the embryo beyond the globular stage for *BIO2* (Patton et al., 1998) and later for *BIO1*. Recently, another auxotrophic mutant impaired in a different metabolic pathway has been characterized (Ishikawa et al., 2003). The *GLOBULAR ARREST 1 (GLA1)* gene encodes a protein with homologies to enzymes that participate in folate biosynthesis. The *GLA1* gene corresponds to the *AtDFA* gene, described previously as encoding a mitochondrial protein (Ravanel et al., 2001). The *gla1* mutant embryos are arrested at the globular stage, hence the name, and can be rescued by addition of 5-formyl THF in the culture medium. More interestingly, overproduction of GLA1 protein in heterozygous *gla1/+* plants leads to the production of *gla1/gla1* embryos arrested predominantly at the torpedo stage. Similar results were observed when heterozygous *bio2* plants were grown on an exogenous source of biotin (Schneider et al., 1989). Taken together, these results demonstrate the role of maternal tissues in supporting early embryo development and that the globular-heart stage marks the transition from heterotrophism to autotrophism. Disruption of the *ACC1* gene, which codes for acetyl-CoA carboxylase 1 (ACCase1) that catalyzes the formation of malonyl-CoA used for fatty acid synthesis and elongation, is lethal for the embryo (Baud et al., 2003). Null mutations lead to cucumber-like structures lacking cotyledons, while the shortened hypocotyl and root exhibit a normal radial pattern organization of the body axis. The mutant embryos do not produce very-long-chain fatty acids (VLCFA) and may be affected in different signaling pathways (Baud et al., 2003).

The function of the *TOR (target of rapamycin)* gene may provide a link between nutrients and cell growth and division. *TOR* encodes a protein kinase, expressed in primary meristems, embryos, and endosperm but not in differentiated cells. Disruption of *TOR* leads to embryonic lethality at the globular stage (Menand et al., 2002). A subset of globular mutants with an abnormal growth of the suspensor has been named *SUSPENSOR (SUS)* mutants (Schwartz et al., 1994). The identity of the proliferating cells of the suspensor is thought to be embryonic (Yadegari, Paiva, et al., 1994). The *twin (twn)* mutants could be described as the most extreme *sus* phenotype with a complete secondary embryo formed from a suspensor cell (Vernon and Meinke, 1994). *SUS2* and *SUS1* have been cloned. Little information is available for *SUS2. SUS2* encodes a protein exhibiting similarities to splicing factors. The SUS1 protein is probably the homologue of the *Drosophila* DICER protein, a multidomain ribonuclease specific for double-stranded RNA (Golden et al., 2002). The *sus1* phenotype exemplified only one aspect of DICER functions in plants, since mutations affecting ovule development (SHORT INTEGUMENTS 1) and flower development (CARPEL FACTORY) are defective in the same gene. The extraordinary phenotype of *twn2* has been found to be due to a deregulation of a putative *valyl-tRNA synthetase* gene (Zhang and Somerville, 1997). It is probably the uneven distribution of Val-tRNA synthetase in the embryo's apical and basal cells due to T-DNA integration in an enhancer region that is responsible for the *twin* phenotype.

Heart-Torpedo Mutants

After the globular stage, the rate of cell division declines. Cells undergo orientated and regionalized cell expansion, producing the shape of the torpedo. These processes require the biogenesis of new membranes and new cell walls. The identification of several mutants supports the importance of these cellular mechanisms at the heart-torpedo stage.

The characterization of *fackel (fk)* (Jang et al., 2000; Schrick et al., 2000), together with the description of the *sterol methyltransferase 1 (smt1)* mutant (Diener et al., 2000), provided the first evidence for a role of sterols in embryogenesis. A third gene, *HYDRA1 (HYD1),* was later identified (Souter et al., 2002). These three genes encode enzymes of the sterol biosynthesis pathway, positioned upstream of the brassinosteroids (BR) branch. *Arabidopsis* produces two main sterols: sitosterol, a major component of membrane lipids, and camposterol, a precursor of BR. Since mutations in BR biosynthesis do not produce embryonic defect, it was deduced that BR signaling is crucial for only postembryonic growth. Therefore, steroid molecules, other than BR, play an essential role in embryogenesis. The early defects of *fk, hyd,* and *smt* can be traced back to the late globular stage but are clearly visible at the heart-torpedo stage. The central cells of the mutant embryos do not divide normally and fail to extend. These defects will result in the production of seedlings with short hypocotyls and roots and malformed cotyledons. It was hypothesized that in mutant embryos, cell division and/or polarity decisions are randomized instead of being organized and controlled (Schrick et al., 2002). The study of the localization of PIN1, PIN3, and AUX1 proteins involved in auxin transport in the roots of the *smt1* mutant further supported this hypothesis (Willemsen et al., 2003). Balanced sterol composition is required for cell polarity and auxin efflux. Individually, mutations in each gene do not lead to embryonic lethality, but the double mutations *smt fk* and *smt hyd* have a more drastic effect on embryo viability (Schrick et al., 2002).

Both *cyt1* and *gcs1/knf* mutant embryos are arrested at the heart stage and display abnormal (swollen) embryo morphology (Boisson et al., 2001; Gillmor et al., 2002; Lukowitz et al., 2001). *CYT1* encodes a mannose-1 phosphate guanyltransferase (Lukowitz et al., 2001) and *GCS1/KNOPF* encodes an alpha-glucosidase-1 (Boisson et al., 2001; Gillmor et al., 2002), both of which are involved in N-glycan biosynthesis. It has been suggested that the phenotype is due to a defect in cellulose biosynthesis found in both mutants. However, several mutations affecting cell wall biogenesis that do not impair seed viability have been identified (Arioli et al., 1998; Fagard et al., 2000). Thus, is it possible that the severe/lethal *knf/gcs1* and *cyt1* phenotypes are due not only to the defect in cellulose synthesis but also to other molecular and cellular processes for which normal N-glycosylation is required. For instance, that *gsc1* embryos present a reduction in storage proteins and protein bodies emphasized the importance of N-glycan trimming in the accumulation of storage proteins (Boisson et al., 2001).

The heart stage marks also the beginning of storage product accumulation. A gross defect in storage may lead to the production of incompletely filled seeds. *Shrunken seed 1 (sse1)* mutant seeds are not viable and altered in storage composition, favoring starch over proteins and lipids (Lin et al., 1999). This striking phenotype is due to a mutation in a protein homologous to the PEX16 protein in yeast. PEX16p is a membrane-associated protein required for the assembly and proliferation of peroxisomes and the trafficking of plasma membranes and cell wall proteins. PEX16p is transiently located in the endoplasmic reticulum (ER). SSE1 could participate in storage protein delivery and the formation of oil bodies since both are ER dependent. Moreover, the study of this mutant suggests that starch accumulation may be a default storage pathway during seed development.

Sugar in plants can act as a carbon source but also as a metabolic signal, controlling many aspects of plant growth. Mutation in the *Trehalose-6-phosphate synthase 1 (TPS1)* gene revealed that trehalose synthesis is essential for embryo maturation (Eastmond et al., 2002). *tps1* mutant embryos are retarded in their development and eventually arrest at the torpedo stage. Trehalose-6-phosphate in normal developing seeds cannot be detected, and, therefore, it is hypothesized that TPS1 functions by responding to the increase in sugar supply at the onset of seed maturation.

Embryo Maturation

During the maturation phase, the embryo growth and cell cycle activities are stopped (Raz et al., 2001). The embryo accumulates nitrogen (proteins) and carbon storage (lipids) compounds that make up to 40 percent of dry matter each (Baud et al., 2002; Mansfield and Briarty, 1992). Lipids accumulate in the form of triacylglycerols (TAG) in cytosolic oil bodies, which occupy about 60 percent of the cell volume in the cotyledons of mature embryos. The pathway of fatty acid biosynthesis in developing seeds involves different subcellular compartments and is well established. The most abundant mRNAs in embryos, however, are those encoding the seed storage proteins, 12S globulin (cruciferin) and 2S albumin (arabin), that are encoded by multigene families. During late maturation, the embryo becomes metabolically quiescent, tolerant to desiccation, and sometimes dormant.

Genes Involved in the Control of Embryo Maturation

A first group of genes that control seed maturation includes *LEC1, LEC2,* and *FUSCA 3 (FUS3)* (Keith et al., 1994; Meinke, 1992; Meinke et al., 1994). The pleiotropic effects of their mutations suggest that they are key regulators of embryo development. These genes function during both early and late embryo development, and the mutants share overlapping phenotypes, such as lack of desiccation tolerance, defect in cotyledon identity, reduced level of storage compounds, lack of growth arrest, and precocious germination (Harada, 2001; Raz et al., 2001). LEC1 and LEC2 have been presented in a previous paragraph, since both are able to induce ectopic embryos when ectopically expressed. *LEC1* encodes a transcription factor homologue, the CCAATT box-binding factor HAP3 subunit (Lotan et al., 1998), and *LEC2* encodes a protein showing similarity to transcription factors with a B3 domain, a DNA binding domain unique to plants that is also found in ABI3 and FUSCA3 proteins (Stone et al., 2001). LEC1 and LEC2 seem to have partially redundant function, and *lec2* mutants have a less extreme phenotype and are less sensitive to desiccation than *lec1* mutants (Meinke et al., 1994). *FUS3* encodes a protein with a conserved B3 domain (Luerssen et al., 1998) that has been shown to control the expression of several embryogenic genes (Keith et al., 1994; Nambara et al., 2000; Parcy et al., 1997; Reidt et al., 2001, 2000; Vicient et al., 2000). The mutations of all three factors lead to the premature appearance of germination-specific transcripts (West et al., 1994) and may be interpreted as a tendency to shortcut the maturation phase. They would act both by conferring embryogenic competence to cells and by suppressing activation of vegetative development. The presence of typical vegetative traits, such as an active apical meristem between cotyledons carrying trichomes, supports this model.

In the *abi3 (abscisic acid insensitive 3)* mutant, similar aspects of seed development are also affected, such as resistance to seed desiccation and accumulation of storage compounds, but they differ from *lec fus3* mutants by the green color of the embryo (Giraudat et al., 1992; Koornneef et al., 1984). The *abi3* mutations are additive to *lec fus3* and double mutants are viviparous (Parcy et al., 1997; Raz et al., 2001). *ABI3* encodes a putative transcription factor with a B3 domain, as do *LEC2* and *FUS3* (Giraudat et al., 1992), but how it may regulate the expression of several genes is still not fully understand (Nambara et al., 2000; Parcy et al., 1994). However, it has been shown that ABI3, LEC1, and FUS3 act synergistically to control different aspects of seed development, and both FUS3 and LEC1 regulate positively the abundance of ABI3 protein (Parcy et al., 1997). Although the characterization of *abi3* and other mutants underlines the importance of ABA for seed maturation (Koornneef et al., 1984; Ooms et al., 1993; Wobus and Weber, 1999), ABI3 may play a regulatory role larger than only ABA signaling in the embryo. For instance, it would be also involved in bud dormancy (Rohde et al., 2000, 2002), auxin signaling, and lateral root development (Brady et al., 2003). Thus, it has been suggested that ABI3 may have a general function in cellular differentiation that has been recruited more specifically for embryo maturation, during plant seed evolution (Rohde et al., 2000, 2002).

ABI4 and ABI5 also have similar functions during seed maturation, although the effect of the null ABI3 mutations are stronger and more pleiotropic (Finkelstein et al., 1998, 2002; Finkelstein and Lynch, 2000). ABI4 protein shows homology to a plant-specific family of transcriptional regulators with an APETALA 2 (AP2) domain (Finkelstein et al., 1998) and ABI5 is a transcription factor of the bZIP family (Finkelstein and Lynch, 2000; Lopez-Molina and Chua, 2000). Analysis of the double mutants suggests that ABI4 and ABI5 act in the same signaling pathway, but downstream of ABI3 (Brocard-Gifford et al., 2003; Lopez-Molina et al., 2002; Soderman et al., 2000). ABI3 has been shown to interact with ABI5 in a yeast two-hybrid assay, suggesting that ABI3 might directly modulate transcriptional activity of ABI5 (Nakamura et al., 2001). Interestingly, it has been recently shown that this control may be antagonistically regulated by another bZIP of the same family, ENHANCED EM LEVEL (EEL) (Bensmihen et al., 2002). ABI4 and ABI5 also interact genetically with both LEC1 and FUS3 to control gene expression, although the interactions vary depending of the responses/genes considered (Carles et al., 2002). However, the ABI transcription factors do not seem to interact directly with FUS3 or LEC1. As with ABI3, both ABI4 and ABI5 are also involved in other functions, such as sugar and salt responses (Finkelstein et al., 2002; Lopez-Molina et al., 2001; Quesada et al., 2000). Furthermore, these factors are involved, to different extents, in the control of the transition from maturation to seed germination (Bentsink and Koornneef, 2002; Brocard-Gifford et al., 2003; Holdsworth et al., 1999; Koornneef et al., 2002). Their specific function in seed development may be conferred by repressors of embryonic fate, such as PICKLE, in vegetative tissues or by activators of embryogenesis, such as LEC1, in embryonic tissue. These results illustrate the complexity of the regulatory network including cross talk, synergistic, and epistatic relationships among the different regulatory proteins.

LIMITATIONS OF THE GENETIC APPROACH AND NEW PROSPECTS

Most of the *Arabidopsis* genes expressed during seed development have still unknown functions, and it is obvious that the classical genetic approach suffers from some limitations, such as genetic redundancy, as exemplified in this chapter. In *Arabidopsis,* about 40 percent of genes appear to belong to families containing at least five members. Furthermore, some mutations affecting essential cellular functions might not be transmitted through both types of gametes. Finally, it appears that many mutations are conditional or quantitative and do not modify phenotypes drastically in many environments (Bouche and Bouchez, 2001). Consequently, it is obvious that gene knockouts are not sufficient for the assignment of function for all genes. These must be integrated into a more global approach for determining biological functions, and new tools are required to better investigate subtle changes in seed phenotypes. Various approaches are being used to work around redundancy. Reverse genetics allows the identification of mutants for putative redundant genes that can be used to generate double (or more) mutants. It may also be appropriate to use strategies based on gain-of-function (including ectopic expression or activation tagging), although one has to be careful in interpreting resulting phenotypes. Indeed, ectopic expression of a protein may give some insights into its biochemical function (e.g., cell division) but needs to be interpreted in the light of its real spatiotemporal location in the seed. More recently, antisense, cosuppression, and RNAi have been used to simultaneously down-regulate the expression of homologous genes (e.g., Nishihama et al., 2002). For mutations that prevent the formation of both male and female gametes, new specific and conditional expression systems may be used (Guo et al., 2003). The availability of the complete genome sequence for *Arabidopsis* provides the opportunity to develop a systems-based approach, allowing the investigation of the function of all genes and how their products interact. This includes systematic analyses of gene expression, protein accumulation and activity, search for partners and targets; structure-function relationships, and metabolic analyses. However, a major limitation arises from the multicellular and multi-tissue/organ structure of the seed (with additional complexity of intracellular compartmentation). It will be important to be able to carry out transcriptome, proteome, or metabolome analyses at the cellular level or, at least, for a given tissue, especially in seeds. These new approaches should further our knowledge and help to take into account other levels of complexity. Probably a combination of both reductionist and systems-based approaches, with the development of bioinformatics to integrate the different data, will allow the assigning of functions to uncharacterized coding regions and, finally, the unraveling of the complex regulatory networks that control seed development.

REFERENCES

Adams, S., Vinkenoog, R., Spielman, M., Dickinson, H.G., Scott, R.J. (2000). Parent-of-origin effects on seed development in *Arabidopsis thaliana* require DNA methylation. *Development* 127: 2493-2502.

Albert, S., Despres, B., Guilleminot, J., Bechtold, N., Pelletier, G., Delseny, M., Devic, M. (1999). The *EMB* 506 gene encodes a novel ankyrin repeat containing protein that is essential for the normal development of *Arabidopsis* embryos. *Plant J.* 17: 169-179.

Angenent, G.C., Franken, J., Busscher, M., van Dijken, A., van Went, J.L., Dons, H., van Tunen, A.J. (1995). A novel class of MADS box genes is involved in ovule development in petunia. *Plant Cell* 7: 1569-1582.

Apuya, N.R., Yadegari, R., Fischer, R.L., Harada, J.H., Goldberg, R.B. (2002). *RASPBERRY3* gene encodes a novel protein important for embryo development. *Plant Physiol.* 129: 691-705.

Apuya, N.R., Yadegari, R., Fischer, R.L., Harada, J.J., Zimmerman, J.L., Goldberg, R.B. (2001). The *Arabidopsis* embryo mutant *schlepperless* has a defect in the chaperonin-60 α gene. *Plant Physiol.* 126: 717-730.

Arabidopsis Genome Initiative (2000). Analysis of the genome sequence of the flowering plant *Arabidopsis thaliana*. *Nature* 408: 796-815.

Arioli, T., Peng, L., Betzner, A.S., Burn, J., Wittke, W., Herth, W., Camilleri, C., Hofte, H., Plazinski, J., Birch, R., et al. (1998). Molecular analysis of cellulose biosynthesis in *Arabidopsis*. *Science* 279: 717-720.

Assaad, F.F. (2001). Plant cytokinesis: Exploring the links. *Plant Physiol.* 126: 509-516.

Baker, S.C., Robinson-Beers, K., Villanueva, J.M., Gaiser, J.C., Gasser, C.S. (1997). Interactions among genes regulating ovule development in *Arabidopsis thaliana*. *Genetics* 145: 1109-1124.

Balasubramanian, S., Schneitz, K. (2002). NOZZLE links proximal-distal and adaxial-abaxial pattern formation during ovule development in *Arabidopsis thaliana*. *Development* 129: 4291-4300.

Balzergue, S., Dubreucq, B., Chauvin, S., Le-Clainche, I., Le Boulaire, F., de Rose, R., Samson, F., Biaudet, V., Lecharny, A., Cruaud, C., et al. (2001). Improved PCR-walking for large-scale isolation of plant T-DNA borders. *Biotechniques* 30: 496-498, 502, 504.

Baud, S., Boutin, J. P., Miquel, M., Lepiniec, L., Rochat, C. (2002). An integrated overview of seed development in *Arabidopsis thaliana* ecotype Ws. *Plant Physiol. Biochem.* 40: 151-160.

Baud, S., Guyon, V., Kronenberger, J., Wuilleme, S., Miquel, M., Caboche, M., Lepiniec, L., Rochat, C. (2003). Multifunctional acetyl-CoA carboxylase 1 is essential for very long chain fatty acid elongation and embryo development in *Arabidopsis. Plant J.* 33: 75-86.

Bechtold, N., Ellis, J., Pelletier, G. (1993). In-planta *Agrobacterium*-mediated gene-transfer by infiltration of adult *Arabidopsis thaliana* plants. *C.R.A.S. Life Sciences* 316: 1194-1199.

Becker, A., Kaufmann, K., Freialdenhoven, A., Vincent, C., Li, M.A., Saedler, H., Theissen, G. (2002). A novel MADS-box gene subfamily with a sister-group relationship to class B floral homeotic genes. *Mol. Genet. Gen.* 266: 942-950.

Beeckman, T., De Rycke, R., Viane, R., Inze, D. (2000). Histological study of seed coat development in *Arabidopsis thaliana. J. Plant Res.* 113: 139-148.

Bensmihen, S., Rippa, S., Lambert, G., Jublot, D., Pautot, V., Granier, F., Giraudat, J., Parcy, F. (2002). The homologous ABI5 and EEL transcription factors function antagonistically to fine-tune gene expression during late embryogenesis. *Plant Cell* 14: 1391-1403.

Bentsink, L., Koornneef, M. (2002). Seed dormancy and germination. In Somerville, C.R., Meyerowitz, E.M. (Eds.), *The Arabidopsis Book.* Rockville, MD: American Society of Plant Biologists. Available online at <http://www.aspb.org/publications/arabidopsis/toc.cfm>.

Berger, F. (2003). Endosperm: The crossroad of seed development. *Curr. Opin. Plant Biol.* 6: 42-50.

Birve, A., Sengupta, A. K., Beuchle, D., Larsson, J., Kennison, J.A., Rasmuson-Lestander, A., Muller, J. (2001). *Su(z)12,* a novel *Drosophila* polycomb group gene that is conserved in vertebrates and plants. *Development* 128: 3371-3379.

Blilou, I., Frugier, F., Folmer, S., Serralbo, O., Willemsen, V., Wolkenfelt, H., Eloy, N.B., Ferreira, P.C.G., Weisbeek, P., Scheres, B. (2002). The *Arabidopsis HOBBIT* gene encodes a CDC27 homolog that links the plant cell cycle to progression of cell differentiation. *Genes Dev.* 16: 2566-2575.

Boisnard-Lorig, C., Colon-Carmona, A., Bauch, M., Hodge, S., Doerner, P., Bancharel, E., Dumas, C., Haseloff, J., Berger, F. (2001). Dynamic analyses of the expression of the HISTONE::YFP fusion protein in *Arabidopsis* show that syncytial endosperm is divided in mitotic domains. *Plant Cell* 13: 495-509.

Boisson, M., Gomord, V., Audran, C., Berger, N., Dubreucq, B., Granier, F., Lerouge, P., Faye, L., Caboche, M., Lepiniec, L. (2001). *Arabidopsis* glucosidase I mutants reveal a critical role of N-glycan trimming in seed development. *EMBO J.* 20: 1010-1019.

Bouche, N., Bouchez, D. (2001). *Arabidopsis* gene knockout: Phenotypes wanted. *Curr. Opin. Plant Biol.* 4: 111-117.

Boutilier, K., Offringa, R., Sharma, V.K., Kieft, H., Ouellet, T., Zhang, L., Hattori, J., Liu, C.-M., van Lammeren, A.A., Miki, B.L., et al. (2002). Ectopic expression of BABY BOOM triggers a conversion from vegetative to embryonic growth. *Plant Cell* 14: 1737-1749.

Bowman, J.L., Drews, G.N., Meyerowitz, E.M. (1991). Expression of the *Arabidopsis* floral homeotic gene *AGAMOUS* is restricted to specific cell-types late in flower development. *Plant Cell* 3: 749-758.

Brady, S.M., Sarkar, S.F., Bonetta, D., McCourt, P. (2003). The *ABSCISIC ACID INSENSITIVE 3 (ABI3)* gene is modulated by farnesylation is involved in auxin signaling and lateral root development in *Arabidopsis*. *Plant J.* 34: 67-75.

Broadhvest, J., Baker, S.C., Gasser, C.S. (2000). *SHORT INTEGUMENTS 2* promotes growth during *Arabidopsis* reproductive development. *Genetics* 155: 899-907.

Brocard-Gifford, I.M., Lynch, T.J., Finkelstein, R.R. (2003). Regulatory networks in seeds integrating developmental, abscisic acid, sugar, light signaling. *Plant Physiol.* 131: 78-92.

Brown, R.C., Lemmon. B.E., Nguyen, H., Olsen, O.-A. (1999). Development of endosperm in *Arabidopsis thaliana*. *Sex. Plant Reprod.* 12: 32-42.

Brunaud, V., Balzergue, S., Dubreucq, B., Aubourg, S., Samson, F., Chauvin, S., Bechtold, N., Cruaud, C., DeRose, R., Pelletier, G., et al. (2002). T-DNA integration into the *Arabidopsis* genome depends on sequences of pre-insertion sites. *EMBO Rep.* 3: 1152-1157.

Canales, C., Bhatt, A.M., Scott, R., Dickinson, H. (2002). EXS, a putative LRR receptor kinase, regulates male germline cell number tapetal identity and promotes seed development in *Arabidopsis*. *Curr. Biol.* 12: 1718-1727.

Carles, C., Bies-Etheve, N., Aspart, L., Leon-Kloosterziel, K.M., Koornneef, M., Echeverria, M., Delseny, M. (2002). Regulation of *Arabidopsis thaliana Em* genes: Role of *ABI5*. *Plant J.* 30: 373-383.

Castle, L.A., Errampalli, D., Atherton, T.L., Franzmann, L.H., Yoon, E.S., Meinke, D.W. (1993). Genetic and molecular characterization of embryonic mutants identified following seed transformation in *Arabidopsis*. *Mol. Gen. Genet.* 241: 504-514.

Chaudhury, A.M., Ming, L., Miller, C., Craig, S., Dennis, E.S., Peacock, W.J. (1997). Fertilization-independent seed development in *Arabidopsis thaliana*. *Proc. Natl. Acad. Sci. USA* 94: 4223-4228.

Chen, J.G., Ullah, H., Young, J.C., Sussman, M.R., Jones, A.M. (2001). ABP1 is required for organized cell elongation and division in *Arabidopsis* embryogenesis. *Genes Dev.* 15: 902-911.

Cheng, W., Taliercio, E.W., Chourey, P.S. (1996). The *miniature1* seed locus of maize encodes a cell wall invertase required for normal development of endosperm and maternal cells in the pedicel. *Plant Cell* 8: 971-983.

Chevalier, D., Sieber, P., Schneitz, K. (2001). The genetic and molecular control of ovule development. In Roberts, J.A. (Ed.), *Plant Reproduction,* Volume 6 (pp. 61-85). Boca Raton: CRC Press.

Choi, Y., Gehring, M., Johnson, L., Hannon, M., Harada, J.J., Goldberg, R.B., Jacobsen, S.E., Fischer, R.L. (2002). DEMETER, a DNA glycosylase domain protein, is required for endosperm gene imprinting and seed viability in *Arabidopsis*. *Cell* 110: 33-42.

Colombo, L., Franken, J., Koetje, E., van Went, J., Dons, H., Angenent, G.C., van Tunen, A.J. (1995). The petunia MADS box gene *FBP11* determines ovule identity. *Plant Cell* 7: 1859-1868.

Colombo, L., Franken, J., Van der Krol, A.R., Wittich, P.E., Dons, H.J., Angenent, G.C. (1997). Downregulation of ovule-specific MADS box genes from petunia results in maternally controlled defects in seed development. *Plant Cell* 9: 703-715.

Conner, J., Liu, Z.C. (2000). LEUNIG, a putative transcriptional corepressor that regulates AGAMOUS expression during flower development. *Pro. Natl. Acad. Sci. USA* 97: 12902-12907.

Debeaujon, I., Leon-Kloosterziel, K.M., Koornneef, M. (2000). Influence of the testa on seed dormancy, germination, and longevity in *Arabidopsis. Plant Physiol.* 122: 403-414.

Debeaujon, I., Nesi, N., Perez, P., Devic, M., Grandjean, O., Caboche, M., Lepiniec, L. (2003). Proanthocyanidin-accumulating cells in *Arabidopsis* testa: Regulation of differentiation and role in seed development. *Plant Cell* 15: 2514-2531.

Debeaujon, I., Peeters, A.J., Leon-Kloosterziel, K.M., Koornneef, M. (2001). The *TRANSPARENT TESTA12* gene of *Arabidopsis* encodes a multidrug secondary transporter-like protein required for flavonoid sequestration in vacuoles of the seed coat endothelium. *Plant Cell* 13: 853-871.

Despres, B., Delseny, M., Devic, M. (2001). Partial complementation of embryo defective mutations: A general strategy to elucidate gene function. *Plant J.* 27: 149-159.

Devic, M., Albert, S., Delseny, M., Roscoe, T.J. (1997). Efficient PCR walking on plant genomic DNA. *Plant Physiol. Biochem.* 35: 331-339.

Diener, A.C., Li, H., Zhou, W., Whoriskey, W.J., Nes, W.D., Fink, G.R. (2000). Sterol methyltransferase 1 controls the level of cholesterol in plants. *Plant Cell* 12: 853-870.

Doelling, J.H., Yan, N., Kurepa, J., Walker, J., Vierstra, R.D. (2001). The ubiquitin-specific protease UBP14 is essential for early embryo development in *Arabidopsis thaliana. Plant J.* 27: 393-405.

Drews, G.N., Lee, D., Christensen, C.A. (1998). Genetic analysis of female gametophyte development and function. *Plant Cell* 10: 5-18.

Drews, G.N., Yadegari, R. (2002). Development and function of the angiosperm female gametophyte. *Ann. Rev. Genet.* 36: 99-124.

Dubreucq, B., Berger, N., Vincent, E., Boisson, M., Pelletier, G., Caboche, M., Lepiniec, L. (2000). The *Arabidopsis AtEPR1* extensin-like gene is specifically expressed in endosperm during seed germination. *Plant J.* 23: 643-652.

Eastmond, P.J., van Dijken, A.J., Spielman, M., Kerr, A., Tissier, A.F., Dickinson, H.G., Jones, J.D., Smeekens, S.C., Graham, I.A. (2002). Trehalose-6-phosphate synthase 1, which catalyses the first step in trehalose synthesis, is essential for *Arabidopsis* embryo maturation. *Plant J.* 29: 225-235.

Edgar, B.A., Lehner, C.F. (1996). Developmental control of cell cycle regulators: A fly's perspective. *Science* 274: 1646-1652.

Elliott, R.C., Betzner, A.S., Huttner, E., Oakes, M.P., Tucker, W.Q., Gerentes, D., Perez, P., Smyth, D.R. (1996). *AINTEGUMENTA*, an *APETALA2*-like gene of *Arabidopsis* with pleiotropic roles in ovule development and floral organ growth. *Plant Cell* 8: 155-168.

Fagard, M., Hofte, H., Vernhettes, S. (2000). Cell wall mutants. *Plant Physiol. Biochem.* 38: 15-25.

Felker, F.C., Peterson, D.M., Nelson, O.E. (1985). Anatomy of immature grains of eight maternal effect shrunken endosperm barley mutants. *Am. J. Bot.* 72: 248-256.

Finkelstein, R.R., Gampala, S.S., Rock, C.D. (2002). Abscisic acid signaling in seeds and seedlings. *Plant Cell* 14(Suppl): S15-45.

Finkelstein, R.R., Lynch, T.J. (2000). The *Arabidopsis* abscisic acid response gene *ABI5* encodes a basic leucine zipper transcription factor. *Plant Cell* 12: 599-609.

Finkelstein, R.R., Wang, M.L., Lynch, T.J., Rao, S., Goodman, H.M. (1998). The *Arabidopsis* abscisic acid response locus *ABI4* encodes an APETALA 2 domain protein. *Plant Cell* 10: 1043-1054.

Finnegan, E.J., Peacock, W.J., Dennis, E.S. (1996). Reduced DNA methylation in *Arabidopsis thaliana* results in abnormal plant development. *Proc. Natl. Acad. Sci. USA* 93: 8449-8454.

Gaiser, J.C., Robinson-Beers, K., Gasser, C.S. (1995). The *Arabidopsis SUPERMAN* gene mediates asymmetric growth of the outer integument of ovules. *Plant Cell* 7: 333-345.

Gallois, J.L., Woodward, C., Reddy, G.V., Sablowski, R. (2002). Combined SHOOT MERISTEMLESS and WUSCHEL trigger ectopic organogenesis in *Arabidopsis. Development* 129: 3207-3217.

Galweiler, L., Guan, C.H., Muller, A., Wisman, E., Mendgen, K., Yephremov, A., Palme, K. (1998). Regulation of polar auxin transport by AtPIN1 in *Arabidopsis* vascular tissue. *Science* 282: 2226-2230.

Garcia, D., Saingery, V., Chambrier, P., Mayer, U., Jurgens, G., Berger, F. (2003). *Arabidopsis haiku* mutants reveal new controls of seed size by endosperm. *Plant Physiol.* 131: 1661-1670.

Gasser, C.S., Broadhvest, J., Hauser, B.A. (1998). Genetic analysis of ovule development. *Annu. Rev. Plant Physiol. Plant Mol. Biol.* 49: 1-24.

Geldner, N., Anders, N., Wolters, H., Keicher, J., Kornberger, W., Muller, P., Delbarre, A., Ueda, T., Nakano, A., Jurgens, G. (2003). The *Arabidopsis* GNOM ARF-GEF mediates endosomal recycling, auxin transport, and auxin-dependent plant growth. *Cell* 112: 219-230.

Gillmor, C.S., Poindexter, P., Lorieau, J., Palcic, M.M., Somerville, C. (2002). Alpha-glucosidase I is required for cellulose biosynthesis and morphogenesis in *Arabidopsis. J. Cell Biol.* 156: 1003-1013.

Giraudat, J., Hauge, B.M., Valon, C., Smalle, J., Parcy, F., Goodman, H.M. (1992). Isolation of the *Arabidopsis ABI3* gene by positional cloning. *Plant Cell* 4: 1251-1261.

Golden, T.A., Schauer, S.E., Lang, J.D., Pien, S., Mushegian, A.R., Grossniklaus, U., Meinke, D.W., Ray, A. (2002). SHORT INTEGUMENTS1/SUSPENSOR1/CARPEL FACTORY, a dicer homolog, is a maternal effect gene required for embryo development in *Arabidopsis. Plant Physiol.* 130: 808-822.

Grini, P.E., Jürgens, G., Hülskamp, M. (2002). Embryo and endosperm development is disrupted in the female gametophytic capulet mutants of *Arabidopsis. Genetics* 162: 1911-1925.

Gross-Hardt, R., Lenhard, M., Laux, T. (2002). WUSCHEL signaling functions in interregional communication during *Arabidopsis* ovule development. *Genes Dev.* 16: 1129-1138.
Grossniklaus, U., Schneitz, K. (1998). The molecular and genetic basis of ovule and megagametophyte development. *Semin. Cell Dev. Biol.* 9: 227-238.
Grossniklaus, U., Vielle-Calzada, J.P., Hoeppner, M.A., Gagliano, W.B. (1998). Maternal control of embryogenesis by *MEDEA*, a polycomb group gene in *Arabidopsis. Science* 280: 446-450.
Guo, H.-S., Fei, J.-F., Xie, Q., Chua, N.-H. (2003). A chemical-regulated inducible RNAi system in plants. *Plant J.* 34: 383-392.
Hamann, T., Benkova, E., Baurle, I., Kientz, M., Jurgens, G. (2002). The *Arabidopsis BODENLOS* gene encodes an auxin response protein inhibiting MONOPTEROS-mediated embryo patterning. *Genes Dev.* 16: 1610-1615.
Harada, J.J. (2001). Role of *Arabidopsis LEAFY COTYLEDON* genes in seed development. *J. Plant Physiol.* 158: 405-409.
Hardtke, C.S., Berleth, T. (1998). The *Arabidopsis* gene *MONOPTEROS* encodes a transcription factor mediating embryo axis formation and vascular development. *EMBO J.* 17: 1405-1411.
Haseloff, J. (1999). GFP variants for multispectral imaging of living cells. *Methods Cell. Biol.* 58: 139-151.
Hauser, B.A., Villanueva, J.M., Gasser, C.S. (1998). *Arabidopsis* TSO1 regulates directional processes in cells during floral organogenesis. *Genetics* 150: 411-423.
Hecht, V., Vielle-Calzada, J.P., Hartog, M.V., Schmidt, E.D., Boutilier, K., Grossniklaus, U., de Vries, S.C. (2001). The *Arabidopsis SOMATIC EMBRYOGENESIS RECEPTOR KINASE 1* gene is expressed in developing ovules and embryos and enhances embryogenic competence in culture. *Plant Physiol.* 127: 803-816.
Hill, L.M., Morley-Smith, E.R., Rawsthorne, S. (2003). Metabolism of sugars in the endosperm of developing seeds of oilseed rape. *Plant Physiol.* 131: 228-236.
Holding, D.R., Springer, P.S. (2002). The *Arabidopsis* gene *PROLIFERA* is required for proper cytokinesis during seed development. *Planta* 214: 373-382.
Holdsworth, M., Kurup, S., McKibbin, R. (1999). Molecular and genetic mechanisms regulating the transition from embryo development to germination. *Trends Plant Sci.* 4: 275-280.
Ishida, T., Aida, M., Takada, S., Tasaka, M. (2000). Involvement of *CUP-SHAPED COTYLEDON* genes in gynoecium and ovule development in *Arabidopsis thaliana. Plant Cell Physiol.* 41: 60-67.
Ishikawa, T., Machida, C., Yoshioka, Y., Kitano, H., Machida, Y. (2003). The *GLOBULAR ARREST1* gene, which is involved in the biosynthesis of folates, is essential for embryogenesis in *Arabidopsis thaliana. Plant J.* 33: 235-244.
Jang, J.C., Fujioka, S., Tasaka, M., Seto, H., Takatsuto, S., Ishii, A., Aida, M., Yoshida, S., Sheen, J. (2000). A critical role of sterols in embryonic patterning and meristem programming revealed by the *fackel* mutants of *Arabidopsis thaliana. Genes Dev.* 14: 1485-1497.

Jenik, P.D., Irish, V.F. (2000). Regulation of cell proliferation patterns by homeotic genes during *Arabidopsis* floral development. *Development* 127: 1267-1276.

Jofuku, K.D., Denboer, B.G.W., Vanmontagu, M., Okamuro, J.K. (1994). Control of *Arabidopsis* flower and seed development by the homeotic gene *Apetala2*. *Plant Cell* 6: 1211-1225.

Johnson, C.S., Kolevski, B., Smyth, D.R. (2002). *TRANSPARENT TESTA GLABRA2*, a trichome and seed coat development gene of *Arabidopsis*, encodes a WRKY transcription factor. *Plant Cell* 14: 1359-1375.

Jorgensen, R.A., Que, Q., Napoli, C.A. (2002). Research note: Maternally-controlled ovule abortion results from cosuppression of dihydroflavonol-4-reductase or flavonoid-3', 5'-hydroxylase genes in *Petunia hybrida*. *Funct. Plant Biol.* 29: 1500-1506.

Jürgens, G., Mayer, U., Busch, M., Lukowitz, W., Laux, T. (1995). Pattern formation in the *Arabidopsis* embryo: A genetic perspective. *Philos. Trans. R. Soc. Lond. Biol. Sci.* 350: 19-25.

Jürgens, G., Mayer, U., Torres-Ruiz, R.A., Berleth, T., Misera, S. (1991). Genetic analysis of pattern formation in the *Arabidopsis* embryo. *Development* 1(Suppl): 27-38.

Kang, B.H., Busse, J.S., Dickey, C., Rancour, D.M., Bednarek, S.Y. (2001). The *Arabidopsis* cell plate-associated dynamin-like protein, ADL1Ap, is required for multiple stages of plant growth and development. *Plant Physiol.* 126: 47-68.

Kaplan, D.R., Cooke, T.J. (1997). Fundamental concepts in the embryogenesis of dicotyledons: A morphological interpretation of embryo mutants. *Plant Cell* 9: 1903-1919.

Karssen, C.M., Brinkhorst-van der Swan, D.L.C., Breekland, A.E., Koornneef, M. (1983). Induction of dormancy during seed development by endogenous abscisic-acid: Studies on abscisic-acid deficient genotypes of *Arabidopsis thaliana* (L.) Heynh. *Planta* 157: 158-165.

Keith, K., Kraml, M., Dengler, N.G., McCourt, P. (1994). *fusca3:* A heterochronic mutation affecting late embryo development in *Arabidopsis*. *Plant Cell* 6: 589-600.

Kieber, J.J., Rothenberg, M., Roman, G., Feldmann, K.A., Ecker, J.R. (1993). CTR1, a negative regulator of the ethylene response pathway in *Arabidopsis*, encodes a member of the raf family of protein kinases. *Cell* 72: 427-441.

Kinoshita, T., Yadegari, R., Harada, J.J., Goldberg, R.B., Fischer, R.L. (1999). Imprinting of the *MEDEA* polycomb gene in the *Arabidopsis* endosperm. *Plant Cell* 11: 1945-1952.

Klucher, K.M., Chow, H., Reiser, L., Fischer, R.L. (1996). The *AINTEGUMENTA* gene of *Arabidopsis* required for ovule and female gametophyte development is related to the floral homeotic gene *APETALA2*. *Plant Cell* 8: 137-153.

Koornneef, M. (1981). The complex syndrome of *ttg* mutants: *Arabidopsis*. *Arabid. Info. Serv.* 18: 45-51.

Koornneef, M., Bentsink, L., Hilhorst, H. (2002). Seed dormancy and germination. *Curr. Opin. Plant Biol.* 5: 33-36.

Koornneef, M., Dellaert, L.W.M., Vanderveen, J.H. (1982). EMS-induced and radiation-induced mutation frequencies at individual loci in *Arabidopsis thaliana* (L.) Heynh. *Mut. Res.* 93: 109-123.

Koornneef, M., Reuling, G., Karssen, C.M. (1984). The isolation and characterization of abscisic-acid insensitive mutants of *Arabidopsis thaliana. Physiol. Plant.* 61: 377-383.

Kwong, R.W., Bui, A.Q., Lee, H., Kwong, L.W., Fischer, R.L., Goldberg, R.B., Harada, J.J. (2003). LEAFY COTYLEDON1-LIKE defines a class of regulators essential for embryo development. *Plant Cell* 15: 5-18.

Lang, J.D., Ray, S., Ray, A. (1994). *sin 1,* a mutation affecting female fertility in *Arabidopsis,* interacts with *mod 1,* its recessive modifier. *Genetics* 137: 1101-1110.

Lauber, M.H., Waizenegger, I., Steinmann, T., Schwarz, H., Mayer, U., Hwang, I., Lukowitz, W., Jürgens, G. (1997). The *Arabidopsis* KNOLLE protein is a cytokinesis-specific syntaxin. *J. Cell Biol.* 139: 1485-1493.

Laux, T., Mayer, K.F.X., Berger, J., Jürgens, G. (1996). The *WUSCHEL* gene is required for shoot and floral meristem integrity in *Arabidopsis. Development* 122: 87-96.

Leon-Kloosterziel, K.M., Keijzer, C.J., Koornneef, M. (1994). A seed shape mutant of *Arabidopsis* that is affected in integument development. *Plant Cell* 6: 385-392.

Liljegren, S.J., Ditta, G.S., Eshed, Y., Savidge, B., Bowman, J.L., Yanofsky, M.F. (2000). *SHATTERPROOF* MADS-box genes control seed dispersal in *Arabidopsis. Nature* 404: 766-770.

Lin, B.-Y. (1984). Ploidy barrier to endosperm development in maize. *Genetics* 107: 103-115.

Lin, Y., Sun, L., Nguyen, L.V., Rachubinski, R.A., Goodman, H.M. (1999). The Pex16p homolog SSE1 and storage organelle formation in *Arabidopsis* seeds. *Science* 284: 328-330.

Liu, C.M., McElver, J., Tzafrir, I., Joosen, R., Wittich, P., Patton, D., Van Lammeren, A.A., Meinke, D. (2002). Condensin and cohesin knockouts in *Arabidopsis* exhibit a titan seed phenotype. *Plant J.* 29: 405-415.

Liu, C.M., Meinke, D.W. (1998). The *titan* mutants of *Arabidopsis* are disrupted in mitosis and cell cycle control during seed development. *Plant J.* 16: 21-31.

Liu, C.M., Xu, Z.H., Chua, N.H. (1993). Auxin polar transport is essential for the establishment of bilateral symmetry during early plant embryogenesis. *Plant Cell* 5: 621-630.

Liu, Z., Franks, R.G., Klink, V.P. (2000). Regulation of gynoecium marginal tissue formation by *LEUNIG* and *AINTEGUMENTA. Plant Cell* 12: 1879-1892.

Liu, Z.C., Meyerowitz, E.M. (1995). *Leunig* regulates *Agamous* expression in *Arabidopsis* flowers. *Development* 121: 975-991.

Liu, Z.C., Running, M.P., Meyerowitz, E.M. (1997). TSO1 functions in cell division during *Arabidopsis* flower development. *Development* 124: 665-672.

Lopez-Molina, L., Chua, N.H. (2000). A null mutation in a bZIP factor confers ABA-insensitivity in *Arabidopsis thaliana. Plant Cell Physiol.* 41: 541-547.

Lopez-Molina, L., Mongrand, S., Chua, N.H. (2001). A postgermination developmental arrest checkpoint is mediated by abscisic acid and requires the ABI5 transcription factor in *Arabidopsis*. *Proc. Natl. Acad. Sci. USA* 98: 4782-4787.

Lopez-Molina, L., Mongrand, B., McLachlin, D.T., Chait, B.T., Chua, N.H. (2002). ABI5 acts downstream of ABI3 to execute an ABA-dependent growth arrest during germination. *Plant J.* 32: 317-328.

Lotan, T., Ohto, M., Yee, K.M., West, M.A., Lo, R., Kwong, R.W., Yamagishi, K., Fischer, R.L., Goldberg, R.B., Harada, J.J. (1998). *Arabidopsis* LEAFY COTYLEDON1 is sufficient to induce embryo development in vegetative cells. *Cell* 93: 1195-1205.

Luerssen, H., Kirik, V., Herrmann, P., Misera, S. (1998). *FUSCA3* encodes a protein with a conserved VP1/AB13-like B3 domain which is of functional importance for the regulation of seed maturation in *Arabidopsis thaliana*. *Plant J.* 15: 755-764.

Lukowitz, W., Mayer, U., Jürgens, G. (1996). Cytokinesis in the *Arabidopsis* embryo involves the syntaxin-related *KNOLLE* gene product. *Cell* 84: 61-71.

Lukowitz, W., Nickle, T.C., Meinke, D.W., Last, R.L., Conklin, P.L., Somerville, C.R. (2001). *Arabidopsis cyt1* mutants are deficient in a mannose-1-phosphate guanylyltransferase and point to a requirement of N-linked glycosylation for cellulose biosynthesis. *Proc. Natl. Acad. Sci. USA* 98: 2262-2267.

Luo, M., Bilodeau, P., Dennis, E.S., Peacock, W.J., Chaudhury, A.M. (2000). Expression and parent-of-origin effects for *FIS2*, *MEA*, and *FIE* in the endosperm and embryo of developing *Arabidopsis* seeds. *Proc. Natl. Acad. Sci. USA* 97: 10637-10642.

Luo, M., Bilodeau, P., Koltunow, A., Dennis, E.S., Peacock, W.J., Chaudhury, A.M. (1999). Genes controlling fertilization-independent seed development in *Arabidopsis thaliana*. *Proc. Natl. Acad. Sci. USA* 96: 296-301.

Maitz, M., Santandrea, G., Zhang, Z., Lal, S., Hannah, L.C., Salamini, F., Thompson, R.D. (2000). *rgf1*, a mutation reducing grain filling in maize through effects on basal endosperm and pedicel development. *Plant J.* 23: 29-42.

Mansfield, S.G., Briarty, L.G. (1990). Development of the free nuclear endosperm in *Arabidopsis thaliana* L. *Arabid. Info. Serv.* 27: 53-64.

Mansfield, S.G., Briarty, L.G. (1991). Early embryogenesis in *Arabidopsis thaliana*. 2. The developing embryo. *Canad. J. Bot.* 69: 461-476.

Mansfield, S.G., Briarty, L.G. (1992). Cotyledon cell development in *Arabidopsis thaliana* during reserve deposition. *Canad. J. Bot.* 70: 151-164.

Mayer, K.F., Schoof, H., Haecker, A., Lenhard, M., Jurgens, G., Laux, T. (1998). Role of WUSCHEL in regulating stem cell fate in the *Arabidopsis* shoot meristem. *Cell* 95: 805-815.

Mayer, U., Buttner, G., Jürgens, G. (1993). Apical-basal pattern formation in the *Arabidopsis* embryo: Studies on the role of the *Gnom* gene. *Development* 117: 149-162.

Mayer, U., Herzog, U., Berger, F., Inze, D., Jürgens, G. (1999). Mutations in the *pilz* group genes disrupt the microtubule cytoskeleton uncouple cell cycle progression from cell division in *Arabidopsis* embryo and endosperm. *Eur. J. Cell Biol.* 78: 100-108.

Mayer, U., Jürgens, G. (1998). Pattern formation in plant embryogenesis: A reassessment. *Semin. Cell Dev. Biol.* 9: 187-193.

Mayer, U., Torres-Ruiz, R.A., Berleth, T., Misera, S., Jürgens, G. (1991). Mutations affecting body organization in the *Arabidopsis* embryo. *Nature* 353: 402-407.

McCallum, C.M., Comai, L., Greene, E.A., Henikoff, S. (2000). Targeting induced local lesions in genomes (TILLING) for plant functional genomics. *Plant Physiol.* 123: 439-442.

McElver, J., Tzafrir, I., Aux, G., Rogers, R., Ashby, C., Smith, K., Thomas, C., Schetter, A., Zhou, Q., Cushman, M.A., et al. (2001). Insertional mutagenesis of genes required for seed development in *Arabidopsis thaliana. Genetics* 159: 1751-1763.

Meinke, D.W. (1991). Perspectives on genetic analysis of plant embryogenesis. *Plant Cell* 3: 857-866.

Meinke, D.W. (1992). A homeotic mutant of *Arabidopsis thaliana* with leafy cotyledons. *Science* 258: 1647-1650.

Meinke, D.W. (1995). Molecular genetics of plant embryogenesis. *Ann. Rev. Plant Physiol. Plant Mol. Biol.* 46: 369-394.

Meinke, D.W., Franzmann, L.H., Nickle, T.C., Yeung, E.C. (1994). Leafy cotyledon mutants of *Arabidopsis. Plant Cell* 6: 1049-1064.

Meinke, D.W., Meinke, L.K., Showalter, T.C., Schissel, A.M., Mueller, L.A., Tzafrir, I. (2003). A sequence-based map of *Arabidopsis* genes with mutant phenotypes. *Plant Physiol.* 131: 409-418.

Meinke, D.W., Sussex, I.M. (1979). Embryo-lethal mutants of *Arabidopsis thaliana:* A model system for genetic analysis of plant embryo development. *Dev. Biol.* 72: 50-61.

Meister, R.J., Kotow, L.M., Gasser, C.S. (2002). SUPERMAN attenuates positive *INNER NO OUTER* autoregulation to maintain polar development of *Arabidopsis* ovule outer integuments. *Development* 129: 4281-4289.

Menand, B., Desnos, T., Nussaume, L., Berger, F., Bouchez, D., Meyer, C., Robaglia, C. (2002). Expression and disruption of the *Arabidopsis TOR* (*target of rapamycin*) gene. *Proc. Natl. Acad. Sci. USA* 99: 6422-6427.

Messing, J., Grossniklaus, U. (1999). Genomic imprinting in plants. *Results Probl. Cell Differ.* 25: 23-40.

Mizukami, Y., Fischer, R.L. (2000). Plant organ size control: *AINTEGUMENTA* regulates growth and cell numbers during organogenesis. *Proc. Natl. Acad. Sci. USA* 97: 942-947.

Modrusan, Z., Reiser, L., Feldmann, K.A., Fischer, R.L., Haughn, G.W. (1994). Homeotic transformation of ovules into carpel-like structures in *Arabidopsis. Plant Cell* 6: 333-349.

Mordhorst, A.P., Hartog, M.V., El Tamer, M.K., Laux, T., de Vries, S.C. (2002). Somatic embryogenesis from *Arabidopsis* shoot apical meristem mutants. *Planta* 214: 829-836.

Mordhorst, A.P., Voerman, K.J., Hartog, M.V., Meijer, E.A., van Went, J., Koornneef, M., de Vries, S.C. (1998). Somatic embryogenesis in *Arabidopsis thaliana* is facilitated by mutations in genes repressing meristematic cell divisions. *Genetics* 149: 549-563.

Nacry, P., Mayer, U., Jurgens, G. (2000). Genetic dissection of cytokinesis. *Plant Mol. Biol.* 43: 719-733.

Nakamura, S., Lynch, T.J., Finkelstein, R.R. (2001). Physical interactions between ABA response loci of *Arabidopsis*. *Plant J.* 26: 627-635.

Nambara, E., Hayama, R., Tsuchiya, Y., Nishimura, M., Kawaide, H., Kamiya, Y., Naito, S. (2000). The role of *ABI3* and *FUS3* loci in *Arabidopsis thaliana* on phase transition from late embryo development to germination. *Dev. Biol.* 220: 412-423.

Nesi, N., Debeaujon, I., Jond, C., Pelletier, G., Caboche, M., Lepiniec, L. (2000). The *TT8* gene encodes a basic helix-loop-helix domain protein required for expression of *DFR* and *BAN* genes in *Arabidopsis* siliques. *Plant Cell* 12: 1863-1878.

Nesi, N., Debeaujon, I., Jond, C., Stewart, A.J., Jenkins, G.I., Caboche, M., Lepiniec, L. (2002). The *TRANSPARENT TESTA16* locus encodes the ARABIDOPSIS BSISTER MADS domain protein and is required for proper development and pigmentation of the seed coat. *Plant Cell* 14: 2463-2479.

Nesi, N., Jond, C., Debeaujon, I., Caboche, M., Lepiniec, L. (2001). The *Arabidopsis TT2* gene encodes an R2R3 MYB domain protein that acts as a key determinant for proanthocyanidin accumulation in developing seed. *Plant Cell* 13: 2099-2114.

Nishihama, R., Soyano, T., Ishikawa, M., Araki, S., Tanaka, H., Asada, T., Irie, K., Ito, M., Terada, M., Banno, H., et al. (2002). Expansion of the cell plate in plant cytokinesis requires a kinesin-like protein/MAPKKK complex. *Cell* 109: 87-99.

Ogas, J., Cheng, J.C., Sung, Z.R., Somerville, C. (1997). Cellular differentiation regulated by gibberellin in the *Arabidopsis thaliana* pickle mutant. *Science* 277: 91-94.

Ogas, J., Kaufmann, S., Henderson, J., Somerville, C. (1999). PICKLE is a CHD3 chromatin-remodeling factor that regulates the transition from embryonic to vegetative development in *Arabidopsis*. *Proc. Natl. Acad. Sci. USA* 96: 13839-13844.

Ohad, N., Margossian, L., Hsu, Y.C., Williams, C., Repetti, P., Fischer, R.L. (1996). A mutation that allows endosperm development without fertilization. *Proc. Natl. Acad. Sci. USA* 93: 5319-5324.

Ohad, N., Yadegari, R., Margossian, L., Hannon, M., Michaeli, D., Harada, J.J., Goldberg, R.B., Fischer, R.L. (1999). Mutations in *FIE*, a WD polycomb group gene, allow endosperm development without fertilization. *Plant Cell* 11: 407-416.

Olsen, O.A. (2001). ENDOSPERM DEVELOPMENT: Cellularization and cell fate specification. *Annu. Rev. Plant Physiol. Plant Mol. Biol.* 52: 233-267.

Ooms, J., Leon-Kloosterziel, K.M., Bartels, D., Koornneef, M., Karssen, C.M. (1993). Acquisition of desiccation tolerance and longevity in seeds of *Arabidopsis thaliana:* A comparative study using abscisic acid-insensitive *abi3* mutants. *Plant Physiol.* 102: 1185-1191.

Otegui, M.S., Mastronarde, D.N., Kang, B.H., Bednarek, S.Y., Staehelin, L.A. (2001). Three-dimensional analysis of syncytial-type cell plates during endosperm cellularization visualized by high resolution electron tomography. *Plant Cell* 13: 2033-2051.

Otegui, M., Staehelin, L.A. (2000). Syncytial-type cell plates: A novel kind of cell plate involved in endosperm cellularization of *Arabidopsis. Plant Cell* 12: 933-947.

Parcy, F., Valon, C., Kohara, A., Misera, S., Giraudat, J. (1997). The *ABSCISIC ACID-INSENSITIVE3, FUSCA3, LEAFY COTYLEDON1* loci act in concert to control multiple aspects of *Arabidopsis* seed development. *Plant Cell* 9: 1265-1277.

Parcy, F., Valon, C., Raynal, M., Gaubier-Comella, P., Delseny, M., Giraudat, J. (1994). Regulation of gene expression programs during *Arabidopsis* seed development: Roles of the *ABI3* locus and of endogenous abscisic acid. *Plant Cell* 6: 1567-1582.

Patton, D.A., Schetter, A.L., Franzmann, L.H., Nelson, K., Ward, E.R., Meinke, D.W. (1998). An embryo-defective mutant of *Arabidopsis* disrupted in the final step of biotin synthesis. *Plant Physiol.* 116: 935-946.

Penfield, S., Meissner, R.C., Shoue, D.A., Carpita, N.C., Bevan, M.W. (2001). MYB61 is required for mucilage deposition and extrusion in the *Arabidopsis* seed coat. *Plant Cell* 13: 2777-2791.

Pinyopich, A., Ditta, G.S., Savidge, B., Liljegren, S.J., Baumann, E., Wisman, E., Yanofsky, M.F. (2003). Assessing the redundancy of MADS-box genes during carpel and ovule development. *Nature* 424: 85-88.

Quesada, V., Ponce, M.R., Micol, J.L. (2000). Genetic analysis of salt-tolerant mutants in *Arabidopsis thaliana. Genetics* 154: 421-436.

Ravanel, S., Cherest, H., Jabrin, S., Grunwald, D., Surdin-Kerjan, Y., Douce, R., Rebeille, F. (2001). Tetrahydrofolate biosynthesis in plants: Molecular and functional characterization of dihydrofolate synthetase and three isoforms of folylpolyglutamate synthetase in *Arabidopsis thaliana. Proc. Natl. Acad. Sci. USA* 98: 15360-15365.

Raz, V., Bergervoet, J.H.W., Koornneef, M. (2001). Sequential steps for developmental arrest in *Arabidopsis* seeds. *Development* 128: 243-252.

Reidt, W., Ellerstrom, M., Kolle, K., Tewes, A., Tiedemann, J., Altschmied, L., Baumlein, H. (2001). FUS3-dependent gene regulation during late embryogenesis. *J. Plant Physiol.* 158: 411-418.

Reidt, W., Wohlfarth, T., Ellerstrom, M., Czihal, A., Tewes, A., Ezcurra, I., Rask, L., Baumlein, H. (2000). Gene regulation during late embryogenesis: The RY motif of maturation-specific gene promoters is a direct target of the *FUS3* gene product. *Plant J.* 21: 401-408.

Reiser, L., Modrusan, Z., Margossian, L., Samach, A., Ohad, N., Haughn, G.W., Fischer, R.L. (1995). The *Bell1* gene encodes a homeodomain protein involved in pattern-formation in the *Arabidopsis* ovule primordium. *Cell* 83: 735-742.

Rerie, W.G., Feldmann, K.A., Marks, M.D. (1994). The *GLABRA2* gene encodes a homeo domain protein required for normal trichome development in *Arabidopsis. Genes Dev.* 8: 1388-1399.

Robinson-Beers, K., Pruitt, R.E., Gasser, C.S. (1992). Ovule development in wild-type *Arabidopsis* and two female-sterile mutants. *Plant Cell* 4: 1237-1249.

Roe, J.L., Durfee, T., Zupan, J.R., Repetti, P.P., McLean, B.G., Zambryski, P.C. (1997). TOUSLED is a nuclear serine/threonine protein kinase that requires a coiled-coil region for oligomerization and catalytic activity. *J. Biol. Chem.* 272: 5838-5845.

Roe, J.L., Rivin, C.J., Sessions, R.A., Feldmann, K.A., Zambryski, P.C. (1993). The tousled gene in *A. thaliana* encodes a protein-kinase homolog that is required for leaf and flower development. *Cell* 75: 939-950.

Rohde, A., Kurup, S., Holdswoth, M. (2000). ABI3 emerges from the seed. *Trends Plant Sci.* 5: 418-419.

Rohde, A., Prinsen, E., De Rycke, R., Engler, G., Van Montagu, M., Boerjan, W. (2002). PtABI3 impinges on the growth and differentiation of embryonic leaves during bud set in poplar. *Plant Cell* 14: 1885-1901.

Rojo, E., Gillmor, C.S., Kovaleva, V., Somerville, C.R., Raikhel, N.V. (2001). *VACUOLELESS1* is an essential gene required for vacuole formation and morphogenesis in *Arabidopsis*. *Dev. Cell* 1: 303-310.

Rounsley, S.D., Ditta, G.S., Yanofsky, M.F. (1995). Diverse roles for MADS box genes in *Arabidopsis* development. *Plant Cell* 7: 1259-1269.

Sagasser, M., Lu, G.H., Hahlbrock, K., Weisshaar, B. (2002). *A. thaliana TRANSPARENT TESTA 1* is involved in seed coat development and defines the WIP subfamily of plant zinc finger proteins. *Genes Dev.* 16: 138-149.

Sakai, H., Krizek, B.A., Jacobsen, S.E., Meyerowitz, E.N. (2000). Regulation of SUP expression identifies multiple regulators involved in *Arabidopsis* floral meristem development. *Plant Cell* 12: 1607-1618.

Samson, F., Brunaud, V., Balzergue, S., Dubreucq, B., Lepiniec, L., Pelletier, G., Caboche, M., Lecharny, A. (2002). FLAGdb/FST: A database of mapped flanking insertion sites (FSTs) of *Arabidopsis thaliana* T-DNA transformants. *Nucleic Acids Res.* 30: 94-97.

Schauer, S.E., Jacobsen, S.E., Meinke, D.W., Ray, A. (2002). *DICER-LIKE1:* Blind men and elephants in *Arabidopsis* development. *Trends Plant Sci.* 7: 487-491.

Schiefthaler, U., Balasubramanian, S., Sieber, P., Chevalier, D., Wisman, E., Schneitz, K. (1999). Molecular analysis of *NOZZLE*, a gene involved in pattern formation and early sporogenesis during sex organ development in *Arabidopsis thaliana*. *Proc. Natl. Acad. Sci. USA* 96: 11664-11669.

Schneider, T., Dinkins, R., Robinson, K., Shellhammer, J., Meinke, D.W. (1989). An embryo-lethal mutant of *Arabidopsis thaliana* is a biotin auxotroph. *Dev. Biol.* 131: 161-167.

Schneitz, K. (1999). The molecular and genetic control of ovule development. *Curr. Opin. Plant Biol.* 2: 13-17.

Schneitz, K., Balasubramanian, S., Schiefthaler, U. (1998). Organogenesis in plants: The molecular and genetic control of ovule development. *Trends Plant Sci.* 3: 468-472.

Schneitz, K., Hülskamp, M., Kopczak, S.D., Pruitt, R.E. (1997). Dissection of sexual organ ontogenesis: A genetic analysis of ovule development in *Arabidopsis thaliana*. *Development* 124: 1367-1376.

Schneitz, K., Hülskamp, M., Pruitt, R.E. (1995). Wild-type ovule development in *Arabidopsis thaliana:* A light-microscope study of cleared whole-mount tissue. *Plant J.* 7: 731-749.

Schrick, K., Mayer, U., Horrichs, A., Kuhnt, C., Bellini, C., Dangl, J., Schmidt, J., Jürgens, G. (2000). FACKEL is a sterol C-14 reductase required for organized cell division and expansion in *Arabidopsis* embryogenesis. *Genes Dev.* 14: 1471-1484.

Schrick, K., Mayer, U., Martin, G., Bellini, C., Kuhnt, C., Schmidt, J., Jürgens, G. (2002). Interactions between sterol biosynthesis genes in embryonic development of *Arabidopsis. Plant J.* 31: 61-73.

Schultz, P., Jensen, W.A. (1971). Capsella embryogenesis: The chalazal proliferating tissue. *J. Cell Sci.* 8: 201-227.

Schwartz, B.W., Yeung, E.C., Meinke, D.W. (1994). Disruption of morphogenesis and transformation of the suspensor in abnormal suspensor mutants of *Arabidopsis. Development* 120: 3235-3245.

Scott, R., Spielman, M., Bailey, J., Dickinson, H. (1998). Parent-of-origin effects on seed development in *Arabidopsis thaliana. Development* 125: 3329-3341.

Shen, W.H., Parmentier, Y., Hellmann, H., Lechner, E., Dong, A., Masson, J., Granier, F., Lepiniec, L., Estelle, M., Genschik, P. (2002). Null mutation of *AtCUL1* causes arrest in early embryogenesis in *Arabidopsis. Mol. Biol. Cell* 13: 1916-1928.

Shevell, D.E., Leu, W.M., Gillmor, C.S., Xia, G., Feldmann, K.A., Chua, N.H. (1994). *EMB30* is essential for normal cell division, cell expansion, and cell adhesion in *Arabidopsis* and encodes a protein that has similarity to Sec7. *Cell* 77: 1051-1062.

Soderman, E.M., Brocard, I.M., Lynch, T.J., Finkelstein, R.R. (2000). Regulation and function of the *Arabidopsis ABA-insensitive4* gene in seed and abscisic acid response signaling networks. *Plant Physiol.* 124: 1752-1765.

Somerville, C., Koornneef, M. (2002). A fortunate choice: The history of *Arabidopsis* as a model plant. *Nat. Rev. Genet.* 3: 883-889.

Sorensen, M.B., Chaudhury, A.M., Robert, H., Bancharel, E., Berger, F. (2001). Polycomb group genes control pattern formation in plant seed. *Curr. Biol.* 11: 277-281.

Sorensen, M.B., Mayer, U., Lukowitz, W., Robert, H., Chambrier, P., Jurgens, G., Somerville, C., Lepiniec, L., Berger, F. (2002). Cellularisation in the endosperm of *Arabidopsis thaliana* is coupled to mitosis and shares multiple components with cytokinesis. *Development* 129: 5567-5576.

Souter, M., Topping, J., Pullen, M., Friml, J., Palme, K., Hackett, R., Grierson, D., Lindsey, K. (2002). *hydra* mutants of *Arabidopsis* are defective in sterol profiles and auxin and ethylene signaling. *Plant Cell* 14: 1017-1031.

Spillane, C., MacDougall, C., Stock, C., Kohler, C., Vielle-Calzada, J.P., Nunes, S.M., Grossniklaus, U., Goodrich, J. (2000). Interaction of the *Arabidopsis* polycomb group proteins FIE and MEA mediates their common phenotypes. *Curr. Biol.* 10: 1535-1538.

Springer, P.S., Holding, D.R., Groover, A., Yordan, C., Martienssen, R.A. (2000). The essential Mcm7 protein PROLIFERA is localized to the nucleus of dividing cells during the G(1) phase and is required maternally for early *Arabidopsis* development. *Development* 127: 1815-1822.

Stacey, M.G., Koh, S., Becker, J., Stacey, G. (2002). AtOPT3, a member of the oligopeptide transporter family, is essential for embryo development in *Arabidopsis. Plant Cell* 14: 2799-2811.

Steinborn, K., Maulbetsch, C., Priester, B., Trautmann, S., Pacher, T., Geiges, B., Kuttner, F., Lepiniec, L., Stierhof, Y.D., Schwarz, H., et al. (2002). The *Arabidopsis PILZ* group genes encode tubulin-folding cofactor orthologs required for cell division but not cell growth. *Genes Dev.* 16: 959-971.

Steinmann, T., Geldner, N., Grebe, M., Mangold, S., Jackson, C.L., Paris, S., Galweiler, L., Palme, K., Jürgens, G. (1999). Coordinated polar localization of auxin efflux carrier PIN1 by GNOM ARF GEF. *Science* 286: 316-318.

Stone, S.L., Kwong, L.W., Yee, K.M., Pelletier, J., Lepiniec, L., Fischer, R.L., Goldberg, R.B., Harada, J.J. (2001). *LEAFY COTYLEDON2* encodes a B3 domain transcription factor that induces embryo development. *Proc. Natl. Acad. Sci. USA* 98: 11806-11811.

Takada, S., Hibara, K., Ishida, T., Tasaka, M. (2001). The *CUP-SHAPED COTYLEDON1* gene of *Arabidopsis* regulates shoot apical meristem formation. *Development* 128: 1127-1135.

Thompson, R.D., Hueros, G., Becker, H., Maitz, M. (2001). Development and functions of seed transfer cells. *Plant Sci.* 160: 775-783.

Tsugeki, R., Kochieva, E.Z., Fedoroff, N.V. (1996). A transposon insertion in the *Arabidopsis SSR16* gene causes an embryo-defective lethal mutation. *Plant J.* 10: 479-489.

Tzafrir, I., Dickerman, A., Brazhnik, O., Nguyen, Q., McElver, J., Frye, C., Patton, D., Meinke, D. (2003). The *Arabidopsis* SeedGenes Project. *Nucleic Acids Res.* 31: 90-93.

Tzafrir, I., McElver, J.A., Liu, C.M., Yang, L.J., Wu, J.Q., Martinez, A., Patton, D.A., Meinke, D.W. (2002). Diversity of TITAN functions in *Arabidopsis* seed development. *Plant Physiol.* 128: 38-51.

Uwer, U., Willmitzer, L., Altmann, T. (1998). Inactivation of a glycyl-tRNA synthetase leads to an arrest in plant embryo development. *Plant Cell* 10: 1277-1294.

Vernon, D.M., Meinke, D.W. (1994). Embryogenic transformation of the suspensor in twin, a polyembryonic mutant of *Arabidopsis. Dev. Biol.* 165: 566-573.

Vicient, C.M., Bies-Etheve, N., Delseny, M. (2000). Changes in gene expression in the *leafy cotyledon1 (lec1)* and *fusca3 (fus3)* mutants of *Arabidopsis thaliana* L. *J. Exp. Bot.* 51: 995-1003.

Vielle-Calzada, J.P., Baskar, R., Grossniklaus, U. (2000). Delayed activation of the paternal genome during seed development. *Nature* 404: 91-94.

Vielle-Calzada, J.P., Thomas, J., Spillane, C., Coluccio, A., Hoeppner, M.A., Grossniklaus, U. (1999). Maintenance of genomic imprinting at the *Arabidopsis medea* locus requires zygotic DDM1 activity. *Genes Dev.* 13: 2971-2982.

Villanueva, J.M., Broadhvest, J., Hauser, B.A., Meister, R.J., Schneitz, K., Gasser, C.S. (1999). *INNER NO OUTER* regulates abaxial-adaxial patterning in *Arabidopsis* ovules. *Genes Dev.* 13: 3160-3169.

Vinkenoog, R., Spielman, M., Adams, S., Fischer, R.L., Dickinson, H.G., Scott, R.J. (2000). Hypomethylation promotes autonomous endosperm development and rescues postfertilization lethality in fie mutants. *Plant Cell* 12: 2271-2282.

Waizenegger, I., Lukowitz, W., Assaad, F., Schwarz, H., Jürgens, G., Mayer, U. (2000). The *Arabidopsis KNOLLE* and *KEULE* genes interact to promote vesicle fusion during cytokinesis. *Curr. Biol.* 10: 1371-1374.

Walker, A.R., Davison, P.A., Bolognesi-Winfield, A.C., James, C.M., Srinivasan, N., Blundell, T.L., Esch, J.J., Marks, M.D., Gray, J.C. (1999). The *TRANSPARENT TESTA GLABRA1* locus, which regulates trichome differentiation and anthocyanin biosynthesis in *Arabidopsis*, encodes a WD40 repeat protein. *Plant Cell* 11: 1337-1350.

Weigel, D., Jürgens, G. (2002). Stem cells that make stems. *Nature* 415: 751-754.

Weijers, D., Franke-van Dijk, M., Vencken, R.J., Quint, A., Hooykaas, P., Offringa, R. (2001). An *Arabidopsis* minute-like phenotype caused by a semi-dominant mutation in a *RIBOSOMAL PROTEIN S5* gene. *Development* 128: 4289-4299.

West, M.A.L., Yee, K.M., Danao, J., Zimmerman, J.L., Fischer, R.L., Goldberg, R.B., Harada, J.J. (1994). LEAFY COTYLEDON1 is an essential regulator of late embryogenesis and cotyledon identity in *Arabidopsis*. *Plant Cell* 6: 1731-1745.

Western, T.L., Burn, J., Tan, W.L., Skinner, D.J., Martin-McCaffrey, L., Moffatt, B.A., Haughn, G.W. (2001). Isolation and characterization of mutants defective in seed coat mucilage secretory cell development in *Arabidopsis*. *Plant Physiol.* 127: 998-1011.

Western, T. L., Haughn, G.W. (1999). *BELL1* and *AGAMOUS* genes promote ovule identity in *Arabidopsis thaliana*. *Plant J.* 18: 329-336.

Western, T.L., Skinner, D.J., Haughn, G.W. (2000). Differentiation of mucilage secretory cells of the *Arabidopsis* seed coat. *Plant Physiol.* 122: 345-356.

Willemsen, V., Friml, J., Grebe, M., Van Den Toorn, A., Palme, K., Scheres, B. (2003). Cell polarity and PIN protein positioning in *Arabidopsis* require *STEROL METHYLTRANSFERASE1* function. *Plant Cell* 15: 612-625.

Windsor, J.B., Symonds, V.V., Mendenhall, J., Lloyd, A.M. (2000). *Arabidopsis* seed coat development: Morphological differentiation of the outer integument. *Plant J.* 22: 483-493.

Winkel-Shirley, B. (2001). Flavonoid biosynthesis: A colorful model for genetics, biochemistry, cell biology, and biotechnology. *Plant Physiol.* 126: 485-493.

Winkel-Shirley, B. (2002). A mutational approach to dissection of flavonoid biosynthesis in *Arabidopsis*. *Rec. Adv. Phytochem.* 36: 96-110.

Wobus, U., Weber, H. (1999). Seed maturation: Genetic programmes and control signals. *Curr. Opin. Plant Biol.* 2: 33-38.

Xie, D.-Y., Sharma, S.B., Paiva, N.L., Ferreira, D., Dixon, R.A. (2003). Role of anthocyanidin reductase, encoded by *BANYULS* in plant flavonoid biosynthesis. *Science* 299: 396-399.

Yadegari, R., Depaiva, G.R., Laux, T., Koltunow, A.M., Apuya, N., Zimmerman, J.L., Fischer, R.L., Harada, J.J., Goldberg, R.B. (1994). Cell-differentiation and morphogenesis are uncoupled in *Arabidopsis raspberry* embryos. *Plant Cell* 6: 1713-1729.

Yadegari, R., Kinoshita, T., Lotan, O., Cohen, G., Katz, A., Choi, Y., Nakashima, K., Harada, J.J., Goldberg, R.B., Fischer, R.L., Ohad, N. (2000). Mutations in the *FIE* and *MEA* genes that encode interacting polycomb proteins cause parent-of-origin effects on seed development by distinct mechanisms. *Plant Cell* 12: 2367-2382.

Yadegari, R., Paiva, G., Laux, T., Koltunow, A.M., Apuya, N., Zimmerman, J.L., Fischer, R.L., Harada, J.J., Goldberg, R.B. (1994). Cell differentiation and morphogenesis are uncoupled in *Arabidopsis raspberry* embryos. *Plant Cell* 6: 1713-1729.

Yang, W.-C., Sundaresan, V. (2000). Genetics of gametophyte biogenesis in *Arabidopsis*. *Curr. Opin. Plant Biol.* 3: 53-57.

Zhang, J.Z., Somerville, C.R. (1997). Suspensor-derived polyembryony caused by altered expression of valyl-tRNA synthetase in the *twn2* mutant of *Arabidopsis*. *Proc. Natl. Acad. Sci. USA* 94: 7349-7355.

Zuo, J., Niu, Q.W., Frugis, G., Chua, N.H. (2002). The *WUSCHEL* gene promotes vegetative-to-embryonic transition in *Arabidopsis*. *Plant J.* 30: 349-359.

SECTION V: PROTEIN FAMILIES

Chapter 21

Plant Transporters

Marcelo Desimone
Reinhard Kunze
Ulf-Ingo Flügge
Wolf B. Frommer

INTRODUCTION

Asymmetrical distribution of metabolites within cell compartments and between organs is a basic requirement for maintenance of cell homeostasis and plant functionality. Unequal distribution is achieved by separating compartments via lipid bilayers, which represent a permeability barrier, in combination with selective transport of ions and metabolites in a vectorial manner. The cellular components responsible for transport facilitation are integral membrane proteins that are characterized by stretches of hydrophobic amino acid sequences spanning the cell membranes (for a detailed review, see Ludewig and Frommer, 2003). These transport proteins can function alone or as part of a protein complex, and, similar to enzymes, work by lowering the activation energy required for transport and have more or less stringent substrate specificity. In addition, transporters play a central role in plant nutrient acquisition and elimination.

Transport processes are also important for biotechnological applications. Crop plants having specific transport features can potentially help to solve agricultural problems (Kunze et al., 2002). Salt stress, for instance, limits agricultural productivity in many regions worldwide. Recently, an interesting way has been suggested to improve salt stress tolerance in tomato

---

This work was supported by the German Bundesministerium für Bildung und Forschung in the framework of the Genomanalyse im biologischen System Pflanze—GABI program. We are grateful to Elisabetta Catoni, Melanie Hilpert, Michael Fitz, Anja Schmidt, Giorgio Gonella, Karin Schumacher, Yan-Hua Su, and Uwe Ludewig from the ZMBP, Tübingen, and to Anja Schneider, Eric van der Graaff, Rainer Schwacke, and Patrycja Niewiadomski from the Botanical Institute, Universität zu Köln, for the excellent teamwork.

(Zhang and Blumwald, 2001). Starting from the finding that the vacuolar H^+/Na^+ antiporter AtNHX1 confers salt stress resistance in *Arabidopsis*, this gene was overexpressed in tomato plants. The transgenic plants were able to grow, flower, and produce fruits under salt stress conditions. Moreover, salt was accumulated in leaves, but not in the fruits, demonstrating the potential use of the transgenic plants in saline soils. This finding exemplifies how manipulation of transport may improve agricultural productivity.

However, to solve other agricultural problems related to transport in plants, a more solid knowledge of transport physiology at the molecular level is required. For example, attempts to optimize transport processes in plants by manipulating a single gene have often failed due to redundant gene functions or to lack of sufficient knowledge of vectoriality and regulation of transport. To unravel the interconnections and interactions of all parts of a biological system is a challenging task, termed *functional genomics,* that requires new experimental tools. New technologies based on known methods are adapted for large-scale studies and allow the simultaneous analysis of many components in a single experiment.

This chapter focuses on current methodological advances in the field of plant transporter functional genomics. Special attention is given to the progress achieved in *Arabidopsis*. However, efforts to obtain the complete genome sequence of other plant species are in progress, and the future impact of these data is discussed at the end of the chapter.

IDENTIFICATION OF NOVEL TRANSPORTERS BY HETEROLOGOUS EXPRESSION

The most important task in functional genomics is to convert genome data into gene function data. In the case of transporters, this means to assign new transport functions to known sequences. Although the structural features of proteins extrapolated from the DNA sequences can reveal important information about transport, e.g., transmembrane domains or other typical sequence motifs, such additional properties as expression regulation, subcellular localization, interaction with other cell components, and transport mechanism define the physiological gene function.

A starting point for the identification of the function of plant transporters has been the experimental determination of the potential substrate(s) by expression in heterologous systems. The different strategies used for this objective will be outlined and advantages and limitations will be briefly discussed.

Expression in Bacterial Systems

Bacterial systems are used for protein expression from diverse organisms, including plants. Heterologous protein expression in mutant *Escherichia coli* strains allowed the identification of several plant genes by complementation of functional deficiencies (Hausmann et al., 1993; Smith et al., 1989). However, attempts to identify transporter genes using this strategy have not been successful in many cases. The main problem seems to be the toxic effect of eukaryotic membrane proteins on *E. coli.* Highly hydrophobic proteins cannot correctly fold and consequently accumulate as inactive aggregates (Frommer and Ninnemann, 1995). Only in a few cases has it been possible to obtain active plant transporters in *E. coli* (Kim et al., 1998; Tjaden et al., 1998; Uozumi et al., 1998). Nevertheless, heterologous expression in *E. coli,* followed by protein isolation and reconstitution in liposomes, allowed the identification and characterization of several members of the mitochondrial carrier family (MCF), including plant transporters (Palmieri, 1994). This strategy represents a potential alternative for the functional identification of intracellular transporters. Indeed, central questions regarding the intracellular compartmentation of metabolism are still unresolved because the key carriers connecting the cell compartments remain largely unknown. In plastids, for example, myriads of metabolites are synthesized and utilized, but only a few carriers have been identified via classical biochemical approaches (Flügge, 2000). Because of the prokaryote-like features of chloroplasts and mitochondria, it was hypothesized that at least a set of proteins localizing in these organelles will yield active transporters in bacterial systems.

Yeast Suppression Cloning

The most widely used and probably most successful approach to identifying plant transporters for a given substrate is suppression cloning in heterologous hosts, especially yeast (for reviews, see Barbier-Brygoo et al., 2001; Frommer and Ninnemann, 1995). The basis of this approach is a yeast strain lacking specific transport capacities. This can occur (1) because yeast does not have such a system, (2) because the endogenous system has been inactivated by the supply of a suppressing agent (e.g., ammonium suppresses the general amino acid permease GAP1), (3) by mutating (in most cases deleting by homologous recombination) the transporter gene(s), or (4) by manipulating the strain in a way that it becomes dependent on a heterologous uptake system. The transport-deficient strain is transformed with a plant cDNA library cloned into a yeast expression vector, and yeast mutant growth is monitored under selective conditions. Selective conditions may be, for example, media with low concentrations of a substance for which the yeast is brady- or auxotrophic. The cells can grow only if the transfected cDNA, expressed under control of a strong host promoter, encodes a transport protein that ameliorates the auxotrophy. From the few viable colonies of millions of transformants, the cDNA-carrying plasmid encoding a putative transporter for the substrate in question is isolated and sequenced. This approach has the enormous advantage that no other information about the transporter is necessary, and the isolated cDNA encodes a functional transporter. This method has been successfully used to isolate plasma membrane (Breitkreuz et al., 1999; Gillissen et al., 2000; Riesmeier et al., 1992) and intracellular plant transporters (Catoni et al., 2003).

In addition to cDNA library transformation, a candidate gene strategy has been followed in some cases (Catoni et al., 2003). Using sequence information, the function of a given gene can be anticipated by similarity with known transporters in other species or by the presence of typical signature sequences. The cDNA of a candidate gene is then cloned in a yeast expression vector and expressed in yeast mutants that are deficient in the assumed function. Considering the experimental information available about transporters in animals and microorganisms, the candidate gene approach represents a strong alternative to the unbiased cDNA library screening.

A large collection of yeast mutants has become available and can be used as selection systems for expression cloning of transporters (<http://www.uni-frankfurt.de/fb15/mikro/euroscarf/index.html> and <http://www.uni-koeln.de/math-nat-fak/botanik/bot2/agflue/HOME/projects/GABI_rkunze/>). In the near future, it is expected that the function elucidation of putative yeast transporters will allow the development of additional selection systems for new transporter substrates.

Other model organisms, such as the genetically well-characterized fungi *Neurospora crassa* and *Aspergillus nidulans* have also been used for complementation of transport deficiencies. Compared with the domesticated baker's yeast, these organisms provide a rich source of mutants and offer in some cases more suitable selection possibilities. Recently, the function of LPE1 (leaf permease 1), a maize homologue of known transporters for xanthine and uric acid, could be identified by complementation of an *A. nidulans* mutant deficient in uptake of oxidized purines (Argyrou et al., 2001). In this case, *A. nidulans* is a more appropriate system than *Saccharomyces cerevisiae,* which cannot transport and utilize xanthine or uric acid.

Expression in Animal Cells

Expression cloning using *Xenopus laevis* oocytes has been extremely successful for identification of mammalian ion channels and transporters (for a review, see Romero et al., 1998). The *X. laevis* oocyte as an expression system for transporters offers several unique features. It is a relatively large cell, making it convenient for RNA microinjection and two-electrode voltage clamp studies. In addition, the cells exhibit low levels of endogenous transport activity and can readily synthesize membrane proteins. Basically, the method starts with injecting high-quality poly(A)$^+$ RNA obtained from fresh tissues into oocytes, adding potential substrates to the medium, and measuring the resulting electrical current across the plasma membrane. If induced currents are observed, the RNAs are size fractionated and tested again. The positive RNA fraction is used to synthesize cDNA and cloned into *E. coli* plasmids. Finally, the obtained library is screened by injecting in vitro transcribed cRNAs into oocytes and isolation of the individual transporter-encoding plasmid(s).

Using plant poly(A)$^+$ RNA as material, this approach has so far led to the cloning of only a syntaxin, which may indirectly affect endogenous ion channels in the oocytes (Leyman et al., 1999). However, many plant transporters could be expressed in oocytes if single cRNAs were injected, indicating that plant transporters can be principally expressed in this system (Boorer et al., 1996; Desimone et al., 2002; Fischer et al., 2002; Zhou et al., 1997). Moreover, the expressed plant transporters have been characterized in detail concerning substrate specificities, kinetic properties, and mechanism of transport. For this reason, transporter expression in oocytes represents an important tool either to identify the function of candidate genes or to characterize transporters identified in other systems, in which detailed studies are not possible.

In a few cases, cultured animal cell lines have been used for expression of plant transporters. For instance, the African green monkey COS-1 cell line has been used to identify heterologously expressed plant aquaporins (Kammerloher et al., 1994) and insect cells were used to express plant K^+ channels (Gaymard et al., 1996). Although these expression systems may be beneficial for special applications, their use has been limited in studies of plant genes.

Expression in Plants and Plant Cell Cultures

Overexpression of plant genes in plant cells would be particularly advantageous for resolving plant-specific problems. Although the methodology is available, it is surprising that little has been undertaken for transporter studies. Potential organisms for overexpression are unicellular algae or cell lines of higher plants with well-established transformation protocols. Using transiently or stably transformed plant cells, potential substrates can be evaluated by screening for a gain-of-function. In addition, plants overexpressing putative transporters in distinct organs, e.g., in roots, could be similarly tested.

One of the most critical factors to be solved may be the endogenous existence of analogous transport systems producing transport background. To this end, inactivation of endogenous transporters by homologous recombination is not feasible in higher plants. Therefore, this methodology would be applicable under the condition that the endogenous transport system can be properly down-regulated or that mutants with the required transport deficiency are available. Alternatively, cell cultures of the moss *Physcomitrella patens,* in which homologous recombination is possible, can be used as an expression system (Egener et al., 2002; Reski, 1999) (see Chapter 9).

Limitations and Perspectives

As described previously, several heterologous expression systems for plant transporters are available. All of them feature technical limitations and can probably be used for functional expression of only a discrete group of transporters. They allow the identification of transporters for a given substrate, but whether this substrate is also the physiological substrate within the plant is not known. In addition, only transporters that are functional as single polypeptides, but not those which are active as multisubunit complexes, will be isolated by that approach. Transporters may even be mistargeted to other membranes in heterologous systems; i.e., a transporter located in the plasma membrane in yeast must not necessarily be a plasma membrane transporter in plants. Correct processing, folding, posttranslational modification, and degradation may be assumed to occur in the host organism in the same or similar way as in the original donor organism. However, for transporters whose trafficking or activity depends on additional cellular components or special conditions, this may not always be the case.

For these reasons, a functional genomic approach ideally should exploit the advantages of each one of them. This task could be facilitated by the generation of ordered expression libraries (Figure 21.1). Conventional expression libraries generated from expressed mRNAs have some disadvantages for screening purposes: (1) they contain only the gene set expressed in the source tissue, (2) the abundance of individual clones can vary dramatically in non-normalized libraries, (3) they usually contain 5'-UTRs of variable length, potentially interfering with promoter activity. These problems can be avoided by the construction of libraries that contain only the coding sequences (ORFs) of all expressed putative membrane proteins. A preselection of the candidates can be conveniently carried out using bioinformatic

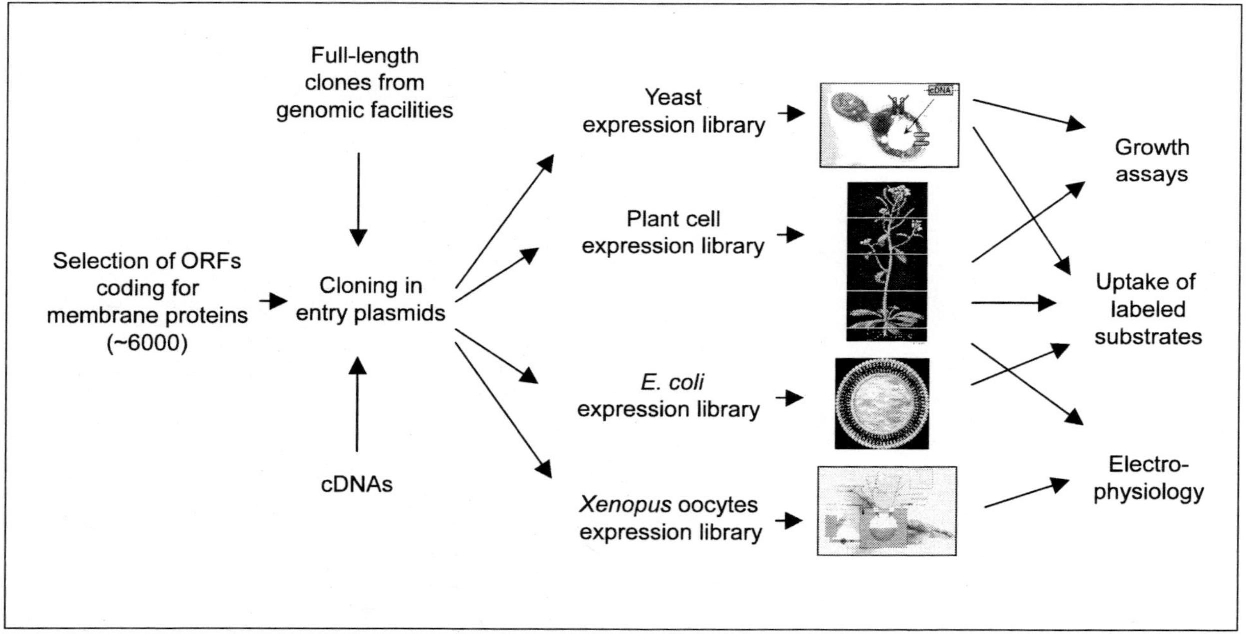

FIGURE 21.1. Genomic approach for functional identification of novel plant transporters.

platforms, such as the *Arabidopsis* membrane protein database ARAMEMNON. To facilitate the transfer of the ORFs into expression plasmids for different host organisms, the library vector should contain in vitro recombination sites, e.g., GATEWAY linkers. In this way, the putative transporter genes can be analyzed under optimal expression conditions in several systems. An additional advantage of such modular expression libraries is the possibility to array the single genes or pools of them in conventional devices, e.g., microtiter plates for cell transformation as well as for analytic purposes. This will be a first step toward automation, which would eventually allow the screening of thousands of transporter genes, expressed in multiple expression systems, with hundreds of possible substrates.

GENE EXPRESSION

Knowledge of temporal, spatial, and conditional expression patterns of membrane proteins is crucially required to assign physiological functions. For example, transporters that are induced or repressed by such external stimuli as environmental conditions may be important factors in future breeding strategies. In plants, many transporters are organized in gene families, and the amino acid sequences of some transporters are extremely conserved. For any transporter functional genomics project, a major challenge will be to untangle the individual functions of these proteins. It is conceivable that different family members have the same or very similar transport specificities but are active in different tissues or cellular compartments or are differentially regulated in response to changing growth conditions.

Ample evidence suggests that transporters are transcriptionally regulated by external stimuli. For example, various transporters respond at the transcription level to high and low nitrate concentrations (e.g., the *Arabidopsis* nitrate transporters NRT1-1 and NRT2-1, calcium antiporters, and the ammonium transporter AMT1-1; Filleur and Daniel-Vedele, 1999; Forde, 2000; Lejay et al., 1999; Wang et al., 2000) (see also Chapter 17). In *Arabidopsis,* ureide permeases are induced by nitrogen deficiency (Desimone et al., 2002), the phosphate transporter PHO1 is weakly up-regulated under phosphate stress (Hamburger et al., 2002), and plasma membrane H^+-ATPases are induced by exogenous sugars (Mito et al., 1996). The *AtNHX1* gene encodes a vacuolar Na^+/H^+-antiporter that is important in salt tolerance and up-regulated by treatment with NaCl, KCl, or ABA (Shi and Zhu, 2002). Recently, for the first time, activation of a sugar carrier, AtSUC2, by a pathogen was reported (Juergensen et al., 2003). A tomato ammonium transporter was found to be induced in tomato roots upon association with N_2-fixing bacteria (Becker et al., 2002). Transporters may also be developmentally controlled, as, for example, plasma membrane H^+-ATPases in *Arabidopsis* and tomato (Mito et al., 1996), or the *Arabidopsis* amino acid transporters AAP1 and AAP2, which are strongly induced in siliques at the heart stage of embryogenesis, shortly before induction of storage protein genes (Hirner et al., 1998).

Multiplex expression profiling using macro- or microarrays has been successfully used to discover conditionally regulated genes in plants (Koo and Ohlrogge, 2002; Ruan et al., 1998; Schaffer et al., 2001; Seki et al., 2001; Sreenivasulu et al., 2002) (see Chapter 4). Expression profiling is also the method of choice to discover novel transporters and other membrane proteins that are up- or down-regulated in response to challenging growth conditions.

For *Arabidopsis,* microarrays carrying specific oligonucleotides of all approximately 26,000 genes are commercially available. However, these arrays are comparatively expensive and the experimental processing requires special equipment. For academic purposes, *Arabidopsis* and other plant microarrays are also available from different research institutions. Links to microarray providers and other useful information concerning multiplex expression profiling can be found on the Web, for example:

<http://genome-www5.stanford.edu/MicroArray/SMD/resources.html>
<www.mpiz-koeln.mpg.de/~weisshaa/Adis/DNA-array-links.html>
<www.biologie.ens.fr/en/genetiqu/puces/microarraysframe.html>

However, as data extraction and evaluation are tedious with full-genome arrays, for large-scale screenings, the use of specialized custom arrays carrying defined subsets of genes may be sufficient and also more economical. Especially for the analysis of *Arabidopsis* membrane protein expression, a set of gene-specific cDNA fragments was developed for all putative membrane proteins with four or more transmembrane spans (approximately 2,000 genes). Primer pairs for the amplification of cDNA segments of 150 to 400 bp, with, if possible, less than 75 percent sequence similarity to any other *Arabidopsis* gene and preferably located at the very 3'-end of the gene, were designed. The amplified cDNA fragments were cloned and verified by sequencing. Probe DNA was amplified on the plasmids and can be used for spotting of macroarrays on membranes or low-density microarrays on glass slides. Taken together, a large number of arrays can be produced that are suitable for screening purposes at affordable cost (E. van der Graaff, R. Kunze, and U.-I. Flügge, unpublished).

The observation that plants with loss-of-function mutations in certain transporters frequently have no obvious phenotype suggests that other transporters may (partially) substitute for the lost function, or a compensation may be achieved by other metabolic changes (e.g., Dennison et al., 2001; Kaiser et al., 2002; Schneider et al., 2002; Szyroki et al., 2001). Therefore, expression profiling is also the preferred technique to identify regulatory networks of transporters and other proteins that are alternatively regulated as a consequence of mutations in specific transport processes.

SUBCELLULAR LOCALIZATION

In each cellular compartment, specific biochemical processes are executed. Therefore, to determine the physiological function of a given transporter, it is necessary to know its membrane localization. It is broadly accepted that subcellular targeting is a strictly regulated process in eukaryotic cells. Each protein is targeted to the respective membrane by a specific targeting signal to allow delivery to the correct compartment, while some proteins may be targeted to more than one organelle (Peeters and Small, 2001). In some specialized cases, even protein transport to adjacent cells may occur, suggesting that the cell-cell-connecting plasmodesmata are able to recognize specific targeting signals.

Plasma membrane proteins contain specific information that directs targeting via endoplasmic reticulum (ER) and Golgi to their destination in the plasma membrane (Goder and Spiess, 2001; Persson and Argos, 1996). Briefly, a hydrophobic signal sequence in a polytopic membrane protein may serve as a signal for the signal recognition particle (SRP). After initiation of translation upon ribosome binding to the transcript, the SRP/ribosome/nascent chain complex (SRP-RNC) docks with the SRP receptor on the cytoplasmic face of the ER. Thereafter, each transmembrane domain (TMD) is translated on the ribosome and may be inserted cotranslationally into the ER membrane. Concurrently and/or subsequently, one or more transmembrane domains at a time presumably move laterally to the periphery of the translocon and into the bilayer. From there, the membrane protein traffics via vesicles budding off from the ER membrane to the Golgi. Loading of the vesicles may involve cargo specifiers (Antonny and Schekman, 2001; Gilstring et al., 1999). Within the Golgi, proteins are sorted into vesicles destined either to the plasma membrane or to the vacuole (Pelham, 2001). Thus, membrane proteins use a pathway similar to those of secreted soluble proteins, but they always remain integrated into the membranes to reach their target membrane. However, in contrast to soluble protein targeting signals, which are normally cleaved within the ER, the signal sequences of polytopic membrane proteins are usually not cleaved. It is often assumed that the N-terminus and/or first transmembrane spans serve as targeting signals (for a review, see Chin et al., 2002). This trafficking is highly dynamic and also involves endocytosis and cycling of membrane proteins in the trafficking system.

Membrane proteins that transport solutes across organellar membranes also use targeting signals for posttranslational sorting. In the case of the well-studied plastidic triosephosphate translocator (TPT), a cleaved signal sequence has been identified (Flügge et al., 1989). A member of the plasma membrane hexose transporter family is responsible for plastidic glucose transport and contains a signal similar to that of the triosephosphate translocator for targeting to plastids (Weber et al., 2000). Similarly, as sulfur metabolism is predominantly found in chloroplasts, a member of the plasma membrane sulfate transporter family containing a signal sequence has been localized to plastids (Takahashi et al., 1999). Some inner mitochondrial membrane proteins also have clear signal sequences, while others, although known to localize to these membranes, do not contain obvious signal sequences.

Various algorithms and programs have been developed, allowing prediction of the destination of a query protein (Schwacke et al., 2003) (see Chapters 14 and 15). Although the prediction for membrane proteins is so far not very reliable (Bannai et al., 2002; Emanuelsson and von Heijne, 2001), such bioinformatics tools are a valuable support in directing functional studies. Experimental subcellular localization of membrane proteins has been performed either by analysis of the proteome of isolated membranes (Ferro et al., 2002) or by expression of proteins carrying a detectable tag.

Combining the use of highly purified and characterized membrane fractions, extraction of the hydrophobic proteins with organic solvents, SDS/PAGE fractionation, and tandem mass spectrometry analysis, it has recently been possible to identify more than 50 proteins residing in the envelope of spinach chloroplasts (Ferro et al., 2002). Twenty-seven of these were new envelope proteins that had not been predicted by bioinformatic tools. Moreover, this study also revealed common features among the putative transporters localized in the inner envelope membrane, e.g., an isoelectric point >8.8. However, proteomics approaches can detect only proteins present at a precise developmental stage. Therefore, localization data obtained, e.g., for the envelope of chloroplasts, cannot necessarily be extrapolated to other plastid types.

A complementary method to membrane proteome analysis is the expression of tagged proteins in plant cells. A simple variant of this approach is to transiently express the protein N- or C-terminally fused to GFP in plant protoplasts under the control of a constitutive promoter, generally the

CaMV 35S promoter. In this way, localization data can be collected for many proteins in a relatively short time. The main problem associated with this methodology is the strong and potentially misleading background induced by gene overexpression, so that the results have to be interpreted with care.

A proteomics version of this approach was achieved by random fusion of GFP to cDNA fragments and analysis of a large number of transgenic plants (Cutler et al., 2000; <http://deepgreen.stanford.edu>). Alternatively, proteins tagged with C- or N-terminal short peptides, e.g., c-myc, have been expressed in plants under control of their own promoter and detected with antibodies against this tag sequence (DeWitt and Sussman, 1995; Okumoto et al., 2003). The disadvantage of this latter approach, at least for functional genomics applications, is that the generation of transgenic plants is very time-consuming and difficult in high-throughput.

The conjunction of computer predictions and experimental data generated both by proteome analysis and by in vivo localization of tagged proteins will provide a reliable map of transporter subcellular localization. This will be valuable not only to relate known transporters with cellular processes but also to hypothesize and experimentally test putative functions of unknown membrane proteins.

PROTEIN INTERACTIONS

Protein-protein interactions constitute essential regulatory steps in intracellular processes. Accordingly, their comprehensive description would contribute considerably to the functional interpretation of fully sequenced genomes (Ito, Chiba, and Yoshida, 2001; Walhout and Vidal, 2001) (see Chapter 6). Due to the extreme variability between protein structures, the determination of physical association between proteins cannot fully be accomplished by any single methodology. A variety of complementary approaches should be undertaken to complete the protein interaction map. Two principally different approaches can be distinguished: (1) the mass spectrometric analysis of native protein complexes purified mainly by affinity capture and (2) the examination of mutual interactions after heterologous expression (e.g., in yeast). In the former, protein complexes are first isolated from the organisms using a known feature of one protein, e.g., one epitope that serves for immunotrapping, and subsequently the other proteins present in the complex are identified (Ho et al., 2002). The associations obtained in this way should reflect the true interactions in vivo. However, this approach is often limited by several factors, including low protein concentration, weak affinity interactions, or unknown conditions promoting interaction. In particular, proteomics studies of highly hydrophobic membrane proteins require the use of special extraction conditions, e.g., inclusion of detergents, that not only can substantially interfere with the stability of protein complexes but also may generate false-positive interactions due to rearrangement of hydrophobic protein domains.

The second approach includes the classical yeast two-hybrid system (Ito, Chiba, Ozawa, et al., 2001; Uetz et al., 2000). However, this approach is unsuitable for membrane proteins since a fusion of the activator domain to a membrane protein will prevent reconstitution of a functional transcription factor in the nucleus. Therefore, an alternative system optimized for membrane protein interaction studies has been developed (Johnsson and Varshavsky, 1994). The recent development and application of the split ubiquitin system to plant membrane proteins provides a new tool to detect and study such interactions (Reinders et al., 2001; Schulze et al., 2003). The plant H^+/sucrose cotransporters represent distant homologues of lacy, lacS, and GLUT1 and are also able to form oligomers. Three SUT paralogues are present in the same cell and can interact with one another in the split ubiquitin system, potentially forming hetero-oligomers. Also, separately expressed halves of one sucrose transporter can interact with the other half or the half of a paralogous sucrose transporter, leading to the reconstitution of functional transporters with new properties. Similar results were obtained for the *Arabidopsis* sucrose transport proteins (Schulze et al., 2003). As in other organisms, *Arabidopsis* membrane proteins are assumed to assemble in homo- or heteromeric complexes, in clusters or "rafts" (Jacobson et al., 1995; Reinders et al., 2001). To systematically identify membrane protein interactions both in yeast and *Arabidopsis,* a large-scale screening is being undertaken by several academic teams in the Sixth Framework Program of the European Union (http://blizzard.ulb.ac.be/Associoport/). The split ubiquitin system will be developed as a high-throughput screening tool and applied to 1,024 selected membrane proteins in yeast and *Arabidopsis*.

ANALYSIS OF LOSS-OF-FUNCTION ("KNOCKOUT") MUTANTS AND "KNOCKDOWN" MUTANTS

To assign the physiological function of a novel transporter, it is important to determine the temporal, spatial, and conditional expression profile; subcellular localization; transport mechanism and specificity; and interaction with other proteins. However, a central requirement remains the phenotypic and biochemical analysis of loss-of-function ("knockout") mutants in order to be able to assign a clear physiological function and to determine the relative role, e.g., within a pathway. In the case of novel candidate genes that were discovered by any of the strategies discussed previously, reverse genetics techniques need to be employed to isolate mutant plants.

In *Arabidopsis,* T-DNA and transposable elements from maize have successfully been used to generate large numbers of mutants. Due to the tremendous improvements in T-DNA transformation efficiency and ease during recent years (Bechtold et al., 1993; Clough and Bent, 1998), hundreds of thousands of T-DNA insertion mutants are available now from various sources. Assuming random integration of the T-DNA into the genome, approximately 120,000 mutants are required to recover a mutation in any gene with a 95 percent chance. For PCR-screening regimes, a primer from the gene of interest is combined with an outward-oriented primer from the left or right T-DNA border in the amplification reaction. To reduce the number of PCR reactions required, genomic template DNA is prepared from pooled mutant plants (Sussman et al., 2000). By arranging the DNA of (pools of) mutants in two- or three-dimensional arrays, a single mutant can be identified among tens of thousands of wild-type plants with only a few dozen reactions (Koes et al., 1995). Seeds of various mutant populations can be obtained from the *Arabidopsis* stock centers (NBRC, <http://nasc.nott.ac.uk/>, and ABRC, <http://www.arabidopsis.org/abrc/>). A screening service for another 133,000 T-DNA insertion mutants is provided by the Arabidopsis Knockout Facility at the University of Wisconsin, Madison (<www.biotech.wisc.edu/NewServicesAndResearch/Arabidopsis/Arabidopsis Home.asp>; Krysan et al., 1999; Sussman et al., 2000). With the advent of flanking sequence tag (FST) databases, the identification of insertion mutants has been significantly simplified. This topic is covered in Chapter 2 of this book.

Since it is reasonable to expect that not all genes are essential for normal growth but may be required under certain conditions, conditional screening in so-called "gauntlets" may be the tool that will allow the observation of a specific phenotype for many of the transporter genes (Sussman et al., 2000). Respective schemes have been set up in many labs for multiparallel screening of many mutants under a wide variety of differing conditions.

In past years, knockout mutants were identified for many plant transporters (e.g., Schneider et al., 2002; Scholz-Starke et al., 2003; Sherson et al., 2000; Streatfield et al., 1999; Taniguchi et al., 2002; Varotto et al., 2002). However, in spite of the high overall number of insertion mutants, for some genes, no knockout mutants have yet been found. These genes either belong to the small fraction of genes (statistically less than 1 percent) that are not yet hit by a T-DNA or transposon or potentially fall into the class of mutants for which loss-of-function mutations are dominantly lethal. To overcome this problem and to study the immediate consequences of mutations in such genes, it may be necessary to employ inducible antisense-RNA or posttranscriptional gene silencing (PTGS) approaches (Wassenegger, 2002). The advantages of antisense or PTGS are that genes can be analyzed in the absence of knockout mutants, the option to restrict repression to certain tissues and/or developmental stages, and the possibility to obtain only partial inactivation ("knockdown" mutants; Koch et al., 2003; Kühn et al., 2003). This, however, can also be problematic, as suppression levels are variable and not predictable (e.g., Chuang and Meyerowitz, 2000).

In *Arabidopsis,* many transporters are organized in gene families whose members can be highly similar and may have partially overlapping specificities and functions. Therefore, knockout mutations in single genes may not exhibit any phenotype under normal growth conditions (e.g., Sherson et al., 2000; Taniguchi et al., 2002; Scholz-Starke et al., 2003), and double or triple mutants may be required. Especially in those cases where the genes occur in tandem arrangements, double mutants are extremely difficult to obtain. Here, PTGS or antisense suppression offers a potential alternative. It is possible to simultaneously suppress several genes, if these contain partially identical regions, or by using a single chimeric transgene produced by fusing partial sense sequences of the individual genes (Abbott et al., 2002; Jones et al., 1998).

THE ARAMEMNON DATABASE: A NOVEL BIOINFORMATICS TOOL FOR SEQUENCE INTERPRETATION AND DATA MINING ON TRANSPORTERS AND MEMBRANE PROTEINS

A prime task of functional genomics programs is to elucidate the participating enzymes and the regulation of metabolic pathways in the context of plant development and response to environmental conditions. In almost all metabolic pathways, membrane passages of solutes are involved; however, biological membranes are essentially impermeable for ions and large and small polar molecules. Thus, membranes contain a multitude of proteins that facilitate passive diffusion of solutes or actively pump molecules across. As these transporters are frequently positioned at the entry or exit points of a metabolic pathway, they potentially control the activity of the whole pathway.

Accordingly, to understand the integration of metabolic pathways and biochemical responses to changing environmental conditions, it is important to gain a complete overview of an organism's membrane transporters and channels. The *Arabidopsis* genome is completely sequenced (Arabidopsis Genome Initiative, 2000) and the building of gene models and gene annotation are close to completion. This, in combination with the relative ease and efficiency of generating transgenic plants, makes *Arabidopsis* the first plant system that is suitable for comprehensive bioinformatic analyses.

The vast majority of transporters and also other integral membrane proteins are anchored within the membranes by alpha-helical transmembrane (TM) domains that pass through or dip into the lipid bilayer. The TM spans are approximately 16 to 24 amino acids in length and contain mostly hydrophobic residues, except toward the inside of the pore, where predominantly hydrophilic residues are located. Consequently, the TM helices typically are amphipathic. These characteristics enabled the development of algorithms to detect potential TM spans in a protein (Eisenberg et al., 1984). However, as the properties of membranes in different organisms and organelles are not identical and the prediction of protein structure is error prone, there is no perfect method to calculate TM spans. Over the past twenty years, more than a dozen algorithms were developed and optimized by different sets of membrane proteins with known structures (almost exclusively prokaryotic proteins). Most of these programs are accessible on the Internet (Table 21.1).

Biological membranes constitute not only a chemical barrier to the extracellular environment; in eukaryotes, they are also responsible for the formation of chemically distinct intracellular compartments (vacuole, mitochondria, plastids, ER/Golgi). To unravel the biological function of a transporter, it is critical to know in which membrane compartment within the cell it is located. In most cases, proteins are directed to mitochondria, plastids, and the secretory system by means of N-terminal signal peptides that differ slightly among these subcellular targets. Various signal sequence prediction programs were developed that calculate a probability for a protein to be targeted to one of these compartments (Table 21.2). However, the character of signal peptides can be very diverse and is sometimes ambiguous (Peeters and Small, 2001), and, thus, the reliability of single target predictions is rather low.

Transporters are often organized in families with sometimes highly similar members and show surprising conservation among plants, fungi, animals, and bacteria (see, e.g., amino acid transport systems; Wipf et al., 2002). The identification of related proteins and analysis of the gene family structures is one important step toward elucidating whether similar proteins fulfill redundant or distinct functions. A couple of classification systems have been developed, such as Pfam (Bateman et al., 2002), the SLC system for human solute carriers (www.gene.ucl.ac.uk/nomenclature), a system comparable to the enzyme nomenclature, i.e., the T numbers (<www.biology.ucsd.edu/~msaier/transport/>; Saier, 1999, 2000) and an integrated documentation resource for protein families, domains, and functional sites (InterPro; Mulder et al., 2002). For *S. cerevisiae,* a transport protein data-

TABLE 21.1. Transmembrane protein prediction programs used in ARAMEMNON.

Program	Creator	Web site address
Alom_v2	National Institute for Basic Biology, Okazaki	<http://psort.nibb.ac.jp/form2.html>
HmmTop_v2	Institute of Enzymology, Hungarian Academy of Sciences	<http://www.enzim.hu/hmmtop/html/submit.html>
MemSat_v1.8	Department of Computer Science, University College, London	<http://bioinf.cs.ucl.ac.uk/psiform.html>
PredTmr_v1	Deptartment of Cell Biology and Biophysics, University of Athens	<http://biophysics.biol.uoa.gr/PRED-TMR/input.html>
SosuiG_v1.1	Department of Biotechnology, Tokyo University of Agriculture and Technology	<http://sosui.proteome.bio.tuat.ac.jp/cgi-bin/sosui.cgi?/sosui_submit.html>
TMap	Department of Medical Biochemistry and Biophysics, Karolinska Institutet, Stockholm	<http://www.mbb.ki.se/tmap/>
TmHMM_v2	Center for Biological Sequence Analysis, Technical University of Denmark	<http://www.cbs.dtu.dk/services/TMHMM/>
TmPred	EMBnet Switzerland	<http://www.ch.embnet.org/software/TMPRED_form.html>
TopPred_v2	Institute Pasteur, Paris	<http://bioweb.pasteur.fr/seqanal/interfaces/toppred.html>

base (DB) has been established (<http://alize.ulb.ac.be/YTPdb/>; Andre, 1995; Van Belle and Andre, 2001). Genomic comparisons of membrane transport systems (www.biology.ucsd.edu/~ipaulsen/transport/) and P-type ATPases (<www.Patbase.kvl.dk>; Axelsen and Palmgren, 2001) have also been published. An *Arabidopsis* library of all transmembrane candidate proteins containing more than one TM span was used to identify novel membrane protein families not known from other organisms (<www.biosci.cbs.umn.edu/arabidopsis>; Ward, 2001). PlantsT is an *Arabidopsis* and yeast transporter DB with a focus on metal ion transporters from *Arabidopsis* (<http://plantst.sdsc.edu>; Mäser et al., 2001).

To gather information about an *Arabidopsis* membrane protein from different TM span and signal sequence predictions, to identify protein families and similar proteins from other organisms, and to find relevant publications

TABLE 21.2. Signal sequence prediction programs used in ARAMEMNON.

Program	Creator	Web site address
ChloroP_v1.1	Center for Biological Sequence Analysis, Technical University of Denmark	<http://www.cbs.dtu.dk/services/ChloroP/>
iPsort	Human Genome Center, Institute of Medical Science, University of Tokyo	<http://www.hypothesiscreator.net/iPSORT/>
MitoProt_v2	Institut für Humangenetik, TU München [Service]	<http://ihg.gsf.de/ihg/mitoprot.html>
PCLR_v0.9	Department of Computer and Information Science, University of Pennsylvania	<http://apicoplast.cis.upenn.edu/pclr/>
Predotar_v1.0	Unité de Recherche en Génomique Végétale (URGV), INRA/CNRS, Évry	<http://genoplante-info.infobiogen.fr/predotar/index.html>
SignalP_HMM_v2	Center for Biological Sequence Analysis, Technical University of Denmark	<http://www.cbs.dtu.dk/services/SignalP-2.0/>
SignalP_NN_v2	Center for Biological Sequence Analysis, Technical University of Denmark	<http://www.cbs.dtu.dk/services/SignalP-2.0/>
TargetP_v1	Center for Biological Sequence Analysis, Technical University of Denmark	<http://www.cbs.dtu.dk/services/TargetP/>

about these proteins presently is a tedious and time-consuming task that requires sequence submission to many different Web-based services. Recently, a novel *Arabidopsis thaliana* membrane protein DB was developed (ARAMEMNON) that integrates multiple features and greatly facilitates the analysis and interpretation of gene/protein sequence data (<http://aramemnon.botanik.uni-koeln.de>; Schwacke et al., 2003).

The complete set of approximately 27,060 nuclear-encoded *Arabidopsis* genes was analyzed by nine TM prediction programs. Based on all TM predictions, the statistical median of the number of TM spans was calculated for each protein. As an example, the TM analysis of the chloroplastic triose phosphate translocator (TPT) is shown in Figure 21.2.

ER targeting sequences of soluble proteins are often characterized by a hydrophobic core that may be recognized by prediction programs as a TM

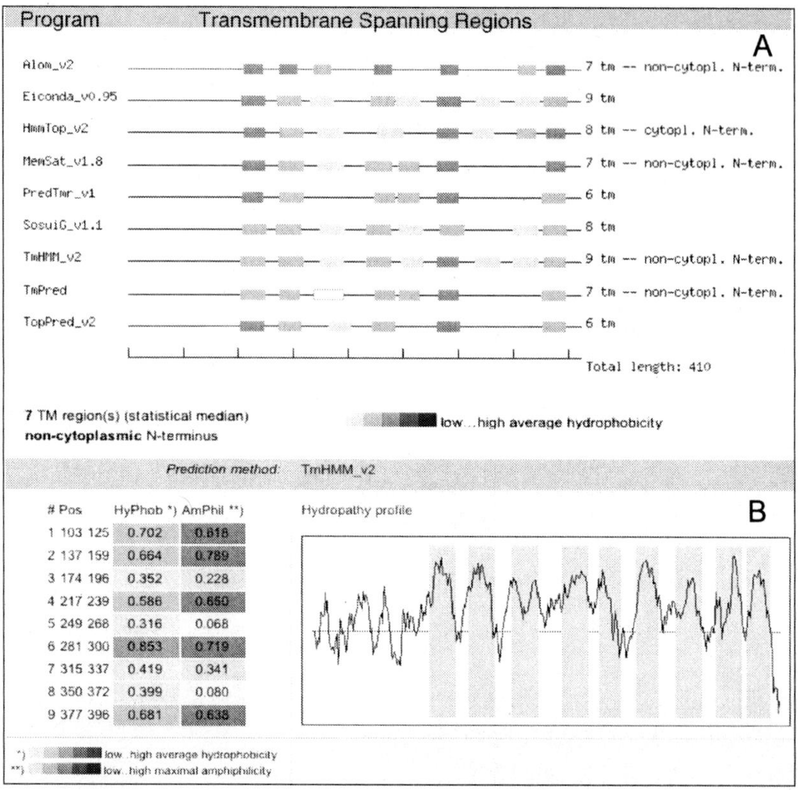

FIGURE 21.2. TM analysis of the triosephosphate translocator (TPT). Upper panel (A) shows the prediction of transmembrane spanning regions by nine different programs. The lower panel (B) shows the protein hydropathy profile and hydrophobicity of TM regions as defined by TmHMM_v2.

span (Nielsen et al., 1997). To exclude such proteins from classification as integral membrane proteins, those proteins which have one predicted transmembrane region at the N-terminus and a cleavable N-terminal signal peptide were removed from the database. Finally, some 6,700 nuclear-encoded proteins (25 percent of all nuclear-encoded proteins) and 80 organellar-encoded proteins were classified as TM proteins.

Plastids have evolved from progenitors of the present cyanobacteria and the majority of their genes have been translocated to the nucleus. Therefore, the knowledge about cyanobacterial orthologues can be helpful in the assignment of function. To facilitate this kind of analysis, all membrane proteins from *Synechocystis* sp., *Anabaena* sp., and *Thermosynechococcus elongatus* were included in the ARAMEMNON database.

The subcellular localization of membrane proteins is critical for their function. Several prediction programs were developed (Table 21.2); however, the ambiguity of individual subcellular localization predictions is rather high (see also Emanuelsson and von Heijne, 2001). To enable a comparison, ARAMEMNON provides the prediction results of eight different programs.

CONNECTING ARABIDOPSIS *FUNCTIONAL GENOMICS RESOURCES WITH CROP PLANT RESEARCH*

Functional genomics in the model plant *Arabidopsis* is only the prelude to an imminent revolution in plant production. Several genome projects of crop plants with economic importance are now in the sequencing process. For instance, the public Rice Genome Project will probably be completed by the end of 2005, representing the second plant genome known so far (see Chapter 11). The genomes of many other plant species, including maize, *Medicago truncatulata,* and tomato, are undergoing large-scale sequencing. These organisms will serve as "reference species" for economically important plant families, such as Poaceae (grasses), Fabaceae (legumes), and Solanaceae (tomato, potato), and the emerging functional genomics projects thereon will provide information for research and crop development for all the related crop species.

In this context, functional genomics in *Arabidopsis* will have a pioneer character. It will serve to develop and optimize high-throughput techniques for plant organisms, will provide a reference system for comparative genomics in plants, and will constitute a fundamental data platform for understanding, in general, how plants function. Gene function information obtained for *Arabidopsis* will facilitate the assignment of functions of orthologous genes in other plants species, thereby providing a basis for the development of effective strategies to manipulate important physiological processes for agricultural purposes. This will include strategies of gain-of-function and loss-of-function using transgenic approaches. In addition, as the resolution of genetic maps in the major crops increases, and as the molecular basis for specific traits or physiological responses becomes clearer, it will be increasingly possible to associate candidate genes, discovered in model species, with corresponding loci in crop plants. Appropriate relational databases will make it possible to freely associate across genomes with respect to gene sequence, putative function, or genetic map position. Once such tools have been implemented, the distinction between quantitative genetics (breeding) and molecular genetics will become blurred. Breeders will routinely use computer models to formulate predictive hypotheses to create phenotypes of interest from complex allele combinations and then construct those combinations by scoring large populations for very large numbers of genetic markers. On the other hand, the vast resource of breeding knowledge gathered over the past several decades will become directly linked to basic plant biology and enhance the ability to elucidate gene function in model organisms. For instance, traits that are poorly defined at the biochemical level, but well-established as a visible phenotype, can be associated by high-resolution mapping with candidate genes. Orthologous genes in a model species, such as *Arabidopsis* or rice, may not have a known association with a quantitative trait similar to that seen in the crop but might have been implicated in a particular pathway or signaling chain by genetic or biochemical experiments. These concepts are schematically summarized in Figure 21.3.

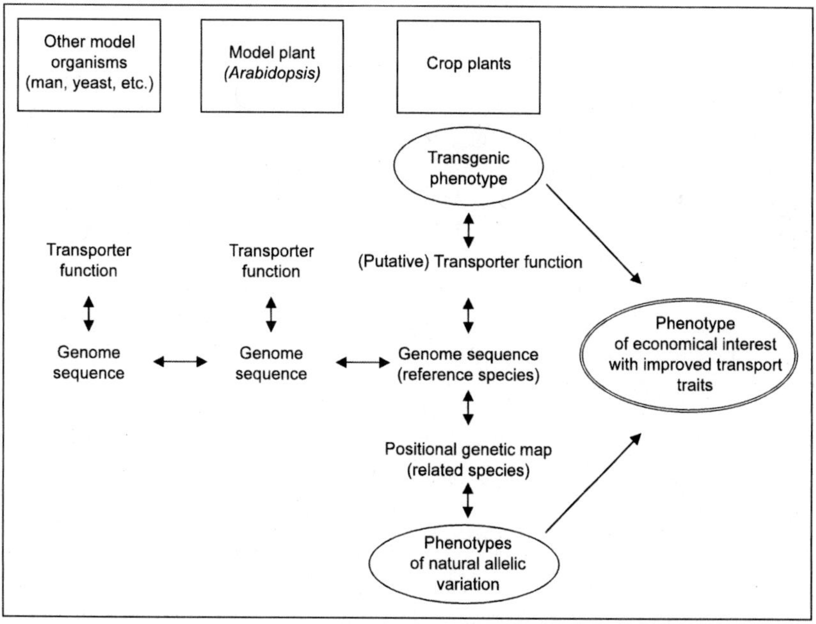

FIGURE 21.3. A model for cross-genome referencing to improve economically interesting transport traits in crop plants.

REFERENCES

Abbott, J.C., Barakate, A., Pincon, G., Legrand, M., Lapierre, C., Mila, I., Schuch, W., Halpin, C. (2002). Simultaneous suppression of multiple genes by single transgenes: Down-regulation of three unrelated lignin biosynthetic genes in tobacco. *Plant Physiol.* 128: 844-853.

Andre, B. (1995). An overview of membrane transport proteins in *Saccharomyces cerevisiae*. *Yeast* 11: 1575-1611.

Antonny, B., Schekman, R. (2001). ER export: Public transportation by the COPII coach. *Curr. Opin. Cell. Biol.* 13: 438-443.

Arabidopsis Genome Initiative (2000). Analysis of the genome sequence of the flowering plant *Arabidopsis thaliana*. *Nature* 408: 796-815.

Argyrou, E., Sophianopoulou, V., Schultes, N., Diallinas, G. (2001). Functional expression of a maize purine transporter by expression in *Aspergillus nidulans*. *Plant Cell* 13: 953-964.

Axelsen, K.B., Palmgren, M.G. (2001). Inventory of the superfamily of P-type ion pumps in *Arabidopsis*. *Plant Physiol.* 126: 696-706.

Bannai, H., Tamada, Y., Maruyama, O., Nakai, K., Miyano, S. (2002). Extensive feature detection of N-terminal protein sorting signals. *Bioinformatics* 18: 298-305.

Barbier-Brygoo, H., Gaymard, F., Rolland, N., Joyard, J. (2001). Strategies to identify transport systems in plants. *Trends Plant Sci.* 6: 577-585.

Bateman, A., Birney, E., Cerruti, L., Durbin, R., Etwiller, L., Eddy, S.R., Griffiths-Jones, S., Howe, K.L., Marshall, M., Sonnhammer, E.L. (2002). The Pfam protein families database. *Nucleic Acids Res.* 30: 276-280.

Bechtold, N., Ellis, J., Pelletier, G. (1993). In planta *Agrobacterium* mediated gene transfer by infiltration of adult *Arabidopsis thaliana* plants. *C.R. Acad. Sci. Paris Life Sci.* 316: 1194-1199.

Becker, D., Stanke, R., Fendrik, I., Frommer, W.B., Vanderleyden, J., Kaiser, W.M., Hedrich, R. (2002). Expression of the NH_4^+-transporter gene *LEAMT1;2* is induced in tomato roots upon association with N_2-fixing bacteria. *Planta* 215: 424-429.

Boorer, K.J., Frommer, W.B., Bush, D.R., Kreman, M., Loo, D.D.F., Wright, E.M. (1996). Kinetics and specificity of a H^+/amino acid transporter from *Arabidopsis thaliana*. *J. Biol. Chem.* 272: 13040-13046.

Breitkreuz, K.E., Shepl, B.J., Fischer, W.N., Schwacke, R., Rentsch, D. (1999). Identification and characterization of GABA, proline and quaternary ammonium compound transporters from *Arabidopsis thaliana*. *FEBS Lett.* 450: 280-284.

Catoni, E., Schwab, R., Hilpert, M., Desimone, M., Schwacke, R., Flügge, U.-I., Schumacher, K., Frommer, W.B. (2003). Identification of an *Arabidopsis* mitochondrial succinate-fumarate translocator. *FEBS Lett.* 534: 87-92.

Chin, C.N., von Heijne, G., de Gier, J.W.L. (2002). Membrane proteins: Shaping up. *Trends Biochem. Sci.* 27: 231-234.

Chuang, C.F., Meyerowitz, E.M. (2000). Specific and heritable genetic interference by double-stranded RNA in *Arabidopsis thaliana*. *Proc. Natl. Acad. Sci. USA* 97: 4985-4990.

Clough, S.J., Bent, A.F. (1998). Floral dip: A simplified method for *Agrobacterium*-mediated transformation of *Arabidopsis thaliana*. *Plant J.* 16: 735-743.

Cutler, S.R., Ehrhardt, D.W., Griffitts, J.S., Somerville, C.R. (2000). Random GFP::cDNA fusions enable visualization of subcellular structures in cells of *Arabidopsis* at a high frequency. *Proc. Natl. Acad. Sci. USA* 28: 3718-3723.

Dennison, K.L., Robertson, W.R., Lewis, B.D., Hirsch, R.E., Sussman, M.R., Spalding, E.P. (2001). Functions of AKT1 and AKT2 potassium channels determined by studies of single and double mutants of *Arabidopsis*. *Plant Physiol.* 127: 1012-1019.

Desimone, M., Catoni, E., Ludewig, U., Hilpert, M., Schneider, A., Kunze, R., Tegeder, M., Frommer, W.B., Schumacher, K. (2002). A novel superfamily of transporters for allantoin and other oxo derivatives of nitrogen heterocyclic compounds in *Arabidopsis*. *Plant Cell* 14: 847-856.

DeWitt, N.D., Sussman, M.R. (1995). Immunocytological localization of an epitope-tagged plasma membrane proton pump (H^+-ATPase) in phloem companion cells. *Plant Cell* 7: 2053-2067.

Egener, T., Granado, J., Guitton, M.C., Hohe, A., Holtorf, H., Lucht, J.M., Rensing, S.A., Schlink, K., Schulte, J., Schween, G., et al. (2002). High frequency of phenotypic deviations in *Physcomitrella patens* plants transformed with a gene-disruption library. *BMC Plant Biol.* 2: 6.

Eisenberg, D., Schwarz, E., Komaromy, M., Wall, R. (1984). Analysis of membrane and surface protein sequences with the hydrophobic moment plot. *J. Mol. Biol.* 179: 125-142.

Emanuelsson, O., von Heijne, G. (2001). Prediction of organellar targeting signals. *Biochim. Biophys. Acta* 1541: 114-119.

Ferro, M., Salvi, D., Riviere-Rolland, H., Vermat, T., Seigneurin-Berny, D., Grundwald, D., Garin, J., Joyard, J., Rolland, N. (2002). Integral membrane proteins of chloroplast envelope: Identification and subcellular localization of new transporters. *Proc. Natl. Acad. Sci. USA* 99: 11487-11492.

Filleur, S., Daniel-Vedele, F. (1999). Expression analysis of a high-affinity nitrate transporter isolated from *Arabidopsis thaliana* by differential display. *Planta* 207: 461-469.

Fischer, W.N., Loo, D.D.F., Koch, W., Ludewig, U., Tegeder, M., Boorer, K.J., Rentsch, D., Wright, E.M., Frommer, W.B. (2002). Low and high affinity amino acid H^+-cotransporters for cellular import of neutral and charged amino acids. *Plant J.* 29: 717-731.

Flügge, U.I. (2000). Transport in and out of plastids: Does the outer envelope membrane control the flow? *Trends Plant Sci.* 5: 135-137.

Flügge, U.I., Fischer, K., Gross, A., Sebald, W., Lottspeich, F., Eckerskorn, C. (1989). The triose phosphate-3-phosphoglycerate-phosphate translocator from spinach chloroplasts: Nucleotide sequence of a full-length cDNA clone and import of the in vitro synthesized precursor protein into chloroplasts. *EMBO J.* 8: 39-46.

Forde, B.G. (2000). Nitrate transporters in plants: Structure, function and regulation. *Biochim. Biophys. Acta* 1465: 219-235.

Frommer, W.B., Ninnemann, O. (1995). Heterologous expression of genes in bacterial, fungal, animal, and plant cells. *Ann. Rev. Plant Phys. Plant Mol. Biol.* 46: 419-444.

Gaymard, F., Cerutti, M., Horeau, C., Lemaillet, G., Urbach, S., Ravallec, M., Devauchelle, G., Sentenac, H., Thibaud, J.B. (1996). The baculovirus/insect cell system as an alternative to *Xenopus* oocytes: First characterization of the AKT1 K^+ channel from *Arabidopsis thaliana*. *J. Biol. Chem.* 271: 22863-22870.

Gillissen, B., Bürkle, L., André, B., Kühn, C., Rentsch, D., Brandl, B., Frommer, W.B. (2000). A new family of high affinity transporters for adenine, cytosine and purine derivatives in *Arabidopsis*. *Plant Cell* 12: 291-300.

Gilstring, C.F., Melin-Larsson, M., Ljungdahl, P.O. (1999). Shr3p mediates specific COPII coatomer-cargo interactions required for the packaging of amino acid permeases into ER-derived transport vesicles. *Mol. Biol. Cell* 10: 3549-3565.

Goder, V., Spiess, M. (2001). Topogenesis of membrane proteins: Determinants and dynamics. *FEBS Lett.* 504: 87-93.

Hamburger, D., Rezzonico, E., MacDonald-Comber Petetot, J., Somerville, C., Poirier, Y. (2002). Identification and characterization of the *Arabidopsis PHO1* gene involved in phosphate loading to the xylem. *Plant Cell* 14: 889-902.

Hausmann, L., Schell, J., Töpfer, R. (1993). Glycerol-3-phosphate dehydrogenase, a candidate for providing the backbone of glycerolipids. *Biol. Chem.* 374: 525-526.

Hirner, B., Fischer, W.N., Rentsch, D., Kwart, M., Frommer, W.B. (1998). Developmental control of H^+/amino acid permease gene expression during seed development of *Arabidopsis*. *Plant J.* 14: 535-544.

Ho, Y., Gruhler, A., Heilbut, A., Bader, G. D., Moore, L., Adams, S.L., Millar, A., Taylor, P., Bennett, K., Boutilier, K., et al. (2002). Systematic identification of protein complexes in *Saccharomyces cerevisiae* by mass spectrometry. *Nature* 415: 180-183.

Ito, T., Chiba, T., Ozawa, R., Yoshida, M., Hattori, M., Sakaki, Y. (2001). A comprehensive two-hybrid analysis to explore the yeast protein interactome. *Proc. Natl. Acad. Sci. USA* 98: 4569-4574.

Ito, T., Chiba, T., Yoshida, M. (2001). Exploring the protein interactome using comprehensive two-hybrid projects. *Trends Biotech.* 10: 23-27.

Jacobson, K., Sheets, E. D., Simson, R. (1995). Revisiting the fluid mosaic model of membranes. *Science* 268: 1441-1442.

Johnsson, N., Varshavsky, A. (1994). Split ubiquitin as a sensor of protein interactions in vivo. *Proc. Natl. Acad. Sci. USA* 91: 10340-10344.

Jones, C.G., Scothern, G.P., Lycett, G.W., Tucker, G.A. (1998). The effect of chimeric transgene architecture on co-ordinated gene silencing. *Planta* 204: 499-505.

Juergensen, K., Scholz-Starke, J., Sauer, N., Hess, P., Van Bel, A.J., Grundler, F.M. (2003). The companion cell-specific *Arabidopsis* disaccharide carrier AtSUC2 is expressed in nematode-induced syncytia. *Plant Physiol.* 131: 61-69.

Kaiser, B.N., Rawat, S.R., Siddiqi, M.Y., Masle, J., Glass, A.D. (2002). Functional analysis of an *Arabidopsis* T-DNA "knockout" of the high-affinity NH_4^+ transporter AtAMT1;1. *Plant Physiol.* 130: 1263-1275.

Kammerloher, W., Fischer, U., Piechottka, G.P., Schäffner, A.R. (1994). Water channels in the plant plasma membrane cloned by immunoselection from a mammalian expression system. *Plant J.* 6: 187-199.

Kim, E.J., Kwak, J.M., Uozumi, N., Schroeder, J.I. (1998). *AtKUP1:* An *Arabidopsis* gene encoding high-affinity potassium transport activity. *Plant Cell* 10: 51-62.

Koch, W., Kwart, M., Laubner, M., Heineke, D., Stransky, H., Frommer, W.B., Tegeder, M. (2003). Reduced amino acid content in transgenic potato tubers due to antisense inhibition of the leaf H^+/amino acid symporter StAAP1. *Plant J.* 33: 211-220.

Koes, R., Souer, E., Vanhouwelingen, A., Mur, L., Spelt, C., Quattrocchio, F., Wing, J., Oppedijk, B., Ahmed, S., Maes, T., et al. (1995). Targeted gene inactivation in petunia by PCR-based selection of transposon insertion mutants. *Proc. Natl. Acad. Sci. USA* 92: 8149-8153.

Koo, A.J., Ohlrogge, J.B. (2002). The predicted candidates of *Arabidopsis* plastid inner envelope membrane proteins and their expression profiles. *Plant Physiol.* 130: 823-836.

Krysan, P.J., Young, J.C., Sussman, M.R. (1999). T-DNA as an insertional mutagen in *Arabidopsis. Plant Cell* 11: 2283-2290.

Kühn, C., Hajirezaei, M.R., Fernie, A.R., Roessner-Tunali, U., Czechowski, T., Hirner, B., Frommer, W.B. (2003). The sucrose transporter StSUT1 localizes to sieve elements in potato tuber phloem and influences tuber physiology and development. *Plant Physiol.* 131: 102-113.

Kunze, R., Frommer, W.B., Flügge, U.-I. (2002). Metabolic engineering of plants: The role of membrane transport. *Metab. Eng.* 4: 57-66.

Lejay, L., Tillard, P., Lepetit, M., Olive, F., Filleur, S., Daniel-Vedele, F., Gojon, A. (1999). Molecular and functional regulation of two NO_3^- uptake systems by N- and C-status of *Arabidopsis* plants. *Plant J.* 18: 509-519.

Leyman, B., Geelen, D., Quintero, F.J., Blatt, M.R. (1999). A tobacco syntaxin with a role in hormonal control of guard cell ion channels. *Science* 283: 537-540.

Ludewig, U., Frommer, W.B. (2003). Metabolite transport. In Somerville, C.R., Meyerowitz, E.M. (Eds.), *The Arabidopsis Book*. Rockville, MD: American Society of Plant Biologists. Available online at <http://.aspb.org/publications/arabidopsis/>.

Mäser, P., Thomine, S., Schroeder, J.I., Ward, J.M., Hirschi, K., Sze, H., Talke, I.N., Amtmann, A., Maathuis, F.J., Sanders, D., et al. (2001). Phylogenetic relationships within cation transporter families of *Arabidopsis. Plant Physiol.* 126: 1646-1667.

Mito, N., Wimmers, L.E., Bennett, A.B. (1996). Sugar regulates mRNA abundance of H^+-ATPase gene family members in tomato. *Plant Physiol.* 112: 1229-1236.

Mulder, N.J., Apweiler, R., Attwood, T.K., Bairoch, A., Bateman, A., Binns, D., Biswas, M., Bradley, P., Bork, P., Bucher, P., et al. (2002). InterPro: An integrated documentation resource for protein families, domains and functional sites. *Brief Bioinform.* 3: 225-235.

Nielsen, H., Engelbrecht, J., Brunak, S., von Heijne, G. (1997). Identification of prokaryotic and eukaryotic signal peptides and prediction of their cleavage sites. *Protein Eng.* 10: 1-6.

Okumoto, S., Schmidt, R., Tegeder, M., Fischer, W.N., Rentsch, D., Frommer, W.B., Koch, W. (2003). High affinity amino acid transporters specifically expressed in xylem parenchyma and developing seeds of *Arabidopsis*. *J. Biol. Chem.* 277: 45338-45346.

Palmieri, F. (1994). Mitochondrial carrier proteins. *FEBS Lett.* 346: 48-54.

Peeters, N., Small, I. (2001). Dual targeting to mitochondria and chloroplasts. *Biochim. Biophys. Acta* 1541: 54-63.

Pelham, H.R. (2001). Traffic through the Golgi apparatus. *J. Cell Biol.* 155: 1099-1101.

Persson, B., Argos, P. (1996). Topology prediction of membrane proteins. *Protein Sci.* 5: 363-371.

Reinders, A., Schulze, W., Thaminy, S., Stagljar, I., Frommer, W.B., Ward, J.M. (2001). Intra- and intermolecular interactions in sucrose transporters at the plasma membrane detected by the split-ubiquitin system and functional assays. *Structure* 10: 763-772.

Reski, R. (1999). Molecular genetics of *Physcomitrella*. *Planta* 208: 301-309.

Riesmeier, J.W., Willmitzer, L., Frommer, W.B. (1992). Isolation and characterization of a yeast carrier cDNA from spinach by functional expression in yeast. *EMBO J.* 11: 4705-4713.

Romero, M.F., Kanai, Y., Gunshin, H., Hediger, M.A. (1998). Expression cloning using *Xenopus laevis* oocytes. *Meth. Enzymol.* 296: 17-52.

Ruan, Y., Gilmore, J., Conner, T. (1998). Towards *Arabidopsis* genome analysis: Monitoring expression profiles of 1400 genes using cDNA microarrays. *Plant J.* 15: 821-833.

Saier, M.H., Jr. (1999). A functional-phylogenetic system for the classification of transport proteins. *J. Cell Biochem.* 32-33(Suppl): l 84-94.

Saier, M.H., Jr. (2000). A functional-phylogenetic classification system for transmembrane solute transporters. *Microbiol. Mol. Biol. Rev.* 64: 354-411.

Schaffer, R., Landgraf, J., Accerbi, M., Simon, V.V., Larson, M., Wisman, E. (2001). Microarray analysis of diurnal and circadian-regulated genes in *Arabidopsis*. *Plant Cell* 13: 113-123.

Schneider, A., Häusler, R.E., Kolukisaoglu, U., Kunze, R., van der Graaff, E., Schwacke, R., Catoni, E., Desimone, M., Flügge, U.I. (2002). An *Arabidopsis thaliana* knock-out mutant of the chloroplast triose phosphate/phosphate translocator is severely compromised only when starch synthesis, but not starch mobilisation is abolished. *Plant J.* 32: 685-699.

Scholz-Starke, J., Buttner, M., Sauer, N. (2003). AtSTP6, a new pollen-specific H^+-monosaccharide symporter from *Arabidopsis*. *Plant Physiol.* 131: 70-77.

Schulze, W., Reinders, A., Ward, J.M., Lalonde, S., Frommer, W.B. (2003). Interactions between co-expressed *Arabidopsis* sucrose transporters in the split-ubiquitin system. *BMC Biochemistry* 4: 3.

Schwacke, R., Schneider, A., van der Graaff, E., Fischer, K., Catoni, E., Desimone, M., Frommer, W.B., Flügge, U.I., Kunze, R. (2003). ARAMEMNON, a novel database for *Arabidopsis* integral membrane proteins. *Plant Physiol.* 131: 16-26.

Seki, M., Narusaka, M., Abe, H., Kasuga, M., Yamaguchi-Shinozaki, K., Carninci, P., Hayashizaki, Y., Shinozaki, K. (2001). Monitoring the expression pattern of 1300 *Arabidopsis* genes under drought and cold stresses by using a full-length cDNA microarray. *Plant Cell* 13: 61-72.

Sherson, S.M., Hemmann, G., Wallace, G., Forbes, S., Germain, V., Stadler, R., Bechtold, N., Sauer, N., Smith, S.M. (2000). Monosaccharide/proton symporter AtSTP1 plays a major role in uptake and response of *Arabidopsis* seeds and seedlings to sugars. *Plant J.* 24: 849-857.

Shi, H., Zhu, J.K. (2002). Regulation of expression of the vacuolar Na^+/H^+ antiporter gene *AtNHX1* by salt stress and abscisic acid. *Plant Mol. Biol.* 50: 543-550.

Smith, J.K., Schloss, J.V., Mazur, B.J. (1989). Functional expression of plant acetolactate synthase genes in *E. coli. Proc. Natl. Acad. Sci. USA* 86: 4197-4183.

Sreenivasulu, N., Altschmied, L., Panitz, R., Hahnel, U., Michalek, W., Weschke, W., Wobus, U. (2002). Identification of genes specifically expressed in maternal and filial tissues of barley caryopses: A cDNA array analysis. *Mol. Genet. Genomics* 266: 758-767.

Streatfield, S.J., Weber, A., Kinsman, E.A., Häusler, R.E., Li, J., Post-Beittenmiller, D., Kaiser, W.M., Pyke, K.A., Flügge, U.I., Chory, J. (1999). The phosphoenolpyruvate/phosphate translocator is required for phenolic metabolism, palisade cell development, and plastid-dependent nuclear gene expression. *Plant Cell* 11: 1609-1622.

Sussman, M.R., Amasino, R.M., Young, J.C., Krysan, P.J., Austin-Phillips, S. (2000). The *Arabidopsis* knockout facility at the University of Wisconsin-Madison. *Plant Physiol.* 124: 1465-1467.

Szyroki, A., Ivashikina, N., Dietrich, P., Roelfsema, M.R., Ache, P., Reintanz, B., Deeken, R., Godde, M., Felle, H., Steinmeyer, R., et al. (2001). KAT1 is not essential for stomatal opening. *Proc. Natl. Acad. Sci. USA* 98: 2917-2921.

Takahashi, H., Asanuma, W., Saito, K. (1999). Cloning of an *Arabidopsis* cDNA encoding a chloroplast localizing sulfate transporter isoform. *J. Exp. Bot.* 50: 1713-1714.

Taniguchi, M., Taniguchi, Y., Kawasaki, M., Takeda, S., Kato, T., Sato, S., Tabata, S., Miyake, H., Sugiyama, T. (2002). Identifying and characterizing plastidic 2-oxoglutarate/malate and dicarboxylate transporters in *Arabidopsis thaliana*. *Plant Cell Physiol.* 43: 706-717.

Tjaden, J., Schwoppe, C., Mohlmann, T., Quick, P.W., Neuhaus, H.E. (1998). Expression of a plastidic ATP/ADP transporter gene in *Escherichia coli* leads to a functional adenine nucleotide transport system in the bacterial cytoplasmic membrane. *J. Biol. Chem.* 273: 9630-9636.

Uetz, P., Giot, L., Cagney, G., Mansfield, T.A., Judson, R.S., Knight, J.R., Lockshon, D., Narayan, V., Srinivasan, M., Pochart, P., et al. (2000). A comprehensive analysis of protein-protein interactions in *Saccharomyces cerevisiae*. *Nature* 403: 623-627.

Uozumi, U., Nakamura, T., Schroeder, J.I., Muto, S. (1998). Determination of transmembrane topology of an inward-rectifying potassium channel from *Arabidopsis thaliana* based on functional expression in *E. coli*. *Proc. Natl. Acad. Sci. USA* 95: 9773-9778.

Van Belle, D., Andre, B. (2001). A genomic view of yeast membrane transporters. *Curr. Opin. Cell Biol.* 13: 389-398.

Varotto, C., Maiwald, D., Pesaresi, P., Jahns, P., Salamini, F., Leister, D. (2002). The metal ion transporter IRT1 is necessary for iron homeostasis and efficient photosynthesis in *Arabidopsis thaliana*. *Plant J.* 31: 589-599.

Walhout, A.J.M., Vidal, M. (2001). Protein interaction maps for model organisms. *Nature Rev.* 2: 55-62.

Wang, R., Guegler, K., LaBrie, S.T., Crawford, N.M. (2000). Genomic analysis of a nutrient response in *Arabidopsis* reveals diverse expression patterns and novel metabolic and potential regulatory genes induced by nitrate. *Plant Cell* 12: 1491-1509.

Ward, J.M. (2001). Identification of novel families of membrane proteins from the model plant *Arabidopsis thaliana*. *Bioinformatics* 17: 560-563.

Wassenegger, M. (2002). Gene silencing. *Int. Rev. Cytol.* 219: 61-113.

Weber, A., Servaites, J.C., Geiger, D.R., Kofler, H., Hille, D., Gröner, F., Hebbeker, U., Flügge, U.I. (2000). Identification, purification and molecular cloning of a plastidic glucose translocator. *Plant Cell* 12: 787-801.

Wipf, D., Ludewig, U., Tegeder, M., Rentsch, D., Koch, W., Frommer, W.B. (2002). Conservation of amino acid transporters in fungi, plants and animals. *Trends Biochem. Sci.* 27: 139-147.

Zhang, H.X., Blumwald, E. (2001). Transgenic salt-tolerant tomato plants accumulate salt in foliage but not in fruit. *Nat. Biotechnol.* 19: 765-768.

Zhou, J.J., Theodoulou, F., Sauer, N., Sanders, D., Miller, A.J. (1997). A kinetic model with ordered cytoplasmic dissociation for SUC1, an *Arabidopsis* H^+/sucrose cotransporter expressed in *Xenopus* oocytes. *J. Membr. Biol.* 159: 113-125.

Chapter 22

Functional Genomics of the Cytochrome P450 Gene Superfamily in *Arabidopsis thaliana*

David W. Galbraith
Søren Bak

WHAT ARE CYTOCHROME P450S?

The wealth of information generated by genome-sequencing projects confronts the modern biologist with the task of identifying the function of genes that are being revealed. One of the largest and most divergent multigene families is that of the cytochrome P450s, being present in all phyla, where they are involved in both primary and secondary metabolism, as well as in detoxification of xenobiotics. For recent reviews on P450s in general, and plant P450s in particular, please see Chapple (1998), Nelson (1999), Kahn and Durst (2001), Kim and Tsukaya (2002), and Schuler and Werck-Reichhart (2003).

P450s are heme-thiolate proteins, which in prokaryotes are found as soluble enzymes, whereas in eukaryotes they are N-terminally membrane anchored. P450s were historically recognized and defined as proteins possessing a diagnostic peak of absorbance at 450 nm in the presence of carbon monoxide (Omura and Sato, 1964). This red-shifted Soret peak is a consequence of the formation of a cysteinate ligand to ferrous iron within the heme in a position *trans* to that of carbon monoxide. CO binding occurs at high affinity, in competition to that of oxygen, resulting in inhibition of P450 activity, but is reversible by illumination at 450 nm. As the catalytic mechanisms of different P450s have been elucidated, it is clear that the

---

The authors gratefully acknowledge support of this work in part by grants from the Human Frontiers Science Program, the National Science Foundation, the U.S. Department of Agriculture, the Danish Agricultural and Veterinary Research Council, and the Danish National Research Foundation.

spectral shift and its photoreversibility remain the only biophysical and biochemical properties common to all cytochrome P450s (Mansuy, 1998). The proposition that heme proteins of the cytochrome P450 superfamily be referred to as heme-thiolate protein P450s reflects the presence of conserved structural folds within which diagnostic sequence motifs are located. These sequence motifs provide not only the means to classify different P450s but also, in some cases, information concerning enzymatic activities.

P450 proteins catalyze many reactions involving dioxygen activation and the transfer of one oxygen atom into a variety of substrates. In many cases, electrons are received from NADH or NADPH via electron transfer proteins. In other cases, catalytic functions have been defined that do not involve permanent oxygen atom transfers, such as carbon-carbon or carbon-nitrogen bond cleavage, reductions, dehydrations, dehydrogenations, isomerizations, and dimerizations.

The catalytic mechanisms of cytochrome P450s can be usefully imagined as similar to that of a "molecular blowtorch" (Schlichting et al., 2000), whereby inherently inert hydrocarbon substrates are subjected to regio- and stereospecific attack at physiological temperatures, reactions that uncatalyzed would require extremely high temperatures. The first steps in the catalytic process are understood in reasonable detail, being a function of the more-conserved portions of the P450 proteins (Mansuy, 1998; Werck-Reichhart and Feyereisen, 2000). The active center of the P450 in its resting state usually comprises a mixture of low-spin ferric state hexacoordinated iron-protoporphyrin IX (heme), with the fifth coordination position occupied by the thiolate of the conserved cysteine, and the sixth occupied by solvent, and of the pentacoordinate high-spin ferric state, with the cysteinate as the sole axial ligand. Next, substrate binding displaces solvent and shifts the equilibrium toward the pentacoordinate state. A one-electron reduction converts the heme to a high-spin ferrous state, which can then bind molecular oxygen. A second one-electron reduction step produces the short-lived activated oxygen species that provide the means for substrate attack. Evidence exists that these electrophilic oxidants are a mixture of iron-peroxo (minor form) and iron-oxo (major form) complexes (Werck-Reichhart and Feyereisen, 2000), formed by protonation of the two-electron reduced form of dioxygen. The subsequent pathways of reaction, which are a function of the less-conserved parts of P450 molecules, are highly diverse and, consequently, are much less well understood.

The sources of electrons employed for these reductive steps vary according to the individual P450s and thereby provide a means for overall classification. Class I proteins receive electrons from NAD(P)H via a FAD-containing reductase and an iron sulfur redoxin. For class II proteins, transfer is effected by a FAD/FMN-containing P450 reductase. Two NADPH-P450 reductases, ATR1 and ATR2, have been identified in *Arabidopsis,* with a third potential one still being tested. In addition to receiving electrons via a NADPH-P450 reductase, class II P450s can receive a second electron from NAD(P)H via an additional carboxy terminally ER-anchored electron donor, cytochrome b_5, which has been shown to enhance the activity of some P450 enzymes (e.g., de Vetten et al., 1999). Class III proteins are self-sufficient, not requiring an electron donor. Finally, class IV proteins receive electrons directly from NAD(P)H.

COMPARATIVE GENOME ANALYSIS

Cytochrome P450s are found in all kingdoms and constitute a large group of polyphyletic sequences. Overall, P450s are highly divergent, in that individual pairs of P450s can have less than 20 percent sequence similarity. A basis for the phylogenetic analysis of large and divergent multigene families, such as the P450s, is the construction of optimal multiple sequence alignments. Only a few domains are highly conserved, including the heme-binding domain in the C-terminus, the I-helix involved in oxygen binding, the proline-rich region in the N-terminus that functions as a membrane hinge, and the E-R-R triad. This triad comprises the conserved Glu and Arg of the K-helix and the Arg of the "PERF" consensus and is involved in locking the heme pocket into position and in securing stabilization of the core structure by forming a salt bridge.

Primary classification of P450s is done according to amino acid similarity. P450s having >40 percent identity at the amino acid level are placed in the same family. Sequences that are more than 55 percent identical are placed in the same subfamily, while sequences between 40 percent and 55 percent are placed in different subfamilies. P450s are named in chronological order of submission to a nomenclature committee (http://drnelson.utmem.edu/CytochromeP450.html). Naming is according to the following convention: the CYP prefix is accompanied by a number indicating the family. Those P450s first described in plants have been given family numbers 71 through 99 and subsequently from 701 and higher. Subfamilies are denoted with a letter and subfamily members with a chronological number. As an example, the CYP51 family in *Arabidopsis* has two members that are 72 percent identical at the amino acid level; thus, according to the nomenclature rules, they are designated CYP51A1 and CYP51A2. It should be noted that this classification groups P450s across species, although not generally across kingdoms. The CYP51 family constitutes an exception to this convention, since this family is present in animals, fungi, and some bacteria. Sterols are essential in eukaryotes, and CYP51 proteins catalyze a key step in sterol biosynthesis. Thus, the broad occurrence of CYP51 orthologues in eukaryotes reflects the fact that sterols are ancient biomolecules pivotal for eukaryotic life-forms. Interestingly, CYP51 is absent from nematodes and insects, and these correspondingly are sterol heterotrophs. This condition appears to have conferred some selective advantage, allowing the gene to be superfluous and consequently extinct over evolutionary time within these organisms. The presence of CYP51 in bacteria such as *Mycobacterium* and *Streptomyces* has been suggested to represent horizontal gene transfer from a eukaryote, most likely from plants (Debeljak et al., 2000; Debeljak et al., 2003), but their functional role within the bacteria remains obscure.

Plant Cytochrome P450s Constitute Ten Clans

A disadvantage of the P450 nomenclature system is that, as more and more sequences are emerging, the various families and subfamilies have a natural tendency to overlap. The explosion in numbers of P450 families and subfamilies, which is not restricted to the plant kingdom, has led to the call for a nomenclature system based at a higher level, to permit grouping of clusters of evolutionary-related sequences. These clusters have been termed *clans* (Nelson, 1998, 1999). Traditionally, plant P450s have been divided into "A" and "non-A" types. The A type P450s are plant specific and of a monophyletic origin, whereas the non-A types comprise a less homogeneous group, the members of which often share greater sequence similarity to nonplant sequences than to other plant P450s (Paquette et al., 2000; Werck-Reichhart et al., 2002). A key signature of the A type cytochrome P450s is the consensus motif surrounding the conserved cysteine in the heme-binding region. Whereas the A types contain the consensus PFGXGRRXCXG, the non-A types have a less-conserved motif: FXXGXRXCXG. Similar differences in consensus motifs between the A and non-A cytochrome P450s can be seen in the so-called PERF region, which reads PER(F/H) for the A types, and PxRX for the non-A types. Multiple sequence alignments of the *Arabidopsis* A type and non-A type cytochrome P450s illustrating these differences in consensus motifs can be found at <http://www.biobase.dk/P450/cyp_Atype_112401.pdf> and <http://www.biobase.dk/P450/cyp_nonA_112401.pdf>. Recently, the plant P450s have been divided into ten clans (<http://drnelson.utmem.edu>; Nelson, 1999; Nelson, personal communication). These clans are expected to be present in all higher plants. The 727 clan is an exception, in that it has not been found in the *Arabidopsis* genome, arguing that it might be a clan specific to monocots. Intron splice sites are generally conserved through evolution (Long et al., 1995; Stoltzfus et al., 1997) and have been used as a diagnostic tool to validate the phylogenetic relationships predicted by the multiple sequence alignments of the *Arabidopsis* P450s (Werck-Reichhart et al., 2002; <http://www.biobase.dk/P450/>). The intron-exon organization concurs with ten clans, representing P450s having a phylogenetic origin.

Cytochrome P450s Across Kingdoms

The availability of complete or nearly complete sequenced genomes for an increasing number of organisms makes possible phylogenetic analysis across and within phyla, while reducing the bias caused by the availability of a limited number of genes that may represent specific subgroups, such as, for example, highly expressed genes. To locate the phylogenetic origin of the plant P450s, a multiorganism tree was constructed (Figure 22.1), including a single member of each plant P450 subfamily, as well as selected P450 families from animals and fungi representing the various clans, and a complete listing of the species chosen to represent the clans and families. The accession number as well as the actual alignment can be found at (http://www.biobase.dk/P450/cyp_allfam_reviewpaper.pdf). This highly divergent set of sequences represents class II P450s as well as the class I P450s found in mitochondria and class III P450s. To date, no class I P450s have been identified in plants. The vast majority of plant P450s are of class II, it being generally accepted that these proteins are N-terminally bound to the endoplasmic reticulum (ER) membrane. The 74 clan represents class III P450s catalyzing conversion of already oxygenated fatty acid hydroperoxides into allene epoxides or volatile cleavage products (Itoh and Howe, 2001; Howe and Schilmiller, 2002). The only P450 conserved across kingdoms is the orthologous CYP51, although substrate specificity varies slightly. For simplicity, only the CYP51A2 sequence from *Arabidopsis* was included and only the plant clans have been bracketed.

The multiorganism tree is in agreement with the general view that the A type/71 clan P450s are monophyletic in origin and have diverged after the bifurcation of the animal/fungi/plant kingdoms (Werck-Reichhart et al., 2002; Paquette et al., 2003). In *Arabidopsis,* 62 percent of the 272 full-length P450s are A type/71 clan (Paquette et al., 2003), reflecting the explosion of this clan in the evolution of terrestrial plants starting 430 million years ago. In comparison to other phyla, plants have cytochrome P450s in abundance. *Saccharomyces cerevisiae* has only three, while higher eukaryotes, such as *Caenorhabditis elegans* and *Drosophila melanogaster,* have 74 and 90 P450s, respectively. The conspicuous proliferation of plant

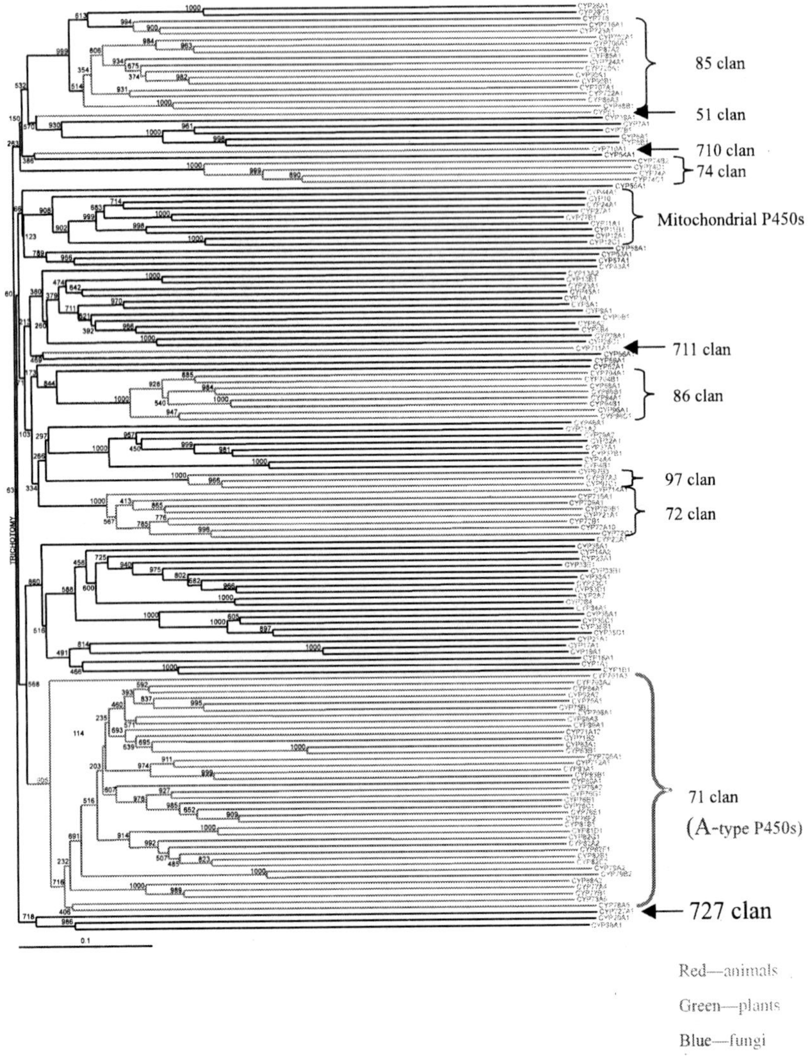

FIGURE 22.1. Phylogenetic tree of cytochrome P450s illustrating the interrelationships within and among P450s of different kingdoms (see corresponding figure in color photo section).

cytochrome P450s can to a large extent be ascribed to the explosion of the A type/71 clan. Plants are committed to a particular location throughout their life span. Accordingly, a wide range of strategies have evolved to respond to abiotic and biotic stresses as well as to secure communication with other plants. The ability to produce more than 200,000 known different plant products is an important part of these strategies. The majority of A type/71 clan P450s are involved in the biosynthesis of such metabolites commonly referred to as secondary metabolites or natural products, which reflects the necessity of plants constantly to produce new defense compounds. These compounds are not only defense compounds, such as phytoalexins and phytoanticipins, but also metabolites pivotal to terrestrial plants, such as lignin monomers and UV protectants. That the A type/71 clan is still subject to rather frequent changes to secure recruitment for novel function is reflected by the number of pseudogenes and by the clusters of multiple subfamily members on the chromosomes, indicative of recent gene-duplication events. The ratio of genes to pseudogenes for the A type/71 clan in *Arabidopsis* is 1:0.124, which is in striking contrast to the overall ratio of 1:0.03 (Arabidopsis Genome Initiative, 2000), being approximately four times higher. This high ratio of pseudogenes is characteristic of gene families that are not stable but are in the process of being recruited for new functions (Walsh, 1987, 1995; Lynch and Conery, 2000). A typical example is in biosynthesis of glucosinolates, a class of amino acid-derived phytoanticipins occurring primarily in the Chapparales (Rodman et al., 1996; Wittstock and Halkier, 2002). The occurrence of glucosinolates is a rather recent evolutionary event. The step thought to dictate the glucosinolate profile is catalyzed by members of the CYP79 family. The CYP79 homologues are highly substrate specific and metabolize amino acids or chain-elongated forms to their corresponding oximes in glucosinolate biosynthesis. Of the 13 CYP79 family members present in the *Arabidopsis* genome (http://www.biobase.dk/P450/), six are pseudogenes. The CYP705 family is another example of an unstable family; eight of the 33 CYP705 genes are pseudogenes. The functions of the CYP705s are currently unknown.

The so-called non-A type P450s comprise nine individual clans of polyphyletic origin that evolved before the radiation of the animal, fungi, and plant kingdoms (Figure 22.1; Werck-Reichhart et al., 2002). Most of the known non-A type P450s possess functions within household metabolism or provide various conserved functions essential for development and signaling. Examples of the former include fatty acid metabolism by the 86 clan and sterol biosynthesis by CYP51 in the 51 clan. Examples of the latter are the biosynthesis of oxylipids by the CYP74 family (Itoh and Howe, 2001) (74 clan) and CYP72B1 in brassinosteroid homeostasis (Neff et al., 1999) (72 clan). As far as we are aware, no function has been assigned to any of the members of the 97, 710, 711, and 727 clans. The latter four clans, as for the 51 clan, are characterized by containing only a single family. This is in contrast to the five other clans and argues that these P450 families encode specific and essential steps in metabolism. Furthermore, since no apparent pseudogenes have been found for these families, at least within the *Arabidopsis* genome, this implies these families are stable and not subject to ongoing recruitment for novel functions.

Cytochrome P450s in Green Algae

Further insight into P450 function and origins can be obtained by comparative study of the unicellular green alga *Chlamydomonas reinhardtii* (Chlorophyta), which is considered a valid representative of the ancestor of terrestrial plants (Willis and McElwain, 2002). The *C. reinhardtii* genome is currently being sequenced, and the first draft release has recently appeared (http://genome.jgi-psf.org/chlre1/chlre1.home.html) (see Chapter 8). In addition, various EST programs are underway. Blast searches indicate around 11 different *Chlamydomonas* P450s. Orthologues of CYP51, CYP97, CYP710, and CYP711 are seen, emphasizing the significance of these families in plants. For the CYP97s, these orthologues are classified as being members of the CYP97A and CYP97B subfamilies.

Two CYP51s are found in the *Arabidopsis* genome, *CYP51A1* and *CYP51A2,* which encode proteins that are only 72 percent identical at the amino acid level. These two genes are differentially expressed, *CYP51A1* being expressed mainly in the root, while *CYP51A2* appears to be evenly expressed in all tissues (Kim et al., 2000). Knockouts of *CYP51A2* are lethal, whereas knockouts of *CYP51A1* do not display obvious physiological phenotypes (Kim et al., 2000). Since the *Arabidopsis CYP51A2* is more related to the genes encoding CYP51s from sorghum, rice, and wheat than is *CYP51A1,* this suggests that *Arabidopsis CYP51A2* represents the *CYP51* orthologue, and *CYP51A1* a paralogue. In agreement with this, *CYP51* from *C. reinhardtii* provides a closer match to *Arabidopsis CYP51A2* than to *CYP51A1*. Clear orthologues of the 71, 72, 74, 85, 86, and 727 clans could not be identified. This does not exclude that they may be present, given the current status of the *C. reinhardtii* genome. The identification of *C. reinhardtii* P450s belonging to the 51, 97, 710, and 711 plant clans validates that these clans are ancient and represent essential and conserved functions in plants, while the apparent lack of representatives of the remaining clans indicates that they are relatively younger clans, having evolved perhaps as a consequence of the multicellular life-form of higher plants to facilitate the emergence of conserved functions essential for development, signaling, and secondary metabolites. Interestingly, some *C. reinhardtii* sequence fragments that showed similarity to P450 sequences could not be assigned to existing known plant P450 families. These fragments may represent P450 families that have been lost in the evolution of land plants, or, alternatively, they may simply represent misaligned contigs or remnants of pseudogenes. Evidently, our understanding will change as more, and better quality, sequence data become available for *C. reinhardtii*.

A SMALL PROPORTION OF THE ARABIDOPSIS *P450S HAVE BEEN ASSIGNED A SPECIFIC CELLULAR FUNCTION*

To date, specific enzymatic functions based on experimental evidence have been assigned to only a small fraction (<15 percent) of the *Arabidopsis* P450s (Werck-Reichhart et al., 2002; Schuler and Werck-Reichhart, 2003; The Arabidopsis Information Resource, http://www.arabidopsis.org/info/genefamily/P450_functions.html). An obstacle to assignment of functionality is that, given that the great majority of P450 genes currently lack assigned functions for their encoded proteins, obtaining the primary sequence of a P450 only rarely reveals its enzymatic function. For P450s that have sequence similarity to unique orthologues, such as CYP51 (sterol biosynthesis) and CYP73 (central phenylpropanoid biosynthesis), prediction of function based on primary sequence is relatively straightforward. However, for P450s belonging to families that comprise paralogous genes that have vastly expanded, a comparison to P450s of known function in other plants is much less helpful. One of the most extreme cases is encountered in the CYP71 family, which currently comprises over 14 subfamilies, many of which appear to be species specific. For example, the *Arabidopsis* CYP71 family comprises 52 members covering only two subfamilies, CYP71A and CYP71B. The CYP71B subfamily alone contains 36 members. While CYP71A subfamily members have been identified in dicots, the CYP71B subfamily appears to be specific for the Brassicaceae. Of the 52 *Arabidopsis* genes encoding CYP71s, the role of only one has been identified to date. This P450 (CYP71B15) is known to be required for biosynthesis of the phytoalexin camalexin, since it corresponds to *PAD3* (Zhou et al., 1999), although its precise enzymatic function remains unknown.

A listing of known functionalities is presented in Table 22.1. For the assignment of functionality to P450s for which the primary sequence is not informative, other methods are required.

TABLE 22.1. Functions described for individual members of the *Arabidopsis thaliana* P450 superfamily.

	Gene name	Activity	Pathway	Comments	Reference
1	51A2	Obtusifoliol 14α-demethylase	Sterols/steroids	Complements yeast *erg11*	Kushiro et al. (2001)
2	72B1	26-hydroxylase for brassinolide	Degradation of brassinolides		Neff et al. (1999)
3	73A5	Cinnamic acid 4-hydroxylase	Phenylpropanoid	Expressed in baculovirus	Mizutani et al. (1997)
4	74A1	Allene oxide synthase	Oxylipins (jasmonic acid)	Expressed in *E. coli*	Laudert et al. (1996)
5	74B2	Hydroperoxide lyase	Oxylipins (degradation of hydroperoxides)	Expressed in *E. coli*	Bate et al. (1998)
6	75B1	Naringenin 3'-hydroxylase, dihydrokaempferol (flavonoid 3'-hydroxylase)	Phenylpropanoid	Complements *tt7*; expressed in yeast	Schoenbohm et al. (2000)
7	79A2	Phenylalanine *N*-hydroxylase	Benzylglucosinolate	Expressed in *E. coli*	Wittstock and Halkier (2000)
8	79B2	Tryptophan *N*-hydroxylase	Indole glucosinolate	Expressed in *E. coli*	Hull et al. (2000); Mikkelsen et al. (2000)
9	79B3	Tryptophan *N*-hydroxylase	Indole glucosinolate	Expressed in *E. coli*	Hull et al. (2000)
10	79F1	Short-chain methionine derivatives *N*-hydroxylase (dihomomethionine, trihomomethionine)	Aliphatic glucosinolate	Substrates different from those of 79F2; expressed in *E. coli* and yeast	Hansen, Wittstock, et al. (2001); Reintanz et al. (2001); Chen et al. (2003)
11	79F2	Long-chain methionine derivatives *N*-hydroxylase	Aliphatic glucosinolate	Substrates different from those of 79F1	Reintanz et al. (2001); Chen et al. (2003)
12	83A1	Oxime *N*-hydroxylase	Glucosinolates	Complements 83B1 knockouts; expressed in yeast	Bak and Feyereisen (2001)
13	83B1	Indole-3-acetaldoxime *N*-hydroxylase	Indole glucosinolate	83B1 knockout rescued by 83A1; expressed in yeast and *E. coli;* 20-fold more efficient on indole-3-acetaldoxime than CYP83A1	Bak et al. (2001); Bak and Feyereisen (2001); Hansen, Du, et al. (2001)
14	84A1	Coniferaldehyde 5-hydroxylase, coniferyl alcohol and ferulic acid (ferulate 5-hydroxylase)	Phenylpropanoid	Complements *fah1;* expressed in yeast; much less efficient on ferulic acid	Meyer et al. (1996); Ruegger et al. (1999); Humphreys et al. (1999)
15	85A1	6-Deoxycastasterone C6-oxidase (and other steroids)	Brassinolide	Expressed in yeast	Shimada et al. (2001)

16	86A1	ω-Hydroxylase for saturated and unsaturated C12 to C18 fatty acids	Fatty acids	Substrates similar to those of 86A8; expressed in yeast	Benveniste et al. (1998)
17	86A8	ω-Hydroxylase for saturated and unsaturated C12 to C18 fatty acids	Fatty acids	Substrates similar to those of 86A1; expressed in yeast	Wellesen et al. (2001)
18	88A3	Multifunctional ent-kaurenoic acid oxidase	Gibberellin	Substrates similar to those of 88A4; expressed in yeast	Helliwell et al. (2001)
19	88A4	Multifunctional ent-kaurenoic acid oxidase	Gibberellin	Substrates similar to those of 88A3; expressed in yeast	Helliwell et al. (2001)
20	90A1	6-Oxo-cathasterone 23 α-hydroxylase	Brassinolide	Complements *cpd*	Szekeres et al. (1996)
21	90B1	6-Oxo-campestanol 22 α-hydroxylase	Brassinolide	Complements *dwf4*	Choe et al. (1998)
22	98A3	*p*-Coumaryl shikimic acid and *p*-coumaryl quinic acid 3'-hydroxylase	Phenylpropanoid	Expressed in yeast	Schoch et al. (2001)
23	701A3	Multifunctional ent-kaurene oxidase	Gibberellin	Complements *ga3;* expressed in yeast	Helliwell et al. (1998, 1999)

Source: Adapted from <http://www.arabidopsis.org/info/genefamily/P450_functions.html>.

A VARIETY OF NEW METHODS ARE BECOMING INCREASINGLY AVAILABLE TO ASSIGN FUNCTION TO P450 FAMILY MEMBERS

Over the past decade, a variety of high-throughput methods have been developed that provide alternative yet complementary ways to assign functions to genes. Many of these have been developed particularly for use in *Arabidopsis thaliana*. These include the analysis of (1) the effects of selective elimination of function of specific P450 genes, primarily based on insertional mutagenesis; (2) ectopic under- and overexpression of P450 genes; (3) the patterns of coregulated expression of specific P450 genes within the context of the entire genome (transcriptional profiling); and (4) the patterns of expression encoded by specific regulatory sequences of selected P450 genes. Less suited to high-throughput, although very valuable for precise definition of enzymatic function, is the heterologous expression of individual P450s within bacteria or other eukaryotes. Finally, molecular modeling will provide increased insight as the number of P450s of defined function increases.

Reverse Genetics

One of the most important and valuable methods for analysis of gene function involves reverse genetics, whereby inactivation of specific genes is followed by analysis of the resultant phenotype. Experimental details of methods of reverse genetics can be found elsewhere in this book. These methods rely on the more or less random insertion of foreign DNA sequences within the genome, and if this insertion occurs at an appropriate position relative to the gene model, it will result in a null allele. The sites of insertion are identified ex post facto, and the resultant phenotype of the null homozygote is then examined. Initial methods were labor intensive, requiring multiple rounds of PCR amplification to detect insertions within genes of interest. Strategies of pooling have been devised to minimize the number of PCR reactions required (Winkler et al., 1998), followed more recently, for *Arabidopsis,* by the use of sequencing to identify the precise site of foreign DNA insertion, which is then displayed in a Web-accessible manner (e.g., <http://signal.salk.edu/cgi-bin/tdnaexpress>) (see Chapter 2). Resources such as this greatly facilitate the identification of homozygous nulls for specific P450 genes, which can then be characterized in terms of potential P450-related phenotypes.

An early example of systematic screening of P450 gene families for functions based on induced null phenotypes was first established at the University of Arizona in 1998 (Winkler et al., 1998). This screen employed a PCR-based approach to comprehensively search for T-DNA inserts within P450 genes. Collections of up to 130,000 random T-DNA insertion lines were screened for insertions in 150 selected P450s (Winkler et al., 1998; Feldmann et al., 2002). Fifty-three P450 gene tags were identified and insertions into 40 different P450s were confirmed by sequencing. One of the initial mutants recovered from the first round of screening was an insertional mutant in *CYP83B1,* designated *rnt1* (Winkler et al., 1998). The mutant was a complete null and was later identified as being allelic to *sur2* (Bak et al., 2001; Barlier et al., 2000). The visual physiological phenotype of the mutant suggested that the gene product was involved in auxin (IAA) homeostasis (Bak et al., 2001; Barlier et al., 2000; Delarue et al., 1998). However, heterologous expression of CYP83B1 in a yeast system unexpectedly revealed that CYP83B1 participates in biosynthesis of indole glucosinolates, a class of tryptophan-derived secondary metabolites found in the Brassicaceae. CYP83B1 catalyzes the conversion of indole-3-acetaldoxime to the corresponding *S*-alkyl-thiohydroximate in the presence of thiol compounds, the first committed step in indole glucosinolate biosynthesis (Bak et al., 2001). Indole-3-acetaldoxime is also an intermediate in IAA biosynthesis; thus, complete elimination of CYP83B1 activity perturbs the control of the flux of the shared intermediate indole-3-acetaldoxime between IAA and indole glucosinolate biosynthesis and results in the observed high IAA phenotype.

Another mutant that emerged from this screen was a null of *CYP74A*, which is involved in the jasmonic acid signaling pathway, specifically having allene oxide synthase (AOS) activity (Park et al., 2002). This mutant cannot make endogenous jasmonic acid or its cyclopentenone precursors and characteristically displays a severe male sterility phenotype. The male sterility could be rescued by spraying with exogenous methyl jasmonate. As expected for a mutant defective in jasmonic acid biosynthesis, the mutant displayed defective wound signal transduction for *AtVSP2* (*Arabidopsis* vegetative storage protein 2) and *lipoxygenase 2* expression, both of which are wound- and jasmonic acid-induced genes in wild-type plants. As already mentioned, null alleles of *CYP51A1* and *CYP51A2* were identified and characterized (Kim et al., 2000).

RNA Interference

A problem with insertion-mediated inactivation is a consequence of the existence of extensive sequence duplications within plant genomes, found even within genomes as small as that of *A. thaliana,* coupled to the observation that individual P450 enzymes can catalyze the conversion of different, structurally related substrates to products, albeit with different turnover numbers and K_m values. This makes it difficult to assert confidently the absolute substrate specificity of a given P450 and underlines the potential for considerable functional redundancy within this multigene family. Insertional mutagenesis does not easily address the problem of redundancy, especially if the redundant genes are closely clustered on the chromosome. One possible solution would be to saturate local regions with multiple insertion events, utilizing local transposon movement. This can also be coupled to production of local deletions, or such deletions can be directly sought based on fast neutron irradiation (Li et al., 2001, 2002) (see Chapter 3).

A highly flexible alternative has recently emerged based on the use of RNA silencing (Waterhouse et al., 1998; Waterhouse and Helliwell, 2003), also termed *posttranscriptional gene silencing* and, more recently, *RNA interference* (RNAi). RNA silencing occurs as a consequence of the intracellular detection of double-stranded RNA (dsRNA), the response to which is the induction of a sequence-specific RNA degradation mechanism. The operation of this mechanism efficiently removes homologous transcripts, thereby silencing the gene. RNAi may have originally evolved as an antiviral defense but has been found to operate on any source of dsRNA. It involves degradation of the dsRNA into about 21-nucleotide fragments, which are incorporated into the RNAi silencing complex (RISC) and thereby direct the endolytic cleavage of complementary mRNA. Thus, if the dsRNA corresponds in sequence to the gene whose silencing is required, for example, specific P450 genes or sequences homologous to specific groups of P450 genes, specific silencing of these genes will occur, following which the resultant phenotype can be scored, assuming silencing is not lethal. Details of methods for induction of RNAi are found elsewhere in this book (see Chapter 3 for more details).

For analysis of P450 functions, the advantages of RNAi appear obvious. Silencing of individual P450s should occur using sequences specific to those P450s, sequence lengths of about 100 bases appearing sufficient for this purpose (Wesley et al., 2001). Silencing of groups of P450s can be envisaged through identification of shared identical sequences, as has recently been shown for the *GluB* genes in rice (Kusaba et al., 2003). To date, we are aware of only one example of the use of RNAi to silence a P450, namely, CYP71D16, which is involved in diterpene metabolism in tobacco trichomes (Wang and Wagner, 2003). It should be noted that 5' and 3' untranslated sequences act as good silencing targets (Wesley et al., 2001). Furthermore, there is evidence that promoter sequences can be employed as targets of effective RNAi, mediated not by RNA destruction, but by promoter methylation. This should be expected to increase considerably the number of potential P450 target sequences.

Analysis of Coordination of Transcriptional Regulation

Coordinated regulation of expression is a feature that is sometimes observed between genes encoding protein products that function within the context of a single metabolic pathway or for subunits of a multiprotein complex. Coordinated regulation is rationalized as a means whereby the organism can efficiently and effectively respond to changing environmental conditions or to the requirements of development. A recent and particularly popular means to characterize transcriptional regulation at the level of entire genomes involves the use of DNA microarrays (Pease et al., 1994; Schena et al., 1995), which serve to characterize, on a genomic scale, the relative abundance of mRNA transcripts within the cells or tissues of interest.

Details of the use of microarrays in gene-expression profiling can be found elsewhere in this book (see Chapter 4). Recent developments include the increasing use of long (50-70 mer), single-stranded, sense 5'-amino modified oligonucleotides as microarray "probes" (Phimister, 1999). The entire *Arabidopsis* genome, currently 29,388 annotated genes, of which 27,170 are predicted to be protein coding and 2,218 pseudogenes (*Arabidopsis* genome annotation, release 4.0, <ftp://ftp.tigr.org/pub/data/a_ thaliana/ath1/Arabidopsis_release.v4.0.README>), can be readily accommodated on a single microarray slide having elements spaced at 160 μm intervals. For long oligonucleotides, specificity is achieved by application of a set of empirical criteria for oligonucleotide design. For the *Arabidopsis* full-genome set, commercially available from Qiagen-Operon, these criteria include that the 70 mer should be within 1,000 bp of the 3'-end of the gene; have identity to the transcript of interest; have a common Tm; and lack sequence motifs confounding to hybridization, including fold-back stem-loops having stems longer than 9 bp and homopolymeric sequences >8 bp in length. Finally, the 70 mers should have <70 percent similarity to other genes and should lack identical sequences longer than 20 bp. These criteria correspond to empirical results (Xu et al., 2001) obtained using *Arabidopsis* P450 cDNAs as microarray probes and a single P450 cDNA as the hybridization target. It should be noted that absence of crosshybridization should not be restricted to comparisons within the *Arabidopsis* P450 superfamily since crosshybridization can occur from any transcript within the genome (Galbraith, 2003).

We have been supplying whole-genome *Arabidopsis* 70 mer microarrays to not-for-profit organizations since mid-2002 (http://www.ag.arizona.edu/microarray), and these comprise single, nonredundant, elements for each gene. Given the ready availability and reasonable cost of full-genome microarrays, the need for smaller, P450-specific microarrays appears less compelling. Although smaller microarrays can accommodate multiple element replications on a single slide surface, the advantage of this type of replication is obviated by the need for between-slide replication, for which the variance is much greater. Smaller arrays are also more difficult to normalize for expression levels, since global methods of normalization become increasingly skewed (and in unpredictable ways) when the number of independent array elements is reduced. An alternative to microarrays is Affymetrix GeneChips, in which the array elements are synthesized in situ, using photolithographic methods, and typically are <25 bp in length. Given that the hybridization performance of these array elements is close to the thermodynamic limit, specificity is achieved by defining "probe sets" for each gene, each comprising up to 20 array elements whose hybridization signals collectively are employed as a measure of abundance of the transcript of that gene. In general, GeneChips provides highly reproducible signals for abundant transcripts but has a lower technical reproducibility than

70-mers deposited microarrays for genes whose transcripts produce low-intensity signals (Galbraith, 2003).

A beginning has been made to providing a compilation of various changes in P450 transcript abundances within different tissues and in response to different treatments, primarily using amplicon-based microarrays and GeneChips (Xu et al., 2001; Schuler and Werck-Reichhart, 2003; Kristensen and Bak, unpublished data). Data of this type can be viewed as providing general clues as to potential P450 functions, for example, by consideration of the types of small molecules synthesized within tissues specifically expressing particular P450 genes, such as flowers. However, given that microarrays can rapidly report genome-wide changes in transcript abundances over different conditions of treatment, over different genotypes, and as a function of different tissue and cell types, a very powerful conceptual approach to discovery of unknown gene function has relied on the observation that the transcript abundances of genes within specific metabolic pathways can be coregulated (Eisen et al., 1998; Hughes et al., 2000; Ideker et al., 2001). Microarray-based identification of coregulation of groups of genes is most frequently achieved using algorithms to "cluster" gene-expression profiles (for a recent review of methods for clustering gene-expression profiles produced by microarrays, see Quackenbush, 2001). Using these methods, the identification of clusters containing genes of P450s of unknown function along with genes encoding protein components of specific metabolic pathways should provide a means to assign possible functions to these unknowns. These might then be validated more directly, such as by heterologous expression, followed by enzymatic assay using defined substrates, following heterologous expression of the P450 cDNA. For example, *CYP98A3* expression was found to cluster tightly with that of *CYP73A5* and of the gene encoding phenylalanine ammonia lyase (Xu, Galbraith, et al., unpublished data), consistent with its enzymatic role in phenylpropanoid metabolism (Schoch et al., 2001).

The use of expression profiling for assignment of function to a previously unknown P450 has not yet been reported. The resolution of expression profiling, and therefore the accuracy with which specific predictions of function might be made, depends on the number of different "cellular states" (Hughes et al., 2000) that are sampled, on the inherent experimental accuracy of the microarray platform, and on statistical considerations that are a direct function of the number of replications (biological and technical) performed. Statistical issues most likely have represented the limiting factor to this approach, given the unavailability, until recently, of whole-genome *Arabidopsis* microarrays of reasonable cost. A second limitation is a consequence of the occurrence, previously mentioned, of crosshybridization between similar gene sequences. This restricts the ultimate resolution of amplicon-based microarrays but can be greatly ameliorated through the use of oligonucleotide-based arrays (Galbraith, 2003). Given the availability of low-cost, whole-genome, long-oligonucleotide microarrays for *Arabidopsis,* accumulation of high-quality information concerning global gene expression should lead to the increasingly successful assignment of function to P450 family members.

Analysis of Overexpression

Transgenic overexpression of specific P450 gene family members is recognized as an important biotechnological goal, in that it can provide alterations in the levels of specific cellular components if the specificity of the P450 is known. To date, little progress has been made in employing analysis of overexpression of specific P450s for the assignment of function. As described in the next paragraph on purification/reconstitution and heterologous expression, the lack of an enzymatic assay for a given P450 makes it difficult to determine whether this P450 has been successfully overexpressed. Complicating the situation is the induction of gene silencing, which can unpredictably reduce or even eliminate transcript levels (Reintanz et al., 2001).

Biochemical Purification and Reconstitution of Activities

Biochemical purification of P450 proteins followed by reconstitution of activities in vitro provides the gold standard for biochemical analysis of the functions of P450s having defined catalytic activities. However, this approach has not been particularly useful in the assignment of catalytic activities to P450s of unknown function. It is not possible to monitor the successful purification and reconstitution of function of an enzyme in the absence of an assay to monitor progress. Availability and instability of potential substrates have also hampered this approach. Adaptation of purification and reconstitution methods to a high-throughput format might be possible, since P450s can be grouped according to their subcellular location, and since one might reasonably expect that experimental methods for purification and reconstitution could emerge that are universally applicable to P450s, or at least to specific families or clans. Thence, multiplexed assays using mixtures of potential substrates could allow identification of those molecules that can bind to the protein (producing diagnostic spectral changes) and/or be catalytically converted into products.

This approach does have some potential advantages over heterologous expression, an experimental approach that otherwise would appear superior. In particular, P450 proteins reconstituted in vitro would not be subject to cellular mechanisms of degradation in vivo. Nonetheless, this method appears both difficult to implement and to have a high risk of failure, not least since purification of large amounts of materials would be required.

Heterologous Expression of P450s

Since most P450 enzymes are membrane bound, and most for proper function require electrons transferred from the appropriate NADPH P450 reductase, heterologous cytochrome P450 expression requires the availability of a system within which the P450 can be synthesized and correctly folded and assembled, plus the availability of a P450 reductase that can transfer electrons to the P450. Eukaryotic heterologous expression systems are of two types, being based on yeast and insect cells (see, for examples, Bak et al., 2001; Mizutani et al., 1997). In the former case, galactose-inducible expression of P450 cDNAs is employed, using either an endogenous yeast NADPH P450 reductase or transgenic strains coexpressing *Arabidopsis* P450 reductases (Urban et al., 1997).

The success of this approach is evidently dependent upon production of sufficient P450 proteins capable of assembling into functional complexes within the heterologous system, as well as identification of a means for assaying the enzymatic activity of the P450 in question. In most cases, heterologous expression has provided a means to confirm the substrate-product conversion properties of the enzyme (see examples in Table 22.1). Occasionally, heterologous expression has allowed the screening of a limited number of potential compounds to identify true substrates. For example, heterologous expression in yeast was employed (Bak et al., 2001; Bak and Feyereisen, 2001) as a means to identify indole-3-acetaldoxime, *p*-hydroxyphenylacetaldoxime, and phenylacetaldoxime as substrates for CYP83A1 and CYP83B1.

To date, strategies have not been described for adapting these methods for high-throughput characterization of P450s for which there are no preexisting clues as to potential substrates and products. One future possibility could involve employing multiplexed NMR- or MS-based techniques for high-throughput analysis of substrate-product conversions using pooled mixtures of biochemicals.

Molecular Modeling

The availability of crystal structures for specific P450s leads to the possibility of employing modeling to establish structures for related P450 sequences, using molecular-threading programs that are continuously being improved (Forster, 2002). From these structures, it can become possible to predict catalytic functions based on the nature of the amino acids that provide the contact points for putative substrates. This type of approach has proved useful in elucidating the enzymatic functions of various gene families (see, for example, *O*-methyl transferases; Gang et al., 2002). One might expect that it should find increasing applicability to the P450 gene family, as more functions are discovered for individual members, whether within *Arabidopsis* or within other species and kingdoms. However, since substrate specificity in P450s is largely dictated by hypervariable regions located in loop structures, this restricts the accuracy with which the folding of such loops can be predicted, thereby also limiting predictions about substrate binding. This may therefore limit the value and utility of this type of approach.

CONCLUSIONS AND PERSPECTIVES

The completion and consequent availability of high-quality sequence data of additional plant genomes is anticipated for species including rice, lotus, and poplar. This will inevitably lead to a better understanding of the evolution of higher-plant P450s. A deeper understanding of P450 complexity in green algae and mosses will facilitate approaches using comparative genomics that, coupled to the understanding of the chemistries in these species, will provide valuable clues as to the evolution of P450s in higher plants. It is clear from the first draft of the *C. reinhardtii* genome that P450s within such families as CYP51, CYP97, CYP710, and CYP711 must encode for functions that are pivotal to all plant forms, not simply for vascular plants. It is particularly intriguing that, of those four families, only the function of CYP51 has been established. *In silico* homology searching of the *Physcomitrella patens* EST database (http://www.moss.leeds.ac.uk/blast.html) has revealed the presence of CYP73 orthologues (cinnamate-4-hydroxylase) in mosses, reflecting the emergence of phenylpropanoids in mosses. Orthologues of phenylalanine ammonia lyase as well as other non-P450 genes involved in phenylpropanoid biosynthesis can also be identified. In plants, the CYP97, CYP710, and CYP711 families appear to be distinct orthologues that generally have not expanded into paralogues, except perhaps for CYP97, and that their function is currently not understood.

The recent simplification of access to insertional mutagenesis lines provides a powerful tool to accelerate the elucidation of P450 functionality. It should be noted that *Arabidopsis* was initially thought to possess a simple genome lacking redundant gene copies. The increasing use of knockout lines has illustrated limitations due to a surprising degree of functional redundancies within P450s, which has the effect of masking the function of a given P450 within a knockout background. It is also important to note that pleiotropic effects may give rise to misleading phenotypes, such as those due to accumulation of intermediates that are further metabolized by, or act as inhibitors of, other pathways. Many examples of this have been seen for mutants identified within the glucosinolate pathway; e.g., the knockout of the *N*-hydroxylase for short-chain methionine derivatives CYP79F1 (Chen et al., 2003; Hansen, Wittstock, et al., 2001) results in a bushy phenotype (Reintanz et al., 2001), and the knockout of the oxime *N*-hydroxylase CYP83A1 results in a decrease in phenylpropanoid pathway-derived products, including sinapoylmalate, sinapoylcholine, and syringyl lignin, in addition to a decrease in aliphatic glucosinolates and an increase in indole glucosinolates (Hemm et al., 2003). A case studied in particular detail is that of CYP83B1, the indole-3-acetaldoxime *N*-hydroxylase involved in indole glucosinolate biosynthesis. Knockouts of CYP83B1 have effects relative to IAA homeostasis (Bak et al., 2001; Barlier et al., 2000; Delarue et al., 1998) and in relation to the regulation of the tryptophan pathway (Smolen and Bender, 2002), as well as their expected effects on indole glucosinolate metabolism (Bak et al., 2001).

In general, to understand fully the role of specific P450s clearly calls for multidisciplinary research, combining the use of genetic analysis and the various "-omics" (genomics, transcriptomics, proteomics, metabolomics, etc.) with biochemical analysis of recombinantly expressed proteins. Comparative genomics and cluster analysis of microarray data should provide hints for physiological roles and function. Heterologous expression followed by kinetic characterization, to confirm the identity of the substrate or substrates and the products, is pivotal for the elucidation of the true in vivo function. Finally, an emerging and important tool in substrate identification will lie in molecular modeling based on structure availabilities. Nevertheless, it appears that the specific task of assigning functions to the ~85 percent of the members of the *Arabidopsis* P450 multigene family for which functions are not yet known will remain challenging for the foreseeable future.

REFERENCES

Arabidopsis Genome Initiative (2000). Analysis of the genome sequence of the flowering plant *Arabidopsis thaliana*. *Nature* 408:796-815.

Bak, S., Feyereisen, R. (2001). The involvement of two P450 enzymes, CYP83B1 and CYP83A1, in auxin homeostasis and glucosinolate biosynthesis. *Plant Physiol.* 127: 108-118.

Bak, S., Tax, F.E., Feldmann, K.A., Galbraith, D.W., Feyereisen, R. (2001). CYP83B1, a cytochrome P450 at the metabolic branch point in auxin and indole glucosinolate biosynthesis in *Arabidopsis*. *Plant Cell* 13: 101-111.

Barlier, I., Kowalczyk, M., Marchant, A., Ljung, K., Bhalerao, R., Bennett, M., Sandberg, G., Bellini, C. (2000). The *SUR2* gene of *Arabidopsis thaliana* encodes the cytochrome P450 CYP83B1, a modulator of auxin homeostasis. *Proc. Natl. Acad. Sci. USA* 97: 14819-14824.

Bate, N.J., Sivasankar, S., Moxon, C., Riley, J.M., Thompson, J.E., Rothstein, S.J. (1998). Molecular characterization of an *Arabidopsis* gene encoding hydroperoxide lyase, a cytochrome P-450 that is wound inducible. *Plant Physiol.* 117: 1393-1400.

Benveniste, I., Tijet, N., Adas, F., Philipps, G., Salaun, J.P., Durst, F. (1998). *CYP86A1* from *Arabidopsis thaliana* encodes a cytochrome P450-dependent fatty acid ω-hydroxylase. *Biochem. Biophys. Res. Commun.* 243: 688-693.

Chapple, C. (1998). Molecular-genetic analysis of plant cytochrome P450-dependent monooxygenases. *Annu. Rev. Plant Physiol. Plant Mol. Biol.* 49: 311-343.

Chen, S., Glawischnig, E., Jorgensen, K., Naur, P., Jorgensen, B., Olsen, C.E., Hansen, C.H., Rasmussen, H., Pickett, J.A., Halkier, B.A. (2003). CYP79F1 and CYP79F2 have distinct functions in the biosynthesis of aliphatic glucosinolates in *Arabidopsis*. *Plant J.* 33: 923-937.

Choe, S.W., Dilkes, B.P., Fujioka, S., Takatsuto, S., Sakurai, A., Feldmann, K.A. (1998). The *DWF4* gene of *Arabidopsis* encodes a cytochrome P450 that mediates multiple 22 α-hydroxylation steps in brassinosteroid biosynthesis. *Plant Cell* 10: 231-243.

de Vetten, N., ter Horst, J., van Schaik, H.P., de Boer, A., Mol, J., Koes, R. (1999). A cytochrome b_5 is required for full activity of flavonoid 3',5'-hydroxylase, a cytochrome P450 involved in the formation of blue flower colors. *Proc. Natl. Acad. Sci. USA* 96: 778-783.

Debeljak, N., Fink, M., Rozman, D. (2003). Many facets of mammalian lanosterol 14 α-demethylase from the evolutionarily conserved cytochrome P450 family CYP51. *Arch. Biochem. Biophys.* 409: 159-171.

Debeljak, N., Horvat, S., Vouk, K., Lee, M., Rozman, D. (2000). Characterization of the mouse lanosterol 14α-demethylase (CYP51), a new member of the evolutionarily most conserved cytochrome P450 family. *Arch. Biochem. Biophys.* 379: 37-45.

Delarue, M., Prinsen, E., Onckelen, H.V., Caboche, M., Bellini, C. (1998). *Sur2* mutations of *Arabidopsis thaliana* define a new locus involved in the control of auxin homeostasis. *Plant J.* 14: 603-611.

Eisen, M.B., Spellman, P.T., Brown, P.O., Botstein, D. (1998). Cluster analysis and display of genome-wide expression patterns. *Proc. Natl. Acad. Sci. USA* 95: 14863-14868.

Feldmann, K.A., Choe, S., Kim, H., Park, J.H. (2002). Functional genomics of cytochromes P450 in plants. *Recent Adv. Phytochem.* 36: 223-248.

Forster, M.J. (2002). Molecular modelling in structural biology. *Micron* 33: 365-384.

Galbraith, D.W. (2003). Global analysis of cell type-specific gene expression. *Comp. Func. Genomics* 4: 208-215.

Gang, D.R., Lavid, N., Zubieta, C., Chen, F., Beuerle, T., Lewinsohn, E., Noel, J.P., Pichersky, E. (2002). Characterization of phenylpropene *O*-methyltransferases from sweet basil: Facile change of substrate specificity and convergent evolution within a plant *O*-methyltransferase family. *Plant Cell* 14: 505-519.

Hansen, C.H., Du, L., Naur, P., Olsen, C.E., Axelsen, K.B., Hick, A.J., Pickett, J.A., Halkier, B.A. (2001). CYP83B1 is the oxime-metabolizing enzyme in the glucosinolate pathway in *Arabidopsis*. *J. Biol. Chem.* 276: 24790-24796.

Hansen, C.H., Wittstock, U., Olsen, C.E., Hick, A.J., Pickett, J.A., Halkier, B.A. (2001). Cytochrome P450 CYP79F1 from *Arabidopsis* catalyzes the conversion of dihomomethionine and trihomomethionine to the corresponding aldoximes in the biosynthesis of aliphatic glucosinolates. *J. Biol. Chem.* 276: 11078-11085.

Helliwell, C.A., Chandler, P.M., Poole, A., Dennis, E.S., Peacock, W.J. (2001). The CYP88A cytochrome P450, ent-kaurenoic acid oxidase, catalyzes three steps of the gibberellin biosynthesis pathway. *Proc. Natl. Acad. Sci. USA* 98: 2065-2070.

Helliwell, C.A., Poole, A., Peacock, W.J., Dennis, E.S. (1999). *Arabidopsis* ent-kaurene oxidase catalyzes three steps of gibberellin biosynthesis. *Plant Physiol.* 119: 507-510.

Helliwell, C.A., Sheldon, C.C., Olive, M.R., Walker, A.R., Zeevaart, J.A.D., Peacock, W.J., Dennis, E.S. (1998). Cloning of the *Arabidopsis* ent-kaurene oxidase gene *GA3*. *Proc. Natl. Acad. Sci. USA* 95: 9019-9024.

Hemm, M.R., Ruegger, M.O., Chapple, C. (2003). The *Arabidopsis ref2* mutant is defective in the gene encoding CYP83A1 and shows both phenylpropanoid and glucosinolate phenotypes. *Plant Cell* 15: 179-194.

Howe, G.A., Schilmiller, A.L. (2002). Oxylipin metabolism in response to stress. *Current Opin. Plant Biol.* 5: 230-236.

Hughes, T.R., Marton, M.J., Jones, A.R., Roberts, C.J., Stoughton, R., Armour, C.D., Bennett, H.A., Coffey, E., Dai, H., He, Y.D., et al. (2000). Functional discovery via a compendium of expression profiles. *Cell* 102: 109-126.

Hull, A.K., Vij, R., Celenza, J.L. (2000). *Arabidopsis* cytochrome P450s that catalyze the first step of tryptophan-dependent indole-3-acetic acid biosynthesis. *Proc. Natl. Acad. Sci. USA* 97: 2379-2384.

Humphreys, J.M., Hemm, M.R., Chapple, C. (1999). New routes for lignin biosynthesis defined by biochemical characterization of recombinant ferulate 5-hydroxylase, a multifunctional cytochrome P450-dependent monooxygenase. *Proc. Natl. Acad. Sci. USA*. 96: 10045-10050.

Ideker, T., Thorsson, V., Ranish, J.A., Christmas, R., Buhler, J., Eng, J.K., Bumgarner, R., Goodlett, D.R., Aebersold, R., Hood, L. (2001). Integrated genomic and proteomic analyses of a systematically perturbed metabolic network. *Science* 292: 929-934.

Itoh, A., Howe, G.A. (2001). Molecular cloning of a divinyl ether synthase: Identification as a CYP74 cytochrome P-450. *J. Biol. Chem.* 276: 3620-3627.

Kahn, R., Durst, F. (2001). Function and evolution of plant cytochrome P450. *Recent. Adv. Phytochem.* 34: 151-189.

Kim, G.T., Tsukaya, H. (2002). Regulation of the biosynthesis of plant hormones by cytochrome P450s. *J. Plant Res.* 115: 169-177.

Kim, H.B., Park, J.H., Qi, C., Ma, J., Van Etten, H., Galbraith, D., Feldmann, K., Feyereisen, R. (2000). Isolation of *Arabidopsis* CYP51 mutants using systematic reverse genetic and molecular genetic analysis. In *Fifth International Symposium on P450 Biodiversity* (Copenhagen), C4.

Kusaba, M., Miyahara, K., Iida, S., Fukuoka, H., Takano, T., Sassa, H., Nishimura, M., Nishio, T. (2003). *Low glutelin content1:* A dominant mutation that suppresses the *Glutelin* multigene family via RNA silencing in rice. *Plant Cell.* 15: 1455-1466.

Kushiro, M., Nakano, T., Sato, K., Yamagishi, K., Asami, T., Nakano, A., Takatsuto, S., Fujioka, S., Ebizuka, Y., Yoshida, S. (2001). Obtusifoliol 14α-demethylase (CYP51) antisense *Arabidopsis* shows slow growth and long life. *Biochem. Biophys. Res. Commun.* 285: 98-104.

Laudert, D., Pfannschmidt, U., Lottspeich, F., Hollander-Czytko, H., Weiler, E.W. (1996). Cloning, molecular and functional characterization of *Arabidopsis thaliana* allene oxide synthase (CYP74), the first enzyme of the octadecanoid pathway to jasmonates. *Plant Mol. Biol.* 31: 323-335.

Li, X., Lassner, M., Zhang, Y. (2002). Deleteagene: A fast neutron deletion mutagenesis-based gene knockout system for plants. *Comp. Funct. Genom.* 3: 158-160.

Li, X., Song, Y., Century, K., Straight, S., Ronald, P., Dong, X., Lassner, M., Zhang, Y. (2001). A fast neutron deletion mutagenesis-based reverse genetics system for plants. *Plant J.* 27: 235-242.

Long, M., Rosenberg, C., Gilbert, W. (1995). Intron phase correlations and the evolution of the intron/exon structure of genes. *Proc. Natl. Acad. Sci. USA* 92: 12495-12499.

Lynch, M., Conery, J.S. (2000). The evolutionary fate and consequences of duplicated genes. *Science* 290: 1151-1155.

Mansuy, D. (1998). The great diversity of reactions catalyzed by cytochromes P450. *Comp. Biochem. Physiol. Part C* 121: 5-14.

Meyer, K., Cusumano, J.C., Somerville, C., Chapple, C.C.S. (1996). Ferulate-5-hydroxylase from *Arabidopsis thaliana* defines a new family of cytochrome P450-dependent monooxygenases. *Proc. Natl. Acad. Sci. USA* 93: 6869-6874.

Mikkelsen, M.D., Hansen, C.H., Wittstock, U., Halkier, B.A. (2000). Cytochrome P450 CYP79B2 from *Arabidopsis* catalyzes the conversion of tryptophan to indole-3-acetaldoxime, a precursor of indole glucosinolates and indole-3-acetic acid. *J. Biol. Chem.* 275: 33712-33717.

Mizutani, M., Ohta, D., Sato, R. (1997). Isolation of a cDNA and a genomic clone encoding cinnamate 4-hydroxylase from *Arabidopsis* and its expression manner *in planta*. *Plant Physiol.* 113: 755-763.

Neff, M.M., Nguyen, S.M., Malancharuvil, E.J., Fujioka, S., Noguchi, T., Seto, H., Tsubuki, M., Honda, T., Takatsuto, S., Yoshida, S., Chory, J. (1999). *BAS1:* A gene regulating brassinosteroid levels and light responsiveness in *Arabidopsis. Proc. Natl. Acad. Sci. USA* 96: 15316-15323.

Nelson, D. (1999). Cytochrome P450 and the individuality of species. *Arch. Biochem. Biophys.* 369: 1-10.

Nelson, D.R. (1998). Metazoan cytochromes P450 evolution. *Comp. Biochem. Physiol. C.* 121: 15-22

Omura, T., Sato, R. (1964). The carbon monoxide binding pigment of liver microsomes. I. Evidence for its hemoprotein nature. *J. Biol. Chem.* 239: 2370-2378.

Paquette, S.M., Bak, S., Feyereisen, R. (2000). Intron-exon organization and phylogeny in a large superfamily, the paralogous cytochrome P450 genes of *Arabidopsis thaliana. DNA Cell Biol.* 19: 307-317.

Paquette, S.M., Møller, B.L., Bak, S. (2003). On the origin of family 1 plant glycosyltransferases. *Phytochemistry* 62: 399-413.

Park, J.H., Halitschke, R., Kim, H.B., Baldwin, I.T., Feldmann, K.A., Feyereisen, R. (2002). A knock-out mutation in allene oxide synthase results in male sterility and defective wound signal transduction in *Arabidopsis* due to a block in jasmonic acid biosynthesis. *Plant J.* 31: 1-12.

Pease, A.C., Solas, D., Sullivan, E.J., Cronin, M.T., Holmes, C.P., Fodor, S.P. (1994). Light-generated oligonucleotide arrays for rapid DNA sequence analysis. *Proc. Natl. Acad. Sci. USA* 91: 5022-5026.

Phimister, B. (1999). Going global. *Nat. Genet.* 21S: 1.

Quackenbush, J. (2001). Computational analysis of microarray data. *Nat. Rev. Genet.* 2: 418-427.

Reintanz, B., Lehnen, M., Reichelt, M., Gershenzon, J., Kowalczyk, M., Sandberg, G., Godde, M., Uhl, R., Palme, K. (2001). *Bus,* a bushy *Arabidopsis* CYP79F1 knockout mutant with abolished synthesis of short-chain aliphatic glucosinolates. *Plant Cell* 13: 351-367.

Rodman, J.E., Karol, K.G., Price, R.A., Sytsma, K.J. (1996). Molecules, morphology, and Dahlgren's expanded order Capparales. *System. Bot.* 21: 289-307.

Ruegger, M., Meyer, K., Cusumano, J.C., Chapple, C. (1999). Regulation of ferulate-5-hydroxylase expression in *Arabidopsis* in the context of sinapate ester biosynthesis. *Plant Physiol.* 119: 101-110.

Schena, M., Shalon, D., Davis, R.W., Brown, P.O. (1995). Quantitative monitoring of gene expression patterns with a complementary DNA microarray. *Science* 270: 467-470.

Schlichting, I., Berendzen, J., Chu, K., Stock, A.M., Maves, S.A., Benson, D.E., Sweet, R.M., Ringe, D., Petsko, G.A., Sligar, S.G. (2000). The catalytic pathway of cytochrome P450cam at atomic resolution. *Science* 287: 1615-1622.

Schoch, G., Goepfert, S., Morant, M., Hehn, A., Meyer, D., Ullmann, P., Werck-Reichhart, D. (2001). CYP98A3 from *Arabidopsis thaliana* is a 3'-hydroxylase of phenolic esters, a missing link in the phenylpropanoid pathway. *J. Biol. Chem.* 276: 36566-36574.

Schoenbohm, C., Martens, S., Eder, C., Forkmann, G., Weisshaar, B. (2000). Identification of the *Arabidopsis thaliana* flavonoid 3'-hydroxylase gene and functional expression of the encoded P450 enzyme. *Biol. Chem.* 381: 749-753.

Schuler, M.A., Werck-Reichhart, D. (2003). Functional genomics of P450s. *Ann. Rev. Plant Biol.* 54: 629-667.

Shimada, Y., Fujioka, S., Miyauchi, N., Kushiro, M., Takatsuto, S., Nomura, T., Yokota, T., Kamiya, Y., Bishop, G.J., Yoshida, S. (2001). Brassinosteroid-6-oxidases from *Arabidopsis* and tomato catalyze multiple C-6 oxidations in brassinosteroid biosynthesis. *Plant Physiol.* 126: 770-779.

Smolen, G., Bender, J. (2002). *Arabidopsis* cytochrome P450 CYP83B1 mutations activate the tryptophan biosynthetic pathway. *Genetics* 160: 323-332.

Stoltzfus, A., Logsdon, J.M. Jr., Palmer, J.D., Doolittle, W.F. (1997). Intron "sliding" and the diversity of intron positions. *Proc. Natl. Acad. Sci. USA* 94: 10739-10744.

Urban, P., Mignotte, C., Kazmaier, M., Delorme, F., Pompon, D. (1997). Cloning, yeast expression, and characterization of the coupling of two distantly related *Arabidopsis thaliana* NADPH-cytochrome P450 reductases with P450 CYP73A5. *J. Biol. Chem.* 272: 19176-19186.

Walsh, J.B. (1987). Sequence-dependent gene conversion: Can duplicated genes diverge fast enough to escape conversion? *Genetics* 117: 543-557.

Walsh, J.B. (1995). How often do duplicated genes evolve new functions? *Genetics* 139: 421-428.

Wang, E., Wagner, G.J. (2003). Elucidation of the functions of genes central to diterpene metabolism in tobacco trichomes using posttranscriptional gene silencing. *Planta* 216: 686-691.

Waterhouse, P.M., Graham, M.W., Wang, M.B. (1998). Virus resistance and gene silencing in plants can be induced by simultaneous expression of sense and antisense RNA. *Proc. Natl. Acad. Sci. USA* 95: 13959-13964.

Waterhouse, P.M., Helliwell, C.A. (2003). Exploring plant genomes by RNA-induced gene silencing. *Nat. Rev. Genet.* 4: 29-38.

Wellesen, K., Durst, F., Pinot, F., Benveniste, I., Nettesheim, K., Wisman, E., Steiner-Lange, S., Saedler, H., Yephremov, A. (2001). Functional analysis of the *LACERATA* gene of *Arabidopsis* provides evidence for different roles of fatty acid ω-hydroxylation in development. *Proc. Natl. Acad. Sci. USA* 98: 9694-9699.

Werck-Reichhart, D., Bak, S., Paquette, S. (2002). Cytochromes P450. In Somerville, C.R., Meyerowitz, E.M., (Eds.), *The Arabidopsis Book.* Rockville, MD: American Society of Plant Biologists. Available online at <http://www.aspb.org/downloads/arabidopsis/werckfinal.pdf>.

Werck-Reichhart, D., Feyereisen, R. (2000). Cytochromes P450: A success story. *Genome Biology* 1: reviews3003.1-3003.9.

Wesley, S.V., Helliwell, C.A., Smith, N.A., Wang, M.B., Rouse, D.T., Liu, Q., Gooding, P.S., Singh, S.P., Abbott, D., Stoutjesdijk, P.A., et al. (2001). Construct design for efficient, effective and high-throughput gene silencing in plants. *Plant J.* 27: 581-590.

Willis, H.J., McElwain, J.C. (2002). *The Evolution of Plants.* Oxford: Oxford University Press.

Winkler, R.G., Frank, M.R., Galbraith, D.W., Feyereisen, R., Feldmann, K.A. (1998). Systematic reverse genetics of Transfer-DNA-Tagged lines of *Arabi-*

dopsis: Isolation of mutations in the cytochrome P450 gene superfamily. *Plant Physiology* 118: 743-750.

Wittstock, U., Halkier, B.A. (2000). Cytochrome P450 CYP79A2 from *Arabidopsis thaliana* L. catalyzes the conversion of L-phenylalanine to phenylacetaldoxime in the biosynthesis of benzylglucosinolate. *J. Biol. Chem.* 275: 14659-14666.

Wittstock, U., Halkier, B.A. (2002). Glucosinolate research in the *Arabidopsis* era. *Trends. Plant. Sci.* 7: 263-270.

Xu, W., Bak, S., Decker, A., Paquette, S.M., Feyereisen, R., Galbraith, D.W. (2001). Microarray-based analysis of gene expression in very large gene families: The cytochrome P450 gene superfamily of *Arabidopsis thaliana*. *Gene* 272: 61-74.

Zhou, N., Tootle, T.L., Glazebrook, J. (1999). Arabidopsis *PAD3*, a gene required for camalexin biosynthesis, encodes a putative cytochrome P450 monooxygenase. *Plant Cell* 11: 2419-2428.

Chapter 23

Functional Genomics of Protein Phosphorylation in *Arabidopsis thaliana*

Claudia Hass
Stefan Weinl
Klaus Harter
Jörg Kudla

INTRODUCTION

Reversible protein phosphorylation plays a cardinal role in regulating and coordinating important aspects of metabolism, gene expression, cell growth, and cell division. In particular, protein phosphorylation is a major currency of signal transduction pathways. Processes that are reversibly controlled by protein phosphorylation require not only a protein kinase (PK) but also the function of an antagonistic protein phosphatase (PP). Therefore, to ensure cellular regulation in an orderly manner, the switching on and off of protein kinases and phosphatases is as crucial for their function as their inherent catalytic activity.

Since the discovery of protein phosphorylation-dephosphorylation as a possible regulatory device in the mid-1950s, there has been a steady expansion of the field. Current conservative estimations predict about one-third of all known eukaryotic proteins to be phosphorylated, underlining the general importance of this protein modification. Purification of the first mammalian protein kinase was reported in 1959 (phosphorylase kinase; Krebs et al., 1959), followed in 1968 by another serine-threonine protein kinase, the cAMP-dependent protein kinase (Walsh et al., 1968). Still it was not before 1989 that the first cDNA sequence of a higher-plant protein kinase was published (Lawton et al., 1989). Since then, the number of identified plant protein kinases has exploded, but so far this development has not been paralleled by a similar successful analysis of kinase gene function.

Nevertheless, the complete sequencing of several eukaryotic genomes provides the first opportunity to ascertain the full complement of kinases and phosphatases encoded in the genomes of these organisms and enables comparative evolutionary studies of protein phosphorylation. Since each of the analyzed species harbors at least several hundred different protein kinases, their genes generally represent a significant proportion of eukaryotic genomes, while protein phosphatases are usually less numerous. Accordingly, the completely sequenced genome of the model plant *Arabidopsis thaliana* encodes 1,085 typical protein kinases (Hrabak et al., 2003) in comparison to 112 phosphatase catalytic subunit sequences that have been identified (Kerk et al., 2002).

The comparison of the amount and types of protein kinases and phosphatases prevalent in different species reveals some rather specific aspects in the evolution of protein phosphorylation in plants. The more than 1,000 kinase genes identified in *Arabidopsis* encompass about 4 percent of the predicted 27,300 genes (Wortman et al., 2003), which is about twice the proportion observed in *Saccharomyces cerevisiae* (Hunter and Plowmann, 1997) and *Caenorhabditis elegans* (Plowmann et al., 1999). Moreover, there are major differences in the types of kinases found in animals and plants. For example, receptor kinases in animals predominantly phosphorylate tyrosine residues, whereas in plants receptor kinases phosphorylate serine and threonine residues. Histidine kinases have been identified in bacteria, fungi, and plants, but not in animals. In addition, two major types of kinases that decode calcium signals in animals (calmodulin-dependent protein kinases [CaMKs] and protein kinase C [PKC]) appear to be underrepresented or missing in plants. Similarly, the calcium/calmodulin-dependent protein phosphatase calcineurin (protein phosphatase type 2B), which regulates a multitude of cellular processes in fungi and animals, has not yet been identified in plants.

Conversely, plants contain a number of kinase families that either are not found in animals and fungi or are highly divergent. Prominent examples are the calcium-dependent protein kinases (CDPKs), which occur in plants, algae, and certain protozoa (Hrabak, 2000), and the calcium decoding CBL-interacting protein kinases (CIPKs), which so far have been identified in only mosses and higher plants (Luan et al., 2002).

Since it is impossible to present comprehensively the genomics of more than 1,000 phosphorylating or dephosphorylating proteins, this chapter will exemplarily discuss the plant histidine kinases and their relationship to two-component signaling as a model to demonstrate our current knowledge and future challenges in the field of genomics of plant phosphorylation. In addition, a separate section of this chapter will briefly summarize the current knowledge about the distribution of specific types of kinases and phosphatases in the model plant *Arabidopsis* and will refer to recent publications for more specific information.

PROTEIN KINASES AND PHOSPHATASES ENCODED IN THE ARABIDOPSIS *GENOME*

Receptor-Like Protein Kinases (RLKs)

A general feature of all kinase types in *Arabidopsis* is their occurrence in gene families. Receptor-like kinases (RLKs) constitute about 60 percent of all kinases in this plant and account for nearly all transmembrane kinases in this species. A recent study identified 610 members of this large gene family in *Arabidopsis,* which represent a monophyletic group structurally defined by a predicted signal sequence, a single transmembrane domain, and a cytoplasmic kinase domain (Shiu and Bleeker, 2001b). Corresponding to this huge number, plant RLKs function in many physiological processes as diverse as self-pollen rejection, stem-cell maintenance and differentiation of the shoot meristem, the response to brassinosteroid hormones, and the innate response to bacterial pathogens (Shiu and Bleecker, 2001a; Tichtinsky et al., 2003). Plant RLKs can be distinguished from animal receptor tyrosine kinases (RTKs) by the finding that all plant RLKs examined to date show serine/threonine kinase specificity, whereas animal receptor kinases, with the exception of transforming growth factor-β (TGF-β, receptor serine/threonine kinases [RSKs]) receptors, are tyrosine kinases. Nevertheless, in an extended analysis, animal RSKs, Raf kinases (ten members in *Arabidopsis*), plant RLKs, and animal receptor tyrosine kinases form a (well supported by high bootstrap values) group sharing a common origin (Shiu and Bleecker, 2001b). Among animal kinase sequences, *Drosophila* Pelle and related cytoplasmic kinases fall within the plant RLK clade. According to this analysis, an early gene duplication event led to the founding of two lineages that diversified into the RTK and Raf families, on one hand, and the RLK/Pelle family, on the other. This diversification seems to have occurred before the divergence of plants and animals. The small number of representatives of the RLK/Pelle kinase family in animals compared to the much larger number in *Arabidopsis* could indicate that the immense expansion of the plant RLKs occurred after the divergence of the plant and animal lineages. Whether this plant-specific gene amplification reflects a function of RLKs in plant-specific physiological processes remains to be established.

Of the 610 *Arabidopsis* RLKs, 193 do not have an obvious receptor configuration, as determined by the absence of the putative signal sequences and/or transmembrane regions. According to the structure of their extracellular domains, the remaining 417 kinases with receptor configuration can be classified into 21 classes, which vary greatly in size. Among these classes, the leucine-rich repeat (LRR)-containing RLKs represent the largest group in *Arabidopsis* with 216 genes. Genomic analyses uncovered that tandem duplications and segmental whole-genome duplications represent two of the major mechanisms for the enormous expansion of the RLK gene family in *Arabidopsis* (Shiu and Bleecker, 2002). More than 33 percent of the RLK genes are found in tandem clusters. In addition, 470 of the RLKs are located within segmentally duplicated regions in the *Arabidopsis* genome, and 268 of them have close relatives in the corresponding region. This situation could significantly interfere with the further functional analyses of these genes by reverse genetics approaches, since the duplicated genes still might exhibit some degree of functional overlap, and, in addition, the tandemly duplicated genes will be hardly amenable for double-mutant generation by T-DNA insertion mutagenesis.

Kinases of the MAPK Protein Phosphorylation Cascade

As RLKs, mitogen-activated protein (MAP) kinases link stimuli that are activated by external sensors to cellular responses. This modular signaling system, consisting of MAPKK kinases (MAPKKKs), MAPK kinases (MAPKKs), and MAP kinases (MAPKs), is conserved in all eukaryotes analyzed so far. Again, though, plants appear to harbor an unusually high number of these signaling components. Whereas yeast has six MAPKs and 13 are found in mammals, 20 MAPKs have been identified in *Arabidopsis* (Ligterink and Hirt, 2000). In all systems, mitogen-activated protein kinase signaling pathways fulfill crucial functions in regulating cell growth and death, the cell cycle, and stress responses (Jonak et al., 2002; Innes, 2001).

MAPKKKs are serine/threonine kinases and activate MAPKKs through phosphorylation of two serine-threonine residues in a conserved S/T-X_{3-5}-S/T motif. In contrast, MAPKKs are dual-specific kinases that phosphorylate MAPKs on threonine and tyrosine residues in the T-X-Y motif. MAPKs are promiscuous serine-threonine kinases that phosphorylate a multitude of substrates, including transcription factors, protein kinases, and cytoskeletal proteins (Garrington and Johnson, 1999; Tena et al., 2001). The phylogenetic analysis of potential MAPK cascades has revealed a bewildering complexity. In an analysis of the *Arabidopsis* genome, 20 MAPKs, 10 MAPKKs, and 60 MAPKKKs were identified (Jonak et al., 2002). These families can be further grouped into distinct subfamilies. The rather low number of MAPKKs as compared to MAPKKKs may suggest that the MAPKK signaling modules represent information bottlenecks or convergence points in signaling cascades, which might be amenable for reverse genetics approaches. On the other hand, the large number of MAPKs argues in favor of alternative experimental approaches to unravel MAPK functions. The recent identification of the first MAPK signaling module for a bacterial elicitor, by a combination of transient expression studies in protoplasts and biochemical techniques, might point to a promising approach to unraveling MAPK functions (Asai et al., 2002). Contrary to the situation in other eukaryotes, these studies revealed that a given plant MAPKK can interact with and activate more than one MAPK. This aspect will have to be considered in future analyses of MAPKK and MAPK functions.

SNF-Like Kinases and cAMP-Dependent Kinases

Outside of the plant kingdom, sucrose nonfermenting (SNF)-like kinases and cAMP-dependent kinases usually form a rather small family of closely related serine/threonine-specific protein kinases. In accordance with the generally occurring expansion of gene families in plants, this SNF-like protein kinase group encompasses a large multigene family in plants. Detailed phylogenetic analyses revealed that besides the 38 SnRKs (sucrose nonfermenting related kinases), which are divided into three subgroups, the *Arabidopsis* genome also encodes 34 phylogenetic-related CDPKs (calcium-dependent protein kinase), eight CRKs (CDPK-related kinase), two PPCKs (phosphoenolpyruvate [PEP] carboxylase kinase), and two PEPRKs (PEP carboxylase kinase-related kinase). These kinases are structurally related but functionally are clearly diverged (Hrabak et al., 2003).

The SnRK1 subgroup of these kinases is most closely related to SNF1 from yeast and AMP-activated kinase from animals. Only three *Arabidopsis* sequences fall into this subgroup. Studies on the biochemistry and physiology of SnRK1s from several plant species have suggested that, similar to SNF1 and AMPK, they may regulate metabolism in response to nutritional or environmental stresses (Hallford et al., 2000, 2003). Not unexpectedly, SnRK1s interact with plant orthologues of proteins known to regulate yeast SNF1. The SnRK2 group appears to be unique in plants. There are 10 SnRK2 genes in plants. Recent analyses revealed insights into the function of one of these genes. Mutants of SnRK2.6 are affected in stomatal closure in response to absisic acid (Mustilli et al., 2002; Yoshida et al., 2002).

Twenty-five *Arabidopsis* kinases form the structurally defined SnRK3 subgroup. These kinases share a two-domain structure being composed of a conserved N-terminal kinase domain and a more variable C-terminal domain. Functionally, these kinases are clearly distinct from SNF kinases, since they represent central mediators of calcium signaling in plants (Shi et al., 1999). Interaction studies as well as functional analyses have established that these kinases are regulated by differential interaction with a group of ten calcineurin B-like (CBL) calcium sensor proteins (Luan et al., 2002). A conserved 21aa protein-protein interaction module, designated as NAF-domain, represents the calcium sensor-kinase interaction interface, which defines this group of proteins as components of calcium-mediated signaling in plants (Albrecht et al., 2001). Elegant forward genetics studies have identified one of these calcium sensors (SOS3/CBL4) and one specifically interacting kinase (SOS2/CIPK24) as major components of the plant salt signaling and adaptation circuit (Liu and Zhu, 1998; Halfter et al., 2000). Moreover, recent reverse genetics analyses established a function of several CBLs and CIPKs in plant stress signaling. For example, knockout mutant analysis of CIPK3 indicated a role of this kinase in regulating absisic acid and cold signal transduction in *Arabidopsis* (Kim et al., 2003). In addition, studies of two CBL1 loss-of-function alleles uncovered that this calcium sensor protein differentially mediates plant signaling and adaptation in response to cold, drought, salt, and osmotic environmental stresses (Cheong et al., 2003; Albrecht et al.). Taken together, these findings suggest that the CBL-CIPK network represents an emerging important signaling system mediating calcium-regulated processes in plants.

A long-established, plant-specific signaling system functioning in calcium responses is represented by the plant calcium-dependent protein kinases (CDPKs), which harbor a calcium-binding domain as well as a catalytic kinase domain (Sanders et al., 1999, 2002). Functional analyses of plant CDPKs has recently provided strong evidence for a crucial function of these kinases in processes such as hormone and stress signaling as well as plant pathogen response (Harmon et al., 2000; Romeis et al., 2001). The *Arabidopsis* genome is predicted to encode 34 different CDPKs. This family of calcium-decoding kinases was identified in the 1980s (Harmon et al., 1987). Nevertheless, there has not been a report on the function of one of these proteins based on reverse genetics. In any case, comprehensive biochemical and physiological studies underline the general importance of this family of kinases in plants (Harmon et al., 2000; Cheng et al., 2002).

Protein Phosphatases

In contrast to the analysis of plant phosphorylating protein activities, much less is known about the genomics and function of plant phosphatases. That these proteins are functionally as important as their kinase counterparts has become evident from the identification of such important signal transductions mutants as, for example, *abscisic acid insensitive (abi)* (Leung et al., 1994; Meyer et al., 1994). Moreover, recent studies place protein phosphatases in various signaling cascades, including pathogen and stress responses as well as developmental processes (Luan, 1998, 2003). The comprehensive analysis of the *Arabidopsis* genome identified 112 phosphatase catalytic subunit sequences (Kerk, 2002). Exploration of this largely unexplored set of plant signaling components will be as important and as exciting as the analyses of the function of plant kinases.

HISTIDINE KINASES AND THE TWO-COMPONENT SIGNALING SYSTEM

The two-component system represents an additional phosphorylation cascade present in prokaryotes, lower eukaryotes, and plants, but not in animals. In its simple architecture, this cascade consists of a histidine kinase that monitors an external stimulus and transmits the information to a second element, a response regulator (Figure 23.1A). In *Arabidopsis,* several genes code for two-component histidine kinases and histidine kinase-like proteins

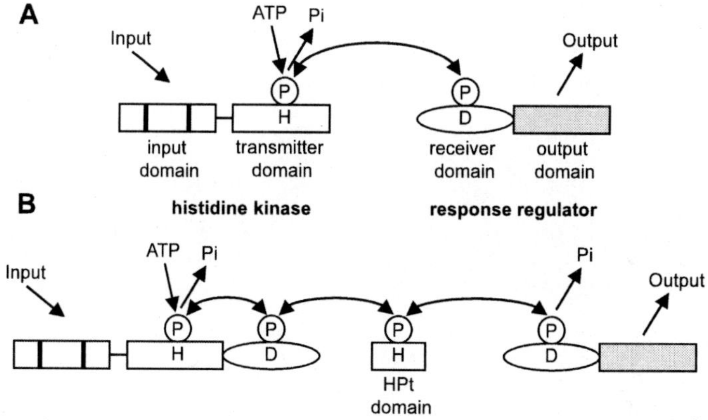

FIGURE 23.1. Schematic representation and basic features of the simple (A) and the multistep (B) two-component signaling systems. Signal perception by the input domain of the histidine kinase induces autophosphorylation of the transmitter domain at a conserved histidine residues (H). The phosphoryl group (Pi) is relayed to a conserved Asp residue (D) that is localized either in the receiver domain of the cognate response regulator (A) or, in the case of hybrid histidine kinases, in an attached receiver domain (B). In the multistep, two-component signaling system, histidine-containing phosphotransfer (HPt) domain proteins serve as phospho-histidine intermediates between the hybrid histidine kinase and the response regulator. The phosphotransfer can be bidirectional. Black vertical bars indicate transmembrane segments. (*Source:* Adapted from Kakimoto, 2003; for further details, see text.)

(Table 23.1; Figure 23.2). Five of them function as ethylene receptors (ETR1, ETR2, EIN4, ERS1, ERS2; Wang et al., 2002) and three as cytokinin receptors (AHK2, AHK3, AHK4/WOL/CRE1; Kakimoto, 2003). CKI1 has recently been shown to be involved in female gametophyte development. The function of AHK5/CKI2 is not clear yet and awaits further analysis (Hwang et al., 2002). The identification of the putative osmosensor histidine kinase AtHK1/AHK1 (Urao et al., 1999) further supports the importance of two-component proteins in plant signal transduction.

The higher-plant red and far-red light-absorbing photoreceptors (phytochromes) have bacterial ancestors that function as light-regulated histidine kinases (see Chapter 7). However, plant phytochromes have significantly

TABLE 23.1. Putative histidine protein kinases and related proteins in *Arabidopsis*.

Name[a]	(AGI) ID[b]	Expression[c]	AA[d]	Function
CKI1	At2g47430	Yes*	1,122	Female gametophyte development
AHK5 (CKI2)	At5g10720	1 EST	950	Unknown
AtHK1 (AHK1)	At2g17820	Yes	1,190	Osmosensor?
AHK2	At5g35750	1 EST	1,173	Cytokinin receptor
AHK3	At1g27320	3 ESTs	1,176	Cytokinin receptor
AHK4 (CRE1/WOL)	At2g01830	11 ESTs	1,057	Cytokinin receptor
ETR1	At1g66340	Yes	738	Ethylene receptor
ERS1	At2g40940	6 ESTs	613	Ethylene receptor
ETR2	At3g23150	2 ESTs	773	Ethylene receptor
EIN4	At3g04580	4 ESTs	766	Ethylene receptor
ERS2	At1g04310	5 ESTs	645	Ethylene receptor

[a]Alternative names are given in parentheses.
[b]Systematic names given to genes by Munich Information Center for Protein Sequence (MIPS) (http://mips.gsf.de/proj/thal/) from AGI sequencing project.
[c]Gene expression based on cloned cDNA or the presence of EST. The number of ESTs in the MIPS database is indicated. An asterisk indicates that there is no EST in the database.
[d]The number of predicted amino acids in the proteins.

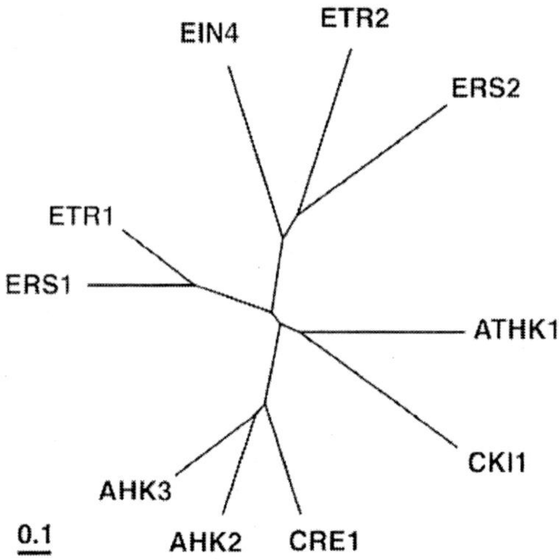

FIGURE 23.2. Phylogenetic relationship of histidine kinases in *Arabidopsis thaliana*. The CLUSTAL W program was used to align the amino acid sequences of the transmitter domain of histidine kinases. The phylogenetic tree was drawn with the TreeView program. (*Source:* Adapted from Urao et al., 2001.)

diverged from bacteriophytochromes and possess no histidine kinase activity but Ser/Thr kinase activity (Yeh and Lagarias, 1998). Therefore, they will be not further discussed here.

Two-Component Signal Transduction

The two-component signaling system serves as a sensing/responding mechanism. In reaction to exogenous or endogenous stimuli, autophosphorylation of the histidine kinase at a conserved histidine residue within the catalytic core of its transmitter domain is induced and, thereby, signaling is initiated. In single-step two-component systems, as is predominantly found in prokaryotes, the phosphoryl group is then transferred to an invariant Asp residue within the receiver domain of the response regulator that results in modulation of its activity (Figure 23.1A). Especially in eukaryotes such as yeast and plants, a more complex type of two-component signaling systems is found, the multistep system, with additional phosphorylation steps (Figure 23.1B). In the multistep two-component system, the histidine kinase is called hybrid histidine kinase and carries an additional receiver domain at the COOH-terminal end. Instead of transferring the phosphoryl group directly to the receiver of a response regulator, it is first transferred to a conserved Asp residue of its receiver domain (Figure 23.1B). Additional proteins with a histidine-containing phosphotransmitter (HPt) domain perceive the phosphoryl group from the receiver domain of the hybrid histidine kinase and transfer it to the receiver of the response regulator (Figure 23.1B). Whether modified via simple or multistep two-component systems, phosphorylation of the response regulator induces a conformational change of its output domain. Because response regulators of eukaryotes act either as transcription factors or modulators of specific target proteins (Hwang et al., 2002; Lohrmann and Harter, 2002; Kakimoto, 2003), this conformational alteration causes a response regulator to become active or, in the case of negative regulation, inactive.

The chemistry of multistep two-component signal transduction is a His-to-Asp phosphorelay and incorporates five phosphoryl transfer reactions and four phosphoprotein intermediates mechanistically linked by phosphoryl transfer events (Figure 23.1B). Though histidine kinases function catalytically similar to Ser/Thr/Tyr protein kinases, the chemistry of His phosphorylation is different. In contrast to Ser/Thr/Tyr kinases, which generate phosphoesters, histidine kinases produce His phosphoramidates. Due to a significantly higher negative free energy of phosphoramidates compared to phosphoesters, the equilibrium for the autophosphorylation reaction prefers the unphosphorylated histidine kinase. Thus, only a minor fraction of the histidine kinase population exists in a phosphorylated state even after application of a phosphorylation-inducing stimulus (Stock et al., 2000).

Phosphorylation of Asp residues produces a high-energy acyl phosphate. The high energy within the acyl phosphate is suspected to drive the long-distance conformational changes observed in phosphorylated receiver domains of hybrid histidine kinases and response regulators. This is in contrast to phospho-Ser/Thr, which alters protein activity mainly by local electrostatic effects. Taken together, it is the flux of phosphoryl groups, rather than stoichiometric phosphorylation, that is relevant for the function of histidine kinases and two-component signaling systems, respectively (Stock et al., 2000).

Sequence analysis of the entire *Arabidopsis* genome has revealed that, in addition to the mentioned 11 histidine kinases (AHKs), there are 6 HPt proteins (AHPs) and 23 response regulators (ARRs), indicating a high complexity and great importance of two-component systems in plant signal transduction (Hwang et al., 2002).

Structure-Function Relationship of Histidine Kinases

In contrast to prokaryotes, histidine kinases without an N-terminal receiver domain are rarely found in eukaryotes, one known exception being not a hybrid histidine kinase is the ethylene receptor ERS1 (Hua et al., 1995). The molecular masses of plant hybrid histidine kinases range from 68 to 136 kDa, and with the exception of AHK5/CKI2, they are thought to be integral membrane proteins (Hwang et al., 2002; Lohrmann and Harter, 2002; Kakimoto, 2003). Similar to their prokaryotic relatives, plant histidine kinases consist of several functional domains for sensing, catalyzing, and auxiliary tasks.

Catalytic Domain (Kinase Core)

As mentioned previously, the activity of a histidine kinase is induced by a specific input signal initiating autophosphorylation. As far as is known from prokaryotic histidine kinases, autophosphorylation is a bimolecular reaction between homodimers, in which one histidine kinase monomer catalyzes the phosphorylation of the conserved histidine residue in the second monomer. The responsible kinase core, composed of a dimerization domain and an ATP/ADP-binding phosphotransfer domain, consists of the five structurally and functionally unique motifs H, N, G1, F, and G2 and is about 350 amino acids in length. The H box contains the conserved His and participates in dimerization, while the contiguously arranged N, G1, F, and G2 boxes form the nucleotide-binding pocket (Figure 23.3). The three-dimensional structure of the catalytic domain of histidine kinases differs strongly from that of known Ser/Thr/Tyr kinases. This is indicative of a mechanism of histidine kinase action that is different from that of prokaryotic and eukaryotic Ser/Thr/Tyr kinases (Stock et al., 2000).

Sensing Domain

The N-terminal sensing domains of the different plant hybrid histidine kinase subfamilies do not show high sequence similarity. This observation suggests that they have evolved for interaction with specific ligands. For some plant histidine kinases, as for instance the ethylene and cytokinin receptors, the ligand is known. However, the exact biochemical mechanism of sensing and the mode of transmission of the signal to the intracellular kinase domain still have to be clarified.

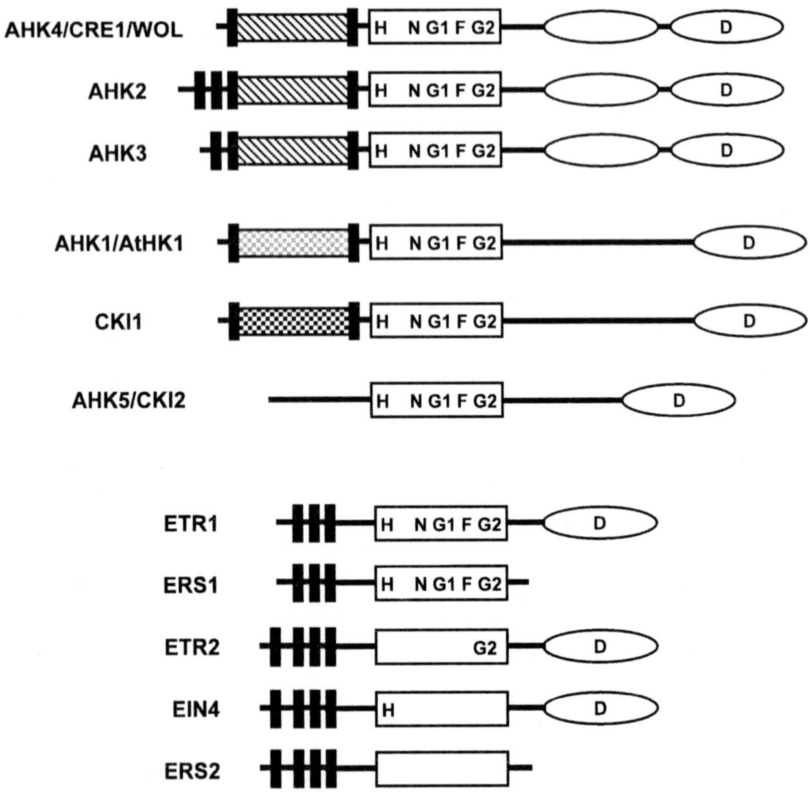

FIGURE 23.3. Structural characteristics of histidine kinases and histidine kinase-related proteins from *Arabidopsis thaliana*. Black rectangles indicate the transmembrane segments; striped rectangles, the extracellular CHASE domains; checker boxes, the extracellular domains between the transmembrane segments of CKI1 and AtHK1; open rectangle, the transmitter domain; H, N, G1, F, G2, the conserved boxes common to the catalytic kinase core of functional histidine kinases; open ovals, the receiver domains with the conserved phosphorylatable Asp (D) residue. (*Source:* Adapted from Kakimoto, 2003.)

Linker Domain

In plant transmembrane histidine kinases, the sensing domain is linked to the cytoplasmic kinase core through a linker (Singh et al., 1998). Although there is no information for plant histidine kinases, these linker regions appear to be essential for proper signal transduction in prokaryotic histidine kinases (Aravind and Ponting, 1999; Stock et al., 2000). Interestingly, the linker regions of the ethylene receptors contain a GAF-related protein-protein interaction domain (Hwang et al., 2002; Wang et al., 2002). However, the functional role of this domain in ethylene perception or signaling is not known.

Receiver Domain

The receiver domain is a regulatory part of either a hybrid histidine kinase or a response regulator and catalyzes the transfer of a phosphoryl group from phosphorylated His to a conserved Asp residue (Figure 23.1 and Figure 23.3). The modification of the phosphorylable Asp residue occurs within the acid pocket of the receiver domain, which is assembled by additional conserved acidic amino acids, and requires Mg^{2+} ions (Stock et al., 2000). Phosphorylation of the Asp residue promotes conformational changes within the receiver domain that facilitate a distinct set of inter- or intramolecular interactions, such as, for instance, dimerization. Structural analysis of the receiver domain of ETR1 suggests that its dimerization depends on phosphorylation. This could affect the interaction of ETR1 with downstream signaling elements (Müller-Dieckmann et al., 1999).

The Ethylene Receptor Family

Molecular and genetic studies revealed that in *Arabidopsis* the phytohormone ethylene is detected by a family of five receptors (ETR1, ETR2, EIN4, ERS1, and ERS2) (Chang et al., 1993; Hua et al., 1995; Sakai et al., 1998; reviewed in Wang et al., 2002). The receptors consist of an N-terminal transmembrane-sensing domain for ethylene binding, the already mentioned GAF-like protein-protein interaction domain of unknown function, and a histidine kinase or histidine kinase-like domain. ETR1, ETR2, and EIN4 also carry COOH-terminal receiver domains (Figure 23.3). All five receptors bind ethylene with similar high affinity in the presence of Cu^{2+}, which is probably provided by the copper transporter-like protein RAN1 (Hirayama et al., 1999; Rodriguez et al., 1999). Based on sequence similarity, the receptors are divided into two subfamilies, ETR1 and ERS1 with 68 percent identity and 80 percent similarity, and ETR2, EIN4, and ERS2 with about 48 percent identity and around 62 percent similarity, respectively (Hwang et al., 2002). ETR1 and ERS1 have all five motifs (H, N, G1, F, G2) typical for the catalytic kinase core of histidine kinases, whereas the other subfamily members lack at least some of these motifs (Figure 23.3). Histidine kinase activity has been demonstrated for ETR1 in vitro (Gamble et al., 1998, 2002), suggesting that two-component elements participate in ethylene signaling. As exemplarily shown for ETR1, the ethylene receptors appear to be predominantly localized to the endoplasmic reticulum (Chen et al., 2002).

Genetic and biochemical analyses performed within the past two decades have provided evidence of how the ethylene receptors actually work. As indicated by the absence of any phenotype in single-receptor mutants and despite structural differences, at least partial functional redundancy or compensation of function exists among the receptors, suggesting that they form multiprotein complexes to initiate ethylene signaling (Hwang et al., 2002; Wang et al., 2002). Interestingly, quadruple loss-of-function receptor mutants exhibited a constitutive ethylene response phenotype and gain-of-function mutations in any of the five receptor genes confer dominant ethylene insensitivity (referred to as "dominant insensitivity"; Urao et al., 2001). These observations led to the conclusion that the ethylene receptors negatively regulate ethylene responses (Hua and Meyerowitz, 1998). Indeed, binding studies in yeast using the gene products encoded by the *etr1-1* and *etr1-4* dominant gain-of-function alleles, which cause ethylene-insensitive plants, did not reveal ethylene binding. These results indicate that the observed dominant insensitivity of the receptors to ethylene is due to the disruption of the ethylene-binding activity. This disruption apparently generates ethylene receptors that are locked in a constitutively active signaling state. In this locked "on" state, the receptors repress the downstream ethylene response pathway regardless of whether ethylene is present (Hall et al., 1999; Gamble et al., 2002). In the wild type, binding of ethylene inactivates the ethylene receptors and de-represses ethylene signal transduction, resulting in hormone-induced responses.

The Cytokinin Receptor Family

The cytokinin receptor subfamily consists of the three hybrid histidine kinases AHK2, AHK3, and AHK4/CRE1/WOL, sharing about 53 percent identity with the entire amino acid sequence. The members of the family contain either one (AHK4/CRE1/WOL), two (AHK3), or three (AHK2) N-terminal transmembrane segments as well as a C-terminal transmitter and receiver domain (Figure 23.3). Noteworthy is that all three cytokinin receptors carry an additional receiver-like motif of unknown function embedded between the transmitter and the receiver domain.

AHK4/CRE1/WOL was initially identified by screening *Arabidopsis* hypocotyl explants for the loss of cytokinin-dependent greening, cell proliferation, and shoot formation. The identified mutant was named *cytokinin response 1 (cre1)*. Loss-of-function mutants for AHK4/CRE1/WOL also displayed a reduced sensitivity to cytokinin (Inoue et al., 2001; Ueguchi et al., 2001). A mutant allele of the AHK4/CRE1/WOL gene is also responsible for the phenotype of *wodden leg (wol),* which was identified as a mutant with abnormal roots. In the *wol* mutant, the vascular system of the root consists of a reduced number of cell files and is composed of only primary xylem. This root phenotype is very likely caused by impaired cell divisions in the embryo that normally produce root vascular initial cells and altered periclinal divisions that increase the vascular cell files after germination (Kakimoto, 2003). These data indicate that AHK4/CRE1/WOL contributes to the regulation of cytokinin-mediated responses, such as cell growth and cell division.

The function of AHK4/CRE1/WOL as hormone-triggered histidine kinase was demonstrated by the cytokinin-dependent complementation of the lethal growth phenotype of a mutant of *Saccharomyces cerevisiae,* in which the only histidine kinase gene, *Sln1,* was disrupted. A similar cytokinin-dependent complementation activity of AHK4/CRE1/WOL was observed in mutants of fission yeast *(Saccharomyces pombe)* and *Escherichia coli,* in which a particular histidine kinase gene has been inactivated (Suzuki et al., 2001). In the yeast system, the conserved His and Asp residues in the catalytic core of the transmitter and receiver domains, respectively, were required for complementation, suggesting that AHK4/CRE1/WOL indeed functions as a cytokinin-regulated receptor histidine kinase in vivo. This was supported by the finding that in *Arabidopsis* protoplasts AHK4/CRE1/WOL enhanced the promoter activity of the cytokinin-inducible reporter gene in a cytokinin-dependent manner (Hwang and Sheen, 2001). Similar cytokinin-dependent histidine kinase activity was also found for AHK2 and AHK3, indicating that these proteins function as hormone-triggered receptor histidine kinases as well (Yamada et al., 2001; Hwang and Sheen, 2001; Kakimoto, 2003).

Binding of cytokinin was proven for AHK4/CRE1/WOL by using microsomal fractions isolated from *S. pombe* extracts expressing the histidine kinase. Interestingly, the extracellular domains of the N-terminal region of the cytokinin receptors reflect the so-called cyclase/histidine kinase-associated sensing extracellular (CHASE) domain. This domain is also found in transmembrane receptors of lower eukaryotes and bacteria and is predicted to be a ligand-binding module (Anantharaman and Aravind, 2001; Mougel and Zhulin, 2001). Because the *wol* mutant has a mutation in the CHASE domain, and because this mutation disrupts the ability of AHK4/CRE1/WOL to bind cytokinin (Yamada et al., 2001) and to confer cytokinin-dependent complementation of the *sln1* mutant in yeast (Kakimoto, 2003), it is very likely that the CHASE motif is the cytokinin-sensing domain of AHK2, AHK3, and AHK4/CRE1/WOL.

In summary, it appears to be established that AHK2, AHK3, and AHK4/CRE1/WOL function as cytokinin-sensing receptor histidine kinases that transfer the hormone signal from the extracellular space to the cytoplasmic side of the plant cell.

CKI1 and AHK5/CKI2

As shown in Figure 23.3, CKI1 has all attributes of a membrane-associated receptor histidine kinase, whereas AHK5/CKI2 lacks any transmembrane segments and is likely localized to the cytoplasm of plant cells. CKI1 and AHK5/CKI2 were originally identified in a screen for cytokinin-independent growth of *Arabidopsis* calli (Kakimoto, 1996). In transgenic plants, overexpression of CKI1 caused cytokinin responses even in the absence of cytokinin (Kakimoto, 1996; Hwang and Sheen, 2001). Therefore, CKI1 and AHK5/CKI2 were discussed to be prime candidates for cytokinin-sensing receptor histidine kinases. However, the function of CKI1 and AHK5/CKI2 as cytokinin receptors is now in question. Based on genetic studies, microscopic examination of developing female gametophytes, and mRNA expression data, Pischke and co-workers (2002) suggested that CKI1 function is required for megagametophyte development. Phenotypic abnormalities in female gametophytes of *cki1* loss-of-function mutants could be observed in the four-nucleate stage. At this point, such important developmental steps as cellularization, vacuole formation, and the establishment of cell identities occur. At the terminal developmental stage of gametophyte development, the embryo sac collapsed and degenerated in *cki1* mutants (Pischke et al., 2002). Hence, CKI1 disruption is lethal to female gametophytes of *Arabidopsis.* Cytokinin is unlikely to play a role in this developmental process and is probably not the ligand for CKI1 (and AHK5/CKI2). Membranes isolated from fission yeast expressing CKI1 do not bind cytokinin (Yamada et al., 2001). Furthermore, expression of CKI1 in *E. coli* lacking a particular histidine kinase and in *Arabidopsis* protoplasts coexpressing a cytokinin-inducible reporter gene has been demonstrated to activate histidine kinase signaling pathways in a constitutive, cytokinin-independent manner (Yamada et al., 2001; Hwang and Sheen, 2001). It has therefore to be proposed that other compounds bind to CKI1 and AHK5/CKI2 and induce their histidine kinase activity. The observed effects of CKI1 and AHK5/CKI2 on cytokinin responses have to be discussed as unexpected interference of the overexpressed proteins with cytokinin receptor-related signal transduction pathways (Pischke et al., 2002; Kakimoto, 2003).

AHK1/AtHK1

Similar to the bacterial and yeast osmosensing pathways, the plant histidine kinase AHK1/AtHK1 has been implicated in osmosensing (Urao et al., 1999). AHK1/AtHK1 has two putative transmembrane segments in the N-terminal half (Figure 23.3) and has structural similarity to the yeast osmosensing hybrid histidine kinase SLN1 (Urao et al., 1999). Expression of AHK1/AtHK1 in a yeast double mutant lacking its two osmosensors suppressed the lethal growth phenotype on high-salinity media and activated the high-osmolarity glycerol response pathway that results in the activation of stress-responsive genes (Urao et al., 1999). This suppressor activity requires an intact catalytic core in the transmitter domain and also depends on the phosphorylable Asp residue in the receiver domain, suggesting that histidine kinase and phosphotransfer activity are necessary for AHK1/AtHK1 function (Urao et al., 1999). Interestingly, AHK1/AtHK1 seems to be active under normal growth conditions and changes from the active to the inactive state in response to increasing osmolarity, as is also the case for yeast SLN1. The *AHK1/AtHK1* transcript is more abundant in roots than in other tissues and accumulates in response to high salinity. These observations suggest that AHK1/AtHK1 senses changes in osmolarity in yeast, and likely also in plant cells, caused, for instance, by dehydration and high salinity, and subsequently transduces the stress signal to the nucleus (Urao et al., 1999). However, the precise molecular function of AHK1/AtHK1 in plants requires further investigation.

EARLY STEPS IN PLANT HISTIDINE KINASE SIGNALING

Ethylene Receptor Signaling

As mentioned previously, analysis of ethylene signal transduction in *Arabidopsis* revealed the presence of a small family of ethylene receptors (namely, ETR1, ERS1, ETR2, EIN4, ERS2), and all members of the family contribute to ethylene-related signaling. However, from these five family members, only ETR1 and ERS1 contain the amino acids essential for histidine kinase activity. These observations opened the discussion of whether the transmission of the ethylene signal occurs through a canonical multistep phosphorelay mechanism. Genetic studies on the triple response of *Arabidopsis* established a different model. In this model, ethylene signaling downstream of the receptors occurs without involvement of two-component elements. Accordingly, the five receptors interact with CTR1, a Raf-related protein kinase, for negative regulation of the ethylene response pathways. CTR1 also represents a negative regulator of ethylene signaling because loss-of-function mutations in CTR1 lead to a constitutive response phenotype in the absence of ethylene (Wang et al., 2002). Recombinant CTR1 protein was recently shown to possess intrinsic Ser/Thr protein kinase activity similar to Raf-1 (Huang et al., 2003) and is associated with ETR1 in an ER-associated ethylene signaling complex (Gao et al., 2003). For association of CTR1 with ETR1, histidine kinase activity of the receptor appears not to be required (Gao et al., 2003). Further molecular and genetic analysis indicates that kinase activity is necessary for CTR1 function and suggests that the ethylene receptors activate by protein-protein interaction the kinase activity of CTR1 in the absence of ethylene. Active, receptor-bound CTR1 may then phosphorylate downstream targets to down-regulate the ethylene response pathway. Thus, association of CTR1 with ethylene receptors is required to keep the ethylene signal transduction in the "off" state (Huang et al., 2003). The observation that phosphorylation of CTR1 may play an important role in ethylene signaling was recently supported by the isolation and molecular characterization of the *Arabidopsis eer1* mutant (Larsen and Cancel, 2003). The *eer1* mutant displays increased sensitivity and increased amplitude of response to ethylene and carries a loss-of-function mutation in the *RCN1* gene, one of three PP2A regulatory subunits in *Arabidopsis*. PP2A-1C, a PP2A catalytic subunit that interacts with RCN1, associates with the kinase domain of CTR1 in vitro. This likely represents a role of PP2A in modulation of the phosphorylation state and activity of CTR1 (Larsen and Cancel, 2003). How the phosphorylation state affects the capacity of CTR1 to modify substrate proteins awaits further analysis.

As indicated by recent data, a MAPK cascade functions downstream of CTR1, which seems to activate the Nramp metal transporter-like protein EIN2. EIN2 in turn appears to initiate, by a hitherto unknown mechanism, a transcription factor cascade that eventually triggers ethylene responses (Ouaked et al., 2003; for further review, see Wang et al., 2002).

The current model and recently published reports (Gamble et al., 2002; Wang et al., 2003) may suggest that histidine kinase activity of ETR1 and ERS1 and, therefore, two-component signaling may not to be required to mediate ethylene signal transduction. However, some evidence suggests that ethylene receptor-dependent two-component mechanisms may contribute to or interfere with CTR1-dependent signaling:

1. The ethylene receptors ETR1 and ERS1 play a predominant role in the regulation of ethylene responses that could be because of the in-

herent histidine kinase activity (Gamble et al., 2002; Wang et al., 2003).

2. In yeast, ETR1 strongly interacts with several *Arabidopsis* HPt proteins, including AHP1 and AHP2 (Urao et al., 2000). These HPt proteins in turn efficiently associate with response regulators, as for instance ARR4 and ARR2 (Urao et al., 2000; Lohrmann et al., 2001; Sweere et al., 2001).
3. *Arabidopsis* plants carrying a loss-of-function allele of the response regulator ARR2 display reduced sensitivity not only to cytokinin but also to ethylene. In contrast, ARR2-overexpressing plants showed enhanced responsiveness to these phytohormones (Hass et al., 2004). Furthermore, overexpression of the HPt protein AHP2 caused increased sensitivity to ethylene in at least one of three analyzed independent transgenic *Arabidopsis* lines (Suzuki et al., 2002).
4. In *Arabidopsis* mesophyll protoplast ARR2 regulates the activity of an ethylene-responsive reporter gene in an ethylene- and phosphorylation-dependent manner (Hass et al., 2004).
5. Efficient Asp phosphorylation of ARR2 depends on the presence of ETR1 in a plant cell-free phosphorelay system (Hass et al., 2004).

Hence, further detailed physiological, molecular, and biochemical analyses are necessary to clarify the functional relationship between the CTR1-dependent and a two-component phosphotransfer mechanism in ethylene signal transduction.

Cytokinin Receptor Signaling

It is now accepted that AHK2, AHK3, and AHK4/CRE1/WOL act as cytokinin-sensing receptor histidine kinases in *Escherichia coli,* yeast, and fission yeast. However, how exactly these cytokinin receptors regulate downstream events including cytokinin-responsive genes in plant cells, awaits a more detailed analysis. After treatment of plant cells with cytokinin, one has to assume that AHK2, AHK3, and AHK4/CRE1/WOL first phosphorylate an HPt protein (AHP; Figure 23.1B) to initiate a multistep phosphorelay. The phosphoryl group is then relayed to at least one response regulator that eventually regulates cellular responses. The principle ability of *Arabidopsis* HPt proteins to perceive a phosphoryl group from AHK4/CRE1/WOL has been suggested by an in vivo interference assay in *E. coli* (Suzuki et al., 2001, 2002). There, expression of AHP2 and AHP4—and to a weaker extent expression of AHP1 and AHP3—negatively interfered with an artificial AHK4/CRE1/WOL → YojN (HPt protein) → RcsB (response regulator) phosphorelay system. The observed negative interference can be explained, assuming that the phosphoryl group on AHK4/CRE1/WOL is acquired more readily by *Arabidopsis* AHPs than by YojN. As shown by in vitro assays, AHP1 and AHP2 can transfer an artificially acquired phosphoryl group to the response regulators ARR3 and ARR4 and ARR6 and ARR10, respectively (Suzuki et al., 1998; Imamura et al., 1999). From these data, it is suggestive that the cytokinin receptors could, in principle, induce a multistep phosphorelay in plant cells.

The existence of a cytokinin-signaling circuitry in plant cells that is based on two-component signaling elements, such as AHK2, AHK3 or AHK4/CRE1/WOL, and AHPs and ARRs, has been proposed by several groups (Hwang and Sheen, 2001; Sakai et al., 2001; Kakimoto, 2003). However, there is an ongoing discussion about the functional relevance of His-to-Asp phosphotransfer events in mediating the cytokinin signal transfer, especially on the level of HPt proteins and response regulators (Hwang et al., 2002; Lohrmann and Harter, 2002).

Despite this discussion, the existing evidence can be combined in a model that proves a phosphorelay mechanism may at least partially function in plant cells that transduce the cytokinin signal from plasmalemma-associated cytokinin receptors to those cellular compartments and molecular events that mediate the expression of plant cytokinin responses (Kakimoto, 2003).

CKI1, AHK5/CKI2, and AHK1/ATHK1 Signaling

As discussed previously, CKI1 function is required for female gametophyte development in *Arabidopsis* (Pischke et al., 2002). In yeast two-hybrid assays, CKI1 interacts with the *Arabidopsis* HPt proteins AHP1 and AHP2 (Urao et al., 2000). In addition, CKI1 acts as a phosphatase in vitro, when coincubated with phosphorylated AHP1 and AHP2 (Nakamura et al., 1999). Hence, the molecular mechanism of how CKI1 modulates female gametophyte development could involve a two-component phosphorelay mechanism. Nothing is known about the molecular function and interaction partners of AHK5/CKI2.

Comparable to CKI1 and AHK5/CKI2, there is only very limited information about signaling events initiated by the osmolarity-sensing histidine kinase AHK1/AtHK1 in plant cells. AHK1/AtHK1 interacts with the *Arabidopsis* HPt protein AHP2 in the yeast two-hybrid system (Urao et al., 2000), suggesting that, similar to yeast cells, the transduction of the stress signal could occur via a multistep phosphorelay.

CONCLUSION

During the past few years, our knowledge of the functional role of protein kinases and phosphatases in plant growth and development, as well as in the plant's communication with the environment, has increased considerably. Despite this progress, and with regard to the more than 1,100 genes encoding protein kinases and phosphatases in *Arabidopsis,* our understanding of which specific protein is involved in which manner in a particular phosphorylation cascade is far from being complete.

For instance, for the vast majority of receptor-like kinases and for several receptor histidine kinases, the ligands have still to be identified. In addition, the molecular mechanisms, how these receptor kinases actually transmit the signal to downstream signaling components, are largely unknown. The same holds true for the substrates of plant kinases and phosphatases.

Cross talk between different phosphorylation cascades is another issue. In the case of two-component signaling, it is an open question whether signaling pathways initiated by different histidine kinases are separated or do converge at a particular crossing point (Kakimoto, 2003). Cross talk appears also to exist between MAPK cascade-depending and CDPK signal transduction in regulation of pathogen response pathways (Romeis et al., 2001; Asai et al., 2002). Emergence and intensity of cross talk strongly depend on the expression levels of the participating kinases and phosphatases. However, for the multitude of plant protein kinases and phosphatases, our knowledge of tissue- and cell-specific expression levels is very limited.

In summary, understanding the complexity of the phosphorylation-dephosphorylation-dependent signaling network in plants will be a challenging task for the future, one that requires combinatorial approaches in functional genomics, cell biology, biochemistry, and physiology.

REFERENCES

Albrecht, V., Ritz, O., Linder, S., Harter, K., Kudla, J. (2001). The NAF domain defines a novel protein-protein interaction module conserved in Ca^{2+}-regulated kinases. *EMBO J.* 20: 1051-1063.

Albrecht, V., Weinl, S., Blazevic, D., D'Angelo, C., Batistic, O., Kolukisaoglu, Ü., Bock, R., Schulz, B., Harter, K., Kudla, J. (2003). The calcium sensor CBL1 integrates plant responses to abiotic stresses. *Plant J.* 36: 457-470.

Anantharaman, V., Aravind, L. (2001). The CHASE domain: A predicted ligand-binding module in plant cytokinin receptors and other eukaryotic and bacterial receptors. *Trends Biochem. Sci.* 26: 579-582.

Aravind, L., Ponting, C.P. (1999). The cytoplasmic helical linker domain of receptor histidine kinase and methyl-accepting proteins is common to many prokaryotic signalling proteins. *FEMS Microbiol. Lett.* 176: 111-116.

Asai, T., Tena, G., Plotnikova, J., Willmann, M.R., Chiu, W.L., Gomez-Gomez, L., Boller, T., Ausubel, F.M., Sheen, J. (2002). MAP kinase signalling cascades in *Arabidopsis* innate immunity. *Nature* 415: 977-983.

Chang, C., Kwok, S.F., Bleecker, A.B., Meyerowitz, E.M. (1993). *Arabidopsis* ethylene-response gene *ETR1:* Similarity of product to two-component regulators. *Science* 262: 539-544.

Chen, Y.-F., Randlett, M.D., Findell, J.L., Schaller, E. (2002). Localization of the ethylene receptor ETR1 to the endoplasmic reticulum of *Arabidopsis. J. Biol. Chem.* 277: 19851-19866.

Cheng, S.H., Willmann, M.R., Chen, H.C., Sheen, J. (2002). Calcium signaling through protein kinases: The *Arabidopsis* calcium-dependent protein kinase gene family. *Plant Physiol.* 129: 469-485.

Cheong, Y.H., Kim, K.-N., Pandey, G.K., Gupta, R., Grant, J.J., Luan, S. (2003). CBL1, a calcium sensor that differentially regulates salt, drought, and cold responses in *Arabidopsis. Plant Cell.* 15: 1833-1845.

Gamble, R.L., Coonfield, M.L., Schaller, G.E. (1998). Histidine kinase activity of the ETR1 ethylene receptor from *Arabidopsis. Proc. Natl. Acad. Sci. USA* 95: 7825-7829.

Gamble, R.L., Qu, X., Schaller, G.E. (2002). Mutational analysis of the ethylene receptor ETR1: Role of the histidine kinase domain in dominant ethylene insensitivity. *J. Biol. Chem.* 128: 1428-1438.

Gao, Z., Chen, Y.-F., Randlett, M.D., Zao, X.-C., Findell, J.L., Kieber, J.J., Schaller, G.E. (2003). Localization of the Raf-like kinase CTR1 to the endoplasmatic reticulum of *Arabidopsis* through participation in ethylene receptor signaling complexes. *J. Biol. Chem.* 278: 34725-34732.

Garrington, T.P., Johnson, G.L. (1999). Organization and regulation of mitogen-activated protein signaling pathways. *Curr. Opin. Cell Biol.* 11: 211-218.

Halfter, U., Ishitani, M., Zhu, J.K. (2000). The *Arabidopsis* SOS2 protein kinase physically interacts with and is activated by the calcium-binding protein SOS3. *Proc. Natl. Acad. Sci. USA* 97: 3735-3740.

Hall, A.E., Chen, Q.G., Findell, J.L., Schaller, G.E., Bleecker, A.B. (1999). The relationship between ethylene binding and dominant insensitivity conferred by mutant forms of the ETR1 ethylene receptor. *Plant Physiol.* 121: 291-299.

Hallford, N.G.., Bouly, J.P., Thomas, M. (2000). SNF1-related kinases (SnRKs) regulators at the heart of the control of carbon metabolism and partitioning. *Adv. Bot. Res.* 32: 405-434.

Hallford, N.G., Hey, S., Jhurreea, D., Laurie, S., McKibbin, R.S., Paul, M., Zhang, Y. (2003). Metabolic signaling and carbon partitioning: Role of SNF1-related (SnRK1) protein kinase. *J. Exp. Bot.* 54: 467-475.

Harmon, A.C., Gribskov, M., Harper, J.F. (2000). CDPKs: A kinase for every Ca^{2+} signal? *Trends Plant Sci.* 5: 154-159.

Harmon, A.C., Putnam-Evenas, C., Cormier, M.J. (1987). A calcium-dependent but calmodulin-independent protein kinase from soybean. *Plant Physiol.* 83: 830-837.

Hass, C., Albrecht, V., Lohrmann, J., Hummel, F., Sweere, U., Grefen, C., Schäfer, E., Hwang, I., Zhu, T., Kudla, J., Harter, K. (2004). ARR2-involving two-component systems mediate hormone and stress signalling in *Arabidopsis* (submitted).

Hirayama, T., Kieber, J.J., Hirayama, N., Kogan, M., Guzman, P., Nourizadeh, S., Alonso, J.M., Dailey, W.P., Dancis, A., Ecker, J.R. (1999). RESPONSIVE-TO-ANTAGONIST1, a Menkes/Wilson disease-related copper transporter, is required for ethylene signaling in *Arabidopsis. Cell* 97: 383-393.

Hrabak, E.M. (2000). Calcium-dependent protein kinases and their relatives. *Adv. Bot. Res.* 32: 185-223.

Hrabak, E.M., Chan, C.W.M., Gribskov, M., Harper, J.F., Choi, J.H., Halford, N., Kudla, J., Luan, S., Nimmo, H.G., Sussman, M.R., et al. (2003). The *Arabidopsis* CDPK-SnRK superfamily of protein kinases. *Plant Physiol.* 132: 666-680.

Hua, J., Chang, C., Sun, Q., Meyerowitz, E.M. (1995). Ethylene insensitivity conferred by *Arabidopsis ERS* gene. *Science* 269: 1712-1714.

Hua, J., Meyerowitz, E.M. (1998). Ethylene responses are negatively regulated by a receptor gene family in *Arabidopsis thaliana. Cell* 94: 261-271.

Huang, Y., Li, H., Hutchison, C.E., Laskey, J., Kieber, J.J. (2003). Biochemical and functional analysis of CTR1, a protein kinase that negatively regulates ethylene signaling in *Arabidopsis. Plant J.* 33: 221-233.

Hunter, T., Plowman, G.D. (1997). The protein kinases of budding yeast: Six score and more. *Trends Biochem. Sci.* 22: 18-22.

Hwang, I., Chen, H.-C., Sheen, J. (2002). Two-component signal transduction pathways in *Arabidopsis. Plant Physiol.* 129: 500-515.

Hwang, I., Sheen, J. (2001). Two-component circuitry in *Arabidopsis* cytokinin signal transduction. *Nature* 413: 383-389.

Imamura, A., Hanaki, N., Nakamura, A., Suzuki, T., Taniguchi, M., Ueguchi, C., Sugiyama, T. Mizuno, T. (1999). Compilation and characterization of *Arabidopsis thaliana* response regulator implicated in His-to-Asp phosphorelay signal transduction. *Plant Cell Physiol.* 40: 733-742.

Innes, R.W. (2001). Mapping out the roles of MAP kinases in plant defense. *Trends Plant Sci.* 6: 392-394.

Inoue, T., Higuchi, M., Hashimoto, Y., Seki, M., Kobayashi, M., Kato, T., Tabata, S., Shinozaki, K., Kakimoto, T. (2001). Identification of CRE1 as a cytokinin receptor from *Arabidopsis. Nature* 409: 1060-1063.

Jonak, C., Ökresz, L., Börge, L., Hirt, H. (2002). Complexity, cross talk and integration of plant MAP kinase signalling. *Curr. Opin. Plant Biol.* 5: 415-424.

Kakimoto, T. (1996). CKI1, a histidine kinase homolog implicated in cytokinin signal transduction. *Science* 274: 982-985.

Kakimoto, T. (2003). Perception and signal transduction of cytokinins. *Annu. Rev. Plant Biol.* 54: 605-627.

Kerk, D., Bulgrien, J., Smith, D.W., Brooke, B., Veretnik, S., Gribskov, M. (2002). The complement of protein phosphatase catalytic subunits encoded in the genome of *Arabidopsis. Plant. Physiol.* 129: 908-925.

Kim, K.N., Cheong, Y.H.., Grant, J.J., Pandey, G.K., Laun, S. (2003). CIPK3, a calcium sensor-associated protein kinase that regulates abscisic acid and cold signal transduction in *Arabidopsis. Plant Cell* 15: 411-423.

Krebs, E.G., Graves, D.J., Fischer, E.H. (1959). Factors affecting the activity of muscle phophorylase b kinase. *J. Biol. Chem.* 234: 2867-2873.

Larsen, P.B., Cancel, J.D. (2003). Enhanced ethylene responsiveness in the *Arabidopsis eer1* mutant results from a loss-of-function mutation in the protein phosphatase 2A A regulatory subunit, *RCN1. Plant J.* 34: 709-718.

Lawton, M.A., Yamamoto, R.T. Hanks, S.K., Lamb, C.J. (1989). Molecular cloning of plant transcripts encoding protein kinase homologs. *Proc. Natl. Acad. Sci. USA* 86: 3140-3144.

Leung, J., Bouvier-Durand, M., Morris, P.C., Guerrier, D., Chefdor, F., Giraudad, J. (1994). *Arabidopsis* ABA gene *ABI1:* Features of a calcium-modulated protein phosphates. *Science* 264: 1448-1452.

Ligterink, W., Hirt, H. (2001). Mitogen-activated protein (MAP) kinase pathways in plants: Versatile signaling tools. *Int. Rev. Cytol.* 201: 209-275.

Lohrmann, J. Harter, K. (2002). Plant two-component signaling systems and the role of response regulators. *Plant Physiol.* 128: 363-369.

Lohrmann, J., Sweere, U., Zabaleta, E., Baeurle, I., Keitel, C., Kozma-Bognar, L., Brennicke, A., Schäfer, E., Kudla, J., Harter, K. (2001). The response regulator ARR2: A pollen-specific transcription factor involved in the expression of nuclear genes for components of mitochondrial complex I in *Arabidopsis. Mol. Genet. Genomics* 265: 2-13.

Liu, J., Zhu, J.-K. (1998). A calcium sensor homolog required for plant salt tolerance. *Science* 280: 1943-1945.

Luan, S. (1998). Protein phosphatases and signaling cascades in higher plants. *Trends Plant Sci.* 3: 271-275.

Luan, S. (2003). Protein phosphatases in plants. *Annu. Rev. Plant. Biol.* 54: 63-92.

Luan, S., Kudla, J., Rodriguez-Concepcion, M., Yalovsky, S., Gruissem, W. (2002). Calmodulins and calcineurin B–like proteins: Calcium sensors for specific signal response coupling in plants. *Plant Cell* 14(Suppl): 389-400.

Meyer, K., Leube, M.P., Grill, E. (1994). A protein phosphatase 2C involved in ABA signal transduction in *Arabidopsis thaliana. Science* 264: 1452-1455.

Müller-Dieckmann, H.-J., Grantz, A.A., Kim, S.-H. (1999). The structure of the signal receiver domain of the *Arabidopsis thaliana* ethylene receptor ETR1. *Structure* 7: 1547-1556.

Mougel, C., Zhulin, I.B. (2001). CHASE: An extracellular sensing domain common to transmembrane receptors from prokaryotes, lower eukaryotes and plants. *Trends Biochem. Sci.* 26: 582-584.

Mustilli, A.C., Merlot, S., Vavasseur, A., Fenzi, F., Giraudad, J. (2002). *Arabidopsis* OST1 protein kinase mediates the regulation of stomatal aperture by abscisic acid and acts upstream of reactive oxygen species production. *Plant Cell* 14: 3089-3099.

Nakamura, A., Kakimoto, T., Imamura, A., Suzuki, T., Ueguchi, C., Mizuno, T. (1999). Biochemical characterization of a putative cytokinin-responsive His-kinase, CKI1, from *Arabidopsis thaliana. Biosci. Biotechnol. Biochem.* 63: 1627-1630.

Ouaked, F., Rozhon, W., Lecourieux, D., Hirt, H. (2003). A MAPK pathway mediates ethylene signalling in plants. *EMBO J.* 22: 1282-1288.

Park, H., Inouye, M. (1997). Mutational analysis of the linker region of EnvZ, an osmosensor in *Escherichia coli. J. Bacteriol.* 179: 4382-4390.

Pischke, M.S., Jones, L.G., Otsuga, D., Fernandez, D.E., Drews, G.N., Sussman, M.R. (2002). An *Arabidopsis* histidine kinase is essential for megagametogenesis. *Proc. Natl. Acad. Sci. USA* 99: 15800-15805.

Plowman, G.D., Sundarsanam, S., Bingham, J, Whyte, D., Hunter, T. (1999). The protein kinases of *Caenorhabditis elegans:* A model for signal transduction in multicellular organisms. *Proc. Natl. Acad. Sci. USA* 96: 13603-13610.

Rodriguez, F.I., Esch, J.J., Hall, A.E., Binder, B.M., Schaller, G.E., Bleecker, A.B. (1999). A copper cofactor for the ethylene receptor ETR1 from *Arabidopsis. Science* 283: 996-998.

Romeis, T., Ludwig, A.A., Martin, R., Jones, J.D. (2001). Calcium-dependent protein kinases play an essential role in a plant defence response. *EMBO J.* 20: 5556-5567.

Sakai, H., Honma, T., Aoyama, T., Sato, S., Kato, T., Tabata, S., Oka, A. (2001). ARR1, a transcription factor for genes immediately responsive to cytokinins. *Science* 294: 1519-1521.

Sakai, H., Hua, J., Chen, Q.G., Chang, C., Medrano, L.J., Bleecker, A.B., Meyerowitz, E.M. (1998). *ETR2* is an *ETR1*-like gene involved in ethylene signaling in *Arabidopsis. Proc. Natl. Acad. Sci. USA* 95: 5812-5817.

Sanders, D., Brownlee, C., Harper, J.F. (1999). Communicating with calcium. *Plant Cell* 11: 691-706.

Sanders, D., Pelloux, J., Brownlee, C., Harper, J.F. (2002). Calcium at the crossroads of signaling. *Plant Cell* 14(Suppl): 401-417.

Shi, J., Kim, K.N., Ritz, O., Albrecht, V., Gupta, R., Harter, K., Luan, S., Kudla, J. (1999). Novel protein kinases associated with calcineurin B-like calcium sensors in *Arabidopsis. Plant Cell* 11: 2393-2405.

Shiu, S.H., Bleecker, A.B. (2001a). Plant receptor-like kinase gene family: Diversity, function and signaling. Available at <www.stke.org/cgi/content/full/OC_sigtrans;2001/113/re22>.

Shiu, S.H., Bleecker, A.B. (2001b). Receptor-like kinases from *Arabidopsis* form a monophyletic gene family related to animal receptor kinases. *Proc. Natl. Acad. Sci. USA* 98: 10763-10768.

Shiu, S.H., Bleecker, A.B. (2002). Expansion of the receptor-like kinase/Pelle gene family and receptor-like proteins in *Arabidopsis. Plant Physiol.* 132: 530-543.

Singh, M., Berger, B., Kim, P.S., Berger, J.M., Cochran, A.G. (1998). Computational learning reveals coiled coil-like motifs in histidine kinase linker domains. *Proc. Natl. Acad. Sci. USA* 95: 2738-2743.

Stock, A.M., Robinson, V.L., Goudreau, P.N. (2000). Two-component signal transduction. *Annu. Rev. Biochem.* 69: 183-215.

Suzuki, T., Imamura, A., Ueguchi, C., Mizuno, T. (1998). Histidine-containing phosphotransfer (HPt) signal transducers implicated in His-to-Asp phosphorelay in *Arabidopsis*. *Plant Cell Physiol.* 39: 1258-1268.

Suzuki, T., Ishikawa, K., Yamashino, T., Mizuno, T. (2002). An *Arabidopsis* histidine-containing phosphotransfer (HPt) factor implicated in phosphorelay signal transduction: Overexpression of AHP2 in plants results in hypersensitiveness to cytokinin. *Plant Cell Physiol.* 43: 123-129.

Suzuki, T., Miwa, K., Ishikawa, K., Yamada, H., Aiba, H., Mizuno, T. (2001). The *Arabidopsis* sensor His-kinase, AHK4, can respond to cytokinins. *Plant Cell Physiol.* 42: 107-113.

Sweere, U., Eichenberg, K., Lohrmann, J., Mira-Rodado, V., Bäurle, I., Kudla, J., Nagy, F., Schäfer, E., Harter, K. (2001). Interaction of the response regulator ARR4 with phytochrome B in modulating red light signalling. *Science* 294: 1108-1111.

Tena, G., Asai, T., Chiu, W.L., Sheen, J. (2001). Plant mitogen-activated protein kinase signaling cascades. *Curr. Opin. Plant Biol.* 4: 392-400.

Tichtinsky, G., Vanoosthuyse, V., Cock, J.M., Gaude, T. (2003). Making inroads into plant receptor kinase signaling pathways. *Trends Plant Sci.* 8: 231-237.

Ueguchi, C., Sato, S., Kato, T., Tabata, S. (2001). The *AHK4* gene involved in the cytokinin-signaling pathway as a direct receptor molecule in *Arabidopsis thaliana*. *Plant Cell Physiol.* 42: 751-755.

Urao, T., Miyata, S., Yamaguchi-Shinozaki, K., Shinozaki, K. (2000). Possible His to Asp phosphorelay signalling in an *Arabidopsis* two-component system. *FEBS Lett.* 478: 227-232.

Urao, T., Yakubov, B., Satoh, R., Yamaguchi-Shinozaki, K., Seki, M., Hirayama, T., Shinozaki, K. (1999). A transmembrane hybrid-type histidine kinase in *Arabidopsis* functions as an osmosensor. *Plant Cell* 11: 1743-1754.

Urao, T., Yamaguchi-Shinozaki, K., Shinozaki, K. (2001). Plant histidine kinases: An emerging picture of two-component signal transduction in hormone and environmental responses. *Science STKE* 109: RE18

Walsh, D.A., Perkins, J.P., Krebs, E.G. (1968). An adenosine 3',5'-monopnosphate-dependent protein kinase from rabbit skeletal muscle. *J. Biol. Chem.* 243: 3763-3765.

Wang K.L.-C., Li, H., Ecker, J.R. (2002). Ethylene biosynthesis and signaling networks. *Plant Cell* 14(Suppl): 131-151.

Wang, W., Hall, A.E., O'Malley, R., Bleecker, A.B. (2003). Canonical histidine kinase activity of the transmitter domain of ETR1 ethylene receptor from *Arabidopsis* is not required for signal transmission. *Proc. Natl. Acad. Sci. USA* 100: 352-357.

Wortman, J.R., Haas, B.J., Hannik, L.I., Smith, R.K., Maiti, R., Ronning, C.M., Chan, A.P., Yu, C., Ayele, M., Whitelaw, C.A., et al. (2003). Annotation of the *Arabidopsis* genome. *Plant Physiol.* 132: 461-468.

Yamada, H., Suzuki, T., Terada, K., Takei, K., Ishikawa, K., Miwa, K., Mizuno, T. (2001). The *Arabidopsis* AHK4 histidine kinase is a cytokinin-binding receptor that transduces cytokinin signals across the membrane. *Plant Cell Physiol.* 42: 1017-1023.

Yeh, K.C., Lagarias, J.C. (1998). Eukaryotic phytochromes: Light-regulated serine/threonine protein kinases with histidine kinase ancestry. *Proc. Natl. Acad. Sci. USA* 95: 13976-13981.

Yoshida, R., Hobo, T., Ichimura, K., Muzoguchi, T., Takahashi, F., Alonso, J., Ecker, J.R., Shinozaki, K. (2002). ABA-activated SnRK2 protein kinase is required for dehydration stress signaling in *Arabidopsis*. *Plant Cell Physiol*. 43: 1473-1483.

Index

Page numbers followed by the letter "f" indicate figures; those followed by the letter "t" indicate tables.